D0206377

Biomedical Instrumentation and Measurements

Biomedical Instrumentation and Measurements

Second Edition

Leslie Cromwell

California State University,
Los Angeles, California

Fred J. Weibell

Veterans Administration
Biomedical Engineering and Computing Center
Sepulveda, California

Erich A. Pfeiffer

Wells Fargo Alarm Services
Engineering Center
Hawthorne, California

Prentice-Hall, Inc., Englewood Cliffs, New Jersey 07632

Library of Congress Cataloging in Publication Data

Cromwell, Leslie.
  Biomedical instrumentation and measurements.

  Bibliography: p.
  Includes index.
  1. Biomedical engineering. 2. Medical instruments and apparatus. 3. Physiological apparatus. I. Weibell, Fred J., joint author. II. Pfeiffer, Erich A., joint author. III. Title. [DNLM: 1. Biomedical engineering—Instrumentation. QT26 C946b]
  R856.C7 1980   610´.28   79-22696
  ISBN 0-13-076448-5

© 1980 by Prentice-Hall, Inc., Englewood Cliffs, N.J. 07632

Printed in the United States of America

10 9 8 7 6 5 4 3 2 1

Editorial/production supervision by *Virginia Huebner*
Page layout by *Rita Schwartz*
Cover design by *James Wall*
Manufacturing buyer: *Gordon Osbourne*

PRENTICE-HALL INTERNATIONAL, INC., *London*
PRENTICE-HALL OF AUSTRALIA PTY. LIMITED, *Sydney*
PRENTICE-HALL OF CANADA, LTD., *Toronto*
PRENTICE-HALL OF INDIA PRIVATE LIMITED, *New Delhi*
PRENTICE-HALL OF JAPAN, INC., *Tokyo*
PRENTICE-HALL OF SOUTHEAST ASIA PTE. LTD., *Singapore*
WHITEHALL BOOKS LIMITED, *Wellington, New Zealand*

To our wives:

IRINA CROMWELL

CAROL WEIBELL

LIANNE PFEIFFER

Contents

PREFACE TO THE FIRST EDITION   xi
PREFACE TO THE SECOND EDITION   xv

1. INTRODUCTION TO BIOMEDICAL INSTRUMENTATION   1

    1.1.   The Age of Biomedical Engineering,   4
    1.2.   Development of Biomedical Instrumentation,   4
    1.3.   Biometrics,   6
    1.4.   Introduction to the Man–Instrument System,   10
    1.5.   Components of the Man–Instrument System,   13
    1.6.   Physiological Systems of the Body,   16
    1.7.   Problems Encountered in Measuring a Living System,   21
    1.8.   Some Conclusions,   24
    1.9.   The Objectives of the Book,   25

2. BASIC TRANSDUCER PRINCIPLES   26

2.1. The Transducer and Transduction Principles,   27
2.2. Active Transducers, 27
2.3. Passive Transducers,   35
2.4. Transducers for Biomedical Applications,   42

3. SOURCES OF BIOELECTRIC POTENTIALS   49

3.1. Resting and Action Potentials,   50
3.2. Propagation of Action Potentials,   53
3.3. The Bioelectric Potentials,   54

4. ELECTRODES   63

4.1. Electrode Theory,   64
4.2. Biopotential Electrodes,   66
4.3. Biochemical Transducers,   76

5. THE CARDIOVASCULAR SYSTEM   84

5.1. The Heart and Cardiovascular System,   85
5.2. The Heart,   89
5.3. Blood Pressure,   93
5.4. Characteristics of Blood Flow,   98
5.5. Heart Sounds,   100

6. CARDIOVASCULAR MEASUREMENTS   105

6.1. Electrocardiography,   106
6.2. Measurement of Blood Pressure,   126
6.3. Measurement of Blood Flow and Cardiac Output,   150
6.4. Plethysmography,   163
6.5. Measurement of Heart Sounds,   169

7. PATIENT CARE AND MONITORING   173

7.1. The Elements of Intensive-Care Monitoring,   174
7.2. Diagnosis, Calibration, and Repairability of
       Patient-Monitoring Equipment,   185
7.3. Other Instrumentation for Monitoring Patients,   187

7.4. The Organization of the Hospital for
Patient-Care Monitoring, 193
7.5. Pacemakers, 195
7.6. Defibrillators, 206

8. MEASUREMENTS IN THE RESPIRATORY SYSTEM 213

8.1. The Physiology of the Respiratory System, 215
8.2. Tests and Instrumentation for the Mechanics of Breathing, 218
8.3. Gas Exchange and Distribution, 232
8.4. Respiratory Therapy Equipment, 237

9. NONINVASIVE DIAGNOSTIC INSTRUMENTATION 243

9.1. Temperature Measurements, 244
9.2. Principles of Ultrasonic Measurement, 255
9.3. Ultrasonic Diagnosis, 263

10. THE NERVOUS SYSTEM 277

10.1. The Anatomy of the Nervous System, 278
10.2. Neuronal Communication, 282
10.3. The Organization of the Brain, 286
10.4. Neuronal Receptors, 290
10.5. The Somatic Nervous System and Spinal Reflexes, 291
10.6. The Autonomic Nervous System, 292
10.7. Measurements from the Nervous System, 292

11. INSTRUMENTATION FOR SENSORY MEASUREMENTS AND THE STUDY OF BEHAVIOR 304

11.1. Psychophysiological Measurements, 305
11.2. Instruments for Testing Motor Responses, 308
11.3. Instrumentation for Sensory Measurements, 309
11.4. Instrumentation for the Experimental Analysis of Behavior, 311
11.5. Biofeedback Instrumentation, 314

12. BIOTELEMETRY 316

12.1. Introduction to Biotelemetry, 317
12.2. Physiological Parameters Adaptable to Biotelemetry, 318

12.3. The Components of a Biotelemetry System,   321
12.4. Implantable Units,   332
12.5. Applications of Telemetry in Patient Care,   337

13. INSTRUMENTATION FOR THE CLINICAL LABORATORY   344

13.1. The Blood,   345
13.2. Tests on Blood Cells,   347
13.3. Chemical Tests,   351
13.4. Automation of Chemical Tests,   357

14. X-RAY AND RADIOISOTOPE INSTRUMENTATION   363

14.1. Generation of Ionizing Radiation,   364
14.2. Instrumentation for Diagnostic X Rays,   369
14.3. Special Techniques,   374
14.4. Instrumentation for the Medical Use of Radioisotopes,   376
14.5. Radiation Therapy,   383

15. THE COMPUTER IN BIOMEDICAL INSTRUMENTATION   384

15.1. The Digital Computer,   386
15.2. Microprocessors,   398
15.3. Interfacing the Computer with Medical Instrumentation
          and Other Equipment,   401
15.4. Biomedical Computer Applications,   409

16. ELECTRICAL SAFETY OF MEDICAL EQUIPMENT   430

16.1. Physiological Effects of Electrical Current,   431
16.2. Shock Hazards from Electrical Equipment,   437
16.3. Methods of Accident Prevention,   439

APPENDICES   449

A. MEDICAL TERMINOLOGY AND GLOSSARY   451

B. PHYSIOLOGICAL MEASUREMENTS SUMMARY   462

C. SI METRIC UNITS AND EQUIVALENCIES   467

D. PROBLEMS AND EXERCISES   468

INDEX   493

Preface
to the
First Edition

As the world's population grows, the need for health care increases. In recent years progress in medical care has been rapid, especially in such fields as neurology and cardiology. A major reason for this progress has been the marriage of two important disciplines: medicine and engineering.

There are similarities between these two disciplines and there are differences, but there is no doubt that cooperation between them has produced excellent results. This fact can be well attested to by the man or woman who has received many more years of useful life because of the help of a prosthetic device, or from careful and meaningful monitoring during a critical illness.

The disciplines of medicine and engineering are both broad. They encompass people engaged in a wide spectrum of activities from the basic maintenance of either the body, or a piece of equipment, to research on the frontiers of knowledge in each field. There is one obvious common denominator: the need for instrumentation to make proper and accurate measurements of the parameters involved.

Personnel involved in the design, use, and maintenance of biomedical instrumentation come from either the life sciences or from engineering and technology, although most probably from the latter areas. Training in the life sciences includes physiology and anatomy, with little circuitry, electronics, or instrumentation. For the engineer or electronics technician the reverse is true, and anything but a meager knowledge of physiology is usually lacking on the biomedical side.

Unfortunately for those entering this new field, it is still very young and few reference books are available. This book has been written to help fill the gap. It has grown out of notes prepared by the various authors as reference material for educational courses. These courses have been presented at many levels in both colleges and hospitals. The participants have included engineers, technicians, doctors, dentists, nurses, psychologists, and many others covering a multitude of professions.

This book is primarily intended for the reader with a technical background in electronics or engineering, but with not much more than a casual familiarity with physiology. It is broad in its scope, however, covering a major portion of what is known as the field of biomedical instrumentation. There is depth where needed, but, in general, it is not intended to be too sophisticated. The authors believe in a down-to-earth approach. There are ample illustrations and references to easily accessible literature where more specialization is required. The presentation is such that persons in the life sciences with some knowledge of instrumentation should have little difficulty using it.

The introductory material is concerned with giving the reader a perspective of the field and a feeling for the subject matter. It also introduces the concept of the man-instrument system and the problems encountered in attempting to obtain measurements from a living body. An overall view of the physiological systems of the body is presented and then is later reinforced by more detailed explanations in appropriate parts of the text. The physiological material is presented in a language that should be readily understood by the technically trained person, even to the extent of using an engineering-type analysis. Medical terminology is introduced early, for one of the problems encountered in the field of biomedical instrumentation is communication between the doctor and the engineer or technician. Variables that are meaningful in describing the body system are discussed, together with the type of difficulties that may be anticipated.

It should also be noted that although reference works on physiology are included for those needing further study, enough fundamentals are presented within the context of this book to make it reasonably self-sufficient.

All measurements depend essentially on the detection, acquisition, display, and quantification of signals. The body itself provides a source of many types of signals. Some of these types—the bioelectric potentials responsible for the electrocardiogram, the electroencephalogram, and the electromyogram—are discussed in Chapter 3. In later chapters the measurement of each of these forms of biopotentials is discussed. One chapter is devoted to electrodes—the transducers for the biopotential signals.

With regard to the major physiological systems of the body, each segment is considered as a unit but often relies on material presented in the earlier chapters. The physiology of each system is first discussed in general, followed by an analysis of those parameters that have clinical importance. The fundamental principles and methods of measurement are discussed, with descriptions of principles of equipment actually in use today. This is done in turn for the cardiovascular, respiratory, and nervous systems. There are certain physical measurements that do not belong to any specific system but could relate to any or all of them. These physical variables, including temperature, displacement, force, velocity and acceleration, are covered in Chapter 9.

One of the novel ideas developed in this book is the fact that, together with a discussion of the nervous system, behavioral measurements are covered as well as the interaction between psychology and physiology.

The latter chapters are devoted to special topics to give the reader a true overall view of the field. Such topics include the use of remote monitoring by radio techniques commonly known as biotelemetry; radiation techniques, including X rays and radioisotopes; the clinical laboratory; and the digital computer as it applies to the medical profession, since this is becoming a widely used tool.

The final chapter is one of the most important. Electrical safety in the hospital and clinic is of vital concern. The whole field becomes of no avail unless this topic is understood.

For a quick reference, a group of appendices are devoted to medical terminology, an alphabetical glossary, a summary of physiological measurements, and some typical values.

The book has been prepared for a multiplicity of needs. The level is such that it could be used by those taking bioinstrumentation in a community college program, but the scope is such that it can also serve as a text for an introductory course for biomedical engineering students. It should also prove useful as a reference for medical and paramedical personnel with some knowledge of instruments who need to know more.

The background material was developed by the authors in courses presented at California State University, Los Angeles, and at various centers and hospitals of the United States Veterans Administration.

In a work of this nature, it is essential to illustrate commercial systems in common use. In many examples there are many manufacturers who produce similar equipment, and it is difficult to decide which to use as illustrative material. All companies have been most cooperative, and we apologize for the fact that it is not possible always to illustrate alternate examples. The authors wish to thank all the companies that were willing to supply illustrative material, as well as the authors of other textbooks for some borrowed descriptions and drawings. All of these are acknowledged in the text at appropriate places.

The authors wish to thank Mrs. Irina Cromwell, Mrs. Elissa J. Schrader, and Mrs. Erna Wellenstein for their assistance in typing the manuscript; Mr. Edward Francis, Miss Penelope Linskey, and the Prentice-Hall Company for their help, encouragement, and cooperation; and Mr. Joseph A. Labok, Jr., for his efforts in encouraging us to write this book.

Los Angeles, California   LESLIE CROMWELL
FRED J. WEIBELL
ERICH A. PFEIFFER
LEO B. USSELMAN

Preface
to the
Second Edition

It is extremely gratifying to write a book in a relatively new field and achieve the wide acceptance enjoyed by the first edition of Biomedical Instrumentation and Measurements. Many major Biomedical Engineering and BMET programs have adopted the book over the last six years both in the U.S.A. and abroad. The reviews by our professional colleagues have also has been most encouraging and the remarks we have had from them and from students with respect to the "readability" have been a stimulant to the authors.

However, this field is dynamic and has progressed tremendously since the book was written. When the authors and publishers decided that a second edition should be prepared much soul searching was necessary to decide on what changes should be made to improve the work. Obviously everything had to be updated. Fortunately the original edition was written in "building block" style so that this could be achieved with relative ease. Also there were the constructive criticisms of our colleagues around the world to consider.

Perhaps the three major impacts on biomedical engineering in recent years are the tremendous expansion of non-invasive techniques, the sophistication built up in special care units and, along with other fields, the greater use of computers and the advent of microprocessors.

Taking all these facts together, the authors re-studied the book and the field and decided on the direction for the new edition. With respect to criticism, it was obvious, even after early adoptions, that the concept and principles of transducers should be presented earlier in the work. The original Chapters 1 and 2 were combined into a new introductory chapter and a new Chapter 2 was written on basic transducers, including some material drawn from the old Chapter 9. Chapter 9 was transformed into a new chapter on non-invasive techniques with the major emphasis on ultrasonics, a field that has developed greatly in recent years. However, some non-invasive techniques not covered in Chapter 9 are more appropriately included in other chapters.

Most of the material on physiology and basic principles has not been changed much, but the illustrative chapters contain many changes. Cardiovascular techniques have progressed considerably as reflected in the changes in Chapter 6. Because intensive care equipment and computers have also changed, Chapters 7 and 15 were virtually re-written. New topics such as echocardiography and computerized axial tomography have been added. Over two thirds of the illustrative photographs are new to reflect the many changes in the field.

This book now includes SI (Systeme Internationale) metric units, although other measurement units have been retained for comparison. Parentheses have been used where two sets of units are mentioned. However, it should be pointed out that the transition to SI metric units in the health care field is far from complete. Whereas some changes, such as linear measurements in centimeters, are now widely accepted others are not. Kilopascals is rarely used as a measurement for blood pressure, mmHg still being preferred. (1 mmHg is equal to 0.133 kilopascals.)

The authors again wish to thank the many manufacturers for their help and photographs, and the hospitals and physicians for their cooperation. These are individually acknowledged in the text. The authors also wish to thank Mrs. Irina Cromwell for typing and assembling the manuscript and Mrs. Elissa Schrader and Mrs. Erna Wellenstein for helping again in many ways in the preparation of this book. We are also indebted to California State University, Los Angeles, and the U.S. Veterans Administration for encouragement and use of facilities.

Los Angeles, California

LESLIE CROMWELL
FRED J. WEIBELL
ERICH A. PFEIFFER

Biomedical
Instrumentation
and Measurements

1

Introduction to Biomedical Instrumentation

Science has progressed through many gradual states. It is a long time since Archimedes and his Greek contemporaries started down the path of scientific discoveries, but a technological historian could easily trace the trends through the centuries. Engineering has emerged out of the roots of science, and since the Industrial Revolution the profession has grown rapidly. Again, there are definite stages that can be traced.

1.1. THE AGE OF BIOMEDICAL ENGINEERING

It is common practice to refer to developmental time eras as "ages." The age of the steam engine, of the automobile, and of radio communication each spanned a decade or so of rapid development. Since World War II there have been a number of overlapping technological ages. Nuclear engineering and aerospace engineering are good examples. Each of these fields reached a peak of activity and then settled down to a routine, orderly progression. The age of computer engineering, with all its ramifications, has been developing rapidly and still has much momentum. The time for the age of biomedical engineering has now arrived.

The probability is great that the 1970s will be known as the decade in which the most rapid progress was made in this highly important field. There is one vital advantage that biomedical engineering has over many of the other fields that preceded it: the fact that it is aimed at keeping people healthy and helping to cure them when they are ill. Thus, it may escape many of the criticisms aimed at progress and technology. Many purists have stated that technology is an evil. Admittedly, although the industrial age introduced many new comforts, conveniences, and methods of transportation, it also generated many problems. These problems include air and water pollution, death by transportation accidents, and the production of such weapons of destruction as guided missiles and nuclear bombs. However, even though biomedical engineering is not apt to be criticized as much for producing evils, some new problems have been created, such as shock hazards in the use of electrical instruments in the hospital. Yet these side effects are minor compared to the benefits that mankind can derive from it.

One of the problems of "biomedical engineering" is defining it. The prefix *bio-*, of course, denotes something connected with life. Biophysics and biochemistry are relatively old "interdisciplines," in which basic sciences have been applied to living things. One school of thought subdivides bioengineering into different engineering areas—for example, biomechanics and bioelectronics. These categories usually indicate the use of that area of engineering applied to living rather than to physical components. *Bioinstrumentation* implies measurement of biological variables, and this field of measurement is often referred to as *biometrics,* although the latter term is also used for mathematical and statistical methods applied to biology.

[Naturally committees have been formed to define these terms; the professional societies have become involved.] The latter include the IEEE Engineering in Medicine and Biology Group, the ASME Biomechanical and Human Factors Division, the Instrument Society of America, and the American Institute of Aeronautics and Astronautics. Many new "cross-disciplinary" societies have also been formed.

A few years ago an engineering committee was formed to define bioengineering. This was Subcommittee B (Instrumentation) of the Engineers Joint Council Committee on Engineering Interactions with Biology and Medicine. Their recommendation was that *bioengineering* be defined as application of the knowledge gained by a cross fertilization of engineering and the biological sciences so that both will be more fully utilized for the benefit of man.

More recently, as new applications have emerged, the field has produced definitions describing the personnel who work in it. A tendency has arisen to define the *biomedical engineer* as a person working in research or development in the interface area of medicine and engineering, whereas the practitioner working with physicians and patients is called a *clinical engineer.*

One of the societies that has emerged in this interface area is the Association for the Advancement of Medical Instrumentation (AAMI). This association consists of both engineers and physicians. In late 1974, they developed a definition that is widely accepted:

> A *clinical engineer* is a professional who brings to health care facilities a level of education, experience, and accomplishment which will enable him to responsibly, effectively, and safely manage and interface with medical devices, instruments, and systems and the use thereof during patient care, and who can, because of this level of competence, responsibly and directly serve the patient and physician, nurse, and other health care professionals relative to their use of and other contact with medical instrumentation.

Most clinical engineers go into the profession through the engineering degree route, but many start out as physicists or physiologists. They must have at least a B.S. degree, and many of them have M.S. or Ph.D. degrees.

Another new term, also coined in recent years, the "biomedical equipment technician" (BMET), is defined as follows:

> A *biomedical equipment technician* (BMET) is an individual who is knowledgeable about the theory of operation, the underlying physiologic principles, and the practical, safe clinical application of biomedical equipment. His capabilities may include installation, calibration, inspection, preventive maintenance and repair of general biomedical and related technical equipment as well as operation or supervision of equipment control, safety and maintenance programs and systems.

This was also an AAMI definition. Typically, the BMET has two years of training at a community college. This person is not to be confused with the *medical technologist.* The latter is usually used in an operative sense, for example in blood chemistry and in the taking of electrocardiograms. The level of sophistication of the BMET is usually higher than that of the technologist in terms of equipment, but lower in terms of the life sciences.

In addition, other titles have been used, such as *hospital engineer* and *medical engineer.* In one hospital the title *biophysicist* is preferred for their biomedical engineers, for reasons best known to themselves.

It is now possible to become professionally registered as a clinical engineer, but unfortunately there are two different agencies who certify clinical engineers, and the requirements have not been standardized. Some states are also considering adding "biomedical" to their professional engineering registration.

These definitions are all noteworthy, but whatever the name, this age of the marriage of engineering to medicine and biology is destined to benefit all concerned. Improved communication among engineers, technicians, and doctors, better and more accurate instrumentation to measure vital physiological parameters, and the development of interdisciplinary tools to help fight the effects of body malfunctions and diseases are all a part of this new field. Remembering that Shakespeare once wrote "A rose by any other name . . .," it must be realized that the name is actually not too important; however, what the field can accomplish *is* important. With this point in mind, the authors of this book use the term *biomedical engineering* for the field in general and the term *biomedical instrumentation* for the methods of measurement within the field.

Another major problem of biomedical engineering involves communication between the engineer and the medical profession. The language and jargon of the physician are quite different from those of the engineer. In some cases, the same word is used by both disciplines, but with entirely different meanings. Although it is important for the physician to understand enough engineering terminology to allow him to discuss problems with the engineer, the burden of bridging the communication gap usually falls on the latter. The result is that the engineer, or technician, must learn the doctor's language, as well as some anatomy and physiology, in order that the two disciplines can work effectively together.

To help acquaint the reader with this special aspect of biomedical engineering, a basic introduction to medical terminology is presented in Appendix A. This appendix is in two parts: Appendix A.1 is a list of the more common roots, prefixes, and suffixes used in the language of medicine, and Appendix A.2 is a glossary of some of the medical terms frequently encountered in biomedical instrumentation.

In addition to the language problem, other differences may affect communication between the engineer or technician and the doctor. Since the physician is often self-employed, whereas the engineer is usually salaried, a different concept of the fiscal approach exists. Thus, some physicians are reluctant to consider engineers as professionals and would tend to place them in a subservient position rather than class them as equals. Also, engineers, who are accustomed to precise quantitative measurements based on theoretical principles, may find it difficult to accept the often imprecise, empirical, and qualitative methods employed by their counterparts.

Since the development and use of biomedical instrumentation must be a joint effort of the engineer or technician and the physician (or nurse), every effort must be exerted to avoid or overcome these "communication" problems. By being aware of their possible existence, the engineer or technician can take steps to avert these pitfalls by adequate preparation and care in establishing his relationship with the medical profession.

1.2. DEVELOPMENT OF BIOMEDICAL INSTRUMENTATION

The field of medical instrumentation is by no means new. Many instruments were developed as early as the nineteenth century—for example, the electrocardiograph, first used by Einthoven at the end of that century. Progress was rather slow until after World War II, when a surplus of electronic equipment, such as amplifiers and recorders, became available. At that time many technicians and engineers, both within industry and on their own, started to experiment with and modify existing equipment for medical use. This process occurred primarily during the 1950s and the results were often disappointing, for the experimenters soon learned that physiological parameters are not measured in the same way as physical parameters. They also encountered a severe communication problem with the medical profession.

During the next decade many instrument manufacturers entered the field of medical instrumentation, but development costs were high and the medical profession and hospital staffs were suspicious of the new equipment and often uncooperative. Many developments with excellent potential seemed to have become lost causes. It was during this period that some progressive companies decided that rather than modify existing hardware, they would design instrumentation specifically for medical use. Although it is true that many of the same components were used, the philosophy was changed; equipment analysis and design were applied directly to medical problems.

A large measure of help was provided by the U.S. government, in particular by NASA (National Aeronautics and Space Administration). The Mercury, Gemini, and Apollo programs needed accurate physiological monitoring for the astronauts; consequently, much research and development money went into this area. The aerospace medicine programs were expanded considerably, both within NASA facilities, and through grants to universities and hospital research units. Some of the concepts and features of patient-monitoring systems presently used in hospitals throughout the world evolved from the base of astronaut monitoring. The use of adjunct fields, such as biotelemetry, also finds some basis in the NASA programs.

Also, in the 1960s, an awareness of the need for engineers and technicians to work with the medical profession developed. All the major engineering technical societies recognized this need by forming "Engineering in Medicine and Biology" subgroups, and new societies were organized.* Along with the medical research programs at the universities, a need developed for courses and curricula in biomedical engineering, and today almost every major university or college has some type of biomedical engineering program. However, much of this effort is not concerned with biomedical instrumentation per se.

1.3. BIOMETRICS

The branch of science that includes the measurement of physiological variables and parameters is known as *biometrics.* Biomedical instrumentation provides the tools by which these measurements can be achieved.

In later chapters each of the major forms of biomedical instrumentation is covered in detail, along with the physiological basis for the measurements involved. The physiological measurements themselves are summarized in Appendix B, which also includes such information as amplitude and frequency range where applicable.

Some forms of biomedical instrumentation are unique to the field of medicine but many are adaptations of widely used physical measurements. A thermistor, for example, changes its electrical resistance with temperature, regardless of whether the temperature is that of an engine or the human body. The principles are the same. Only the shape and size of the device might be different. Another example is the strain gage, which is commonly used to measure the stress in structural components. It operates on the principle that electrical resistance is changed by the stretching of a wire or a piece of semiconductor material. When suitably excited by a source of constant voltage, an electrical output can be obtained that is proportional to the amount of the strain. Since pressure can be translated into strain by various means, blood pressure can be measured by an adaptation of this device. When the transducer is connected into a typical circuit, such as a bridge configuration, and this circuit is excited from a source of constant input voltage, the changes in resistance are reflected in the output as voltage changes. For a thermistor, the temperature is indicated on a voltmeter calibrated in degrees Celsius or Fahrenheit.

In the design or specification of medical instrumentation systems, each of the following factors should be considered.

*An example is the Biomedical Engineering Society.

1.3.1 Range

The *range* of an instrument is generally considered to include all the levels of input amplitude and frequency over which the device is expected to operate. The objective should be to provide an instrument that will give a usable reading from the smallest expected value of the variable or parameter being measured to the largest.

1.3.2. Sensitivity

The *sensitivity* of an instrument determines how small a variation of a variable or parameter can be reliably measured. This factor differs from the instrument's range in that sensitivity is not concerned with the absolute levels of the parameter but rather with the minute changes that can be detected. The sensitivity directly determines the *resolution* of the device, which is the minimum variation that can accurately be read. Too high a sensitivity often results in nonlinearities or instability.Thus, the optimum sensitivity must be determined for any given type of measurement. Indications of sensitivity are frequently expressed in terms of scale length per quantity to be measured—for example, inches per microampere in a galvanometer coil or inches per millimeter of mercury.* These units are sometimes expressed reciprocally. A sensitivity of 0.025 centimeter per millimeter of mercury (cm/mm Hg) could be expressed as 40 millimeters of mercury per centimeter.*

1.3.3. Linearity

The degree to which variations in the output of an instrument follow input variations is referred to as the *linearity* of the device. In a *linear system* the sensitivity would be the same for all absolute levels of input, whether in the high, middle, or low portion of the range. In some instruments a certain form of nonlinearity is purposely introduced to create a desired effect, whereas in others it is desirable to have linear scales as much as possible over the entire range of measurements. Linearity should be obtained over the most important segments, even if it is impossible to achieve it over the entire range.

1.3.4. Hysteresis

Hysteresis (from the Greek, *hysterein,* meaning "to be behind" or "to lag") is a characteristic of some instruments whereby a given value of the measured variable results in a different reading when reached in an ascend-

*1 mm Hg is 133.3 pascals in the SI metric system; 1 mm Hg is also equivalent to 1 torr.

ing direction from that obtained when it is reached in a descending direction. Mechanical friction in a meter, for example, can cause the movement of the indicating needle to lag behind corresponding changes in the measured variable, thus resulting in a hysteresis error in the reading.

1.3.5. Frequency Response

The *frequency response* of an instrument is its variation in sensitivity over the frequency range of the measurement. It is important to display a waveshape that is a faithful reproduction of the original physiological signal. An instrument system should be able to respond rapidly enough to reproduce all frequency components of the waveform with equal sensitivity. This condition is referred to as a "flat response" over a given range of frequencies.

1.3.6. Accuracy

Accuracy is a measure of systemic error. Errors can occur in a multitude of ways. Although not always present simultaneously, the following errors should be considered:

1. Errors due to tolerances of electronic components.
2. Mechanical errors in meter movements.
3. Component errors due to drift or temperature variation.
4. Errors due to poor frequency response.
5. In certain types of instruments, errors due to change in atmospheric pressure or temperature.
6. Reading errors due to parallax, inadequate illumination, or excessively wide ink traces on a pen recording.

Two additional sources of error should not be overlooked. The first concerns correct instrument zeroing. In most measurements, a zero, or a baseline, is necessary. It is often achieved by balancing the Wheatstone bridge or a similar device. It is very important that, where needed, balancing or zeroing is done prior to each set of measurements. Another source of error is the effect of the instrument on the parameter to be measured, and vice versa. This is especially true in measurements in living organisms and is further discussed later in this chapter.

1.3.7. Signal-to-Noise Ratio

It is important that the *signal-to-noise ratio* be as high as possible. In the hospital environment, power-line frequency noise or interference is common and is usually picked up in long leads. Also, interference due to electromagnetic, electrostatic, or diathermy equipment is possible. Poor grounding is often a cause of this kind of noise problem.

Such "interference noise," however, which is due to coupling from other energy sources, should be differentiated from thermal and shot noise, which originate within the elements of the circuit itself because of the discontinuous nature of matter and electrical current. Although thermal noise is often the limiting factor in the detection of signals in other fields of electronics, interference noise is usually more of a problem in biomedical systems.

It is also important to know and control the signal-to-noise ratio in the actual environment in which the measurements are to be made.

1.3.8. Stability

In control engineering, *stability* is the ability of a system to resume a steady-state condition following a disturbance at the input rather than be driven into uncontrollable oscillation. This is a factor that varies with the amount of amplification, feedback, and other features of the system. The overall system must be sufficiently stable over the useful range. *Baseline stability* is the maintenance of a constant baseline value without drift.

1.3.9. Isolation

Often measurements must be made on patients or experimental animals in such a way that the instrument does not produce a direct electrical connection between the subject and ground. This requirement is often necessary for reasons of electrical safety (see Chapter 16) or to avoid interference between different instruments used simultaneously. *Electrical isolation* can be achieved by using magnetic or optical coupling techniques, or radio telemetry. Telemetry is also used where movement of the person or animal to be measured is essential, and thus the encumbrance of connecting leads should be avoided (see Chapter 12).

1.3.10. Simplicity

All systems and instruments should be as simple as possible to eliminate the chance of component or human error.

Most instrumentation systems require calibration before they are actually used. Each component of a measurement system is usually calibrated individually at the factory against a standard. When a medical system is assembled, it should be calibrated as a whole. This step can be done external to the living organism or in situ (connected to or within the body). This point is discussed in later chapters. Calibration should always be done by using error-free devices of the simplest kind for references. An example would be that of a complicated, remote blood-pressure monitoring system, which is calibrated against a simple mercury manometer.

1.4. INTRODUCTION TO THE MAN–INSTRUMENT SYSTEM

A classic exercise in engineering analysis involves the measurement of outputs from an unknown system as they are affected by various combinations of inputs. The object is to learn the nature and characteristics of the system. This unknown system, often referred to as a *black box,* may have a variety of configurations for a given combination of inputs and outputs. The end product of such an exercise is usually a set of input–output equations intended to define the internal functions of the box. These functions may be relatively simple or extremely complex.

One of the most complex black boxes conceivable is a living organism, especially the living human being. Within this box can be found electrical, mechanical, acoustical, thermal, chemical, optical, hydraulic, pneumatic, and many other types of systems, all interacting with each other. It also contains a powerful computer, several types of communication systems, and a great variety of control systems. To further complicate the situation, upon attempting to measure the inputs and outputs, an engineer would soon learn that none of the input–output relationships is deterministic. That is, repeated application of a given set of input values will not always produce the same output values. In fact, many of the outputs seem to show a wide range of responses to a given set of inputs, depending on some seemingly relevant conditions, whereas others appear to be completely random and totally unrelated to any of the inputs.

The living black box presents other problems, too. Many of the important variables to be measured are not readily accessible to measuring devices. The result is that some key relationships cannot be determined or that less accurate substitute measures must be used. Furthermore, there is a high degree of interaction among the variables in this box. Thus, it is often impossible to hold one variable constant while measuring the relationship between two others. In fact, it is sometimes difficult to determine which are the inputs and which are the outputs, for they are never labeled and almost inevitably include one or more feedback paths. The situation is made even worse by the application of the measuring device itself, which often affects the measurements to the extent that they may not represent normal conditions reliably.

At first glance an assignment to measure and analyze the variables in a living black box would probably be labeled ''impossible'' by most engineers; yet this is the very problem facing those in the medical field who attempt to measure and understand the internal relationships of the human body. The function of medical instrumentation is to aid the medical clinician and researcher in devising ways of obtaining reliable and meaningful measurements from a living human being.

Still other problems are associated with such measurements: the process of measuring must not in any way endanger the life of the person on whom the measurements are being made, and it should not require the subject to endure undue pain, discomfort, or any other undesirable conditions. This means that many of the measurement techniques normally employed in the instrumentation of nonliving systems cannot be applied in the instrumentation of humans.

Additional factors that add to the difficulty of obtaining valid measurements are (1) safety considerations, (2) the environment of the hospital in which these measurements are performed, (3) the medical personnel usually involved in the measurements, and (4) occasionally even ethical and legal considerations.

Because special problems are encountered in obtaining data from living organisms, especially human beings, and because of the large amount of interaction between the instrumentation system and the subject being measured, it is essential that the person on whom measurements are made be considered an integral part of the instrumentation system. In other words, in order to make sense out of the data to be obtained from the black box (the human organism), the internal characteristics of the black box must be considered in the design and application of any measuring instruments. Consequently, the overall system, which includes both the human organism and the intrumentation required for measurement of the human is called the *man–instrument system.*

An instrumentation system is defined as the set of instruments and equipment utilized in the measurement of one or more characteristics or phenomena, plus the presentation of information obtained from those measurements in a form that can be read and interpreted by man. In some cases. the instrumentation system includes components that provide a stimulus or drive to one or more of the inputs to the device being measured. There may also be some mechanism for automatic control of certain processes within the system, or of the entire system. As indicated earlier, the complete man–instrument system must also include the human subject on whom the measurements are being made.

The basic objectives of any instrumentation system generally fall into one of the following major categories:

1. ***Information gathering:*** In an information-gathering system, instrumentation is used to measure natural phenomena and other

variables to aid man in his quest for knowledge about himself and the universe in which he lives. In this setting, the characteristics of the measurements may not be known in advance.

2. *Diagnosis:* Measurements are made to help in the detection and, hopefully, the correction of some malfunction of the system being measured. In some applications, this type of instrumentation may be classed as "troubleshooting equipment."

3. *Evaluation:* Measurements are used to determine the ability of a system to meet its functional requirements. These could be classified as "proof-of-performance" or "quality control" tests.

4. *Monitoring:* Instrumentation is used to monitor some process or operation in order to obtain continuous or periodic information about the state of the system being measured.

5. *Control:* Instrumentation is sometimes used to automatically control the operation of a system based on changes in one or more of the internal parameters or in the output of the system.

The general field of biomedical instrumentation involves, to some extent, all the preceding objectives of the general instrumentation system. Instrumentation for biomedical research can generally be viewed as information-gathering instrumentation, although it sometimes includes some monitoring and control devices. Instrumentation to aid the physician in the diagnosis of disease and other disorders also has widespread use. Similar instrumentation is used in evaluation of the physical condition of patients in routine physical examinations. Also, special instrumentation systems are used for monitoring of patients undergoing surgery or under intensive care.

Biomedical instrumentation can generally be classified into two major types: clinical and research. *Clinical instrumentation* is basically devoted to the diagnosis, care, and treatment of patients, whereas *research instrumentation* is used primarily in the search for new knowledge pertaining to the various systems that compose the human organism. Although some instruments can be used in both areas, clinical instruments are generally designed to be more rugged and easier to use. Emphasis is placed on obtaining a limited set of reliable measurements from a large group of patients and on providing the physician with enough information to permit him to make clinical decisions. On the other hand, research instrumentation is normally more complex, more specialized, and often designed to provide a much higher degree of accuracy, resolution, and so on. Clinical instruments are used by the physician or nurse, whereas research instruments are generally operated by skilled technologists whose primary training is in the operation of such instruments. The concept of the man–instrument system applies to both clinical and research instrumentation.

Measurements in which biomedical instrumentation is employed can also be divided into two categories: in vivo and in vitro. An *in vivo* measurement is one that is made on or within the living organism itself. An example would be a device inserted into the bloodstream to measure the pH of the blood directly. An *in vitro* measurement is one performed outside the body, even though it relates to the functions of the body. An example of an in vitro measurement would be the measurement of the pH of a sample of blood that has been drawn from a patient. Literally, the term in vitro means "in glass," thus implying that in vitro measurements are usually performed in test tubes. Although the man–instrument system described here applies mainly to in vivo measurements, problems are often encountered in obtaining appropriate samples for in vitro measurements and in relating these measurements to the living human being.

1.5. COMPONENTS OF THE MAN-INSTRUMENT SYSTEM

A block diagram of the man–instrument system is shown in Figure 1.1. The basic components of this system are essentially the same as in any instrumentation system. The only real difference is in having a living human being as the subject. The system components are given below.

1.5.1. The Subject

The *subject* is the human being on whom the measurements are made. Since it is the subject who makes this system different from other instrumentation systems, the major physiological systems that constitute the human body are treated in much greater detail in Section 1.6.

1.5.2 Stimulus

In many measurements, the response to some form of external *stimulus* is required. The instrumentation used to generate and present this stimulus to the subject is a vital part of the man–instrument system whenever responses are measured. The stimulus may be visual (e.g., a flash of light), auditory (e.g., a tone), tactile (e.g., a blow to the Achilles tendon), or direct electrical stimulation of some part of the nervous system.

1.5.3. The Transducer

In general, a *transducer* is defined as a device capable of converting one form of energy or signal to another. In the man–instrument system, each

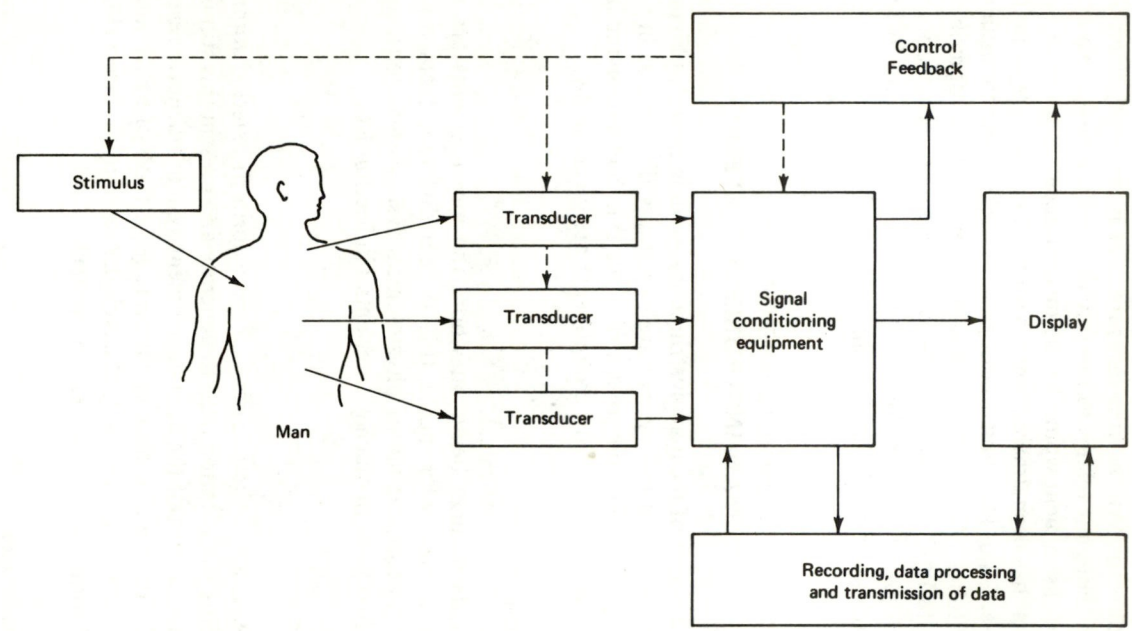

14

Figure 1.1. Block diagram—
the man-instrument system.

transducer is used to produce an electric signal that is an analog of the phenomenon being measured. The transducer may measure temperature, pressure, flow, or any of the other variables that can be found in the body, but its output is always an electric signal. As indicated in Figure 1.1, two or more transducers may be used simultaneously to obtain relative variations between phenomena.

1.5.4. Signal-Conditioning Equipment

The part of the instrumentation system that amplifies, modifies, or in any other way changes the electric output of the transducer is called *signal-conditioning* (or sometimes *signal-processing*) equipment. Signal-conditioning equipment is also used to combine or relate the outputs of two or more transducers. Thus, for each item of signal-conditioning equipment, both the input and the output are electric signals, although the output signal is often greatly modified with respect to the input. In essence, then, the purpose of the signal-conditioning equipment is to process the signals from the transducers in order to satisfy the functions of the system and to prepare signals suitable for operating the display or recording equipment that follows.

1.5.5. Display Equipment

To be meaningful, the electrical output of the signal-conditioning equipment must be converted into a form that can be perceived by one of man's senses and that can convey the information obtained by the measurement in a meaningful way. The input to the *display* device is the modified electric signal from the signal-conditioning equipment. Its output is some form of visual, audible, or possibly tactile information. In the man–instrumentation system, the display equipment may include a graphic pen recorder that produces a permanent record of the data.

1.5.6. Recording, Data-Processing, and Transmission Equipment

It is often necessary, or at least desirable, to *record* the measured information for possible later use or to *transmit* it from one location to another, whether across the hall of the hospital or halfway around the world. Equipment for these functions is often a vital part of the man–instrument system. Also, where *automatic storage* or *processing* of data is required, or where computer control is employed, an on-line analog or digital computer may be part of the instrumentation system. It should be noted that the term *recorder* is used in two different contexts in biomedical instrumentation. A graphic pen recorder is actually a display device used to produce a paper record of analog waveforms, whereas the recording equipment referred to in this paragraph includes devices by which data can be recorded for future playback, as in a magnetic tape recorder.

1.5.7. Control Devices

Where it is necessary or desirable to have *automatic control* of the stimulus, transducers, or any other part of the man–instrument system, a control system is incorporated. This system usually consists of a feedback loop in which part of the output from the signal-conditioning or display equipment is used to control the operation of the system in some way.

1.6.   PHYSIOLOGICAL SYSTEMS OF THE BODY

From the previous sections it should be evident that, to obtain valid measurements from a living human being, it is necessary to have some understanding of the subject on which the measurements are being made. Within the human body can be found electrical, mechanical, thermal, hydraulic, pneumatic, chemical, and various other types of systems, each of which communicates with an external environment, and internally with the other systems of the body. By means of a multilevel control system and communications network, these individual systems are organized to perform many complex functions. Through the integrated operation of all these systems, and their various subsystems, man is able to sustain life, learn to perform useful tasks, acquire personality and behavioral traits, and even reproduce himself.

Measurements can be made at various levels of man's hierarchy of organization. For example, the human being as a whole (the highest level of organization) communicates with his environment in many ways. These methods of communicating could be regarded as the inputs and outputs of the black box and are illustrated in Figure 1.2. In addition, these various inputs and outputs can be measured and analyzed in a variety of ways. Most are readily accessible for measurement, but some, such as speech, behavior, and appearance, are difficult to analyze and interpret.

Next to the whole being in the hierarchy of organization are the major functional systems of the body, including the nervous system, the cardiovascular system, the pulmonary system, and so on. Each major system is discussed later in this chapter, and most are covered in greater detail in later chapters. Just as the whole person communicates with his environment, these major systems communicate with each other as well as with the external environment.

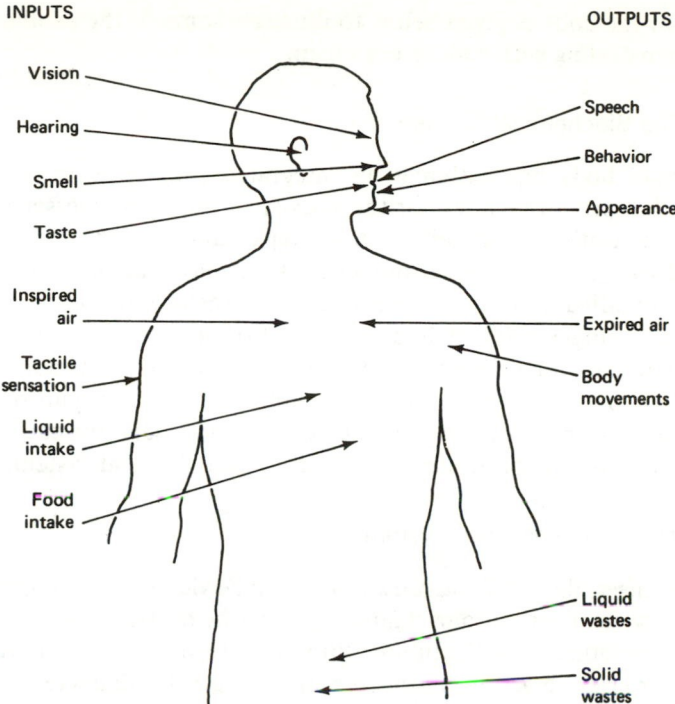

Figure 1.2. Communication of man with his environment.

These functional systems can be broken down into subsystems and organs, which can be further subdivided into smaller and smaller units. The process can continue down to the cellular level and perhaps even to the molecular level. The major goal of biomedical instrumentation is to make possible the measurement of information communicated by these various elements. If all the variables at all levels of the organization heirarchy could be measured, and all their interrelationships determined, the functions of the mind and body of man would be much more clearly understood and could probably be completely defined by presently known laws of physics, chemistry, and other sciences. The problem is, of course, that many of the inputs at the various organizational levels are not accessible for measurement. The interrelationships among elements are sometimes so complex and involve so many systems that the "laws" and relationships thus far derived are inadequate to define them completely. Thus, the models in use today contain so many assumptions and constraints that their application is often severely limited.

Although each of the systems is treated in much more detail in later chapters, a brief engineering-oriented description of the major physiological systems of the body is given below to illustrate some of the problems to be expected in dealing with a living organism.

1.6.1. The Biochemical System

The human body has within it an integrated conglomerate of *chemical systems* that produce energy for the activity of the body, messenger agents for communication, materials for body repair and growth, and substances required to carry out the various body functions. All operations of this highly diversified and very efficient chemical factory are self-contained in that from a single point of intake for fuel (food), water, and air, all the source materials for numerous chemical reactions are produced within the body. Moreover, the chemical factory contains all the monitoring equipment needed to provide the degree of control necessary for each chemical operation, and it incorporates an efficient waste disposal system.

1.6.2. The Cardiovascular System

To an engineer, the *cardiovascular system* can be viewed as a complex, closed hydraulic system with a four-chamber pump (the heart), connected to flexible and sometimes elastic tubing (blood vessels). In some parts of the system (arteries, arterioles), the tubing changes its diameter to control pressure. Reservoirs in the system (veins) change their volume and characteristics to satisfy certain control requirements, and a system of gates and variable hydraulic resistances (vasoconstrictors, vasodilators) continually alters the pattern of fluid flow. The four-chamber pump acts as two synchronized but functionally isolated two-stage pumps. The first stage of each pump (the atrium) collects fluid (blood) from the system and pumps it into the second stage (the ventricle). The action of the second stage is so timed that the fluid is pumped into the system immediately after it has been received from the first stage. One of the two-stage pumps (right side of the heart) collects fluid from the main hydraulic system (systemic circulation) and pumps it through an oxygenation system (the lungs). The other pump (left side of the heart) receives fluid (blood) from the oxygenation system and pumps it into the main hydraulic system. The speed of the pump (heart rate) and its efficiency (stroke volume) are constantly changed to meet the overall requirements of the system. The fluid (blood), which flows in a laminar fashion, acts as a communication and supply network for all parts of the system. Carriers (red blood cells) of fuel supplies and waste materials are transported to predetermined destinations by the fluid. The fluid also contains mechanisms for repairing small system punctures and for rejecting foreign elements from the system (platelets and white blood cells, respectively). Sensors provided to detect changes in the need for supplies, the buildup of waste materials, and out-of-tolerance pressures in the system are known as *chemoreceptors,* P_{co_2} *sensors,* and *baroreceptors,* respectively. These and other mechanisms control the pump's speed and efficiency, the fluid flow pattern through the system, tubing diameters, and other factors. Because part of the system is required to work against gravity at times, special one-way valves are provided to prevent gravity from pulling fluid against the direction of flow between pump cycles. The variables of prime importance in this system are the pump (cardiac) output and the pressure, flow rate, and volume of the fluid (blood) at various locations throughout the system.

1.6.3. The Respiratory System

Whereas the cardiovascular system is the major hydraulic system in the body, the *respiratory system* is the pneumatic system. An air pump (diaphragm), which alternately creates negative and positive pressures in a sealed chamber (thoracic cavity), causes air to be sucked into and forced out of a pair of elastic bags (lungs) located within the compartment. The bags are connected to the outside environment through a passageway (nasal cavities, pharynx, larynx, trachea, bronchi, and bronchioles), which at one point is in common with the tubing that carries liquids and solids to the stomach. A special valving arrangement interrupts the pneumatic system whenever liquid or solid matter passes through the common region. The passageway divides to carry air into each of the bags, wherein it again subdivides many times to carry air into and out of each of many tiny air spaces (pulmonary alveoli) within the bags. The dual air input to the system (nasal cavities) has an alternate vent (the mouth) for use in the event of nasal blockage and for other special purposes. In the tiny air spaces of the bags is a membrane interface with the body's hydraulic system through which certain gases can diffuse. Oxygen is taken into the fluid (blood) from the incoming air, and carbon dioxide is transferred from the fluid to the air, which is exhausted by the force of the pneumatic pump. The pump operates with a two-way override. An automatic control center (respiratory center of the brain) maintains pump operation at a speed that is adequate to supply oxygen and carry off carbon dioxide as required by the system. Manual control can take over at any time either to accelerate or to inhibit the operation of the pump. Automatic control will return, however, if a condition is created that might endanger the system. System variables of primary importance are respiratory rate, respiratory airflow, respiratory volume, and concentration of CO_2 in the expired air. This system also has a number of relatively fixed volumes and capacities, such as tidal volume (the volume inspired or expired during each normal breath), inspiratory reserve volume (the additional volume that can be inspired after a normal inspiration), expiratory reserve volume (the additional amount of air that can be forced out of the lungs after normal expiration), residual volume (amount of air remaining in the lungs after all possible air has been forced out), and vital capacity (tidal volume, plus inspiratory reserve volume, plus expiratory reserve volume).

1.6.4. The Nervous System

The nervous system is the communication network for the body. Its center is a self-adapting central information processor or computer (the brain) with memory, computational power, decision-making capability, and a myriad of input–output channels. The computer is self adapting in that if a certain section is damaged, other sections can adapt and eventually take over (at least in part) the function of the damaged section. By use of this computer, a person is able to make decisions, solve complex problems, create art, poetry, and music, "feel" emotions, integrate input information from all parts of the body, and coordinate output signals to produce meaningful behavior. Almost as fascinating as the central computer are the millions of communication lines (afferent and efferent nerves) that bring sensory information into, and transmit control information out of the brain. In general, these lines are not single long lines but often complicated networks with many interconnections that are continually changing to meet the needs of the system. By means of the interconnection patterns, signals from a large number of sensory devices, which detect light, sound, pressure, heat, cold, and certain chemicals, are channeled to the appropriate parts of the computer, where they can be acted upon. Similarly, output control signals are channeled to specific motor devices (motor units of the muscles), which respond to the signals with some type of motion or force. Feedback regarding every action controlled by the system is provided to the computer through appropriate sensors. Information is usually coded in the system by means of electrochemical pulses (nerve action potentials) that travel along the signal lines (nerves). The pulses can be transferred from one element of a network to another in one direction only, and frequently the transfer takes place only when there is the proper combination of elements acting on the next element in the chain. Action by some elements tends to inhibit transfer by making the next element less sensitive to other elements that are attempting to actuate it. Both serial and parallel coding are used, sometimes together in the same system. In addition to the central computer, a large number of simple decision-making devices (spinal reflexes) are present to control directly certain motor devices from certain sensory inputs. A number of feedback loops are accomplished by this method. In many cases, only situations where important decision making is involved require that the central computer be utilized.

1.7. PROBLEMS ENCOUNTERED IN MEASURING A LIVING SYSTEM

The previous discussions of the man–instrument system and the physiological systems of the body imply measurements on a human subject. In some cases, however, animal subjects are substituted for humans in order to permit measurements or manipulations that cannot be performed without some risk. Although ethical restrictions sometimes are not as severe with animal subjects, the same basic problems can be expected in attempting measurements from any living system. Most of these problems were introduced in earlier sections of the chapter. However, they can be summarized as follows.

1.7.1. Inaccessibility of Variables to Measurement

One of the greatest problems in attempting measurements from a living system is the difficulty in gaining access to the variable being measured. In some cases, such as in the measurement of dynamic neurochemical activity in the brain, it is impossible to place a suitable transducer in a position to make the measurement. Sometimes the problem stems from the required physical size of the transducer as compared to the space available for the measurement. In other situations the medical operation required to place a transducer in a position from which the variable can be measured makes the measurement impractical on human subjects, and sometimes even on animals. Where a variable is inaccessible for measurement, an attempt is often made to perform an indirect measurement. This process involves the measurement of some other related variable that makes possible a usable estimate of the inaccessible variable under certain conditions. In using indirect measurements, however, one must be constantly aware of the limitations of the substitute variable and must be able to determine when the relationship is not valid.

1.7.2. Variability of the Data

Few of the variables that can be measured in the human body are truly deterministic variables. In fact, such variables should be considered as stochastic processes. A *stochastic process* is a time variable related to other variables in a nondeterministic way. Physiological variables can never be viewed as strictly deterministic values but must be represented by some kind of statistical or probabilistic distribution. In other words, measurements taken under a fixed set of conditions at one time will not necessarily be the same as similar measurements made under the same conditions at another time. The variability from one subject to another is even greater. Here, again, statistical methods must be employed in order to estimate relationships among variables.

1.7.3. Lack of Knowledge About Interrelationships

The foregoing variability in measured values could be better explained if more were known and understood about the interrelationships within the body. Physiological measurements with large tolerances are often accepted by the physician because of a lack of this knowledge and the resultant inability to control variations. Better understanding of physiological relationships would also permit more effective use of indirect measurements as substitutes for inaccessible measures and would aid engineers or technicians in their job of coupling the instrumentation to the physiological system.

1.7.4. Interaction Among Physiological Systems

Because of the large number of feedback loops involved in the major physiological systems, a severe degree of interaction exists both within a given system and among the major systems. The result is that stimulation of one part of a given system generally affects all other parts of that system in some way (sometimes in an unpredictable fashion) and often affects other systems as well. For this reason, "cause-and-effect" relationships become extremely unclear and difficult to define. Even when attempts are made to open feedback loops, collateral loops appear and some aspects of the original feedback loop are still present. Also, when one organ or element is rendered inactive, another organ or element sometimes takes over the function. This situation is especially true in the brain and other parts of the nervous system.

1.7.5. Effect of the Transducer on the Measurement

Almost any kind of measurement is affected in some way by the presence of the measuring transducer. The problem is greatly compounded in the measurement of living systems. In many situations the physical presence of the transducer changes the reading significantly. For example, a large flow transducer placed in a bloodstream partially blocks the vessel and changes the pressure-flow characteristics of the system. Similarly, an attempt to measure the electrochemical potentials generated within an individual cell requires penetration of the cell by a transducer. This penetration can easily kill the cell or damage it so that it can no longer function normally. Another problem arises from the interaction discussed earlier. Often the presence of a transducer in one system can affect responses in other systems. For example, local cooling of the skin, to estimate the circulation in the area, causes feedback that changes the circulation pattern as a reaction to the cooling. The psychological effect of the measurement can also affect the results. Long-term recording techniques for measuring blood pressure have shown that some individuals who would otherwise have normal pressures show an elevated pressure reading whenever they are in the physician's office. This is a fear response on the part of the patient, involving the autonomic nervous system. In designing a measurement system, the biomedical instrumentation engineer or technician must exert extreme care to ensure that the effect of the presence of the measuring device is minimal. Because of the limited amount of energy available in the body for many physiological variables, care must also be taken to prevent the measuring system from "loading" the source of the measured variable.

1.7.6. Artifacts

In medicine and biology, the term *artifact* refers to any component of a signal that is extraneous to the variable represented by the signal. Thus, random noise generated within the measuring instrument, electrical interference (including 60-Hz pickup), cross-talk, and all other unwanted variations in the signal are considered artifacts. A major source of artifacts in the measuring of a living system is the movement of the subject, which in turn results in movement of the measuring device. Since many transducers are sensitive to movement, the movement of the subject often produces variations in the output signal. Sometimes these variations are indistinguishable from the measured variable; at other times they may be sufficient to obscure the desired information completely. Application of anesthesia to reduce movement may itself cause unwanted changes in the system.

1.7.7. Energy Limitations

Many physiological measurement techniques require that a certain amount of energy be applied to the living system in order to obtain a measurement. For example, resistance measurements require the flow of electric current through the tissues or blood being measured. Some transducers generate a small amount of heat due to the current flow. In most cases, this energy level is so low that its effect is insignificant. However, in dealing with living cells, care must continually be taken to avoid the possibility of energy concentrations that might damage cells or affect the measurements.

1.7.8. Safety Considerations

As previously mentioned, the methods employed in measuring variables in a living human subject must in no way endanger the life or normal functioning of the subject. Recent emphasis on hospital safety requires that extra caution must be taken in the design of any measurement system to protect the patient. Similarly, the measurement should not cause undue pain, trauma, or discomfort, unless it becomes necessary to endure these conditions in order to save the patient's life.

1.8. SOME CONCLUSIONS

From the foregoing discussion is should be quite obvious that obtaining data from a living system greatly increases the complexity of instrumentation problems. Fortunately, however, new developments resulting in improved, smaller, and more effective measuring devices are continually being announced, thereby making possible measurements that had previously been considered impossible. In addition, greater knowledge of the physiology of the various systems of the body is emerging as man progresses in his monumental task of learning about himself. All of this will benefit the engineer, the technician, and the physician as time goes on by adding to the tools at their disposal in overcoming instrumentation problems.

When measurements are made on human beings, one further aspect must be considered. During its earlier days of development biomedical apparatus was designed, tested, and marketed with little specific governmental control. True, there were the controls governing hospitals and a host of codes and regulations such as those described in Chapter 16, but today a number of new controls exist, some of which are quite controversial. On the other hand, there is little control on the effectiveness of devices or their side effects. Food and drugs have long been subject to governmental control by a U.S. government agency, the Food and Drug Administration (FDA). In 1976 a new addition, the Medical Devices Amendments (Public Law 94-295), placed all medical devices from the simple to the complex under the jurisdiction of the FDA. Since then, panels and committees have been formed and symposia have been held by both physicians and engineers. Regulations have been issued which include "Good Laboratory Practices" and "Good Manufacturing Practices." Although some control is essential, unfortunately many of the new regulations are tied up with so much red tape that producing new devices may be hazardous to one's economic health!

Engineers in this field should understand that they are subject to legal, moral, and ethical considerations in their practice since they deal with people's health. They should always be fully conversant with what is going on and be aware of issues and regulations that are brought about by technological, economic and political realities.

1.9. THE OBJECTIVES OF THIS BOOK

The purpose of this book is to relate specific engineering and instrumentation principles to the task of obtaining physiological data.

Each of the major body systems is discussed by presenting physiological background information. Then the variables to be measured are considered, followed by the principles of the instrumentation that could be used. Finally, applications to typical medical, behavioral, and biological use are given.

The subject matter is presented in such a way that it could be extended to classes of instruments that will be used in the future. Thus the material can be used as building blocks for the health-care instrumentation systems of tomorrow.

2

Basic Transducer Principles

A major function of medical instrumentation is the measurement of physiological variables. A *variable* is any quantity whose value changes with time. A variable associated with the physiological processes of the body is called a *physiological variable*. Examples of physiological variables used in clinical medicine are body temperature, the electrical activity of the heart (ECG), arterial blood pressure, and respiratory airflow. The physiological systems from which these variables originate were introduced in Chapter 1. The principal physiological variables and their methods of measurement are summarized in Appendix B and discussed in detail in various chapters of this book.

Physiological variables occur in many forms: as ionic potentials and currents, mechanical movements, hydraulic pressures and flows, temperature variations, chemical reactions, and many more. As stated in Chapter 1, a transducer is required to convert each variable into an electrical signal which can be amplified or otherwise processed and then converted into some form of display. Electrodes, which convert ionic potentials into electrical signals, are discussed in Chapter 4. Transducers for other types of variables are covered in this chapter. The fundamental principles involved in both active and passive transducers are presented, after which several basic types of transducers used in medical instrumentation are discussed.

2.1. THE TRANSDUCER AND TRANSDUCTION PRINCIPLES

The device that performs the conversion of one form of variable into another is called a *transducer.* In this book the primary concern is the conversion of all other forms of physiological variables into electrical signals. In this way a transducer is a component which has a nonelectrical variable as its input and an electrical signal as its output. To conduct its function properly, one (or more) parameters of the electrical output signal (say, its voltage, current, frequency, or pulse width) must be a nonambiguous function of the nonelectrical variable at the input. Ideally, the relationship between output and input should be linear with, for example, the voltage at the output of a pressure transducer being proportional to the applied pressure. A linear relationship is not always possible. For example, the relationship between input and output may follow a logarithmic function or a square law. As long as the transduction function is nonambiguous it is possible to determine the magnitude of the input variable from the electrical output signal, at least in principle. Certain other variables may interfere with the transduction process and can influence the accuracy of the measurement system, such as the hysteresis error, frequency response and baseline drift, which were discussed in Chapter 1.

Two quite different principles are involved in the process of converting nonelectrical variables into electrical signals. One of these is *energy conversion;* transducers based on this principle are called *active transducers.* The other principle involves control of an *excitation voltage* or *modulation of a carrier signal.* Transducers based on this principle are called *passive transducers.* In practical applications, the fact that a transducer is of the active or passive type is not usually significant. Occasionally, it is not even obvious to which group a transducer belongs. The two transducer types will nevertheless be described separately in the following sections.

2.2. ACTIVE TRANSDUCERS

In theory *active transducers* can utilize every known physical principle for converting nonelectrical energy. However, not all principles are of practical importance in the design of actual transducers, especially for biomedical applications. It is a characteristic of active transducers that frequently, but not always, the same transduction principle used to convert from a nonelectrical form of energy can also be used in the reverse direction to convert electrical energy into nonelectrical forms. For example, a magnetic loudspeaker can also be used in the opposite direction as a microphone. Sometimes different names are used to refer to essentially the same effect when used in opposite directions because the two applications were discovered by different people. Table 2.1 shows these conversion principles. These principles (with the exception of the Volta effect and electrical polarization, both of which are treated in Chapter 4) are described in later sections of this chapter.

TABLE 2.1. SOME METHODS OF ENERGY CONVERSION
USED IN ACTIVE TRANSDUCERS

Energy Form	Transduced Form	Device or Effect	Reversible
Mechanical	Electrical	Magnetic induction	Yes
		Electric induction	
Pressure	Electrical	Piezoelectric	Yes
Thermal	Electrical	Thermoelectric	Yes
		Seebeck	No
Electrical	Thermal	Peltier	No
Light radiation	Electrical	Photoelectric	No
Electrical	Light	Light-emitting diodes	No
		Injection laser	
Chemical	Electrical	Volta	No
Electrical	Chemical	Electrical polarization	No
Sound	Electrical	Microphone	Yes
Electrical	Sound	Loudspeaker	Yes

2.2.1. Magnetic Induction

If an electrical conductor is moved in a magnetic field in such a way that the magnetic flux through the conductor is changed, a voltage is induced which is proportional to the rate of change of the magnetic flux. Conversely, if a current is sent through the same conductor, a mechanical force is exerted upon it proportional to the current and the magnetic field. The result, which depends on the polarities of voltage and current on the electrical side or the directions of force and motion on the mechanical side, is a conversion from electrical to mechanical energy, or vice versa. All electrical motors and generators and a host of other devices, such as solenoids and loudspeakers, utilize this principle.

Two basic configurations for transducers that use the principle of *magnetic induction* for the measurement of linear or rotary motion are shown in Figure 2.1(a) and (b). The output voltage in each case is propor-

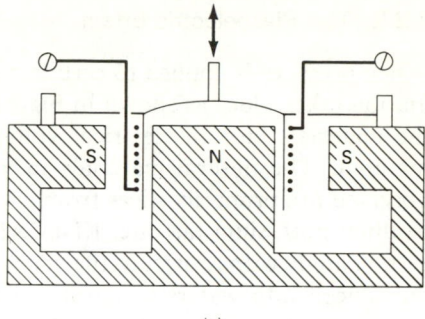

(a)

Figure 2.1. Inductive transducers (a) for linear motion; (b) for rotary motion.

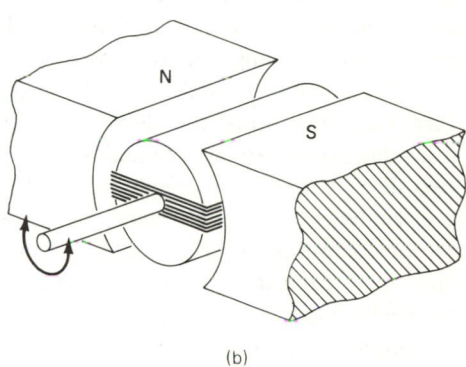

(b)

tional to the linear or angular velocity. The most important biomedical applications are heart sound microphones, pulse transducers, and electromagnetic blood-flow meters, all described in Chapter 6.

Magnetic induction also plays an important role at the output of many biomedical instrumentation systems. Analog meters using d'Arsonval movements, light-beam galvanometers in photographic recorders, and pen motors in ink or thermal recorders are all based on the principle of magnetic induction and closely resemble the basic transducer configuration shown in Figure 2.1(b).

It might be mentioned in passing that the principle of magnetic induction has an electrostatic equivalent called *electric induction*. Microphones based on this principle (condensor microphones) are now finding increasing use in audio applications because of their wide frequency response and high sensitivity. These microphones use an *electret* to create an electrostatic field between two capacitor plates. Electrets—which are the electrostatic equivalent of magnets—are normally in the form of foils of a special plastic material that have been heat-treated while being exposed to a strong electric field. It is conceivable that the principle of the electret microphone could also be applied advantageously to biomedical transducers.

2.2.2. The Piezoelectric Effect

When pressure is applied to certain nonconductive materials so that deformation takes place as shown in Figure 2.2(a) a charge separation occurs in the materials and an electrical voltage, V_P, can be measured across the material. The natural materials in which this *piezoelectric effect* can be observed are primarily slices from crystals of quartz (SiO_2) or Rochelle salt (sodium-potassium tartrate, $KNaC_4H_4O_6 \bullet 4H_2O$) which have been cut at a certain angle with respect to the crystal axis. Piezoelectric properties can be introduced into wafers of barium titanate (a ceramic material that is frequently used as a dielectric in disk-type capacitors) by heat-treating them in the presence of a strong electric field. The piezoelectric process is reversible. If an electric field is applied to a slab of material that has piezoelectric properties, it changes its dimensions.

By cutting the slab from the crystal at a different angle (or by a different application of the electrical field in the case of the barium titanate) the same effect can be obtained when a bending force is applied. Frequently, two slices, with proper orientation of the polarity of the piezoelectric voltages, are sandwiched between layers of conductive metal foil, thus forming the *bimorph* configuration shown in Figure 2.2(b). The electrically equivalent circuit of a piezoelectric transducer, shown in Figure 2.2(c), is that of a voltage source having a voltage, V_P, proportional to the applied mechanical force connected in series with a capacitor, which represents the conductive plates separated by the insulating piezoelectric material. The capacitive properties of the piezoelectric transducer interacting with the input impedance of the amplifier to which they are connected affect the response of the transducer. This effect is shown in Figure 2.3. The top trace shows the force applied to the transducer, which, after time T, is removed again. While the electrical field generated by the piezoelectric effect and the internal transducer voltage, V_P, of Figure 2.2(c) follow the applied force, the voltage, V_A, measured at the input of the amplifier depends on the values of the transducer capacitance, C, and the amplifier input impedance, R, with respect to the duration of the force (time T). If the product of R and C is much larger than T, the effect of the voltage division between these two components can be neglected and the measured voltage is proportional to the applied mechanical force as shown in trace 2. To meet this condition, even for large values of T, it may be necessary to make the amplifier input impedance very large. In some applications, electrometer amplifiers or charge amplifiers with extremely high input impedances have to be used. As an alternative, an external capacitor can be connected in parallel with the amplifier input. This effectively increases the capacity of the transducer but also reduces its sensitivity. Because the output voltages of piezoelectric transducers can be very high (they have occasionally even been used as high-voltage generators for ignition purposes),

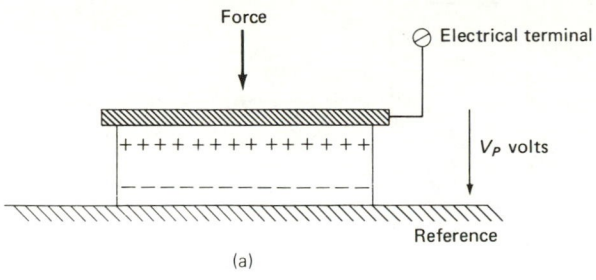

(a)

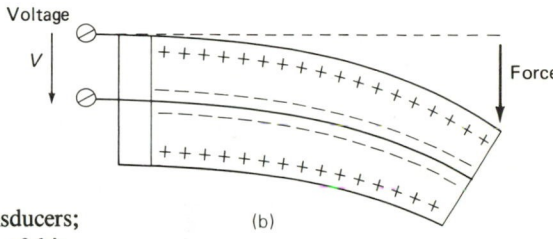

Figure 2.2. Piezoelectric transducers; (a) principle; (b) transducer of biomorph type; (c) equivalent circuit of a piezoelectric transducer connected to an amplifier.

(b)

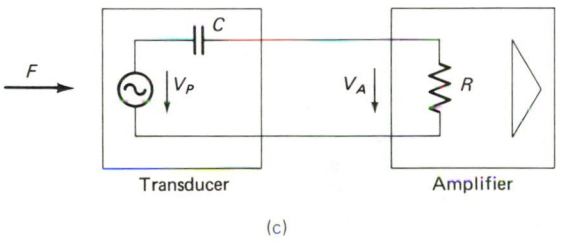

(c)

this approach may be permissible in certain applications. Changes of the input capacity, and thus the sensitivity, can also be caused by the mechanical movement of attached shielded or coaxial cables which can introduce motion artifacts. Special types of shielded cable that reduce this effect are available for piezoelectric transducers.

If the product of resistance and capacitance is made much smaller than T, the voltage at the amplifier input is proportional to the time derivative of the force at the transducer (or proportional to the rate at which the applied force changes) as shown in trace 3 of Figure 2.3. If the product of R and C *is of the same order of magnitude as* T, the resulting voltage is a compromise between the extremes in the two previous traces, as shown in trace 4. Because any mechanical input signal will contain various frequencies (corresponding to different times, $T,$ in the time domain), a distortion of

31

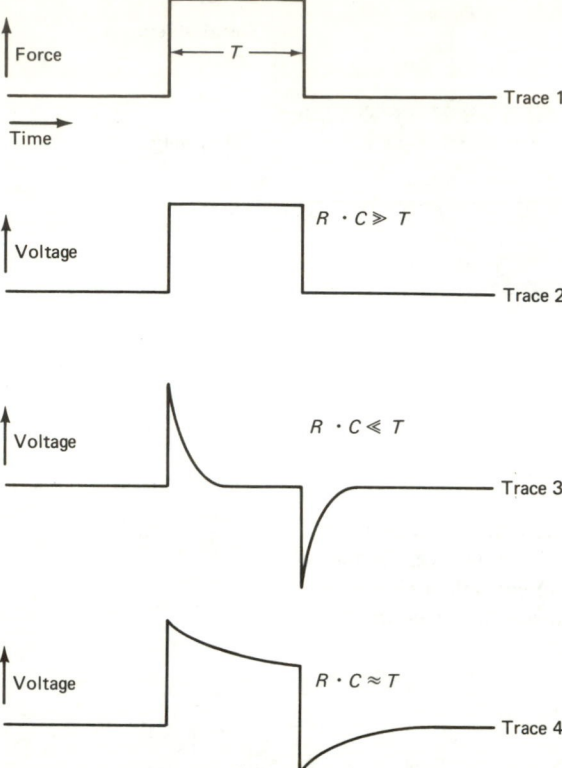

Figure 2.3. Output signal of a piezoelectric transducer under different conditions. Trace 1: Force at the input of the transducer. Trace 2: Output signal when the product of R and C is much larger than T; the output voltage is proportional to the force. Trace 3: Output signal if the product of R and C is much smaller than T; the output voltage is proportional to the rate of change of the force. Trace 4: Output signal if the product of R and C is approximately equal to T; the output signal is a combination of the two other cases.

the waveform of the resulting signal can occur if these relationships are not taken into consideration.

The piezoelectric principle is occasionally used in microphones for heart sounds or other acoustical signals from within the body. A more important application of piezoelectric transducers in biomedical instrumentation is in ultrasonic instruments, where a piezoelectric transducer is used to both transmit and receive ultrasonic signals. Principles of ultrasound and biomedical applications are covered in Chapter 9.

2.2.3. The Thermoelectric Effect

If two wires of dissimilar metals (e.g., iron and copper) are connected so that they form a closed conductive loop as shown in Figure 2.4(a), a voltage can be observed at any point of interruption of the loop which is proportional to the difference in temperature between the two junctions between the metals. The polarity depends on which of the two junctions is warmer. The device formed in this fashion is called a *thermocouple,* shown in Figure 2.4(a). The sensitivity of a thermocouple is small and amounts to only 40 microvolts per degree Celsius (μV/°C) for a copper-constantan and 53 μV/°C for an iron-constantan pair (constantan is an alloy of nickel and copper).

The principle of active transducers requires that any electrical energy delivered at the output of the transducer be obtained from the nonelectrical variable at the input of the transducer. In the case of the thermocouple it might not be quite obvious how the thermal energy is converted. Actually, the delivery of electrical energy causes the transfer of heat from the hotter to the colder junction; the hotter junction gets cooler while the colder junction gets warmer. In most practical applications of thermocouples this effect can be neglected. Because the thermocouple measures a temperature difference rather than an absolute temperature, one of the junctions must be kept at a known reference temperature, usually at the freezing point of water (0°C or 32°F). Frequently, instead of an icebath for the reference

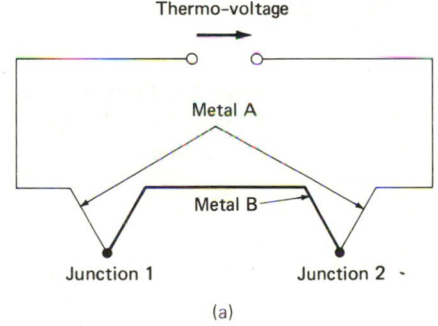

Figure 2.4. Thermocouple (a) principle; (b) thermocouple with double reference junction to connect to measurement circuit using copper wire.

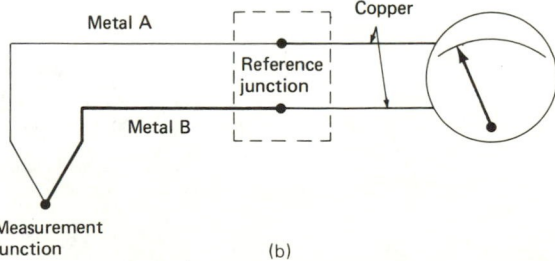

junction, an electronic compensating circuit is used. The inconvenience of having to make the whole circuit from the two metals used in the thermo-couple can be overcome by using a double reference junction that connects to copper conductors as shown in Figure 2.4(b).

Because of their low sensitivity, thermocouples are seldom used for the measurement of physiological temperatures, where the temperature range is so limited. Instead, one of the passive transducers described later is usually preferred. Thermocouples have an advantage at very high tempera-tures where passive transducers might not be usable or sometimes where transducers of minute size are required.

The use of the thermoelectric effect to convert from thermal to elec-trical energy is called the *Seebeck effect.* In the reverse direction it is called the *Peltier effect,* where the flow of current causes one junction to heat and the other to cool. The Peltier effect is occasionally used to cool parts of in-struments (e.g., a microscopic stage).

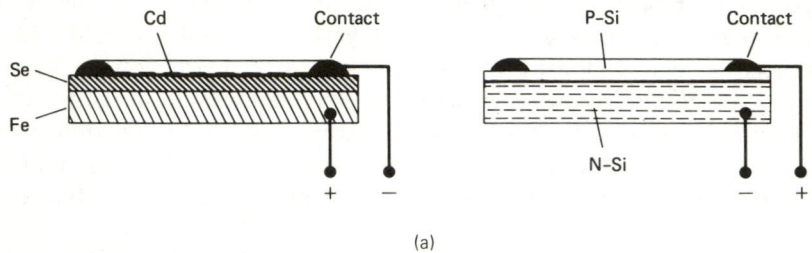

(a)

Figure 2.5. Photoelectric cells (a) selenium cell (left) and silicon (solar) cell (right). (b) Spectral sensitivity of the two cell types.

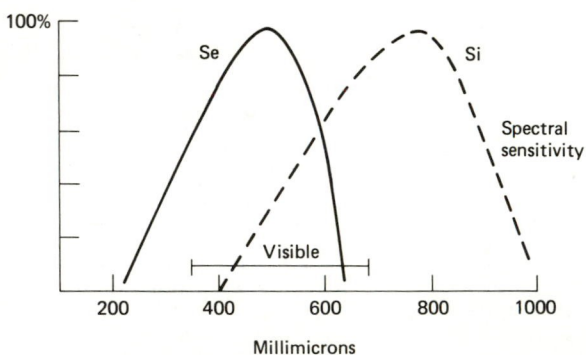

(b)

2.2.4. The Photoelectric Effect

The *selenium cell,* shown in Figure 2.5(a), has long been used to measure the intensity of light in photographic exposure meters or the light absorption of chemical solutions. The silicon photoelectric cell, better known as the *solar cell,* has a much higher efficiency than the selenium cell. Its spectral sensitivity peaks in the infrared, however, while that of the selenium cell is maximum in the visible light range. When operated into a small load resistance the current delivered by either cell is proportional to the intensity of the incident light. The voltage of these cells cannot exceed a certain value (about 0.6 V for the silicon cell); if the light intensity or the load resistance is such that the output voltage approaches this value, it becomes nonlinear.

2.3. PASSIVE TRANSDUCERS

Passive transducers utilize the principle of controlling a dc excitation voltage or an ac carrier signal. The actual transducer consists of a usually passive circuit element which changes its value as a function of the physical variable to be measured. The transducer is part of a circuit, normally an arrangement similar to a Wheatstone bridge, which is powered by an ac or dc excitation signal. The voltage at the output of the circuit reflects the physical variable. There are only three passive circuit elements that can be utilized as passive transducers: resistors, capacitors, and inductors. It should be noted that active circuit elements, vacuum tubes and transistors, are also occasionally used. This terminology might seem confusing since the terms ''active'' and ''passive'' have different meanings when they are applied to transducers than when they are applied to circuit elements. Unlike active transducers, passive transducers cannot be operated in the reverse direction (i.e., to convert an electrical signal into a physical variable) since a different basic principle is involved.

2.3.1. Passive Transducers Using Resistive Elements

Any resistive element that changes its resistance as a function of a physical variable can, in principle, be used as a transducer for that variable. An ordinary potentiometer, for example, can be used to convert rotary motion or displacement into a change of resistance. Similarly, the special linear potentiometers shown in Figure 2.6 can be used to convert linear displacement into a resistance change.

The resistivity of conductive materials is a function of temperature. In resistors this characteristic is a disadvantage; however, in resistive temperature transducers it serves a useful purpose. Temperature transducers are described in more detail in Chapter 9.

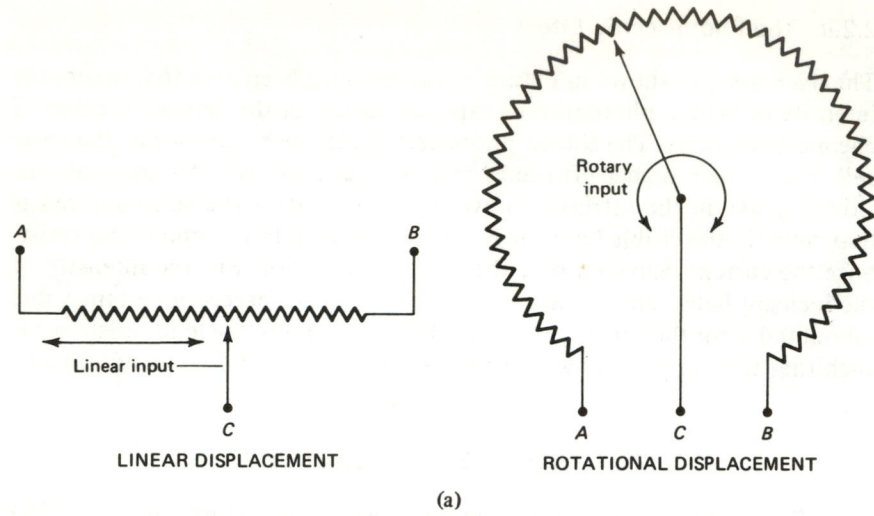

LINEAR DISPLACEMENT

ROTATIONAL DISPLACEMENT

(a)

Figure 2.6. Linear potentiometer (a) principle; (b) view of the device. (Courtesy of Bourns, Inc., Riverside, CA.)

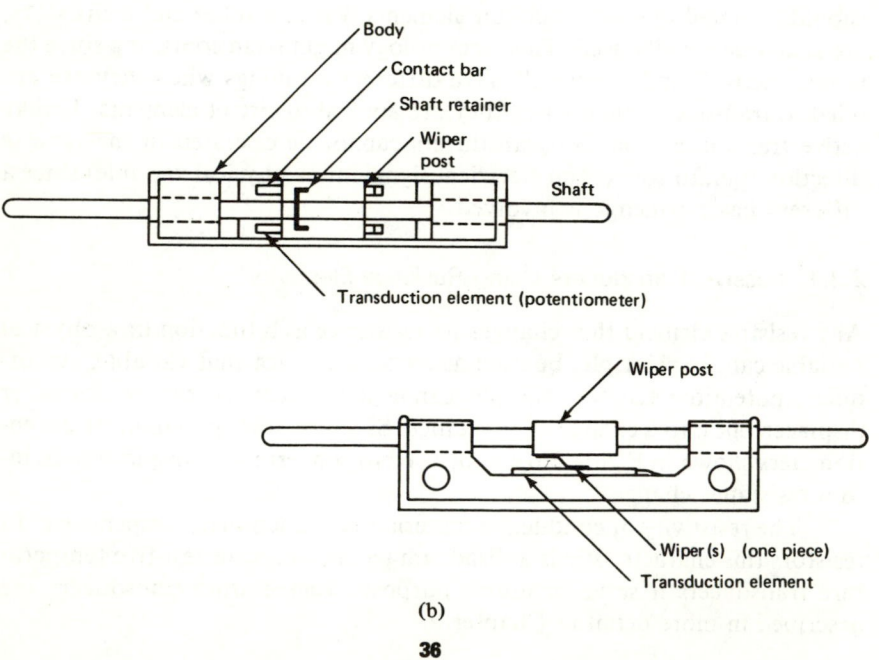

(b)

In certain semiconductor materials the conductivity is increased by light striking the material. This effect which occurs as a surface effect in certain polycrystalline materials such as cadmium sulfide, is used in *photoresistive cells,* a form of photoelectric transducer. This type of transducer is very sensitive, but has a somewhat limited frequency response. A different type of photoelectric transducer is the *photo diode,* which utilizes charge carriers generated by incident radiation in a reverse-biased diode junction. Although less sensitive than the photoresistive cell, the photodiode has improved frequency response. A photo diode can also be used as a photoelectric transducer without a bias voltage. In this case it operates as an active transducer. The *photoemissive cell* (either vacuum or gas-filled) is only of historical interest because it has generally been replaced by photoelectric transducers of the semiconductor type.

Most transducers used for mechanical variables utilize a resistive element called the *strain gage.* The principle of a strain gage can easily be understood with the help of Figure 2.7. Figure 2.7(a) shows a cylindrical resistor element which has length, *L,* and cross-sectional area, *A.* If it is made of a material having a resistivity of *r* ohm-cm, its resistance is

$$R = r \bullet L/A \text{ ohms } (\Omega).$$

If an axial force is applied to the element to cause it to stretch, its length increases by an amount, ΔL, as shown (exaggerated) in Figure 2.7(b). This stretching, on the other hand, causes the cross-sectional area of the cylinder to decrease by an amount ΔA. Either an increase in *L* or a decrease in *A* results in an increase in resistance. The ratio of the resulting resistance change $\Delta R/R$ to the change in length $\Delta L/L$ is called the *gage factor, G.* Thus;

$$G = \frac{\Delta R/R}{\Delta L/L}$$

The gage factor for metals is about 2, whereas the gage factor for silicon (a crystalline semiconductor material) is about 120.

Figure 2.7. Principle of strain gage; (a) cylindrical conductor with length, L, and cross sectional area, A. (b) Application of an axial force has increased the length by L while the cross sectional area has been reduced by A.

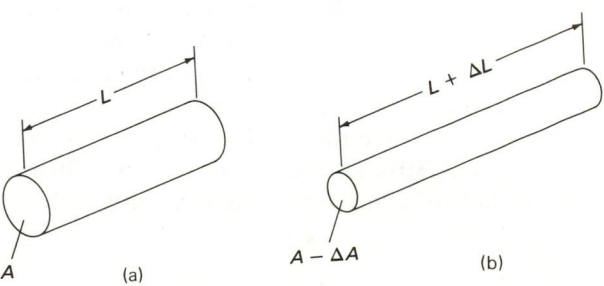

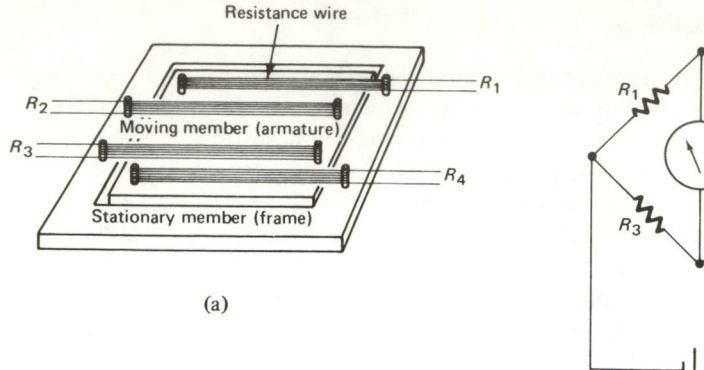
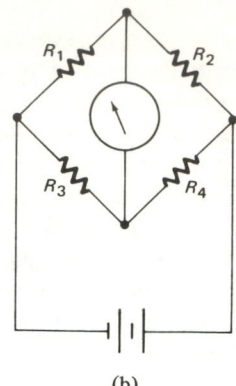

(a)

(b)

Figure 2.8. Unbonded strain gage transducer. (From D. Bartholomew, *Electrical Measurement and Instruments.* Allyn & Bacon, Inc., Boston, MA., by permission.)

The basic principle of the strain gage can be utilized for transducers in a number of different ways. In the *mercury strain gage* the resistive material consists of a column of mercury enclosed in a piece of silicone rubber tubing. The use of this type of strain gage for the measurement of physiological variables (the diameter of blood vessels) was first described by Whitney. Mercury strain gages are, therefore, sometimes called *Whitney gages.* An application of this type of transducer, the *mercury strain gage plethysmograph,* is described in Chapter 6. Because the silicone rubber yields easily to stretching forces, mercury strain gages are frequently used to measure changes in the diameter of body sections or organs. A disadvantage is that for practical dimensions the resistance of the mercury columns is inconveniently low (usually only a few ohms). This problem can be overcome by substituting an electrolyte solution for the mercury. However, silicone rubber is permeable to water vapor, so elastomers other than silicone rubber have to be used as the enclosures for gages containing electrolytes.

When metallic strain gages are used rather than mercury, the possible amount of stretching and the corresponding resistance changes are much more limited. Metal strain gages can be of two different types: unbonded and bonded. In the *unbonded strain gage,* thin wire is stretched between insulating posts as shown in Figure 2.8(a). In order to obtain a convenient resistance (120 Ω is a common value), several turns of wire must be used. Here the moving part of the transducer is connected to the stationary frame by four unbonded strain gages, R_1 through R_4. If the moving member is forced to the right, R_2 and R_3 are stretched and their resistance increases while the stress in R_1 and R_4 is reduced, thus decreasing the resistance of these strain gage wires. By connecting the four strain gages into a bridge circuit as shown in Figure 2.8(b), all resistance changes influence the output voltage in the same direction, increasing the sensitivity of the transducer by a factor of 4. At the same time, resistance changes of the strain gage due to changing temperatures tend to compensate each other. In the form shown, the unbonded strain gage is basically a force transducer. The same principle is also utilized in transducers for other variables. For example, the blood pressure transducers shown in Chapter 6 employ unbonded strain gages as the transducer elements.

The principle of the *bonded strain gage* is shown in Figure 2.9. A thin wire shaped in a zigzag pattern is cemented between two paper covers or is cemented to the surface of a paper carrier. This strain gage is then cemented to the surface of a structure. Any changes in surface dimensions of the structure due to mechanical strain are transmitted to the resistance wire, causing an increase or decrease of its length and a corresponding resistance change. The bonded strain gage, therefore, is basically a transducer for surface strain.

Related to the bonded wire strain gage is the *foil gage.* In this gage the conductor consists of a foil pattern on a substrate of plastic which is manufactured by the same photoetching techniques as those used in printed circuit boards. This process permits the manufacture of smaller gages with more complicated gage patterns (rosettes), which allow the measurement of different strain components.

In *semiconductor strain gages* a small slice of silicon replaces the wire or foil pattern as a conductor. Because of the crystalline nature of the silicon, these strain gages have a much larger gage factor than metal strain gages. Typical values are as high as 120. By varying the amount of im-

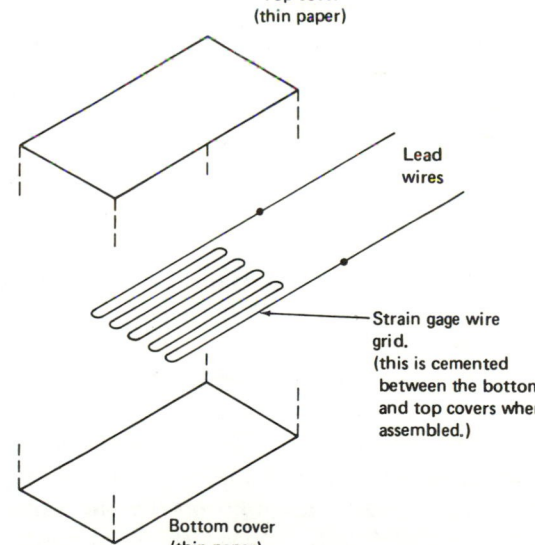

Figure 2.9. Typical bonded strain gage configuration.

purities in the silicon its conductivity can be controlled. With modern manufacturing techniques developed for semiconductor components, silicon strain gages can be made even smaller than the smallest foil gages. If the structure whose surface strain is to be measured is also made of silicon (e.g, in the shape of a beam or diaphragm), the size of the strain gage can be reduced even further by manufacturing it as a resistive pattern on the silicon surface. Such patterns can be obtained using the photolithographic and diffusion techniques developed for the manufacture of integrated circuits. The gages are isolated from the silicon substrate by reverse-biased diode junctions.

As with the unbonded gage, the resistance of a bonded strain gage is influenced by a change in temperature. In semiconductor strain gages, these changes are even more pronounced. Therefore, at least two strain-gage elements are usually used, with the second element either employed strictly for temperature compensation, or as part of a bridge in an arrangement similar to that shown in Figure 2.8(b) to increase the transducer sensitivity at the same time.

2.3.2. Passive Transducers Using Inductive Elements

In principle, the inductance of a coil can be changed either by varying its physical dimensions or by changing the effective permeability of its magnetic core. The latter can be achieved by moving a core having a permeability higher than air through the coil as shown in Figure 2.10. This arrangement appears to be very similar to that of an inductive transducer. However, in the inductive transducer the core is a permanent magnet which when moved induces a voltage in the coil. In this passive transducer the core is made of a soft magnetic material which changes the inductance of the coil when it is moved inside. The inductance can then be measured using an ac signal.

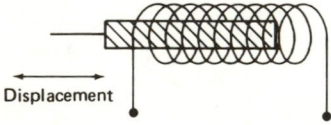

Displacement

Figure 2.10. Example of variable inductance displacement transducer.

Another passive transducer involving inductance is the *variable reluctance transducer,* in which the core remains stationary but the air gap in the magnetic path of the core is varied to change the effective permeability. This principle is also used in active transducers in which the magnetic path includes a permanent magnet.

The inductance of the coil in these types of transducers is usually not related linearly to the displacement of the core or the size of the air gap, especially if large displacements are encountered. The *linear variable differential transformer* (LVDT), shown in Figure 2.11, overcomes this limita-

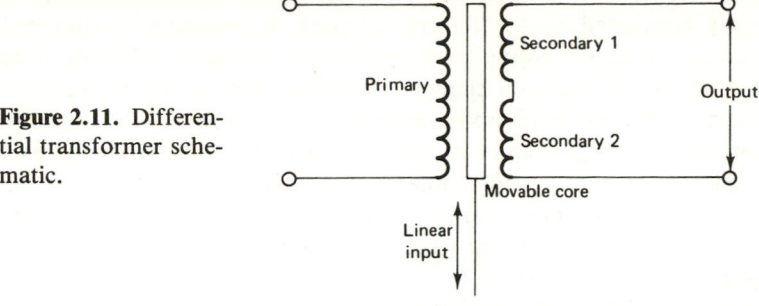

Figure 2.11. Differential transformer schematic.

tion. It consists of a transformer with one primary and two secondary windings. The secondary windings are connected so that their induced voltages oppose each other. If the core is in the center position, as shown in the figure, the voltages in the two secondary windings are equal in magnitude and the resulting output voltage is zero. If the core is moved upward as indicated by the arrow, the voltage in secondary 1 increases while that in secondary 2 decreases. The magnitude of the output voltage changes with the amount of displacement of the core from its central or neutral position. Its phase with respect to the voltage at the primary winding depends on the direction of the displacement. Because nonlinearities in the magnitudes of the voltages induced in the two output coils tend to compensate each other, the output voltage of the differential transducer is proportional to core movement even with fairly large displacements.

2.3.3. Passive Transducers Using Capacitive Elements

The capacitance of a plate capacitor can be changed by varying the physical dimensions of the plate structure or by varying the dielectric constant of the medium between the capacitor plates. Both effects have occasionally been used in the design of transducers for biomedical applications. The capacitance plethysmograph shown in Chapter 6 is an example. As with the transducers using an inductive element, it is sometimes not apparent whether a capacitive transducer is of the passive type or is actually an active transducer utilizing the principle of electric induction. If there is doubt, an examination of the carrier signal can help in the classification. Passive transducers utilize ac carriers, whereas a dc bias voltage is used in transducers based on the principle of electric induction.

2.3.4. Passive Transducers Using Active Circuit Elements

The distinction between "active" and "passive" when used for circuit elements is based on a different principle than that which is used for transducers. Active circuit elements are those which provide power gain for a signal (i.e., vacuum tubes and transistors). Such circuit elements have occasionally been used as transducers. Because, as transducers they employ the principle of carrier modulation (the carrier being the plate or collector voltage), these active circuit elements are nevertheless passive transducers, by definition. A variable-transconductance vacuum tube in which the distance between the control grid and cathode of a vacuum tube was changed by the displacement of a mechanical connection is an early example of this type of transducer. More recently, transistors have been manufactured in which a mechanical force applied to the base region of the planar transistor causes a change in the current gain.

The most important application of active circuit elements in passive transducers is in the area of photoelectric transducers. The *photomultiplier* consists of a photoemissive cathode of the type used in photoemissive cells. When struck by photons, the electrons emitted by the cathode are amplified by several stages of secondary emission electrodes called dynodes. The photomultiplier is still the most sensitive light detector. One of its applications for biomedical purposes is in the scintillation detector for nuclear radiation described in Chapter 14.

The sensitivity of a photo diode can be increased if the reverse-biased diode is incorporated into a transistor as the collector–base junction to form a *photo transistor.* In this device, the photo-diode current is essentially amplified by the transistor and appears at the collector, multiplied by the current gain. In the *photo Darlington,* a photo transistor is connected to a second transistor on the same substrate, with the two transistors forming a Darlington circuit. This effectively multiplies the photo current of the collector-base junction of the first transistor by the product of the current gains of both transistors. This arrangement makes the photo Darlington a very sensitive transducer.

Another semiconductor transducer element is the *Hall generator,* which provides an output voltage that is proportional to both the applied current and any magnetic field in which it is placed.

2.4. TRANSDUCERS FOR BIOMEDICAL APPLICATIONS

Several basic physical variables and the transducers (active or passive) used to measure them are listed in Table 2.2. It should be noted that many variables of great interest in biomedical applications, such as pressure and fluid or gas flow, are not included. These and many other variables of interest can be measured, however, by first converting each of them into one of the variables for which basic transducers are available. Some very ingenious methods have been developed to convert some of the more elusive quantities for measurement by one of the transducers described.

Table 2.2. BASIC TRANSDUCERS

Physical Variable	Type of Transducer
Force (or pressure)	Piezoelectric
	Unbonded strain gage
Displacement	Variable resistance
	Variable capacitance
	Variable inductance
	Linear variable differential transducer
	Mercury strain gage
Surface strain	Strain gage
Velocity	Magnetic induction
Temperature	Thermocouple
	Thermistor
Light	Photovoltaic
	Photoresistive
Magnetic field	Hall effect

[a]In medical applications the basic physiological variables is first transformed into one of the physical variables listed. Examples would be measurement of blood pressure using strain gages and blood flow by magnetic induction.

2.4.1. Force Transducers

A design element frequently used for the conversion of physical variables is the *force-summing member*. One possible configuration of this device is shown in Figure 2.12(a). In this case, the force-summing member is a leaf spring. When the spring is bent downward, it exerts an upward-directed force that is proportional to the displacement of the end of the spring. If a force is applied to the end of the spring in a downward direction, the spring bends until its upward-directed force equals the downward-directed applied force, or, expressed differently, until the vector sum of both forces equals zero. From this it derives its name "force-summing member." In the configuration shown, the force-summing member can be used to convert a force into a variable for which transducers are more readily available. The bending of the spring, for example, results in a surface strain that can be measured by means of bonded strain gages as shown in Figure 2.12(b).

The transducers shown in Figure 2.13 utilize this principle. The photographs illustrate that force and displacement transducers are closely related. Sometimes, the terms *isotonic* and *isometric* are used to describe the characteristics of these transducers. Ideally a force transducer would be isometric; that is, it would not yield (change its dimensions) when a force is applied. On the other hand, a displacement transducer would be isotonic and offer zero or a constant resistance to an applied displacement. In reality, almost all transducers combine the characteristics of both ideal transducer

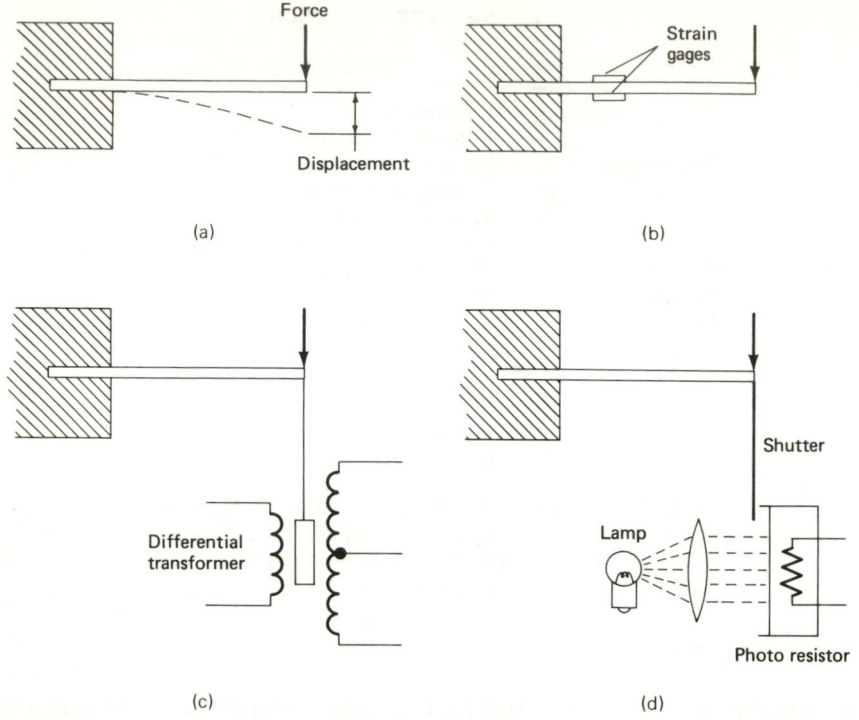

Figure 2.12. Force transducers using various transduction principles. (a) The force summing member, here in the form of a leaf spring. (b) Force transducer with bonded strain gages. (c) Force transducer using a differential transformer. (d) Force transducer using a lamp and photo resistor to measure the displacement of the force summing member.

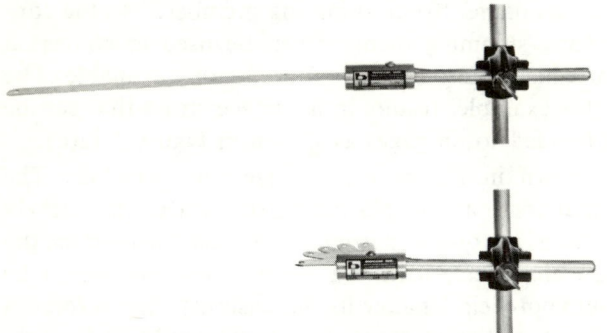

Figure 2.13. Force-displacement transducer with bonded strain gage. (Courtesy of Biocom, Inc., Culver City, CA.)

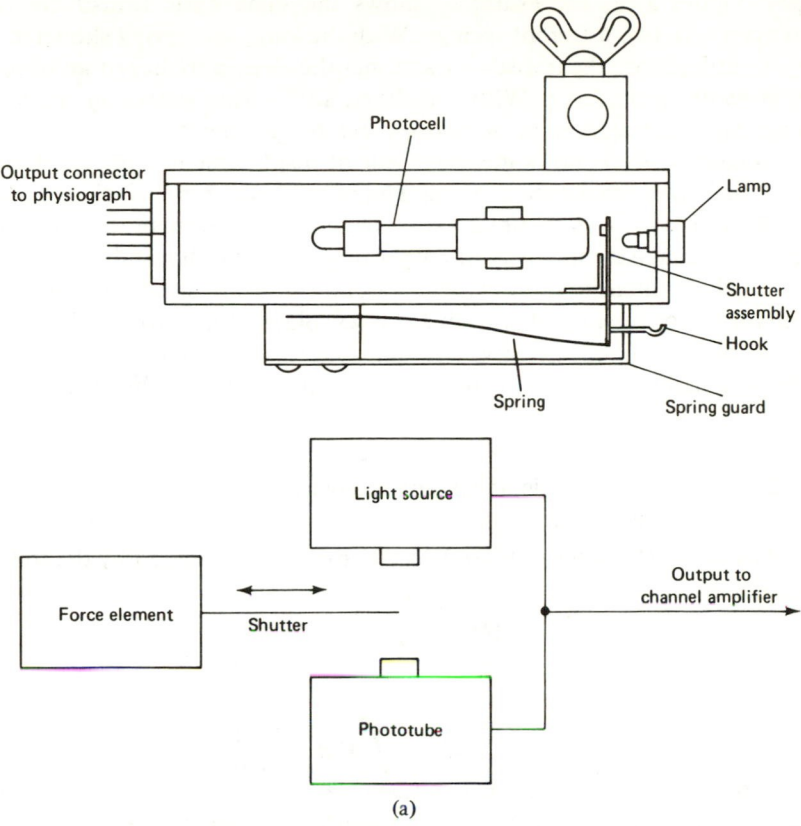

Figure 2.14. Photoelectric displacement transducer: (a) block diagram; (b) photograph. (Courtesy of Narco BioSystems, Houston, TX.)

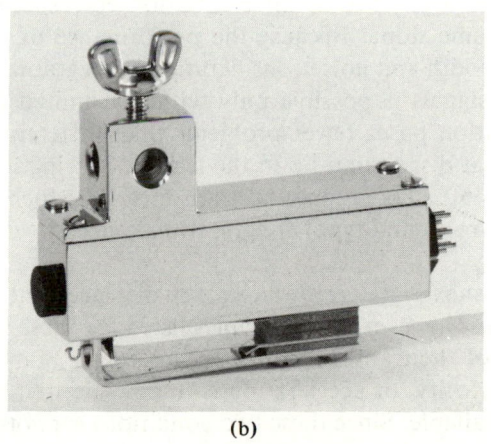

(b)

types. Figure 2.13, for example, shows the same basic transducer type equipped with two different springs. With the long, soft spring shown in the upper photograph, the transducer assumes the characteristics of an *isotonic displacement transducer.* With the short, stiff spring shown in the lower photograph, it becomes an *isometric force transducer.*

Figure 2.12(c) shows measurement of displacement using a differential transformer transducer. A less frequently used type of displacement transducer is shown in Figure 2.12(d). Here the displacement of a spring is used to modulate the intensity of a light beam via a mechanical shutter. The resulting light intensity is measured by a photoresistive cell. In this example, a multiple conversion of variables takes place: force to displacement, displacement to light intensity, and light intensity to resistance. This principle is actually employed in the commercial transducer shown in Figure 2.14.

2.4.2. Transducers for Displacement, Velocity, and Acceleration

Displacement, *D,* velocity, *V,* and acceleration, *A,* are linked by the following relationships:

$$V = \frac{dD}{dt} \qquad A = \frac{dV}{dt} = \frac{d^2D}{dt^2}$$

and the inverse:

$$V = \int A \, dt \qquad D = \int V \, dt = \iint A \, (dt)^2$$

If any one of the three variables can be measured, it is possible—at least in principle—to obtain the other two variables by integration or differentiation. Both operations can readily be performed by electronic methods operating on either analog or digital signals. Expressed in the frequency domain, the integration of a signal corresponds to a lowpass filter with a slope of 6 dB/octave, whereas differentiation corresponds to a highpass filter with the same slope. Because the performance of analog circuits is limited by bandwidth and noise considerations, integration and differentiation of analog signals is possible only within a limited frequency range. Usually, integration poses fewer problems than differentiation. It should also be noted that discontinuities in the transducer characteristic (e.g., the finite resolution of a potentiometric transducer in which the resistive element is of the wire-wound type) are greatly enhanced by the differentiation process.

Table 2.2 shows that transducers for displacement and velocity are readily available. However, the principles listed for these measurements require that part of the transducer be attached to the body structure whose displacement, velocity, or acceleration is to be measured, and that a reference point be available. Since these two conditions cannot always be met in biomedical applications, indirect methods sometimes have to be used. Contactless methods for measuring displacement and velocity, based on optical or magnetic principles, are occasionally used. Magnetic methods usually require that a small magnet or piece of metal be attached to the body structure. Ultrasonic methods, described in Chapter 9, are used more frequently.

2.4.3. Pressure Transducers

Pressure transducers are closely related to force transducers. Some of the force-summing members used in pressure transducers are shown in Figure 2.15. Pressure transducers utilizing flat diaphragms normally have bonded or semiconductor strain gages attached directly to the diaphragms. The small implantable pressure transducer shown in Chapter 6 is of this design. Even smaller dimensions are possible if the diaphragm is made directly from a thin silicon wafer with the strain gages diffused into its surface. The corrugated diaphragm lends itself to the design of pressure transducers using unbonded strain gages or a differential transformer as the transducer element. The LVDT blood pressure transducer shown in Chapter 6 uses these principles. Flat or corrugated diaphragms have also occasionally been used in transducers which employ the variable reluctance or variable capacitance principles. Although diaphragm-type pressure transducers can be designed for a wide range of operating pressures, depending on the diameter and stiffness of the diaphragm, *Bourdon tube* transducers are usually used for high pressure ranges.

It should be noted that the amount of deformation of the force-summing member in a pressure transducer actually depends on the difference in the pressure between the two sides of the diaphragm. If absolute pressure is to be measured, there must be a vacuum on one side of the diaphragm. It is much more common to measure the pressure relative to atmospheric pressure by exposing one side of the diaphragm to the atmosphere. In *differential pressure transducers* the two pressures are applied to opposite sides of the diaphragm.

Figure 2.15. Force-summing members used in pressure transducers; (a) flat diaphragm; (b) corrugated diaphragm; (c) Bourdon tube. (Dashed line shows new position by motion.)

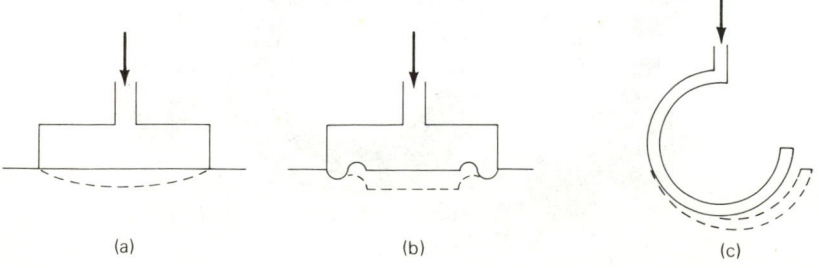

(a)   (b)   (c)

2.4.4. Flow Transducers

The flow rate of fluids or gases is a very elusive variable and many different methods have been developed to measure it. These methods are described in detail in Chapter 6 for blood flow and cardiac output, and in Chapter 8 for the measurement of gas flow as used in measurements in the respiratory system.

2.4.5. Transducers with Digital Output

Increasingly, biomedical instrumentation systems are utilizing digital methods for the processing of data, which require that any data entered into the system be in digital rather than in analog form. Analog-to-digital converters, described in Chapter 15, can be used to convert an analog transducer output into digital form. It is often desirable to have a transducer whose output signal originates in digital form. Although such transducers are very limited in their application, they are available for measurement of linear or rotary displacement. These transducers contain encoding disks or rulers with digital patterns (see Figure 2.16) photographically etched on glass plates. A light source and an array of photodetectors, usually made up of photos diodes or photo transistors, are used to obtain a digital signal in parallel format that indicates the position of the encoding plate, and thereby represents the displacement being measured.

Figure 2.16. Digital shaft encoder patterns. (Courtesy of Itek, Wayne George Division, Newborn, MA.)

3

Sources of Bioelectric Potentials

In carrying out their various functions, certain systems of the body generate their own monitoring signals, which convey useful information about the functions they represent. These signals are the bioelectric potentials associated with nerve conduction, brain activity, heartbeat, muscle activity, and so on. Bioelectric potentials are actually ionic voltages produced as a result of the electrochemical activity of certain special types of cells. Through the use of transducers capable of converting ionic potentials into electrical voltages, these natural monitoring signals can be measured and results displayed in a meaningful way to aid the physician in his diagnosis and treatment of various diseases.

The idea of electricity being generated in the body goes back as far as 1786, when an Italian anatomy professor, Luigi Galvani, claimed to have found electricity in the muscle of a frog's leg. In the century that followed several other scientists discovered electrical activity in various animals and in man. But it was not until 1903, when the Dutch physician Willem

Einthoven introduced the string galvanometer, that any practical application could be made of these potentials. The advent of the vacuum tube and amplification and, more recently, of solid-state technology has made possible better representation of the bioelectric potentials. These developments, combined with a large amount of physiological research activity, have opened many new avenues of knowledge in the application and interpretation of these important signals.

3.1. RESTING AND ACTION POTENTIALS

Certain types of cells within the body, such as nerve and muscle cells, are encased in a semipermeable membrane that permits some substances to pass through the membrane while others are kept out. Neither the exact structure of the membrane nor the mechanism by which its permeability is controlled is known, but the substances involved have been identified by experimentation.

Surrounding the cells of the body are the body fluids. These fluids are conductive solutions containing charged atoms known as *ions*. The principal ions are sodium (Na^+), potassium (K^+), and chloride (C^-). The membrane of excitable cells readily permits entry of potassium and chloride ions but effectively blocks the entry of sodium ions. Since the various ions seek a balance between the inside of the cell and the outside, both according to concentration and electric charge, the inability of the sodium to penetrate the membrane results in two conditions. First, the concentration of sodium ions inside the cell becomes much lower than in the intercellular fluid outside. Since the sodium ions are positive, this would tend to make the outside of the cell more positive than the inside. Second, in an attempt to balance the electric charge, additional potassium ions, which are also positive, enter the cell, causing a higher concentration of potassium on the inside than on the outside. This charge balance cannot be achieved, however, because of the concentration imbalance of potassium ions. Equilibrium is reached with a potential difference across the membrane, negative on the inside and positive on the outside.

This membrane potential is called the *resting potential* of the cell and is maintained until some kind of disturbance upsets the equilibrium. Since measurement of the membrane potential is generally made from inside the cell with respect to the body fluids, the resting potential of a cell is given as negative. Research investigators have reported measuring membrane potentials in various cells ranging from -60 to -100 mV. Figure 3.1 illustrates in simplified form the cross section of a cell with its resting potential. A cell in the resting state is said to be *polarized*.

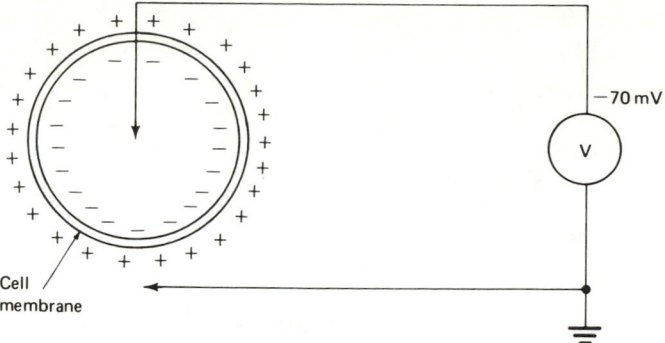

Figure 3.1. Polarized cell with its resting potential.

When a section of the cell membrane is excited by the flow of ionic current or by some form of externally applied energy, the membrane changes its characteristics and begins to allow some of the sodium ions to enter. This movement of sodium ions into the cell constitutes an ionic current flow that further reduces the barrier of the membrane to sodium ions. The net result is an avalanche effect in which sodium ions literally rush into the cell to try to reach a balance with the ions outside. At the same time potassium ions, which were in higher concentration inside the cell during the resting state, try to leave the cell but are unable to move as rapidly as the sodium ions. As a result, the cell has a slightly positive potential on the inside due to the imbalance of potassium ions. This potential is known as the *action potential* and is approximately +20 mV. A cell that has been excited and that displays an action potential is said to be *depolarized;* the process of changing from the resting state to the action potential is called *depolarization.* Figure 3.2 shows the ionic movements associated with depolarization, and Figure 3.3 illustrates the cross section of a depolarized cell.

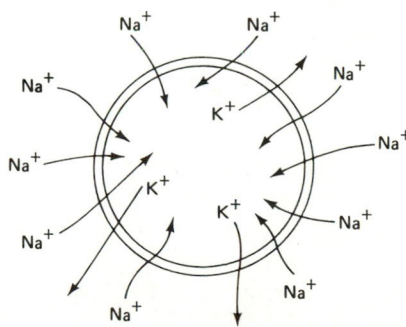

Figure 3.2. Depolarization of a cell. Na^+ ions rush into the cell while K^+ ions attempt to leave.

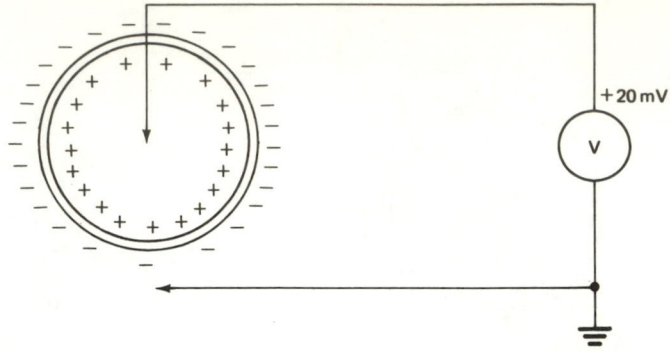

Figure 3.3. Depolarized cell during an action potential.

Once the rush of sodium ions through the cell membrane has stopped (a new state of equilibrium is reached), the ionic currents that lowered the barrier to sodium ions are no longer present and the membrane reverts back to its original, selectively permeable condition, wherein the passage of sodium ions from the outside to the inside of the cell is again blocked. Were this the only effect, however, it would take a long time for a resting potential to develop again. But such is not the case. By an active process, called a *sodium pump,* the sodium ions are quickly transported to the outside of the cell, and the cell again becomes polarized and assumes its resting potential. This process is called *repolarization.* Although little is known of the exact chemical steps involved in the sodium pump, it is quite generally believed that sodium is withdrawn against both charge and concentration gradients supported by some form of high-energy phosphate compound. The rate of pumping is directly proportional to the sodium concentration in the cell. It

Figure 3.4. Waveform of the action potential. (Time scale varies with type of cell.)

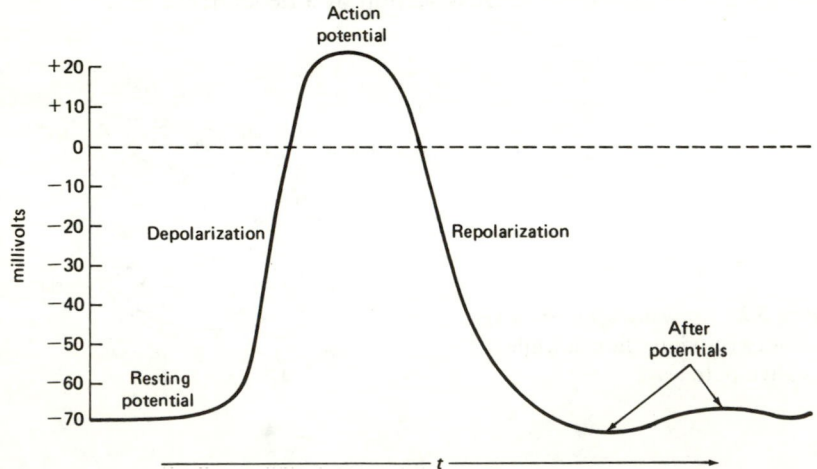

is also believed that the operation of this pump is linked with the influx of potassium into the cell, as if a cyclic process involving an exchange of sodium for potassium existed.

Figure 3.4 shows a typical action-potential waveform, beginning at the resting potential, depolarizing, and returning to the resting potential after repolarization. The time scale for the action potential depends on the type of cell producing the potential. In nerve and muscle cells, repolarization occurs so rapidly following depolarization that the action potential appears as a spike of as little as 1 msec total duration. Heart muscle, on the other hand, repolarizes much more slowly, with the action potential for heart muscle usually lasting from 150 to 300 msec.

Regardless of the method by which a cell is excited or the intensity of the stimulus (provided it is sufficient to activate the cell), the action potential is always the same for any given cell. This is known as the *all-or-nothing* law. The *net height* of the action potential is defined as the difference between the potential of the depolarized membrane at the peak of the action potential and the resting potential.

Following the generation of an action potential, there is a brief period of time during which the cell cannot respond to any new stimulus. This period, called the *absolute refractory period,* lasts about 1 msec in nerve cells. Following the absolute refractory period, there occurs a *relative refractory period,* during which another action potential can be triggered, but a much stronger stimulation is required. In nerve cells, the relative refractory period lasts several milliseconds. These refractory periods are believed to be the result of after-potentials that follow an action potential.

3.2. PROPAGATION OF ACTION POTENTIALS

When a cell is excited and generates an action potential ionic currents begin to flow. This process can, in turn, excite neighboring cells or adjacent areas of the same cell. In the case of a nerve cell with a long fiber, the action potential is generated over a very small segment of the fiber's length but is propagated in both directions from the original point of excitation. In nature, nerve cells are excited only near their "input end" (see Chapter 10 for details). As the action potential travels down the fiber, it cannot reexcite the portion of the fiber immediately upstream, because of the refractory period that follows the action potential.

The rate at which an action potential moves down a fiber or is propagated from cell to cell is called the *propagation rate.* In nerve fibers the propagation rate is also called the *nerve conduction rate,* or *conduction velocity.* This velocity varies widely, depending on the type and diameter of the nerve fiber. The usual velocity range in nerves is from 20 to 140 meters per second (m/sec). Propagation through heart muscle is slower, with an average rate from 0.2 to 0.4 m/sec. Special time-delay fibers between the atria and ventricles of the heart cause action potentials to propagate at an even slower rate, 0.03 to 0.05 m/sec.

3.3. THE BIOELECTRIC POTENTIALS

To measure bioelectric potentials, a transducer capable of converting ionic potentials and currents into electric potentials and currents is required. Such a transducer consists of two *electrodes,* which measure the ionic potential difference between their respective points of application. Electrodes are discussed in detail in Chapter 4.

Although measurement of individual action potentials can be made in some types of cells, such measurements are difficult because they require precise placement of an electrode inside a cell. The more common form of measured biopotentials is the combined effect of a large number of action potentials as they appear at the surface of the body, or at one or more electrodes inserted into a muscle, nerve, or some part of the brain.

The exact method by which these potentials reach the surface of the body is not known. A number of theories have been advanced that seem to explain most of the observed phenomena fairly well, but none exactly fits the situation. Many attempts have been made, for example, to explain the biopotentials from the heart as they appear at the surface of the body. According to one theory, the surface pattern is a summation of the potentials developed by the electric fields set up by the ionic currents that generate the individual action potentials. This theory, although plausible, fails to explain a number of the characteristics indicated by the observed surface patterns. A closer approximation can be obtained if it is assumed that the surface pattern is a function of the summation of the first derivatives (rates of change) of all the individual action potentials, instead of the potentials themselves. Part of the difficulty arises from the numerous assumptions that must be made concerning the ionic current and electric field patterns throughout the body. The validity of some of these assumptions is considered somewhat questionable. Regardless of the method by which these patterns of potentials reach the surface of the body or implanted measuring electrodes, they can be measured as specific bioelectric signal patterns that have been studied extensively and can be defined quite well.

The remainder of this chapter is devoted to a description of each of the more significant bioelectric potential waveforms. The designation of the waveform itself generally ends in the suffix *gram,* whereas the name of the instrument used to measure the potentials and graphically reproduce the waveform ends in the suffix *graph.* For example, the *electrocardiogram* (the name of the waveform resulting from the heart's electrical activity) is measured on an *electrocardiograph* (the instrument). Ranges of amplitudes and frequency spectra for each of the biopotential waveforms described below are included in Appendix B.

3.3.1. The Electrocardiogram (ECG)

The biopotentials generated by the muscles of the heart result in the *electrocardiogram,* abbreviated *ECG (sometimes EKG, from the German electrokardiogram).* To understand the origin of the ECG, it is necessary to have some familiarity with the anatomy of the heart. Figure 3.5 shows a cross section of the interior of the heart. The heart is divided into four chambers. The two upper chambers, the left and right *atria,* are synchronized to act together. Similarly, the two lower chambers, the *ventricles,* operate together. The right atrium receives blood from the veins of the body and pumps it into the right ventricle. The right ventricle pumps the blood through the lungs, where it is oxygenated. The oxygen-enriched blood then enters the left atrium, from which it is pumped into the left ventricle. The left ventricle pumps the blood into the arteries to circulate throughout the body. Because the ventricles actually pump the blood through the vessels (and therefore do most of the work), the ventricular muscles are much larger and more important than the muscles of the atria. For the cardiovascular system to function properly, both the atria and the ventricles must operate in a proper time relationship.

Each action potential in the heart originates near the top of the right atrium at a point called the *pacemaker* or *sinoatrial* (SA) *node.* The pacemaker is a group of specialized cells that spontaneously generate action potentials at a regular rate, although the rate is controlled by innervation. To initiate the heartbeat, the action potentials generated by the pacemaker propagate in all directions along the surface of both atria. The wavefront of activation travels parallel to the surface of the atria toward the junction of the atria and the ventricles. The wave terminates at a point near the center of the heart, called the *atrioventricular* (AV) *node.* At this point, some special fibers act as a "delay line" to provide proper timing between the action of the atria and the ventricles. Once the electrical excitation has passed through the delay line, it is rapidly spread to all parts of both ventricles by the *bundle of His* (pronounced "hiss"). The fibers in this bundle, called *Purkinje fibers,* divide into two branches to initiate action potentials simultaneously in the powerful musculature of the two ventricles. The wavefront in the ventricles does not follow along the surface but is perpendicular to it and moves from the inside to the outside of the ventricular wall, terminating at the tip or *apex* of the heart. As indicated earlier, a wave of repolarization follows the depolarization wave by about 0.2 to 0.4 second. This repolarization, however, is not initiated from neighboring muscle cells but occurs as each cell returns to its resting potential independently.

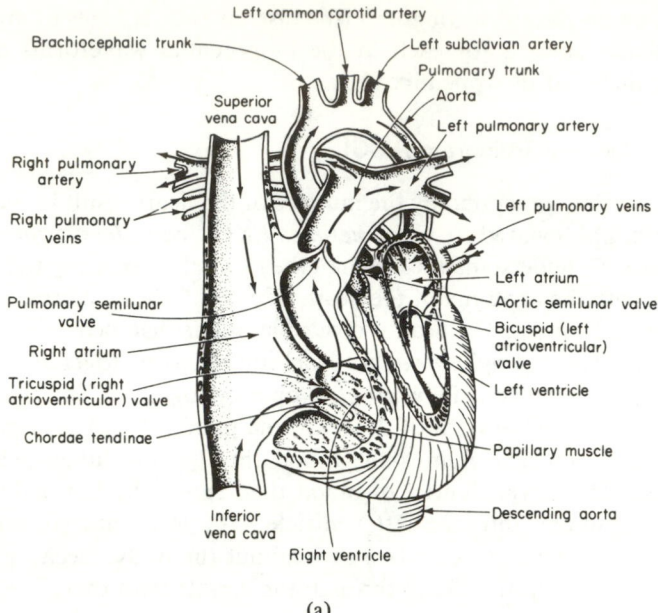

(a)

Figure 3.5. The heart: (a) internal structure; (b) conducting system. (From W.F. Evans, *Anatomy and Physiology, The Basic Principles,* Englewood Cliffs, N.J., Prentice-Hall, Inc., 1971, by permission.)

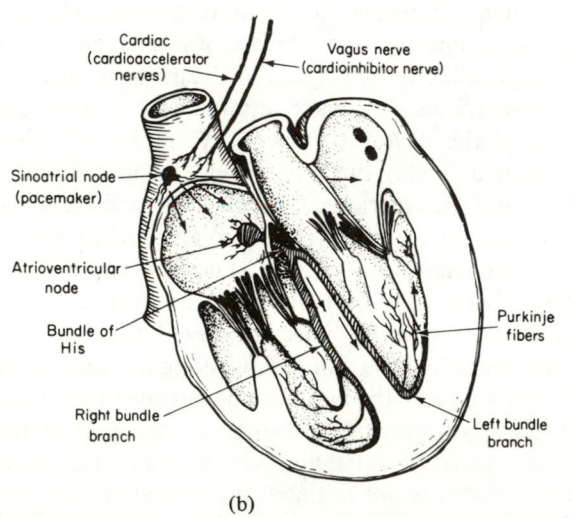

(b)

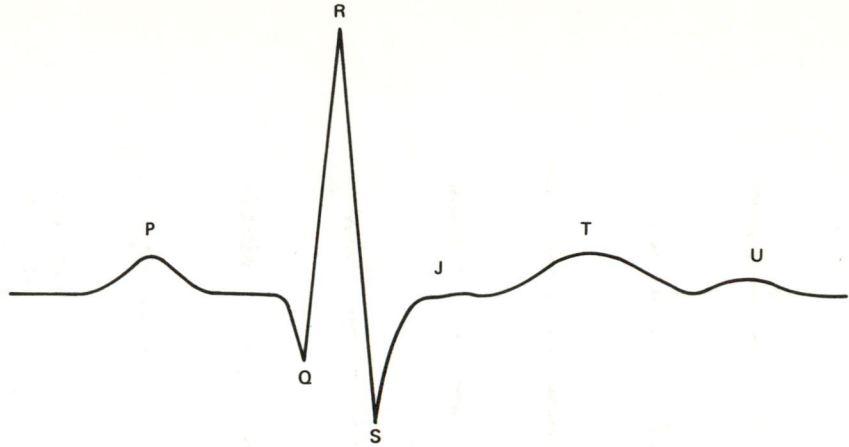

Figure 3.6. The electrocardiogram waveform.

Figure 3.6 shows a typical ECG as it appears when recorded from the surface of the body. Alphabetic designations have been given to each of the prominent features. These can be identified with events related to the action potential propagation pattern. To facilitate analysis, the horizontal segment of this waveform preceding the P wave is designated as the *baseline* or the *isopotential line*. The *P wave* represents depolarization of the atrial musculature. The *QRS complex* is the combined result of the repolarization of the atria and the depolarization of the ventricles, which occur almost simultaneously. The *T wave* is the wave of ventricular repolarization, whereas the *U wave,* if present, is generally believed to be the result of after-potentials in the ventricular muscle. The *P-Q interval* represents the time during which the excitation wave is delayed in the fibers near the AV node.

The shape and polarity of each of these features vary with the location of the measuring electrodes with respect to the heart, and a cardiologist normally bases his diagnosis on readings taken from several electrode locations. Measurement of the electrocardiogram is covered in more detail in Chapter 6.

3.3.2. The Electroencephalogram (EEG)

The recorded representation of bioelectric potentials generated by the neuronal activity of the brain is called the *electroencephalogram,* abbreviated EEG. The EEG has a very complex pattern, which is much more difficult to recognize than the ECG. A typical sample of the EEG is shown in Figure 3.7. As can be seen, the waveform varies greatly with the location of the measuring electrodes on the surface of the scalp. EEG potentials, measured at the surface of the scalp, actually represent the combined effect of potentials from a fairly wide region of the cerebral cortex and from various points beneath.

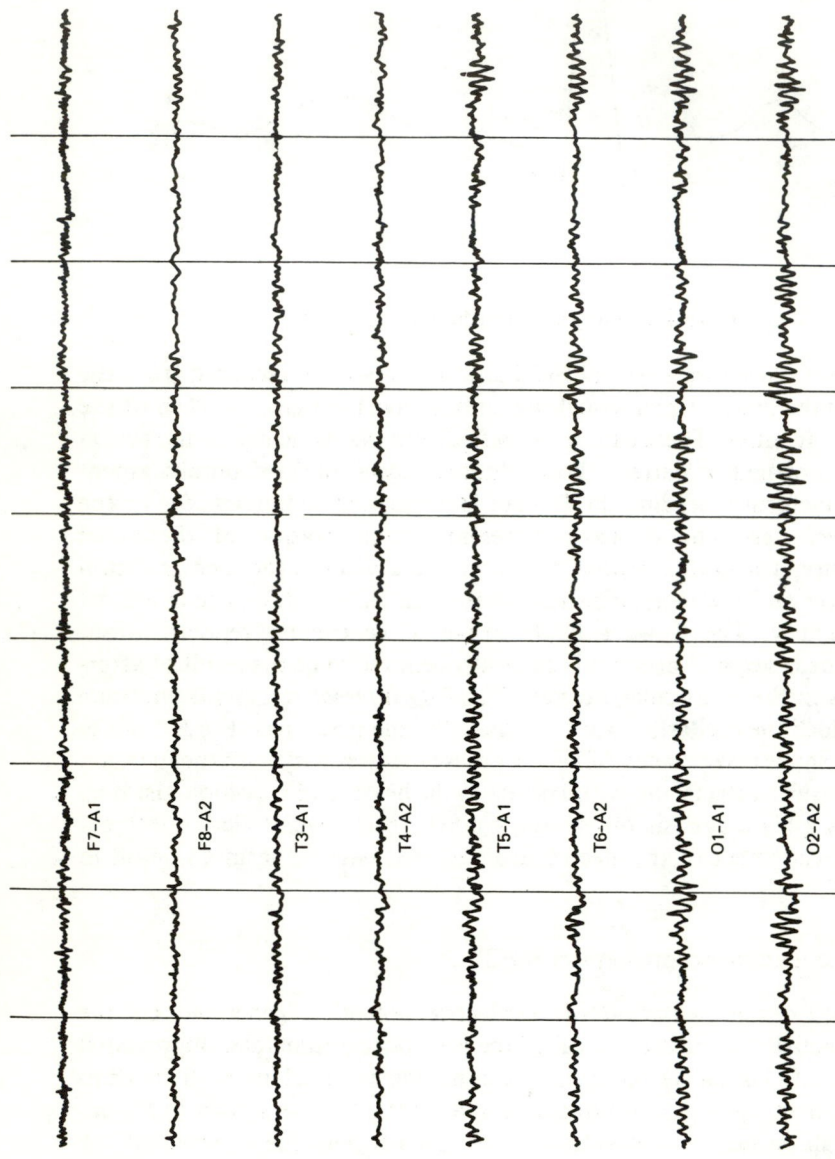

F7-A1

F8-A2

T3-A1

T4-A2

T5-A1

T6-A2

O1-A1

O2-A2

Figure 3.7. Typical human electroencephalogram. The eight tracings indicate regions of the scalp from which each channel of EEG was measured with respect to one of two reference ear electrodes (A1 and A2). Figure 10.12 shows the layout of the electrodes.) (Courtesy Veterans Administration Hospital, Sepulveda, CA.)

Experiments have shown that the frequency of the EEG seems to be affected by the mental activity of a person. The wide variation among individuals and the lack of repeatability in a given person from one occasion to another make the establishment of specific relationships difficult. There are, however, certain characteristic EEG waveforms that can be related to epileptic seizures and sleep. The waveforms associated with the different stages of sleep are shown in Figure 3.8. An alert, wide-awake person usually displays an unsynchronized high-frequency EEG. A drowsy person, particularly one whose eyes are closed, often produces a large amount of rhythmic activity in the range 8 to 13 Hz. As the person begins to fall asleep, the amplitude and frequency of the waveform decrease; and in light sleep, a large-amplitude, low-frequency waveform emerges. Deeper sleep generally results in even slower and higher-amplitude waves. At certain times, however, a person, still sound asleep, breaks into an unsynchronized high-frequency EEG pattern for a time and then returns to the low-frequency sleep pattern. The period of high-frequency EEG that occurs during sleep is called *paradoxical sleep,* because the EEG is more like that of an awake, alert person than of one who is asleep. Another name is *rapid eye movement* (REM) sleep, because associated with the high-frequency EEG is a large amount of rapid eye movement beneath the closed eyelids. This phenomenon is often associated with dreaming, although it has not been shown conclusively that dreaming is related to REM sleep.

The various frequency ranges of the EEG have arbitrarily been given Greek letter designations because frequency seems to be the most prominent feature of an EEG pattern. Electroencephalographers do not agree on the exact ranges, but most classify the EEG frequency bands or rhythms approximately as follows:

Below 3½ Hz	delta
From 3½ Hz to about 8 Hz	theta
From about 8 Hz to about 13 Hz	alpha
Above 13 Hz	beta

Portions of some of these ranges have been given special designations, as have certain subbands that fall on or near the stated boundaries. Most humans seem to develop EEG patterns in the alpha range when they are relaxed with their eyes closed. This condition seems to represent a form of synchronization, almost like a "natural" or "idling" frequency of the brain. As soon as the person becomes alert or begins "thinking," the alpha rhythm disappears and is replaced with a "desynchronized" pattern, generally in the beta range. Much research is presently devoted to attempts to learn the physiological sources in the brain responsible for these phenomena, but so far nothing conclusive has resulted.

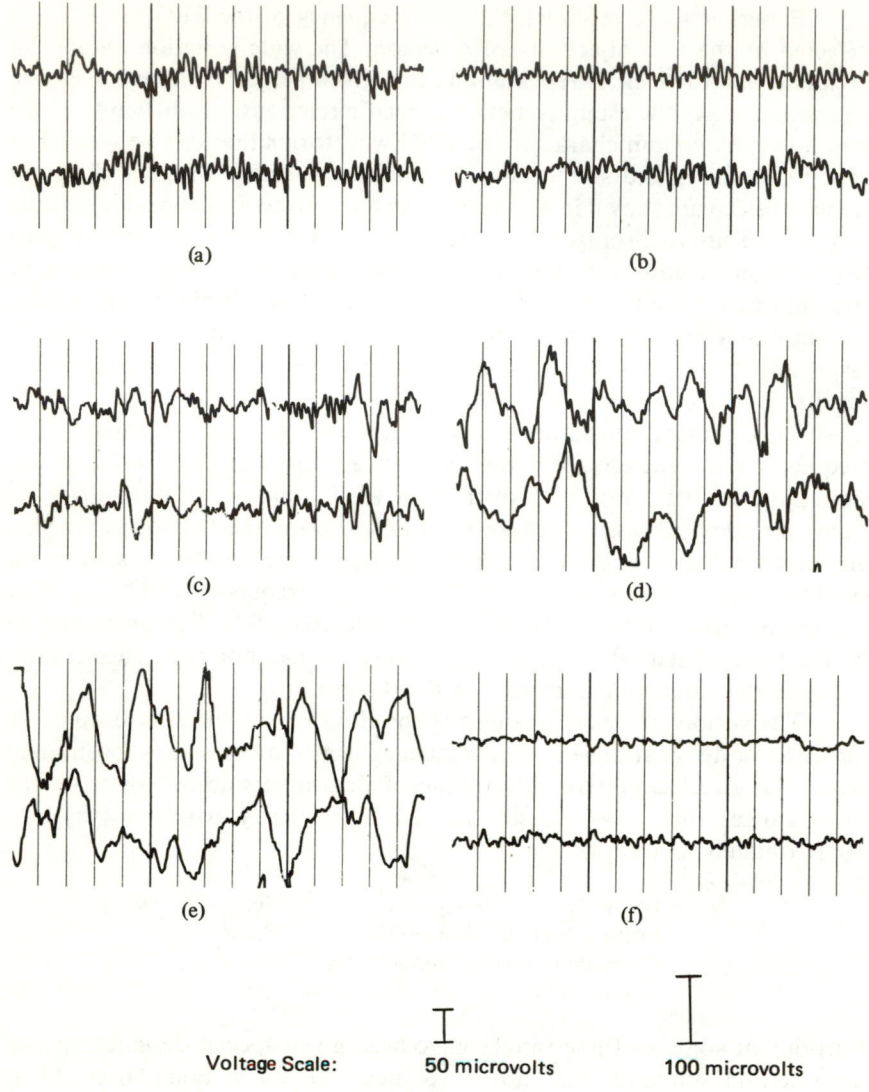

Figure 3.8. Typical human EEG patterns for different stages of sleep. In each case the upper record is from the left frontal region of the brain and the lower tracing is from the right occipital region. (a) Awake and alert—mixed EEG frequencies; (b) Stage 1—subject is drowsy and produces large amount of alpha waves; (c) Stage 2—light sleep; (e) Stage 4—deeper slow wave sleep; (f) Paradoxical or rapid eye movement (REM) sleep. (Courtesy Veterans Administration Hospital, Sepulveda, CA.)

Experiments in biofeedback have shown that under certain conditions, people can learn to control their EEG patterns to some extent when information concerning their EEG is fed back to them either visibly or audibly. The reader is referred to the section on biofeedback in Chapter 11.

As indicated, the frequency content of the EEG pattern seems to be extremely important. In addition, phase relationships between similar EEG patterns from different parts of the brain are also of great interest. Information of this type may lead to discoveries of EEG sources and will, hopefully, provide additional knowledge regarding the functioning of the brain.

Another form of EEG measurement is the *evoked response*. This is a measure of the "disturbance" in the EEG pattern that results from external stimuli, such as a flash of light or a click of sound. Since these "disturbance" responses are quite repeatable from one flash or click to the next, the evoked response can be distinguished from the remainder of EEG activity, and from the noise, by averaging techniques. These techniques, as well as other methods of measuring EEG, are covered in Chapter 10.

3.3.3. Electromyogram (EMG)

The bioelectric potentials associated with muscle activity constitute the *electromyogram,* abbreviated EMG. These potentials may be measured at the surface of the body near a muscle of interest or directly from the muscle by penetrating the skin with needle electrodes. Since most EMG measurements are intended to obtain an indication of the amount of activity of a given muscle, or group of muscles, rather than of an individual muscle fiber, the pattern is usually a summation of the individual action potentials from the fibers constituting the muscle or muscles being measured. As with the EEG, EMG electrodes pick up potentials from all muscles within the range of the electrodes. This means that potentials from nearby large muscles may interfere with attempts to measure the EMG from smaller muscles, even though the electrodes are placed directly over the small muscles. Where this is a problem, needle electrodes inserted directly into the muscle are required.

As stated in Section 3.1, the action potential of a given muscle (or nerve fiber) has a fixed magnitude, regardless of the intensity of the stimulus that generates the response. Thus, in a muscle, the intensity with which the muscle acts does not increase the net height of the action potential pulse but does increase the rate with which each muscle fiber fires and the number of fibers that are activated at any given time. The amplitude of the measured EMG waveform is the instantaneous sum of all the action potentials generated at any given time. Because these action potentials occur in both positive and negative polarities at a given pair of electrodes, they sometimes add and sometimes cancel. Thus, the EMG waveform appears very much like a random-noise waveform, with the energy of the signal a function of the amount of muscle activity and electrode placement. Typical EMG waveforms are shown in Figure 3.9. Methods and instrumentation for measuring EMG are described in Chapter 10.

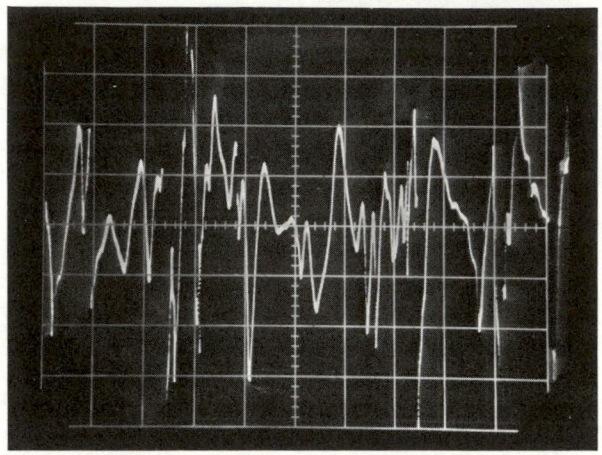

Figure 3.9. Typical electromyogram waveform. EMG of normal "interference pattern" with full strength muscle contraction producing obliteration of the baseline. Sweep speed is 10 milliseconds per cm; amplitude is 1 millivolt per cm. (Courtesy of the Veterans Administration Hospital, Portland, OR.)

3.3.4. Other Bioelectric Potentials

In addition to the three most significant bioelectric potentials (ECG, EEG, and EMG), several other electric signals can be obtained from the body, although most of them are special variations of EEG, EMG, or nerve-firing patterns. Some of the more prominent ones are the following:

1. *Electroretinogram (ERG):*   A record of the complex pattern of bioelectric potentials obtained from the retina of the eye. This is usually a response to a visual stimulus.
2. *Electro-oculogram (EOG):*   A measure of the variations in the corneal-retinal potential as affected by the position and movement of the eye.
3. *Electrogastrogram (EGG):*   The EMG patterns associated with the peristaltic movements of the gastrointestinal tract.

4

Electrodes

In observing the measurement of the electrocardiogram (ECG) or the result of some other form of bioelectric potentials as discussed in Chapter 3, a conclusion could easily be reached that the measurement electrodes are simply electrical terminals or contact points from which voltages can be obtained at the surface of the body. Also, the purpose of the electrolyte paste or jelly often used in such measurements might be assumed to be only the reduction of skin impedance in order to lower the overall input impedance of the system. These conclusions, however, are incorrect and do not satisfy the theory that explains the origin of bioelectric potentials. It must be realized that the bioelectric potentials generated in the body are ionic potentials, produced by ionic current flow. Efficient measurement of these ionic potentials requires that they be converted into electronic potentials before they can be measured by conventional methods. It was the realization of this fact that led to the development of the modern noise-free, stable measuring devices now available.

Devices that convert *ionic* potentials into electronic potentials are called *electrodes*. The theory of electrodes and the principles that govern their design are inherent in an understanding of the measurement of bioelectric potentials. This same theory also applies to electrodes used in chemical transducers, such as those used to measure pH, P_{O_2}, and P_{CO_2} of the blood. This chapter deals first with the basic theory of electrodes and then with the various types used in biomedical instrumentation.

4.1. ELECTRODE THEORY

The interface of metallic ions in solution with their associated metals results in an electrical potential that is called the *electrode potential*. This potential is a result of the difference in diffusion rates of ions into and out of the metal. Equilibrium is produced by the formation of a layer of charge at the interface. This charge is really a double layer, with the layer nearest the metal being of one polarity and the layer next to the solution being of opposite polarity. Nonmetallic materials, such as hydrogen, also have electrode potentials when interfaced with their associated ions in solution. The electrode potentials of a wide variety of metals and alloys are listed in Table 4.1.

It is impossible to determine the absolute electrode potential of a single electrode, for measurement of the potential across the electrode and its ionic solution would require placing another metallic interface in the solution. Therefore all electrode potentials are given as relative values and must be stated in terms of some reference. By international agreement, the normal hydrogen electrode was chosen as the reference standard and arbitrarily assigned an electrode potential of zero volts. All the electrode potentials listed in Table 4.1 are given with respect to the hydrogen electrode. They represent the potentials that would be obtained across the stated electrode and a hydrogen electrode if both were placed in a suitable ionic solution.

Another source of an electrode potential is the unequal exchange of ions across a membrane that is semipermeable to a given ion when the membrane separates liquid solutions with different concentrations of that ion. An equation relating the potential across the membrane and the two concentrations of the ion is called the *Nernst equation* and can be stated as follows:

$$E = -\frac{RT}{nF} \ln \frac{C_1 \, f_1}{C_2 \, f_2}$$

where R = gas constant (8.315×10^7 ergs/mole/degree Kelvin)
  T = absolute temperature, degrees Kelvin

n = valence of the ion (the number of electrons added or removed to ionize the atom)

F = Faraday constant (96,500 coulombs)

C_1, C_2 = two concentrations of the ion on the two sides of the membrane

f_1, f_2 = respective activity coefficients of the ion on the two sides of the membrane

Unfortunately, the gas constant, $R = 8.315 \times 10^7$, is in electromagnetic cgs units, whereas the Faraday constant, $F = 96,500$, is in absolute coulombs. These units are not compatible. To solve the Nernst equation in electromagnetic cgs units, F must be divided by 10 (there are 10 absolute coulombs in each electromagnetic cgs unit). This calculation gives

Table 4.1. ELECTRODE POTENTIALS[a]

Electrode Reaction	*E_0 (volts)*	*Electrode Reaction*	*E_0 (volts)*
Li $\rightleftarrows$ Li+	-3.045	V $\rightleftarrows$ V^3+	-0.876
Rb $\rightleftarrows$ Rb+	-2.925	Zn $\rightleftarrows$ Zn^2+	-0.762
K $\rightleftarrows$ K+	-2.925	Cr $\rightleftarrows$ Cr^2+	-0.74
Cs $\rightleftarrows$ Cs+	-2.923	Ga $\rightleftarrows$ Ga^2+	-0.53
Ra $\rightleftarrows$ Ra^2+	-2.92	Fe $\rightleftarrows$ Fe^2+	-0.440
Ba $\rightleftarrows$ Ba^2+	-2.90	Cd $\rightleftarrows$ Cd^2+	-0.402
Sr $\rightleftarrows$ Sr^2+	-2.89	In $\rightleftarrows$ In^2+	-0.342
Ca $\rightleftarrows$ Ca^2+	-2.87	Tl $\rightleftarrows$ Tl+	-0.336
Na $\rightleftarrows$ Na+	-2.714	Mn $\rightleftarrows$ Mn^3+	-0.283
La $\rightleftarrows$ La^3+	-2.52	Co $\rightleftarrows$ Co^2+	-0.277
Mg $\rightleftarrows$ Mg^2+	-2.37	Ni $\rightleftarrows$ Ni^2+	-0.250
Am $\rightleftarrows$ Am^3+	-2.32	Mo $\rightleftarrows$ Mo^3+	-0.2
Pu $\rightleftarrows$ Pu^3+	-2.07	Ge $\rightleftarrows$ Ge^4+	-0.15
Th $\rightleftarrows$ Th^4+	-1.90	Sn $\rightleftarrows$ Sn^2+	-0.136
Np $\rightleftarrows$ Np^3+	-1.86	Pb $\rightleftarrows$ Pb^2+	-0.126
Bc $\rightleftarrows$ Bc^2+	-1.85	Fe $\rightleftarrows$ Fe^3+	-0.036
U $\rightleftarrows$ U^3+	-1.80	D$_2$ $\rightleftarrows$ D+	-0.0034
Hf $\rightleftarrows$ Hf^4+	-1.70	H$_2$ $\rightleftarrows$ H+	0.000
Al $\rightleftarrows$ Al^3+	-1.66	Cu $\rightleftarrows$ Cu^2+	$+0.337$
Ti $\rightleftarrows$ Ti^2+	-1.63	Cu $\rightleftarrows$ Cu+	$+0.521$
Zr $\rightleftarrows$ Zr^4+	-1.53	Hg $\rightleftarrows$ Hg$_2$^{2}+	$+0.789$
U $\rightleftarrows$ U^4+	-1.50	Ag $\rightleftarrows$ Ag+	$+0.799$
Np $\rightleftarrows$ Np^4+	-1.354	Rh $\rightleftarrows$ Rh^3+	$+0.80$
Pu $\rightleftarrows$ Pu^4+	-1.28	Hg $\rightleftarrows$ Hg^2+	$+0.857$
Ti $\rightleftarrows$ Ti^3+	-1.21	Pd $\rightleftarrows$ Pd^2+	$+0.987$
V $\rightleftarrows$ V^2+	-1.18	Ir $\rightleftarrows$ Ir^3+	$+1.000$
Mn $\rightleftarrows$ Mn^2+	-1.18	Pt $\rightleftarrows$ Pt^2+	$+1.19$
Nb $\rightleftarrows$ Nb^3+	-1.1	Au $\rightleftarrows$ Au^3+	$+1.50$
Cr $\rightleftarrows$ Cr^2+	-0.913	Au $\rightleftarrows$ Au+	$+1.68$

[a]Reproduced by permission from Brown, J. H. V., J. E. Jacobs, and L. Stark, *Biomedical Engineering*, F. A. Davis Company, Philadelphia, 1971.

the membrane potential in abvolts, the electromagnetic cgs unit for potential. However, 1 standard volt equals 10^8 abvolts; therefore, to convert the membrane potential into standard volts, the entire equation must be multiplied by a constant 10^{-8}.

The activity coefficients, f_1 and f_2, depend on such factors as the charges of all ions in the solution and the distance between ions. The product, $C_1 f_1$, of a concentration and its associated activity coefficient is called the *activity* of the ion responsible for the electrode potential. From the Nernst equation it can be seen that the electrode potential across the membrane is proportional to the logarithm of the ratio of the activities of the subject ion on the two sides of the membrane. In a very dilute solution the activity coefficient f approaches unity, and the electrode potential becomes a function of the logarithm of the ratio of the two concentrations.

In electrodes used for the measurement of bioelectric potentials, the electrode potential occurs at the interface of a metal and an electrolyte, whereas in biochemical transducers both membrane barriers and metal-electrolyte interfaces are used. The sections that follow describe electrodes of both types.

4.2.   BIOPOTENTIAL ELECTRODES

A wide variety of electrodes can be used to measure bioelectric events, but nearly all can be classified as belonging to one of three basic types:

1. *Microelectrodes:* Electrodes used to measure bioelectric potentials near or within a single cell.
2. *Skin surface electrodes:* Electrodes used to measure ECG, EEG, and EMG potentials from the surface of the skin.
3. *Needle electrodes:* Electrodes used to penetrate the skin to record EEG potentials from a local region of the brain or EMG potentials from a specific group of muscles.

All three types of biopotential electrodes have the metal–electrolyte interface described in the previous section. In each case, an electrode potential is developed across the interface, proportional to the exchange of ions between the metal and the electrolytes of the body. The double layer of charge at the interface acts as a capacitor. Thus, the equivalent circuit of biopotential electrode in contact with the body consists of a voltage in series with a resistance–capacitance network of the type shown in Figure 4-1.

Since measurement of bioelectric potentials requires two electrodes, the voltage measured is really the difference between the instantaneous potentials of the two electrodes, as shown in Figure 4-2. If the two electrodes are of the same type, the difference is usually small and depends

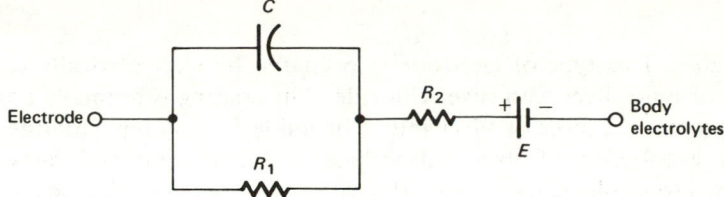

Figure 4.1. Equivalent circuit of biopotential electrode interface.

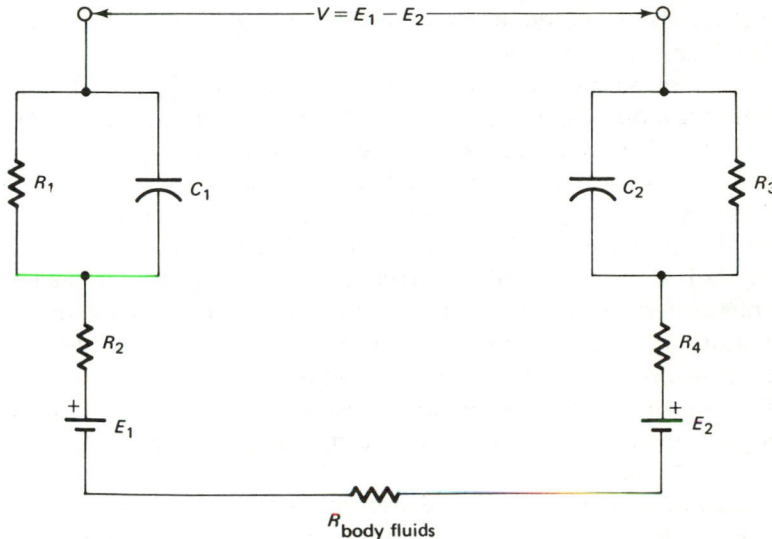

Figure 4.2. Measurement of biopotentials with two electrodes—equivalent circuit.

essentially on the actual difference of ionic potential between the two points of the body from which measurements are being taken. If the two electrodes are different, however, they may produce a significant dc voltage that can cause current to flow through both electrodes as well as through the input circuit of the amplifier to which they are connected. The dc voltage due to the difference in electrode potentials is called the *electrode offset voltage.* The resulting current is often mistaken for a true physiological event. Even two electrodes of the same material may produce a small electrode offset voltage.

In addition to the electrode offset voltage, experiments have shown that the chemical activity that takes place within an electrode can cause voltage fluctuations to appear without any physiological input. Such variations may appear as noise on a bioelectric signal. This noise can be reduced by proper choice of materials or, in most cases, by special treatment, such as coating the electrodes by some electrolytic method to improve stability. It has been found that, electrochemically, the *silver–silver chloride electrode* is very stable. This type of electrode is prepared by electrolytically coating a piece of pure silver with silver chloride. The coating is normally done by placing a cleaned piece of silver into a bromide-free sodium chloride solution. A second piece of silver is also placed in the solution, and the two are connected to a voltage source such that the electrode to be chlorided is made positive with respect to the other. The silver ions combine with the chloride ions from the salt to produce neutral silver chloride molecules that coat the silver electrode. Some variations in the process are used to produce electrodes with specific characteristics.

The resistance-capacitance networks shown in Figures 4.1 and 4.2 represent the impedance of the electrodes (one of their most important characteristics) as fixed values of resistance and capacitance. Unfortunately, the impedance is not constant. The impedance is frequency-dependent because of the effect of the capacitance. Furthermore, both the electrode potential and the impedance are varied by an effect called *polarization.*

Polarization is the result of direct current passing through the metal-electrolyte interface. The effect is much like that of charging a battery with the polarity of the charge opposing the flow of current that generates the charge. Some electrodes are designed to avoid or reduce polarization. If the amplifier to which the electrodes are connected has an extremely high input impedance, the effect of polarization or any other change in electrode impedance is minimized.

Size and type of electrode are also important in determining the electrode impedance. Larger electrodes tend to have lower impedances. Surface electrodes generally have impedances of 2 to 10 k Ω, whereas small needle electrodes and microelectrodes have much higher impedances. For best results in reading or recording the potentials measured by the electrodes, the input impedance of the amplifier must be several times that of the electrodes.

4.2.1. Microelectrodes

Microelectrodes are electrodes with tips sufficiently small to penetrate a single cell in order to obtain readings from within the cell. The tip must be small enough to permit penetration without damaging the cell. This action is usually complicated by the difficulty of accurately positioning an electrode with respect to a cell.

Microelectrodes are generally of two types: metal and micropipet. Metal microelectrodes are formed by electrolytically etching the tip of a fine tungsten or stainless-steel wire to the desired size. Then the wire is coated almost to the tip with an insulating material. Some electrolytic processing can also be performed on the tip to lower the impedance. The metal–ion interface takes place where the metal tip contacts the electrolytes either inside or outside the cell.

The micropipet type of microelectrode is a glass micropipet with the tip drawn out to the desired size [usually about 1 micron (now more commonly called micrometer, μm) in diameter]. The micropipet is filled with an electrolyte compatible with the cellular fluids. This type of microelectrode has a dual interface. One interface consists of a metal wire in contact with the electrolyte solution inside the micropipet, while the other is the interface between the electrolyte inside the pipet and the fluids inside or immediately outside the cell.

A commercial type of microelectrode is shown in Figure 4.3. In this electrode a thin film of precious metal is bonded to the outside of a drawn glass microelectrode. The manufacturer claims such advantages as lower impedance than the micropipet electrode, infinite shelf life, repeatable and reproducible performance, and easy cleaning and maintenance. The metal-electrolyte interface is between the metal film and the electrolyte of the cell.

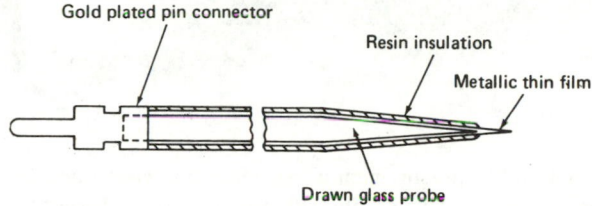

Figure 4.3. Commercial microelectrode with metal film on glass. (Courtesy of Transidyne General Corporation, Ann Arbor, MI.)

Microelectrodes, because of their small surface areas, have impedances well up into the megohms. For this reason, amplifiers with extremely high impedances are required to avoid loading the circuit and to minimize the effects of small changes in interface impedance.

4.2.2. Body Surface Electrodes

Electrodes used to obtain bioelectric potentials from the surface of the body are found in many sizes and forms. Although any type of surface electrode can be used to sense ECG, EEG, or EMG potentials, the larger electrodes are usually associated with ECG, since localization of the measurement is not important, whereas smaller electrodes are used in EEG and EMG measurements.

The earliest bioelectric potential measurements used *immersion electrodes,* which were simply buckets of saline solution into which the subject placed his hands and feet, one bucket for each extremity. As might be expected, this type of electrode (Figure 4.4) presented many difficulties, such as restricted position of the subject and danger of electrolyte spillage.

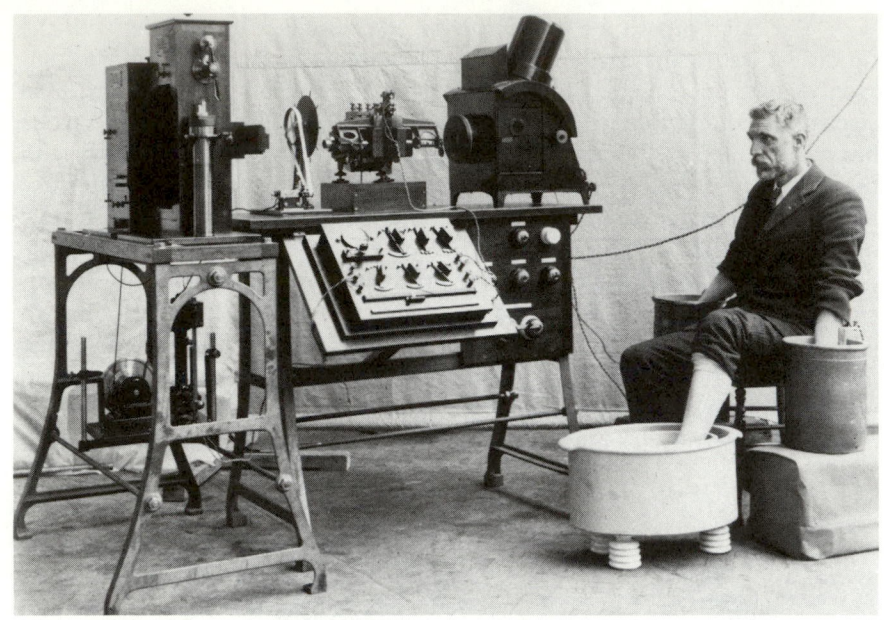

Figure 4.4. ECG measurement using immersion electrodes. Original Cambridge electrocardiograph (1912) built for Sir Thomas Lewis. Produced under agreement with Prof. Willem Einthoven, the father of electrocardiography. (Courtesy of Cambridge Instruments, Inc., Cambridge, MA.)

A great improvement over the immersion electrodes were the plate electrodes, first introduced about 1917. Originally, these electrodes were separated from the subject's skin by cotton or felt pads soaked in a strong saline solution. Later a conductive jelly or paste (an electrolyte) replaced the soaked pads and metal was allowed to contact the skin through a thin coat of jelly. Plate electrodes of this type are still in use today. An example is shown in Figure 4.5.

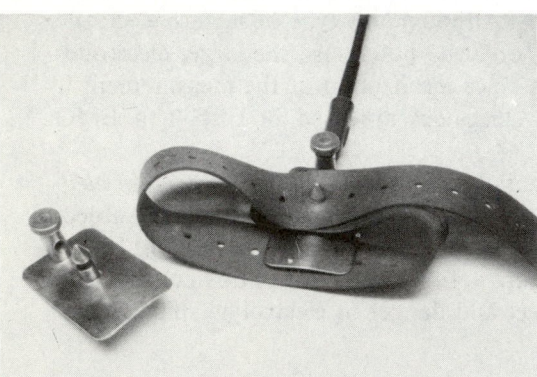

Figure 4.5. Metal plate electrode. These plates are usually made of, or plated with, silver, nickel, or some similar alloy.

Figure 4.6. Suction cup electrode.

Another fairly old type of electrode still in use is the suction-cup electrode shown in Figure 4.6. In this type, only the rim actually contacts the skin.

One of the difficulties in using plate electrodes is the possibility of electrode slippage or movement. This also occurs with the suction-cup electrode after a sufficient length of time. A number of attempts were made to overcome this problem, including the use of adhesive backing and a surface resembling a nutmeg grater that penetrates the skin to lower the contact impedance and reduce the likelihood of slippage.

All the preceding electrodes suffer from a common problem. They are all sensitive to movement, some to a greater degree than others. Even the slightest movement changes the thickness of the thin film of electrolyte between metal and skin and thus causes changes in the electrode potential and impedance. In many cases, the potential changes are so severe that they completely block the bioelectric potentials the electrodes attempt to measure. The adhesive tape and ''nutmeg grater'' electrodes reduce this movement artifact by limiting electrode movement and reducing interface impedance, but neither is satisfactorily insensitive to movement.

Later, a new type of electrode, the *floating electrode,* was introduced in varying forms by several manufacturers. The principle of this electrode is to practically eliminate movement artifact by avoiding any direct contact of the metal with the skin. The only conductive path between metal and skin is the electrolyte paste or jelly, which forms an electrolyte bridge. Even with the electrode surface held at a right angle with the skin surface, performance is not impaired as long as the electrolyte bridge maintains contact with both the skin and the metal. Figure 4.7 shows a cross section of a floating electrode, and Figure 4.8 shows a commercially available configuration of the floating electrode.

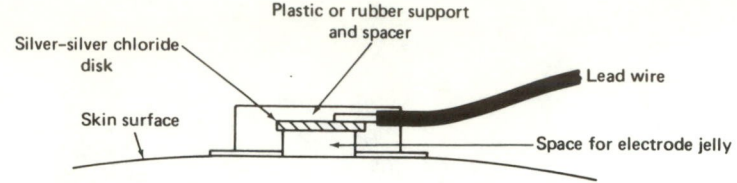

Figure 4.7. Diagram of floating type skin surface electrode.

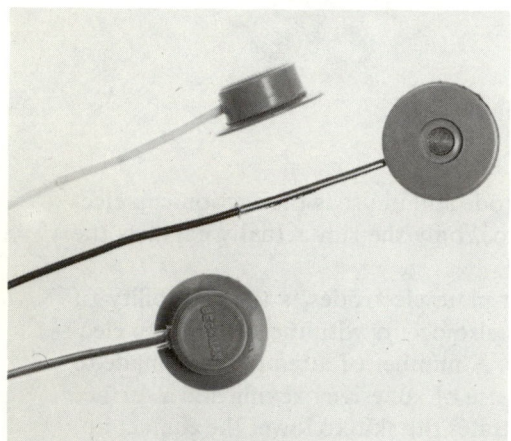

Figure 4.8. Floating skin surface electrode. (Courtesy of Beckman Instruments, Inc., Fullerton, CA.)

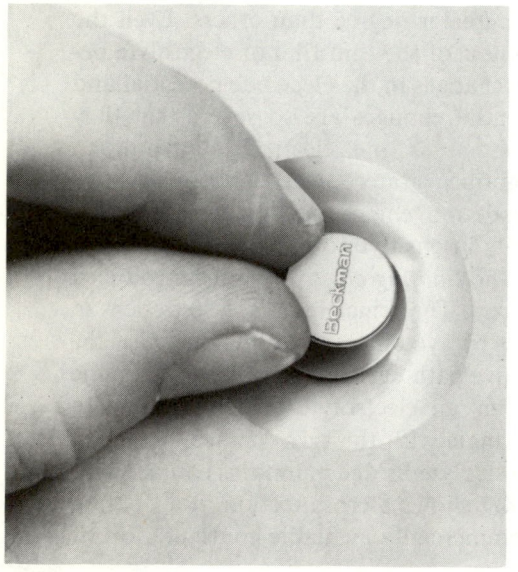

Figure 4.9. Application of floating type skin surface electrode. (Courtesy of Beckman Instruments, Inc., Fullerton, CA.)

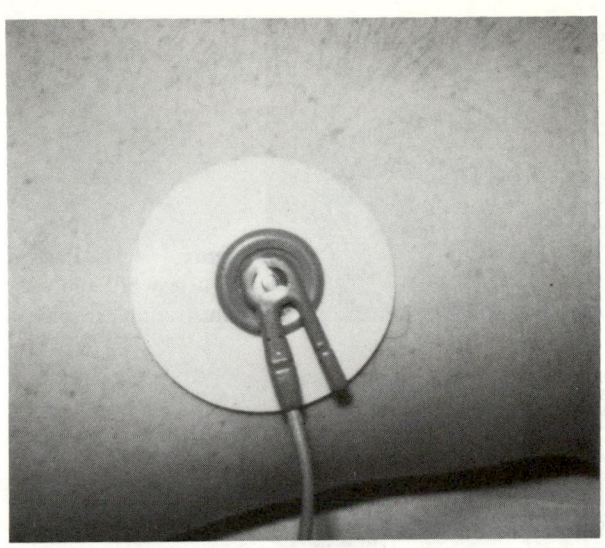

Figure 4.10. Disposable electrodes.

Floating electrodes are generally attached to the skin by means of two-sided adhesive collars (or rings), which adhere to both the plastic surface of the electrode and the skin. Figure 4.9 shows an electrode in position for biopotential measurement.

Special problems encountered in the monitoring of the ECG of astronauts during long periods of time, and under conditions of perspiration and considerable movement, led to the development of *spray-on electrodes,* in which a small spot of conductive adhesive is sprayed or painted over the skin, which had previously been treated with an electrolyte coating.

Various types of *disposable electrodes* have been introduced in recent years to eliminate the requirement for cleaning and care after each use. An example is shown in Figure 4.10. Primarily intended for ECG monitoring, these electrodes can also be used for EEG and EMG as well. In general, disposable electrodes are of the floating type with simple snap connectors by which the leads, which are reusable, are attached. Although some disposable electrodes can be reused several times, their cost is usually low enough that cleaning for reuse is not warranted. They come pregelled, ready for immediate use.

Special types of surface electrodes have been developed for other applications. For example, a special *ear-clip electrode* (Figure 4.11) was developed for use as a reference electrode for EEG measurements. Scalp *surface electrodes* for EEG are usually small disks about 7 mm in diameter or small solder pellets that are placed on the cleaned scalp, using an electrolyte paste. This type of electrode is shown in Figure 4.12.

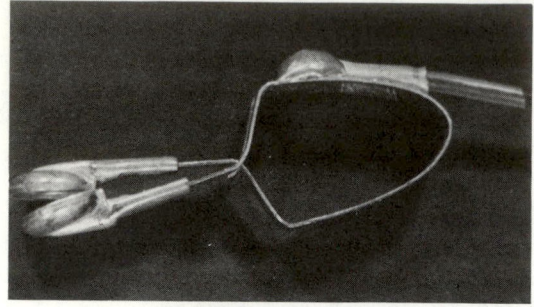

Figure 4.11. Ear-clip electrode. (Courtesy of Sepulveda Veterans Administration Hospital.)

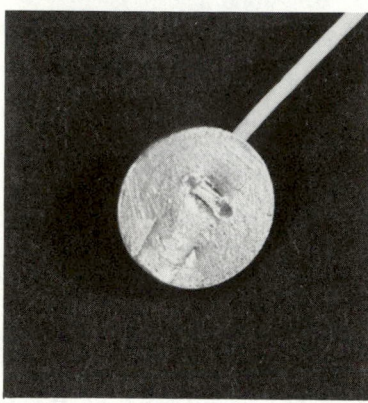

Figure 4.12. EEG scalp surface electrode. (Courtesy of Sepulveda Veterans Administration Hospital.)

4.2.3. Needle Electrodes

To reduce interface impedance and, consequently, movement artifacts, some electroencephalographers use small subdermal needles to penetrate the scalp for EEG measurements. These needle electrodes, shown in Figure 4.13, are not inserted into the brain; they merely penetrate the skin. Generally, they are simply inserted through a small section of the skin just beneath the surface and parallel to it.

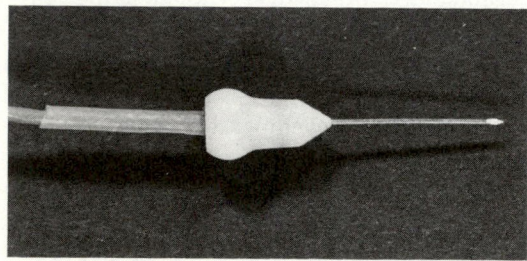

Figure 4.13. Subdermal needle electrode for EEG. (Courtesy of Sepulveda Veterans Administration Hospital.)

In animal research (and occasionally in man) longer needles are actually inserted into the brain to obtain localized measurement of potentials from a specific part of the brain. This process requires longer needles precisely located by means of a map or atlas of the brain. Sometimes a special instrument, called a *stereotaxic instrument,* is used to hold the animal's head and guide the placement of electrodes. Often these electrodes are implanted to permit repeated measurements over an extended period of time. In this case, a connector is cemented to the animal's skull and the incision through which the electrodes were implanted is allowed to heal.

In some research applications, simultaneous measurement from various depths in the brain along a certain axis is required. Special multiple-depth electrodes have been developed for this purpose. This type of electrode usually consists of a bundle of fine wires, each terminating at a different depth or each having an exposed conductive surface at a specific, but different, depth. These wires are generally brought out to a connector at the surface of the scalp and are often cemented to the skull.

Needle electrodes for EMG consist merely of fine insulated wires, placed so that their tips, which are bare, are in contact with the nerve, muscle, or other tissue from which the measurement is made. The remainder of the wire is covered with some form of insulation to prevent shorting. Wire electrodes of copper or platinum are often used for EMG pickup from specific muscles. The wires are either surgically implanted or introduced by means of a hypodermic needle that is later withdrawn, leaving the wire electrode in place. With this type of electrode, the metal–electrolyte interface takes place between the uninsulated tip of the wire and the electrolytes of the body, although the wire is dipped into an electrolyte paste before insertion in some cases. The hypodermic needle is sometimes a part of the electrode configuration and is not withdrawn. Instead, the wires forming the electrodes are carried inside the needle, which creates the hole necessary for insertion, protects the wires, and acts as a grounded shield. A single wire inside the needle serves as a *unipolar electrode,* which measures the potentials at the point of contact with respect to some indifferent reference. If two wires are placed inside the needle, the measurement is called *bipolar* and provides a very localized measurement between the two wire tips.

Electrodes for measurement from beneath the skin need not actually take the form of needles, however. Surgical clips penetrating the skin of a mouse or rat in the spinal region provide an excellent method of measuring the ECG of an essentially unrestrained, unanesthetized animal. Conductive catheters permit the recording of the ECG from within the esophagus or even from within the chambers of the heart itself.

Needle electrodes and other types of electrodes that create an interface beneath the surface of the skin seem to be less susceptible to movement artifacts than surface electrodes, particularly those of the older types. By making direct contact with the subdermal tissue or the intercellular fluids, these electrodes also seem to have lower impedances than surface electrodes of comparable interface area.

4.3. BIOCHEMICAL TRANSDUCERS

At the beginning of this chapter it was stated that an electrode potential is generated either at a metal–electrolyte interface or across a semipermeable membrane separating two different concentrations of an ion that can diffuse through the membrane. Both methods are used in transducers designed to measure the concentration of an ion or of a certain gas dissolved in blood or some other liquid. Also, as stated earlier, since it is impossible to have a single electrode interface to a solution, a second electrode is required to act as a reference. If both electrodes were to exhibit the same response to a given change in concentration of the measured solution, the potential measured between them would not be related to concentration and would, therefore, be useless as a measurement parameter. The usual method of measuring concentrations of ions or gases is to use one electrode (sometimes called the *indicator* or *active electrode*) that is sensitive to the substance or ion being measured and to choose the second, or *reference electrode,* of a type that is insensitive to that substance.

4.3.1. Reference Electrodes

As stated in Section 4.1, the hydrogen gas/hydrogen ion interface has been designated as the reference interface and was arbitrarily assigned an electrode potential of zero volts. For this reason, it would seem logical that the hydrogen electrode should actually be used as the reference in biochemical measurements. Hydrogen electrodes can be built and are available commercially. These electrodes make use of the principle that an inert metal, such as platinum, readily absorbs hydrogen gas. If a properly treated piece of platinum is partially immersed in the solution containing hydrogen ions and is also exposed to hydrogen gas, which is passed through the electrode, an electrode potential is formed. The electrode lead is attached to the platinum.

Unfortunately, the hydrogen electrode is not sufficiently stable to serve as a good reference electrode. Furthermore, the problem of maintaining the supply of hydrogen to pass through the electrode during a measurement limits its usefulness to a few special applications. However, since measurement of electrochemical concentrations simply requires a change of potential proportional to a change in concentration, the electrode potential

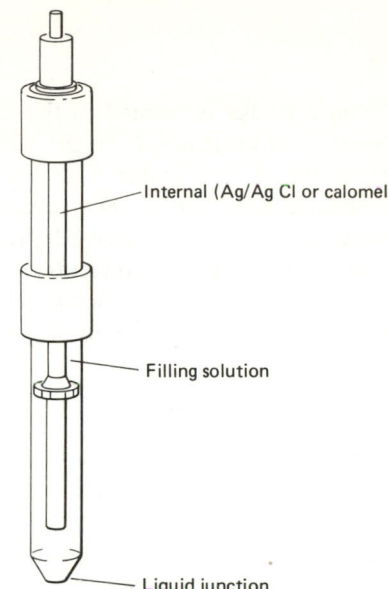

Figure 4.14. Reference electrode—basic configuration. (Courtesy Beckman Instruments, Inc., Fullerton, CA.)

Internal (Ag/Ag Cl or calomel)

Filling solution

Liquid junction

of the reference electrode can be any amount, as long as it is stable and does not respond to any possible changes in the composition of the solution being measured. Thus, the search for a good reference electrode is essentially a search for the most stable electrode available. Two types of electrodes have interfaces sufficiently stable to serve as reference electrodes—the silver–silver chloride electrode and the calomel electrode. Their basic configurations are shown in Figure 4.14.

The *silver–silver chloride electrode* used as a reference in electrochemical measurements utilizes the same type of interface described in Section 4.2 for bioelectric potential electrodes. In the chemical transducer, the ionic (silver chloride) side of the interface is connected to the solution by an electrolyte bridge, usually a dilute potassium chloride (KC1) filling solution which forms a liquid junction with the sample solution. The electrode can be successfully employed as a reference electrode if the KC1 solution is also saturated with precipitated silver chloride. The electrode potential for the silver–silver chloride reference electrode depends on the concentration of the KC1. For example, with a 0.01-mole*-solution, the potential is 0.343 V, whereas for a 1.0-mole solution the potential is only 0.236 V.

An equally popular reference electrode is the *calomel electrode.* Calomel is another name for mercurous chloride, a chemical combination of mercury and chloride ions. The interface between mercury and mercurous chloride generates the electrode potential. By placing the calomel side of the interface in a potassium chloride (KC1) filling solution, an elec-

---

*A 0.01-mole solution of a substance is defined as 0.01 mole of the substance dissolved in 1 liter of solution. A mole is the quantity of the substance that has a weight equal to its molecular weight, usually in grams.

trolytic bridge is formed to the sample solution from which the measurement is to be made. Like the silver–silver chloride electrode, the calomel electrode is very stable over long periods of time and serves well as a reference electrode in many electrochemical measurements. Also, like the silver–silver chloride electrode, the electrode potential of the calomel electrode depends on the concentration of KCl. An electrode with a 0.01-mole solution of KCl has an electrode potential of 0.388 V, whereas a saturated KCl solution (about 3.5 moles) has a potential of only 0.247 V.

4.3.2. The pH Electrode

Perhaps the most important indicator of chemical balance in the body is the pH of the blood and other body fluids. The pH is directly related to the hydrogen ion concentration in a fluid. Specifically, it is the logarithm of the reciprocal of the H^+ ion concentration. In equation form,

$$pH = -\log_{10}[H^+] = \log_{10} \frac{1}{[H^+]}$$

The pH is a measure of the acid–base balance of a fluid. A neutral solution (neither acid nor base) has a pH of 7. Lower pH numbers indicate acidity, whereas higher pH values define a basic solution. Most human body fluids are slightly basic. The pH of normal arterial blood ranges between 7.38 and 7.42. The pH of venous blood is 7.35, because of the extra CO_2.

Because a thin glass membrane allows passage of only hydrogen ions in the form of H_3O^+, a glass electrode provides a "membrane" interface for hydrogen. The principle is illustrated in Figure 4.15. Inside the glass bulb is a highly acidic buffer solution. Measurement of the potential across the glass interface is achieved by placing a silver–silver chloride electrode in the solution inside the glass bulb and a calomel or silver–silver chloride reference electrode in the solution in which the pH is being measured. In the measurement of pH and, in fact, any electrochemical measurement, each of the two electrodes required to obtain the measurement is called a *half-cell*. The electrode potential for a half-cell is sometimes called the *half-cell potential*. For pH measurement, the glass electrode with the silver–silver chloride electrode inside the bulb is considered one half-cell, while the calomel reference electrode constitutes the other half-cell.

To facilitate the measurement of the pH of a solution, combination electrodes of the type shown in Figure 4.16 are available, with both the pH glass electrode and reference electrode in the same enclosure.

The glass electrode is quite adequate for pH measurements in the physiological range (around pH 7), but may produce considerable error at the extremes of the range (near pH of zero or 13 to 14). Special types of pH

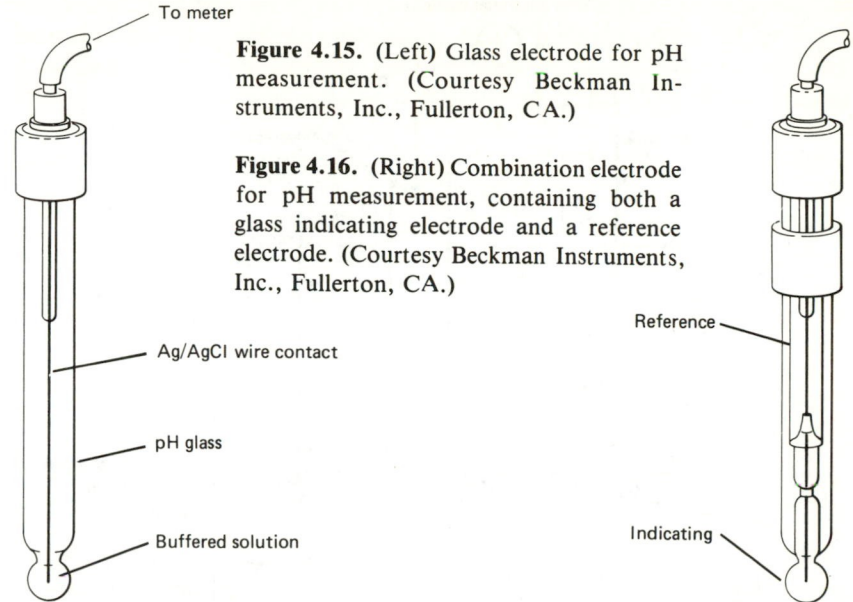

Figure 4.15. (Left) Glass electrode for pH measurement. (Courtesy Beckman Instruments, Inc., Fullerton, CA.)

Figure 4.16. (Right) Combination electrode for pH measurement, containing both a glass indicating electrode and a reference electrode. (Courtesy Beckman Instruments, Inc., Fullerton, CA.)

electrodes are available for the extreme ranges. Glass electrodes are also subject to some deterioration after prolonged use but can be restored repeatedly by etching the glass in a 20 percent ammonium bifluoride solution.

The type of glass used for the membrane has much to do with the pH response of the electrode. Special hydroscopic glass that readily absorbs water provides the best pH response.

Modern pH electrodes have impedances ranging from 50 to 500 megohms ($M\Omega$). Thus, the input of the meter that measures the potential difference between the glass electrode and the reference electrode must have an extremely high input impedance. Most pH meters employ electrometer inputs.

4.3.3. Blood Gas Electrodes

Among the more important physiological chemical measurements are the partial pressures of oxygen and carbon dioxide in the blood. The partial pressure of a dissolved gas is the contribution of that gas to the total pressure of all dissolved gases in the blood. The partial pressure of a gas is proportional to the quantity of that gas in the blood. The effectiveness of both the respiratory and cardiovascular systems is reflected in these important parameters.

The partial pressure of oxygen, P_{O_2}, often called *oxygen tension,* can be measured both in vitro and in vivo. The basic principle is shown in Figure

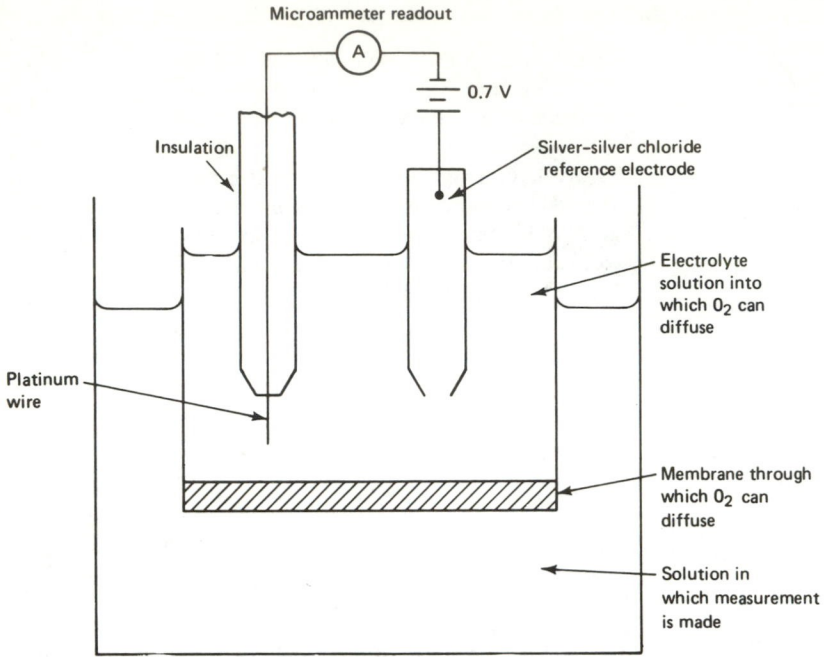

Figure 4.17. Diagram of P_{O_2} electrode with platinum cathode showing principle of operation.

4.17. A fine piece of platinum or some other noble metal wire, embedded in glass for insulation purposes, with only the tip exposed, is placed in an electrolyte into which oxygen is allowed to diffuse. If a voltage of about 0.7 V is applied between the platinum wire and a reference electrode (also placed into the electrolyte), with the platinum wire negative, reduction of the oxygen takes place at the platinum cathode. As a result, an oxidation–reduction current proportional to the partial pressure of the diffused oxygen can be measured. The electrolyte is generally sealed into the chamber that holds the platinum wire and the reference electrode by means of a membrane across which the dissolved oxygen can diffuse from the blood.

The platinum cathode and the reference electrode can be integrated into a single unit (the *Clark electrode*). This electrode can be placed in a cuvette of blood for in vitro measurements, or a micro version can be placed at the tip of a catheter for insertion into various parts of the heart or vascular system for direct in vivo measurements.

One of the problems inherent in this method of measuring P_{O_2} is the fact that the reduction process actually removes a finite amount of the oxygen from the immediate vicinity of the cathode. By careful design and use of proper procedures, modern P_{O_2} electrodes have been able to reduce this potential source of error to a minimum. Another apparent error in P_{O_2} measurement is a gradual reduction of current with time, almost like the polarization effect described for skin surface electrodes in Section 4.2.2. This effect, generally called *aging,* has also been minimized in modern P_{O_2} electrodes.

The measurement of the partial pressure of carbon dioxide, P_{CO_2}, makes use of the fact that there is a linear relationship between the logarithm of the P_{CO_2} and the pH of a solution. Since other factors also influence the pH, measurement of P_{CO_2} is essentially accomplished by surrounding a pH electrode with a membrane selectively permeable to CO_2. A modern, improved type of P_{CO_2} electrode is called the *Severinghaus electrode.* In this type of electrode, the membrane permeable to the CO_2 is made of Teflon, which is not permeable to other ions that might affect the pH. The space between the Teflon and the glass contains a matrix consisting of thin cellophane, glass wool, or sheer nylon. This matrix serves as the support for an aqueous bicarbonate layer into which the CO_2 gas molecules can diffuse. One of the difficulties with older types of CO_2 electrodes is the length of time required for the CO_2 molecules to diffuse and thus obtain a reading. The principal advantage of the Severinghaus-type electrode is the more rapid reading that can be obtained because of the improved membrane and bicarbonate layer.

In some applications, measurements of P_{O_2} and P_{CO_2} are combined into a single electrode that also includes a common reference half-cell. Such a combination electrode is shown in diagram form in Figure 4.18.

Figure 4.18. Combination of P_{CO_2} and P_{O_2} electrode. (Courtesy of J.W. Severinghaus, M.D.)

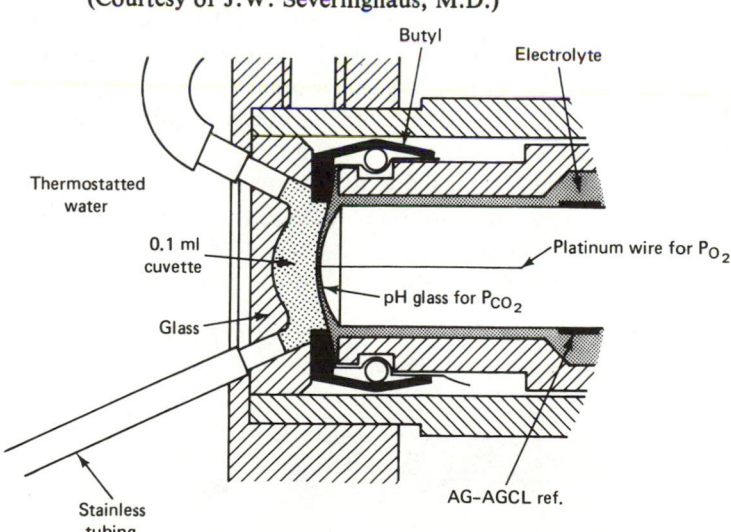

4.3.4. Specific Ion Electrodes

Just as the glass electrode provides a semipermeable membrane for the hydrogren ion in the pH electrode (see Section 4.3.2), other materials can be used to form membranes that are semipermeable to other specific ions. In each case, measurement of the ion concentration is accomplished by measurement of potentials across a membrane that has the correct degree of permeability to the specific ion to be measured. The permeability should be sufficient to permit rapid establishment of the electrode potential. Both liquid and solid membranes are used for specific ions. As in the case of the pH electrode, a silver–silver chloride interface is usually provided on the electrode side of the membrane, and a standard reference electrode serves as the other half-cell in the solution.

Figure 4.19 shows a solid-state electrode of the type used for measurement of fluoride ions. Figure 4.20 shows three specific ion electrodes along with a pH glass electrode. The sodium electrode in Figure 4.20(a) is commonly used to determine sodium ion activity in blood and other physiological solutions. The cationic electrode (b) is used when studying alkaline metal ions or enzymes. The ammonia electrode (d) is designed for determinations of ammonia dissolved in aqueous solutions. Its most popular application is in determining nitrogen as free ammonia or total Kjeldahl nitrogen.

Figure 4.21 is a diagram showing the construction of a flow-through type of electrode. This is a liquid-membrane, specific-ion electrode. One of the difficulties encountered in the measurement of specific ions is the effect of other ions in the solution. In cases where more than one type of membrane could be selected for measurement of a certain ion, the choice of membrane actually used might well depend on other ions that may be expected. In fact, some specific-ion electrodes can be used in measurement of a given ion only in the absence of certain other ions.

For measurement of divalent ions, a liquid membrane is often used for ion exchange. In this case, the exchanger is usually a salt of an organophosphoric acid, which shows a high degree of specificity to the ion being measured. A calcium chloride solution bridges the membrane to the silver–silver chloride electrode. Electrodes with membranes of solid materials are also used for measurement of divalent ions.

Figure 4.19. Electrode for measurement of fluoride ions. (Courtesy Beckman Instruments, Inc., Fullerton, CA.)

Figure 4.20. Specific ion electrodes with pH glass electrode. (a) Sodium ion electrode; (b) cationic electrode; (c) pH glass electrode; (d) ammonia electrode. (Courtesy Beckman Instruments, Inc., Fullerton, CA.)

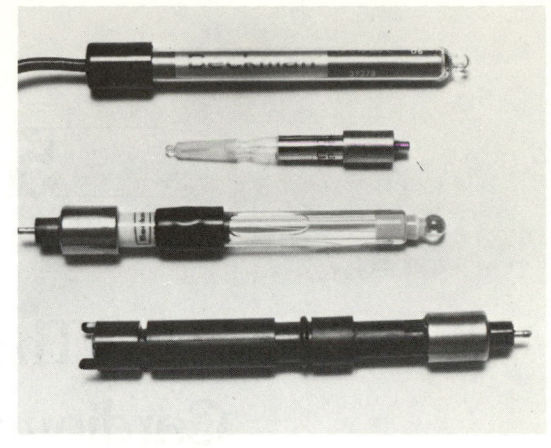

Figure 4.21. Diagram showing construction of flow-through liquid membrane specific ion electrode. (Courtesy of Orion Research, Inc., Cambridge, MA.)

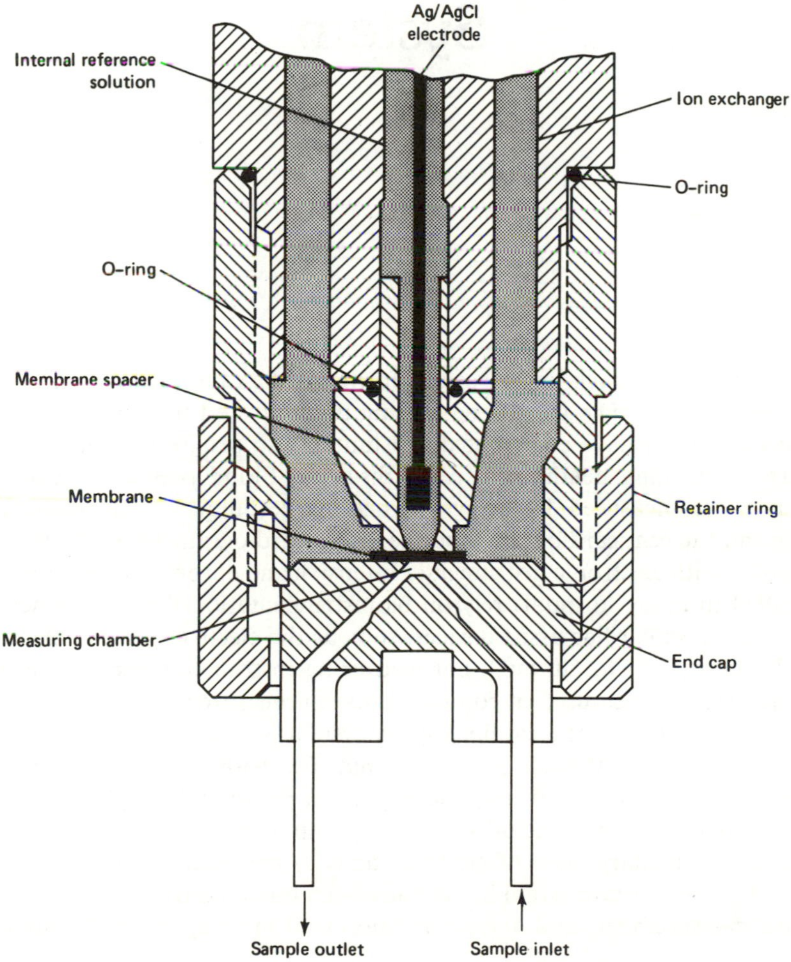

Ag/AgCl electrode

Internal reference solution

Ion exchanger

O-ring

O-ring

Membrane spacer

Membrane

Retainer ring

Measuring chamber

End cap

Sample outlet

Sample inlet

5

The Cardiovascular System

The heart attack, in its various forms, is the cause of many deaths in the world today. The use of engineering methods and the development of instrumentation have contributed substantially to progress made in recent years in reducing death from heart diseases. Blood pressure, flow, and volume are measured by using engineering techniques. The electrocardiogram, the echocardiogram, and the phonocardiogram are measured and recorded with electronic instruments. Intensive and coronary care units now installed in many hospitals rely on bioinstrumentation for their function. There are also cardiac assist devices, such as the electronic pacemaker and defibrillator, which, although not measuring instruments per se, are electronic devices often used in conjunction with measurement systems.

In this chapter the cardiovascular system is discussed, not only from the point of view of basic physiology but also with the idea that it is an engineering system. In this way the important parameters can be illustrated in correct perspective. Included are the pump and flow characteristics, as well as the ancillary ideas of electrical activity and heart sounds.

The electrocardiogram has already been introduced in Chapter 3. The actual measurements and devices are discussed in Chapters 6, 7, and 9.

5.1. THE HEART AND CARDIOVASCULAR SYSTEM

The heart may be considered as a two-stage pump, physically arranged in parallel but with the circulating blood passing through the pumps in a series sequence. The right half of the heart, known as the *right heart,* is the pump that supplies blood to the rest of the system. The circulatory path for blood flow-through the lungs is called the *pulmonary circulation,* and the circulatory system that supplies oxygen and nutrients to the cells of the body is called the *systemic circulation.*

From an engineering standpoint, the systemic circulation is a high-resistance circuit with a large pressure gradient between the arteries and veins. Thus, the pump constituting the left heart may be considered as a pressure pump. However, in the pulmonary circulation system, the pressure difference between the arteries and the veins is small, as is the resistance to flow, so the right heart may be considered as a volume pump. The muscle contraction of the left heart is larger and stronger than that of the right heart because of the greater pressures required for the systemic circulation. The volume of blood delivered per unit of time by the two sides is the same when measured over a sufficiently long interval. The left heart develops a pressure head sufficient to cause blood to flow to all the extremities of the body.

The pumping action itself is performed by contraction of the heart muscles surrounding each chamber of the heart. These muscles receive their own blood supply from the *coronary arteries,* which surround the heart like a crown (corona). The *coronary arterial system* is a special branch of the systemic circulation.

The analogy to a pump and hydraulic piping system should not be used too indiscriminately. The pipes, the arteries and the veins, are not rigid but flexible. They are capable of helping and controlling blood circulation by their own muscular action and their own valve and receptor system. Blood is not a pure Newtonian fluid; rather, it possesses properties that do not always comply with the laws governing hydraulic motion. In addition, the blood needs the help of the lungs for the supply of oxygen, and it interacts with the lymphatic system. Furthermore, many chemicals and hormones affect the operation of the system. Thus, oversimplication could lead to error if carried too far.

The actual physiological system for the heart and circulation is illustrated in Figure 5.1, with the equivalent engineering type of piping diagram shown in Figure 5.2. Referring to these figures, the operation of the circulatory system can be described as follows. Blood enters the heart on the right side through two main veins: the *superior vena cava,* which leads from the body's upper extremities, and the *inferior vena cava* leading from the body's organs and extremities below the heart. The incoming blood fills

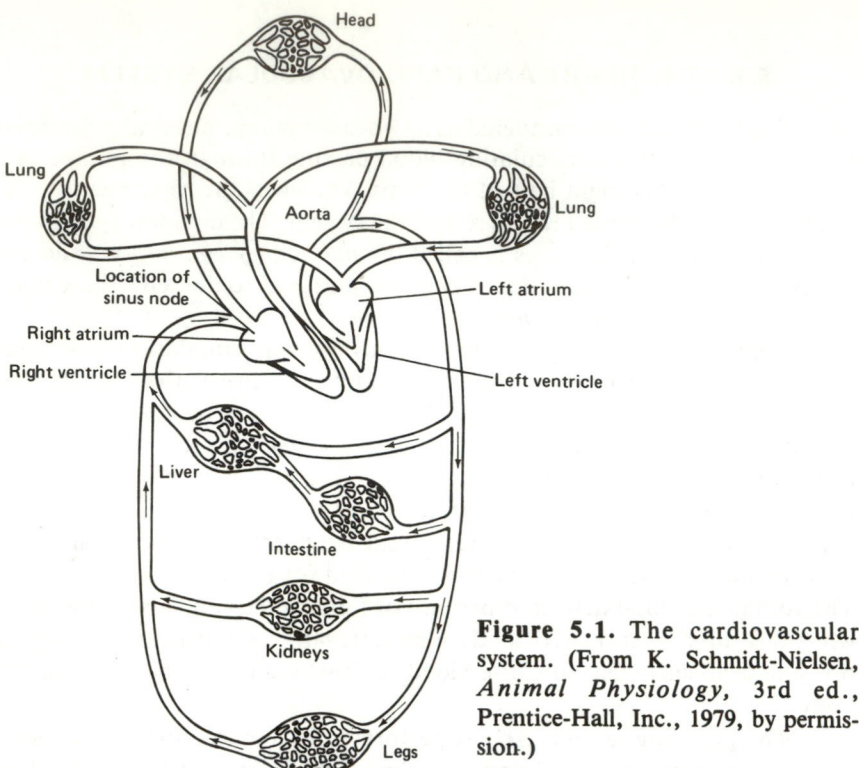

Figure 5.1. The cardiovascular system. (From K. Schmidt-Nielsen, *Animal Physiology,* 3rd ed., Prentice-Hall, Inc., 1979, by permission.)

the storage chamber, the *right atrium.* In addition to the two veins mentioned, the *coronary sinus* also empties into the right atrium. The coronary sinus contains the bood that has been circulating through the heart itself via the coronary loop.

When the right atrium is full, it contracts and forces blood through the *tricuspid valve* into the *right ventricle,* which then contracts to pump the blood into the pulmonary circulation system. When ventricular pressure exceeds atrial pressure, the tricuspid valve closes and the pressure in the ventricle forces the semilunar *pulmonary valve* to open, thereby causing blood to flow into the pulmonary artery, which divides into the two lungs.

In the *alveoli* of the lungs, an exchange takes place. The red blood cells are recharged with oxygen and give up their carbon dioxide. Not shown on the diagram are the details of this exchange. The pulmonary artery *bifurcates* (divides) many times into smaller and smaller arteries, which become arterioles with extremely small cross sections. These arterioles supply blood to the alveolar capillaries, in which the exchange of oxygen and carbon dioxide takes place. On the other side of the lung mass is a similar construction in which the capillaries feed into tiny veins, or *venules.* The latter combine to form larger veins, which in turn combine until ultimately all the oxygenated blood is returned to the heart via the pulmonary vein.

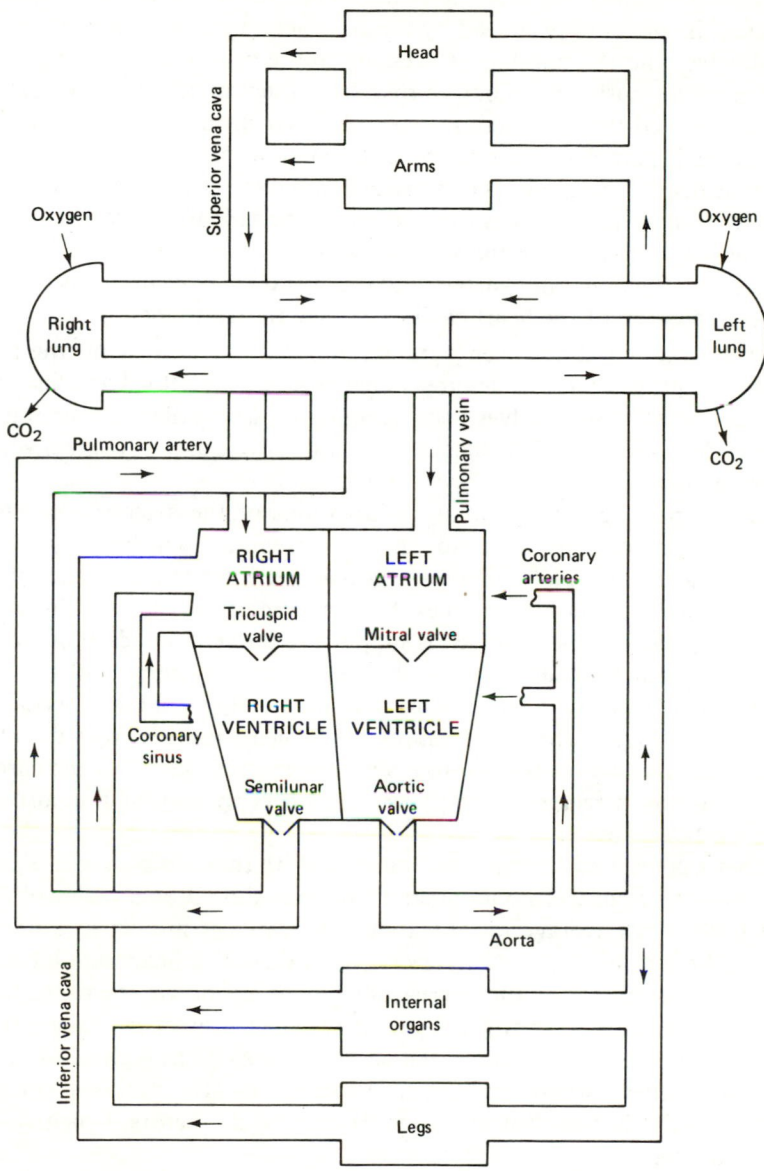

Figure 5.2. Cardiovascular circulation.

The blood enters the *left atrium* from the pulmonary vein, and from there it is pumped through the *mitral,* or *bicuspid valve,* into the left ventricle by contraction of the atrial muscles. When the left ventricular muscles contract, the pressure produced by the contraction mechanically closes the mitral valve, and the buildup of pressure in the ventricle forces the *aortic valve* to open, causing the blood to rush from the ventricle into the *aorta.* It should be noted that this action takes place synchronously with the right ventricle as it pumps blood into the pulmonary artery.

The heart's pumping cycle is divided into two major parts: systole and diastole. *Systole* (sĭs′tō•lē) is defined as the period of contraction of the heart muscles, specifically the ventricular muscle, at which time blood is pumped into the pulmonary artery and the aorta. *Diastole* (dī•ăs′tō•lē) is the period of dilation of the heart cavities as they fill with blood.

Once the blood has been pumped into the arterial system, the heart relaxes, the pressure in the chambers decreases, the outlet valves close, and in a short time the inlet valves open again to restart the diastole and initiate a new cycle in the heart. The mechanism of this cycle and its control will be discussed later.

After passing through many bifurcations of the arteries, the blood reaches the vital organs, the brain, and the extremities. The last stage of the arterial system is the gradual decrease in cross section and the increase in the number of arteries until the smallest type (arterioles) is reached. These feed into the capillaries, where oxygen is supplied to the cells and carbon dioxide is received from the cells. In turn, the capillaries join into venules, which become small veins, then larger veins, and ultimately form the superior and inferior vena cavae. The blood supply to the heart itself is from the aorta through the coronary arteries into a similar capillary system to the cardiac veins. This blood returns to the heart chambers by way of the coronary sinus, as stated above.

Since continued reference has been made to the cardiovascular system in engineering terms, some numbers of interest should be mentioned. The heart beats at an average rate of about 75 beats per minute in a normal adult, although this figure can vary considerably. The heart rate increases when a person stands up and decreases when he sits down, the range being from about 60 to 85. On the average, it is higher in women and generally decreases with age. In an infant, the heart rate may be as high as 140 beats per minute under normal conditions. The heart rate also increases with heat exposure and other physiological and psychological factors, which will be discussed later.

The heart pumps about 5 liters of blood per minute, and since the volume of blood in the average adult is about 5 to 6 liters, this corresponds to a complete turnover every minute during rest. With heavy exercise, the circulation rate is increased considerably. At any given time, about 75 to 80 percent of the blood volume is in the veins, about 20 percent in the arteries, and the remainder in the capillaries.

Systolic (maximum) blood pressure in the normal adult is in the range of 95 to 140 mm Hg, with 120 mm Hg being average.* These figures are subject to much variation with age, climate, eating habits, and other factors. Normal diastolic blood pressure (lowest pressure between beats) ranges from 60 to 90 mm Hg, 80 mm Hg being about average. This pressure is usually measured in the brachial artery in the arm. For comparison purposes with pressures of 130/75 in the aorta, 130/5 can be expected in the left ventricle, 9/5 in the left atrium, 25/0 in the right ventricle, 3/0 in the right atrium, and 25/12 in the pulmonary artery. These values are given as:

systolic pressure/diastolic pressure

5.2 THE HEART

The general behavior of the heart as the pump used to force the blood through the cardiovascular system has been discussed. A more detailed analysis of the anatomy of the heart, plus a discussion of the electrical excitation system necessary to produce and control the muscular contractions, should help to round out the background material needed for an understanding of cardiac dynamics. The electrocardiogram, as a record of biopotential events, has already been discussed in Chapter 3, but some repetition is necessary in order to consider the system as a whole and the relationships that exist between the electrical and mechanical events of the heart. Figure 5.3 is an illustration of the heart.

The heart is contained in the *pericardium,* a membranous sac consisting of an external layer of dense fibrous tissue and an inner serous layer that surrounds the heart directly. The base of the pericardium is attached to the central tendon of the diaphragm, and its cavity contains a thin serous liquid. The two sides of the heart are separated by the *septum,* or dividing wall of tissue. The septum also includes the *atrioventricular node* (AV node), which, as will be explained later, plays a role in the electrical conduction through the cardiac muscles.

Each of the four chambers of the heart is different from the others because of its functions. The *right atrium* is elongated and lies between the inferior (lower) and superior (upper) vena cava. Its interior is complex, the anterior (front) wall being very rough, whereas the posterior (rear) wall

---

*Clinically, the *mm Hg* is still the unit used for blood pressure measurements. To convert to SI metric unit *kilopascals* (kPa), multiply the mm Hg figure by 0.133. Figure 5.5 has both scales, for comparison purposes. The *torr* is also a measure of pressure and is equivalent to the mm Hg.

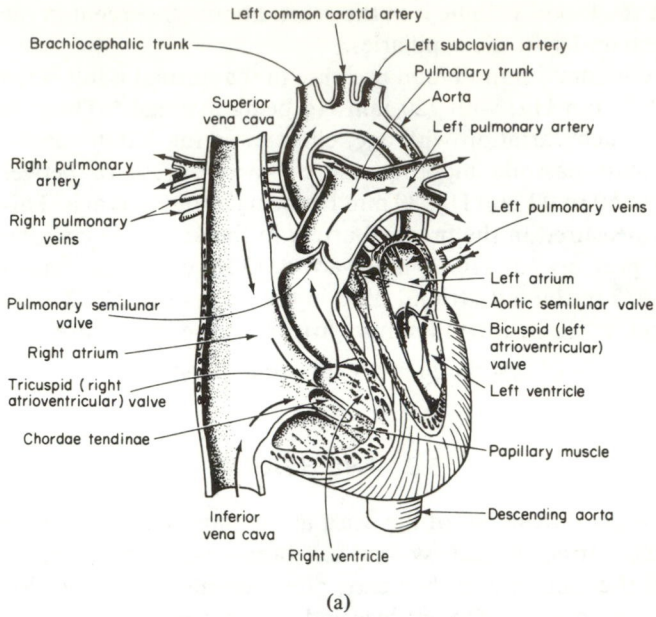

(a)

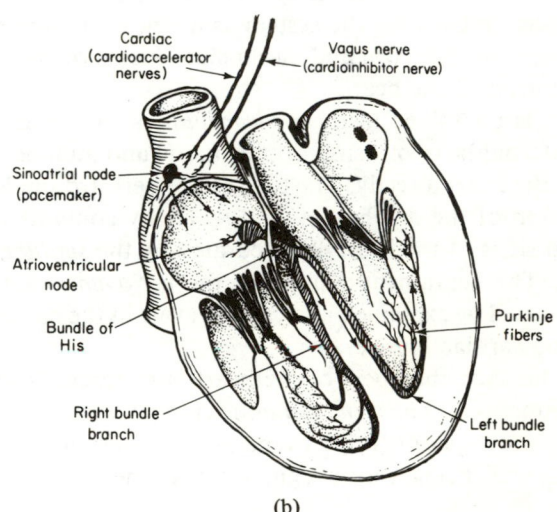

(b)

Figure 5.3. The heart: (a) internal structure; (b) conduction system. (From W.F. Evans, *Anatomy and Physiology, The Basic Principles,* Englewood Cliffs, N.J., Prentice-Hall, Inc., 1971, by permission.)

(which forms a part of the septum) and the remaining walls are smooth. At the junction of the right atrium and the superior vena cava is situated the *sinoatrial node* (SA node), which is the *pacemaker* or initiator of the electrical impulses that excite the heart. The *right ventricle* is situated below and to the left of the right atrium. They are separated by a fatty structure in which is contained the right branch of the coronary artery. This fatty separation is incapable of conducting electrical impulses; communication between the atria and the ventricles is accomplished only via the AV node and delay line.

Since the ventricle has to perform a more powerful pumping action, its walls are thicker than those of the atrium and its surfaces are ridged. Between the anterior wall of the ventricle and the septum is a muscular ridge that is part of the heart's electrical conduction system, known as the *bundle of His,* described in Section 3.3.1. At the junction of the right and left atrium and the right ventricle on the septum is another node, the atrioventricular node. The bundle of His is attached to this node.

The right atrium and right ventricle are joined by a fibrous tissue known as the *atrioventricular ring,* to which are attached the three cusps of the tricuspid valve, which is the connecting valve between the two chambers.

The *left atrium* is smaller than the right atrium. Entry to it is through four pulmonary veins. The walls of the chamber are fairly smooth. It is joined to the left ventricle through the *mitral valve,* sometimes called the *bicuspid* valve since it consists of two cusps.

The *left ventricle* is considered the most important chamber, for this is the power pump for all the systemic circulation. Its walls are approximately three times as thick as the walls of the right ventricle because of this function. Conduction to the left ventricle is through the left bundle branch, which is in the ventricular muscle on the septum side.

As mentioned earlier, the outputs from the ventricles are through the aortic and pulmonary valves, respectively, for left and right ventricles.

Some aspects of the electrical activity of the heart have already been discussed in Section 3.3, but certain details will be elaborated on in the present context.

Unlike most other muscle innervations, excitation of the heart does not proceed directly from the central nervous system but is initiated in the sinoatrial (SA) node, or pacemaker, a special group of excitable cells. The electrical events that occur within the heart are reflected in the electrocardiogram.

The SA node creates an impulse of electrical excitation that spreads across the right and left atria; the right atrium receives the earlier excitation because of its proximity to the SA node. This excitation causes the atria to contract and, a short time later, stimulates the atrioventricular (AV) node.

The activated AV node, after a brief delay, initiates an impulse into the ventricles, through the bundle of His, and into the bundle branches that connect to the *Purkinje fibers* in the myocardium. The contraction resulting in the myocardium supplies the force to pump the blood into the circulatory systems.

The heart rate is controlled by the frequency at which the SA node generates impulses. However, nerves of the sympathetic nervous system and the vagus nerve of the parasympathetic nervous system (see Chapter 10) cause the heart rate to quicken or slow down, respectively. Anatomically, the fibers of the sympathetic and vagus nerves enter the heart at the cardiac plexus under the arch of the aorta and are distributed quite profusely at and near both the SA and AV nodes. Vagal fibers are mostly found distributed in the atria, the bundle of His and branches, whereas the sympathetic fibers are found within the muscular walls of the atria and ventricles.

Although the effects of the sympathetic and vagus nerves are in opposition to each other, if they both occur together in opposite directions, the result is additive. That is, heart rate will increase from a combination of increased sympathetic activity concurrent with decreased vagal activity. The action of these nerves is called their *tone;* and by the various activities of each type of nerve, the rate of the heart, its coronary blood supply, and its contractability may be affected. The nerves affecting the rate of the heart in this way originate from the medullary centers in the brain and are controlled both by cardiac acceleration and by inhibition centers, each being sensitive to stimulation from higher centers of the brain. It is in this sequence that the heart rate is affected when a person is anxious, frightened, or excited. Heart attacks may be caused by this type of stimulation. The heart rate can also be affected, but in a more indirect way, by overeating, respiration problems, extremes of body temperature, and blood changes.

One other effect that should be mentioned is that of the *pressoreceptors* or *baroreceptors* situated in the arch of the aorta and in the carotid sinus. Their function is to alter the vagal tone whenever the blood pressure within the aorta or carotid sinus changes. When blood pressure rises, vagal tone is increased and the heart rate slows; when blood pressure falls, vagal tone is decreased and the heart rate increases.

A good engineering analog is illustrated in Figure 5.4, which shows the physiological system as a pump model. The pump is initially set to operate under predetermined conditions, as are the valves representing the resistance in the various organs. The pressure transducers sense the pressure continuously. With the pressure head set at some normal level, if one of the valves opens farther to obtain greater flow in that branch, the pressure head will decrease. This is picked up as a lower pressure by the sensors, which feed a signal to the controller, which, in turn, closes other valves, speeds up the pump, or does both in order to try to maintain a constant pressure head.

CONTROL OF ARTERIAL BLOOD PRESSURE

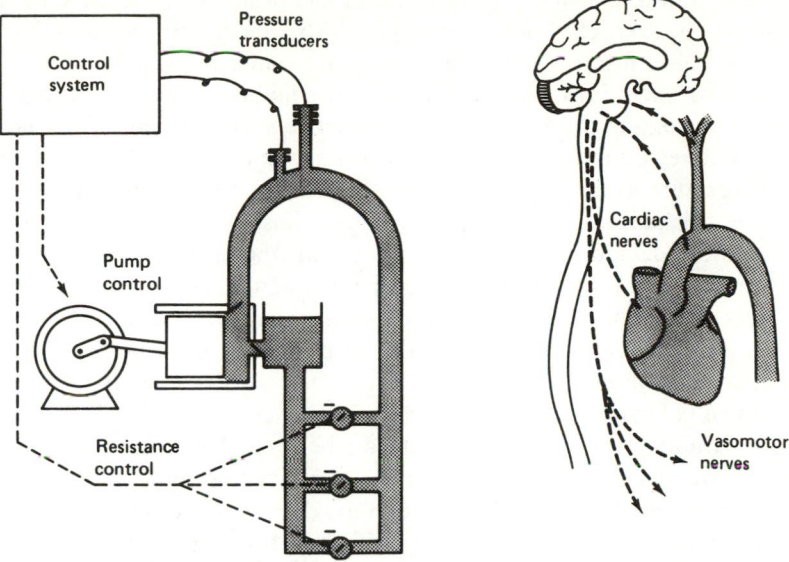

Figure 5.4 Control of arterial blood pressure. (From R.F. Rushmer, *Cardiovascular Dynamics,* 3rd ed., W.B. Saunders Co., 1970, by permission.)

5.3. BLOOD PRESSURE

In the arterial system of the body, the large pressure variations from systole to diastole are smoothed into a relatively steady flow through the peripheral vessels into the capillaries. This system, with some modifications, obeys the simple physical laws of hydrodynamics. As an analog, the potential (blood pressure) acting through the resistance of the arterial vascular pathways causes flow throughout the system. The resistance must not be so great as to impede flow, so that even the most remote capillaries receive suffcient blood and are able to return it into the venous system. On the other hand, the vessels of the system must be capable of damping out any large pressure fluctuations.

Since the system must be capable of maintaining an adequate pressure head while controlling flow, monitoring and feedback control loops are required. Demand on the system comes from various sources, such as from the gastrointestinal tract after a large meal or from the skeletal muscles during exercise. The result is vascular dilation at these particular points. If sufficient demands were to occur simultaneously so that increased blood flow were needed in many parts of the body, the blood pressure would drop. In this way, flow to the vital regions of heart and brain might be affected. Fortunately, however, the body is equipped with a monitoring system that can sense systemic arterial blood pressure and can compensate in the cardiovascular operations. Pressure is therefore maintained within a relatively narrow range, and the flow is kept within the normal range of the heart.

With regard to measurements, the events in the heart that relate to the blood pressure as a function of time should be understood. Figure 5.5 illustrates this point. The two basic stages of diastole and systole are shown with a more detailed time scale of phases of operation below. The blood pressure waves for the aorta, the left atrium, and the left ventricle are drawn to show time and magnitude relationships. Also, the correlated electrical events are shown at the bottom in the form of the electrocardiogram, and the basic relationship of the heart sounds, which are discussed in Section 5.5, are shown in Figure 5.8.

Examining the aortic wave, it can be seen that during systole, the ejection of blood from the left ventricle is rapid at first. As the rate of pressure change decreases, the rounded maximum of the curve is obtained. The peak aortic pressure during systole is a function of left ventricular stroke volume, the peak rate of ejection, and the distensibility of the walls of the aorta. In a diseased heart, ventricular contractability and rigid atherosclerotic arteries produce unwanted rises in blood pressure.

When the systolic period is completed, the aortic valve is closed by the back pressure of blood (against the valve). This effect can be seen on the pressure pulse waveform as the *dicrotic notch*. When the valve is closed completely, the arterial pressure gradually decreases as blood pours into the countless peripheral vascular networks. The rate at which the pressure falls is determined by the pressure achieved during the systolic interval, the rate of outflow through the peripheral resistances, and the diastolic interval.

· The form of the arterial pressure pulse changes as it passes through the arteries. The walls of the arteries cause damping and reflections; and as the arteries branch out into smaller arteries with smaller cross-sectional areas, the pressures and volumes change, hence the rate of flow also changes. The peak systolic pressure gets a little higher and the diastolic pressure flatter. The mean pressure in some arteries (e.g., the brachial artery) can be as much as 20 mm Hg higher than that in the aorta.

As the blood flows into the smaller arteries and arterioles, the pressure decreases and loses its oscillatory character. Pressure in the arterioles can vary from about 60 mm Hg down to 30 mm Hg. As the blood enters the venous system after flowing through the capillaries, the pressure is down to about 15 mm Hg.

In the venous system, the pressure in the venules decreases to approximately 8 mm Hg, and in the veins to about 5 mm Hg. In the vena cava, the

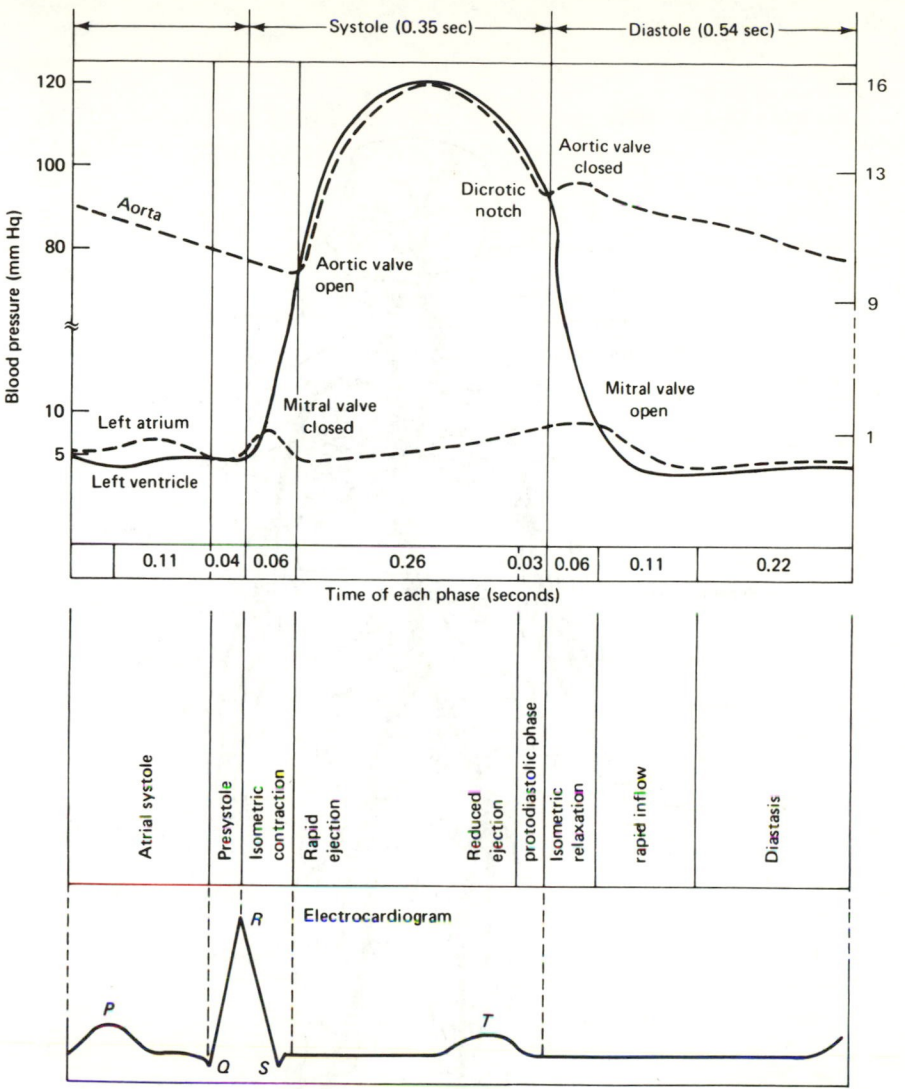

Figure 5.5. Blood pressure variations as a function of time. (Note: S.I. metric scale on right ordinate.)

pressure is only about 2 mm Hg. Because of these differences in pressure, measurement of arterial blood pressure is quite different from that of venous pressure. For example, a 2-mm Hg error in systolic pressure is only of the order of 1.5 percent. In a vein, however, this would be a 100 percent error. Also because of these pressure differences, the arteries have thick walls, while the veins have thin walls. Moreover, the veins have larger internal diameters. Since about 75 to 80 percent of the blood volume is contained in the venous system, the veins tend to serve as a reservoir for the body's blood supply.

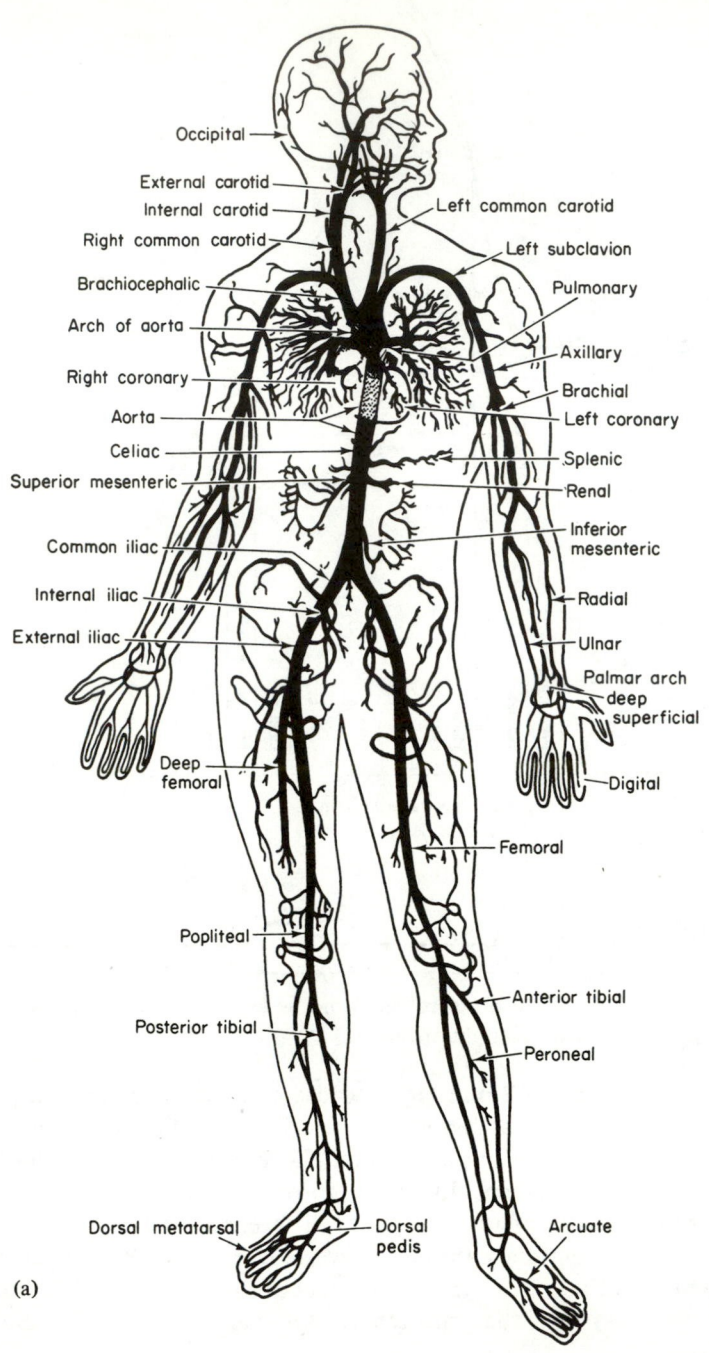

(a)

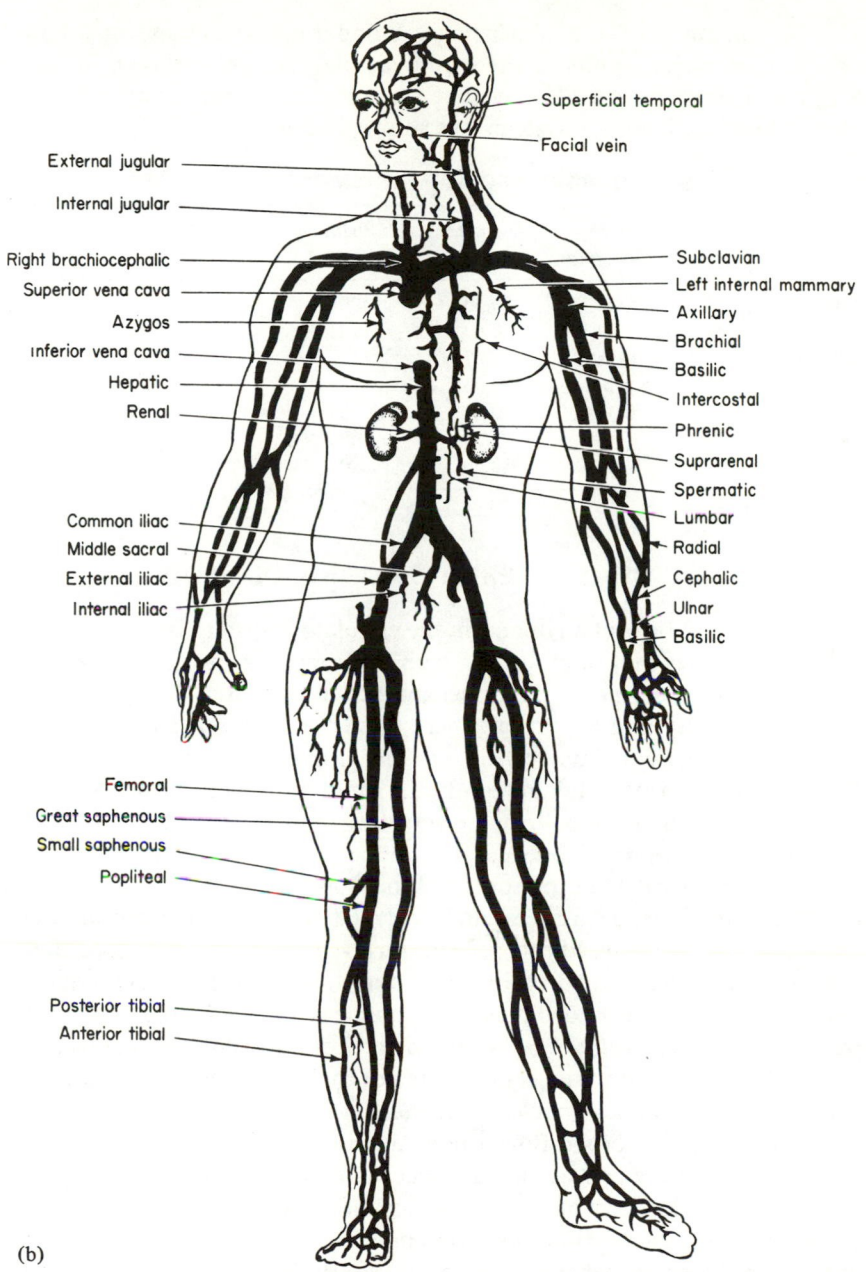

Figure 5.6. (a) Major arteries of the body; (b) major veins of the body. (From W.F. Evans, *Anatomy and Physiology, The Basic Principles,* Englewood Cliffs, N.J., Prentice-Hall, Inc., 1971, by permission.)

A summary of the dimensions, blood flow velocities, and blood pressures at major points in the cardiovascular system is shown in Table 5.1. The figures are typical or average for comparative reference. The complete arterial and venous systems are shown in Figure 5.6.

Table 5.1. CARDIOVASCULAR SYSTEM-TYPICAL VALUES

Vessel	Number (thousands)	Diameter (mm)	Length (mm)	Mean Velocity (cm/sec)	Pressure (mm Hg)
Aorta	—	10.50	400	40.0	100
Terminal arteries	1.8	0.60	10	<10.0	40
Arterioles	40,000	0.02	2	0.5	40–25
Capillaries	> million	0.008	1	<0.1	25–12
Venules	80,000	0.03	2	<0.3	12–8
Terminal veins	1.8	1.50	100	1.0	<8
Vena cava	—	12.50	400	20.0	3–2

5.4. CHARACTERISTICS OF BLOOD FLOW

The blood flow at any point in the circulatory system is the volume of blood that passes that point during a unit of time. It is normally measured in milliliters per minute or liters per minute. Blood flow is highest in the pulmonary artery and the aorta, where these blood vessels leave the heart. The flow at these points, called *cardiac output,* is between 3.5 and 5 liters/min in a normal adult at rest. In the capillaries, on the other hand, the blood flow can be so slow that the travel of individual blood cells can be observed under a microscope.

From the cardiac output or the blood flow in a given vessel, a number of other characteristic variables can be calculated. The cardiac output divided by the number of heartbeats per minute gives the amount of blood that is ejected during each heartbeat, or the *stroke volume.* If the total amount of blood in circulation is known, and this volume is divided by the cardiac output, the *mean circulation time* is obtained. From the blood flow through a vessel, divided by the cross-sectional area of the vessel, the *mean velocity* of the blood at the point of measurement can be calculated.

In the arteries, blood flow is pulsatile. In fact, in some blood vessels a reversal of the flow can occur during certain parts of the heartbeat cycle. Because of the elasticity of their walls, the blood vessels tend to smooth out the pulsations of blood flow and blood pressure. Both pressure and flow are greatest in the aorta, where the blood leaves the heart.

Blood flow is a function of the blood pressure and flow resistance of the blood vessels in the same way as electrical current flow depends on voltage and resistance. The flow resistance of the capillary bed, however, can vary over a wide range. For instance, when exposed to low temperatures or under the influence of certain drugs (e.g., nicotine), the body reduces the blood flow through the skin by *vasoconstriction* (narrowing) of the capillaries. Heat, excitement, or local inflammation, among other things, can cause *vasodilation* (widening) of the capillaries, which increases the blood flow, at least locally. Because of the wide variations that are possible in the flow resistance, the determination of blood pressure alone is not sufficient to assess the status of the circulatory system.

The velocity of blood flowing through a vessel is not constant throughout the cross section of the vessel but is a function of the distance from the wall surfaces. A thin layer of blood actually adheres to the wall, resulting in zero velocity at this place, whereas the highest velocity occurs at the center of the vessel. The resulting "velocity profile" is shown in Figure 5.7. Some blood flow meters do not actually measure the blood flow but measure the mean velocity of the blood. If, however, the cross-sectional area of the blood vessel is known, these devices can be calibrated directly in terms of blood flow.

If the local blood velocity exceeds a certain limit (as may happen when a blood vessel is constricted), small eddies can occur, and the *laminar flow* of Figure 5.7 changes to a *turbulent flow* pattern, for which the flow rate is more difficult to determine.

The proper functioning of all body organs depends on an adequate blood supply. If the blood supply to an organ is reduced by a narrowing of the blood vessels, the function of that organ can be severely limited. When the blood flow in a certain vessel is completely obstructed (e.g., by a blood clot or *thrombus*), the tissue in the area supplied by this vessel may die. Such an obstruction in a blood vessel of the brain is one of the causes of a *cerebrovascular accident* (CVA) or *stroke*. An obstruction of part of the

Figure 5.7. Laminar flow in a blood vessel.

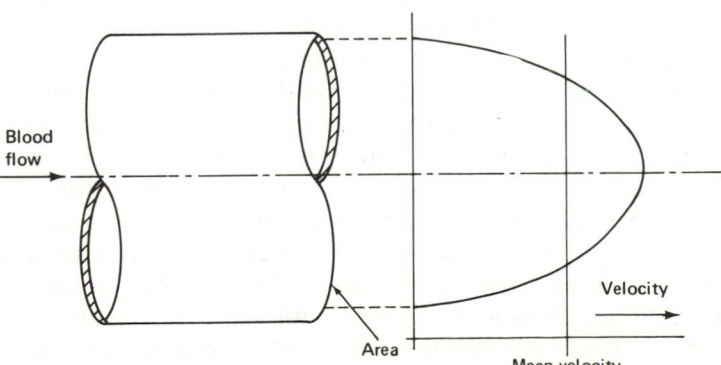

coronary arteries that supply blood for the heart muscle is called a *myocardial* (or *coronary*) *infarct* or *heart attack,* whereas merely a reduced flow in the coronary vessels can cause a severe chest pain called *angina pectoris.* A blood clot in a vessel in the lung is called an *embolism.* Blood clots can also afflict the circulation in the lower extremities (*thrombosis*). Although the foregoing events afflict only a limited, although often vital, area of the body, the total circulatory system can also be affected. Such is the case if the cardiac output, the amount of blood pumped by the heart, is greatly reduced. This situation can be due to a mechanical malfunction such as a leaking or torn heart valve. It can also occur as *shock* — for example, after a severe injury when the body reacts with vasoconstriction of the capillaries, which reduces the blood loss but also prevents the blood from returning to the heart.

Most of these events have severe, and often fatal, results. Therefore, it is of great interest to be able to determine the blood flow in such cases to provide an early diagnosis and begin treatment before irreparable tissue damage has occurred.

5.5. HEART SOUNDS

For centuries the medical profession has been aided in its diagnosis of certain types of heart disorders by the sounds and vibrations associated with the beating of the heart and the pumping of blood. The technique of listening to sounds produced by the organs and vessels of the body is called *auscultation,* and it is still in common use today. During his training the physician learns to recognize sounds or changes in sounds that he can associate with various types of disorders.

In spite of its widespread use, however, auscultation is rather subjective, and the amount of information that can be obtained by listening to the sounds of the heart depends largely on the skill, experience, and hearing ability of the physician. Different physicians may hear the same sounds differently, and perhaps interpret them differently.

The heart sounds heard by the physician through his stethoscope actually occur at the time of closure of major valves in the heart. This timing could easily lead to the false assumption that the sounds which are heard are primarily caused by the snapping together of the vanes of these valves. In reality, this snapping action produces almost no sound, because of the cushioning effect of the blood. The principal cause of heart sounds seems to be vibrations set up in the blood inside the heart by the sudden closure of the valves. These vibrations, together with eddy currents induced in the blood as it is forced through the closing valves, produce vibrations in the walls of the heart chambers and in the adjoining blood vessels.

With each heartbeat, the normal heart produces two distinct sounds that are audible in the stethoscope—often described as "lub-dub." The "lub" is caused by the closure of the *atrioventricular valves,* which permit flow of blood from the atria into the ventricles but prevent flow in the reverse direction. Normally, this is called the *first heart sound,* and it occurs approximately at the time of the QRS complex of the electrocardiogram and just before ventricular systole. The "dub" part of the heart sounds is called the *second heart sound* and is caused by the closing of the *semilunar valves,* which release blood into the pulmonary and systemic circulation systems. These valves close at the end of systole, just before the atrioventricular valves reopen. This second heart sound occurs about the time of the end of the T wave of the electrocardiogram.

A *third heart sound* is sometimes heard, especially in young adults. This sound, which occurs from 0.1 to 0.2 sec. after the second heart sound, is attributed to the rush of blood from the atria into the ventricles, which causes turbulence and some vibration of the ventricular walls. This sound actually precedes atrial contraction, which means that the inrush of blood to the ventricles causing this sound is passive, pushed only by the venous pressure at the inlets to the atria. Actually, about 70 percent of blood flow into the ventricles occurs before atrial contraction.

An *atrial heart sound,* which is not audible but may be visible on a graphic recording, occurs when the atria actually do conract, squeezing the remainder of the blood into the ventricles. The inaudibility of this heart sound is a result of the low amplitude and low frequency of the vibrations.

Figure 5.8 shows the time relationships between the first, second, and third heart sounds with respect to the electrocardiogram, and the various pressure waveforms. Opening and closing times of valves are also shown. This figure should also be compared with Figure 5.5.

In abnormal hearts additional sounds, called *murmurs,* are heard between the normal heart sounds. Murmurs are generally caused either by improper opening of the valves (which requires the blood to be forced through a small aperture) or by regurgitation, which results when the valves do not close completely and allow some backward flow of blood. In either case, the sound is due to high-velocity blood flow through a small opening. Another cause of murmurs can be a small opening in the septum, which separates the left and right sides of the heart. In this case, pressure differences between the two sides of the heart force blood through the opening, usually from the left ventricle into the right ventricle, bypassing the systemic circulation.

Normal heart sounds are quite short in duration, approximately one-tenth of a second for each, while murmurs usually extend between the normal sounds. Figure 5.9 shows a record of normal heart sounds and several types of murmurs.

There is also a difference in frequency range between normal and ab-

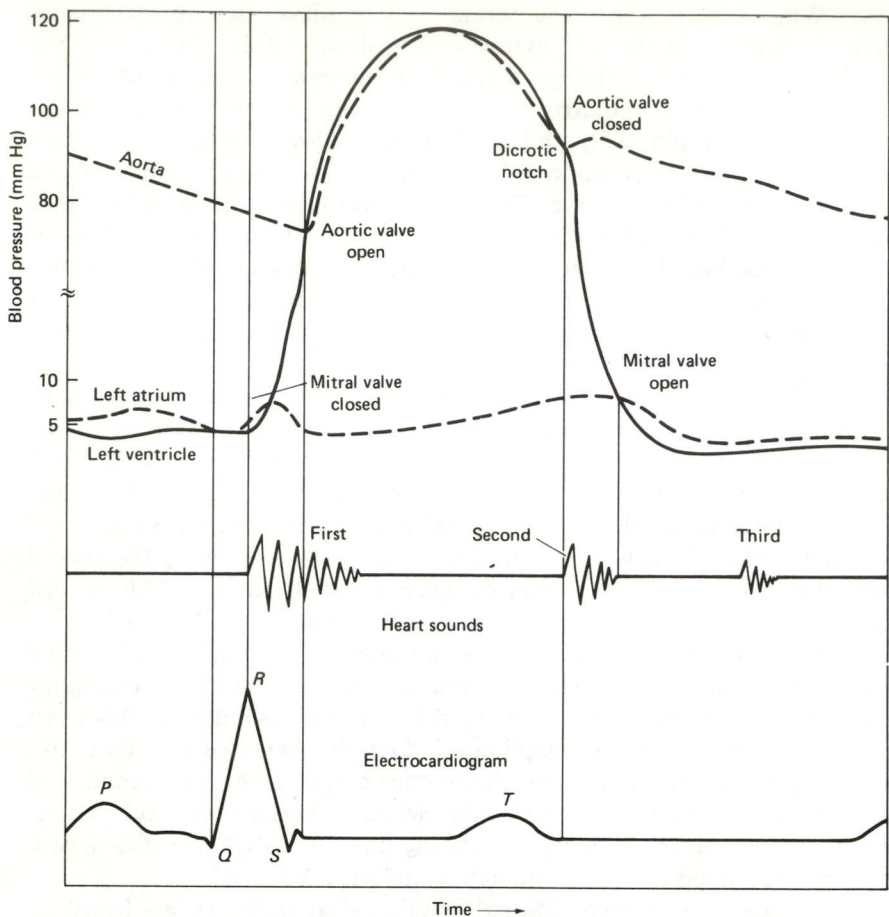

Figure 5.8. Relationship of heart sounds to function of the cardiovascular system.

normal heart sounds. The first heart sound is composed primarily of energy in the 30- to 45-Hz range, with much of the sound below the threshold of audibility. The second heart sound is usually higher in pitch than the first, with maximum energy in the 50- to 70-Hz range. The third heart sound is an extremely weak vibration, with most of its energy at or below 30 Hz. Murmurs, on the other hand, often produce much higher pitched sounds. One particular type of regurgitation, for example, causes a murmur in the 100-to 600-Hz range.

Although auscultation is still the principal method of detecting and analyzing heart sounds, other techniques are also often employed. For example, a graphic recording of heart sounds, such as shown in Figure 5.8, is called a *phonocardiogram*. Even though the phonocardiogram is a graphic record like the electrocardiogram, it extends to a much higher frequency range.

An entirely different waveform is produced by the vibrations of the heart against the thoracic cavity. The vibrations of the side of the heart as it thumps against the chest wall form the *vibrocardiogram,* whereas the tip or apex of the heart hitting the rib cage produces the *apex cardiogram.*

Sounds and pulsations can also be detected and measured at various locations in the systemic arterial circulation system where major arteries approach the surface of the body. The most common one is the *pulse,* which can be felt with the fingertips at certain points on major arteries. The waveform of this pulse can also be measured and recorded. In addition, when an artery is partially occluded so that the blood velocity through the constriction is increased sufficiently, identifiable sounds can be heard downstream through a stethoscope. These sounds, called *Korotkoff sounds,* are used in the common method of blood pressure measurements and are discussed in detail in Chapter 6.

Another cardiovascular measurement worthy of note is the *ballistocardiogram.* Although not a heart sound or vibration measurement

Figure 5.9. Normal and abnormal heart sounds. (From A.C. Guyton, *Textbook of Medical Physiology,* 4th ed., W.B. Saunders Co., 1971, by permission.)

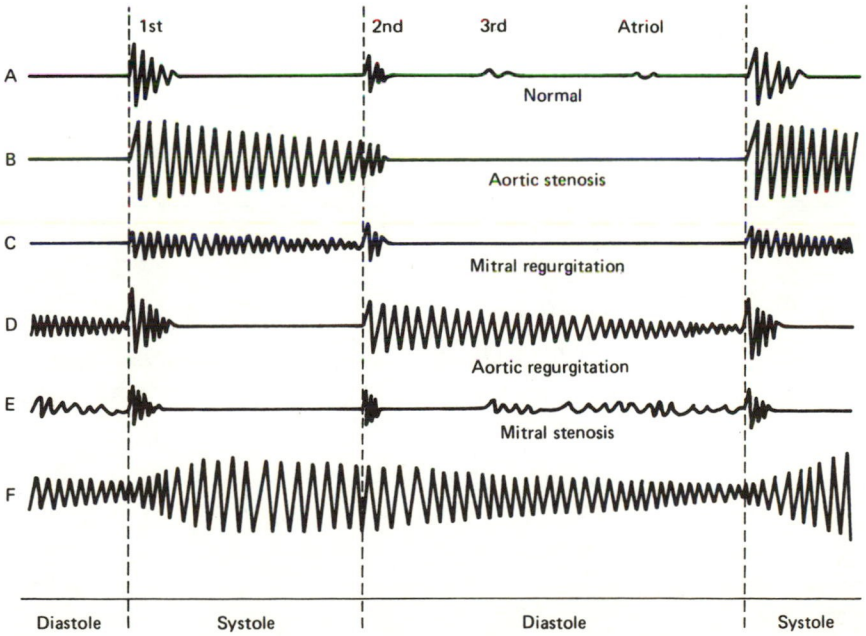

of the type described earlier, the ballistocardiogram is related to these measures in that it is a direct result of the dynamic forces of the heart as it beats and pumps blood into the major arteries. The beating heart exerts certain forces on the body as it goes through its sequence of motions. As in any situation involving forces, the body responds, but because of the greater mass of the body, these responses are generally not noticeable. However, when a person is placed on a platform that is free to move with these small dynamic responses, the motions of the body due to the beating of the heart and the corresponding blood ejection can be measured and recorded to produce the ballistocardiogram. Like the vibrocardiogram and the apex cardiogram, the ballistocardiogram provides information about the heart that cannot be obtained by any other measurement.

6

Cardiovascular Measurements

To this point the reader has been exposed to the overall concepts of the biomedical instrumentation system, some basic descriptions of component parts, and a description of the cardiovascular physiology. The next step is to combine this information and study measurement of a major body system.

It is not by accident that the cardiovascular system has been chosen as the first of the major physiological measurement groupings to be studied. Instrumentation for obtaining measurements from this system has contributed greatly to advances in medical diagnosis.

Since such instrumentation includes devices to measure various types of physiological variables, such as electrical, mechanical, thermal, fluid, and auditory, this chapter is intended to provide a basis for studying all types of biomedical instrumentation. Each type of measurement is considered separately, beginning with the measurement of biopotentials that result in the electrocardiogram. Then the various methods, both direct and indirect, of measuring blood pressure, blood flow, cardiac output, and blood volume (plethysmography) are discussed. The final section is concerned with the measurement of heart sounds and vibrations.

It should be noted that some of the methods discussed in this chapter involve *noninvasive techniques,* measurements that can be made without "invading" the body. Some of these techniques are included in this chapter under cardiovascular methods to keep them in that perspective. Others are covered in Chapter 9.

6.1. ELECTROCARDIOGRAPHY

The *electrocardiogram* (ECG or EKG) is a graphic recording or display of the time-variant voltages produced by the myocardium during the cardiac cycle. Figure 6.1 shows the basic waveform of the normal electrocardiogram. The P, QRS, and T waves reflect the rhythmic electrical depolarization and repolarization of the myocardium associated with the contractions of the atria and ventricles. The electrocardiogram is used clinically in diagnosing various diseases and conditions associated with the heart. It also serves as a timing reference for other measurements.

A discussion of the ECG waveform has already been presented in Section 3.2 and will not be repeated here, except in the concept of measurement details. To the clinician, the shape and duration of each feature of the ECG are significant. The waveform, however, depends greatly upon the lead configuration used, as discussed below. In general, the cardiologist looks critically at the various time intervals, polarities, and amplitudes to arrive at his diagnosis.

Some normal values for amplitudes and durations of important ECG parameters are as follows:

Amplitude:	P	wave	0.25 mV
	R	wave	1.60 mV
	Q	wave	25% of R wave
	T	wave	0.1 to 0.5 mV
Duration:	P-R	interval	0.12 to 0.20 sec
	Q-T	interval	0.35 to 0.44 sec
	S-T	segment	0.05 to 0.15 sec
	P	wave interval	0.11 sec
	QRS	interval	0.09 sec

For his diagnosis, a cardiologist would typically look first at the heart rate. The normal value lies in the range of 60 to 100 beats per minute. A slower rate than this is called *bradycardia* (slow heart) and a higher rate, *tachycardia* (fast heart). He would then see if the cycles are evenly spaced. If

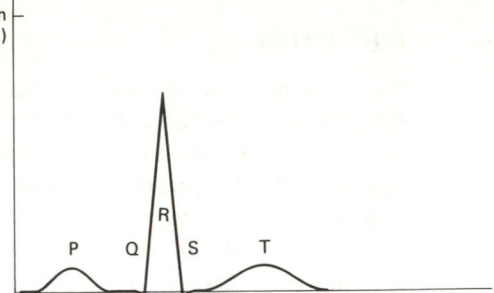

Figure 6.1. The electrocardiogram in detail.

not, an *arrhythmia* may be indicated. If the P-R interval is greater than 0.2 second, it can suggest blockage of the AV node. If one or more of the basic features of the ECG should be missing, a heart block of some sort might be indicated.

In healthy individuals the electrocardiogram remains reasonably constant, even though the heart rate changes with the demands of the body. It should be noted that the position of the heart within the thoracic region of the body, as well as the position of the body itself (whether erect or recumbent), influences the "electrical axis" of the heart. The *electrical axis* (which parallels the anatomical axis) is defined as the line along which the greatest electromotive force is developed at a given instant during the cardiac cycle. The electrical axis shifts continually through a repeatable pattern during every cardiac cycle.

Under pathological conditions, several changes may occur in the ECG. These include (1) altered paths of excitation in the heart, (2) changed origin of waves (ectopic beats), (3) altered relationships (sequences) of features, (4) changed magnitudes of one or more features, and (5) differing durations of waves or intervals.

As mentioned earlier, an instrument used to obtain and record the electrocardiogram is called an *electrocardiograph*. The electrocardiograph was the first electrical device to find widespread use in medical diagnostics, and it still remains the most important tool for the diagnosis of cardiac disorders. Although it provides invaluable diagnostic information, especially in the case of arrhythmias and myocardial infarction, certain disorders—for instance, those involving the heart valves—cannot be diagnosed from the electrocardiogram. Other diagnostic techniques, however, such as angiography (Chapter 14) and echocardiography (Chapter 9), can provide the information not available in the electrocardiogram. The first electrocardiographs appeared in hospitals around 1910, and while ECG machines have benefited from technological innovations over the years, little has actually changed in the basic technique. Most of the terminology and several of the methods still employed date back to the early days of electrocardiography and can be understood best in an historical context.

6.1.1 History

The discovery that muscle contractions involve electrical processes dates to the eighteenth century. At that time, however, the technology was not advanced enough to allow a quantitative study of the electrical voltages generated by the contracting heart muscle. It was not until 1887 that the first electrocardiogram was recorded by Waller, who used the *capillary electrometer* introduced by Lippman in 1875. This device consisted of a mercury-filled glass capillary immersed in dilute sulfuric acid. The position of the meniscus, which formed the dividing line between the two fluids, changed when an electrical voltage was applied between the mercury and acid. This movement was very small, but it could be recorded on a moving piece of light-sensitive paper or film with the help of a magnifying optical projection system. The capillary electrometer, however, was cumbersome to operate and the inertia of the mercury column limited its frequency range.

The string galvanometer, which was introduced to electrocardiography by Einthoven in 1903, was a considerable improvement. It consisted of an extremely thin platinum wire or a gold-plated quartz fiber, about 5 μm thick, suspended in the air gap of a strong electromagnet. An electrical current flowing through the string caused movement of the string perpendicular to the direction of the magnetic field. The magnitude of the movement was small but could be magnified several hundred times by an optical projection system for recording on a moving film or paper. The small mass of the moving fiber resulted in a frequency response sufficiently high for the faithful recording of the electrocardiogram. The sensitivity of the galvanometer could be adjusted by changing the mechanical tension on the string. To measure the sensitivity of the galvanometer, a standardization switch allowed a calibration voltage of 1 mV to be connected to the galvanometer terminals. Modern electrocardiographs, although they have a calibrated sensitivity, still retain this feature. The string galvanometer had dc response, and a difference in the contact potentials of the electrodes could easily drive the string off scale. A compensation voltage, adjustable in magnitude and polarity, was provided to center the shadow of the string on a ground-glass screen prior to recording the electrocardiogram. To facilitate measurement of the time differences between the characteristic parts of the ECG waveform, time marks were provided on the film by a wheel with five spokes driven by a constant-speed motor.

String galvanometer electrocardiographs like the one shown in Figure 4.4 were used until about 1920, when they were replaced by devices incorporating electronic amplification. This allowed the use of less sensitive and more rugged recording devices. Early ECG machines incorporating amplification used the Dudell oscillograph as a recorder. This oscillograph was similar in design to the string galvanometer but had the single string replaced by a hairpin-shaped wire stretched between two fixed terminals and a spring-loaded support pulley. A small mirror cemented across the two legs of the hairpin wire was rotated when a current (the amplified ECG signal) flowed through the wire. The mirror was used to deflect a narrow light beam, throwing a small light spot on a moving film. While recording systems of this type are mechanically more rugged than the fragile string galvanometers, they still require photosensitive film or paper which has to be processed before the electrocardiogram can be read.

This disadvantage was overcome with the introduction of direct-writing recorders (about 1946), which used ink or the transfer of pigment from a ribbon to record the ECG trace on a moving paper strip, where it was immediately visible without processing. Later, a special heat-sensitive paper was developed. This type of paper is now used almost exclusively as a recording medium for electrocardiograms. Basically, the pen motor of such a recorder has a meter movement with a writing tip at the end of the indicator. Because this type of indicator naturally moves in a circular path, special measures are required to convert this motion to a straight line when a *rectilinear* rather than a *curvilinear* recording is desired.

The higher mass of moving parts used in direct-writing pen motors makes their frequency response inherently inferior to that of optical recording systems. Despite this handicap, modern direct-writing electrocardiographs have a frequency range extending to over 100 Hz, which is completely adequate for clinical ECG recordings. An improvement in performance over that of older direct-writing recorders can be partially attributed to the use of servo techniques in which the actual position of the pen is electrically sensed and the pen motor is included as part of a servo loop. For these reasons optical recording methods are seldom used in modern electrocardiographs.

6.1.2. ECG Amplifiers

The early string galvanometer had the advantage that it could easily be isolated from ground. Thus, the potential difference between two electrodes on the patient could be measured with less electrical interference than can be done with a grounded system. Electronic amplifiers, however, are normally referenced to ground through their power supplies. This creates an interference problem (unless special measures are taken) when such amplifiers are used to measure small bioelectric potentials. The technique usually employed, not only in electrocardiography but also in the measurement of other bioelectric signals, is the use of a *differential amplifier*. The principle of the differential amplifier can be explained with the help of Figure 6.2.

A differential amplifier can be considered as two amplifiers with separate inputs [Figure 6.2 (a)], but with a common output terminal, which

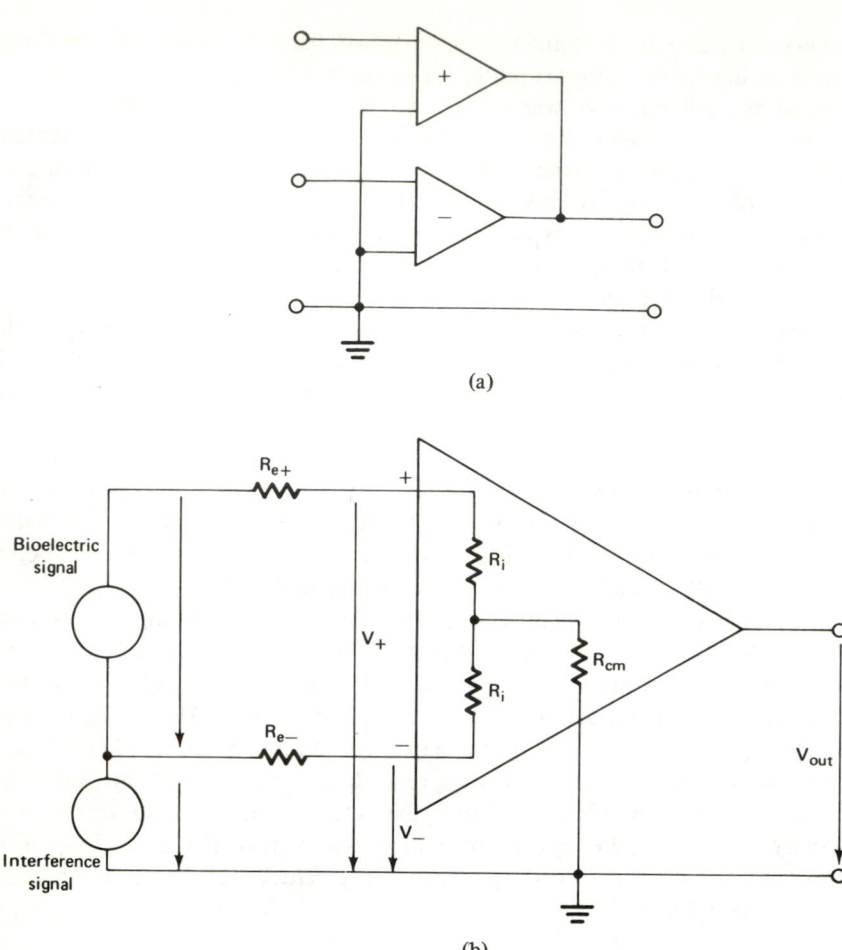

Figure 6.2. The differential amplifier: (a) represented as two amplifiers with separate inputs and common output; (b) as used for amplification of bioelectric signals (see text for explanation.)

delivers the sum of the two amplifier output voltages. Both amplifiers have the same voltage gain, but one amplifier is inverting (output voltage is $180°$ out of phase with respect to the input) while the other is noninverting (input and output voltages are in phase). If the two amplifier inputs are connected to the same input source, the resulting *common-mode gain* should be zero, because the signals from the inverting and the noninverting amplifiers cancel each other at the common output. However, because the gain of the two amplifiers is not exactly equal, this cancellation is not complete. Rather, a small residual common-mode output remains. When one of the amplifier inputs is grounded and a voltage is applied only to the other amplifier input, the input voltage appears at the output amplified by the gain of the amplifier. This gain is called the *differential gain* of the differential amplifier. The ratio of the differential gain to the common-mode gain is called the *common-mode rejection ratio* of the differential amplifier, which in modern amplifiers can be as high as 1,000,000:1.

When a differential amplifier is used to measure bioelectric signals that occur as a potential difference between two electrodes, as shown in Figure 6.2(b), the bioelectric signals are applied between the inverting and noninverting inputs of the amplifier. The signal is therefore amplified by the differential gain of the amplifier. For the interference signal, however, both inputs appear as though they were connected together to a common input source. Thus, the common-mode interference signal is amplified only by the much smaller common-mode gain.

Figure 6.2(b) also illustrates another interesting point. The electrode impedances, R_{e+} and R_{e-}, each form a voltage divider with the input impedance of the differential amplifier. If the electrode impedances are not identical, the interference signals at the inverting and noninverting inputs of the differential amplifier may be different, and the desired degree of cancellation does not take place. Because the electrode impedances can never be made exactly equal, the high common-mode rejection ratio of a differential amplifier can only be realized if the amplifier has an input impedance much higher than the impedance of the electrodes to which it is connected. As indicated in the figure, this input impedance may not be the same for the differential signal as it is for the common-mode signal. The use of a differential amplifier also requires a third connection for the reference or grouped input.

6.1.3. Electrodes and Leads

To record an electrocardiogram, a number of electrodes, usually five, are affixed to the body of the patient. The electrodes are connected to the ECG machine by the same number of electrical wires. These wires and, in a more general sense, the electrodes to which they are connected are usually called *leads*. The electrode applied to the right leg of the patient, for example, is called the RL lead. For the recording of the electrocardiogram, two electrodes or one electrode and an interconnected group of electrodes are selected and connected to the input of the recording amplifier. It is somewhat confusing that the particular electrodes selected and the way in which they are connected are also referred to as a lead. To avoid this ambiguity, in this book the term *lead* will be used only to indicate a particular group of electrodes and the way in which they are connected to the amplifier. For the individual lead wire, as well as the physical connection to the body of the patient, the term *electrode* will be used. The reader, however, should be aware of the double meaning that the term "lead" can have in normal usage.

The voltage generated by the pumping action of the heart is actually a vector whose magnitude, as well as spatial orientation, changes with time. Because the ECG signal is measured from electrodes applied to the surface of the body, the waveform of this signal is very dependent on the placement of the electrodes. Figure 6.1 shows a typical ECG waveform. Some of the segments of this trace may, however, almost disappear for certain electrode placements, whereas others may show up clearly on the recording. For this reason, in a normal electrocardiographic examination, the electrocardiogram is recorded from a number of different leads, usually 12, to ensure that no important detail of the waveform is missed. Placement of electrodes and names and configurations of the leads have become standardized and are used the same way throughout the world.

6.1.3.1. Electrodes. The placement of the electrodes, as well as the color code used to identify each electrode, is shown in Figure 6.3. In his experiments Einthoven had found it advantageous to record the electrocardiogram from electrodes placed vertically as well as horizontally on the body. As shown in Figure 4.4, he had his patients place not only both arms but also one leg into the earthenware crocks used as immersion electrodes.

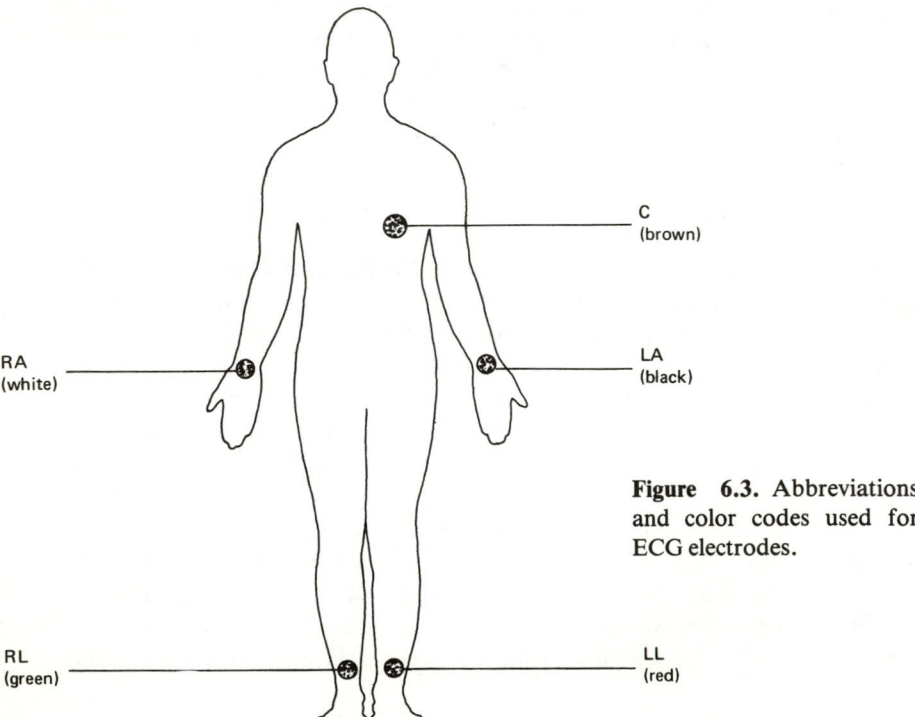

Figure 6.3. Abbreviations and color codes used for ECG electrodes.

The leg selected was the left one, probably because it terminates vertically below the heart. The early electrocardiograph machines thus employed three electrodes, of which only two were used at one time. With the introduction of the electronic amplifier, an additional connection to the body was needed as a ground reference. Although an electrode could have been positioned almost anywhere on the body for this purpose, it became a convention to use the "free" right leg.

Chest or *precordial electrodes* were introduced later. Although plate electrodes are normally used for the electrodes at the extremities, the chest electrode is often the suction type shown in Figure 4.6. It should be noted that abbreviations referring to the extremities are used to identify the electrodes even when they are actually placed on the chest, as in the case of the patient-monitoring applications described in Chapter 7.

6.1.3.2. Leads. In the normal electrode placement shown in Figure 6.3, four electrodes are used to record the electrocardiogram; the electrode on the right leg is only for ground reference. Because the input of the ECG recorder has only two terminals, a selection must be made among the available active electrodes. The 12 standard leads used most frequently are shown in Figure 6.4. The three *bipolar limb lead* selections first introduced by Einthoven, shown in the top row of the figure, are as follows:

> Lead I: Left Arm (LA) and Right Arm (RA)
> Lead II: Left Leg (LL) and Right Arm (RA)
> Lead III: Left Leg (LL) and Left Arm (LA)

These three leads are called *bipolar* because for each lead the electrocardiogram is recorded from two electrodes and the third electrode is not connected.

In each of these lead positions, the QRS of a normal heart is such that the R wave is positive.

In working with electrocardiograms from these three basic limb leads, Einthoven postulated that at any given instant of the cardiac cycle, the frontal plane representation of the electrical axis of the heart is a two-dimensional vector. Further, the ECG measured from any one of the three basic limb leads is a time-variant single-dimensional component of that vector. Einthoven also made the assumption that the heart (the origin of the vector) is near the center of an equilateral triangle, the apexes of which are the right and left shoulder and the crotch. By assuming that the ECG potentials at the shoulders are essentially the same as the wrists and that the potentials at the crotch differ little from those at either ankle, he let the points of this triangle represent the electrode positions for the three limb leads. This triangle, known as the *Einthoven triangle,* is shown in Figure 6.5.

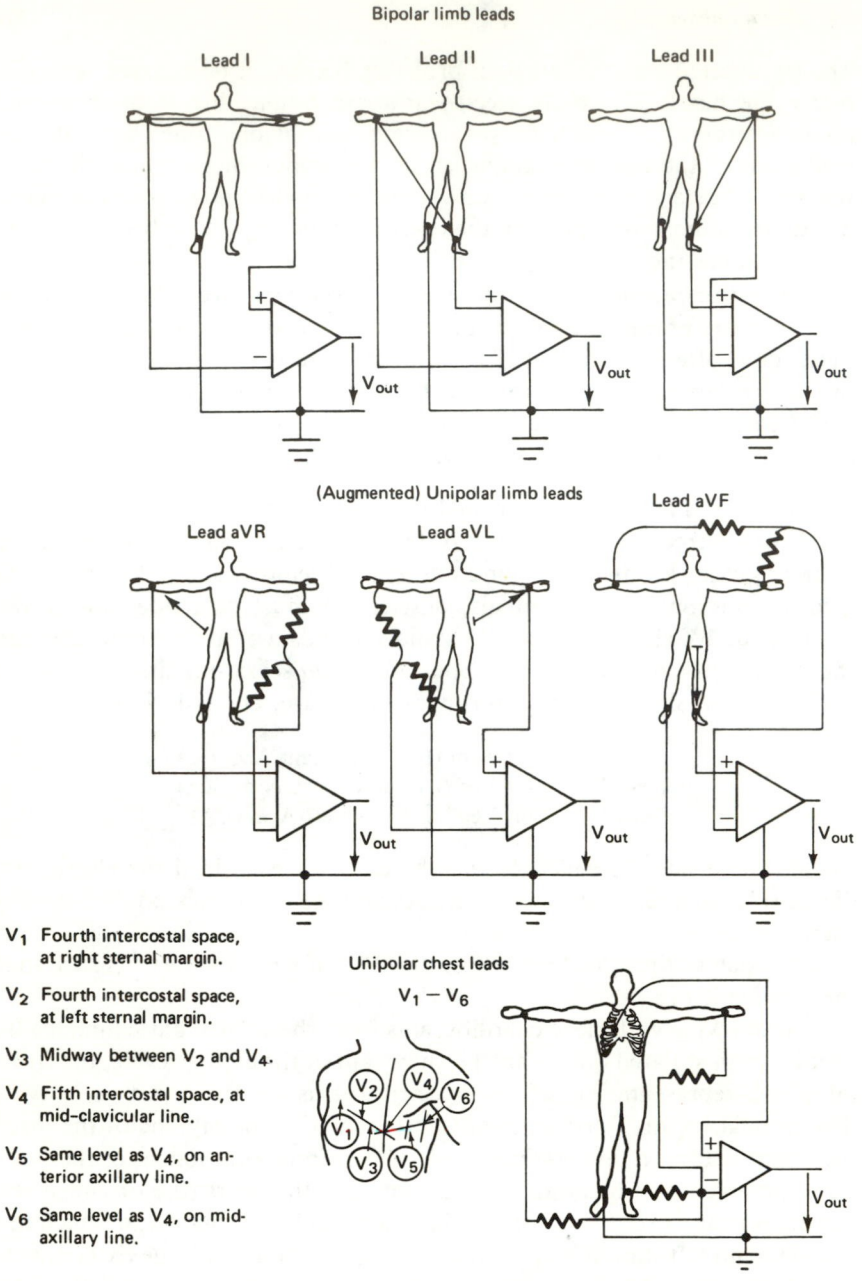

Figure 6.4. ECG lead configurations.

The sides of the triangle represent the lines along which the three projections of the ECG vector are measured. Based on this, Einthoven showed that the instantaneous voltage measured from any one of the three limb lead positions is approximately equal to the algebraic sum of the other two, or

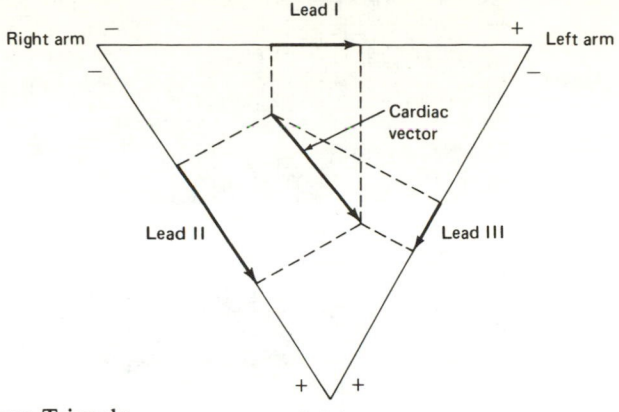

Figure 6.5. The Einthoven Triangle.

that the vector sum of the projections on all three lines is equal to zero. For these statements to actually hold true, the polarity of the lead II measurement must be reversed.

Of the three limb leads, lead II produces the greatest R-wave potential. Thus, when the amplitudes of the three limb leads are measured, the R-wave amplitude of lead II is equal to the sum of the R-wave amplitudes of leads I and III.

The other leads shown in Figure 6.4 are of the *unipolar* type, which was introduced by Wilson in 1944. For unipolar leads, the electrocardiogram is recorded between a single *exploratory electrode* and the *central terminal,* which has a potential corresponding to the center of the body. This central terminal is obtained by connecting the three active limb electrodes together through resistors of equal size. The potential at the connection point of the resistors corresponds to the mean or average of the potentials at the three electrodes. In the *unipolar limb leads,* one of the limb electrodes is used as an exploratory electrode as well as contributing to the central terminal. This double use results in an ECG signal that has a very small amplitude. In *augmented unipolar limb leads,* the limb electrode used as an exploratory electrode is not used for the central terminal, thereby increasing the amplitude of the ECG signal without changing its waveform appreciably. These leads are designated aVR, aVL, and aVF (F as in foot).

For the *unipolar chest leads,* a single chest electrode (exploring electrode) is sequentially placed on each of the six predesignated points on the chest. These chest positions are called the *precordial unipolar* leads and are designated V_1 through V_6. These leads are diagrammed in the lower part of Figure 6.4. All three active limb electrodes are used to obtain the central terminal, while a separate chest electrode is used as an exploratory electrode.

The electrocardiograms recorded from these 12 lead selections are shown in Figure 6.6. It can be seen that the trace from lead selection I or II resembles most closely the idealized waveform of Figure 6.1; some of the other traces are quite different in appearance.

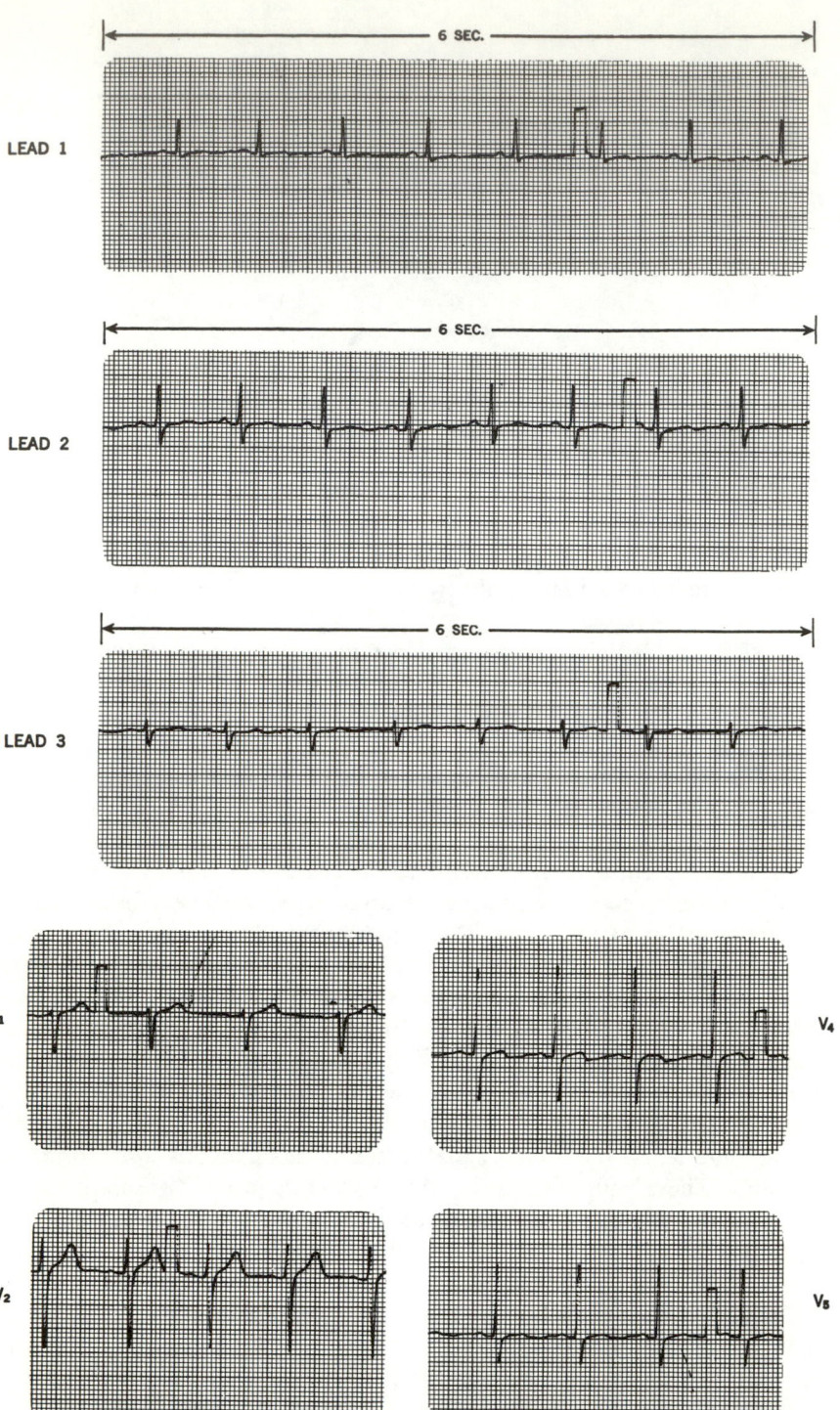

Figure 6.6. Typical patient ECG.

116

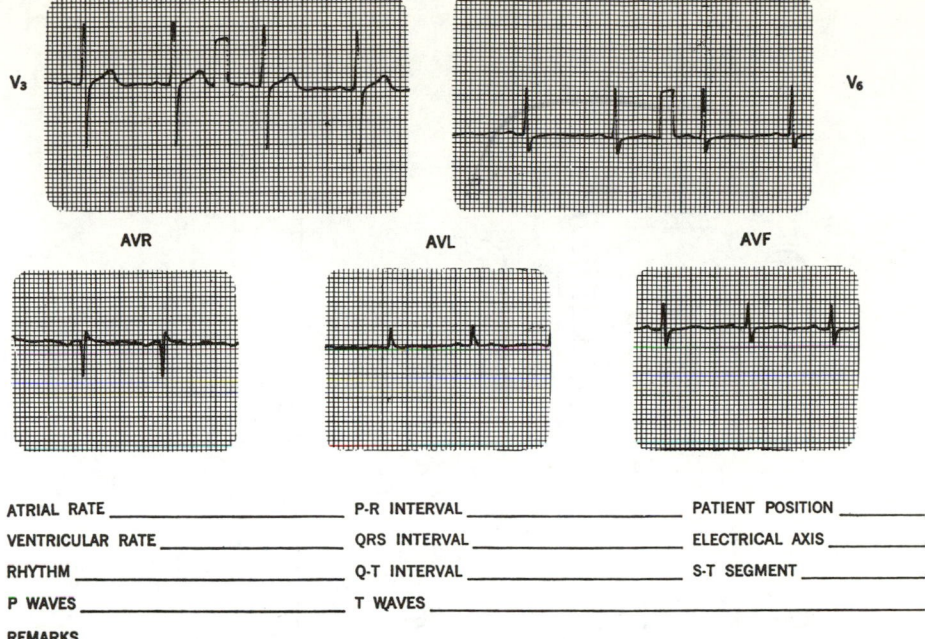

ATRIAL RATE _____	P-R INTERVAL _____	PATIENT POSITION _____
VENTRICULAR RATE _____	QRS INTERVAL _____	ELECTRICAL AXIS _____
RHYTHM _____	Q-T INTERVAL _____	S-T SEGMENT _____
P WAVES _____	T WAVES _____
REMARKS		

Figure 6.6. Continued

In addition to the lead systems already discussed, there are certain additional lead modifications that are of considerable use in the coronary care unit. The most widely used modification for ongoing ECG monitoring is the *modified chest lead I* (MCL$_1$) also called the *Marriott lead,* named after its inventor. This lead system simulates the V$_1$ position with electrode placement as follows: positive electrode, fourth intercostal space, right sternal border; negative electrode just below the outer portion of the left clavicle, with ground just about anywhere, but usually below the right clavicle. The monitor is set on lead I for this bipolar tracing. Recordings obtained in this way are very useful in differentiating left ventricular ectopic rhythms from aberrant right ventricular or supraventricular rhythms. The former situation usually necessitates prompt therapeutic action; the latter is of less clinical significance.

6.1.4. ECG Recorder Principles

The principal parts or building blocks of an ECG recorder are shown in Figure 6.7. Also shown are the controls usually found on ECG recorders; the dashed lines indicate the building block with which each control interacts.

The connecting wires for the patient electrodes originate at the end of a *patient cable,* the other end of which plugs into the ECG recorder. The wires from the electrodes connect to the *lead selector switch,* which also incorporates the resistors necessary for the unipolar leads. A pushbutton

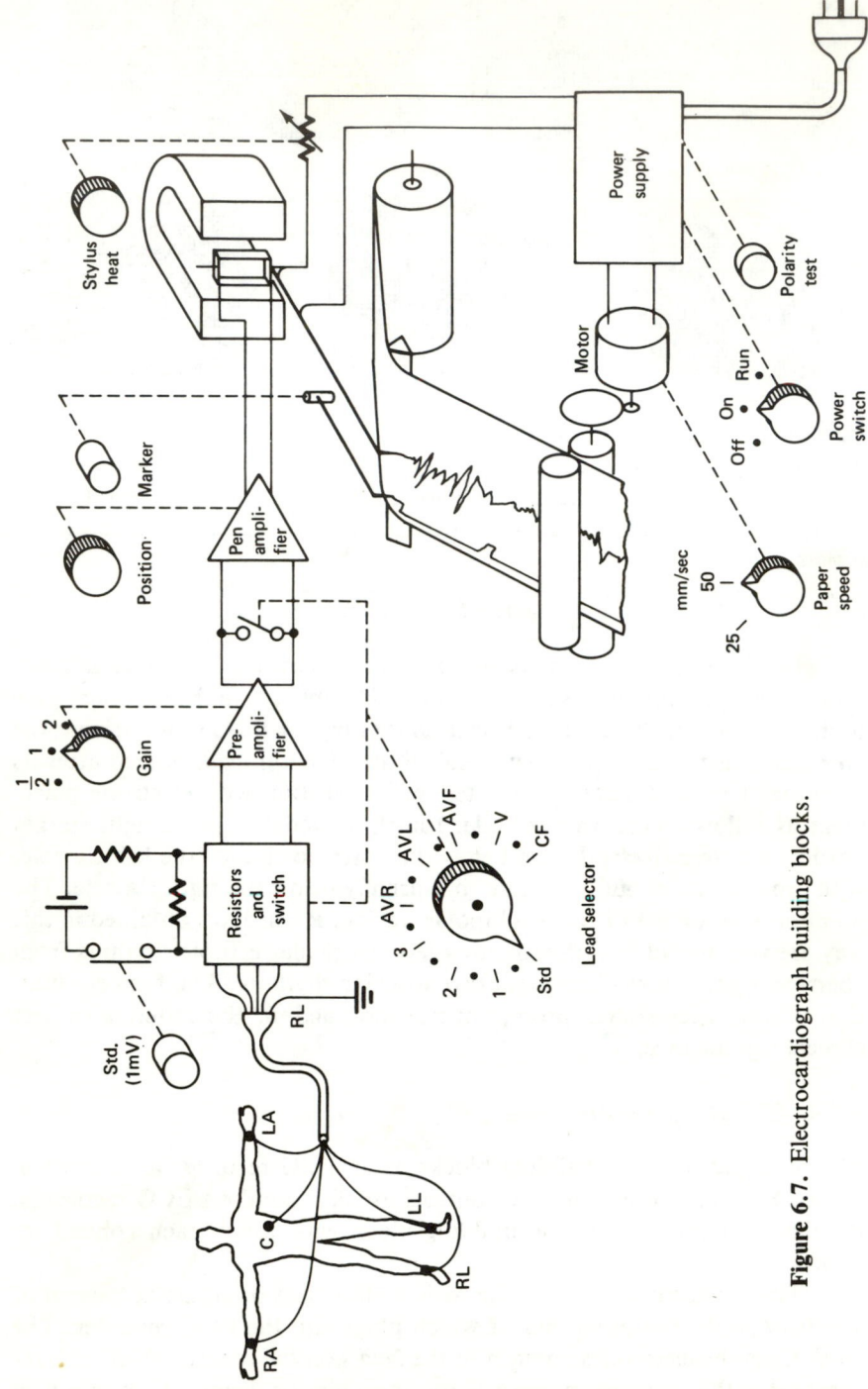

Figure 6.7. Electrocardiograph building blocks.

allows the insertion of a *standardization* voltage of 1 mV to standardize or calibrate the recorder. Although modern recorders are stable and their sensitivity does not change with time, the ritual of inserting the standardization pulse before or after each recording when recording a 12-lead ECG is still followed. Changing the setting of the lead selector switch introduces an artifact on the recorded trace. A special contact on the lead selector switch turns off the amplifier momentarily whenever this switch is moved and turns it on again after the artifact has passed. From the lead selector switch the ECG signal goes to a *preamplifier,* a differential amplifier with high common-mode rejection. It is ac-coupled to avoid problems with small dc voltages that may originate from polarization of the electrodes. The preamplifier also provides a switch to set the *sensitivity* or *gain.* Older ECG machines also have a continuously variable sensitivity adjustment, sometimes marked *standardization adjustment.* By means of this adjustment, the sensitivity of the ECG recorder can be set so that the standardization voltage of 1 mV causes a pen deflection of 10 mm. In modern amplifiers the gain usually remains stable once adjusted, so the continuously variable gain control is now frequently a screwdriver adjustment at the side or rear of the ECG recorder.

The preamplifier is followed by a dc amplifier called the *pen amplifier,* which provides the power to drive the *pen motor* that records the actual ECG trace. The input of the pen amplifier is usually accessible separately, with a special *auxiliary input* jack at the rear or side of the ECG recorder. Thus, the ECG recorder can be used to record the output of other devices, such as the *electromotograph,* which records the Achilles reflex. A *position* control on the pen amplifier makes it possible to center the pen on the recording paper. All modern ECG recorders use heat-sensitive paper, and the pen is actually an electrically heated *stylus,* the temperature of which can be adjusted with a *stylus heat control* for optimal recording trace. Beside the recording stylus, there is a *marker* stylus that can be actuated by a pushbutton and allows the operator to mark a coded indication of the lead being recorded at the margin of the electrocardiogram. Normally, electrocardiograms are recorded at a paper speed of 25 mm/s, but a faster speed of 50 mm/s is provided to allow better resolution of the QRS complex at very high heart rates or when a particular waveform detail is desired.

The *power switch* of an ECG recorder has three positions. In the *ON* position the power to the amplifier is turned on, but the paper drive is not running. In order to start the paper drive, the switch must be placed in the *RUN* position. In some ECG machines the lead selector switch has auxiliary positions (between the lead positions) in which the paper drive is stopped. In older ECG machines a pushbutton or metal "finger contact" allows the operator to check whether the recorder is connected to the power line with the right polarity. Because the improper connection of older machines can create a shock hazard for the patient, this test must be performed prior to

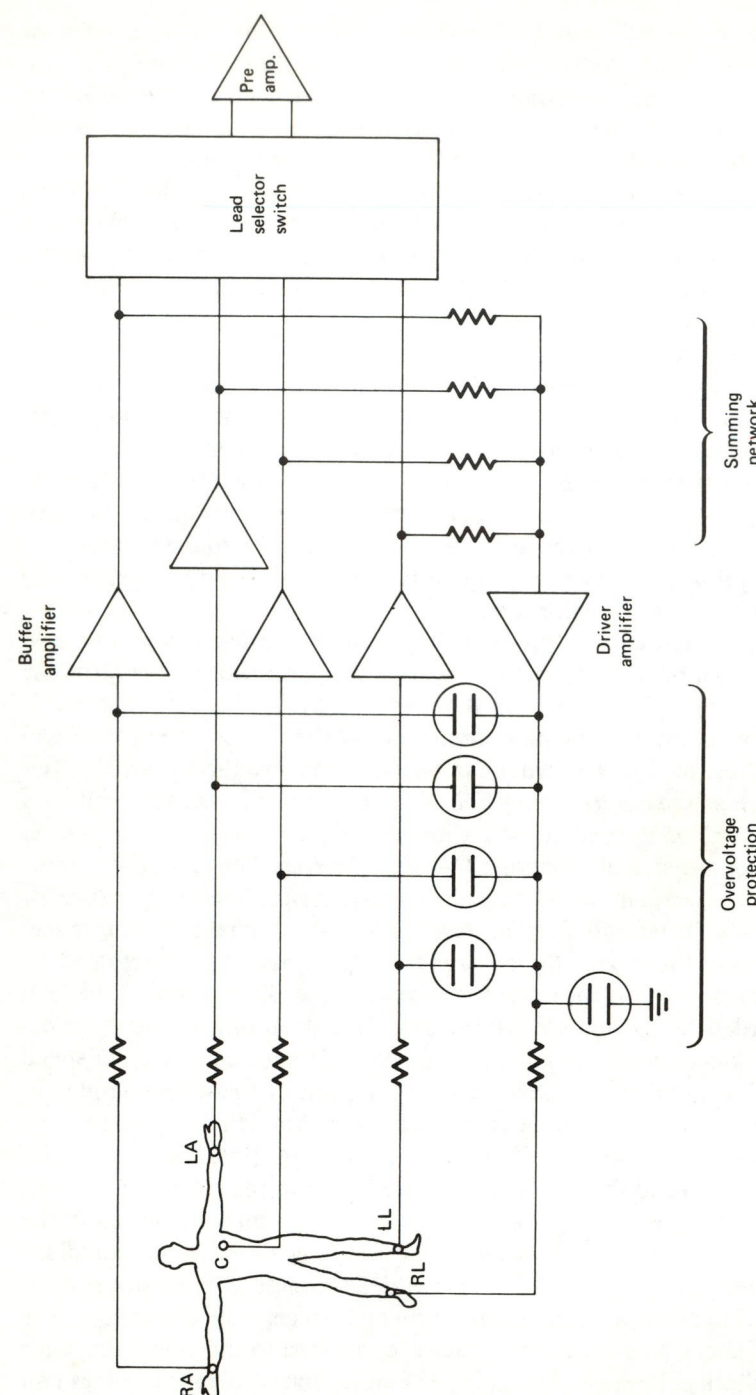

Figure 6.8. Input circuit of modern ECG machine with buffer amplifier; driven right-leg lead and over-voltage protection.

connecting the electrodes to the patient. Modern ECG machines, which have line plugs with grounding pins, do not require such a polarity test.

Although Figure 6.7 shows the principal building blocks of an electrocardiograph, it does not show the circuit details that are found in modern devices of this type. Some of these features are shown in Figure 6.8. To increase the input impedance and thus reduce the effect of variations in electrode impedance, these instruments usually include a *buffer amplifier* for each patient lead. The transistors in these amplifiers are often protected by a network of resistors and neon lamps from overvoltages that may occur when the electrocardiograph is used during surgery in conjunction with high-frequency devices for cutting and coagulation.

A more severe problem is the protection of the electrocardiograph from damage during defibrillation. The voltages that may be encountered in this case can reach several thousand volts. Thus, special measures must be incorporated into the electrocardiograph to prevent burnout of components and provide fast recovery of the trace so as to permit the success of the countershock to be judged.

Some modern devices do not connect the right leg of the patient to the chassis, but utilize a "driven right leg lead." This involves a summing network to obtain the sum of the voltages from all other electrodes and a driving amplifier, the output of which is connected to the right leg of the patient. The effect of this arrangement is to force the reference connection at the right leg of the patient to assume a voltage equal to the sum of the voltages at the other leads. This arrangement increases the common-mode rejection ratio of the overall system and reduces interference. It also has the effect of reducing the current flow in the right leg electrode. Increased concern for the safety aspect of electrical connections to the patient have caused modern ECG designs to abandon the principle of a ground reference altogether and use isolated or floating-input amplifiers, as described in Chapter 16.

6.1.5. Types of ECG Recorders

There are numerous types of ECG recorders. Many of these are portable units, while others are part of permanent installations. In this section the most commonly used types are discussed. The reader will find further examples in Chapter 7 in the context of the coronary-care unit and in Chapter 12 in the discussion of emergency care. The use of the computer in connection with electrocardiography is covered in Chapter 15.

6.1.5.1. Single-channel recorders. The most frequently used type of EGG recorder is the portable single-channel unit illustrated in Figure 6.9. For hospital use this recorder is usually mounted on a cart so that it can be wheeled to the bedside of a patient with relative ease.

If the electrocardiogram of a patient is recorded in the 12 standard lead configurations, the resulting paper strip is from 3 to 6 ft long. Even if folded in accordion fashion, the strip is still inconvenient to read and store. Therefore, it is usually cut up, and sections of the recordings from the 12 leads are mounted as shown in Figure 6.6. Because it is easy to mix up the cut sections, the lead for each trace is encoded at the margin of the paper, using the marker pen, during the recording process. The code markers consist of short marks (dots) and long marks (dashes) and look similar to Morse code. No standard code has been established for this purpose, however.

The cut sections of the electrocardiogram can be mounted by inserting them in pockets of a special folder with cutouts to make the trace visible.

Figure 6.9. Single-channel portable ECG machine. (Courtesy of Hewlett-Packard Co., Andover, MA.)

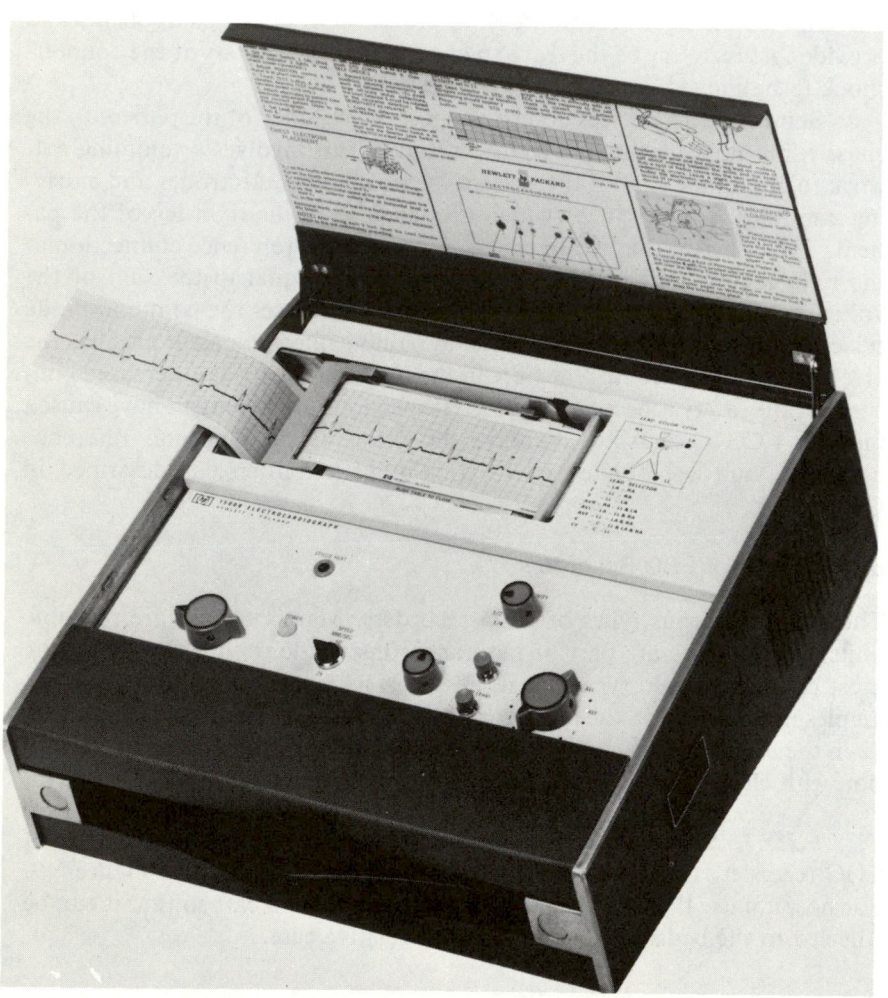

This is the way in which the electrocardiogram in Figure 6.6 was mounted. It should be noted that the recordings from the three limb leads are longer than those from the other lead selections in order to show several QRS complexes; they are called *rhythm strips*. Commercial systems are available to simplify the mounting by die-cutting the paper strip and using mounting cards with adhesive pads. With the automatic three-channel recorders described in the next section, the mounting is greatly simplified.

6.1.5.2. Three-channel recorders. Where large numbers of electrocardiograms are recorded and mounted daily, substantial savings in personnel can be achieved by the use of *automatic three-channel recorders.* These devices not only record three leads simultaneously on a three-channel recorder, but they also switch automatically to the next group of three leads. An electrocardiogram with the 12 standard leads, therefore, can be recorded automatically as a sequence of four groups of three traces. The time required for the actual recording is only about 10 seconds. The groups of leads recorded and the time at which the switching occurs are automatically identified by code markings at the margin of the recording paper. At the end of the recording, standardization pulses are inserted in all three recording channels. Although the actual recording time is reduced substantially compared to single-channel recorders, more time is required to apply the electrodes to the patient because separate electrodes must necessarily be used for each chest position. The mounting of the electrocardiogram, however, is simplified substantially, for no cutting or mounting of the individual lead selections is required. A modern recorder of this type is shown in Figure 6.10.

6.1.5.3. Vector electrocardiographs (vectorcardiographs). As noted in Section 6.1.3, the voltage generated by the activity of the heart can be described as a vector whose magnitude and spatial orientation change with time. In the type of electrocardiography described thus far, only the magnitude of the voltage is recorded. *Vectorcardiography,* on the other hand, presents an image of both the magnitude and the spatial orientation of the heart vector. The heart vector, however, is a three-dimensional variable, and three "views" or projections on orthogonal planes are necessary to describe the variable fully in two-dimensional figures. Special lead placement systems must be used to pick up the ECG signals for vector electrocardiograms, the *Frank system* being the one most frequently employed. The vectorcardiogram is usually displayed on a cathode-ray tube, similar to those used for patient monitors. Each QRS complex is displayed as a sequence of "loops" on this screen, which is then photographed with a Polaroid camera. Vectorcardiographs that use computer techniques to slow down the ECG signals and allow the recording of the vectorcardiogram with a mechanical X-Y recorder are also available.

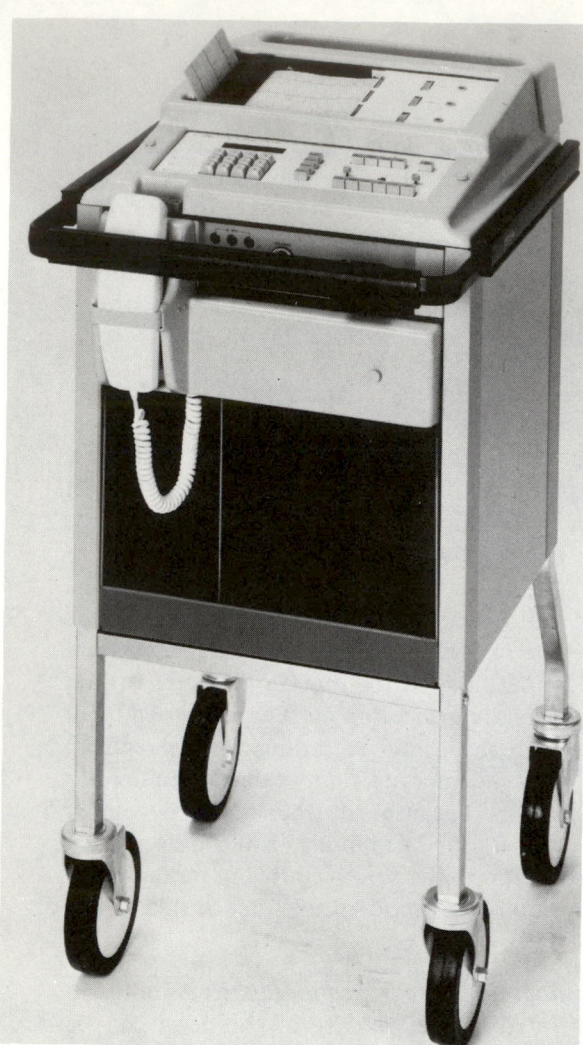

Figure 6.10. Automatic three-channel ECG recorder with keyboard for entering patient data and telephone coupler to send ECG and data to a remote computer via telephone lines. (Courtesy of Hewlett-Packard Co., MA.)

6.1.5.4. Electrocardiograph systems for stress testing. Coronary insufficiency frequently does not manifest itself in the electrocardiogram if the recording is taken during rest. In the *Masters test* or *two-step exercise test,* a physiological stress is imposed on the cardiovascular system by letting the patient repeatedly walk up and down a special pair of 9-inch high steps prior to recording his electrocardiogram. Based on the same principle is the *exercise stress test,* in which the patient walks at a specified speed on a treadmill whose inclination can be changed. While the Masters test is normally conducted using a regular single-channel electrocardiograph, special systems are available for the exercise stress test. These systems, however, are usually made up of a number of individual instruments which are described in this book.

An exercise stress test system typically consists of the following parts:

1. A treadmill which may incorporate an automatic programmer to change the speed and inclination in order to apply a specific physiological stress.
2. An ECG radiotelemetry system to allow recording of the ECG without artifacts while the patient is on the treadmill.
3. An ECG monitor with a cathode-ray-tube display and heart rate meter.
4. An ECG recorder.
5. An automatic or semiautomatic sphygmomanometer for the indirect measurement of blood pressure.

Because the exercise stress test involves a certain risk for patients with known or suspected cardiac disorders, a dc defibrillator is usually kept available while the test is performed.

6.1.5.5. Electrocardiographs for computer processing. The automatic analysis of electrocardiograms by computers is used increasingly (see Chapter 15). This technique requires that the ECG signal from the standard leads be transmitted sequentially to the computer by some suitable means, together with additional information on the patient. The automatic three-channel recorders can frequently be adapted for this purpose. The ECG signals can either be recorded on a tape for later computer entry or can be directly transmitted to the computer through special lines or regular telephone lines using a special acoustical coupler (see Chapter 12). Information regarding the patient is entered with thumbwheel switches or from a keyboard and is transmitted along with the ECG signal. During the transmission of the signal, the electrocardiogram is simultaneously recorded to verify that the transmitted signals are free of artifacts.

6.1.5.6. Continuous ECG recording (Holter recording). Because a normal electrocardiogram represents only a brief sample of cardiac activity, arrhythmias which occur intermittently or only under certain conditions, such as emotional stress, are frequently missed. The technique of continuous ECG recording, which was introduced by Norman Holter, makes it possible to capture these kinds of arrhythmias. To obtain a continuous ECG, the electrocardiogram of a patient is recorded during his normal daily activity by means of a special magnetic tape recorder. The smallest device of this type can actually be worn in a shirt pocket and allows recordings of the ECG for four hours. Other recorders, about the size of a camera case, are worn over the shoulder and can record the electrocardiogram for up to 24 hours. The recorded tape is analyzed using a special scanning device which plays back the tape at a higher speed than that used for recording. By this method a 24-hour tape can be reviewed in as little as 12 minutes. During the playback, the beat-to-beat interval of the electrocardiogram is displayed on a cathode-ray tube as a picket-fence-like pattern in which arrhythmia ep- siodes are clearly visible. Once such an episode has been discovered, the tape is backed up and slowed down to obtain a normal electrocardiogram strip for the time interval during which the arrhythmias occurred. A special time clock is synchronized by the tape drive to correlate the onset of the ep- isode with the activity of the patient.

6.2. MEASUREMENT OF BLOOD PRESSURE

As one of the physiological variables that can be quite readily measured, blood pressure is considered a good indicator of the status of the cardiovascular system. A history of blood pressure measurements has saved many a person from an untimely death by providing warnings of dangerously high blood pressure (*hypertension*) in time to provide treat- ment.

In routine clinical tests, blood pressure is usually measured by means of an indirect method using a *sphygmomanometer* (from the Greek word, *sphygmos,* meaning pulse). This method is easy to use and can be automated. It has, however, certain disadvantages in that it does not pro- vide a continuous recording of pressure variations and its practical repeti- tion rate is limited. Furthermore, only systolic and diastolic arterial pressure readings can be obtained, with no indication of the details of the pressure waveform. The indirect method is also somewhat subjective, and often fails when the blood pressure is very low (as would be the case when a patient is in shock).

Methods for direct blood pressure measurement, on the other hand, do provide a continuous readout or recording of the blood pressure waveform and are considerably more accurate than the indirect method. They require, however, that a blood vessel be punctured in order to in- troduce the sensor. This limits their use to those cases in which the condition of the patient warrants invasion of the vascular system.

This section is divided into three parts. First, indirect or noninvasive methods are discussed. Since there has been much progress in the automating of indirect techniques, automated methods are covered in a separate section. Finally, direct or invasive blood pressure measurements are discussed.

6.2.1. Indirect Measurements.

As stated earlier, the familiar indirect method of measuring blood pres- sure involves the use of a sphygmomanometer and a stethoscope. The

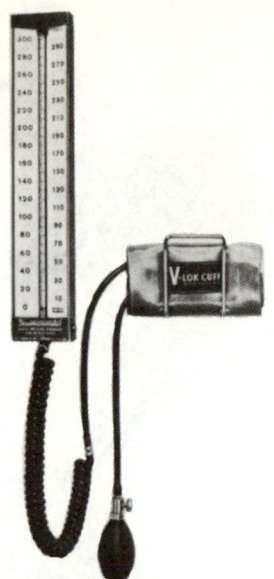

Figure 6.11. Wall-mounted sphygmomanometer. (Courtesy of W.A. Baum, Inc., Copiague, NY.)

sphygmomanometer consists of an inflatable pressure cuff and a mercury or aneroid manometer to measure the pressure in the cuff. The cuff consists of a rubber bladder inside an inelastic fabric covering that can be wrapped around the upper arm and fastened with either hooks or a Velcro fastener. The cuff is normally inflated manually with a rubber bulb and deflated slowly through a needle valve. The stethoscope is described in detail in Section 6.5. A wallmounted sphygmomanometer is shown in Figure 6.11. These devices are also manufactured as portable units.

The sphygmomanometer works on the principle that when the cuff is placed on the upper arm and inflated, arterial blood can flow past the cuff only when the arterial pressure exceeds the pressure in the cuff. Furthermore, when the cuff is inflated to a pressure that only partially occludes the brachial artery, turbulence is generated in the blood as it spurts through the tiny arterial opening during each systole. The sounds generated by this turbulence, *Korotkoff sounds,* can be heard through a stethoscope placed over the artery downstream from the cuff.

To obtain a blood pressure measurement with a sphygmomanometer and a stethoscope, the pressure cuff on the upper arm is first inflated to a pressure well above systolic pressure. At this point no sounds can be heard through the stethoscope, which is placed over the brachial artery, for that artery has been collapsed by the pressure of the cuff. The pressure in the cuff is then gradually reduced. As soon as cuff pressure falls below systolic pressure, small amounts of blood spurt past the cuff and Korotkoff sounds begin to be heard through the stethoscope. The pressure of the cuff that is indicated on the manometer when the first Korotkoff sound is heard is recorded as the systolic blood pressure.

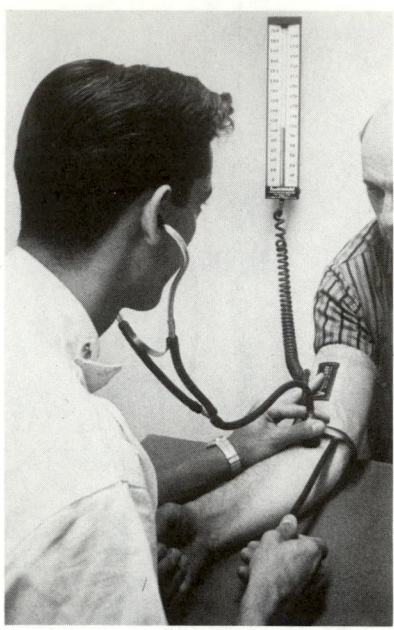

Figure 6.12. Measurement of blood pressure using sphygmomanometer. (Courtesy of W.A. Baum, Inc., Copiague, NY.)

As the pressure in the cuff continues to drop, the Korotkoff sounds continue until the cuff pressure is no longer sufficient to occlude the vessel during any part of the cycle. Below this pressure the Korotkoff sounds disappear, marking the value of the diastolic pressure.

This familiar method of locating the systolic and diastolic pressure values by listening to the Korotkoff sounds is called the *auscultatory method* of sphygmomanometry. An alternative method, called the *palpatory method,* is similar except that the physician identifies the flow of blood in the artery by feeling the pulse of the patient downstream from the cuff instead of listening for the Korotkoff sounds. Although systolic pressure can easily be measured by the palpatory method, diastolic pressure is much more difficult to identify. For this reason, the auscultatory method is more commonly used. Figure 6.12 shows a blood pressure measurement using the auscultatory method.

6.2.2. Automated Indirect Methods

Because of the trauma imposed by direct measurement of blood pressure (described below) and the lack of a more suitable method for indirect measurement, attempts have been made to automate the indirect procedure. As a result, a number of automatic and semiautomatic systems have been developed. Most devices are of a type that utilizes a pressure transducer connected to the sphygmomanometer cuff, a microphone placed beneath the cuff (over the artery), and a standard physiological recording system on which cuff pressure and the Korotkoff sounds are recorded. The basic procedure essentially parallels the manual method. The pressure cuff is automatically inflated to about 220 mm Hg and allowed to deflate slowly. The microphone picks up the Korotkoff sounds from the artery near the surface, just below the compression cuff. One type of instrument either superimposes the signal of the Korotkoff sounds on the voltage recording representing the falling cuff pressure or records the two separately. The pressure reading at the time of the first sound represents the systolic pressure; the diastolic pressure is the point on the falling pressure curve where the signal representing that last sound is seen. This instrument is actually only semiautomatic because the recording thus obtained must still be interpreted by the observer. False indications—caused, for instance, by motion artifacts—can often be observed on the recording. Fully automated devices use some type of signal-detecting circuit to determine the occurrence of the first and last Korotkoff sounds and retain and display the cuff pressure reading for these points, either electronically or with mercury manometers that are cut off by solenoid valves. One of the recent innovations of these techniques is the "do-it-yourself" blood pressure machine to be found in some supermarkets. These devices, by necessity, are more susceptible to false indications caused by artifacts.

Many of the commercially available automatic blood pressure meters work well when demonstrated on a quiet, healthy subject but fail when used to measure blood pressure during activity or when used on patients in circulatory shock. Methods other than those utilizing the Korotkoff sounds have been tried in detecting the blood pulse distal to the occlusion cuff. Among them is impedance plethysmography (see Section 6.4), which indicated directly the pulsating blood flow in the artery, and ultrasonic Doppler methods, which measure the motions of the arterial walls. An early example of an automatic blood pressure meter is the *programmed electrosphygmomanometer PE-300*, illustrated in Figures 6.13 and 6.14 in block diagram and pictorial form. This instrument is designed for use in conjunction with an occluding cuff, microphone, or pulse transducer, and a recorder for the automatic measurement of indirect systolic and diastolic blood pressures from humans and many animal subjects.

The PE-300 incorporates a transducer-preamplifier that provides two output signals, a voltage proportional to the cuff pressure, and the amplified Korotkoff sounds or pulses. These signals can be monitored individually or with the sounds or pulses superimposed on the calibrated cuff-recorder. The combined signal can be recorded on a graphic pen recorder.

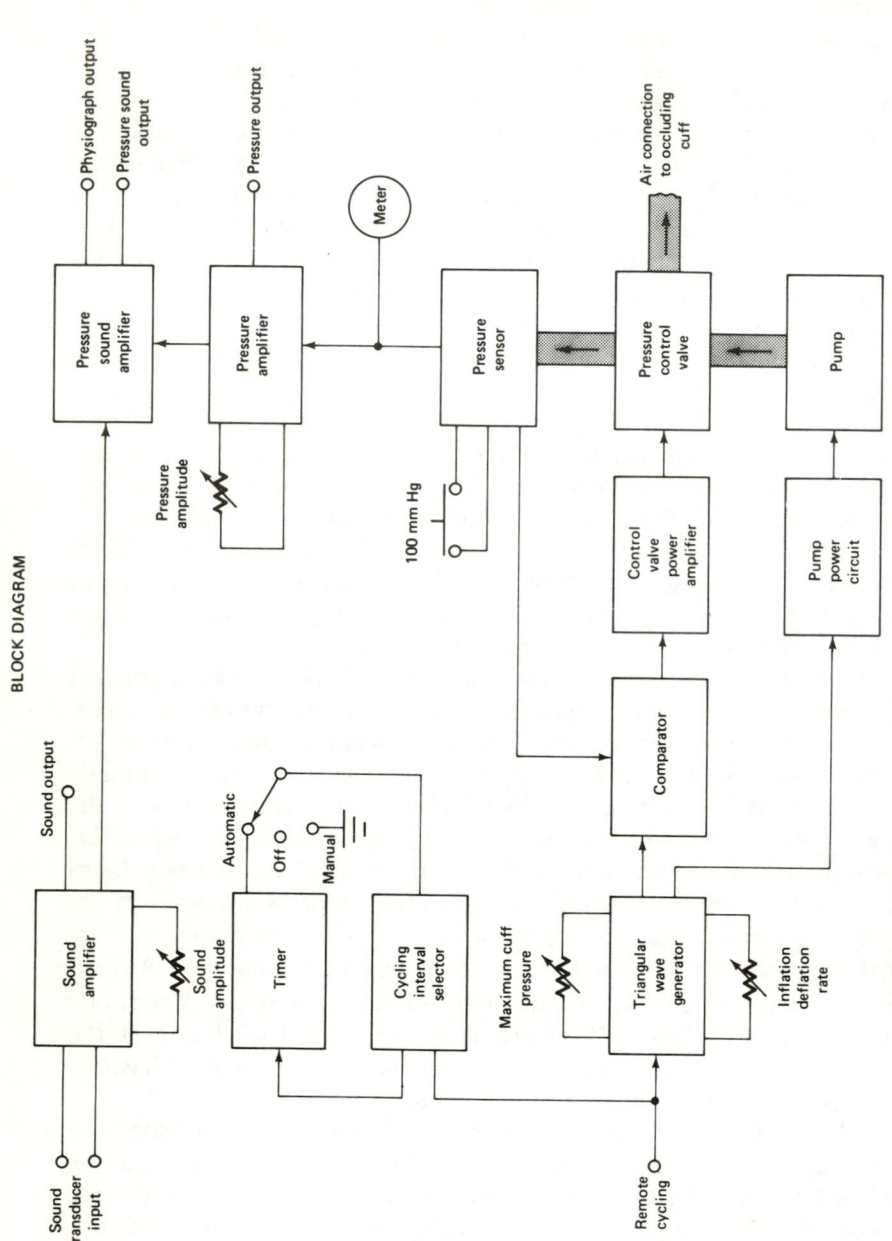

Figure 6.13. Programmed electrosphygmomanometer (block diagram). (Courtesy of Narco BioSystems, Inc., Houston, TX.)

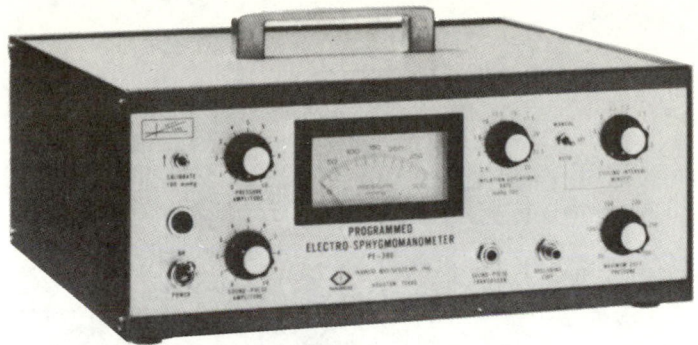

Figure 6.14. Electrosphygmomanometer. (Courtesy of Narco BioSystems, Inc., Houston, TX.)

The self-contained cuff inflation system can be programmed to inflate and deflate an occluding cuff at various rates and time intervals. Equal and linear rates of cuff inflation and deflation permit two blood pressure determinations per cycle. The PE-300 can be programmed for repeat cycles at adjustable time intervals for monitoring of blood pressure over long periods of time. Single cycles may be initiated by pressing a panel-mounted switch. Provision is also made for remote control via external contact closure. The maximum cuff pressure is adjustable, and the front-panel meter gives a continuous visual display of the cuff pressure.

The electronic sphygmomanometer shown in Figure 6.15 is an example of a device incorporating a large gage for easy reading. The scale is 6 in. (over 15 cm) in diameter. In this example the cuff has to be inflated manually and so this instrument is sometimes called "semiautomatic." The Korotkoff sounds are automatically detected by microphones and a light flashes at systolic pressure, which stops at the diastolic value.

While the clinical diagnostic value of systolic and diastolic blood pressure has been clearly established, the role of *mean arterial pressure* (MAP) as an indication of blood pressure trend has become more widely accepted with the expanded use of direct pressure monitoring using arterial cannulae with electronic transducers and displays. Most electrical monitors now provide both diagnostic systolic/diastolic waveform information and the added option of a single-value MAP indication. It is now generally recognized that MAP is a direct indication of the pressure available for tissue perfusion, and that a continuously increasing or decreasing MAP can ultimately result in a hypertensive or hypotensive crisis.

Mean arterial pressure is a weighted average of systolic and diastolic pressure. Generally, MAP falls about one-third of the way between the diastolic low and the systolic peak. A simple formula for calculating MAP is: MAP = 1/3(systolic - diastolic) + diastolic.

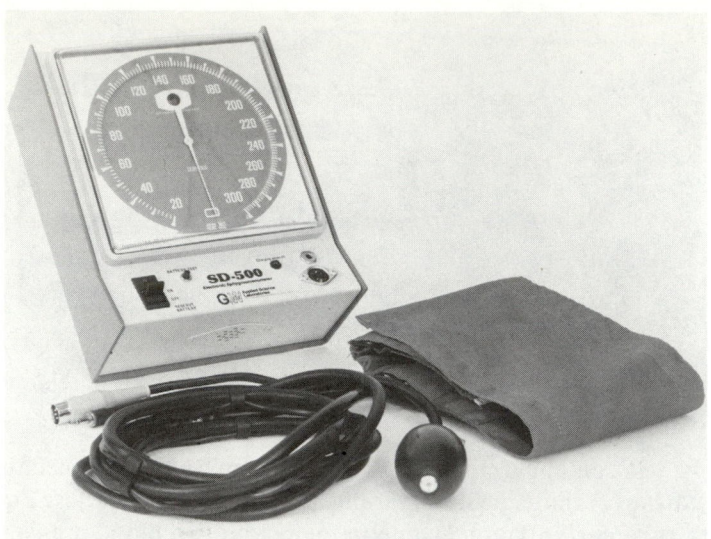

Figure 6.15. Electronic sphygmomanometer. (Courtesy of Applied Science Laboratories, Waltham, MA.)

In the Dinamap illustrated in Figure 6.16, the mean arterial blood pressure is determined by the oscillometric method. The pressure pulsations or oscillations introduced within a cuff bladder are sensed by a solid-state pressure transducer located within the enclosure. As the air pressure in the cuff is decreased, pressure oscillations increase in amplitude and reach a maximum as the cuff pressure passes through the pressure equal to the mean arterial blood pressure. This phenomenon can also be observed as it produces pulsations in the mercury column of a conventional sphygmomanometer or the needle oscillations in an aneroid sphygmomanometer.

Figure 6.16. Dinamap for automatic detection of mean arterial blood pressure. (Courtesy of Applied Medical Research, Tampa, FL.)

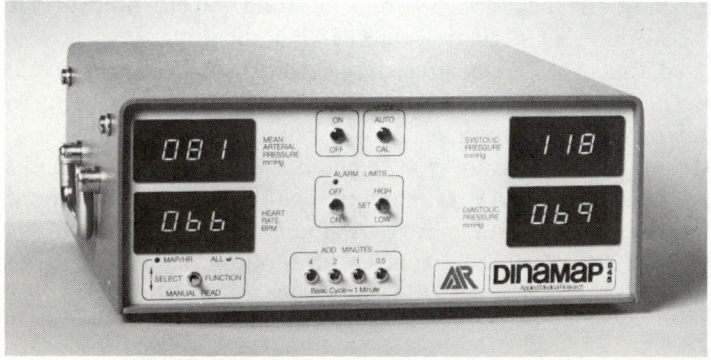

Since the indications associated with the indirect measurement of mean arterial pressure pass through a maximum at the point of interest, as opposed to the phenomenon for the indirect determination of systolic and diastolic pressure, which is at a minimum at the point of interest, the Dinamap can determine arterial pressures over a wide range of physiological states. It works effectively on most patients on whom systolic and diastolic indirect pressures are impossible to determine or are subject to such variabilities as to be of questionable clinical value.

An internal microprocessor rejects most artifacts caused by patient movement or external interference. Under program control, all normal operating variables, such as inflation pressures, deflation rates, alarm limits, and abnormal operation alarms, are automatically determined and controlled without operator adjustments. Adjustable alarm limits and measurement cycle times are simply set via front panel switches by the user as required for varying clinical conditions.

Because, in most clinical situations, systolic and diastolic pressures correlate with each other and MAP is determined from the systolic/diastolic pair, critical patient arterial pressure trends can be monitored by observation of MAP. Adjustable alarm limits, as required for a given clinical situation, can then alert staff of possible patient problems. These units can be used in operating rooms, recovery rooms, and intensive-care units.

The ability to measure blood pressure automatically with portable equipment makes it possible to take measurements while the patient is pursuing his or her normal activities. A system that does this is shown in Figure

Figure 6.17. Ambulatory automatic blood pressure monitor. (Courtesy of Del Mar Avionics, Irvine, CA.)

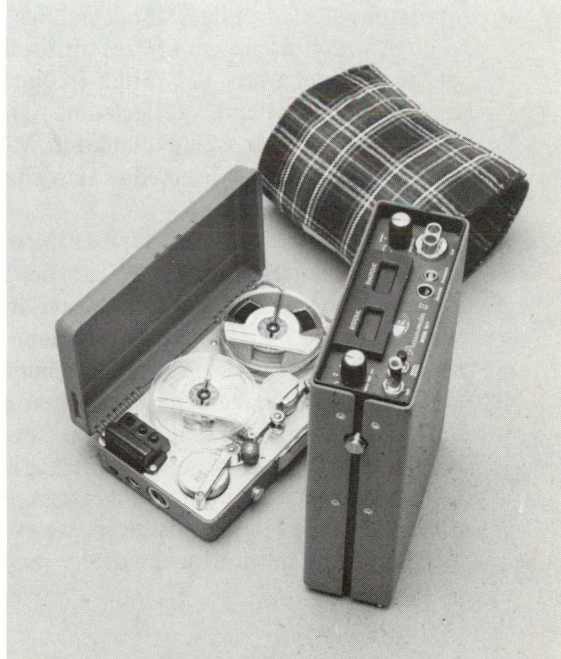

6.17. This device is a 24-hour automatic noninvasive blood pressure and Holter (see Section 6.1.5.6) monitoring system with trend writeouts. The three major components of the ambulatory system are shown in the photograph. The Pressurometer II Model 1977 at the front is the blood pressure measuring device. The recording unit is the Model 446A Electrocardiocorder, which records blood pressure and ECG on a 24-hour basis. The patient wears the units with belts and straps. The cuff is attached comfortably and securely on the left arm.

The Model 1977 utilizes a standard pneumatic cuff. A transducer for the detection of the Korotkoff sounds is held in place with an adhesive disk. Cuff pressure is applied automatically and the patient's systolic and diastolic pressures are measured in excess of 100 preprogrammed intervals in a 24-hour period. The preselected intervals can be overridden by means of manual switching and the unit can be operated manually for checking and calibration. When used with the 446A recorder, the Korotkoff sounds are gated by ECG R-wave signals. The recordings can be fed into a companion trend computer, which gives a digital readout of the data on the recording. The entire system is powered by a portable rechargeable nickelcadmium battery pack which permits up to 26 hours of recording. The computer is a plug-in module and chart-paper documentation is also available.

An important feature of the system is that it can be used with an Electrocardioscanner, which scans all the data at a rate 120 times as fast as they were recorded (the entire 24-hour record can be scanned in 12 minutes) quantitating ectopic beats and total heart beats and printing hourly trends on heart rate and other quantities. It can rewind itself and will search out operator-selected patient abnormalities.

Another approach utilizes ultrasound to measure the pulsatile motion of the brachial artery wall. High-frequency sound energy is transmitted into the patient's arm and is reflected back from the arterial walls. By means of the Doppler effect (see explanation in Chapter 9), the movement of the arterial walls can be detected as they snap open and closed with each pulsation of blood.

An advantage of this type of instrument is that results closer to direct measurements (Section 6.2.3) can be obtained. Also, it can be used for patients under shock and in intensive care units, when direct measurement would not be suitable for the patient. Because vessel-wall movement is sensed, blood flow is not a requirement for measurement.

One such instrument is the Arteriosonde, which has a cloth cuff and air bag with an electric air pump to supply the pressure. The pump can be regulated by a front-panel control. The cuff is placed on the arm in the same fashion as the sphygmomanometer except that there is a transducer array under the cuff. These transducers are arranged as alternate transmitters and receivers. The motion of the artery produces a *Doppler shift*, which identifies the instant the artery is opened and closed with each beat between systolic and diastolic pressure.

The first signal is used to stop the mercury fall in the first of two manometers to indicate systolic pressure. The second manometer is stopped at the point of disappearance of the pulses to indicate diastolic pressure. Hence, both readings can be noted simultaneously.

6.2.3. Direct Measurements

In 1728, Hales inserted a glass tube into the artery of a horse and crudely measured arterial pressure. Poiseuille substituted a mercury manometer for the piezometer tube of Hales, and Ludwig added a float and devised the *kymograph,* which allowed continuous, permanent recording of the blood pressure. It is only quite recently that electronic systems using strain gages as transducers have replaced the kymograph.

Regardless of the electrical or physical principles involved, direct measurement of blood pressure is usually obtained by one of three methods:

1. Percutaneous insertion.
2. Catheterization (vessel cutdown).
3. Implantation of a transducer in a vessel or in the heart.

Other methods, such as clamping a transducer on the intact artery, have also been used, but they are not common.

Figure 6.18 should give a general idea of both methods. Typically, for percutaneous insertion, a local anesthetic is injected near the site of invasion. The vessel is occluded and a hollow needle is inserted at a slight angle toward the vessel. When the needle is in place, a catheter is fed through the hollow needle, usually with some sort of a guide. When the catheter is securely in place in the vessel, the needle and guide are withdrawn. For some measurements, a type of needle attached to an airtight tube is used, so that the needle can be left in the vessel and the blood pressure sensed directly by attaching a transducer to the tube. Other types have the transducer built into the tip of the catheter. This latter type is used in both percutaneous and full catheterization models.

Catheterization was first developed in the late 1940s and has become a major diagnostic technique for analyzing the heart and other components of the cardiovascular system. Apart from obtaining blood pressures in the heart chambers and great vessels, this technique is also used to obtain blood samples from the heart for oxygen-content analysis and to detect the location of abnormal blood flow pathways. Also, catheters are used for investigations with injection of radiopaque dyes for X-ray studies, colored dyes for indicator dilution studies, and of vasoactive drugs directly into the heart and certain vessels. Essentially, a *catheter* is a long tube that is in-

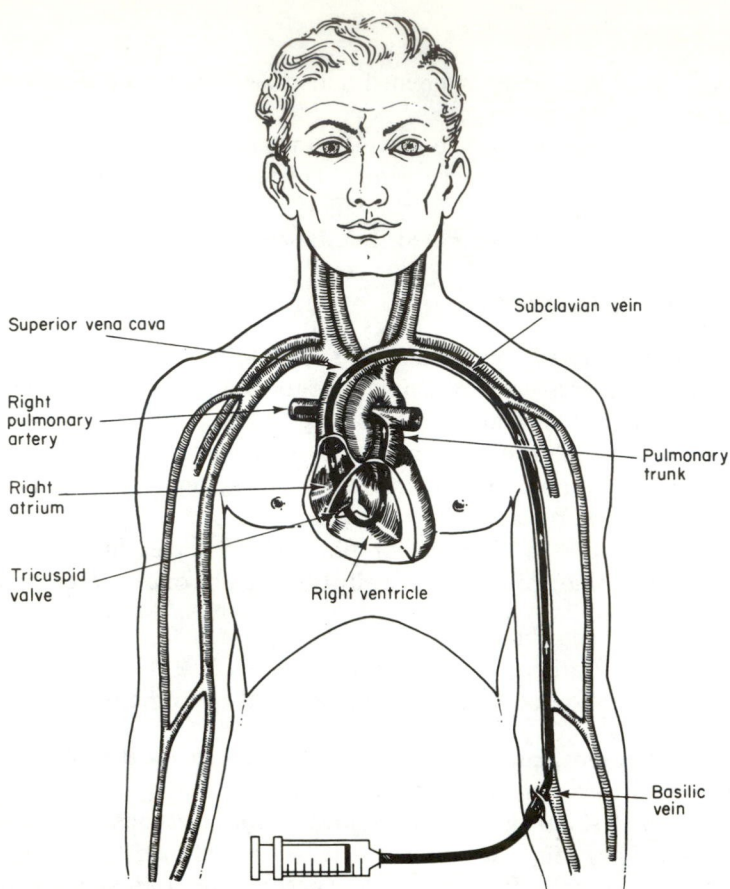

Figure 6.18. Cardiac catheterization. The tube is shown entering the basilic vein in this case. (From W.F. Evans, *Anatomy and Physiology, The Basic Principles,* Englewood Cliffs, NJ., Prentice-Hall, Inc., 1971, by permission.)

troduced into the heart or a major vessel by way of a superficial vein or artery. The sterile catheter is designed for easy travel through the vessels.

Measurement of blood pressure with a catheter can be achieved in two ways. The first is to introduce a sterile saline solution into the catheter so that the fluid pressure is transmitted to a transducer outside the body (extracorporeal). A complete fluid pressure system is set up with provisions for checking against atmospheric pressure and for establishing a reference point. The frequency response of this system is a combination of the frequency response of the transducer and the fluid column in the catheter. In the second method, pressure measurements are obtained at the source. Here, the transducer is introduced into the catheter and pushed to the point at which the pressure is to be measured, or the transducer is mounted at the tip of the catheter. This device is called a *catheter-tip blood pressure transducer.* For mounting at the end of a catheter, one manufacturer uses an unbonded resistance strain gage in the transducer, whereas another uses a variable inductance transducer (see Chapter 2). Each will be discussed later.

Implantation techniques involve major surgery and thus are normally employed only in research experiments. They have the advantage of keeping the transducer fixed in place in the appropriate vessel for long periods of time. The type of transducer employed in that procedure is also described later in this section.

Transducers can be categorized by the type of circuit element used to sense the pressure variations, such as capacitive, inductive, and resistive. Since the resistive types are most frequently used, the other two types are discussed only briefly.

In the *capacitance manometer*, a change in the distance between the plates of a capacitor changes its capacitance. In a typical application, one of the plates is a metal membrane separated from a fixed plate by some one-thousandth of an inch of air. Changes in pressure that change the distance between the plates thereby change the capacitance. If this element is contained in a high-frequency resonant circuit, the changes in capacitance vary the frequency of the resonant circuit to produce a form of frequency modulation. With suitable circuitry, blood pressure information can be obtained and recorded as a function of time.

An advantage of this type of transducer is that its total contour can be long and thin so that it can be easily introduced into the bloodstream without deforming the contour of the recorded pressure waveform. Because of the stiff structure and the small movement of the membrane when pressure is applied, the volume displacement is extremely small (in the region of 10^{-6} cm^3/100 mm Hg of applied pressure).

Disadvantages of this type of transducer are instability and a proneness to variations with small changes in temperature. Also, lead wires introduce errors in the capacitance, and this type of transducer is more difficult to use than resistance types.

A number of different devices use inductance effects. They measure the distortion of a membrane exposed to the blood pressure. In some of these types, two coils are used—a primary and secondary. When a spring-loaded core that couples the coils together magnetically is moved back and forth, the voltage induced into the secondary changes in proportion to the pressure applied.

A better-known method employs a *differential transformer,* described in detail in Chapter 2. In this device two secondary coils are wound oppositely and connected in series. If the spring-loaded core is symmetrically positioned, the induced voltage across one secondary coil opposes the voltage of the other. Movement of the core changes this symmetry, and the result is a signal developed across the combined secondary coils. The core can be spring-loaded to accept pressure from one side, or it can accept pressure from both sides simultaneously, thus measuring the difference of pressure between two different points.

The *physiological resistance transducer* is a direct adaptation of the strain gages used in industry for many years. The principle of a strain gage is that if a very fine wire is stretched, its resistance increases. (A detailed discussion of strain gages is given in Chapter 2.) If voltage is applied to the resistance, the resulting current changes with the resistance variations according to Ohm's law. Thus, the forces responsible for the strain can be recorded as a function of current. The method by which the blood pressure produces the strain is discussed in Section 6.2.4.

To obtain the degree of sensitivity required for blood pressure transducers, two or four strain gages are mounted on a diaphragm or membrane, and these resistances are connected to form a bridge circuit. Figure 6.19 shows such a circuit configuration.

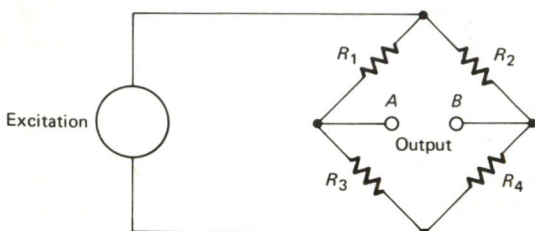

Figure 6.19. Resistance strain-gage bridge.

In general, the four resistances are initially about equal when no pressure or strain is applied. The gages are attached to the pressure diaphragm in such a way that as the pressure increases, two of them stretch while the other two contract. An excitation voltage is applied as shown. When pressure changes unbalance the bridge a voltage appears between terminals A and B proportional to the pressure. Excitation can be either direct current or alternating current, depending on the application.

Resistance-wire strain gages can be *bonded* or *unbonded* (see Chapter 2). In the bonded type, the gage is "bonded" to the diaphragm and stretches or contracts with bending. The unbonded type consists of two pairs of wires, coiled and assembled in such a way that displacement of a membrane connected to them causes one pair to stretch and the other to relax. The two pairs of wires are not bonded to the diaphragm material but are attached only by retaining lugs. Because the wires are very thin, it is possible to obtain relatively large signals from the bridge with small movement of the diaphragm.

Development of semiconductors that change their resistance in much the same manner as wire gages has led to the *bonded silicon element* bridge. Only small displacements (on the order of a few micro meters) of the pressure-sensing diaphragm, are needed for sizable changes of output voltage with low-voltage excitation. For example, with 10 V excitation, a range of 300 mm Hg is obtained with a 3-μm deflection, producing a 30-mV signal.

Semiconductor strain-gage bridges are often temperature-sensitive, however, and have to be calibrated for baseline and true zero. Therefore, it is usually necessary to incorporate external resistors and potentiometers to balance the bridge initially, as well as for periodic correction.

In Chapter 2 the gage factor for a strain gage is defined as the amount of resistance change produced by a given change in length. Wire strain gages have gage factors on the order of 2 to 4, whereas semiconductor strain gages have gage factors ranging from 50 to 200. For silicon, the gage factor is typically 120. The use of semiconductors is restricted to those configurations that lend themselves to this technique.

When strain gages are incorporated in pressure transducers, the sensitivity of the transducer is expressed, not as a gage factor, but as a voltage change that results from a given pressure change. For example, the sensitivity of a pressure transducer can be given in microvolts per (applied) volt per millimeter of mercury.

6.2.4. Specific Direct Measurement Techniques

In Section 6.2.3 methods of direct blood pressure were classified in two ways, first by the clinical method by which the measuring device was coupled to the patient and, second, by the electrical principle involved. In the following discussion, the first category is expanded, with the electrical principles involved being used as subcategories where necessary. The four categories are as follows:

1. A catheterization method involving the sensing of blood pressure through a liquid column. In this method the transducer is external to the body, and the blood pressure is transmitted through a saline solution column in a catheter to this transducer. This method can use either an unbonded resistance strain gage to sense the pressure or a linear variable differential transformer. Externally, these two devices are quite similar in appearance.

2. A catheterization method involving the placement of the transducer through a catheter at the actual site of measurement in the bloodstream (e.g., to the aorta), or by mounting the transducer on the tip of the catheter.

3. Percutaneous methods in which the blood pressure is sensed in the vessel just under the skin by the use of a needle or catheter.

4. Implantation techniques in which the transducer is more permanently placed in the blood vessel or the heart by surgical methods.

The most important aspects of these methods are discussed separately.

6.2.4.1. Liquid-column methods. A typical liquid-column blood pressure transducer, the Gould Statham P 23 ID, is illustrated in Figure 6.20. Figure 6.21 is a cutaway drawing to show the interior construction and the isolation features of the same transducer, which is considered a standard size in hospital practice. The heart of the P 23 transducer is the unbonded strain gage, which is connected in a standard Wheatstone bridge configuration. The metal sensing diaphragm can be seen on the left side. It is a precision-made part that must deflect predictably with a given fluid pressure. When the diaphragm is deflected downward by the pressure of the liquid being measured, the tension on two of the bridge wires is relaxed and the tension on the other two wires is tightened, changing the resistance of the gage. For negative pressures, the opposite wires are stretched and relaxed.

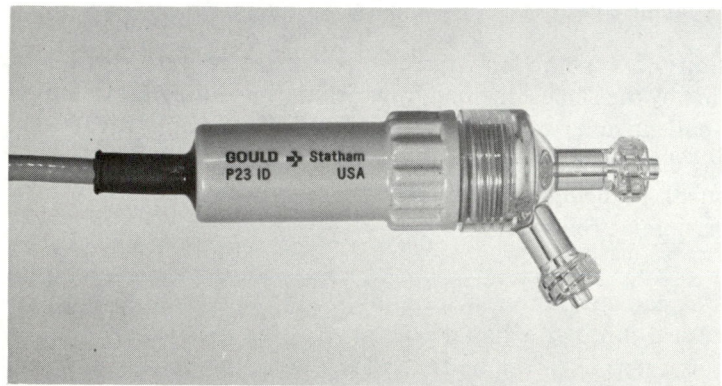

Figure 6.20. Fluid-column blood pressure transducer. (Courtesy of Gould, Inc., Measurement Systems Division, Oxnard, CA.)

The transducer is connected through the cable to an instrument which contains zero-balance and range controls, amplifier circuits, and a readout. The shielded cable is attached to the case through a liquid-tight seal that permits immersion of the transducer for cleaning. The transducer case is vented through the cable so that measurements are always referenced to atmospheric pressure.

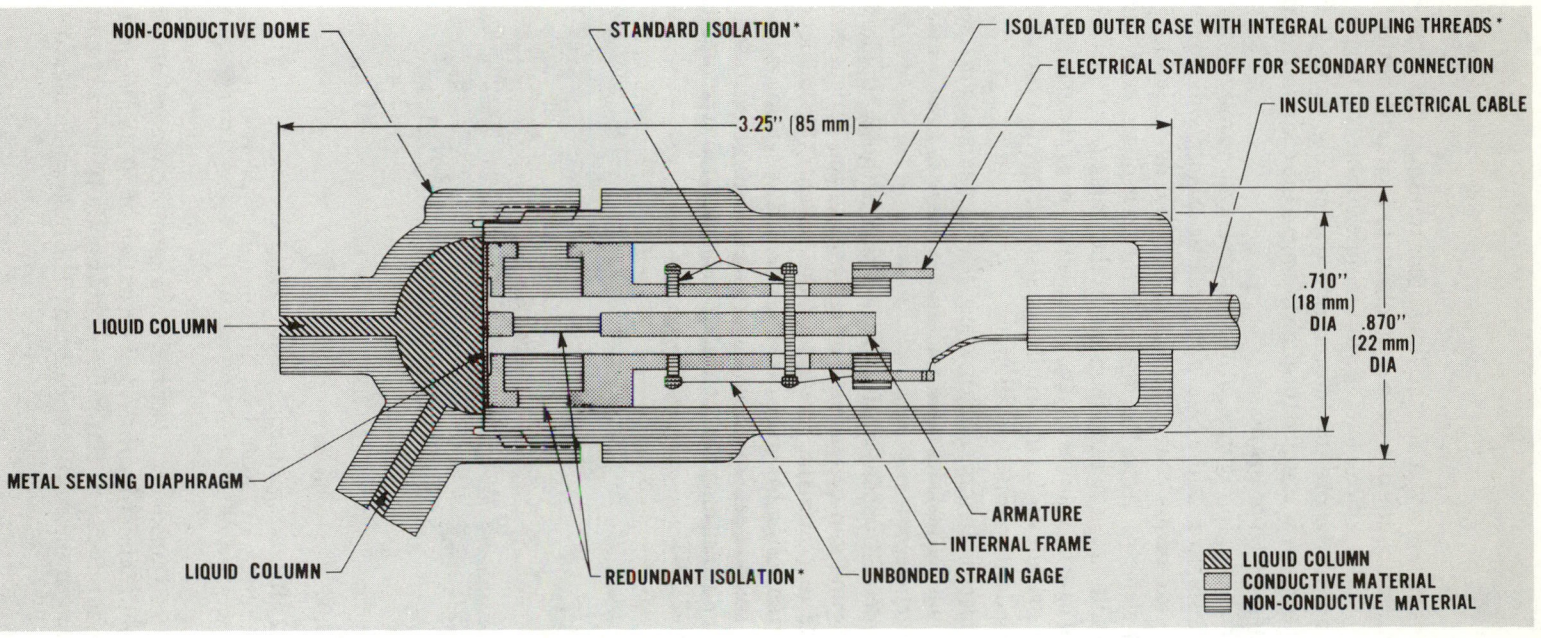

Figure 6.21. Interior construction of P23ID Blood Pressure Transducer. (Courtesy of Gould Inc., Measurement Systems Division, Oxnard, CA.)

The dome is the reservoir for the liquid that transmits blood pressure to the diaphragm. It is made of transparent plastic to facilitate the detection and removal of bubbles, since even the most minute bubble can degrade the frequency response of the pressure-monitoring system. The dome is fitted with two ports. One port is coupled through tubing to the cannula; the other is used for venting air from the dome.

It should be noted in Figure 6.21 that there are three modes of isolation: (1) external isolation of the case with a plastic sheath, which provides protection from extraneous voltages; (2) standard internal isolation of the sensing (bridge) elements from the inside of the transducer case and the frame; and (3) additional isolation (internal) of the frame from the case and the diaphragm in case of wire breakage. Thus, isolation of the patient/fluid column from electrical excitation circuitry is assured, even in the event of failure of the standard internal isolation.

This transducer is 56 mm (2.21 in.) long, with a maximum diameter at the base of 18 mm (0.71 in.). Its rated excitation voltage is 7.5 V which may be dc or an ac carrier.

Another type of transducer of smaller design is the P 50, shown in Figure 6.22. This unit can be mounted or attached to the patient near the measurement site (e.g., on the forearm of the patient). To achieve this miniaturization, the design has to be quite different. The sensing element of the P 50 is a tiny silicon beam upon which strain elements are diffused. This is, therefore, a bonded strain gage instead of the unbonded type used in the P 23 (see Section 2.3.1).

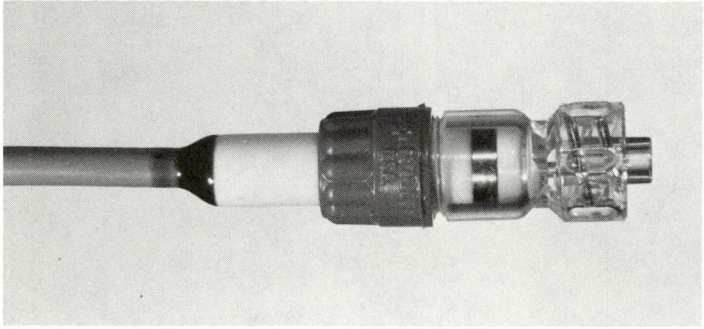

Figure 6.22. Bonded type blood pressure transducer for attaching to patient. (Courtesy of Gould Inc., Measurement Systems Division, Oxnard, CA.)

Both types of transducer described have pressure ranges of -50 to $+300$ mm Hg, with sensitivity for 7.5-V excitation of 50 μV/V/cm Hg.

These transducers can be flushed to remove air bubbles and to prevent blood from clotting at the end of the catheter. The fluid column used with a transducer of this type has a natural frequency or resonance of its own that can affect the frequency response of the system. Care must be taken in selecting a transducer and catheter to be used together so that the frequency response of the complete system will be adequate. There are general-purpose models for arterial pressure (0 to 330 mm Hg) and venous pressure (0 to 50 mm Hg), and special models with differing sensitivities, volume displacement characteristics, and mechanical arrangements.

Pressure transducers are normally mounted on a suitable manifold near the patient's bed. It is important to keep the transducer at the same height as the point at which the measurements are to be made in order to avoid errors due to hydrostatic pressure differences. If a *differential pressure* is desired, two transducers of this type may be used at two different points, and the difference in pressure may be obtained as the difference of their output signals. Figure 6.23 shows a typical infusion manifold incorporating a transducer, a flushing system, and syringes for blood specimen withdrawal.

The signal-conditioning and display devices for these transducers are available in a variety of forms. However, each must provide a method of excitation for the strain-gage bridge, a means of zeroing or balancing the bridge, necessary amplification of the output signal, and a display device, such as a monitor scope, a recorder, panel meter, or digital readout device. Most modern systems permit many possible combinations.

Another type of blood pressure transducer is the *linear variable differential transformer* (LVDT) device, shown in an exploded view in Figure 6.24. Superficially, these transducers look similar to the unbonded strain-gage type. Indeed, with respect to the plastic dome used for visibility, the two pressure fittings for attachment to the catheter and for flushing, and the cable coming out of the bottom, they are similar. Such transducers also come in a variety of models with a range of characteristics for venous or arterial pressure, for different sensitivities, and for alternative volume displacements. The various models also have different natural frequencies and frequency responses.

It should be noted from the exploded view that these units disassemble into three subassemblies—the dome and pressure fittings subassembly, the center portion consisting of a stainless-steel diaphragm and core assembly, and the LVDT subassembly. There are two basic diaphragm and core assemblies with appropriate domes that are interchangeable in the coil-connector assembly. The first is used for venous and general-purpose clinical measurements and has a standard-size diaphragm with an internal fluid volume between the dome and diaphragm of less than 0.5 cm³. The second design, with higher frequency response characteristics for arterial pressure contours, has a reduced diaphragm area and an internal volume of approximately 0.1 cm³.

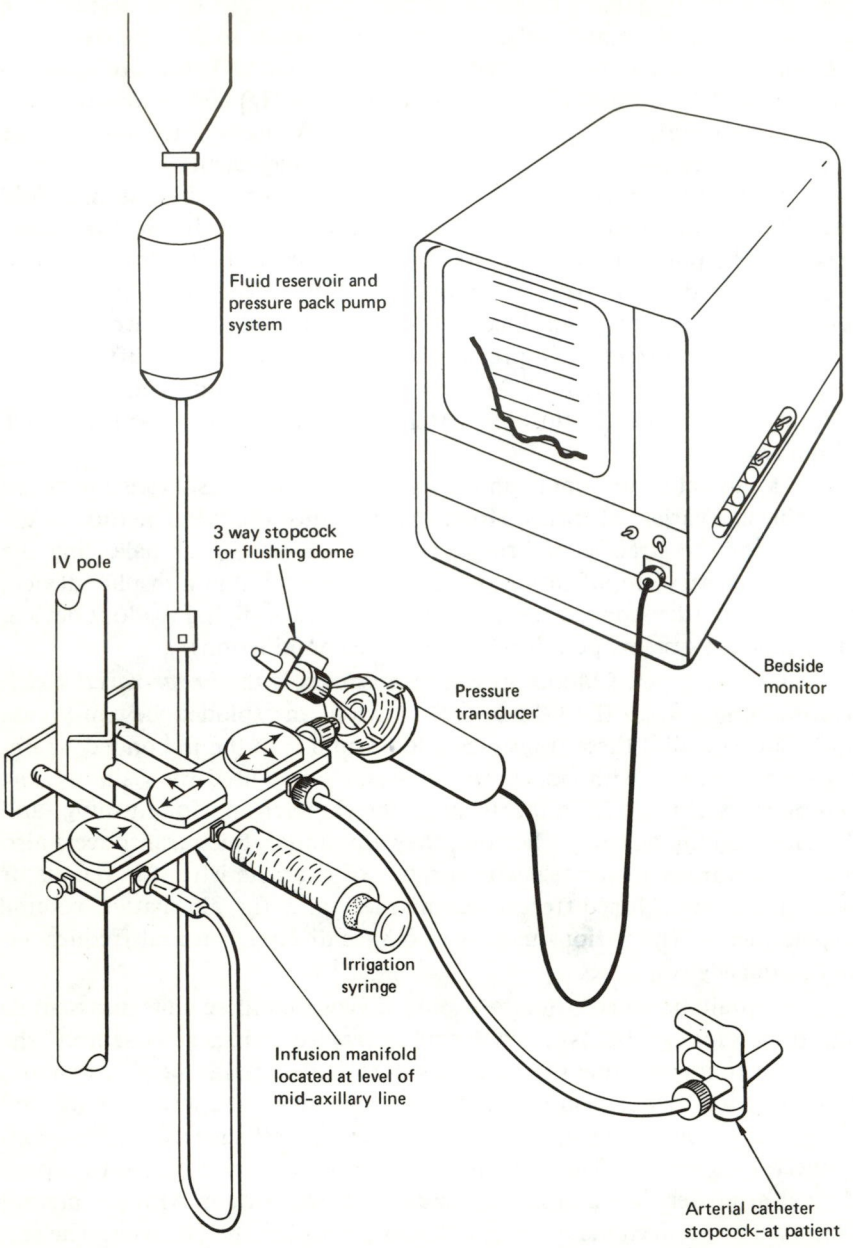

Figure 6.23. Infusion manifold with transducer, flushing system and syringe. (Courtesy of Michael Tomeo, UCLA Medical Center.)

Figure 6.24. LVDT blood pressure transducer—exploded view. (Courtesy of Biotronex Laboratory, Inc., Silver Springs, MD.)

The Biotronex BL-9630 transducer is a linear variable differential transformer in which the primary coil is excited by an ac carrier (5 to 20 V peak to peak) in the range of 1500 to 15,000 Hz. Axial displacement of a movable iron core, attached to the diaphragm, cuts the magnetic lines of flux generated by the primary coil. Voltages induced in the secondary sensing coils are returned to the carrier amplifier, where they are differentially amplified and demodulated to remove the carrier frequency. The output of the carrier amplifier is a dc voltage proportional to diaphragm displacement. Linearity of the gage is better than ±1 percent of full range. Ordinary jarring and handling will not harm the gage. A positive mechanical stop is provided to prevent damage by as much as a 100 per-cent overpressure. The LVDT offers much higher signal levels than do conventional strain-gage transducers for a given excitation voltage.

6.2.4.2. Measurement at the site. To avoid the problems inherent in measuring blood pressure through a liquid column, a "catheter-tip" manometer can be fed through the catheter to the site at which the blood pressure is to be measured. This process requires a small-diameter transducer that is fairly rigid but flexible.

One such transducer makes use of the variable-inductance effect mentioned earlier. The tip is placed directly in the bloodstream so that the blood presses on a membrane surrounded by a protective cap. The membrane is connected to a magnetic slug that is free to move within a coil assembly and thus changes the inductance of the coil as a function of the pressure on the membrane.

Another type has a bonded strain-gage sensor built into the tip of a cardiac catheter. The resistance changes in the strain gage are a result of pressure variations at the site itself rather than through a fluid column. This gage can also be calibrated with a liquid-system catheter at the same location.

6.2.4.3. Floatation catheter. The construction of specialized, multiple-lumen "floatation" catheters has made insertion and continuous monitoring of pulmonary artery pressures feasible in most clinical settings. This type of catheter was designed by Drs. Swan and Ganz of the Cedars–Sinai Medical Center in Los Angeles and bears their names. Although specialized models are available, the basic catheter is approximately 110 cm in length and consists of a double-lumen tube with an inflatable balloon tip (Figure 6.25).

The catheter may be inserted percutaneously or via a venous cutdown. By using continuous pressure and electrocardiographic (ECG) monitoring,

Figure 6.25. (a) Swan-Ganz monitoring catheter for measurement of pulmonary artery and pulmonary capillary wedge pressures, full view; (b) balloon. (Courtesy of Edwards Laboratories, Division of American Hospital Supply Corporation, Santa Ana, CA.)

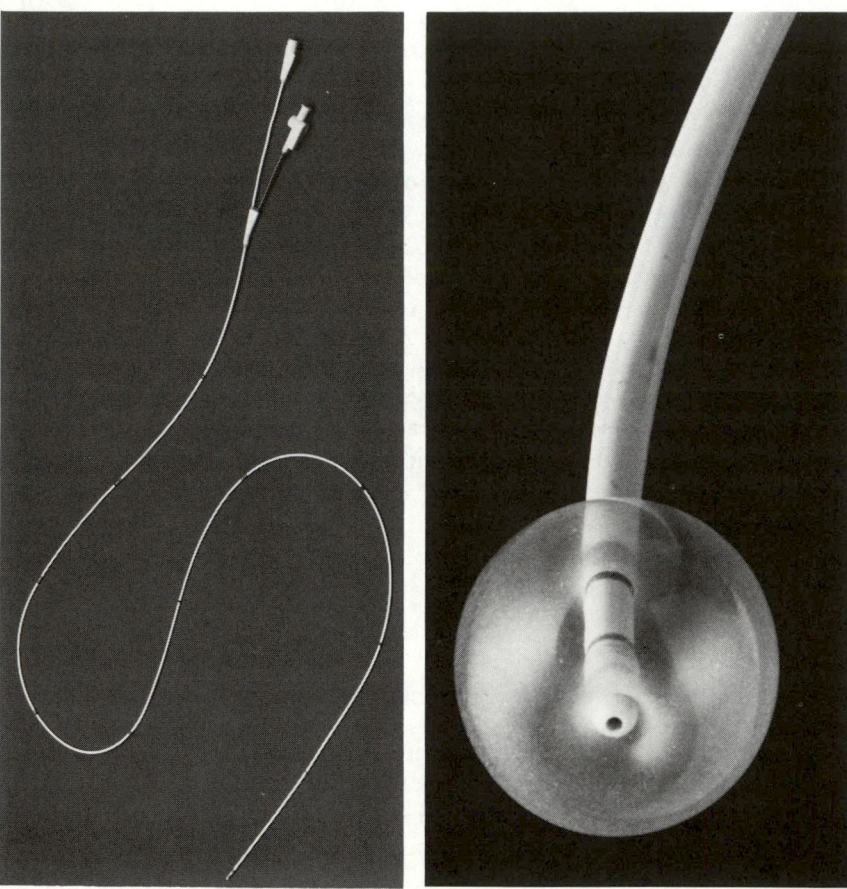

the catheter is threaded into the subclavian vein with the balloon deflated. At this point, the balloon is partially inflated to half capacity (0.4 to 0.6 cm^3 of CO_2 or air) and carried downstream to the right atrium by the flow of blood. The balloon is then fully inflated (0.8 cm^3) and advanced again so that the blood flow propels it through the tricuspid valve into the right ventricle. From there it is carried through the pulmonary valve into the pulmonary artery, where the balloon wedges in a distal artery branch. The position of the catheter is verified by the pressure tracing, which shifts from a pulmonary artery pressure indicator to the "wedged" pressure waveform position. Under ideal circumstances it should take the physician no more than 1 minute to float the catheter from full balloon inflation in the right atrium to the wedge position. During insertion, the fully inflated balloon covers the hard tip of the catheter, distributing pressure forces evenly across a broad area of the endocardium.

6.2.4.4. *Percutaneous transducers.*

6.2.4.4. Percutaneous transducers. An example of a percutaneous blood pressure transducer is shown in Figure 6.26. It shows a transducer connected to a hypodermic needle that has been placed in a vessel of the arm. The three-way stopcock dome permits flushing of the needle, administering of drugs, and withdrawing of blood samples. This transducer can measure arterial or venous pressures, or the pressures of other physiological fluids, by direct attachment to a needle at the point of measurement. It can be used with a continuously self-flushing system without degradation of signal. The transparent plastic dome permits observation of air-bubble formation and consequent ejection. It is designed for use with a portable blood pressure monitor, which provides bridge excitation, balancing, and amplification. The meter scale is calibrated directly into millimeters of mercury. This transducer also has the advantage that it can be connected to a standard intravenous infusion bottle.

6.2.4.5. Implantable transducers. Figure 6.27 shows a type of transducer that can be implanted into the wall of a blood vessel or into the wall of the heart itself. This transducer is particularly useful for long-term investigations in animals.

The transducer's body is made of titanium, which has excellent corrosion-resistance characteristics, a relatively low thermal coefficient of expansion, and a low modulus of elasticity, which results in greater strain per unit stress. Four semiconductor strain gages are bonded to the inner surface of the pressure-sensing diaphragm. Transducers of this type come in a number of sizes (from 3 to 7 mm in diameter) for blood pressure measurement. A popular size is 4.5 mm in diameter. Larger sizes are available for pleural pressure. The thickness of the body is 1.2 to 1.3 mm in the various models.

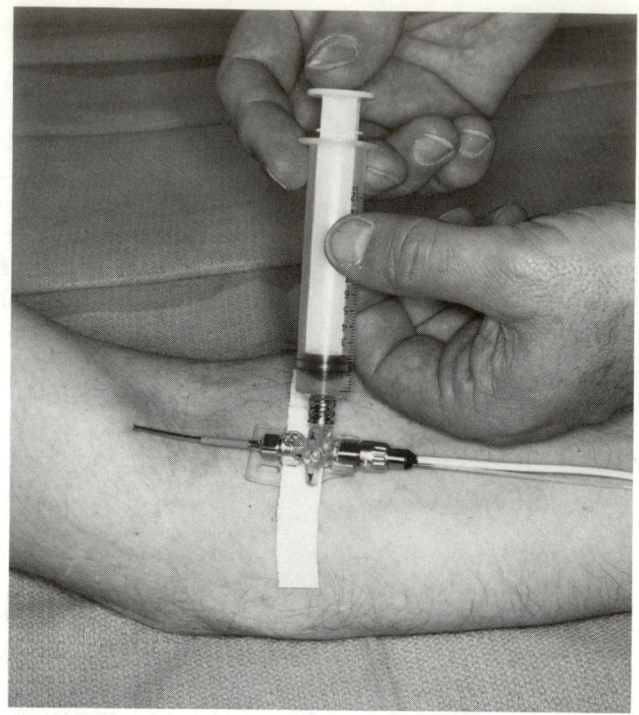

Figure 6.26. Percutaneous blood pressure measurement. Transducer in arm with three-way stopclock dome for administering drugs and withdrawing blood samples. (Courtesy of Gould Inc., Measurement Systems Division, Oxnard, CA.)

Figure 6.27. Implantable pressure transducer. (Courtesy of Konigsberg Instruments, Pasadena, CA.)

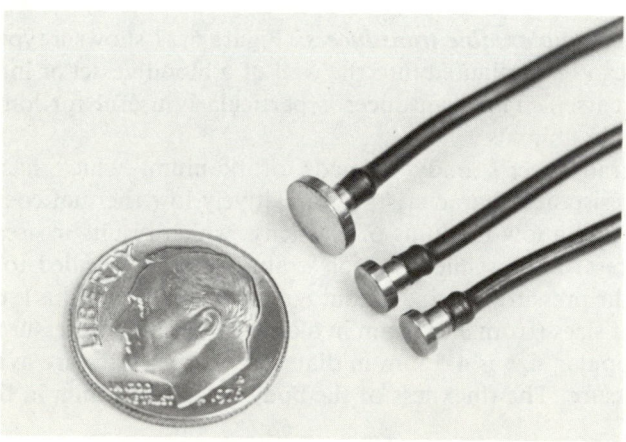

The four semiconductors are connected in bridge fashion as shown in Figure 6.19. As blood pressure increases on the diaphragm, the inner surface is stressed. The strain gages are located so that two of them are strained in tension while two are in compression. When the bridge is excited, an output voltage proportional to the blood pressure can be obtained.

Additional resistors, connected externally to the bridge, provide temperature compensation, although these bridges are not extremely sensitive to temperature. Since they operate in the bloodstream at a fairly constant 37 °C, the temperature effects are not serious.

These transducers can be excited with ac or dc and easily lend themselves to telemetry application. In service, they have proven very reliable. Cases of chronic implants (in excess of 2 years) have been reported with no detrimental effect on the animal, the gage, or the wires. The wires are usually insulated with a plastic compound, polyvinylchloride, which is fairly impervious to body fluids.

There are many examples of the use of this type of transducer in animal research, including the implantation in both ventricles of the heart, the aorta, the carotid artery, and the femoral artery. In addition to blood pressures, they have also been used for measuring abdominal, esophageal, thoracic, intrauterine and intracranial pressures.

To implant a transducer in an artery, a longitudinal incision is made; the transducer is inserted with its housing in intimate contact with the arterial walls. The wound is closed with interrupted sutures. For cardiac implants, a stab wound in the ventricle permits ready insertion, with the transducer placed free of both the myocardium and (in the left ventricle) the chordae tendonae. A technique used for long-term studies of the blood pressure in the aorta is to insert the transducer from the opposite side and use a small intercostal artery to bring the wire through. This creates a

Figure 6.28. Transducer implanted in the aorta.

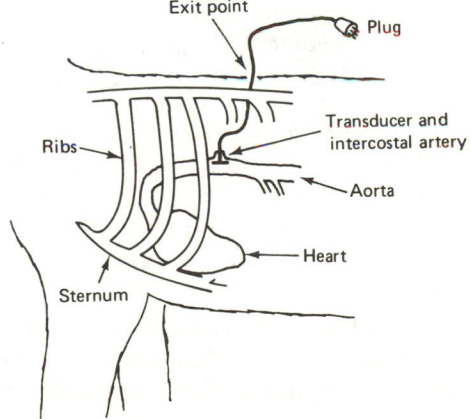

stronger bind in active animals. The wire is held to the artery by a purse-string suture. Figure 6.28 shows such a preparation. In this case the plug was inserted into a biotelemetry transmitter so that the blood pressure data could be received remotely. The use of these transducers and telemetry have been useful in gathering information on exercise, the effect of drugs and extreme environments, and acceleration and impact studies (see Chapter 12).

6.3. MEASUREMENT OF BLOOD FLOW AND CARDIAC OUTPUT

An adequate blood supply is necessary for all organs of the body; in fact, an impaired supply of blood is the cause of various diseases. The ability to measure blood flow in the vessel that supplies a particular organ would therefore be of great help in diagnosing such diseases. Unfortunately, blood flow is a rather elusive variable that cannot be measured easily.

The rate of flow of a liquid or gas in a pipe is expressed as the volume of the substance that passes through the pipe in a given unit of time. Flow rates are therefore usually expressed in liters per minute or milliliters per minute (cm^3/min).

Methods used in industry for flow measurements of other liquids, like the turbine flowmeter and the rotameter, are not very suitable for the measurement of blood flow because they require cutting the blood vessel. These methods also expose the blood to sharp edges, which are conducive to blood-clot formation.

Practically all blood flow meters currently used in clinical and research applications are based on one of the following physical principles:

1. Electromagnetic induction.
2. Ultrasound transmission or reflection.
3. Thermal convection.
4. Radiographic principles.
5. Indicator (dye or thermal) dilution.

Magnetic and *ultrasonic blood flow meters* actually measure the *velocity* of the bloodstream. Because these techniques require that a transducer surround an excised blood vessel, they are mainly used during surgery. Ultrasound, however, can be used transcutaneously to detect obstructions of blood vessels where quantitative blood flow measurements are not required.

A *plethysmograph,* which actually indicates volume changes in body segments, can be used to measure the flow of blood in the limbs. The principle of plethysmography is discussed in Section 6.4.

6.3.1. Magnetic Blood Flow Meters.

Magnetic blood flow meters are based on the principle of magnetic induction. When an electrical conductor is moved through a magnetic field, a
voltage is induced in the conductor proportional to the velocity of its motion. The same principle applies when the moving conductor is not a wire,
but rather a column of conductive fluid that flows through a tube located in
the magnetic field. Figure 6.29 shows how this principle is used in magnetic
blood flow meters. A permanent magnet or electromagnet positioned
around the blood vessel generates a magnetic field perpendicular to the
direction of the blood flow. The voltage induced in the moving blood column is measured with stationary electrodes located on opposite sides of the
blood vessel and perpendicular to the direction of the magnetic field.

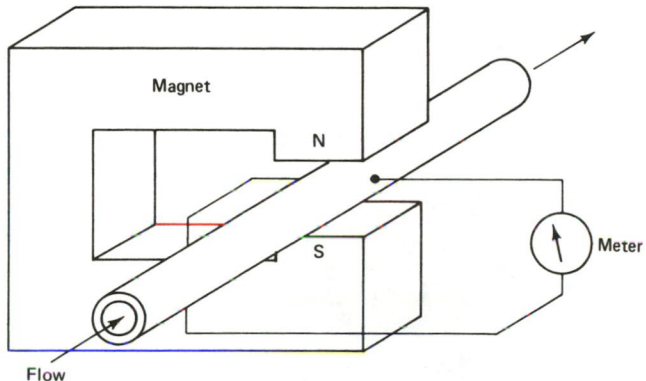

Figure 6.29. Magnetic blood flow meter, principle.

The most commonly used types of implantable magnetic blood flow
probes are shown in Figures 6.30 through 6.32. The *slip-on* or *C type* is applied by squeezing an excised blood vessel together and slipping it through
the slot of the probe. In some transducer models the slot is then closed by
inserting a keystone-shaped segment of plastic, as shown. Contact is provided by two slightly protruding platinum disks that touch the wall of the
blood vessel. For proper operation, the orifice of the probe must fit tightly
around the vessel. For this reason, probes of this type are manufactured in
sets, with diameters increasing in steps of 0.5 or 1 mm from about 2 to 20
mm. The probes shown in Figure 6.30 can be implanted for chronic use. In
contrast, Figure 6.31 shows a model with a long handle for use during
surgery.

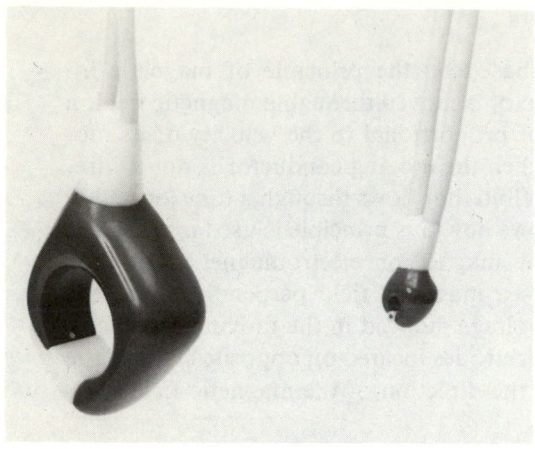

Figure 6.30. Samples of large and small lumen diameter blood flow probes. (Courtesy of Micron Instruments, Los Angeles, CA.)

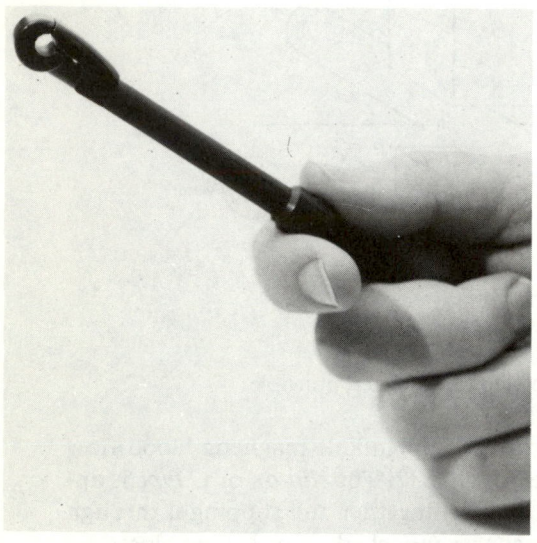

Figure 6.31. Blood flow probe—clip-on type for use during surgery. (Courtesy of Biotronex, Silver Springs, MD.)

In the *cannula-type transducer,* the blood flows through a plastic cannula around which the magnet is arranged. The contacts penetrate the walls of the cannula. This type of transducer requires that the blood vessel be cut and its ends slipped over the cannula and secured with a suture. A similar type of transducer (Figure 6.32) is also used to measure the blood flow in extracorporeal devices, such as dialyzers. Magnetic blood flow meters actually measure the mean blood velocity. Because the cross-sectional area at the place of velocity measurement is well defined with either type of transducer, these transducers can be calibrated directly in units of flow.

Figure 6.32. Extracorporeal blood flow probe. (Courtesy of Biotronex, Silver Springs, MD.)

Magnetic blood flow transducers are also manufactured as catheter-tip transducers. For this type, the normal transducer design is essentially turned "inside out," with the electromagnet being located inside the catheter, which has the electrodes at the outside. Catheter transducers cannot be calibrated in flow units, however, because the cross section of the blood vessel at the place of measurement is not defined.

The output voltage of a magnetic blood flow transducer is very small, typically in the order of a few microvolts. In early blood flow meters, a constant magnetic field was used, which caused difficulties with electrode polarization and amplifier drift. To overcome these problems, all contemporary magnetic blood flow meters use electromagnets that are driven by alternating currents. Doing this, however, creates another problem: the change of the magnetic field causes the transducer to act like a transformer and induces error voltages that often exceed the signal levels by several orders of magnitude. Thus, for recovering the signal in the presence of the error voltage, amplifiers with large dynamic range and phase-sensitive or gated detectors have to be used. To minimize the problem, several different waveforms have been advocated for the magnet current, as shown in Figure 6.33. With a sinusoidal magnet current, the induced voltage is also

Figure 6.33. Waveforms used in magnectic blood flow meters and error signals induced by the current: (a) sine wave; (b) square wave; (c) trapezoidal wave.

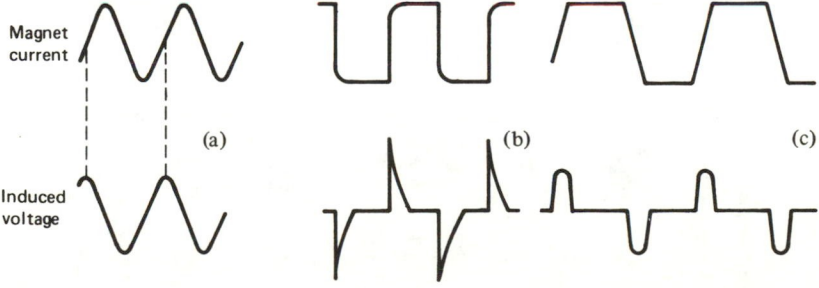

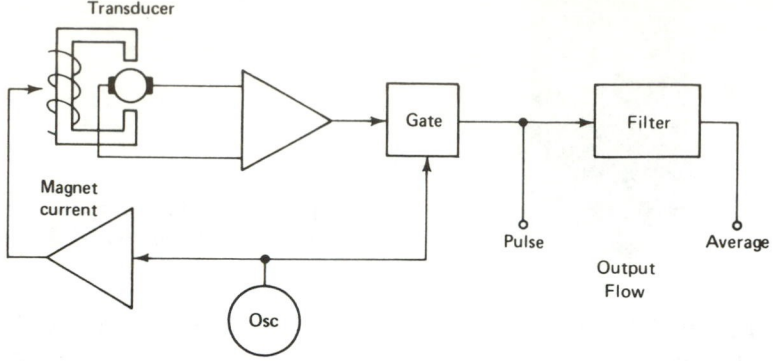

Figure 6.34. Magnetic blood flow meter, block diagram.

sinusoidal but is 90° out of phase with the flow signal. With a suitable circuit, similar to a bridge, the induced voltage can be partially balanced out. With the magnet current in the form of a square wave, the induced voltage should be zero once the spikes from the polarity reversal have passed. In practice, however, these spikes are often of extremely high amplitude, and the circuitry response tends to extend their effect. A compromise is the use of a magnet current having a trapezoidal waveform. None of the three waveforms used seems to have demonstrated a definite superiority.

The block diagram of a magnetic blood flow meter is shown in Figure 6.34. The oscillator, which drives the magnet and provides a control signal for the gate, operates at a frequency of between 60 and 400 Hz. The use of a gated detector makes the polarity of the output signal reverse when the flow direction reverses. The frequency response of this type of system is usually high enough to allow the recording of the flow pulses, while the mean or average flow can be derived by use of a low-pass filter. Figure 6.35 shows a single-channel magnetic blood flow meter that can be used with a variety of different transducers.

Figure 6.35. Magnetic blood flow meter. (Courtesy of Micron Instruments, Los Angeles, CA.)

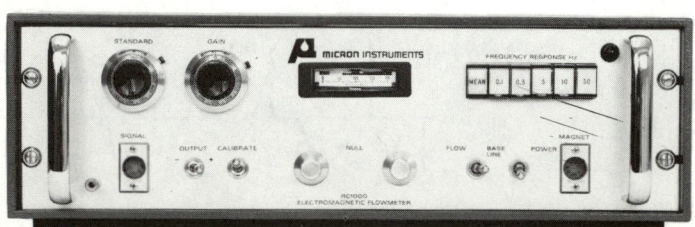

6.3.2. Ultrasonic Blood Flow Meters.

In an *ultrasonic blood flow meter,* a beam of ultrasonic energy is used to measure the velocity of flowing blood. This can be done in two different ways. In the *transit time ultrasonic flow meter,* a pulsed beam is directed through a blood vessel at a shallow angle and its transit time is then measured. When the blood flows in the direction of the energy transmission, the transit time is shortened. If it flows in the opposite direction, the transit time is lengthened.

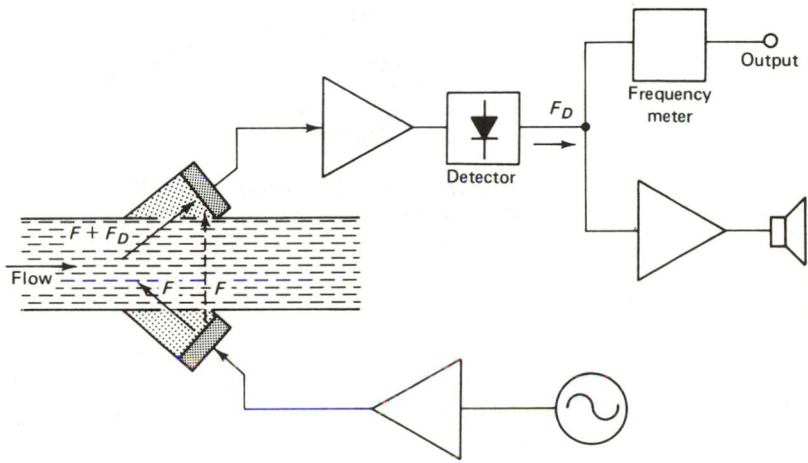

Figure 6.36. Ultrasonic blood flow meter, Doppler type.

More common are ultrasonic flow meters based on the Doppler principle (Figure 6.36). An oscillator, operating at a frequency of several megahertz, excites a piezoelectric transducer (usually made of barium titanate). This transducer is coupled to the wall of an exposed blood vessel and sends an ultrasonic beam with a frequency F into the flowing blood. A small part of the transmitted energy is scattered back and is received by a second transducer arranged opposite the first one. Because the scattering occurs mainly as a result of the moving blood cells, the reflected signal has a different frequency due to the Doppler effect. Its frequency is either $F + F_D$ or $F - F_D$, depending on the direction of the flow. The Doppler component F_D is directly proportional to the velocity of the flowing blood. A fraction of the transmitted ultrasonic energy, however, reaches the second transducer directly, with the frequency being unchanged. After amplification of the composite signal, the Doppler frequency can be obtained at the output of a detector as the difference between the direct and the scattered signal components.

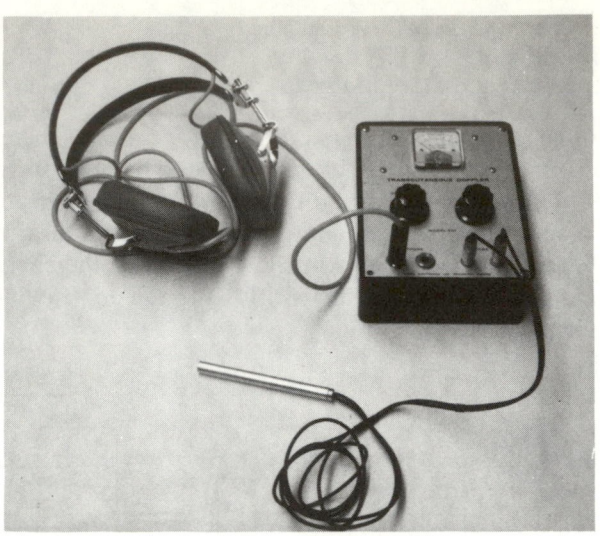

Figure 6.37. Percutaneous Doppler device with probe and earphones. (Courtesy of Veterans Administration Biomedical Engineering and Computing Center, Sepulveda, CA.)

With blood velocities in the range normally encountered, the Doppler signal is typically in the low audio frequency range. Because of the velocity profile of the flowing blood, the Doppler signal is not a pure sine wave, but has more the form of narrow-band noise. Therefore, from a loudspeaker or earphone, the Doppler signal of the pulsating blood flow can be heard as a characteristic "swish—swish—." When the transducers are placed in a suitable mount (which defines the area of the blood vessel), a frequency meter used to measure the Doppler frequency can be calibrated directly in flow-rate units. Unfortunately, Doppler flow meters of this simple design cannot discriminate the direction of flow. More complicated circuits, however, which use the insertion of two quadrature components of the carrier, are capable of indicating the direction of flow.

Transducers for ultrasonic flow meters can be implanted for chronic use. Some commercially available flow meters of this type incorporate a telemetry system to measure the blood flow in unrestrained animals.

Figure 6.37 is an illustration of a simple Doppler device, with the two transducers mounted in a hand-held probe which is now widely used to trace blood vessels close to the surface and to determine the location of vascular obstructions. In order to facilitate transmission of ultrasonic energy, the probe must be coupled to the skin with an aqueous jelly. Such devices can be used to detect the motion of internal structures in the body—for example, the fetal heart. (See Chapter 9 for further discussion of ultrasonic diagnosis.)

6.3.3. Blood Flow Measurement by Thermal Convection

A hot object in a colder-flowing medium is cooled by thermal convection. The rate of cooling is proportional to the rate of the flow of the medium. This principle, often used to measure gas flow, has also been applied to the measurement of blood velocity. In one application, a thermistor in the bloodstream is kept at a constant temperature by a servo system. The electrical energy required to maintain this constant temperature is a measure of the flow rate. In another method an electric heater is placed between two thermocouples or thermistors that are located some distance apart along the axis of the vessel. The temperature difference between the upstream and the downstream sensor is a measure of the blood velocity. A device of the latter type is sometimes called *a thermostromuhr* (literally, from the German "heat current clock"). Thermal convection methods for blood flow determination, although among the oldest ones used for this purpose, have now been widely replaced by the other methods described in this chapter.

6.3.4. Blood Flow Determination by Radiographic Methods

Blood is not normally visible on an X-ray image because it has about the same radio density as the surrounding tissue. By the injection of a contrast medium into a blood vessel (e.g., an iodated organic compound), the circulation pattern can be made locally visible. On a sequential record of the X-ray image (either photographic or on a videotape recording), the progress of the contrast medium can be followed, obstructions can be detected, and the blood flow in certain blood vessels can be estimated. This technique, known as *cine* (or *video*) *angiography,* can be used to assess the extent of damage after a stroke or heart attack.

Another method is the injection of a radioactive isotope into the blood circulation, which allows the detection of vascular obstructions (e.g., in the lung) with an imaging device for nuclear radiation, such as a scanner or gamma camera (see Chapter 14).

Vascular obstructions in the lower extremities can sometimes be detected by measuring differences in the skin temperature caused by the reduced circulation. This can be accomplished by one of the various methods of skin surface temperature measurement described in Chapter 9.

6.3.5. Measurement by Indicator Dilution Methods

The indicator or dye dilution methods are the only methods of blood flow measurement that really measure the blood flow and not the blood velocity. In principle, any substance can be used as an indicator if it mixes readily with blood and its concentration in the blood can be easily determined after mixing. The substance must be stable but should not be retained by the body. It must have no toxic side effects.

An indocyanine dye, Cardiogreen, used in an isotonic solution was long favored as a indicator. Its concentration was detemined by measuring the light absorption with a densitometer (colorimeter). Radioactive isotopes (radioiodited serum albumen) have also been employed for this purpose. The indicator most frequently used today, however, is isotonic saline, which is injected at a temperature lower than the body temperature. The concentration of the saline after mixing with the blood is determined with a sensitive thermistor thermometer (see Chapter 9).

The principle of the dilution method is shown in Figure 6.38. The upper left drawing shows a model of a part of the blood circulation under the (very simplified) assumption that the blood is not recirculated. The indicator is injected into the flow continuously, beginning at time t, at a constant infusion rate I (grams per minute). A detector measures the concentration downstream from the injection point. Figure 6.38 (a) shows the output of a recorder that is connected to the detector. At a certain time after the injection, the indicator begins to appear, the concentration increases, and, finally, it reaches a constant value, C_0 (milligrams per liter). From the measured concentration and the known injection rate, I (in milligrams per minute), the flow can be calculated as

$$F \text{ (liters per minute)} = \frac{I \text{ (milligrams per minute)}}{C_0 \text{ (milligrams per liter)}}$$

The earliest method for determining cardiac output, the *Fick method,* is based on this simple model. The indicator is the oxygen of the inhaled air that is "injected" into the blood in the lungs. The "infusion rate" is determined by measuring the oxygen content of the exhaled air and subtracting it from the known oxygen content of the inhaled room air. The oxygen metabolism only approximately resembles the model of the open circulation, because only part of the oxygen is consumed in the systemic circulation and the returning venous blood still contains some oxygen. Therefore, the oxygen concentration in the returning venous blood has to be determined and subtracted from the oxygen concentration in the arterial blood leaving the lungs. The measurements are averaged over several minutes to reduce the influence of short-term fluctuations. An automated system is available that measures the oxygen concentration (by colorimetry) and the oxygen consumption, and continuously calculates the cardiac output from these measurements.

When a dye or isotope is used as an indicator, the concentration does not assume a steady-state value but increases in steps whenever the recirculated indicator again passes the detector [points R in Figure 6.38 (b)].

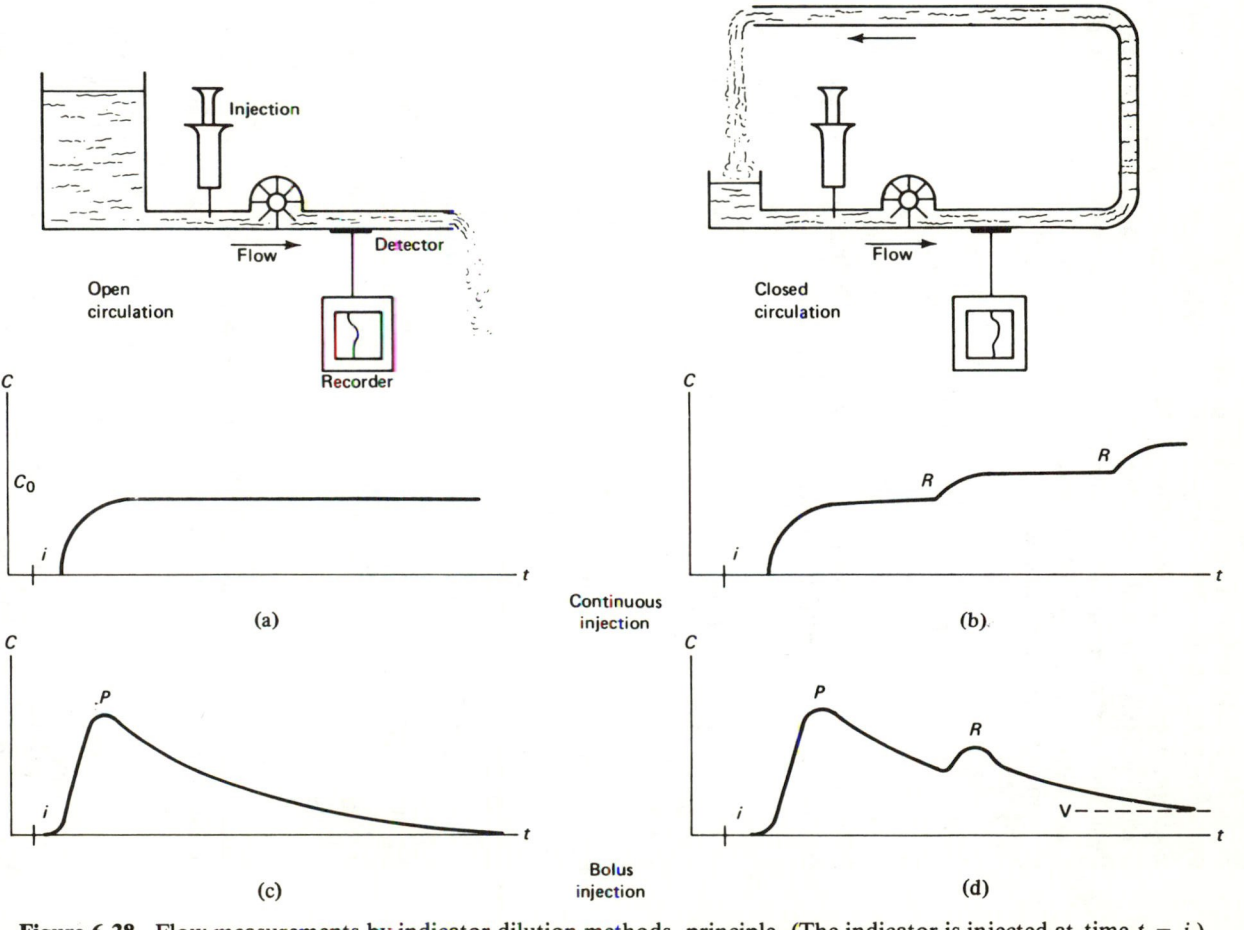

Figure 6.38. Flow measurements by indicator dilution methods, principle. (The indicator is injected at time $t = i$.)

The recirculation often occurs before the concentration has reached a plateau. Consequently, a slightly different method is usually used, and the indicator is injected as a bolus instead of being infused at a constant rate. Figure 6.38(c) shows the concentration for this case, again under the assumption that it is an open system. The concentration increases at first, reaches a peak, P, and then decays as an exponential function. This "washout curve" is mainly a result of the velocity profile of the blood, which causes a "spread" of the bolus. To calculate the flow, the area under the concentration curve has to be determined. This is given by the integral:

$$\int_{t=i}^{t=\infty} C\,dt \quad \left(\frac{\text{milligrams}}{\text{liter}} \times \text{minutes}\right)$$

From the value of this integral, and from the amount B of the injected indicator (in milligrams), the flow can be calculated:

$$F\left(\frac{\text{liters}}{\text{minute}}\right) = \frac{B}{\int_{t=i}^{t=\infty} C\,dt} \quad \left(\frac{\text{milligrams}}{\text{milligrams/liter} \times \text{minutes}}\right)$$

Because of the recirculation, the concentration does not show a monotonous decay as in Figure 6.38(b); instead, after some time, a "hump" (R in the figure) occurs. Normally, the portion of the curve preceding the recirculation hump is exponential, and the decay curve can be determined, despite the recirculation, by exponential extrapolation (*Hamilton method*). This was originally done by manually replotting the curve on semilogarithmic paper, which resulted in a straight line for the exponential part of the curve. This line was then extended, the extended part replotted on the original plot, and the extrapolated curve integrated with a planimeter.

To simplify this complicated and time-consuming operation, various special-purpose analog computers were used which employed several different algorithms or determining the area under the curve despite the recirculation distortion. The use of cold saline as an indicator avoids these problems. Because the volume injected is normally only 10 ml, the circulating blood is rewarmed rapidly and no measurable recirculation occurs. Injection of the cool saline and measurement of the blood temperature are frequently performed with a single catheter of special design. This device is called the *Swan-Ganz catheter,* the principle of which is shown in Figure 6.39. It is a special adaptation of the pulmonary artery floatation catheter described in Section 6.2.4.3. This catheter contains four separate lumens. One lumen terminates about 30 cm (12 in.) from the tip and is used to inject

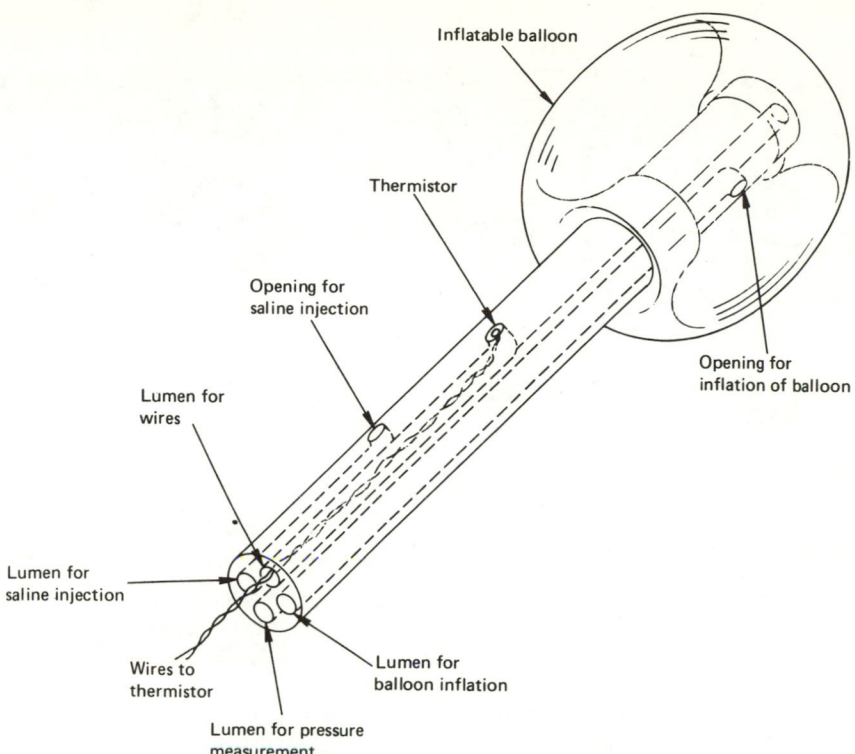

Labels in figure:
Inflatable balloon
Thermistor
Opening for saline injection
Lumen for wires
Lumen for saline injection
Wires to thermistor
Lumen for pressure measurement
Lumen for balloon inflation
Opening for inflation of balloon

Figure 6.39. Swan-Ganz catheter for the measurement of cardiac output by the thermal dilution method. (Not drawn to scale.) The opening for saline injection is actually located 30 cm distal from the catheter tip and the lumens vary in diameter.

the cooled saline. The second lumen contains two thin wires that lead to a tiny electrical temperature sensor close to the top of the catheter. The third lumen ends at the catheter tip and can be used to measure the blood pressure at this point with one of the pressure transducers described in Section 6.2. The fourth lumen is used to inflate a small rubber ballon at the tip of the catheter. Once the catheter has been inserted into a vein, the ballon is inflated and the returning venous blood carries the catheter until its tip is positioned in the pulmonary artery. The position of the catheter can be checked by measuring the pressure at its tip. Thus, the catheter can, if necessary, be inserted without fluoroscopic control.

The thermistor is connected into a bridge circuit which permits measurement and recording of the blood temperature during injection. A relatively simple analog computer, which consists essentially of an electronic integrator and necessary controls, permits direct reading of the cardiac output. The complete thermodilution catheter and the cardiac output computer are illustrated in Figure 6.40.

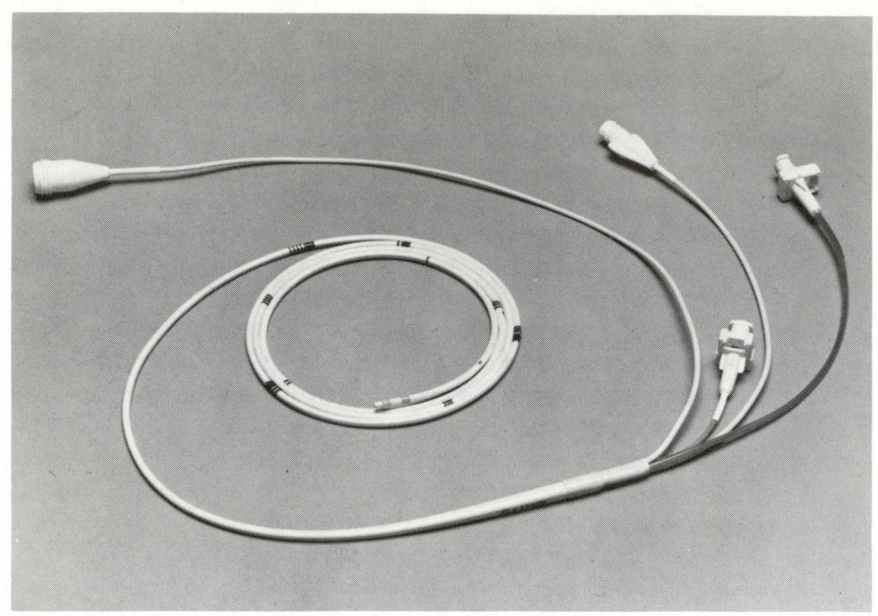

Figure 6.40. Cardiac output catheter and computer: (a) complete thermodilution Swan-Ganz floatation catheter; (b) cardiac output computer. (Courtesy of Edwards Laboratories, Division of American Hospital Supply Corporation, Santa Ana, CA.)

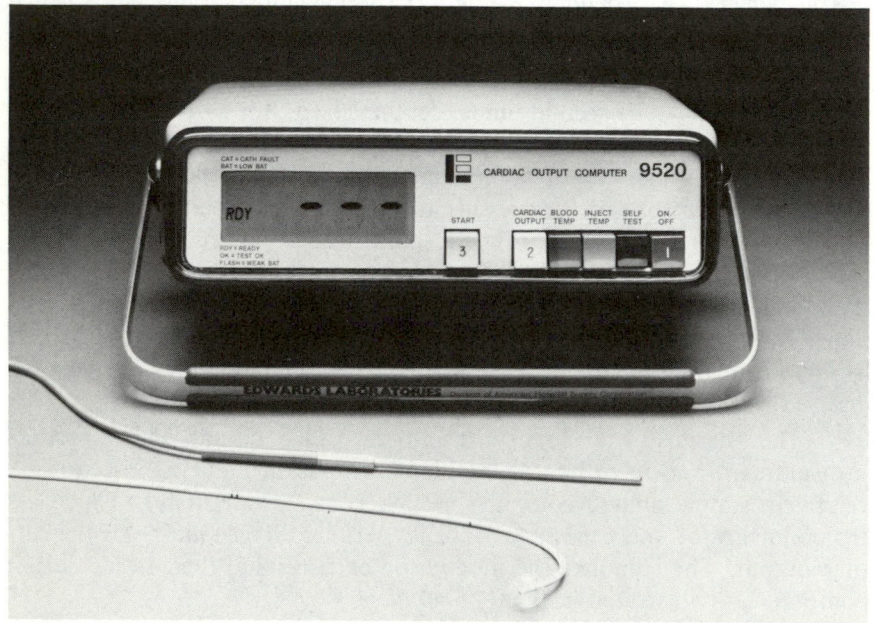

6.4. PLETHYSMOGRAPHY

Related to the measurement of blood flow is the measurement of volume changes in any part of the body that result from the pulsations of blood occuring with each heartbeat. Such measurements are useful in the diagnosis of arterial obstructions as well as for pulse-wave velocity measurements. Instruments measuring volume changes or providing outputs that can be related to them are called *plethysmographs*, and the measurement of these volume changes, or phenomena related thereto, is called *plethysmography*.

A "true" plethysmograph is one that actually responds to changes in volume. Such an instrument consists of a rigid cup or chamber placed over the limb or digit in which the volume changes are to be measured, as shown in Figure 6.41. The cup is tightly sealed to the member to be measured so that any changes of volume in the limb or digit reflect as pressure changes inside the chamber. Either fluid or air can be used to fill the chamber.

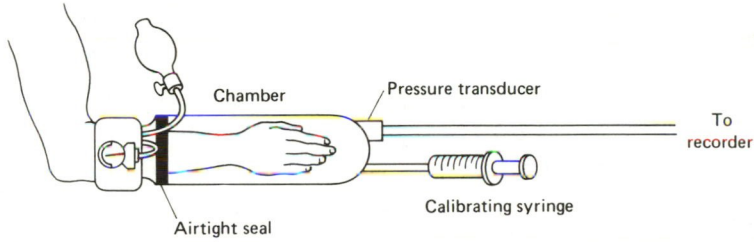

Figure 6.41. Plethysmograph. (Redrawn from A.C. Guyton, *Textbook of Medical Physiology,* 4th ed., W.B. Saunders Co., 1971, by permission.)

Plethysmographs may be designed for constant pressure or constant volume within the chamber. In either case, some form of pressure or displacement transducer must be included to respond to pressure changes within the chamber and to provide a signal that can be calibrated to represent the volume of the limb or digit. (See the description of the pressure transducers in Section 6.2.2. and displacement transducers in Chapter 2.) The baseline pressure can be calibrated by use of a calibrating syringe.

This type of plethysmograph can be used in two ways (see Figue 6.41). If the cuff, placed upstream from the seal, is not inflated, the output signal is simply a sequence of pulsations proportional to the individual volume changes with each heartbeat.

The plethysmograph illustrated in Figure 6.41 can also be used to measure the total amount of blood flowing into the limb or digit being measured. By inflating the cuff (placed slightly upstream from the seal) to a pressure just above venous pressure, arterial blood can flow past the cuff, but venous blood cannot leave. The result is that the limb or digit increases its volume with each heartbeat by the volume of the blood entering during that beat. The output tracing for this measurement is shown in Figure 6.42. The slope of a line along the peaks of these pulsations represents the overall rate at which blood enters the limb or digit. Note, however, that after a few seconds the slope tends to level off. This is caused by a back pressure that builds up in the limb or digit from the accumulation of blood that cannot escape.

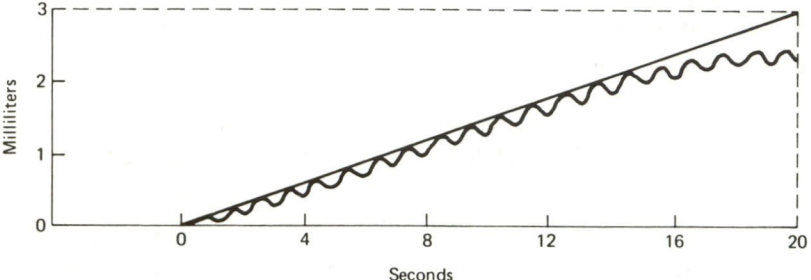

Figure 6.42. Blood volume record from plethysmograph. (From A.C. Guyton, *Textbook of Medical Physiology*, 4th ed., W.B. Saunders Co., 1971, by permission.)

Another device that quite closely approximates a "true" plethysmograph is the *capacitance plethysmograph* shown in Figure 6.43. In this device, which is generally used on either the arm or leg, the limb in which the volume is being measured becomes one plate of a capacitor. The other plate is formed by a fixed screen held at a small distance from the limb by an insulating layer. Often a second screen surrounds the outside plate at a fixed distance to act as a shield for greater electrical stability. Pulsations of the blood in the arm or leg cause variations in the capacitance, because the distance between the limb and the fixed screen varies with these pulsations. Some form of capacitance-measuring device is then used to obtain a continuous measure of these variations. Since the length of the cuff is fixed, the variations in capacitance can be calibrated as volume variations. The device can be calibrated by using a special cone of known volume on which the diameter can be adjusted to provide the same capacitance reading as the limb on which measurements are made. Since the capacitance plethysmograph essentially integrates the diameter changes over a segment of the limb, its readings are reasonably close to those of a "true" plethysmograph. Also, as with the "true" plethysmograph, estimates of the total volume of blood entering an arm or leg over a given period of time can be made by placing an occluding cuff just upstream from the capacitance device and by

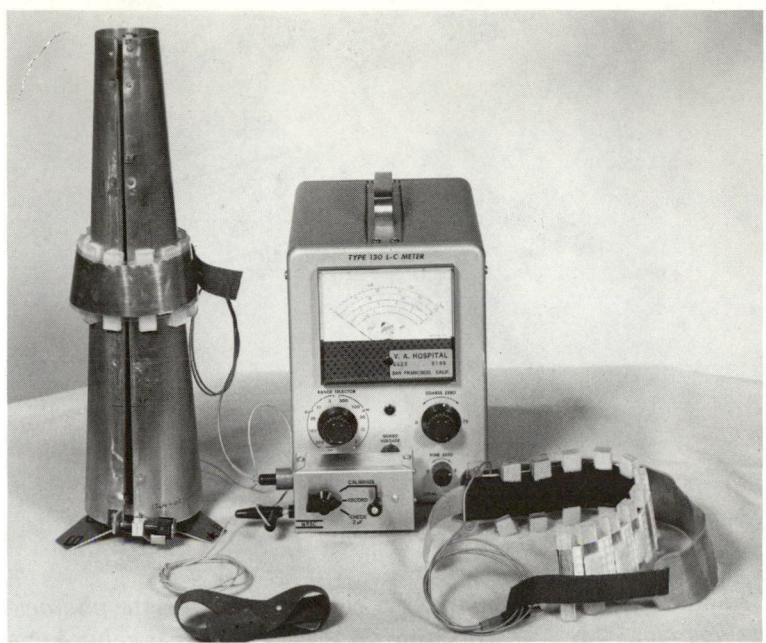

Figure 6.43. Capacitance plethysmograph. (Designed by one of the authors for V.A. Hospital, San Francisco, CA.)

pressurizing the cuff to a pressure greater than venous pressure but below arterial pressure.

Several devices, called plethysmographs, actually measure some variable related to volume rather than volume itself. One class of these "pseudo-plethysmographs" measures changes in diameter at a certain cross section of a finger, toe, arm, leg, or other segment of the body. Since volume is related to diameter, this type of device is sufficiently accurate for many purposes.

A common method of sensing diameter changes is through the use of a *mercury strain gage,* which consists of a segment of small-diameter elastic tubing, just long enough to wrap around the limb or digit being measured. When the tube is filled with mercury, it provides a highly compliant strain gage that changes its resistance with changes in diameter. With each pulsation of blood that increases the diameter of the limb or digit, the strain gage elongates and, in stretching, becomes thinner, thus increasing its resistance. The major difficulty in using the mercury strain gage is its extremely low impedance. This drawback necessitates the use of a low-impedance bridge to measure small resistance variations and convert them into voltage changes that can be recorded. A mercury strain gage plethysmograph is shown in Figure 6.44. A difficulty that is common to all diameter-measuring pseudoplethysmographs is that of interpreting single-point diameter changes as volume changes.

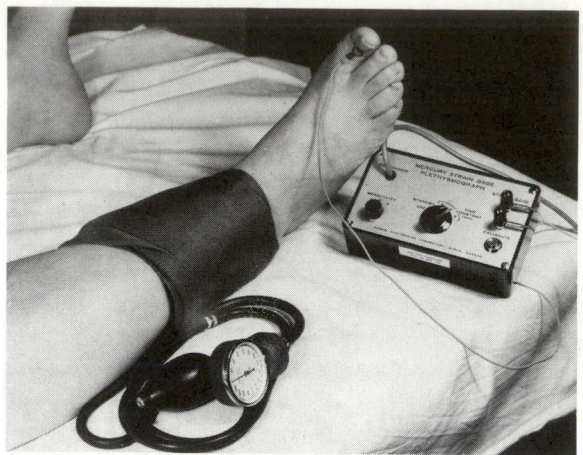

Figure 6.44. Mercury strain gage plethysmograph. (Courtesy of Parks Electronics Laboratory, Beaverton, OR.)

Another step away from the true plethysmograph is the *photoelectric plethysmograph*. This device operates on the principle that volume changes in a limb or digit result in changes in the optical density through and just beneath the skin over a vascular region. A photoelectric plethysmograph is shown in Figure 6.45. A light source in an opaque chamber illuminates a small area of the fingertip or other region to which the transducer is applied. Light scattered and transmitted through the capillaries of the region is picked up by the photocell, which is shielded from all other light. As the capillaries fill with blood (with each pulse), the blood density increases, thereby reducing the amount of light reaching the photocell. The result causes resistance changes in the photocell that can be measured on a Wheatstone bridge and recorded. Pulsations recorded in this manner are

Figure 6.45. Photoelectric plethysmograph. (Courtesy of Narco BioSystems, Inc., Houston, TX.)

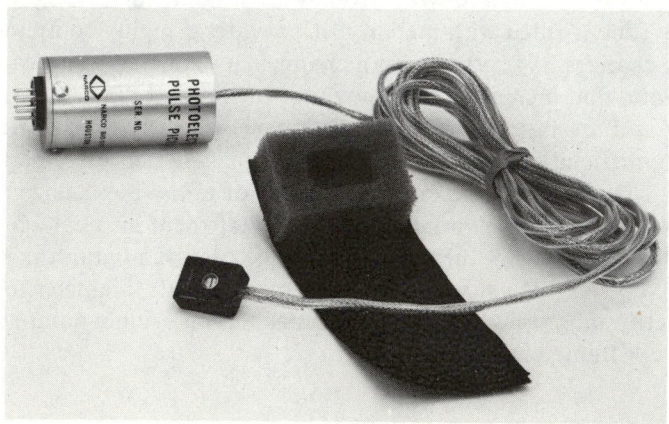

somewhat similar to those obtained by a true plethysmograph, but the photocell device cannot be calibrated to reflect absolute or even relative volumes. As a result, this type of measurement is primarily limited to detecting the fact that there are pulsations into the finger, indicating heart rate and determining the arrival time of the pulses. One serious difficulty experienced with this type of device is the fact that even the slightest movement of the finger with respect to the photocell or light source results in a severe amount of movement artifact. Furthermore, if the light source produces heat, the effect of the heat may change local circulation beneath the light source and photocell.

A more reliable device is the *impedance plethysomgraph,* in which volume changes in a segment of a limb or digit are reflected as impedance changes. These impedance changes are due primarily to changes in the conductivity of the current path with each pulsation of blood. Impedance plethysmographic measurements can be made using a two-electrode or a four-electrode system. The electrodes are either conductive bands wrapped around the limb or digit to be measured or simple conductive strips of tape attached to the skin. In either case, the electrodes contact the skin through a suitable electrolyte jelly or paste to form an electrode interface and to remove the effect of skin resistance. In a two-electrode system, a constant current is forced through the tissue between the two electrodes, and the resulting voltage changes are measured. In the four-electrode system, the constant current is forced through two outer, or current electrodes, and the voltage between the two inner, or measurement, electrodes is measured. The internal body resistances between the electrodes form a physiological voltage divider. The advantage of the four-electrode system is a much smaller amount of current through the measuring electrodes, thus reducing the possiblility of error due to changes in electrode resistance. Currents used for impedance plethysmography are commonly limited to the low-microampere range. The driving current is ac, sometimes a square wave, and usually of a high-enough frequency (around 10 kHz or higher) to reduce the effect of skin resistance. At these frequencies the capacitive component of the skin electrode interface becomes a significant factor.

Several theories attempt to explain the actual cause of the measured impedance changes. One is that the mere presence of additional blood filling a segment of the body lowers the impedance of that segment. Tests reported by critics of this method, however, claim that the actual impedance difference between the blood-filled state and more "empty" state is not significant.

A second theory is that the increase in diameter due to additional blood in a segment of the body increases the cross-sectional area of the segment's conductive path and thereby lowers the resistance of the path. This may be true to some extent, but again the percentage of area change is very small.

Critics of impedance plethysmography argue that the measured impedance changes are actually changes in the impedance of the skin-electrode interface, caused by pressure changes on the electrodes that occur with each blood pulsation.

Whatever the reason, however, impedance plethysmography does produce a measure that closely approximates the output of a true plethysmograph. Its main difficulty is the problem of relating the output resistance to any absolute volume measurement. As with the photocell plethysmograph, detection of the presence of arterial pulsations, measurment of pulse rate, and determination of time of arrival of a pulse at any given point in the peripheral circulation can all be satisfactorily handled by impedance plethysmography. Also, the impedance plethysmograph can measure time-variant changes in blood volume.

A special form of impedance plethysmography is *rheoencephalography,* the measurement of impedance changes between electrodes positioned on the scalp. Although primarily limited to research applications, this technique provides information related to cerebral blood flow and is sometimes used to detect circulatory differences between the two sides of the head. Theoretically, such information might help in locating blockages in the internal carotid system, which supplies blood to the brain.

Another special type of plethysmograph is the *oculo pneumo plethysmograph,* shown in Figure 6.46. As the name implies, this instrument measures every minute volume changes that occur in the eye with each

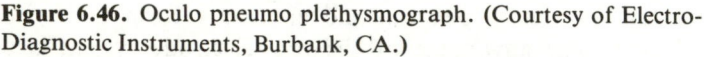

Figure 6.46. Oculo pneumo plethysmograph. (Courtesy of Electro-Diagnostic Instruments, Burbank, CA.)

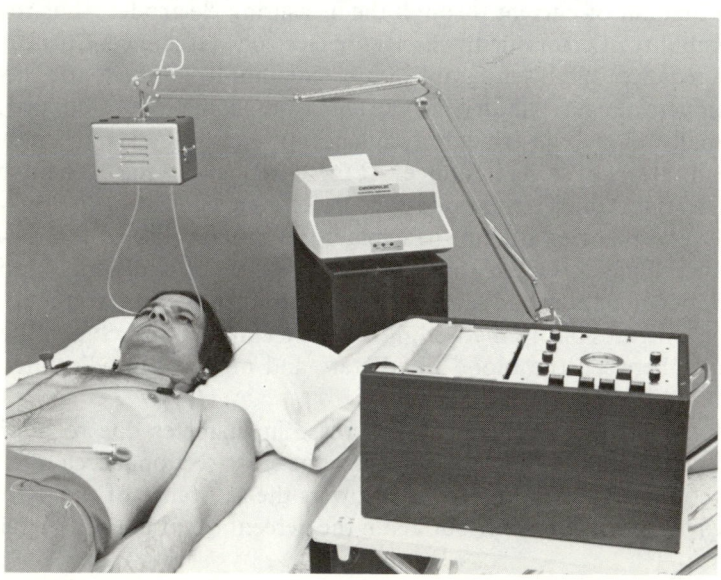

arterial blood pulsation. A small eye cup is placed over the sclera of each eye and is connected to a transducer positioned over the patient's head by a short section of flexible tubing. A vacuum, which can be varied from zero to -300 mm Hg, is applied to hold the eye cups in place. Pulsations are recorded on two channels of a three-channel pen recorder, one for each eye. The third channel is used to record the vacuum. By periodically allowing the vacuum to build up to -300 mm Hg and deplete to zero, the instrument can also be used as a recording suction *ophthalmodynamometer,* an instrument for measuring arterial blood pressure within the eye. Ocular plethysmography and blood pressure measurements are of particular interest because the eye provides a site for noninvasive access to the cerebral circulation system. Occlusion of one of the internal carotid arteries or other interference in the blood supply to one hemisphere of the brain can be detected by nonsymmetric pressures and pulsations at the eyes.

In certain rodents the tail is a convenient region for measurement of circulatory factors. For these measurements, a *caudal plethysmograph* is used. Caudal plethysmographs can utilize any of the previously described methods of sensing volume changes or the presence of blood pulsations. The same limitations encountered in human plethysmographic procedures are also found in caudal plethysmography. In addition, a special physiological factor must be considered in measuring blood pulsations from the tail of a rodent. Many animals use their tails as radiators in the control of body temperature. At low temperatures, very little blood actually flows through the vessels of the tail, and plethysmographic measurements become very difficult. If the animal is heated to a temperature at which the tail is used for cooling, however, sufficient blood flow for good plethysmographic measurements is usually found. Sometimes the necessary temperature for good caudal measurements is so near the point of overheating that traumatic effects are encountered.

6.5. MEASUREMENT OF HEART SOUNDS

In the early days of auscultation a physician listened to heart sounds by placing his ear on the chest of the patient, directly over the heart. It was probably during the process of treating a well-endowed, but bashful young lady that someone developed the idea of transmitting heart sounds from the patient's chest to the physician's ear via a section of cardboard tubing. This was the forerunner of the stethoscope, which has become a symbol of the medical profession.

The *stethoscope* (from the Greek word, *stethos,* meaning chest, and *skopein,* meaning "to examine") is simply a device that carries sound energy from the chest of the patient to the ear of the physician via a column of air. There are many forms of stethoscopes, but the familiar configuration has two earpieces connected to a common bell or chest piece. Since the system is strictly acoustical, there is no amplification of sound, except for any that might occur through resonance and other acoustical characteristics.

Unfortunately, only a small portion of the energy in heart sounds is in the audible frequency range. Thus, since the dawning of the age of electronics, countless attempts have been made to convince the medical profession of the advantage of amplifying heart sounds, with the idea that if the sound level could be increased, a greater portion of the sound spectrum could be heard and greater diagnostic capability might be achieved. In addition, high-fidelity equipment would be able to reproduce the entire frequency range, much of which is missed by the stethoscope. In spite of these apparent advantages, the *electronic stethoscope* has never found favor with the physician. The principal argument is that doctors are trained to recognize heart defects by the way they sound through an ordinary stethoscope, and any variations therefrom are foreign and confusing. Nevertheless, a number of electronic stethoscopes are available commercially.

Instruments for graphically recording heart sounds have been more successful. As stated in Chapter 5, a graphic record of heart sounds is called a *phonocardiogram.* The instrument for producing this recording is called a *phonocardiograph.* Although instruments specifically designed for phonocardiography are rare, components suitable for this purpose are readily available.

The basic transducer for the phonocardiogram is a microphone having the necessary frequency response, generally ranging from below 5 Hz to above 1000 Hz. An amplifier with similar response characteristics is required, which may offer a selective lowpass filter to allow the high-frequency cutoff to be adjusted for noise and other considerations. In one instance, where the associated pen recorder is inadequate to reproduce higher frequencies, an integrator is employed and the envelope of frequencies over 80 Hz is recorded along with actual signals below 80 Hz.

The readout of a phonocardiograph is either a high-frequency chart recorder or an oscilloscope. Because most pen galvanometer recorders have an upper-frequency limitation of around 100 or 200 Hz, photographic or light-galvanometer recorders are required for faithful recording of heart sounds. Although normal heart sounds fall well within the frequency range of pen recorders, the high-frequency murmurs that are often important in diagnosis require the greater response of the photographic device.

Some manufacturers of multiple-channel physiological recording systems claim the phonocardiogram as one of the measurements they offer. They have available as part of their system a microphone and amplifier suitable for the heart sounds, the amplifier often being the same one used for EMG (see Chapter 10). Some of these systems, however, have only a pen recorder output, which limits the high-frequency response of the recorded signal to about 100 or 200 Hz.

The presence of higher frequencies (murmurs) in the phonocardiogram indicates a possible heart disorder. For this reason, a spectral analysis of heart sounds can provide a useful diagnostic tool for discriminating between normal and abnormal hearts. This type of analysis, however, requires a digital computer with a high-speed analog-to-digital conversion capability and some form of Fourier-transform software. A typical spectrum of heart sounds is shown in Figure 6.47.

Microphones for phonocardiograms are designed to be placed on the chest, over the heart. However, heart sounds are sometimes measured from other vantage points. For this purpose, special microphone transducers are placed at the tips of catheters to pick up heart sounds from within the chambers of the heart or from the major blood vessels near the heart. Frequency-response requirements for these microphones are about the same as for phonocardiograph microphones. However, special requirements dictated by the size and configuration of the catheter must be considered in their construction. As might be expected, the difference in acoustical paths makes these heart-sound patterns appear somewhat different from the usual phonocardiogram patterns.

The *vibrocardiograph* and the *apex cardiograph,* which measure the vibrocardiogram and apex cardiogram, respectively, also use microphones

Figure 6.47. Frequency spectrum of heart sounds. (Courtesy of Computer Medical Science Corporation, Tomball, TX.)

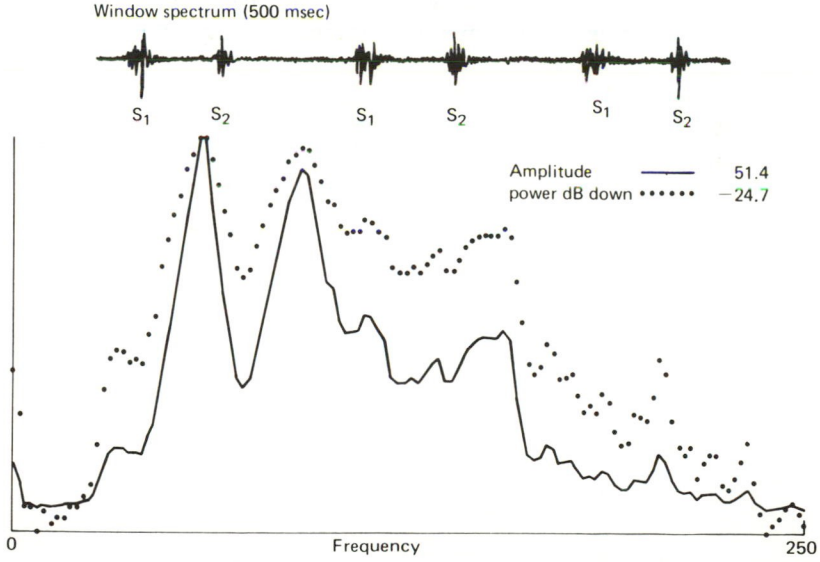

as transducers. However, since these measurements involve the low-frequency vibrations of the heart against the chest wall, the measurement is normally one of displacement or force rather than sound. Thus, the microphone must be a good force transducer, with suitable low-frequency coupling from the chest wall to the microphone element. For the apex cardiogram, the microphone must be coupled to a point between the ribs. A soft rubber or plastic cone attached to the element of the microphone gives good results for this purpose.

Because the vibrocardiogram and the apex cardiogram do not contain the high-frequency components of the heart sounds, these signals can be handled by the same type of amplifiers and recorders as the electrocardiogram (see Section 6.1). Often, these signals are recorded along with a channel of ECG data to maintain time reference. In this case, one channel of a multichannel ECG recorder is devoted to the heart vibration signal.

For recording the Korotkoff sounds from a partially occluded artery (see Chapter 5 and Section 6.2), a microphone is usually placed beneath the occluding cuff or over the artery immediately downstream from the cuff. The waveform and frequency content of these sounds are not as important as the simple identification of their presence, so these sounds generally do not require the high-frequency response specified for the phonocardiogram. Circuitry for identification of these sounds is included in certain automated, indirect blood pressure measuring devices (see Section 6.2.2).

Measurement of the ballistocardiogram requires a platform mounted on a set of extremely flexible springs. When a person lies on the platform, the movement of his body in response to the beating of his heart and the ejection of blood causes similar movement of the platform. The amount of movement can be measured by any of the displacement or velocity transducers or accelerometers described in Chapter 2.

7

Patient Care
and Monitoring

One area of biomedical instrumentation that is becoming increasingly familiar to the general public is that of patient monitoring. Here electronic equipment provides a continuous watch over the vital characteristics and parameters of the critically ill. In the coronary care and other intensive-care units in hospitals, thousands of lives have been saved in recent years because of the careful and accurate monitoring afforded by this equipment. Public awareness of this type of instrumentation has also been greatly increased by its frequent portrayal in television programs, both factual and fictional.

Essentially, patients are monitored because they have an unbalance in their body systems. This can be caused by a heart attack or stroke, for example, or it may be the result of a surgical operation, which can drastically disturb these systems. By continual monitoring, the patient problems can be detected as they occur and remedies taken before these problems get out of hand.

In hospitals that have engineering or electronics departments, patient-monitoring units, both fixed and portable, form a substantial part of the workload of the biomedical engineer or technician. Engineers and technicians are usually involved in the design of facilities for coronary or other intensive-care units, and they work closely with the medical staff to ensure that the equipment to be installed meets the needs of that particular hospital. Ensuring the safety of patients who may have conductive catheters or other direct electrical connections to their hearts is another function of these biomedical engineers and technicians. They also work with the contractors in the installation of the monitoring equipment and, when this job is completed, supervise equipment operation and maintenance. In addition, the engineering staff participates in the planning of improvements and additions, for there are many cases in which an intensive-care unit, even after careful design and installation, fails in some respect to meet the special needs of the hospital, and an in-house solution is required.

Since the first edition of this book was written, much of the equipment then in use has become obsolete except for some basic components. The trend has been toward more automation and computerization. At first, large-scale computers were used and still are, but recent innovations are in the use of microcomputers. Although some of the illustrations in this chapter involve computerized equipment, the reader is referred to Chapter 15 for a discussion of the elements of the computer in biomedical instrumentation.

The first sections of this chapter are concerned with the various elements that compose a patient-monitoring system and the system itself. Some of the problems often encountered are discussed. Two additional topics are included which, although an integral part of the coronary care unit, do have uses in other areas of medical care: the pacemaker and the defibrillator. Many patients have a need for both of these devices while in coronary care, and many leave the unit with implanted pacemakers.

7.1. THE ELEMENTS OF INTENSIVE-CARE MONITORING

The need for intensive-care and patient monitoring has been recognized for centuries. The 24-hour nurse for the critically ill patient has, over the years, become a familiar part of the hospital scene. But only in the last few years has equipment been designed and manufactured that is reliable enough and sufficiently accurate to be used extensively for patient monitoring. Nurses are still there, but roles have changed somewhat, for they now have powerful tools at their disposal for acquiring and assimilating information about the patients under their care. They are therefore able to render better service to a larger number of patients and are better able to react promptly and properly to an emergency situation. With the capability of providing an immediate alarm in the event of certain abnormalities in the behavior of a patient's heart, monitoring equipment makes it possible to summon a physician or nurse in time to administer emergency aid, often before permanent damage can occur. With prompt warning and by providing such information as the electrocardiogram record just prior to, during, and after the onset of cardiac difficulty, the monitoring system enables the physician to give a patient the correct drug rapidly. In some cases, even this process can be automated.

Physicians do not always agree among themselves as to which physiological parameters should be monitored. The number of parameters monitored must be carefully weighed against the cost, complexity, and reliability of the equipment. There are, however, certain parameters that provide vital information and can be reliably measured at relatively low cost. For example, nearly all cardiac-monitoring units continuously measure the electrocardiogram from which the heart rate is easily derived. The electrocardiogram waveform is usually displayed and often recorded. Temperature is also frequently monitored.

On the other hand, there are some variables, such as blood pressure, in which the benefit of continuous monitoring is debatable in light of the problems associated with obtaining the measurement. Since continuous direct blood pressure monitoring requires catheterization of the patient, the traumatic experience of being catheterized may be more harmful to the patient than the lack of continuous pressure information. In fact, intermittent blood pressure measurements by means of a sphygmomanometer, either manual or automatic, might well provide adequate blood pressure information for most purposes. It is not the intent of this book to pass judgment on which measurements should be included but, rather, to familiarize the reader with the instrumentation used in patient care and monitoring.

Since patient-monitoring equipment is usually specified as a system, each manufacturer and each hospital staff has its own ideas as to what should be included in the unit. Thus, a wide variety of configurations can be found in hospitals and in the manufacturers' literature. Since cardiac monitoring is the most extensively used type of patient monitoring today, it provides an appropriate example to illustrate the more general topic of patient monitoring.

The concept of intensive coronary care had little practicality until the development of electronic equipment that was capable of reliably measuring and displaying the electrical activity of the heart on a continuous basis. With such cardiac monitors, instant detection of potentially fatal arrhythmias finally became feasible. Combined with stimulatory equipment to reactivate the heart in the event of such an arrhythmia, a full system of equipment to prevent sudden death in such cases is now available.

In the intensive coronary-care area, monitoring equipment is installed beside the bed of each patient to measure and display the electrocardiogram, heart rate, and other parameters being monitored from that patient. In addition, information from several bedside stations is usually displayed on a central console at the nurses' station.

There are a multiplicity of systems available today, but only a few are illustrated, to give the reader a general idea of their operation. Figure 7.1 shows the bedside unit of an Alpha 9 unit currently in use. The central nurses' station is illustrated in Figure 7.2. To demonstrate some of the changes in design, Figures 7.3 and 7.4 show two earlier types of nurses' stations.

As might be expected, many different room and facility layouts for intensive-coronary-care units are in use. One type that is quite popular is a U-shaped design in which six or eight cubicles or rooms with glass windows surround the nurses' central monitoring station. Although the optimum number of stations per central console has not been established, a group of six or eight seems most efficient. For larger hospitals, monitoring of 16 to 24 beds can be accomplished by two or three central stations. The exact number depends on the individual hospital, its procedures, and the physical layout of the patient-care area. In certain areas in which recruitment of trained nurses is difficult, this factor could also be considered in the selecting of the best design.

Although patient-monitoring systems vary greatly in size and configuration, certain basic elements are common to nearly all of them. A cardiac-care unit, for example, generally includes the following components:

1. Skin electrodes to pick up the ECG potentials. These electrodes are described in Chapter 4.

2. Amplification equipment similar to that described in Chapter 6 for the electrocardiograph.

3. A cathode-ray-tube (CRT) display that permits direct observation of the ECG waveforms. The bedside monitors usually contain fairly small cathode-ray-tube screens (2 to 5 in. in diameter), and each displays the ECG waveform from one patient. The central nurses' station generally has a larger screen on which electrocardiograms from several patients are displayed simultaneously.

4. A rate meter used to indicate the average number of heartbeats per minute and to provide a continuous indication of the heart rate. On most units, an audible beep or flashing light (or both) occurs with each heartbeat.

5. An alarm system, actuated by the rate meter, to alert the nurse or other observer by audible or visible signals whenever the heart rate falls below or exceeds some adjustable preset range (e.g., 40 to 150 beats per minute).

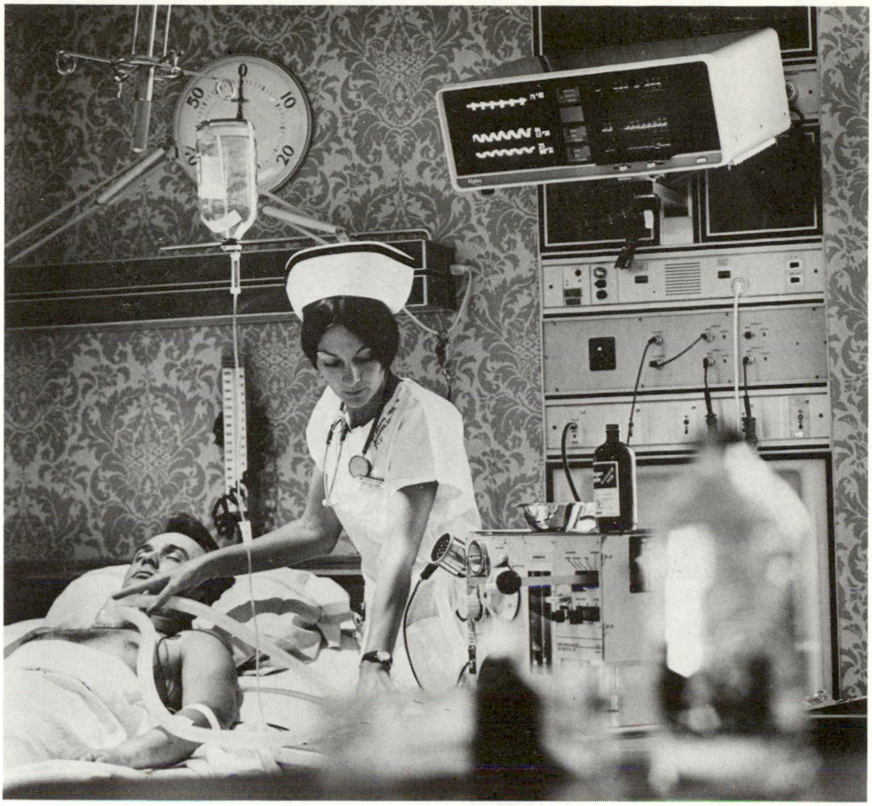

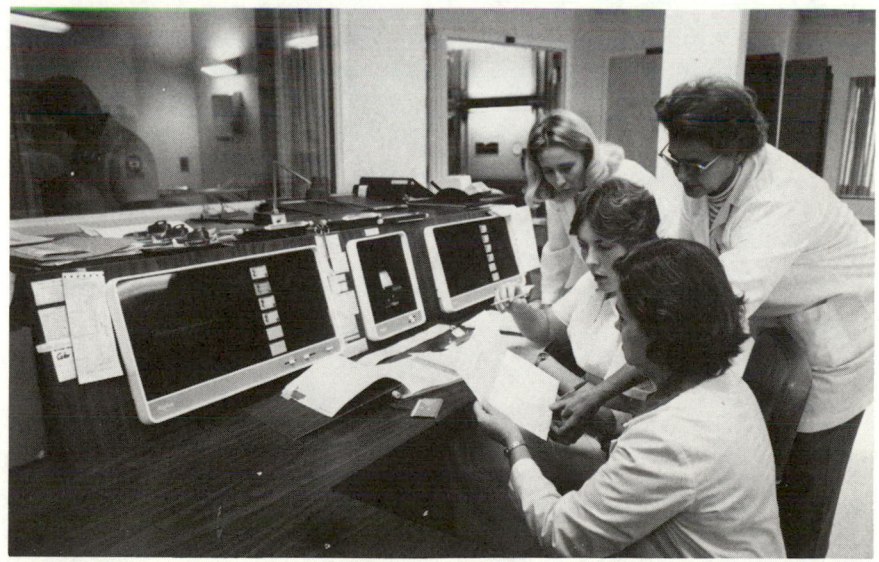

Figure 7.1. Alpha 9 bedside monitor (Courtesy of Spacelabs Inc. Chatsworth, CA.).

Figure 7.2. Nurses' central monitoring station (Courtesy of Spacelabs Inc., Chatsworth, CA.).

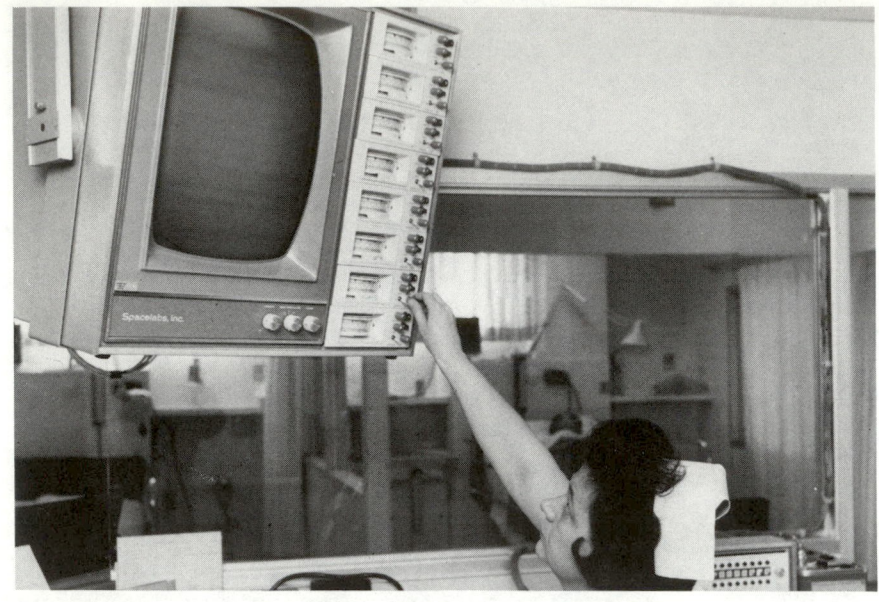

Figure 7.3. Eight channel nurses' station. (Courtesy of Spacelabs, Inc., Chatsworth, CA.)

Figure 7.4. Multiple central monitoring unit. (Courtesy of Spacelabs, Inc., Chatsworth, CA.)

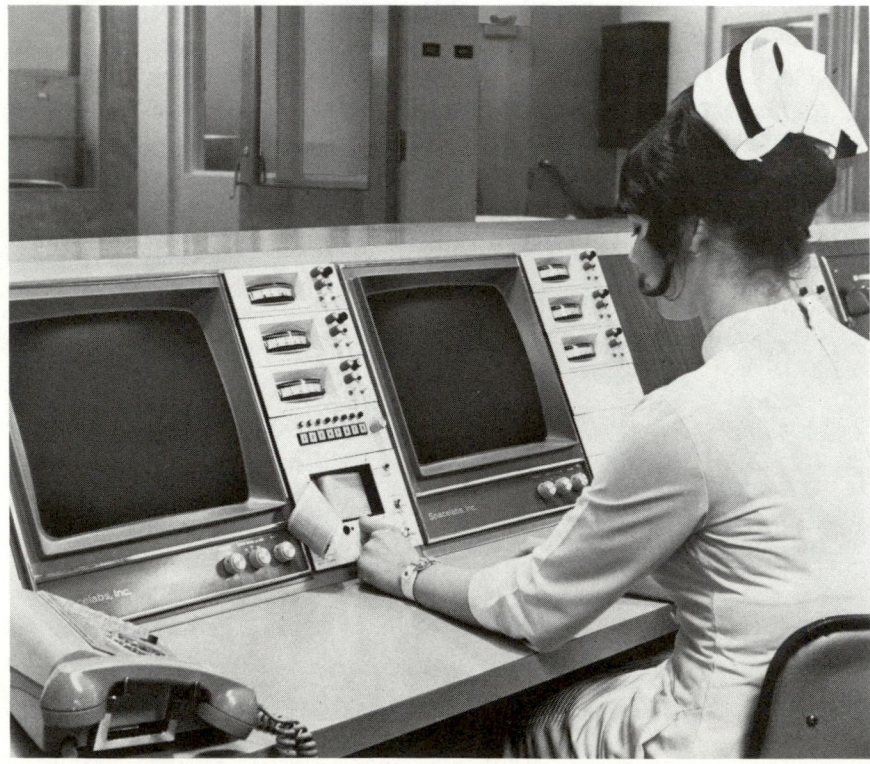

In addition to these basic components, the following elements are useful and are often found in cardiac-monitoring systems:

1. A direct-writeout device (an electrocardiograph) to obtain, on demand or automatically, a permanent record of the electrocardiogram seen on the oscilloscope. Such documentation is valuable for comparative purposes and it usually required in the event of an alarm condition. In combination with the tape loop described below, this written record provides a valuable diagnostic tool.

2. A memory-tape loop to record and play back the electrocardiogram for the 15 to 60 seconds just prior to an alarm condition. Recording of the ECG may continue until the system is reset. In this way, the electrical events associated with the heart immediately before, during, and following an alarm situation can be displayed if a nurse or other observer was not present at the time of the occurrence.

3. Additional alarm systems triggered by ECG parameters other than the heart rate. These alarms may be activated by premature ventricular contractions or by widening of the QRS complex in the ECG (see Chapter 3). Either situation may provide advance indication of a more serious problem.

4. Electrical circuits to indicate that an electrode has become disconnected or that a mechanical failure has occurred somewhere else in the monitoring system. Such a lead failure alarm permits instrumentation problems to be distinguished from true clinical emergencies.

Although not usually considered to be a part of the biomedical instrumentation system for patient monitoring, closed-circuit television is also used in some intensive-care areas to provide visual coverage in addition to monitoring the patients' vital parameters. Where television is employed, a camera is focused on each patient. The nurses' central station has either a bank of monitors, one for each patient, or a single monitor, which can be switched to any camera as desired.

Experience has shown that in spite of its value and increasing popularity, patient-monitoring equipment is not without its problems or limitations. Although many of the difficulties originally encountered in the development of such systems have been corrected, a number of significant problems remain. A few examples are given below.

Example 1. Noise and movement artifacts have always been a problem in the measurement of the electrocardiogram (see Chapter 6). Since heart-rate meters, and subsequently alarm devices, are usually triggered by the R wave of the ECG, and many systems cannot distinguish between the R wave and a noise spike of the same amplitude, movement or muscle interference may be counted as additional heartbeats. As a result, the rate meter shows a higher heart rate than that of the patient, and a high-rate false alarm is actuated. Unfortunately, repeated false alarms tend to cause the staff of the cardiac-care unit to lose confidence in the patient-monitoring equipment and therefore to ignore the alarms or turn them off altogether. Better electrodes, which reduce patient movement artifacts, and more careful placement of electrodes to avoid areas of muscle activity can help to some extent. Electronic filtering to reduce the response of the system at frequencies at which interference might be expected is also partially effective. Some of the more sophisticated systems include circuitry that identifies additional characteristics of the ECG other than simply the amplitude of the R wave, thus further reducing the possibility of mistaking noise for the ECG signal. In spite of all these measures, the possibility of false alarm signals due to movement or muscle artifact is still a real problem.

Example 2.   The low-rate alarm can be falsely activated if the R wave of the ECG is of insufficient amplitude to trigger the rate meter. This can happen if the contact between the electrodes and the skin becomes disturbed because of improper application of the electrodes, excessive patient sweating, or drying of the electrode paste or jelly. In some cases, the indication on the oscilloscope approximates that which might appear if the heart stopped beating. Even though a false low-rate alarm indicates an equipment problem that requires attention, the danger of mistaking failure of the electrode connections for cardiac standstill can have serious consequences. To prevent this possibility, lead-failure alarms have been designed and are built into some patient-monitoring systems.

Example 3.   Because the ECG electrodes must remain attached to the skin for long periods of time during patient monitoring, inflammatory reactions at electrode locations are common. Special skin care and proper application of the electrodes can help minimize this problem.

Special electrode placement patterns are often used in patient-monitoring applications. These patterns are generally intended to approximate the standard limb lead signals (see Chapter 6) while avoiding the actual placement of electrodes on the patient's arms and legs. Instead, the RA, LA, and RL electrodes are placed at appropriate positions on the patient's chest. Because many possible chest placement patterns provide suitable approximations of the three limb leads, no standard pattern has been determined, although some hospitals and manufacturers of patient monitoring equipment may specify a particular arrangement.

7.1.1. Patient-Monitoring Displays

An important feature of any patient-monitoring system is its ability to display the physiological waveforms being monitored. Clear, faithful reproductions of the ECG, arterial blood pressure, and other variables enable the medical staff to periodically check a patient's progress and make vital decisions at times of crisis. Although paper-chart recordings are often used to provide a permanent record of the data, the principal display device for patient monitoring is the cathode-ray tube (CRT). Ranging from a small single- or dual-channel display at the patient's bedside to a large multichannel unit at the nurses' station, the CRT provides a continuous, current presentation of one or more waveforms from a given patient or simultaneous waveforms from several patients. In addition, computerized patient-monitoring systems may permit display of calculated parameters and graphic information, including trend plots and comparisons of current measurement results with past data.

Two types of CRT displays are found in patient-monitoring systems, the conventional or "bouncing-ball" display and the more recent nonfade display. Both utilize the same basic cathode-ray tube, but incorporate different methods of presenting information on the screen.

The *conventional* or *bouncing-ball display* is nothing more than an oscilloscope with the horizontal sweep driven by a slow-speed sweep generator that causes the electron beam to move from left to right at a predetermined rate, selectable from a front-panel control. Sweep rates of 25 and 50 mm/sec are often included to correspond to standard ECG chart speeds. Multiple traces are obtained by an electronic chopper that causes the electron beam to sequentially jump from trace to trace, sharing time among the various channels of data. Eight or more channels can be presented simultaneously in this manner, all of which are presented in time relationship with each other. Because of the high speed with which the beam jumps from trace to trace, the display appears to the viewer as a number of continuous patterns traced horizontally across the screen, each apparently traced by a dot of light that moves up and down to follow the waveform as all of the dots move simultaneously across the screen from left to right. This type of display is often called a bouncing-ball display because each spot of light seems to bounce up and down as it traces the ECG or other waveform being presented.

As the electron beam moves across and writes a pattern on the face of the CRT, the earlier portion of the trace begins to fade away and finally disappears. The ability of the trace to remain visible on the face of the CRT is known as *persistence*. The duration of this persistence is determined by the phosphor (coating) inside the CRT. The longest persistence tubes that are available allow the trace to be visible for approximately 1 second. In the case of patterns like the ECG or blood pressure waveforms, for example, which may occur at a rate of 60 events per minute, the persistence of the tube will allow the viewer to see only one cycle of the waveform. In addition, this displayed waveform will not be uniformly bright. The early portion of the display will always be dimmer than the more current portion. Such a "temporary display" has great limitations for the observer, who may need to evaluate the waveform for diagnostic purposes. Doing so becomes an almost impossible task. When it is necessary to perform any analysis on such waveforms, the waveforms must be permanently recorded on paper. Any event that might have occurred unseen may be lost within a second and cannot be documented on paper even with a time lag or delay between the CRT and writeout device.

For years the bouncing-ball display was the only available type of waveform display, and it was useful despite its limitations. A few of the less expensive systems still use this method, although most modern patient monitors feature nonfade displays.

Nonfade displays also use the cathode-ray tube, but in an entirely different way. In the nonfade method, the electron beam rapidly scans the entire surface of the CRT screen in a television-like raster pattern, but with the brightness level so low that the background raster is not visible. The beam is brightened only when a brightening signal is applied to the CRT by a method called *Z-axis modulation*. This brightening signal is applied only when the electron beam passes a location that is to contain a part of the displayed waveform, at which time it produces a dot on the screen. Each time the entire screen is scanned, each of the traces appears as a series of dots similar to the ECG pattern shown in Figure 7.5. However, the dots are so close together and the scan is so rapid that they appear as a continuous trace.

Fig. 7.5. ECG pattern made up of dots.

The brightening signal is produced from a digital memory in which several seconds of data from each channel of the patient monitor are stored. The memory is constantly renewed so that the oldest data are continually replaced by the most recent. The memory may be an independent digital memory controlled by hardwired logic circuitry or it may be part of a microprocessor or computer incorporated into the patient monitor (see Chapter 15).

The displayed waveforms usually contain several cycles of ECG or respiratory data, all of which are of uniform brightness. The waveforms may either slowly move across the screen from right to left, with the most recent data at the extreme right, or they may remain stationary with the newer data "erasing" the older information as it is written from left to right. Either way, the display may be stopped at any time in order to allow detailed observation of any part of the pattern. Also, in many units, a stored trace may be expanded at any time to permit more detailed observation.

In addition the nonfade display makes possible the presentation of digital numerical readouts on the face of the monitor screen. Thus, the ongoing heart rate, systolic and diastolic blood pressure, and the patient's temperature may be displayed along with the ECG and blood pressure waveform.

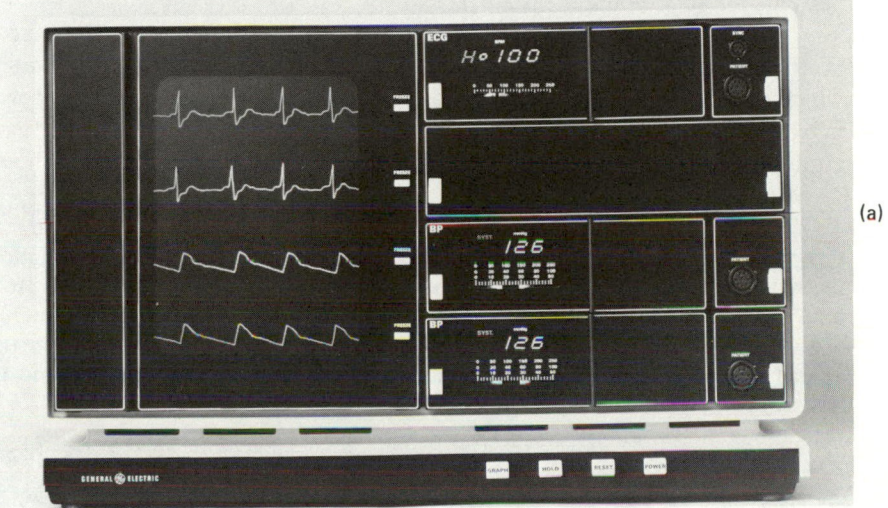

(a)

Figure 7.6. PDS 3000 monitoring system. (a) Bedside unit. (b) Nurses' unit. (Courtesy of General Electric Co., Medical Systems Div., Milwaukee, WI.)

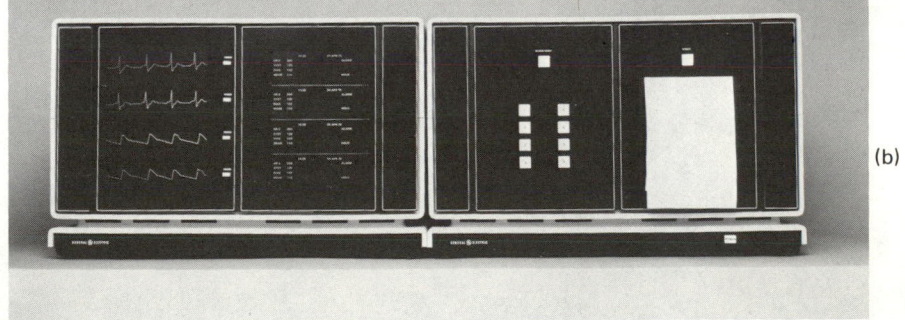

(b)

An example of an intensive-care patient-monitoring system using a nonfade display is the Patient Data System PDS 3000, shown in Figures 7.6 and 7.7. Figure 7.6(a) and (b) show the bedside assembly and central monitoring unit, respectively. The bedside unit displays four nonfade waveforms, together with digital numeric readouts and transmits analog waveforms and digital data to other portions of the system. The central-station unit is equipped with a dual-channel graph. It is a distributed-intelligence system with a microcomputer (see Chapter 15) integrated into each patient's bedside monitor. The bedside microcomputers communicate with another microcomputer at the nurses' central station. The interconnection of the microcomputers form a "star" network with the central nurses' station at the center and the bedside units functioning as satellites. The communications channel from the central station to each bedside is bi-directional so that data and control signals can flow freely in each direction between microcomputers.

It should be noted that the nurses' control station accepts data from up to eight patients; provides single- or dual-channel strip-chart recordings which can be either delayed or produced in real time; identifies all recordings by bed number, time of day, and date; and provides heart rate, lead configuration, and (if monitored) blood pressure values. Figure 7.7 shows a typical hospital installation.

Figure 7.7. This view from the General Electric Patient Data System nurses' station shows two of nine beds in the Intensive Care Unit of Eisenhower Memorial Hospital's expanded facilities. The 185-bed hospital was recently expanded from 138 beds by means of a 78,000-square-foot, three-story addition in the shape of a cross. The addition also features six Coronary Care Unit beds, five new surgeries and 11 recovery beds. (Courtesy of the Eisenhower Medical Center, Palm Desert, CA.)

7.2. DIAGNOSIS, CALIBRATION, AND REPAIRABILITY OF PATIENT-MONITORING EQUIPMENT

Just as the physician must diagnose the patient, engineers or technicians may be required to diagnose the patient-monitoring equipment in the hospital. In connection with the typical intensive-care unit, there is often an associated maintenance and design group composed of engineers and technicians who usually have access to a wide variety of diagnostic devices for electronic equipment.

Not only is it necessary to keep medical instrumentation in good repair, but it is equally important to keep all equipment calibrated accurately. Many pieces of medical electronics have built-in calibration devices. For example, all electrocardiographs have a 1-mV calibration voltage available internally. Many patient monitors have built-in calibration features. In recent years, manufacturers of electronic servicing equipment have designed many items especially for the hospital and other medical applications. An excellent example of such a device is the Medical Instrumentation Calibration System illustrated in Figure 7.8.

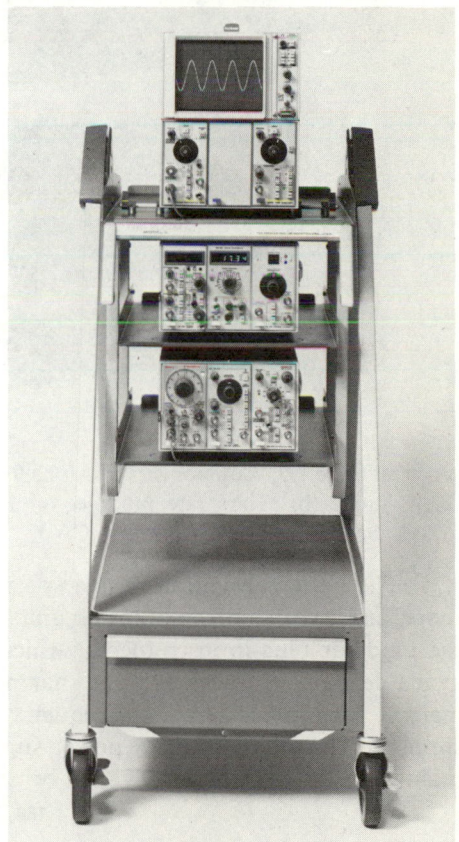

Figure 7.8. Medical instrumentation calibration system. (Courtesy of Tektronix Inc., Beaverton, OR.)

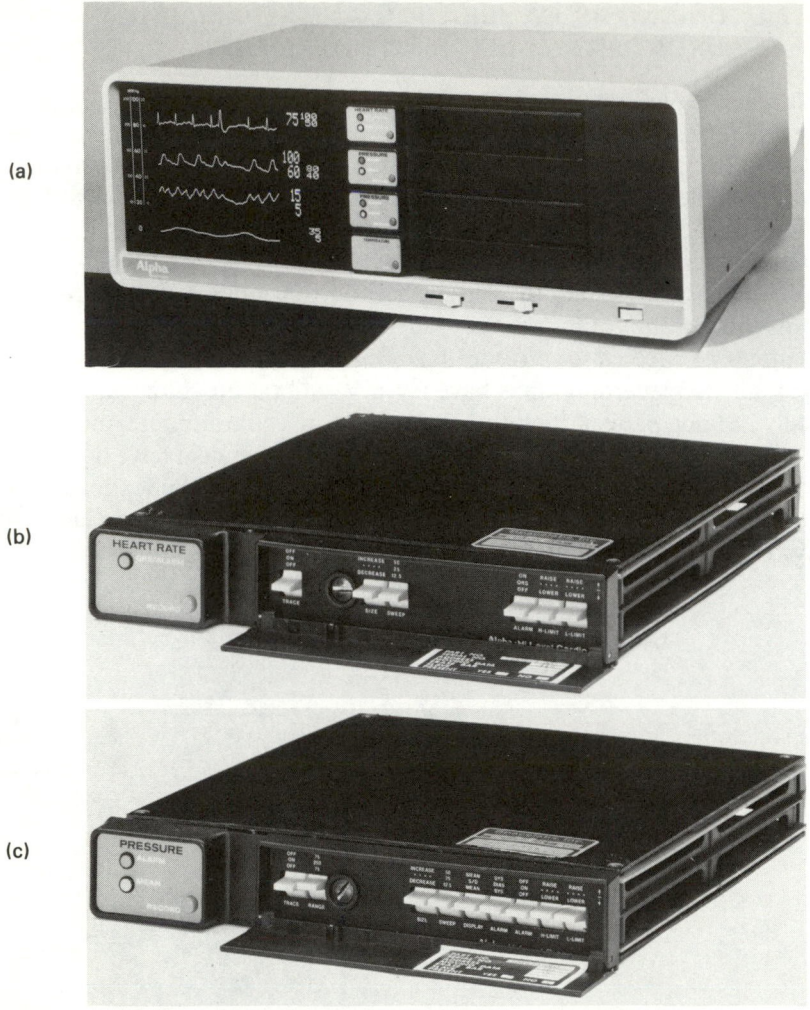

Figure 7.9. Components of Alpha 9 monitoring system. (a) Monitor unit. (b) Heart rate module. (c) Pressure module. (Courtesy of Spacelabs Inc., Chatsworth, CA.)

The basic configuration is a special mobile cart carrying an oscilloscope and two mainframe power units. These mainframes accommodate the modular plug-in instruments, which can be any of the more than 30 available units. Typical plug-ins that might be selected are digital multimeters, function generators, frequency counters, amplifiers, and power supplies. The major primary power supply components are located in the mainframe units, where they can be shared by the plug-in instruments.

Signal interconnections between plug-ins can be made through the mainframe. All units share a common ground.

This unit is capable of repair or calibration of most types of electronic equipment used in medical applications today, including ECGs, EEGs, complete patient monitors, X-ray and cardiac-unit control systems, ultrasound systems, radio-frequency heating and diathermy equipment, and radio-frequency and direct telemetry.

It is important to have all parts of the patient-monitoring equipment designed for easy replacement or repair or both. To achieve the former, most intensive-care equipment is of a modular design, so that individual component groups, such as amplifiers, can be removed and replaced easily. An example of this feature is illustrated in Figure 7.9. These three components are all part of the bedside unit shown in Figure 7.1.

Figure 7.9(a) shows the entire monitor unit. Figure 7.9(b) shows the blood pressure unit and part (c) shows the heart rate unit. Both can be removed and replaced very rapidly in the case of malfunction.

7.3. OTHER INSTRUMENTATION FOR MONITORING PATIENTS

In addition to its use at the bedside, as discussed in Section 7.1 and illustrated in Figures 7.1 through 7.7, patient-monitoring equipment is often found in other applications in the hospital. An important example is in the operating room. Figure 7.10 shows one type of unit used during surgery. The main features of this type of system are the large multichannel oscilloscope, the capability of obtaining a permanent ECG record on the chart recorder, and plug-in signal-conditioner modules that provide versatility and choice of measurement parameters.

The chart recorder has eight channels, to be used as dictated by the specific requirements of the surgical team. The fluid writing system is pressurized and writes dry. The frequency response of such a unit is up to 40 Hz full scale. The trace is rectilinear, and the channel span is 40 mm graduated in 50 divisions. Two event channels are provided to relate information on the chart to specific events. The recorder has a large number of chart speeds selected by pushbuttons, ranging from 0.05 to 200 mm/sec.

As mentioned earlier, a variety of plug-in modules are available for use with the unit. These biomedical signal conditioners include a universal unit for various bioelectric signals, an ECG unit, an EEG unit, a biotachometer, a transducer unit, an integrator, a differentiator, and an impedance unit. These units are all compatible. Figure 7.11 shows the front plates of some of them, and Figure 7.12 shows a complete biotachometer unit.

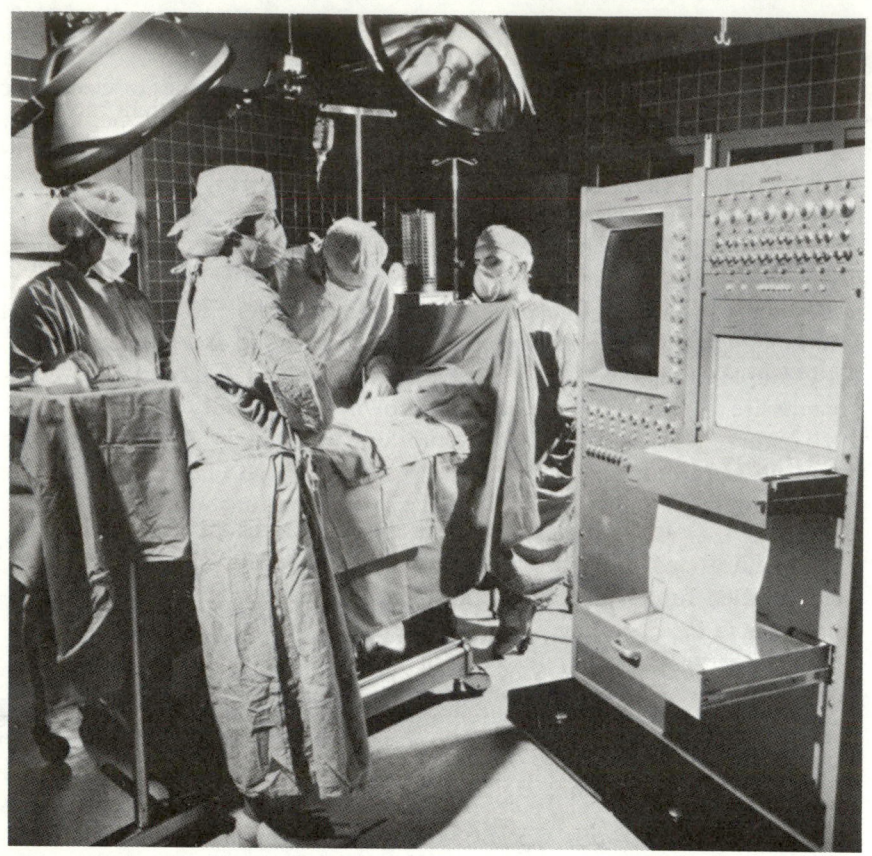

Figure 7.10. Surgical monitoring system. (Courtesy of Gould, Inc., Brush Instruments Division, Cleveland, OH.)

Each of the signal conditioners includes two separate submodules: a coupler and a medical amplifier. The coupler contains the circuitry and controls that are essential to its nameplate function. The amplifier or "back end" contains circuitry that is exactly the same for each channel. This "back end" can be obtained separately, thus reducing a possible investment in idle circuitry. In the last few years many manufacturers of biomedical monitoring equipment have improved their systems to include such versatility.

The amplifier units are designed for broad-band amplification from dc to 10 kHz. The amplifier is a ground-isolation type that eliminates a potential shock hazard to the patient by isolating him to the extent that current through his body cannot exceed $2 \mu A$ (see Chapter 16).

Each amplifier provides a buffered output drive signal for peripheral monitoring equipment, such as tape recorders, oscilloscopes, and computers.

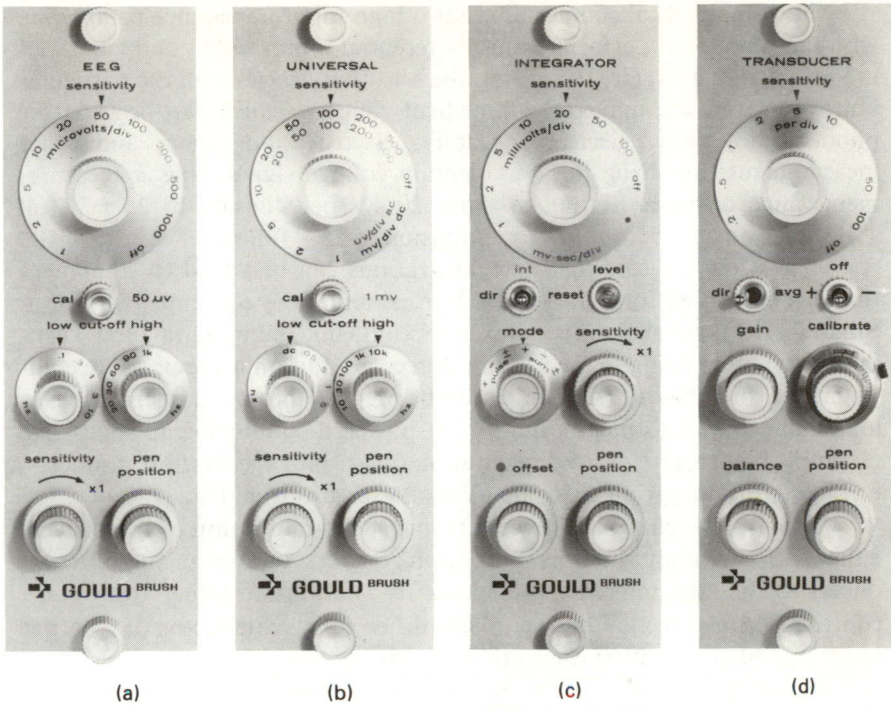

(a) (b) (c) (d)

Figure 7.11. Front panel of signal conditioner units: (a) EEG; (b) universal; (c) integrator; (d) transducer. (Courtesy of Gould, Inc., Brush Instruments Division, Cleveland, OH.)

Figure 7.12. View of biotachometer module. (Courtesy of Gould, Inc., Brush Instruments Division, Cleveland, OH.)

Each biomedical coupler unit has a high input impedance compatible with the function it performs, plus differential inputs with good common-mode rejection at 60 Hz. In general, the sensitivity varies with the particular unit. However, the universal biomedical coupler, which can be used for phonocardiography, electromyography, electrocardiography, and other measurements involving low- or medium-voltage signals, has a measurement range on an ac setting of from 20 μV per division to 25 mV full scale, and, on dc, of from 1 mV per division to 25 V full scale.

It is often the case that new innovations can be added to equipment that has been used for many years. For example, the equipment shown in Figures 7.10 through 7.12 is many years old, but the manufacturer keeps improving and adding to the system. A recent addition is the pressure computer, shown in Figure 7.13, which can be plugged into the existing frame as a modular unit. This plug-in unit is self-powered, employing all solid-state circuitry. It supplies $\pm$ 2.5 Vdc excitation for strain-gage-based transducers having sensitivities from 50 to 500 μV/volt excitation/cm Hg.

Front-panel controls allow electronic shunt calibration of transducers either with or without built-in calibration resistors, and balancing of transducers to the zero reference pressure after calibration. Other front-panel controls permit selection of overall amplifier gain, scale factor of recorder outputs, and pen position for graphic recording.

An eight-position mode switch permits the complete waveform or any derived parameter to be fed to a graphic recorder or other analog device such as an oscilloscope, FM tape recorder, or remote monitor. In addition, all derived parameters are available simultaneously (independent of the mode-switch position) at fixed gain for digital display. The digital display update rate is internally adjustable from 6 to 30 per minute.

The unit displays the complete dynamic waveform in direct mode and simultaneously derives the following parameters from the waveform: systolic pressure, diastolic pressure, pulse pressure, average (mean) pressure, rate of change of pressure (dp/dt), and pulse rate. It has a range of -30 to $+ 500$ mm Hg for systolic and diastolic pressure and 5 to 300 mm Hg for pulse pressure. The pressure derivative range is from 100 mm Hg/sec/division to 15,000 mm Hg/sec full scale. Pulse rate is measured from 20 to 300 beats per minute.

Another special type of monitoring device is the *arterial diagnostic unit* (ADU) shown in Figure 7.14. This mobile cart combines several commonly used instruments for peripheral arterial evaluation. The ADU provides automated pressure-cuff inflation for rapidly determining segmental pressures and post-exercise pressure trends in noninvasive peripheral arterial evaluations. The strip chart can be used for recording Doppler-flow pulse waveforms, ECG traces, or other physiological waveforms. It has a heated stylus strip-chart recorder, a bidirectional Doppler-flow meter with external

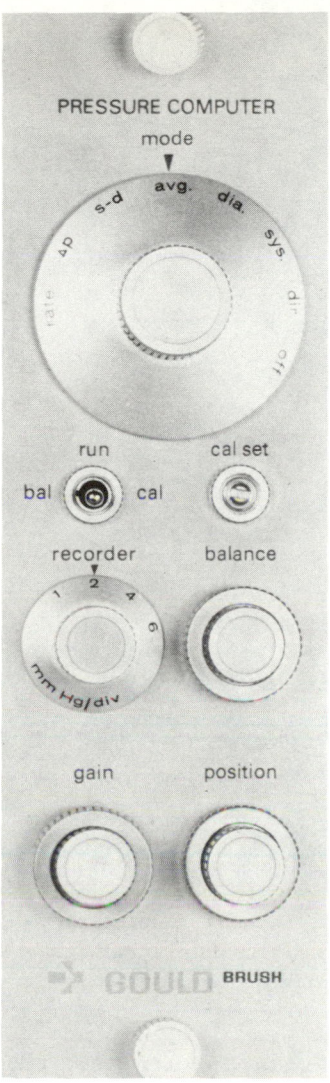

Figure 7.13. Blood pressure computer unit. (Courtesy of Gould Inc., Instrument Systems Division, Cleveland, OH.)

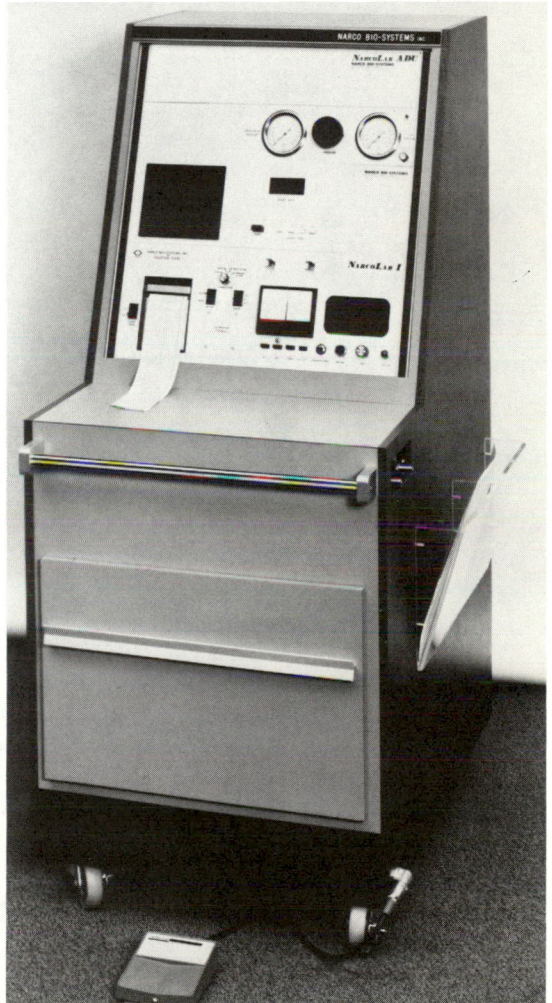

Figure 7.14. Arterial diagnostic unit (Courtesy of Narco Bio-Systems, Houston, TX.)

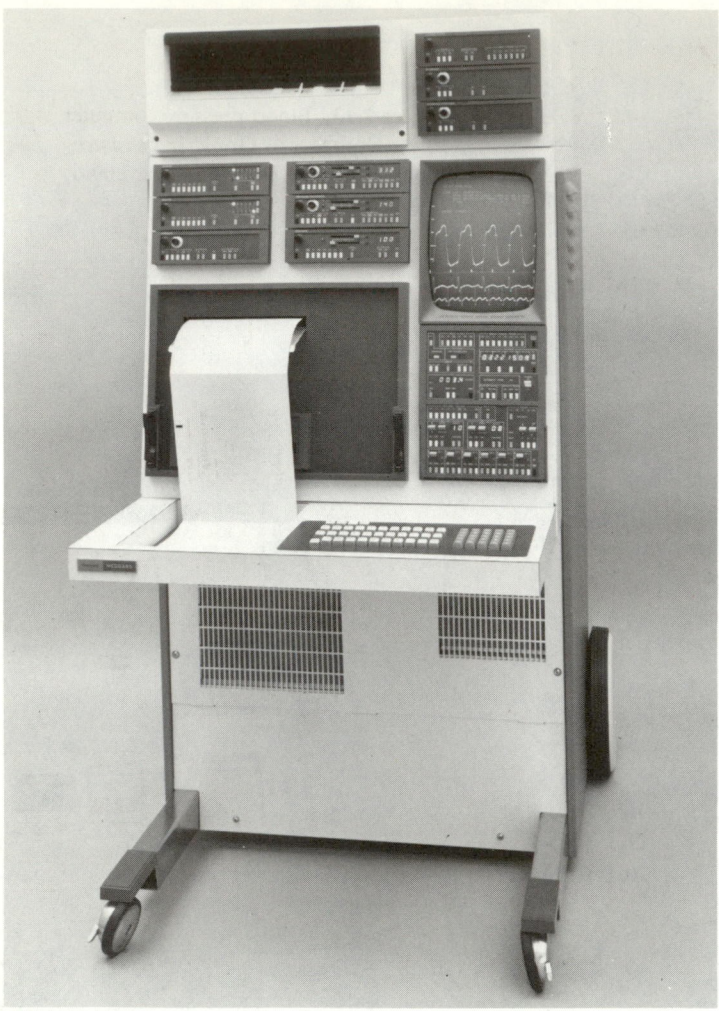

Figure 7.15. Meddars catheterization laboratory recording systems. (Courtesy of Honeywell Inc., Test Instruments Division, Denver, CO.)

loudspeaker and headphones, an ECG telemetry transmitter, a nonfade two-channel oscilloscope with freeze capability, a compressor with regulator and pressure gages, a foot-switch activator for cuffs and recorder, a digital heart rate tachometer, and a noninvasive determination of common femoral artery pressure.

Another place in the hospital where portability can be an advantage is in the catheterization laboratory. The technique of catheterization has been described in a number of places in Chapter 6 (see Figure 6.18 for an illustration of the process). Essentially the *cath. lab* is the room in which cardiologists perform diagnostic catheterizations. If a patient is suspected of having a blockage, say, of one of the coronary arteries, he or she would be brought into the cath. lab to undergo analysis to determine if there is such a blockage and its extent, if it exists.

The catheterization technique is to introduce a catheter into the heart by the method shown in Figure 6.18 and through this catheter to inject a radiopaque dye into the cardiac chamber. The passage of the dye as it traverses the arteries is monitored fluoroscopically on a monitor. In this way, any blockage can be seen and a motion picture can be taken simultaneously.

Catheterization can be a dangerous procedure for the patient, even though fatalities are infrequent. Therefore, the patient must be monitored continuously. Many units have extensive built-in monitoring, but since cath. labs are usually small, there is an advantage to having a smaller mobile unit available. Such a unit is illustrated in Figure 7.15. This is a computerized unit capable of monitoring all the variables usually needed in the cath. lab, such as cardiovascular pressures, cardiac output, and ECG. Patient files may be reviewed by instant recall. Six channels of analog data can be recorded on continuous strip charts, and computer-generated results can be shown on the same page as the waveforms. Test results are eventually incorporated in a comprehensive final report, which includes calculated results, derived results, comments, summary, and a pictorial representation of the heart, showing pressures, oxygen saturation, and so on. During the procedure the patient's name, physiological data in digital form, blood pressure, and ECG waveforms are also displayed on the cathode-ray tube as a nonfade display. The console has a keyboard which allows an operator instant access to any information that he or she wishes to identify. There is also a 12-digit, 10-function calculator adjacent to the keyboard.

7.4. THE ORGANIZATION OF THE HOSPITAL FOR PATIENT-CARE MONITORING

The engineer or technician should be familiar with the overall organization of the hospital with respect to monitoring equipment. In the previous sections of this chapter, many types of equipment and services have been described. The following summary is provided to give an overall view.

A broad classification for patients in the hospital is to categorize them as surgical or nonsurgical. The heart of the surgical facilities is the *operating room,* where the surgery is actually performed. The monitoring equipment in this room usually includes measurement of heart rate, venous and arterial blood pressures, ECG and EEG (see Chapter 10 for a description of EEG methods), and various respiratory therapy devices (see Chapter 8). It is also necessary to have emergency equipment on hand, such as defibrillators and pacemakers (see later sections of this chapter), resuscitation devices, and stimulation equipment.

The *intensive-care unit* (ICU) can be used for postsurgical follow-up or for medical patients with very serious problems. It is usually provided with equipment similar to that of the operating room, the only difference being that there is usually only one set of equipment in the operating room, while in the ICU there are bedside monitoring units and nurses' central consoles to monitor many patients simultaneously.

Most heart-attack victims are placed in a *coronary care unit* (CCU). In some hospitals this is called a *cardiac care unit*. The monitoring equipment in these units center around blood pressure, heart rate, and ECGs.

As the heart attack patient recovers, he or she is usually moved into another unit, where monitoring is not as critical. This unit, typically called the *intermediate coronary care unit* (ICCU), may also contain telemetry equipment to monitor ambulatory patients (see Chapter 12). There are also special groups of rooms in many hospitals that do not have bedside units installed but have portable units available in the corridors for immediate use for short periods of time. These do not necessarily have special names, but *cardiac observation unit* is one name that has been used.

Another important area in the hospital is the *emergency room,* in which emergency care is provided. Here every patient is a possible crisis. For this reason, emergency rooms require equipment of the same types described for the other facilities, but usually of the portable variety. Typically, a rapid diagnosis is made and the patient is rushed off to some other facility, such as the operating room or an intensive-care unit, where the critical measurements are made. In addition to on-the-spot measurements, all the units described above must have provision for taking samples of body fluids, such as urine and blood, where indicated. These samples are usually taken by a medical technologist, technician, or nurse and sent to the laboratory, which is usually situated elsewhere in the hospital. Instrumentation for the laboratory is described in Chapter 13.

Finally, some consideration should be given to what happens to a potential patient if he or she should have a heart attack at home or in the street. Connected with most hospitals today there is a *paramedic service.* Many of these are privately operated or they may be operated by the city or county, typically by the fire department. Paramedic units are described in more detail in Chapter 12, but an introduction is appropriate in the context of this chapter. The need to provide immediate around-the-clock medical care to heart attack and accident victims in the community has resulted in the use of *mobile emergency care units.* Manned by personnel trained to administer first aid as well as emergency cardiopulmonary resuscitation techniques, these vehicles are equipped with instruments and medication similar to that used in the special care units of hospitals. Because these units are in constant radio contact with community organizations (police and fire departments) and with hospitals, they are able to reach an accident scene or the location of a stricken citizen in a short period of time. Typically, on arrival, a portable electrocardiograph (often with an oscilloscope display) is quickly applied to obtain an evaluation of the patient's ECG and heart rate. An indication of cardiac standstill or pulmonary failure will initiate emergency cardiopulmonary resuscitation procedures. After airway clearance and assisted breathing are ensured, defibrillation or cardioversion may be required by a portable defibrillator. The victim's heart may then be temporarily paced by a portable pacer. Some instruments, such as the defibrillator shown in Figure 7.22, are capable of performing the three functions— monitoring of ECG display, defibrillation (or cardioversion), and pacing. The condition of the patient may then be radioed to personnel in the special care unit of a nearby hospital while the ECG is simultaneously transmitted to the unit's display and recording equipment via telemetry (see Chapter 12). In a short time, after a detailed diagnosis is made, instructions returned by radio may prescribe specific medication or additional resuscitation techniques. If the patient is in standstill and cannot breathe, adequate blood circulation can be maintained by means of a mechanical cardiopulmonary resuscitation unit. When the patient is able to be transported to a hospital, vital signs are monitored by appropriate equipment inside the vehicle. In this way, a seriously stricken individual is able to receive timely special care away from the hospital.

7.5. PACEMAKERS

More than a half-million people fall victim to heart attacks in the United States every year and thousands more are critically injured in accidents. Taking care of these patients in special care units of hospitals involves the use of several types of specialized equipment, among which are cardiac pacemakers and defibrillators. Defibrillators and cardiopulmonary resuscitation equipment are also required away from the hospital, in an ambulance or at the scene of an emergency.

In the past few years electronic pacemaker systems have become extremely important in saving and sustaining the lives of cardiac patients whose normal pacing functions have become impaired. Depending on the exact nature of a cardiac dysfunction, a patient may require temporary artificial pacing during the course of treatment or permanent pacing in order to lead an active, productive life after treatment.

This section deals with the various types of cardiac pacemakers. In addition to describing each device, its basic purpose, the physiological conditions under which it is required, and the ways in which it is used are also discussed. Section 7.6 covers defibrillators.

The heart's electrical activity is described in Chapters 3 and 5, but a brief review at this point will be helpful in understanding the need for artificial cardiac pacing. The rhythmic action of the heart is initiated by regularly recurring action potentials (electrochemical impulses) originating at the natural cardiac pacemaker, located at the sinoatrial (SA) node. Each pacing impulse is propagated throughout the myocardium, spreading over the surface of the atria to the atrioventricular (AV) node—which is located within the septum, adjacent to the atrioventricular valves—and depolarizing the atria. After a brief delay at the AV node, the impulse is rapidly conducted to the ventricles to depolarize the ventricular musculature.

A normal sinus rhythm (NSR) depends on the continuous, periodic performance of the pacemaker and the integrity of the neuronal conducting pathways. Any change in the NSR is called an *arrhythmia* (abnormal rhythm). Should the SA node temporarily or permanently fail because of disease (SA node disease) or a congenital defect, the pacing function may be taken over by pacemaker-like cells located near the AV node. However, under certain conditions, cells in the conduction system (an *idioventricular focus*) may pace the ventricles instead. Similarly, an area in the excitable ventricular musculature may try to control the heartbeat. Unfortunately, under these conditions the heart is paced at a much slower rate than normal, ranging between 30 and 50 beats per minute (BPM). The result is a condition called *bradycardia* (slow heart), in which the heart cannot provide sufficient blood circulation to meet the body's physical demands. During the transition period from an NSR to a slow rhythm, dizziness and loss of consciousness (syncope) may occur because of diminished cardiac output.

Heart block occurs whenever the conduction system fails to transmit the pacing impulses from the atria to the ventricles properly. In *first-degree block* an excessive impulse delay at the AV junction occurs that causes the P-R interval to exceed 0.2 second for normal adults. *Second-degree block* results in the complete but intermittent inhibition of the pacing impulse, which may also occur at the AV node. Total and continuous impulse blockage is called *third-degree block*. It may occur either at the AV node or elsewhere in the conduction system. In this case, the ventricles usually continue to contract but at a sharply reduced rate (40 BPM) because of the establishment of an idioventricular escape rhythm or because of impulses that only periodically originate from the atria. In all these conditions, an artificial method of pacing is generally required to ensure that the heart beats at a rate that is sufficient to maintain proper circulation.

7.5.1. Pacemaker Systems

A device capable of generating artificial pacing impulses and delivering them to the heart is known as a *pacemaker system* (commonly called a

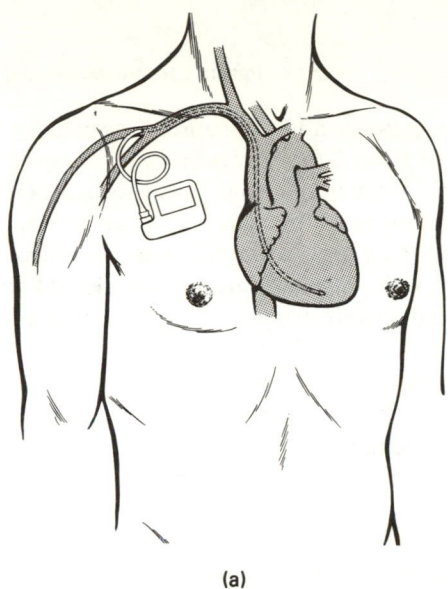

(a)

Figure 7.16. (a) Implanted standby pacemaker with catheter electrodes inserted through the right cephalic vein. (b) Pacing electrodes attached to the myocardium; (c) Myocardial electrodes with pacemaker generator implanted in abdomen.

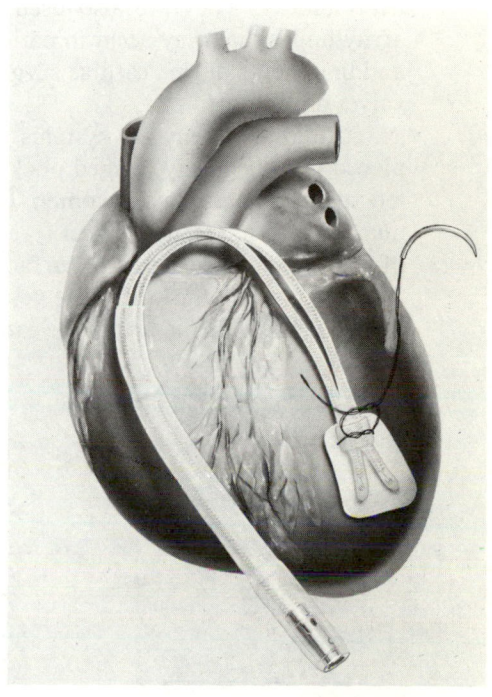

(b)

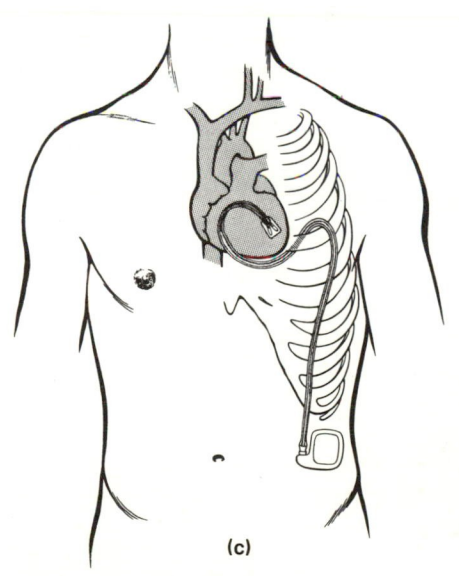

(c)

pacemaker) and consists of a *pulse generator* and appropriate *electrodes*. Pacemakers are available in a variety of forms. *Internal pacemakers* may be permanently implanted in patients whose SA nodes have failed to function properly or who suffer from permanent heart block because of a heart attack. An internal pacemaker is defined as one in which the entire system is inside the body. In contrast, an *external pacemaker* usually consists of an externally worn pulse generator connected to electrodes located on or within the myocardium.

External pacemakers are used on patients with temporary heart irregularities, such as those encountered in the coronary patient, including heart blocks. They are also used for temporary management of certain arrhythmias that may occur in patients during critical postoperative periods and in patients during cardiac surgery, especially if the surgery involves the valves or septum.

Internal pacemaker systems are implanted with the pulse generator placed in a surgically formed pocket below the right or left clavicle, in the left subcostal area, or, in women, beneath the left or right major pectoralis muscle. Internal leads connect to electrodes that directly contact the inside of the right ventricle or the surface of the myocardium (see Figure 7.16). The exact location of the pulse generator depends primarily on the type of electrode used, the nature of the cardiac dysfunction, and the method (mode)

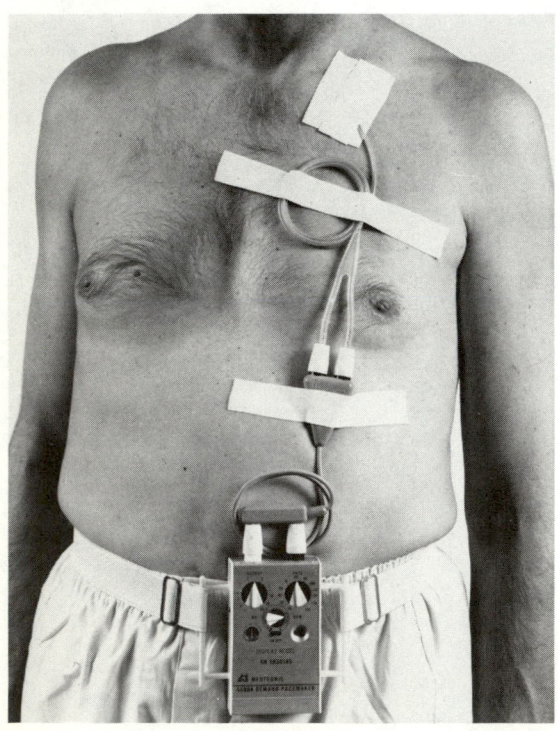

Figure 7.17. Portable external pacemaker. Patient is being temporarily paced with an external demand pacemaker and transvenous pacing catheter. (Courtesy of Medtronic, Inc., Minneapolis, MN.)

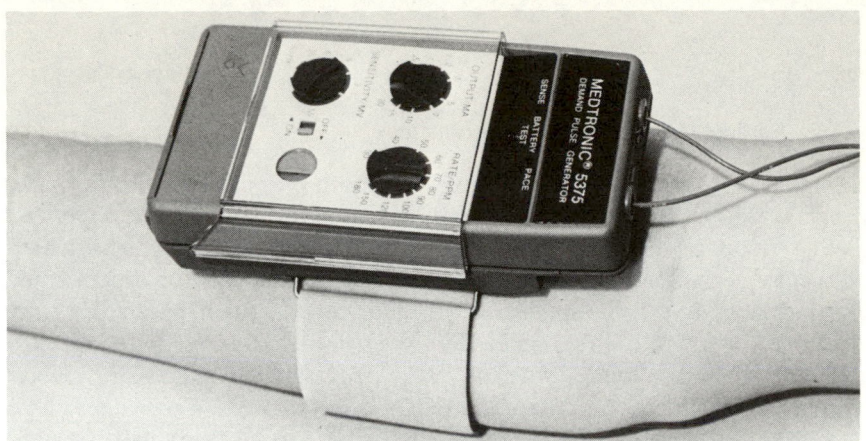

Figure 7.18. Portable external pacemaker, strapped on arm. (Courtesy of Medtronic, Inc., Minneapolis, MN.)

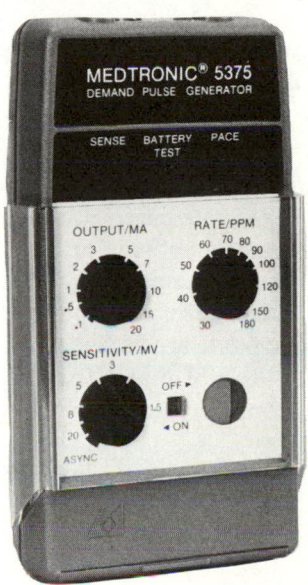

Figure 7.19. Detail view of external demand pacemaker showing adjustment controls. (Courtesy of Medtronic, Inc., Minneapolis, MN.)

of pacing that may be prescribed. Pacing electrodes and modes are described later in this chapter. Since there are no external connections for applying power, the pulse generator must be completely self-contained, with a power source capable of continuously operating the unit for a period of years.

External pacemakers, which include all types of pulse generators located outside the body, are normally connected through wires introduced into the right ventricle via a cardiac catheter, as shown in Figure 7.17. The pulse generator may be strapped to the lower arm of a patient who is confined to bed, or worn at the midsection of an ambulatory patient, as shown in Figures 7.17 and 7.18. A detailed view of the external pulse generator appears in Figure 7.19. Figure 7.17 depicts an older model replaced by that in Figures 7.18 and 7.19, but the idea remains the same.

7.5.2. Pacing Modes and Pulse Generators

Several pacing techniques are possible with both internal and external pace-makers. They can be classed as either *competitive* and *noncompetitive* pacing modes as shown in Figure 7.20. The noncompetitive method, which uses pulse generators that are either *ventricular programmed* or programmed by the atria, is more popular. Ventricular-programmed pacemakers are designed to operate either in a *demand* (R-wave-inhibited) or *standby* (R-wave-triggered) mode, whereas *atrial-programmed pacers* are always synchronized with the P wave of the ECG.

The first (and simplest) pulse generators were *fixed-rate* or *asynchronous* (not synchronized) devices that produced pulses at a fixed rate (set by the physician or nurse) and were independent of any natural cardiac activity. Asynchronous pacing is called *competitive* pacing because the fixed-rate impulses may occur along with natural pacing impulses generated by the heart and would therefore be in competition with them in controlling the heartbeat. This competition is largely eliminated through use of ventricular- or atrial-programmed pulse generators.

Fixed-rate pacers are sometimes installed in elderly patients whose SA nodes cannot provide proper stimuli. They are also used temporarily to determine the amplitude of impulses needed to pace or *capture* the heartbeat of a patient prior to or during the implantation of a more permanent unit. The amplitude at which capture occurs is referred to as the pacing *threshold*. While the implantable fixed-rate units tend to fail less frequently than the more sophisticated demand or standby pacers, their battery life (if the

Figure 7.20. Types of pacing modes.

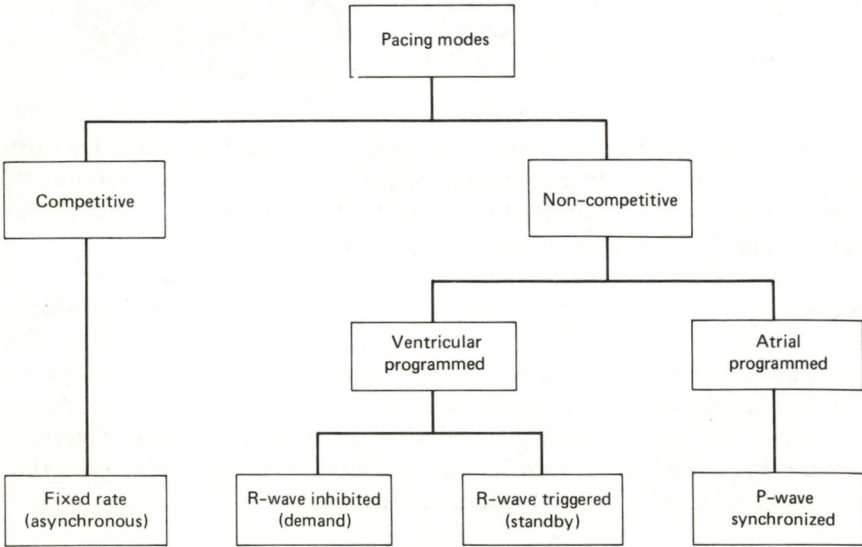

batteries are not rechargeable) is generally shorter because they are in constant operation.

The problems of shorter battery life and competition for control of the heart led, in part, to the development of ventricular-programmed (demand or standby) pulse generators. The models shown in Figures 7.19 and 7.21 are of the demand type. Either type of ventricular-programmed pulse generator, when connected to the ventricles via electrodes, is able to sense the presence (or absence) of a naturally occurring R wave. The output of an R-wave-inhibited (demand) unit is suppressed (no output pulses are produced) as long as natural (intrinsic) R waves are present. Thus, its output is held back or inhibited when the heart is able to pace itself. However, should standstill occur, or should the intrinsic rate fall below the preset rate of the pacer (around 70 BPM), the unit will automatically provide an output to pace the heart after an escape interval at the designated rate. In this way, ventricular-inhibited pacers are able to pace on demand. Some external demand-mode pacers may be adjusted to operate in a fixed-rate mode by means of an accessible mode control of the type shown on the unit in Figure 7.19. Other controls allow the setting of the pacer's rate anywhere between 30 and 180 BPM, as well as the amplitude of output pacing pulses between 0.1 and 20 mA. Some external demand pacers have a sense-pace indicator that deflects for each detected R wave or pacer-initiated impulse. The ON–OFF switch of some external pacers is provided with an interlock mechanism to prevent the unit from being accidentally turned off.

A demand pacer, in the absence of R waves, automatically reverts to a fixed-rate mode of operation. For testing purposes at the time of implantation and for evaluation later, implanted demand pacers are purposely placed in a fixed-rate mode, usually by means of a magnet provided by the manufacturer. When placed over the skin layer covering the pacer, the magnet activates a magnetically operated switch that prevents the pacer from sensing R-wave activity. This process causes the pacer to operate in a fixed-rate mode at a slightly higher rate (about 10 BPM higher than the demand-mode pacing rate that had been preset). For a patient with a normal sinus rhythm, this procedure is used to ensure that an implanted demand pacer whose output is normally inhibited is capable of providing pacing pulses when needed. Evidence of the presence of pacing impulses is obtained from the electrocardiogram. Pacing impulses appear as *pacing artifacts* or *spikes*. Occasionally, they may seriously distort the recorded QRS complex.

When required, the basic pacing rate of some of the earlier implanted pacers (both fixed-rate and demand types) may be changed with the use of a needle-like screwdriver (a Keith surgical skin needle) that is inserted transcutaneously to alter the rate control in the pulse generator. The amplitude of the impulses may also be adjusted in some earlier pacers by using the same type of needle in the appropriate control. In a newer type of pacer, these adjustments are accomplished by means of coded impulses that are magnetically coupled to the implanted pulse generator from the skin surface, thus eliminating the need to puncture the skin. To adjust this pacemaker, a special programming device with an attached coil is placed over the implanted pulse generator. Appropriate controls on the programmer allow the unit to transmit coded signals that cause the pacer to change its basic rate and vary the amplitude of its impulses. The basic rate and impulse amplitude of other recent implantable pulse generators are fixed by the manufacturer and cannot be changed, however.

As explained earlier, R-wave-triggered pulse generators, like the R-wave-inhibited units, sense each intrinsic R wave. However, this pacer emits an impulse with the occurrence of each sensed R wave. Thus, the unit is triggered rather than inhibited by each R wave. The pacing impulses are transmitted to the myocardium during its absolute refractory period, however, so they will have no effect on normal heart activity. Should the intrinsic heart rate fall below the preset rate of the pacer, the pacer will automatically operate synchronously at its preset rate to pace the heart. Thus, this pacemaker stands by to pace when needed. Ventricular-triggered pacing is used less frequently than inhibited-mode pacing. Evidence of pacing impulses from this type of pacer is present on the patient's ECG, although some monitoring modes that utilize greater filtering may distort and even block the pacer artifact. In this case, one should document the ventricular complex following a pacer spike and compare it to the complex in question.

In cases of complete heart block where the atria are able to depolarize but the impulse fails to depolarize the ventricles, atrial synchronous pacing may be used. Here the pulse generator is connected through wires and electrodes to both the atria and the ventricles. The atrial electrode couples atrial impulses to the pulse generator, which then emits impulses to stimulate the ventricles via the ventricular electrode. In this way, the heart is paced at the same rate as the natural pacemaker. When the SA node rate changes because of vagus or sympathetic neuronal control, the ventricle will change its rate accordingly but not above some maximum rate (about 125 per minute).

Pulses applied directly to the heart are usually rectangular in shape with a duration of from 0.15 to 3 msec, depending on the type of pulse generator used and the needs of the patient. Depending on the value of impulse current required to capture, pulse amplitudes may range from 5 to 15 mA for adults, while infants and children require less. If, in an emergency, pacing must be done through the intact chest wall, amplitudes 10 times as great are required. These higher values of current are often painful and may cause burns and contractions of the chest muscles and diaphragm. The amplitude of impulse required to capture the heartbeat of a patient is affected by the duration of the pulse. For example, an impulse of 2-msec duration may capture when its amplitude is only 3 mA. On the other hand, a 0.8-msec pulse may reach 6 mA before capture occurs.

The ability to capture and hence the threshold value of a pacer impulse that has a given amplitude and duration also depend on the electrical quality of the contact between the electrode and the heart. Capture will occur at a higher threshold value for a poor contact than for a good electrical contact at the electrode–heart muscle interface.

The quality of the electrode–heart muscle contact also affects a demand pacer's inhibition capability or *sensitivity*. A good contact will permit the pacer's output to remain inhibited for smaller values of sensed R waves.

The performance of the pulse generators can be checked with the use of a special tester. In one type of tester, pacing impulses are indicated by a lamp that blinks at the pacing rate. In another type, the pacer's pulse rate, amplitude, width, and interval are displayed in digital form. This type of tester is also able to generate impulses used to check the inhibition capability of a demand pacer.

Typical internal pulse generators are shown in Figure 7.21(a). The Xyrel Models 5972 and 5973 pulse generators are of the ventricular-inhibited (demand) type. Programmed from the QRS complex, they deliver their impulses only when the patient's ventricular rate falls below the basic pacing rate of the pulse generator. Rate is preset during manufacture at a typical 72 pulses per minute (ppm). The Model 5972 is a bipolar pulse generator. The 5973 is a unipolar pulse generator. *Unipolar electrodes* have one electrode placed on or in the heart and the other (reference) electrode located somewhere away from the heart, whereas *bipolar electrodes* have both electrodes on or in the heart.

The pulse generators are powered by a hermetically sealed lithium–iodine power source and utilize hermetically sealed hybrid electronic circuitry. To further protect the components of the pulse generator from intrusion of body fluids the electronics assembly and power source are encapsulated and hermetically sealed within a titanium shield.

Nominal dimensions of the circular-shaped pulse generators are 56 mm (2.2 in.) in diameter by 18 mm (0.71 in.) in thickness. Weight is a nominal 95 grams.

Both pulse generators have a self-sealing connector assembly with a corrosion-resistant titanium-alloy body and socket setscrew(s). To help prevent potential migration or rotational complications, the pulse generators have a suture pad which enables the physician to secure the pulse generator within the pocket.

The power source is rated at 5.6 V with 1.1-Ahour capacity and is expected to last for 7 to 10 years with continuous pacing at 72 pulses per minute. Two power-source-depletion indicators are programmed into the circuitry of the pulse generators—a rate decrease and a pulse duration increase. The decrease in rate occurs when voltage has been depleted to about 4.0 V. At this point, replacement of the pulse generator is indicated. The increase in pulse duration, which serves as a secondary power-source-depletion indicator, is gradual and occurs simultaneously with the depletion of the lithium–iodine power source.

(a)

Figure 7.21. Internal pacemaker. (a) Photograph of two units. (b) Block diagram. (Courtesy of Medtronic, Inc., Minneapolis, MN.)

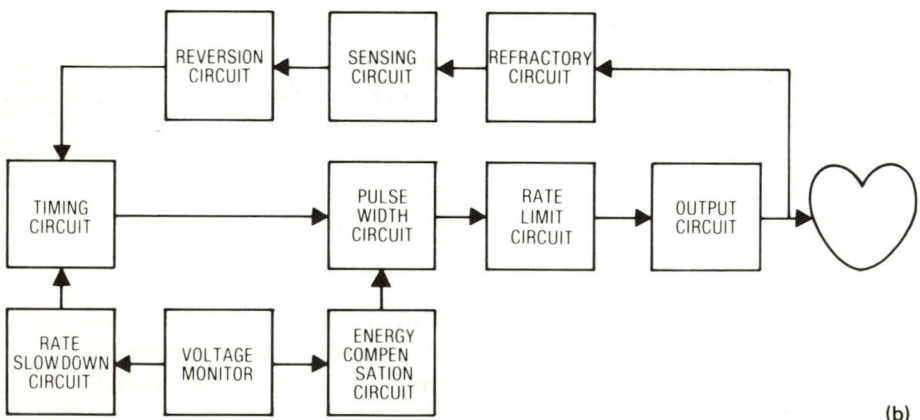

(b)

Models 5972 and 5973 have a rate-limit circuit which prevents the rate from going above 120 ppm for most single-component failures. In the presence of strong continuous interference, the pulse generators are designed to revert to asynchronous operation. The reversion rate is approximately the same as the basic pacing rate.

The pacing function of the pulse generators can be verified during periods of sinus rhythm (when the pulse generator's output is suppressed) by means of a magnet held against the skin over the implanted pulse generator. The rate with the application of the magnet can be slightly higher than the basic pacing rate.

Radiopaque identification permits positive determination of model and series number at all times. With standard X-ray procedures, the five-character code inside the titanium shield appears as black letters and numerals on a white background.

Figure 7.21(b) is a block diagram showing components of the circuitry. The timing circuit which consists of an RC network, a reference voltage source, and a comparator determines the basic pacing rate of the pulse generator. Its output signal feeds into a second RC network, the pulse width circuit, which determines the stimulating pulse duration. A third RC network, the rate-limiting circuit, disables the comparator for a preset interval and thus limits the pacing rate to a maximum of 120 pulses per minute for most single-component failures. The output circuit provides a voltage pulse to stimulate the heart. The voltage monitor circuit senses cell depletion and signals the rate slowdown circuit and energy compensation circuit of this event. The rate slowdown circuit shuts off some of the current to the basic timing network to cause the rate to slow down 8 ± 3 beats per minute when cell depletion has occurred. The energy-compensation circuit causes the pulse duration to increase as the battery voltage decreases, to maintain nearly constant stimulation energy to the heart.

There is also a feedback loop from the output circuit to the refractory circuit, which provides a period of time following an output pulse or a sensed R-wave during which the amplifier will not respond to outside signals. The sensing circuit detects a spontaneous R wave and resets the oscillator timing capacitor. The reversion circuit allows the amplifier to detect a spontaneous R wave in the presence of low-level continuous wave interference. In the absence of an R wave, this circuit allows the oscillator to pace at its preset rate ± 1 beat per minute.

7.5.3. Power Sources and Electromagnetic Interference

The type of power source used for a pulse generator depends on whether the unit is an external or an implantable type. Today most of the manufactured external pulse generators are battery-powered, although earlier units that receive power from the ac power line are still in use. Because of the need to electrically isolate patients with direct-wire connections to their hearts from any possible source of power-line leakage current (see Chapter 16), and for portability, battery-powered units are preferred. Implantable pulse generators commonly use mercury batteries whose life span ranges between 2 and 3 years, after which a new pulse generator must be installed. Recognizing the need to develop longer-lasting batteries for pacing use, the industry has developed the lithium–iodine battery, which has an estimated life expectancy of 5 years. A pulse generator with rechargeable batteries whose life span is estimated at 10 years is now available.

For a short period of time once each week, the patient dons a vest, thus ensuring the correct positioning of a charging head over the implanted pulse generator. Through magnetic coupling between the charging head and the pacer, the pacer's batteries can be recharged. Afterward, the pacer signals the charging unit that the process is completed. The weekly charge provides a pacing safety margin of approximately 6 weeks.

Another technological advance in implantable power sources has been the introduction of nuclear-powered pulse generators. In these devices, heat generated by the decay of radioactive plutonium is converted into direct current that is used to power the pacemaker. These units have an estimated useful life of at least 10 years, with negligible radiation danger to the patient.

Sources of electromagnetic energy, such as microwave ovens, diathermy, electrosurgical units, and auto ignition systems, may affect the operating mode of implanted or external pacemakers. Under certain circumstances such electrical noise signals may be strong enough to mimic the R wave in demand-mode pacers, thus inhibiting their outputs. Some implantable units are shielded to minimize the effects of extraneous noise. Nevertheless, patients with demand pacers should be warned about approaching microwave ovens or other obvious sources of electrical interference.

7.6. DEFIBRILLATORS

As discussed earlier in this chapter, the heart is able to perform its important pumping function only through precisely synchronized action of the heart muscle fibers. The rapid spread of action potentials over the surface of the atria causes these two chambers of the heart to contract together and pump blood through the two atrioventricular valves into the ventricles. After a critical time delay, the powerful ventricular muscles are synchronously activated to pump blood through the pulmonary and systemic circulatory systems. A condition in which this necessary synchronism is lost is known as *fibrillation*. During fibrillation the normal rhythmic contractions of either the atria or the ventricles are replaced by rapid irregular twitching of the muscular wall. Fibrillation of atrial muscles is called *atrial fibrillation;* fibrillation of the ventricles is known as *ventricular fibrillation.*

Under conditions of atrial fibrillation, the ventricles can still function normally, but they respond with an irregular rhythm to the nonsynchronized bombardment of electrical stimulation from the fibrillating atria. Since most of the blood flow into the ventricles occurs before atrial contraction, there is still blood for the ventricles to pump. Thus, even with atrial fibrillation circulation is still maintained, although not as efficiently. The sensation produced, however, by the fibrillating atria and irregular ventricular action can be quite traumatic for the patient.

Ventricular fibrillation is far more dangerous, for under this condition the ventricles are unable to pump blood; and if the fibrillation is not corrected, death will usually occur within a few minutes. Unfortunately, fibrillation, once begun, is not self-correcting. Hence, a patient susceptible to ventricular fibrillation must be watched continuously so that the medical staff can respond immediately if an emergency occurs. This is one of the reasons for cardiac monitoring, which was discussed earlier.

Although mechanical methods (heart massage) for defibrillating patients have been tried over the years, the most successful method of defibrillation is the application of an electric shock to the area of the heart. If sufficient current to stimulate all musculature of the heart simultaneously is applied for a brief period and then released, all the heart muscle fibers enter their refractory periods together, after which normal heart action may resume. The discovery of this phenomenon led to the rather widespread use of defibrillation by applying a brief (0.25 to 1 sec) burst of 60-Hz ac at an intensity of around 6 A to the chest of the patient through appropriate electrodes. This application of an electrical shock to resynchronize the heart is sometimes called *countershock.* If the patient does not respond, the burst is repeated until defibrillation occurs. This method of countershock was known as *ac defibrillation.*

There are a number of disadvantages in using ac defibrillation, however. Successive attempts to correct ventricular fibrillation are often required. Moreover, ac defibrillation cannot be successfully used to correct atrial defibrillation. In fact, attempts to correct atrial fibrillation by this method often result in the more serious ventricular fibrillation. Thus, ac defibrillation is no longer used.

About 1960, a number of experimenters began working with direct-current defibrillation. Various schemes and waveforms were tried until, in late 1962, Bernard Lown of the Harvard School of Public Health and Peter Bent Brigham Hospital developed a new method of *dc defibrillation* that has found common use today. In this method, a capacitor is charged to a high dc voltage and then rapidly discharged through electrodes across the chest of the patient.

It was found that dc defibrillation is not only more successful than the ac method in correcting ventricular fibrillation, but it can also be used successfully for correcting atrial fibrillation and other types of arrhythmias. The dc method requires fewer repetitions and is less likely to harm the patient. A dc defibrillator is shown in Figure 7.22 with a typical dc defibrillator circuit shown in Figure 7.23.

Depending on the defibrillator energy setting, the amount of electrical energy discharged by the capacitor may range between 100 and 400 W-sec, or joules. The duration of the effective portion of the discharge is approximately 5 msec. The energy delivered is represented by the typical waveform shown in Figure 7.24 as a time plot of the current forced to flow through the thoracic cavity. The area under the curve is proportional to the energy delivered. It can be seen that the peak value of current is nearly 20 A and that the wave is essentially *monophasic,* since most of its excursion is above the baseline. An inductor in the defibrillator is used to shape the wave in order to eliminate a sharp, undesirable current spike that would otherwise occur at the beginning of the discharge.

Figure 7.22. DC defibrillator with paddles. This portable unit incorporates a defibrillator, electrocardioscope and pacemaker. (Courtesy of Gould Medical Systems, Sunnyvale, CA.)

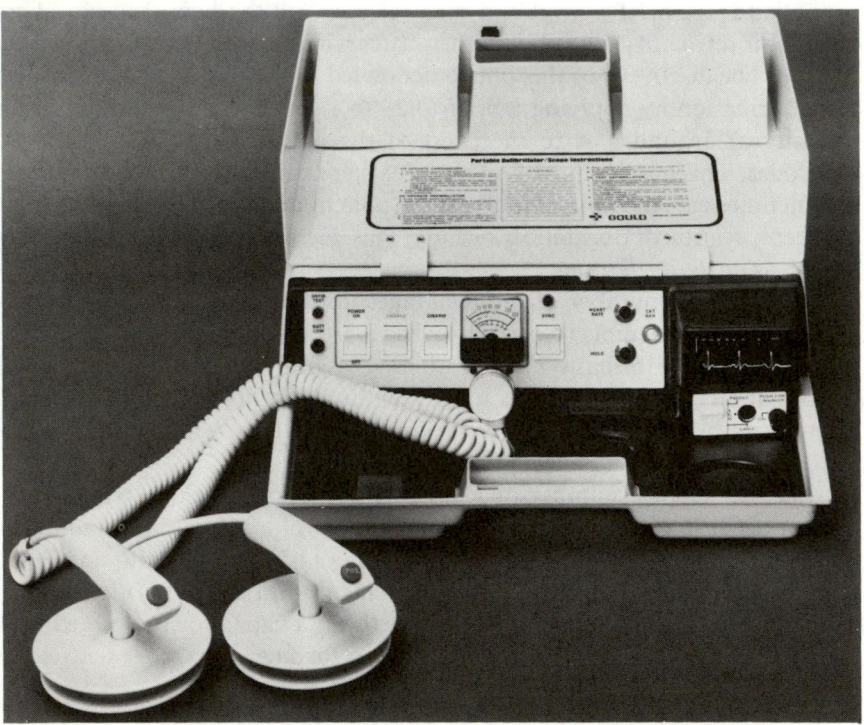

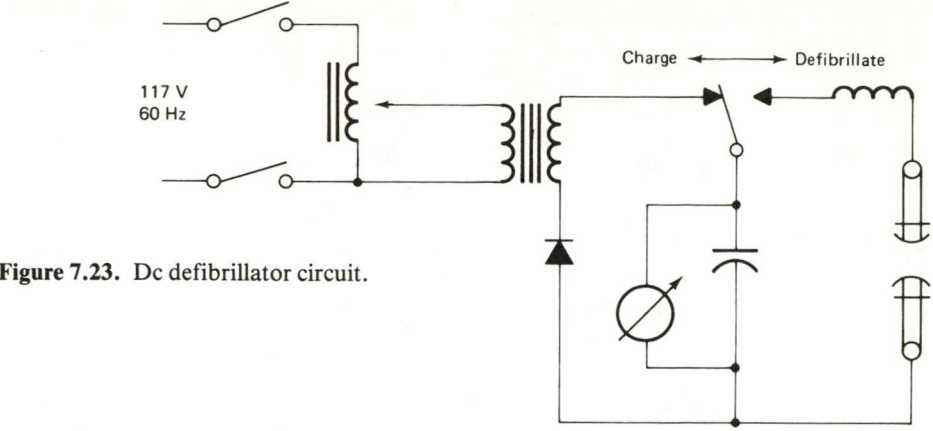

Figure 7.23. Dc defibrillator circuit.

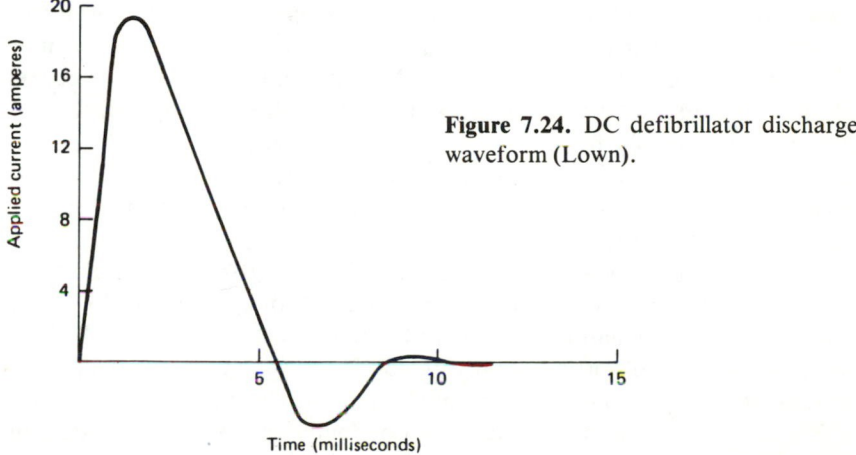

Figure 7.24. DC defibrillator discharge waveform (Lown).

Even with dc defibrillation, there is danger of damage to the myocardium and the chest walls because peak voltages as high as 6000 V may be used. To reduce this risk, some defibrillators produce dual-peak waveforms of longer duration (approximately 10 msec) at a much lower voltage. When this type of waveform is used, effective defibrillation can be achieved in adults with lower levels of delivered energy (between 50 and 200 W-sec). A typical dual-peak waveform is shown in Figure 7.25.

Effective defibrillation at the desirable lower-voltage levels is also possible with the *truncated* waveform shown in Figure 7.26. The amplitude of this waveform is relatively constant, but its duration may be varied to obtain the amount of energy required. To properly deliver a large current discharge applied through the skin large electrodes are used. These electrodes, called *paddles,* have metal disks that usually measure from 8 to 10 cm (3 to

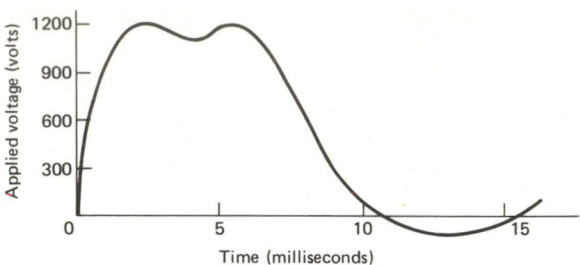

Figure 7.25. Dual-peak monophasic defibrillator discharge waveform.

4 in.) in diameter for external (transthoracic) use. For internal use (direct contact with the heart) or for use on infants, smaller paddles are applied. In external use, a pair of electrodes is firmly pressed against the patient's chest. Conductive jelly or a saline-soaked gauze pad (the latter is preferred) is applied between each paddle surface and the skin to prevent burning. However, if conductive jelly is applied to the paddles prior to electrode placement, care must be taken that when the paddles are applied, the jelly does not accidentally form a conductive bridge between the paddles. If it does, the defibrillation attempt may not be successful. With either of the preceding conductive materials, care must be taken that they will not dry out with repeated discharges.

To protect the person applying the electrodes from accidental electric shock, special insulated handles are provided. A thumb switch, located in one (or both) of the handles, is generally used to discharge the defibrillator when the paddles are properly positioned. This device prevents the patient, or someone else, from receiving a shock prematurely. In earlier equipment, a foot switch was used instead. The possibility of someone accidentally stepping on the foot switch in the excitement of an emergency, before the paddles are in place, makes the thumb switches in the handles preferable.

Figure 7.26. Truncated defibrillator discharge waveform.

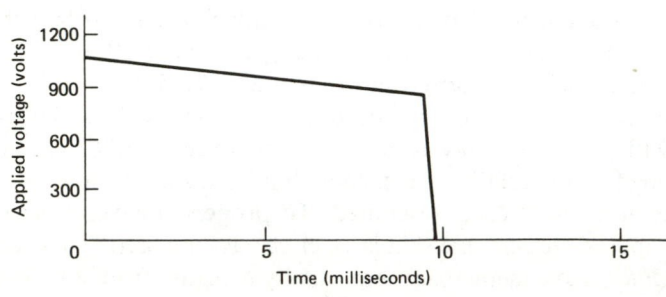

The method by which defibrillators are programmed to become charged (or recharged after use) varies widely. For example, in some defibrillators the charging process is accomplished by means of a charge switch (or push-button) located on the front panel of the unit. A newer model, however, has the charge switch located in the handle of one of its paddles. In a few defribrillators the charging process begins automatically (and immediately) after discharge. Whatever the method, it is important that the person using the defibrillator follows the manufacturer's instructions. Additionally, to ensure the safety of the medical team that is immediately caring for the patient, the user should verbally indicate that the defibrillator is about to be discharged.

The two defibrillator electrodes applied to the thoracic walls are called either *anterior-anterior* or *anterior-posterior* paddles. With anterior-anterior paddles, both paddles are applied to the chest. Anterior-posterior paddles are applied to both the patient's chest wall and back so that the energy is delivered through the heart. This method of paddle application offers better control over arrhythmias that occur as a result of atrial activity. A pair of anterior-posterior paddles consists of the anterior paddle already described and a flat posterior paddle that has a larger electrode diameter than the anterior paddle. Internal paddles, as mentioned above, may be applied directly to the myocardium (during open-chest surgery), or they may be applied to the chest of an infant. Such paddles are flat and are usually available in several sizes, with diameters ranging from 5 to 10 cm. In these applications, the energy levels required for defibrillation may range from 10 to 50 W-sec. Special *pediatric paddles* are available with diameters ranging from 2 to 6 cm. Internal paddles can be either gas-sterilized or autoclaved.

Most defibrillators include watt-second (or joule) meters to indicate the amount of energy stored in the capacitor prior to discharge. For some defibrillators, however, this indication does not assure that the amount of energy to which the unit is set by the user will, in fact, be delivered to a patient. Some of the energy indicated on the meter is lost or dissipated as heat in components (mainly inductors) inside the unit and, to a lesser extent, at the electrode–skin interface. As a result, the patient always receives less energy than the amount indicated on the meter. For example, a user may set a defibrillator to deliver 300 W-sec of energy to a patient (as shown by the meter). The actual amount of energy delivered, however, may be only 240 W-sec, which represents a 20-percent loss of energy. In some defibrillators, the loss may reach 40 percent or more. This inherent drawback of some defibrillators makes it difficult to determine accurately the amount of energy needed for various countershock procedures. For this type of defibrillator, a calibration chart must be prepared as an aid in setting the unit to accurate levels of delivered energy.

Within the last few years, defibrillators whose delivered energy levels essentially equal their preset levels have become available. Their output waveforms are of types shown in Figures 7.25 and 7.26.

Because of the large amount of energy released into the body, an implanted pacemaker pulse generator located immediately beneath a defibrillator paddle could be damaged during a discharge. Furthermore, the lump beneath the skin may reduce the effective skin contact area of the paddle and increase the danger of burns. Thus, care should be taken to avoid placement of a paddle over or near the pulse generator.

Defibrillators are also used to convert other potentially dangerous arrhythmias to one that is easily managed. This process is referred to as *cardioversion*. For this procedure, anterior-posterior paddles are generally used. For example, a defibrillator discharge may be used to convert a *tachycardia* (fast heart) arrhythmia to a normal rhythm. Unlike the ECG for a heart in ventricular fibrillation, the electrocardiogram for a fast heart contains QRS complexes. To avoid the possibility of ventricular fibrillation resulting from the application of the dc pulse in cardioversion, the discharge must be synchronized with the electrocardiogram. The optimum time for discharge is during or immediately after the downward slope of the R wave when the heart is in its absolute refractory period (see Chapter 3). This synchronization will ensure that the countershock is not delivered during the middle of the T wave, which is called the heart's *vulnerable period*. During this time, since it is partially refractory, the heart is susceptible to ventricular fibrillation by the introduction of artificial stimuli.

Most modern defibrillators include a provision for synchronizing the discharge pulse with the patient's ECG. The ECG signal is fed to an amplifier from either a patient monitor or an electrocardiograph. In some cases, the patient's ECG electrodes are connected directly to the amplifier. When properly programmed, the defibrillator will discharge only at the desired portion of the ECG waveform. The closing of the thumb switches on the paddles applied to the patient allows the defibrillator to discharge at the next occurrence of the R wave.

8

Measurements in the Respiratory System

The exchange of gases in any biological process is termed *respiration*. To sustain life, the human body must take in oxygen, which combines with carbon, hydrogen, and various nutrients to produce heat and energy for the performance of work. As a result of this process of *metabolism,* which takes place in the cells, a certain amount of water is produced along with the principal waste product, carbon dioxide (CO_2). The entire process of taking in oxygen from the environment, transporting the oxygen to the cells, removing the carbon dioxide from the cells, and exhausting this waste product into the atmosphere must be considered within the definition of respiration.

In the human body, the tissue cells are generally not in direct contact with their external environment. Instead, the cells are bathed in fluid. This tissue fluid can be considered as the *internal environment* of the body. The cells absorb oxygen from this fluid. The circulating blood is the medium by which oxygen is brought to the internal environment. Carbon dioxide is carried from the tissue fluids by the same mechanism. The exchange of gases between the blood and the external environment takes place in the *lungs* and is termed *external respiration*.

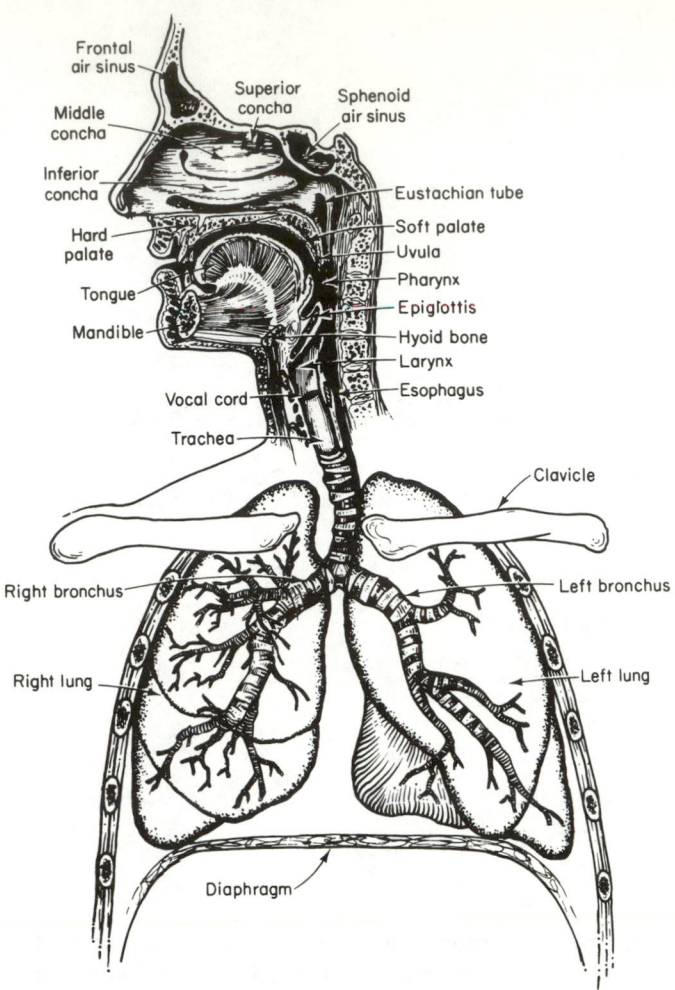

Figure 8.1. The respiratory tract. (From W.F. Evans, *Anatomy and Physiology, The Basic Principles,* Prentice-Hall, Inc., 1971, by permission.)

The function of the lungs is to oxygenate the blood and to eliminate carbon dioxide in a controlled manner. During inspiration fresh air enters the respiratory tract, becomes humidified and heated to body temperature, and is mixed with the gases already present in the region comprising the trachea and bronchi (see Figure 8.1). This gas is then mixed further with the gas residing in the alveoli as it enters these small sacs in the walls of the lungs. Oxygen diffuses from the alveoli to the pulmonary capillary blood supply, whereas carbon dioxide diffuses from the blood to the alveoli. The oxygen is carried from the lungs and distributed among the various cells of the body by the blood circulation system, which also returns the carbon dioxide to the lungs. The entire process of inspiring and expiring air, exchange of gases, distribution of oxygen to the cells, and collection of CO_2 from the cells forms what is known as the *pulmonary function*. Tests for assessing the various components of the process are called *pulmonary function tests*.

Unfortunately, no single laboratory test or even a simple group of tests is capable of completely measuring pulmonary function. In fact, the field of instrumentation for obtaining pulmonary measurements is quite complex. However, tests and instrumentation for the measurement of respiration can be divided into two categories. The first includes tests designed to measure the mechanics of breathing and the physical characteristics of the lungs; the second category is involved with diffusion of gases in the lungs, the distribution of oxygen, and the collection of carbon dioxide.

This chapter begins with a brief presentation of the physiology of the respiratory system; then the tests and instrumentation associated with each of the two categories of measurements described above are covered. Because of the complexity of the field, it is almost impossible to cover all tests or all types of instrumentation used in either category. However, an attempt has been made to include the most meaningful ones, as well as those with which the biomedical engineer or technician is most likely to become associated.

The chapter closes with a section on respiratory therapy equipment, which is used to assist patients who are unable to maintain normal respiration by natural processes.

8.1. THE PHYSIOLOGY OF THE RESPIRATORY SYSTEM

Air enters the lungs through the air passages, which include the *nasal cavities, pharynx, larynx, trachea, bronchi,* and *bronchioles,* as shown in Figure 8.1.

The lungs are elastic bags located in a closed cavity, called the *thorax* or *thoracic cavity*. The right lung consists of three lobes (upper, middle, and lower), and the left lung has two lobes (upper and lower).

The *larynx,* sometimes called the "voice box" (because it contains the vocal cords), is connected to the bronchi through the *trachea,* sometimes called the "windpipe." Above the larynx is the *epiglottis,* a valve that closes whenever a person swallows, so that food and liquids are directed to the esophagus (tube leading to the stomach) and into the stomach rather than into the larynx and trachea.

The trachea is about 1.5 to 2.5 cm in diameter and approximately 11 cm long, extending from the larynx to the upper boundary of the chest. Here it bifurcates (forks) into the right and left main stem *bronchi*. Each bronchus enters into the corresponding lung and divides like the limbs of a tree into smaller branches. The branches are of unequal length and at different angles, with over 20 of these nonsymmetrical bifurcations normally present in the human body. Farther along these branchings, where the diameter is reduced to about 0.1 cm, the air-conducting tubes are called *bronchioles*. As they continue to decrease in size to about 0.05 cm in diameter, they form the *terminal bronchioles,* which branch again into the *respiratory bronchioles,* where some alveoli are attached as small air sacs in the walls of the lung. After some additional branching, these air sacs increase in number, becoming the *pulmonary alveoli*. The alveoli are each about 0.02 cm in diameter. It is estimated that, all told, some 300 million alveoli are found in the lungs (see Figure 8.2).

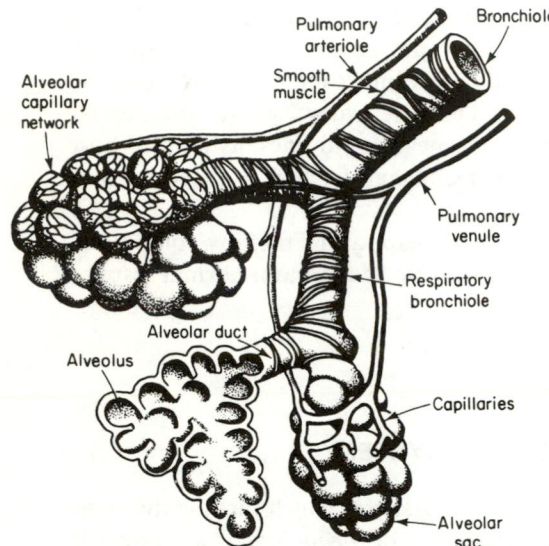

Figure 8.2. Alveoli and capillary network (From W.F. Evans, *Anatomy and Physiology, The Basic Principles,* Prentice-Hall, Inc., 1971, by permission.)

Beyond about the tenth stage of branching, the bronchioles are embedded within alveolar lung tissue; and with the expansion and relaxation of the lung, their diameters are greatly affected by the lung size or lung volume. Up to this point, the diameter of the air sacs is more affected by the *pleural pressure,* the pressure inside the thorax.

The lungs are covered by a thin membrane called the *pleura,* which passes from the lung at its root onto the interior of the chest wall and upper surface of the diaphragm. The two membranous sacs so formed are called the *pleural cavities,* one on each side of the chest, between the lungs and the thoracic boundaries. These "cavities" are potential only, for the pleura covering the lung and that lining the chest are in contact in the healthy condition. Fluid or blood, as well as air, may collect in this potential space to create an actual space in certain diseases. The part of the pleural membrane lining the thoracic wall is called the *parietal pleura,* whereas that portion covering and firmly adherent to the surface of the lungs themselves is called the *pulmonary pleura* or *visceral pleura.* A small amount of fluid, just wetting the surfaces between the pleura, allows the lungs and the lobes of the lungs to slide over each other and on the chest wall easily with breathing.

Breathing is accomplished by musculature that literally changes the volume of the thoracic cavity and, in so doing, creates negative and positive pressures that move air into and out of the lungs. Two sets of muscles are involved: those in and near the diaphragm that cause the diaphragm to move up and down, changing the size of the thoracic cavity in the vertical direction, and those that move the rib cage up and down to change the lateral diameter of the thorax.

The *diaphragm* is a special dome- or bell-shaped muscle located at the bottom of the thoracic cavity, which, when contracted, pulls downward to enlarge the thorax. This action is the prinicpal force involved in inspiration. At the same time as the diaphragm moves downward, a group of external intercostal muscles lifts the rib cage and sternum. Because of the shape of the rib cage, this lifting action also increases the effective diameter of the thoracic cavity. The resultant increase in thoracic volume creates a negative pressure (vacuum) in the thorax. Since the thorax is a closed chamber and the only opening to the outside is from the inside of the lungs, the negative pressure is relieved by air entering the lungs. The lungs themselves are passive and expand only because of the internal pressure of air in the lungs, which is greater than the pressure in the thorax outside the lungs.

Normal expiration is essentially passive, for, on release of the inspiratory muscles, the elasticity of the lungs and the rib cage, combined with the tone of the diaphragm, reduces the volume of the thorax, thereby developing a positive pressure that forces air out of the lungs. In forced expiration a set of abdominal muscles pushes the diaphragm upward very powerfully while the internal intercostal muscles pull the rib cage downward and apply pressure against the lungs to help force air out.

During normal inspiration the pressure inside the lungs, the *intraalveolar pressure,* is about —3 mm Hg, whereas during expiration the pressure becomes about + 3 mm Hg. The ability of the lungs and thorax to expand during breathing is called the *compliance,* which is expressed as the volume increase in the lungs per unit increase in intra-alveolar pressure. The resistance to the flow of air into and out of the lungs is called *airway resistance.*

As described in Chapter 5, blood from the body tissues and their capillaries is brought via the superior and inferior vena cava into the right atrium of the heart, which in turn empties into the right ventricle. The right ventricle pumps the blood into and through the lungs in a pulsating fashion, with a systolic pressure of about 20 mm Hg and a diastolic pressure of 1 to 4 mm Hg. By perfusion, the blood passes through the pulmonary capillaries, which are in the walls of the air sacs, wherein oxygen is taken up by the red blood cells and hemoglobin. The compound formed by the oxygen and the hemoglobin is called *oxyhemoglobin.* At the same time, carbon dioxide is removed from the blood into the alveoli.

From the pulmonary capillaries, the blood is carried through the pulmonary veins to the left atrium. From here it enters the left ventricle, which pumps the blood out into the aorta at pressures of 120/80 mm Hg. It is then distributed to all the organs and muscles of the body. In the tissues, the oxyhemoglobin gives up its oxygen, while carbon dioxide diffuses into the blood from the tissue and surrounding fluids. The blood then flows from the capillaries into the venous system back into the superior and inferior vena cava.

The interchange of the oxygen from the lungs to the blood and the diffusion of carbon dioxide from the blood to the lungs take place in the capillary surfaces of the alveoli. The alveolar surface area is about 80 m^2, of which more than three-fourths is capillary surface.

In order to understand some of the terminology used in conjunction with the tests and instrumentation involved in respiratory measurements, definition of a few medical terms is necessary. Additional definitions are included in the glossary in Appendix A.

Hypoventilation is a condition of insufficient ventilation by an individual to maintain his normal P_{CO_2} level, whereas *hyperventilation* refers to abnormally prolonged, rapid, or deep breathing. Hyperventilation is also the condition produced by overbreathing. *Dyspnea* is the sensation of inadequate or distressful respiration, a condition of abnormal breathlessness. *Hypercapnia* is an excess amount of CO_2 in the system, and *hypoxia* is a shortage of oxygen. Both hypercapnia and hypoxia can result from inadequate ventilation.

8.2. TESTS AND INSTRUMENTATION FOR THE MECHANICS OF BREATHING

The mechanics of breathing concern the ability of a person to bring air into his lungs from the outside atmosphere and to exhaust air from the lungs. This ability is affected by the various components of the air passages, the diaphragm and associated muscles, the rib cage and associated musculature, and the characteristics of the lungs themselves. Tests can be performed to assess each of these factors, but no one measurement has been devised that can adequately and completely evaluate the performance of the breathing mechanism. This section describes a number of the most prominent measurements and tests that are used clinically and in research in connection with the mechanics of breathing. In addition, the instrumentation required for these tests and measurements is described and discussed. In some cases, one instrument can be used for the performance of several tests.

8.2.1. Lung Volumes and Capacities

Among the basic pulmonary tests are those designed for determination of lung volumes and capacities. These parameters, which are a function of an individual's physical characteristics and the condition of his breathing mechanism, are given in Figure 8.3.

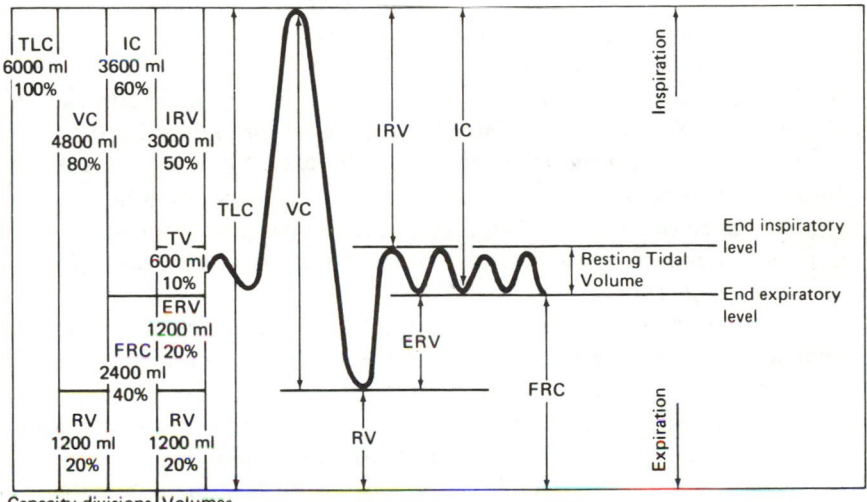

Figure 8.3. Lung volumes and capacities. (From W.F. Evans, *Anatomy and Physiology, The Basic Principles,* Prentice-Hall, Inc., 1971, by permission.)

The *tidal volume* (TV), or normal depth of breathing, is the volume of gas inspired or expired during each normal, quiet, respiration cycle.

Inspiratory reserve volume (IRV) is the extra volume of gas that a person can inspire with maximal effort after reaching the normal end inspiratory level. The *end inspiratory level* is the level reached at the end of a normal, quiet inspiration.

The *expiratory reserve volume* (ERV) is that extra volume of gas that can be expired with maximum effort beyond the end expiratory level. The *end expiratory level* is the level reached at the end of a normal, quiet expiration.

The *residual volume* (RV) is the volume of gas remaining in the lungs at the end of a maximal expiration.

The *vital capacity* (VC) is the maximum volume of gas that can be expelled from the lungs by forceful effort after a maximal inspiration. It is actually the difference between the level of maximum inspiration and the residual volume, and it is measured without respect to time. The vital capacity is also the sum of the tidal volume, inspiratory reserve volume, and expiratory reserve volume.

The *total lung capacity* (TLC) is the amount of gas contained in the lungs at the end of a maximal inspiration. It is the sum of the vital capacity and residual volume. Total lung capacity is also the sum of the tidal volume, inspiratory reserve volume, expiratory reserve volume, and residual volume.

The *inspiratory capacity* (IC) is the maximum amount of gas that can be inspired after reaching the end expiratory level. It is the sum of the tidal volume and the inspiratory reserve volume.

The functional residual capacity, often referred to by its abbreviation, FRC, is the volume of gas remaining in the lungs at the end expiratory level. It is the sum of the residual volume and the expiratory reserve volume. The FRC can also be calculated as the total lung capacity minus the inspiratory capacity, and it is often regarded as the baseline from which other volumes and capacities are determined, for it seems to be more stable than the end inspiratory level.

In addition to the static volumes and capacities given above, several dynamic measures are used to assess the breathing mechanism. These measures are important because breathing is, in fact, a dynamic process, and the rate at which gases can be exchanged with the blood is a direct function of the rate at which air can be inspired and expired.

A measure of the overall output of the respiratory system is the *respiratory minute volume.* This is a measure of the amount of air inspired during 1 minute at rest. It is obtained by multiplying the tidal volume by the number of respiratory cycles per minute.

A number of forced breathing tests are used to assess the muscle power associated with breathing and the resistance of the airway. Among them is the *forced vital capacity* (FVC), which is really a vital capacity measurement taken as quickly as possible. By definition, the FVC is the total amount of air that can forcibly be expired as quickly as possible after taking the deepest possible breath. If the measurement is made with respect to the time required for the maneuver, it is called a *timed vital capacity* measurement. A measure of the maximum amount of gas that can be expelled in a given number of seconds is called the *forced expiratory volume* (FEV). This is usually given with a subscript indicating the number of seconds over which the measurement is made. For example, FEV_1 indicates the amount of air that can be blown out in 1 second following a maximum inspiration, while FEV_3 is the maximum amount of air that can be expired in 3 seconds. FEV is sometimes given as a percentage of the forced vital capacity.

Since forced vital capacity measurements are often encumbered by patient hesitation and the inertia of the instrument, a measure of the *maximum midexpiratory flow* rate may be taken. This is a flow measurement over the middle half of the forced vital capacity (from the 25 percent level to the 75 percent level). The corresponding FEV measurement is called $FEV_{25\%-75\%}$.

Another important flow measurement is the *maximal expiration flow* (MEF) rate, which is the rate during the first liter expired after 200 ml has been exhausted at the beginning of the FEV. It differs from the *peak flow,* which is the maximum rate of airflow attained during a forced expiration.

Another useful measurement for assessing the integrity of the breathing mechanism is the *maximal breathing capacity* (MBC) or *maximal voluntary ventilation* (MVV). This is a measure of the maximum amount of air that can be breathed in and blown out over a sustained interval, such as 15 or 20 seconds. A ratio of the maximal breathing capacity to the vital capacity is also of clinical interest.

In detecting obstruction of the small airways in the lungs, a procedure involving measurement of the *closing volume* is often used. The closing volume level is the volume at which certain zones within the lung cease to ventilate, presumably as the result of airway closure.

The results of many of the preceding tests are generally reported as percentages of predicted normal values. In the presentation of various respiratory volumes, the term *BTPS* is often used, indicating that the measurements were made at body temperature and ambient pressure, with the gas saturated with water vapor. Sometimes, in order to use these values in the reporting of metabolism, they must be converted to standard temperature and pressure and dry measurement conditions, indicated by the term *STPD*.

With each breath, most of the air enters the lungs to fill the alveoli. However, a certain amount of air is required to fill the various cavities of the air passages. This air is called the *dead-space air,* and the space it occupies is called the *dead space.* The amount of air that actually reaches the alveolar interface with the bloodstream with each breath is the tidal volume minus the volume of the dead space. The respiratory minute volume can be broken down into the alveolar *ventilation per minute* and the *dead space ventilation per minute.*

8.2.2. Mechanical Measurements

The volume and capacity measurements just described, particularly the forced measurements, are a good indication of the compliance of the lungs and rib cage and the resistance of the air passages. However, direct measurement of these parameters is also possible and is often used in the measurement of pulmonary function.

Determination of *compliance,* which has been defined as the volume increase in the lungs per unit increase in lung pressure, requires measurement of an inspired or expired volume of gas and of intrathoracic pressure. Compliance is actually a static measurement. However, in practice, two types of compliance measurement, static and dynamic, are made. *Static compliance* is determined by obtaining a ratio of the difference in lung volume at two different volume levels and the associated difference in intra-alveolar pressure. To measure dynamic compliance, tidal volume is used as the volume measurement, while intrathoracic pressure measurements are taken during the instants of zero airflow that occur at the end inspiratory and expiratory levels with each breath (refer to Figure 8.3). The lung compliance varies with the size of the lungs; a child has a smaller compliance than an adult. Furthermore, the volume-pressure curve is not linear. Hence, compliance does not remain constant over the breathing cycle but tends to decrease as the lungs are inflated. Fortunately, over the tidal volume range in which dynamic compliance measurements are usually performed, the relationship is approximately linear and a constant compliance is assumed. Compliance values are given as liters per centimeter H_2O.

Resistance of the air passages is generally called *airway resistance,* which is a pneumatic analog of hydraulic or electrical resistance and, as such, is a ratio of pressure to flow. Thus, for the determination of airway resistance, intra-alveolar pressure and airflow measurements are required. As was the case with compliance, airway resistance is not constant over the respiratory cycle. As the pressure in the thoracic cavity becomes more negative, the airways are widened and the airway resistance is lowered. Conversely, during expiration, when the pressure in the thorax becomes positive, the airways are narrowed and resistance is increased. The intra-alveolar pressure is given in centimeters H_2O and the flow in liters per second; the airway resistance is expressed in centimeters H_2O per liter per second. Most airway resistance measurements are made at or near the functional residual capacity (end expiratory) level.

From the preceding discussion it can be seen that to obtain compliance and airway resistance determinations, volume, intra-alveolar pressure, intrathoracic pressure, and instantaneous airflow measurements are required. The methods for measurement of volume for these determinations are no different from those used for the volume and capacity measurements discussed earlier.

8.2.3. Instrumentation for Measuring the Mechanics of Breathing

As shown in previous sections, all the parameters dealing with the mechanics of breathing can be derived from measurement of lung volumes at various levels and conditions of breathing, pressures within the lungs and the thorax with respect to outside air pressure, and instantaneous airflow. The complexity of pulmonary measurements lies not in the variety required but rather in gaining access to the sources of these measurements and in providing suitable conditions to make them meaningful.

The most widely used laboratory instrument for respiratory volume measurements is the *recording spirometer,* an example of which is shown in Figure 8.4. All lung volumes and capacities that can be determined by measuring the amount of gas inspired or expired under a given set of conditions or during a specified time interval can be obtained by use of the spirometer. Included are the timed vital capacity and forced expiratory volume measurements. The only volume and capacity measurements that cannot be obtained with a spirometer are those requiring measurement of the gas that cannot be expelled from the lungs under any conditions. Such measurements include the residual volume, functional residual capacity, and total lung capacity.

The standard spirometer consists of a movable bell inverted over a chamber of water. Inside the bell, above the water line, is the gas that is to be breathed. The bell is counterbalanced by a weight to maintain the gas inside at atmospheric pressure so that its height above the water is proportional to the amount of gas in the bell. A breathing tube connects the mouth of the patient with the gas under the bell. Thus, as the patient breathes into the tube, the bell moves up and down with each inspiration and expiration

Figure 8.4. Spirometer. (Courtesy of Warren E. Collins, Inc., Braintree, MA.)

in proportion to the amount of air breathed in or out. Attached to the bell or the counterbalancing mechanism is a pen that writes on an adjacent drum recorder, called a *kymograph*. As the kymograph rotates, the pen traces the breathing pattern of the patient.

Various bell volumes are available, but 9 and 13.5 liters are most common. A well-designed spirometer offers little resistance to airflow, and the bell has little inertia. Various paper speeds are available for the kymograph, with 32, 160, 300, and 1920 mm/min most common. The compact spirometer shown in Figure 8.4 is a widely used instrument for pulmonary function testing. It is used both in the physician's office and in the hospital ward. Its 9-liter capacity is often considered adequate for recording the largest vital capacities, for extended-period oxygen-uptake determinations, and even for spirography during mild exercise. However, many physicians prefer the larger size (13.5 liters) because of the extra capacity. The principle of operation is similar for both. Easily removable flutter valves and a CO_2 absorbent container permit minimized breathing resistance during tests for maximal respiratory flow rates. This instrument is equally suitable for clinical spirography, for cardiopulmonary function testing, and for metabolism determinations. The instrument directly records basal minute volume, exercise ventilation, or maximum breathing capacity. The ventilation equivalent for oxygen may be calculated directly from the spirogram slope lines for ventilation and oxygen uptake.

In addition to the type of spirometer just described, and illustrated in Figure 8.4, several other types are available. For example, *waterless spirometers,* which are also used clinically, operate on a principle similar to that of the spirometer just described. One type, called the *wedge spirometer,* is shown in Figure 8.5. In this instrument the air to be breathed is held in a chamber enclosed by two parallel metal pans hinged to each other along one edge. The space between the two pans is enclosed by a flexible bellows (like a fireplace bellows) to form the chamber. One of the pans, which contains an inlet tube, is fixed to a stand and the other swings freely with respect to it. As air is introduced into the chamber or withdrawn from it, the moving pan changes its position to compensate for the volume changes. Construction is such that the pan moves in response to very slight changes in volume. A well-designed wedge spirometer imposes an almost undetectable amount of air pressure on the patient's lungs. The instrument provides electrical outputs proportional to both volume and airflow, from which the required determinations can be obtained.

In a similar type of waterless spirometer, the volume of the chamber is varied by means of a lightweight piston that moves freely in a cylinder as air is withdrawn and replaced in breathing. A Silastic rubber seal between the piston and the cylinder wall keeps the chamber airtight. Instruments of this type have characteristics similar to those of the wedge spirometer.

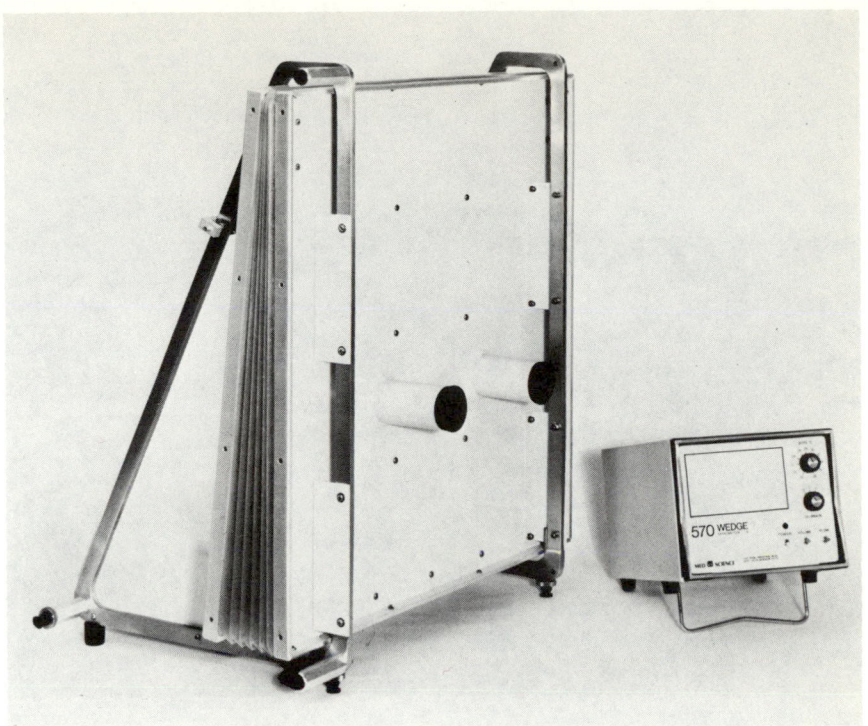

Figure 8.5. Wedge spirometer. (Courtesy of Med. Science Electronics, St. Louis, MO.)

Another group of instruments, sometimes called *electronic spirometers,* measures airflow and, by use of electronic circuitry, calculates the various volumes and capacities. Such a device is shown in Figure 8.6. This instrument provides both a graphic output similar to that of a standard spirometer and a digital readout of the desired parameters. Various types of airflow transducers are used, utilizing such devices as small breath-driven turbines and heated wires that are cooled by the breath.

A *bronchospirometer* is a dual spirometer that measures the volumes and capacities of each lung individually. The air-input device is a double-lumen tube that divides for entry into the airway to each lung and thus provides isolation for differential measurement. The main function of the bronchospirometer is the preoperative evaluation of oxygen consumption of each lung.

The usual output of a spirometer is the *spirogram.* An example is shown in Figure 8.7. The recording is read from right to left. In this particular example, inspiration moves the pen toward the bottom of the chart and expiration toward the top. Some spirometers, however, provide spirograms with inspiration toward the top.

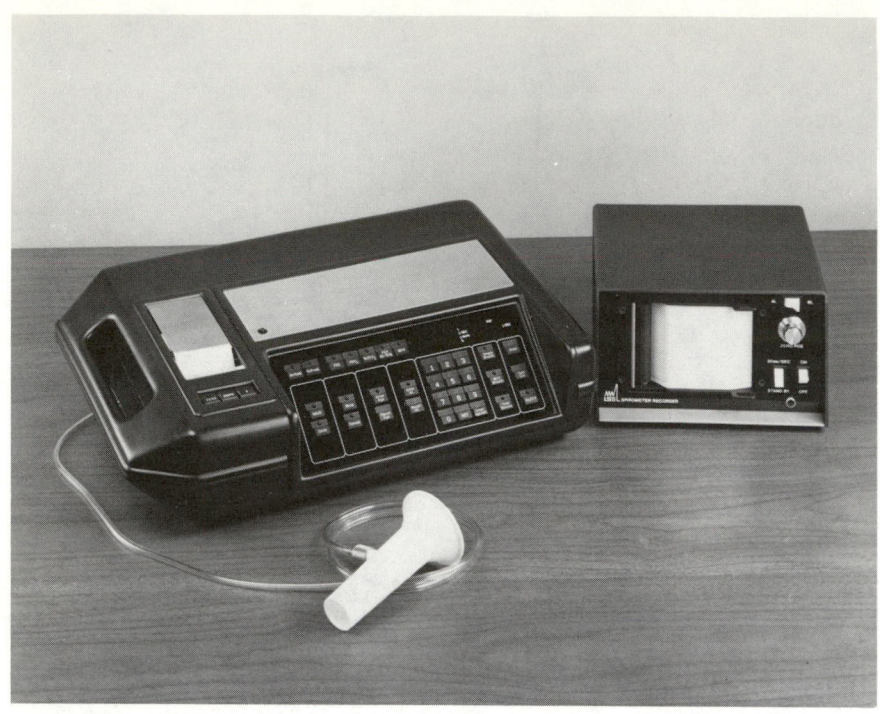

Figure 8.6. Electronic spirometer with digital readout, printed tape, and computer interface capabilities. (Courtesy of Life Support Equipment Co., Woburn, MA.)

Figure 8.7. Typical spirogram. Read right to left. (See text for explanation.)

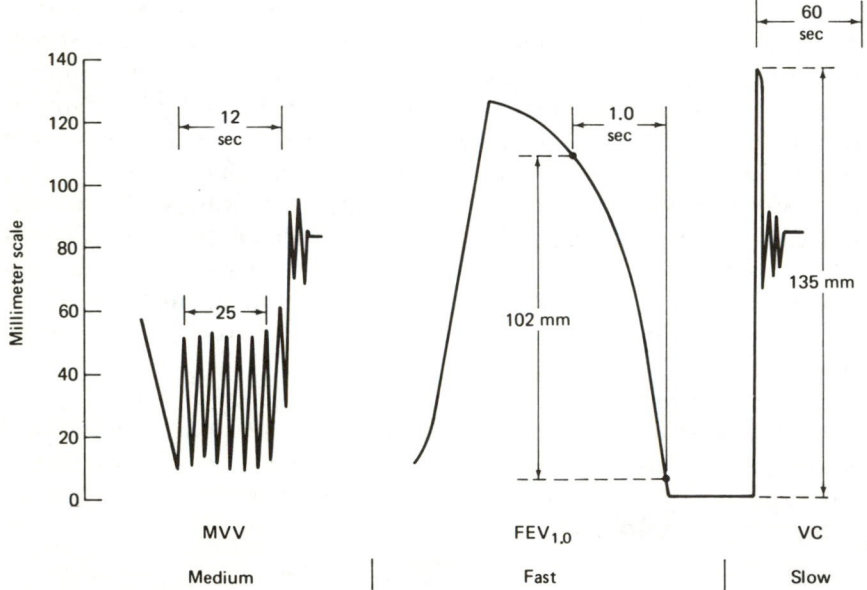

In order to produce a spirogram, the patient is instructed to breathe through the mouthpiece of the spirometer. His nose is blocked with a clip so that all breathing is through the mouth. The recorder is first set to a slow speed to measure vital capacity (typically, 32 mm/min). To produce the spirogram shown in the figure, the patient breathed quietly for a short time at rest so as to provide a baseline. He was then instructed to exhale completely and then to inhale as much as he could. This process produced the vital capacity record at the extreme right of the figure. With his lungs at the maximal inspirational level, the patient held his breath a short time while the recorder was shifted to a higher chart speed (e.g., 1920 mm/min). The patient was then instructed to blow out all the air he could as quickly as possible to produce the FEV_1 curve on the record. To calculate the FEV_1, a 1-second interval was measured from the beginning of the maximum slope. Sometimes it is necessary to determine the beginning point by extending the maximum slope to the level of maximum inspiration. This step ensures that the initial friction and inertia of the spirometer have been overcome and compensates for error on the part of the patient in performing the test as instructed.

The spirogram in Figure 8.7 also shows a maximal voluntary ventilation (MVV) record. For this determination, the recorder is set at an intermediate speed. After a short rest, a few cycles of resting respiration were recorded. The patient was then instructed to breathe in and out as rapidly as possible for about 10 seconds, producing the MVV record in the figure.

Most spirometry tests are repeated two or three times, and the maximum values are used to ensure that the patient performed the test to the best of his ability. Although some instruments are calibrated for direct readout, others require that the height of the tracings be converted to liters by use of a calibration factor for the instrument, called the *spirometer factor*. This calibration factor can be obtained from a table or chart.

Although the usual output for a spirometer is the spirogram, other types of output, including digital readouts, are available, particularly from the waterless and electronic types of spirometers. Some instruments even have built-in computational capability to calculate automatically the required volumes and capacities from the basic measurements.

Incorporation of microprocessors (see Chapter 15) has resulted in instruments that not only calculate all required parameters, but also print additional information and compare the measured results with normal data based on the patient's sex, height, and weight. Figure 8.8 shows a microprocessor-based system for measurement of forced vital capacity (FVC), forced expiratory volume (FEV), forced expiratory flow (FEF), and maximal voluntary ventilation (MVV). When used in conjunction with a wedge spirometer, this instrument provides a digital readout of patient data, test results, and a pulmonary volume-flow loop, which involves both

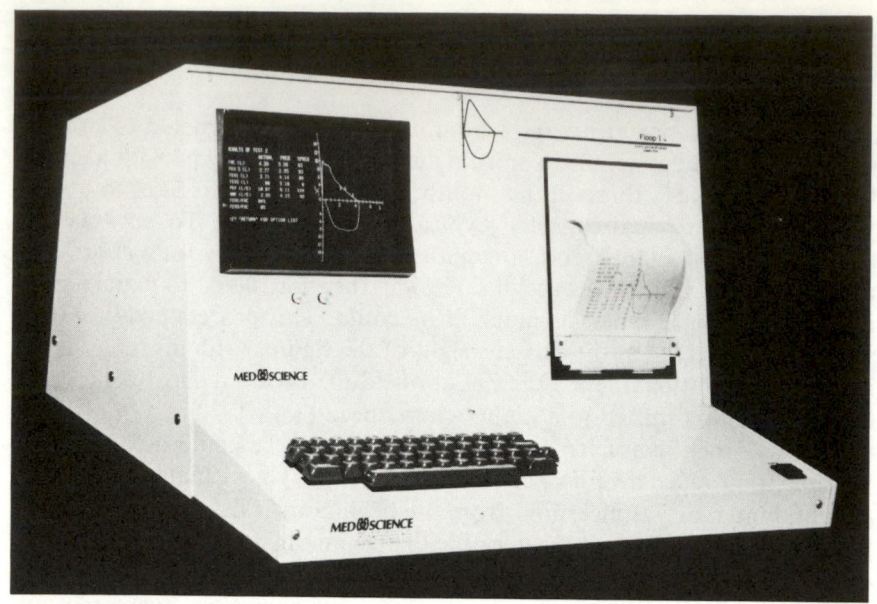

Figure 8.8. Pulmonary function studies system. (Courtesy of Med. Science Electronics, St. Louis, MO.)

compliance and airway resistance. A volume-flow loop is shown on the screen in photograph. Measured results are automatically compared with predicted normal values, based on the sex, height, and weight of the patient. In addition, the instrument is able to correct for ambient temperature and barometric pressure.

8.2.3.1. Measurement of residual volume. From the spirogram and the outputs of some of the other instruments described above, all the lung volumes and capacities can be determined except those that require measurement of the air still remaining in the lungs and airways after maximum expiration. These parameters, which include the residual volume, FRC, and total lung capacity, can be measured through the use of foreign gas mixtures. A gas analyzer is required for these tests. A description of several types of gas analyzers is presented in Section 8.3.1, which is concerned with gas distribution and diffusion.

The *closed-circuit technique* involves rebreathing from a spirometer charged with a known volume and concentration of a marker gas, such as hydrogren or helium. Helium is usually used. After several minutes of breathing, complete mixing of the spirometer and pulmonary gases is assumed, and the residual volume is calculated by a simple proportion of gas volumes and concentrations.

The *open-circuit* or *nitrogen washout method* involves the inspiration of pure oxygen and expiration into an oxygen-purged spirometer. If the patient has been breathing air, the gas remaining in his lungs is 78 percent nitrogen. As he begins to breathe the pure oxygen, it will mix with the gas still in his lungs, and a certain amount of nitrogen will "wash out" with each breath. By measuring the amount of nitrogen in each expired breath, a washout curve is obtained from which the volume of air initially in the lungs can readily be calculated. The preferred breathing level for beginning this measurement is the end expiratory level.

The functional residual capacity (FRC) (from which residual volume can also be calculated by subtracting the expiratory reserve volume) can be measured by using a *body plethysmograph*. This instrument, shown in Figure 8-9, is an airtight box in which the patient is seated. Utilizing Boyle's law (at constant temperature, the volume of gas varies inversely with the pressure), the ratio of the change in lung volume to change in mouth pressure is used to determine the thoracic gas volume. The patient breathes

Figure 8.9. Body plethysmograph. (Courtesy of Warren E. Collins, Inc., Braintree, MA.)

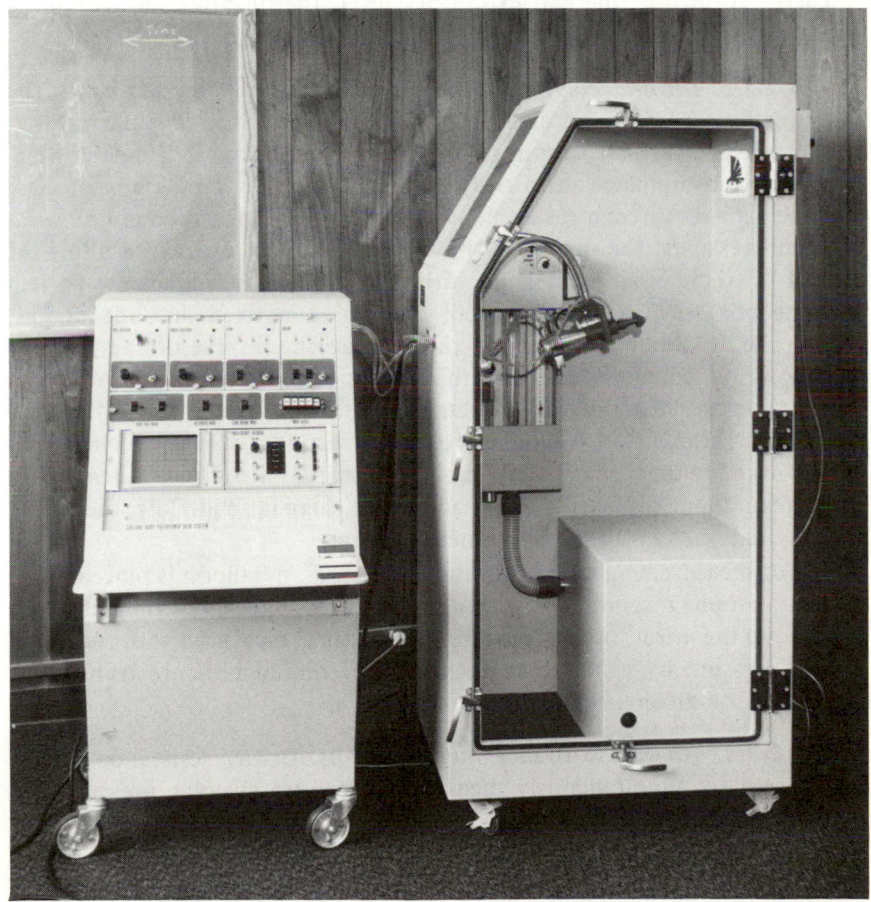

air from within the box through a tube containing an airflow transducer and a shutter to close off the tube for certain portions of the test. Pressure transducers measure the air pressure in the breathing tube on the patient's side of the shutter and inside the box. The amount of air in the box, including that in the patient's lungs, remains constant, since there is no way for air to enter or escape. However, when the patient compresses the air in his lungs during expiration, his total body volume is reduced, thus reducing the pressure in the box. Conversely, when the patient inhales by reducing the pressure in his thoracic region, his body volume increases and increases the box pressure. The FRC is measured with the shutter in the breathing tube closed. With no air allowed to flow, the mouth pressure (sensed by the transducer in the tube) can be assumed to equal the alveolar pressure. The patient is instructed to pant at a slow rate against the closed shutter. As he does so, he alternately expands and compresses the air in his lungs. By measuring the changes in mouth pressure and corresponding changes in intrathoracic volume (Equal and opposite to changes in box volume outside the patient), it is possible to calculate the intrathoracic volume. If the test is performed at the end expiratory level, the intrathoracic volume is equal to the FRC.

8.2.3.2. Intra-alveolar and intra-thoracic pressure measurements. The body plethysmograph can also be used to measure intra-alveolar and intrathoracic pressures. These measurements are important in the determination of both compliance and airway resistance, since inaccessibility of these chambers makes direct measurement impossible. For measurement of intra-alveolar pressures, the shutter in the breathing tube is opened to allow the patient to freely breathe air from within the closed box. Since the patient and the box form a closed system containing a fixed amount of gas, pressure and volume variations in the box are the inverse of the pressure variations in the lungs as the gas within the lungs expands and is compressed due to the positive and negative pressures in the lungs. For calibration, the patient's breathing tube is blocked for a few seconds, during which the patient is asked to the "pant" while mouth pressure is measured. Since mouth pressure and lung pressure are the same when there is no airflow, these data can be used in calibration of the measurement.

For measurement of intrathoracic pressures, a balloon is placed in the patient's esophagus, which is within the thoracic cage. Since the balloon is exposed to the intrathoracic pressure, its pressure, measured with respect to mouth pressure by using some form of differential pressure transducer, represents the difference between pressures.

8.2.3.3. Airway resistance measurements. Airway resistance can be determined by simultaneously measuring the intra-alveolar pressure and airflow in the body plethysmograph and by dividing the difference between the intra-alveolar pressure and the atmospheric pressure by the flow.

$$R = \frac{P_{ia} - P_A}{f}$$

where R = airway resistance
P_{ia} = intra-alveolar pressure
P_A = *atmospheric pressure*
f = airflow

A variety of instruments can be used to measure airflow. One of the most widely used is the *pneumotachometer,* often called the *pneumotachograph,* shown in Figure 8.10. This device utilizes the principle that air flowing through an orifice produces a pressure difference across the orifice that is a function of the velocity of the air. In the more common pneumotachometer, the orifice consists of a set of capillaries or a metal screen. Since the cross section of the orifice is fixed, the pressure difference can be calibrated to represent flow. Two pressure transducers or a differential pressure transducer can be used to measure the pressure difference.

Another method of measuring airflow is a transducer in which a heated wire is cooled by the flow of air, and the resistance change due to the cooling is measured as representative of airflow. Because the cooling effect is the same regardless of the direction of airflow, this transducer is insensitive to direction, whereas the pneumotachograph described above indicates not only the amount of flow but also the direction.

Ultrasonic airflow-measuring devices utilizing the Doppler effect (see Chapters 6 and 9) have been developed. Since flow is the first derivative or rate of change of volume, some volume-measuring devices also produce a measurement of flow. Also in use is a small breath-driven turbine which operates a miniature electrical generator that produces an output voltage proportional to the air velocity.

Figure 8.10. Pneumotachometer. Air inlet and outlet are at left and right. Connections at top are for pressure transducer. Black "knob" is a heating element. (Courtesy of Veterans Administration Hospital, Sepulveda, CA.)

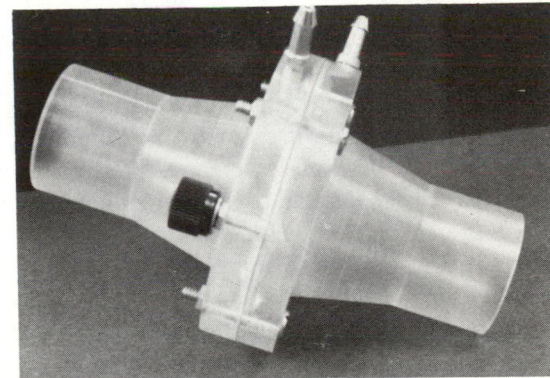

In some applications, the actual flow or volume of respiration is not required, but a measure of *respiration rate* (number of breaths per minute) is needed. Respiration rate can, of course, be obtained from any instrument that records the volume changes during the respiratory cycle. There are, however, other instruments that are difficult to calibrate for volume changes but that well serve the purpose of measuring respiration rate. Such instruments are much simpler and easier to use than the spirometer or other devices intended for volume measurements. These instruments include a mercury plethysmograph of the type described in Chapter 2 and an impedance pneumograph in which impedance changes due to respiration can be measured across the chest.

8.2.3.4. Measurement of closing volume. In measuring the *closing volume,* two techniques can be used. In the *bolus method,* a bolus of a marker gas (usually argon, xenon, or helium) is inspired at the residual volume level. The patient is instructed to inhale air until his or her maximal inspirational level is reached and then to *slowly* expire as much as possible. A person with average lung volume should complete expiration in about 8 to 10 seconds. During expiration, the concentration of the marker gas is monitored at the mouthpiece and plotted against the lung volume level.

In the second method of measuring closing volume, the residual nitrogen in the lungs is used as the marker gas. To perform the measurement, the patient fills his or her lungs with pure oxygen and exhausts all the air possible. The nitrogen concentration of the exhausted air is plotted against lung volume level. At the closing volume level, the nitrogen concentration suddenly begins to increase at a more rapid rate. The *closing volume* is the difference between that level and the residual volume level and is usually expressed as a percentage of the patient's vital capacity.

8.3.   GAS EXCHANGE AND DISTRIBUTION

Once air is in the lungs, oxygen and carbon dioxide must be exchanged between the air and the blood in the lungs and between the blood and the cells in the body tissues. In addition, the gases must be transported between the lungs and the tissue by the blood. The physiological processes involved in this overall task were presented briefly in Section 8.1. A number of tests have been devised to determine the effectiveness with which these processes are carried out. Some of these tests and the instrumentation required for their performance are described and discussed in this section. The tests connected with the exchange of gases are treated first, after which measurements pertaining to the transport of oxygen and CO_2 in the blood are covered.

8.3.1. Measurements of Gaseous Exchange and Diffusion

The mixing of gases within the lungs, the ventilation of the alveoli, and the exchange of oxygen and carbon dioxide between the air and blood in the lungs all take place through a process called *diffusion*. Diffusion is the movement of gas molecules from a point of higher pressure to a point of lower pressure to equalize the pressure difference. This process can occur when the gas is unequally distributed in a chamber or wherever a pressure difference exists in the gas on two sides of a membrane permeable to that gas.

Measurements required for determining the amount of diffusion involve the partial pressures of oxygen and carbon dioxide, P_{O2} and P_{CO2}, respectively. There are many methods by which these measurements can be obtained, including some chemical analysis methods and measurements of diffusing capacity.

8.3.1.1. Chemical analysis methods. The original gas analyzers developed by Haldane, and modified by Scholander, were of the chemical type. In these devices, a gas sample of approximately 0.5 ml is introduced into a reaction chamber by use of a transfer pipet at the upper end of the reaction chamber capillary. An indicator droplet in this capillary allows the sample to be balanced against a trapped volume of air in the thermobarometer. Absorbing fluids for CO_2 and O_2 can be transferred in from side arms without causing any change in the total volume of the system. The micrometer is adjusted so as to put mercury into the system in place of the gases being absorbed. The volume of the absorbed gases is read from the micrometer barrel calibration.

8.3.1.2. Diffusing capacity using CO infrared analyzer. To determine the efficiency of perfusion of the lungs by blood and the diffusion of gases, the most important tests are those that measure O_2, CO_2, pH, and bicarbonate in arterial blood. In trying to measure the diffusion rate of oxygen from the alveoli into the blood, it is usually assumed that all alveoli have an equal concentration of oxygen. Actually, this condition does not exist because of the unequal distribution of ventilation in the lung; hence, the terms *diffusing capacity* or *transfer factor* (rather than *diffusion*) are used to describe the transfer of oxygen from the alveoli into the pulmonary capillary blood.

Carbon monoxide (CO) resembles oxygen in its solubility and molecular weight and also combines with hemoglobin reversibly. Its affinity for hemoglobin is about 200 to 300 times that of oxygen, however. Carbon monoxide can thus be used as a tracer gas in measuring the diffusing capacity of the lung. It passes from the alveolar gas into the alveolar walls, then into the plasma, from which it enters the red blood cells, where it combines with hemoglobin.

A relationship may be obtained that is a function of both the diffusing capacity of the alveolar membrane and the rate at which CO combines with hemoglobin in the alveolar capillaries. This relationship may be expressed as follows:

$$\frac{1}{TF} = \frac{1}{D_m} + \frac{1}{\theta V_c} \quad \text{mm Hg/ml/min}$$

where TF = diffusing capacity for the lung for CO
D_m = diffusing capacity for the alveolar membrane
V_c = volume of blood in the capillaries
θ = reaction rate of CO with oxyhemoglobin

TF, the *diffusing capacity for the whole lung,* in normal adults ranges from 20 to 38 ml/min/mm Hg. It varies with depth of inspiration, increases during exercise, and decreases with anemia or low hemoglobin.

The principal methods of measuring diffusing capacity involve the inhalation of low concentrations of carbon monoxide. The concentration is less than 0.25 percent and usually ranges from 0.05 to 0.1 percent. The concentration of CO in the alveoli and the rate of its uptake into the blood per minute are measured by either the steady-state method or the single-breath method, both of which are described below. In either method, uptake of carbon monoxide is calculated by measuring the concentration and the volume of the air–CO mixture. Since the concentration of CO fluctuates throughout the respiratory cycle, end-tidal expired air is collected and the CO in the air is measured.

In the single-breath method, the last 75 to 100 ml of the expired air is collected so that enough end-tidal air containing CO is available for the measurement. CO in the alveolar gas is measured. In the steady-state method, the patient rebreathes the gas until equilibrium is reached.

The small amount of CO in the blood is negligible, for it combines with the hemoglobin in the red blood cells and exerts no significant back pressure. By estimating the P_{CO} in the blood by the rebreathing method, the diffusing capacity can be calculated as

$$TF \text{ or diffusing capacity} = \frac{\text{ml CO taken up/min}}{P_{CO} \text{ in alveoli (mm/Hg)}}$$

For this measurement, as well as for all methods requiring carbon monoxide determination, a carbon monoxide analyzer or a gas chromatograph is used. The commonly used carbon monoxide analyzer utilizes an infrared energy source, a beam chopper, sample and reference cells, plus a detector and amplifier. A milliammeter or a digital meter may be used for display.

Two infrared beams are generated, one directed through the sample and the other through the reference. The CO gas mixture flowing through the sample cell absorbs more infrared energy than does the reference gas. The two infrared beams are each measured by a differential infrared detector. The output signal is proportional to the amount of monitored gas in the sample cell. The signal is amplified and presented to the output display meter or to a recorder.

8.3.1.3. Gas chromatograph. The quantities of various gases in the expired air can also be determined by means of a *gas chromatograph*, an instrument in which the gases are separated as the air passes through a column containing various substances that interact with the gases. The reactions cause different gases to pass through the column at different rates so that they leave the column at different times. The quantity of each gas is measured as it emerges. To identify the gases in the expired air other than oxygen, nitrogen, or CO_2, a *mass spectrometer* is used in conjunction with the gas chromatograph. The mass spectrometer identifies the ions according to their mass/charge ratio.

8.3.2. Measurements of Gas Distribution

The distribution of oxygen from the lungs to the tissues and carbon dioxide from the tissues to the lungs takes place in the blood. The process by which each gas is transported, however, is quite different. As mentioned earlier, oxygen is carried by the hemoglobin of the red blood cells. On the other hand, carbon dioxide is carried through chemical processes in which CO_2 and water combine to produce carbonic acid, which is dissolved in the blood. The amount of carbonic acid in the blood, in turn, affects the pH of the blood. In assessing the performance of the blood in its ability to transport respiratory gases, then, measurements of the partial pressures of oxygen (P_{O_2}) and carbon dioxide (P_{CO_2}) in the blood, the percent of oxygenation of the hemoglobin, and the pH of the blood are most useful.

Electrodes for measurement of P_{O_2}, P_{CO_2}, and pH are described in detail in Chapter 4. These electrodes, together with amplification and readouts, provide a fairly simple method for this type of analysis. Measurements both in vitro and in vivo are possible with these electrodes. A blood gas analyzer that utilizes such electrodes and provides a digital output of the pH, P_{CO_2}, and P_{O_2} readings is shown in Figure 8.11. This device provides continuous, automatic calibration as well as checking of critical system components and reagent conditions. It can also measure respiratory gases, and a printed readout option is available. All measured and calculated results are displayed in digital form, along with calibration values.

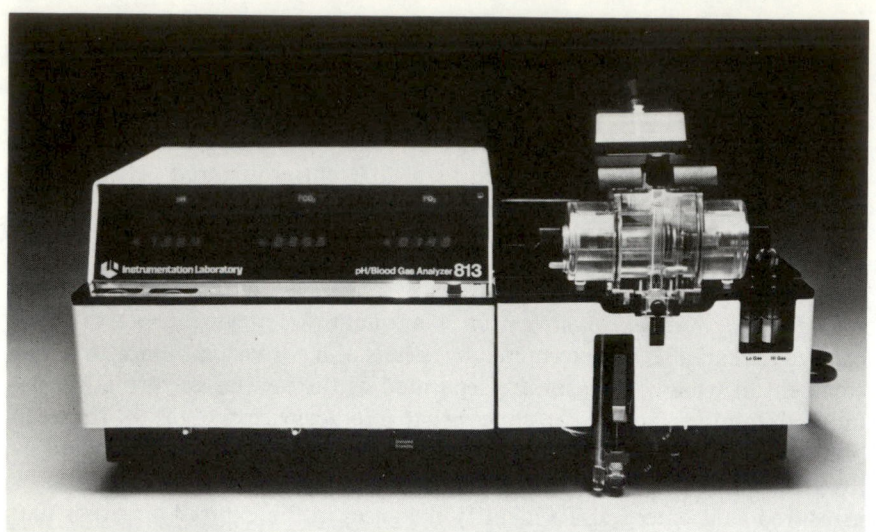

Figure 8.11. Automated digital blood gas analyzer. (Courtesy of Instrumentation Laboratory, Inc., Lexington, MA.)

Another in vitro method for analyzing both P_{O_2} and P_{CO_2} utilizes the *Van Slyke apparatus.* In this device, a measured quantity of blood is used and the O_2 and CO_2 are extracted by vacuum. The quantity of these two gases is measured manometrically, after which the CO_2 is absorbed. The quantity is measured again, the oxygen is absorbed, and the remaining gas, which is nitrogen, is measured. The amount of O_2 and CO_2 may be calculated from these measurements as a percentage of the total gas.

Another method involving the measurement of pH as part of the blood gas determination is called the *Astrup technique* and utilizes a semilogarithmic paper with a special nomogram. In this method a pH determination is made on a heparinized microsample of blood. Two other pH determinations are made on the same sample after it has been equilibrated with two known CO_2 tensions, obtained from cylinders accompanying the apparatus. These three points are plotted on special graph paper and connected by a straight line. The slope of the line is an index of the buffering capacity of the blood, which is calculated using this nomogram.

When hemoglobin is oxygenated, its light-absorption properties change as a function of the percentage of oxygen saturation. At a wavelength of 6500 Å (angstrom units), the difference in absorption between oxygenated and nonoxygenated blood is greatest, whereas at 8050 Å the absorption is the same. Thus, by measuring the absorption of a sample of blood at both wavelengths on a special photometer, the percentage of oxygenation can be determined.

A similar principle can be used to measure the percentage of oxygenation of the blood in vivo. Here an instrument called an *ear oximeter* is used. The ear oximeter is composed of an ear clip that holds a light source on one side of the earlobe and two sensors on the opposite side, so that the light passing through the earlobe is picked up by both of the sensors. As the blood in the capillaries of the earlobe changes color, these changes are reflected in the amount of light transmitted through the ear at each of the two aforementioned wavelengths. Since each of the sensors receives and filters transmitted light so that its maximum response is at one of the two wavelengths, variations in the percentage of oxygenation can be measured. This method should only be used to measure differences in oxyhemoglobin saturation rather than exact oxygen blood level or exact percentage of oxygenation.

8.4. RESPIRATORY THERAPY EQUIPMENT

When a patient is incapable of adequate ventilation by natural processes, mechanical assistance must be provided so that sufficient oxygen is delivered to the organs and tissues of the body and excessive levels of carbon dioxide are not permitted to accumulate. The procedures and instrumentation involved in providing mechanical assistance in respiration and in supplying hypoxic patients with higher-than-normal concentrations of oxygen or other therapeutic gases or medications constitute a field known as *respiratory therapy*. Until the past few years, this field was known as *inhalation therapy*, but since it covers much more than inhalation, the more encompassing term is preferred. Instruments for respiratory therapy include such devices as inhalators, ventilators, respirators, resuscitators, positive-pressure breathing apparatus, humidifiers, and nebulizers. Many of these instruments, however, have overlapping functions, and the name used for a particular device may vary among manufacturers.

8.4.1. Inhalators

The term *inhalator* generally indicates a device used to supply oxygen or some other therapeutic gas to a patient who is able to breathe spontaneously without assistance. As a rule, inhalators are used when a concentration of oxygen higher than that of air is required. The inhalator consists of a source of the therapeutic gas, equipment for reducing the pressure and controlling the flow of the gas, and a device for administering the gas. Devices for administering oxygen to patients include nasal cannulae and catheters, face masks that cover the nose and mouth, and, in certain settings, such as pediatrics, oxygen tents. The oxygen concentration presented to the patient is controlled by adjusting the flow of gas into the mask.

8.4.2. Ventilators and Respirators

The terms *ventilator* and *respirator* are used interchangeably to describe equipment that may be employed continuously or intermittently to improve ventilation of the lungs and to supply humidity or aerosol medications to the pulmonary tree. Most ventilators in clinical settings use positive pressure during inhalation to inflate the lungs with various gases or mixtures of gases (air, oxygen, carbon dioxide, helium, etc.). Expiration is usually passive, although under certain conditions pressure may be applied during the expiratory phase as well, in order to improve arterial oxygen tension. Only under rare circumstances is negative airway pressure utilized during expiration.

Most respirators in common use are classified as assistor-controllers, and can be operated in any of three different modes. These modes differ in the method by which inspiration is initiated.

1. In the *assist* mode inspiration is triggered by the patient. A pressure sensor responds to the slight negative pressure that occurs each time the patient attempts to inhale and triggers the apparatus to begin inflating the lungs. Thus, the respirator helps the patient inspire when he wants to breathe. A sensitivity adjustment is provided to select the amount of patient effort required to trigger the machine. The assist mode is used for patients who are able to control their breathing but are unable to inhale a sufficient amount of air without assistance or for whom breathing requires too much effort.
2. In the *control* mode breathing is controlled by a timer set to provide the desired respiration rate. Controlled ventilation is required for patients who are unable to breathe on their own. In this mode the respirator has complete control over the patient's respiration and does not respond to any respiratory effort on the part of the patient.
3. In the *assist-control* mode the apparatus is normally triggered by the patient's attempts to breathe, as in the assist mode. However, if the patient fails to breathe within a predetermined time, a timer automatically triggers the device to inflate the lungs. Thus, the patient controls his own breathing as long as he can, but if he should fail to do so, the machine is able to take over for him. This mode is most frequently used in critical care settings.

In addition to the three modes described, many respirators can be triggered manually by means of a control on the panel.

Once inspiration has been triggered, inflation of the lungs continues until one of the following conditions occurs:

1. The delivered gas reaches a predetermined pressure in the proximal or upper airways. A ventilator that operates primarily in this manner is said to be *pressure-cycled.*
2. A predetermined volume of gas has been delivered to the patient. This is the primary mode of operation of *volume-cycled ventilators.*
3. The air or oxygen has been applied for a predetermined period of time. This is the characteristic mode of operation for *time-cycled ventilators.*

The various types of ventilators in clinical use can be categorized by two basic types. The first is a *pressure-cycled,* positive-pressure *assistor-controller.* An example of this type of respirator is shown in Figure 8.12. The device is powered pneumatically from a source of gas and requires no electrical power. Devices in this category may contain an electrically powered compressor or can be used with a separate compressor to permit ventilation with ambient air.

Although a ventilator of the type shown in Figure 8.12 is quite small, it includes all the necessary equipment to control the flow of gas, mix air and oxygen, sense the patient's effort to inspire, terminate the inspiration when the desired pressure is reached, permit adjustment of the sensitivity of the triggering mechanism and the desired pressure level, and even generate a negative pressure to assist expiration on some devices. A special type of valve that incorporates a magnet senses the small negative pressure created by a patient when he attempts to inhale. Timing for operation in the controlled mode is accomplished by filling a chamber with gas and letting it bleed off through an adjustable needle valve. In the prescribed time, the pressure drops to a level at which a spring-loaded valve can operate. One widely used respirator in this category includes three pneumatic timing devices of a somewhat different type to provide time cycling as well as pressure cycling.

A form of volume-controlled respiration is possible with the type of pneumatically-operated respirator that permits time cycling. This flexibility is based on the premise that a given amount of airflow for a specified time duration results in a controlled volume.

The second category of respirator is the *volume-cycled ventilator,* often called a *volume respirator.* This type of device shown in Figure 8.13 uses either a piston or bellows to dispense a precisely controlled volume for each breath. In the critical care setting where patients have pulmonary abnormalities and require predictable volumes and concentrations of gas, this type of ventilator is preferred. It is much larger than the pneumatically-operated units, and most units stand on the floor beside the patient's bed. Volume respirators are electrically operated and provide a much greater degree of control over the ventilation than the pressure-cycled types.

Most devices of this type have adjustable pressure limits and alarms for safety. Also, their provision for adjusting pressure limits and both inspiratory and expiratory times can be used in conjunction with the volume setting to ensure therapeutic pulmonary function in the patient who needs it most.

Volume-cycled ventilators used in critical patient care are always supplied with a spirometer to permit accurate monitoring of the patient's ventilation. Other available features include a heated humidifier and optional capabilities for negative pressure and positive end expiratory pressure (PEEP).

Figure 8.12. Mark 7 respirator, and example of a pressure-cycled, positive-pressure, assistor-controller. (Courtesy of Bird Corporation, Palm Springs, CA.)

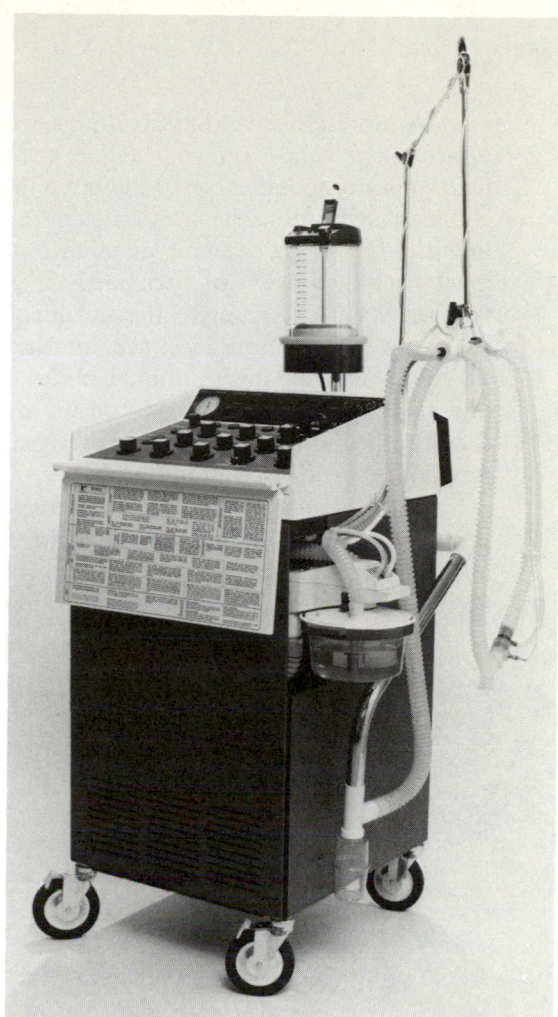

Figure 8.13. MA-2 ventilator. (Courtesy of Puritan-Bennet Corporation, Kansas City, MO.)

8.4.3. Humidifiers, Nebulizers, and Aspirators

In order to prevent damage to the patient's lungs, the air or oxygen applied during respiratory therapy must be humidified. Thus, virtually all inhalators, ventilators, and respirators include equipment to humidify the air, either by heat vaporization (steam) or by bubbling an air stream through a jar of water.

When therapy requires that water or some type of medication be suspended in the inspired air as an aerosol, a device called a *nebulizer* is used. In a nebulizer the water or medication is picked up by a high-velocity jet of oxygen (or some other gas) and thrown against one or more baffles or other surfaces to break the substance into controllable-sized droplets or particles, which are then applied to the patient via a respirator.

A more effective (but also more expensive) type of nebulizer is the *ultrasonic nebulizer,* shown in Figure 8.14. This electronic device produces high-intensity sound energy well above the audible range. When applied to water or medication, the ultrasonic energy vibrates the substance with such intensity that a high volume of minute particles is produced. Such equipment usually consists of two parts, a generator that produces a radio-frequency current to drive the ultrasonic transducer, and the nebulizer itself, in which the transducer generates the ultrasound energy and applies it to the water or medication. Unlike the conventional nebulizer, the ultrasonic unit does not depend on the breathing gas for operation. Thus, the therapeutic agent can be administered during oxygen therapy or a mechanical ventilation procedure.

Aspiration and other types of *suction apparatus* are often included as part of a ventilator or inhalator to remove mucus and other fluids from the airways. Where the aspirator is not provided as part of the respiratory therapy equipment, a separate suction device may be utilized.

Figure 8.14. Ultrasonic nebulizer. (Courtesy of the DeVilbiss Company, Medical Products Division, Somerset, PA.)

9

Noninvasive Diagnostic Instrumentation

In the previous chapters many methods of medical measurements have been discussed that involve getting inside the body, or "invading" it. To say the least, such procedures are usually traumatic for the patient and sometimes result in faulty data or detrimental side effects. As these techniques have become more sophisticated, it has been realized that sometimes equally suitable results can be obtained without invasion of the body. As a result, considerable emphasis has been devoted to developing methods of *noninvasive* testing. Some noninvasive methods, like the indirect method of taking blood pressure, have been around for years. Others have just recently been developed, and many new techniques await development of instrumentation that will make them possible.

In presenting material in a broad textbook such as this, it is often difficult to decide where to place certain material. In the case of noninvasive methods, this is certainly true. Does the material pertaining to a given technique belong in the context of the measurement in the body system involved, or should it be treated as a separate topic? A decision was made to use both approaches. For example, probably the best known noninvasive methods involve the use of X rays. While it is true that X rays are noninvasive in the sense that no physical contact or cutting is involved, the body is nevertheless "invaded" by radiation. Therefore, this type of measurement technique is discussed in its own context of ionizing radiation in Chapter 14. Conversely, the newer technique of the use of ultrasound to obtain information similar to that obtained by X-ray techniques is covered in this chapter. An example can be taken from obstetrics. Prior to the extensive use of ultrasonics, expectant mothers were sometimes X-rayed to determine position of the fetus when there was a possibility of problems during delivery which might necessitate a caesarian section. However, the radiation could have effects on both the mother and the fetus. As far as is presently known, using ultrasound to determine pelvic structure and the like has no known effects that could be detrimental. Ultrasonics is considered as one of the main areas of noninvasive testing.

Another example is in cardiology. In Chapter 6 the traumatic procedure of catheterization was discussed. Some of the results can be obtained today by the use of ultrasound methods. In this case the appropriate measurement technique, echocardiography, is discussed in this chapter. For the brain, one of the latest methods of visualization is computerized axial tomography, but since this procedure involves computers, it is discussed in Chapter 15. There are, of course, cross references for all these topics.

All forms of noninvasive testing are based on the fundamental concepts of physics. Throughout the book there are examples of the use of heat, light, sound, electricity, magnetism, and mechanics. This chapter concentrates on two of these areas, the use of heat and temperature measurements and the application of ultrasound to medicine. Each of these topics is discussed from the point of view of its basic principles, after which the measurement techniques, application, and diagnostic methods are explored. Ultrasonic techniques are covered in greater depth, since this material is not usually as available in broader-based textbooks.

9.1. TEMPERATURE MEASUREMENTS

Body temperature is one of the oldest known indicators of the general well-being of a person. Techniques and instruments for the measurement of temperature have been commonplace in the home for years and throughout all kinds of industry, as well as in the hospital. Except for the narrow range required for physiological temperature measurements and the size and shape of the sensing element, instrumentation for measurement of temperature in the human body differs very little from that found in various industrial applications.

Two basic types of temperature measurements can be obtained from the human body: systemic and skin surface measurements. Both provide valuable diagnostic information, although the systemic temperature measurement is much more commonly used.

Systemic temperature is the temperature of the internal regions of the body. This temperature is maintained through a carefully controlled balance between the heat generated by the active tissues of the body, mainly the muscles and the liver, and the heat lost by the body to the environment. Measurement of systemic temperature is accomplished by temperature-sensing devices placed in the mouth, under the armpits, or in the rectum. The normal oral (mouth) temperature of a healthy person is about 37 °C (98.6 °F). The underarm temperature is about 1 degree lower, whereas the rectal temperature is about 1 degree higher than the oral reading. The systemic body temperature can be measured most accurately at the tympanic membrane in the ear, which is believed to approximate the temperature at the "inaccessible" temperature control center in the brain. For some still unknown reason, the body temperature, even in a healthy person, does not remain constant over a 24-hour period but is often 1 to 1½ degrees lower in the early morning than in late afternoon. Although strenuous muscular exercise may cause a temporary rise in body temperature from about 0.5 to 2 °C (about 0.9 to 3.6 °F), the systemic temperature is not affected by the ambient temperature, even if the latter drops to as low as − 18 °C (0 °F) or rises to over 38 °C (100 °F). This balance is upset only when the metabolism of the body cannot produce heat as rapidly as it is lost or when the body cannot rid itself of heat fast enough.

The temperature-control center for the body is located deep within the brain (in the forepart of the hypothalamus) (see Chapter 10). Here the temperature of the blood is monitored and its control functions are coordinated. In warm, ambient temperatures, cooling of the body is aided by production of perspiration due to secretion of the sweat glands and by increased circulation of the blood near the surface. In this manner, the body acts as a radiator. If the external temperature becomes too low, the body conserves heat by reducing blood flow near the surface to the minimum required for maintenance of the cells. At the same time, metabolism is increased. If these measures are insufficient, additional heat is produced by increasing the tone of skeletal muscles and sometimes by involuntary contraction of skeletal muscles (shivering) and of the arrector muscles in the skin (gooseflesh).

In addition to the central "thermostat" for the body, temperature sensors at the surface of the skin permit some degree of local control in the event a certain part of the body is exposed to local heat or cold. Cooling or heating is accomplished by control of the surface blood flow in the region affected.

The only deviation from normal temperature control is a rise in temperature called "fever," experienced with certain types of infection. The onset of fever is caused primarily by a delicate shutdown of the mechanisms for heat elimination. The body temperature increases as though the "thermostat" in the brain were suddenly turned "up," thus causing additional metabolism because the increased temperature accelerates the chemical reactions of the body. At the beginning of a fever the skin is often pale and dry and shivering usually takes place, for the blood that normally keeps the surface areas warm is shut off, and the skin and muscles react to the coolness. At the conclusion of the fever, as the body temperature is lowered to normal, increased sweating ("breaking of the fever") is often noted as the means by which the additional body heat is eliminated.

Surface or *skin temperature* is also a result of a balance, but here the balance is between the heat supplied by blood circulation in a local area and the cooling of that area by conduction, radiation, convection, and evaporation. Thus, skin temperature is a function of the surface circulation, environmental temperature, air circulation around the area from which the measurement is to be taken, and perspiration. To obtain a meaningful skin temperature measurement, it is usually necessary to have the subject remain with no clothing covering the region of measurement in a fairly cool ambient temperature [approximately 21 °C (70 °F)]. Care must be taken, however, to avoid chilling and the reactions relative to chilling. If a surface measurement is to include the reaction to the cooling of a local region, it should be recognized that the cooling of the skin increases surface circulation, which in turn causes some local warming of adjacent areas. Heat transferred into the site of measurement from adjacent areas of the body must also be accounted for.

9.1.1. Measurement of Systemic Body Temperature

Since the internal or systemic body temperature is a good indicator of the health of a person, measurement of this temperature is considered one of the vital signs of medicine. For this reason, temperature measurement constitutes one of the more important physiological measurements. Although a high degree of accuracy is not always important, methods of temperature measurement must be reliable and easy to perform. In the case of continuous monitoring, the temperature measurement must not cause discomfort to the patient.

Where continuous recording of temperature is not required, the *mercury thermometer* is still the standard method of measurement. Since these devices are inexpensive, easy to use, and sufficiently accurate, they will undoubtedly remain in common use for many years to come. Even so, electronic thermometers, such as that shown in Figure 9.1, are available as

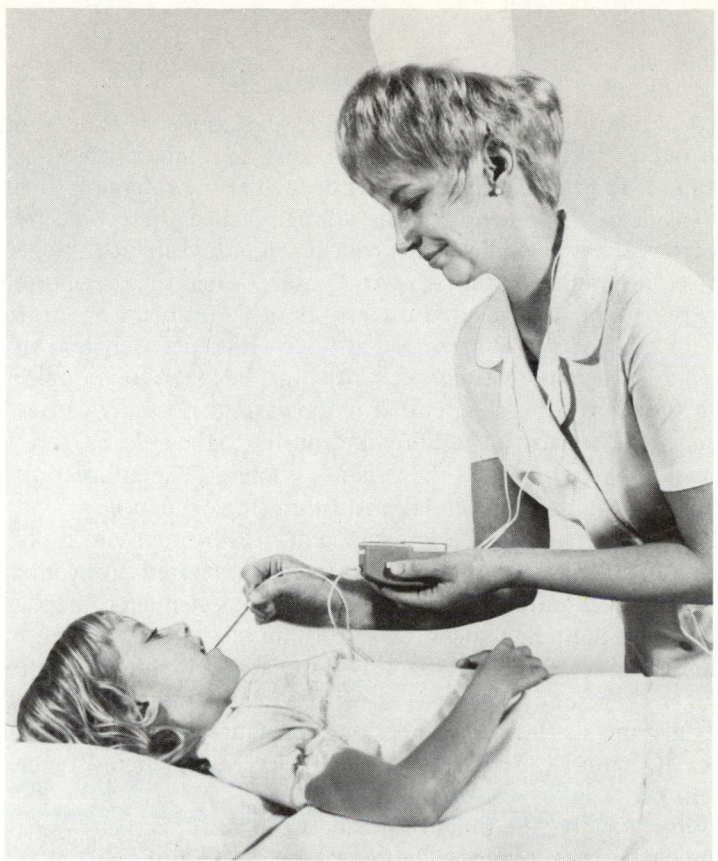

Figure 9.1. Oral temperature measurement using electronic thermometer. (Courtesy of Diagnostic, Inc. Indianopolis, IN.)

replacement for mercury thermometers. With disposable tips, these instruments require much less time for a reading and are much easier to read than the conventional thermometer. Where continuous recording of the temperature is necessary, or where greater accuracy is needed than can be obtained with the mercury thermometer or its electronic counterpart, more sophisticated measuring instruments must be used.

Two types of electronic temperature-sensing devices are found in biomedical applications. They are the *thermocouple,* a junction of two dissimilar metals that produces an output voltage nearly proportional to the temperature at that junction with respect to a reference junction, and the *thermistor,* a semiconductor element whose resistance varies with temperature. Both types are available for medical temperature measurements, although thermistors are used more frequently than thermocouples. This preference is primarily because of the greater sensitivity of the thermistor in the temperature range of interest and the requirement for a reference junction for the thermocouple.

To obtain a voltage proportional to variations in temperature in a thermocouple, the reference junction must be maintained at a known temperature. In practice, the circuit is opened at the reference junction for measurement of the potential. This voltage, called the *contact potential,* ranges from a very few microvolts to a few hundred microvolts per degree centigrade, depending on the two metals used. Generally, the output voltage of a thermocouple is measured directly by using a meter or measured indirectly by comparing the measured voltage with a precisely known voltage obtained by using a potentiometer. Care must be taken to minimize current through the thermocouple circuit, for the current not only causes heating at the junctions but also an additional error due to the *Peltier effect,* wherein one junction is warmed and the other is cooled. (The connections of the leads to the two dissimilar metals constitute a single junction.)

Thermistors are variable resistance devices formed into disks, beads, rods, or other desired shapes. They are manufactured from mixtures of oxides (sometimes sulfates or silicates) of various elements, such as nickel, copper, magnesium, manganese, cobalt, titanium, and aluminum. After the mixture is compressed into shape, it is sintered at a high temperature into a solid mass. The result is a resistor with a large temperature coefficient. Where most metals show an increase of resistance of about 0.3 to 0.5 percent per °C temperature rise, thermistors decrease their resistance by 4 to 6 percent per °C rise.

Unfortunately, the relationship between resistance change and temperature change is nonlinear. The resistance R_{t_1} of a thermistor at a given temperature T_1 can be determined by the following equation:

$$R_{t_1} = R_{t_0}\, e^{\beta(1/T_1 - 1/T_0)}$$

where R_{t_1} = resistance at temperature T_1

R_{t_0} = resistance at a reference temperature T_0

e = base of the natural logarithms (approximately 2.718)

β = temperature coefficient of the material, usually in the range of about 3000 to 4000

T_1 = temperature at which the measurement is being made, (degrees Kelvin)

T_0 = reference temperature, (degrees Kelvin)

To overcome the nonlinear characteristics of thermistors, the instrumentation in which the resistance is measured often incorporates special linearizing circuits. Some such circuits employ pairs of matched thermistors as part of the linearizing network.

In addition to nonlinearity, the use of thermistors can result in other problems, such as the danger of error due to self-heating, the possibility of hysteresis, and the changing of characteristics because of aging. The effect of self-heating can be reduced by limiting the amount of current used in

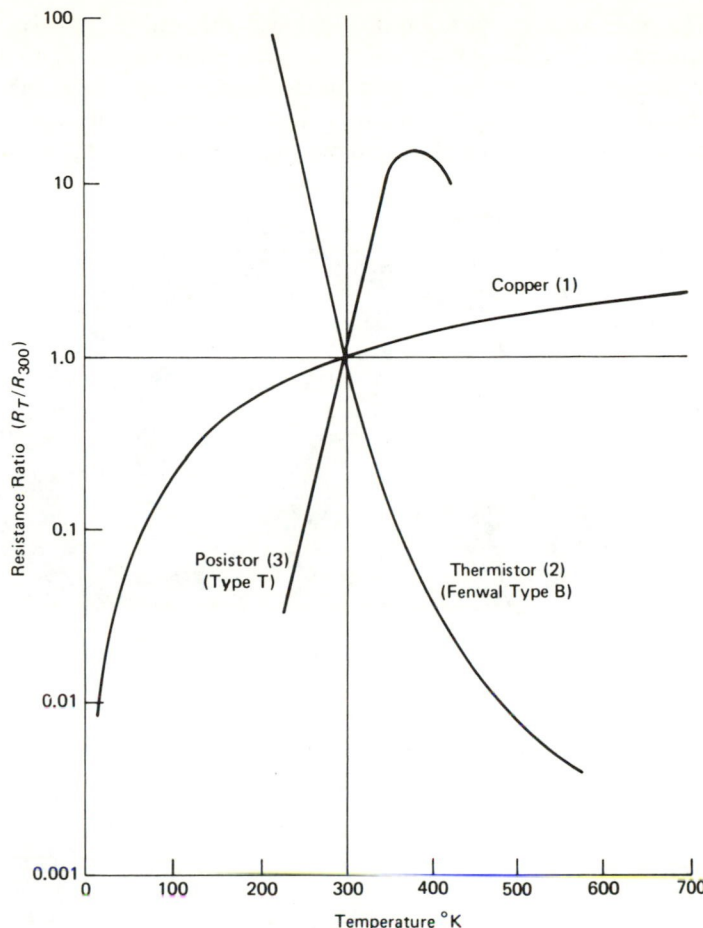

Figure 9.2. Resistance-temperature relationship of copper, thermistor and positor. (From L.A. Geddes and L.E. Baker, *Principles of Applied Biomedical Instrumentation.* John Wiley & Sons, Inc., 1969, by permission.)

measuring the resistance of the thermistor. If the power dissipation of the thermistor can be kept to about a milliwatt, the error should not be excessive, even when temperature differences as small as 0.01 °C are sought.

Semiconductor devices with positive temperature coefficients have been developed but are not commonly used. A comparison of resistance versus temperature curves for copper, a thermistor, and the Posistor (one of the positive coefficient devices) is given in Figure 9.2.

The most important characteristics to consider in selecting a thermistor probe for a specific biomedical application are the following:

1. The physical configuration of the thermistor probe. This is the interface with the site from which the temperature is to be measured. The configuration includes the size, shape, flexibility, and any special features required for the measurement. Commercial probes are available for almost any biomedical application,

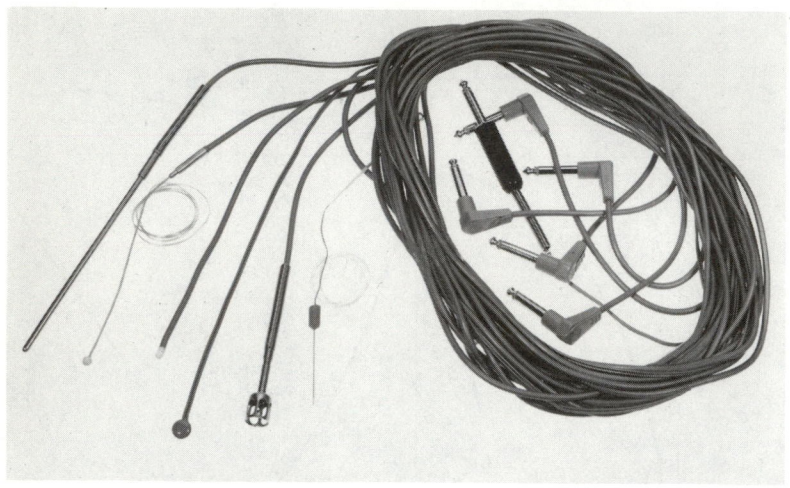

Figure 9.3. Thermistor probes. (Courtesy of Yellow Springs Instruments Company, Yellow Springs, OH.)

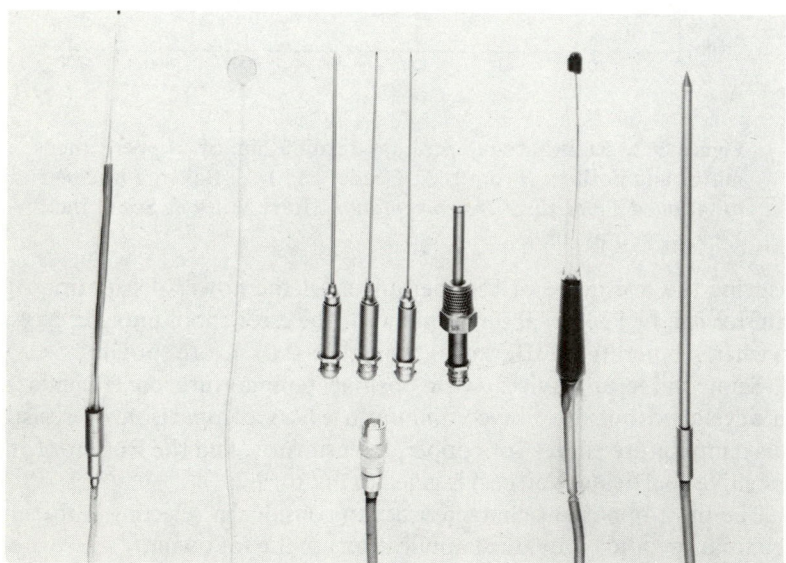

particularly for measurement of oral and rectal temperatures. Some of these probes are shown in Figure 9.3.

2. The sensitivity of the device. This is its ability to measure accurately small changes in temperature, but it can also be interpreted as the resistance change produced by a given temperature change. Usually, overall sensitivity is a function of both the thermistor probe and the circuitry used to measure the resistance, but the limiting factor is the resistance-temperature characteristic of the thermistor (see Figure 9.2).

3. The absolute temperature range over which the thermistor is designed to operate. This is usually no problem with body temperature measurements, for the temperature range to be measured is so limited, but often, if a general-type temperature measuring instrument is used, the range is so wide that the desired resolution is not attainable.

4. Resistance range of the probe. Thermistor probes are available with resistances from a few hundred ohms to several megohms. A probe should be selected with a suitable resistance range corresponding to the temperature range of interest to match the impedance of the bridge or other type of circuit used to measure the resistance.

Although the resistance of a thermistor can be measured by use of an ohmmeter, most thermistor thermometers use a Wheatstone bridge or similar circuit to obtain a voltage output proportional to temperature variations. Generally, the bridge is balanced at some reference temperature and calibrated to read variations above and below that reference. Either ac or dc excitation can be used for the bridge. If the temperature difference between two measurement sites is desired, thermistors at the two locations are placed in adjacent legs of the bridge.

9.1.2. Skin Temperature Measurements

Although the systemic temperature remains very constant throughout the body, skin temperatures can vary several degrees from one point to another. The range is usually from about 30 to 35 °C (85 to 95 °F). Exposure to ambient temperatures, the covering of fat over capillary areas, and local blood circulation patterns are just a few of the many factors that influence the distribution of temperatures over the surface of the body. Often, skin temperature measurements can be used to detect or locate defects in the circulatory system by showing differences in the pattern from one side of the body to the other.

Skin temperature measurements from specific locations on the body are frequently made by using small, flat thermistor probes taped to the skin (Figure 9.3). The simultaneous readings from a number of these probes provide a means of measuring changes in the spatial characteristics of the circulatory pattern over a time interval or with a given stimulus.

Although the effect is insignificant in most cases, the presence of the thermistor on the skin slightly affects the temperature at that location. Other methods of measuring skin temperature that draw less heat from the point of measurement are available. The most popular of these methods involve the measurement of infrared radiation.

The human skin has been found to be an almost perfect emitter of infrared radiation. That is, it is able to emit infrared energy in proportion to the surface temperature at any location of the body. If a person is allowed to remain in a room at about 21°C (70°F) without clothing over the area to be measured, a device sensitive to infrared radiation can accurately read the surface temperature. Such a device, called an *infrared thermometer,* is shown in Figure 9.4. Infrared thermometers in the physiological temperature range are available commercially and can be used to locate breast cancer and other unseen sources of heat. They can also be used to detect areas of poor circulation and other sources of coolness and to measure skin temperature changes that reflect the effects of circulatory changes in the body.

An extension of this method of skin temperature measurement is the *Thermograph,* shown in Figure 9.5(a). This device is an infrared thermometer incorporated into a scanner so that the entire surface of a body, or some portion of the body, is scanned in much the same way that a television camera scans an image, but much slower. While the scanner scans the body, the infrared energy is measured and used to modulate the intensity of a light beam that produces a map of the infrared energy on photographic

Figure 9.4. Infrared thermometer. Barnes model MT-3 noncontact thermometer provides fast, accurate measurements of skin temperature. (Courtesy of Barnes Engineering Company, Stamford, CT.)

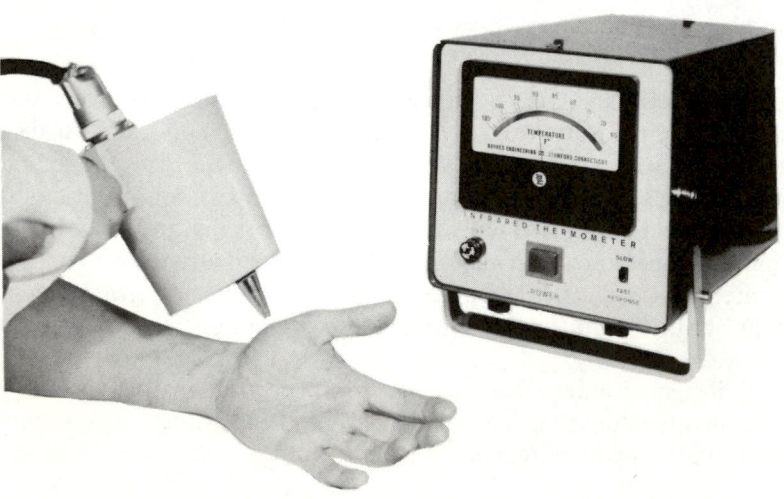

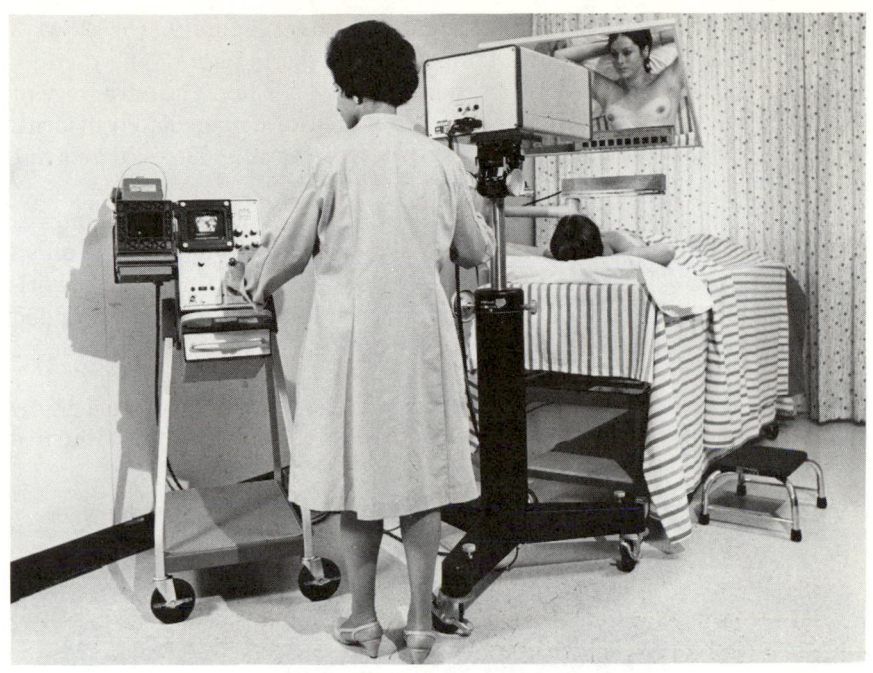

Figure 9.5. Thermography: (a) high resolution thermograph; (b) thermogram (see explanation in text). (Courtesy of Barnes Engineering Company, Stamford, CT.)

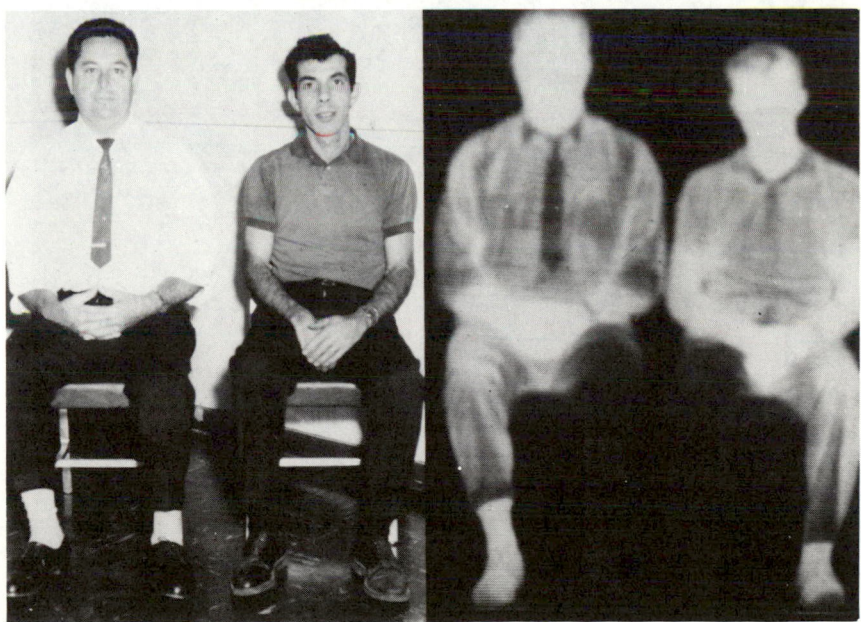

paper. This presentation is called a *thermogram.* Figure 9.5(b) shows a photograph of two men and a corresponding thermogram. The thermogram shows that each of the two men has an artificial leg. The advantage of this method is that relatively warm and cool areas are immediately evident. By calibrating the instrument against known temperature sources, the picture can be read quantitatively.

A similar device, called *Thermovision,* has a scanner that operates at a rate sufficiently high to permit the image to be shown in real time on an oscilloscope. The raster has about 100 vertical lines per frame, and the horizontal resolution is also about 100 lines, which seems to be adequate for good representation. The intensity of the measured infrared radiation is reproduced

Figure 9.6. Thermovision system: (a) Thermovision 680 Medical, camera and display unit; (b) Thermovision 680 Medical with accessories. (Courtesy of AGA Infrared Systems AB, Sweden.)

(a)

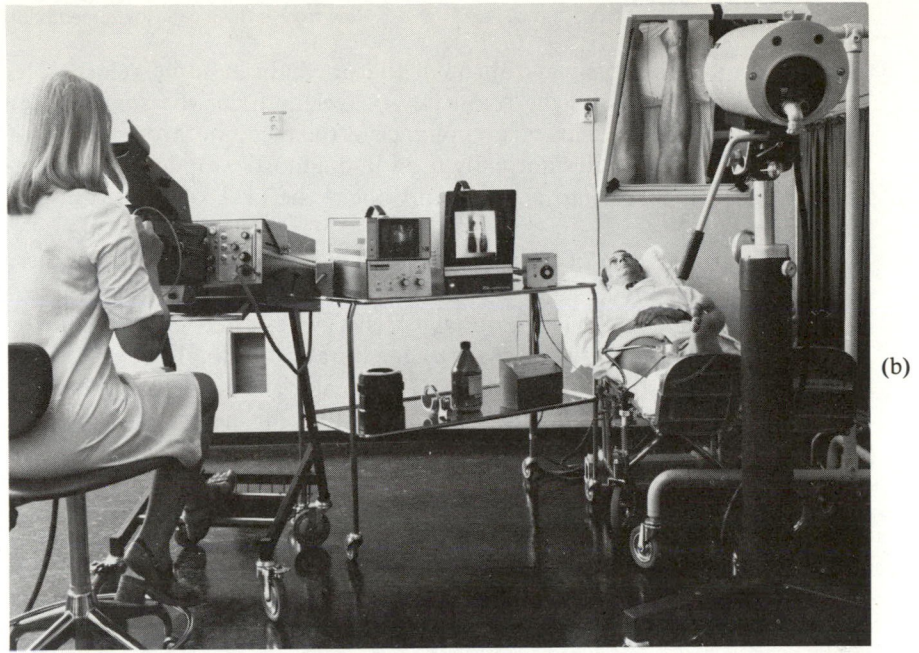
(b)

Figure 9.6. *Continued.*

by Z-axis modulation (brightness variation) of the oscilloscope beam. One advantage of this system is that certain portions of the gray scale can be enhanced to bring out specific features of the picture. Also, the image can be changed so that warm spots appear dark instead of light, as they usually do. All these enhancement measures can be performed while the subject is being scanned. A Thermovision system is shown in Figure 9.6.

9.2. PRINCIPLES OF ULTRASONIC MEASUREMENT

Recently, many of the innovations of medicine have taken place because of the use of ultrasound. By definition, *ultrasound* is sonic energy at frequencies above the audible range (greater than 20 kHz). Its use in medical diagnosis dates back to the period following World War II and is a direct outgrowth of the military development of sonar, in which pulsed ultrasound was used in the detection of submarines and other underwater objects by reflection of the ultrasonic waves.

9.2.1. Properties of Ultrasound

Like other forms of sonic energy, ultrasound exists as a sequence of alternate compressions and rarefactions of a suitable medium (air, water, bone, tissue, etc.) and is propagated through that medium at some velocity. Its behavior also depends on the frequency (wavelength) of the sonic energy and the density and mechanical compliance of the medium through which it travels. At the frequencies normally used in diagnostic applications, ultrasound can be focused into a beam and obeys the laws of reflection and refraction.

Whenever a beam of ultrasound passes from one medium to another, a portion of the sonic energy is reflected and the remainder is refracted, as shown in Figure 9.7. The amount of energy reflected depends on the difference in density between the two media and the angle at which the transmitted beam strikes the medium. The greater the difference in media, the greater will be the amount reflected. Also, the nearer the angle of incidence between the beam and the interface is to 90° the greater will be the reflected portion.

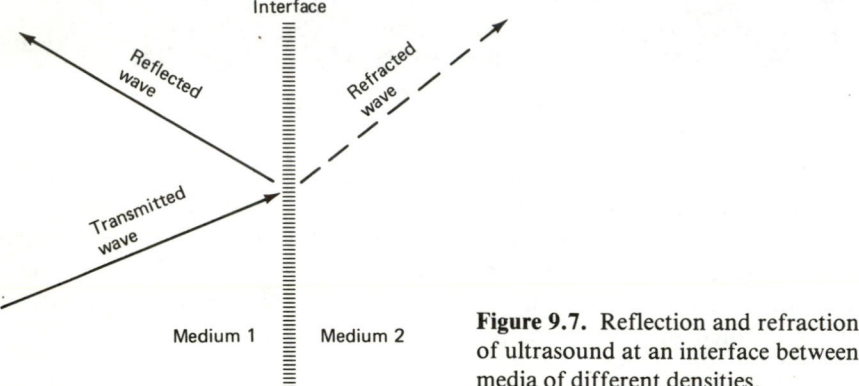

Figure 9.7. Reflection and refraction of ultrasound at an interface between media of different densities.

At interfaces of extreme difference in media, such as between tissue and bone or tissues and a gas, almost all the energy will be reflected and practically none will continue through the second medium. For this reason, the propagation path for ultrasound into or through the body must not include bone or any gaseous medium, such as air. In applying ultrasound to the body, an airless contact is usually produced through use of an aqueous gel or a water bag between the transducer and the skin.

Table 9.1 lists the density and other properties of various materials, including several of biological interest. The temperature and ultrasonic frequency are given for most of the measurements. Note that the density of water and most body fluids and tissues is approximately 1.00 g/cm³. Benzene has a density of 0.88, whereas the density of bone is almost twice as great (1.77 g/cm³).

Table 9.1. ULTRASONIC CHARACTERISTICS OF MATERIALS[a]

Material	Temperature (°C)	Density (g/cm³)	Velocity (m/sec)	Characteristic Impedance × 10⁶ (kg/m²/sec)	β	f (MHz)	α (per cm)
					Attenuation Constant, $\alpha = cf^{\beta}$		
Water	40	0.992	1529	1.517	1	2	0.00025
Saline, 0.9% normal	40	0.998	1539	1.537			
Castor oil	40	0.941	1411	1.328	1	1.67	0.037
Brain, average	37	1.03	1510	1.56	1	1	0.11
					5		0.44
Cortical gray matter	37	1.03			1	1	0.08
Cortical white matter	37	1.03			1	1	0.14
Muscle, skeletal	37	1.07	1570	1.68	1	1	0.13
Fat	37	0.97	1440	1.40	1	1	0.05
Bone, skull	37	1.77	3360	6.00	0.5	1.7	0.37
					1		1.2
					1.5		2.5
					2		4.0
					2.5		5.9
					3		8.1
					3.5		10.5
Skin				1.63			
Liver		1.08	1510	1.63	2		0.19
Blood		1.01	1550	1.56	2		0.04
Lucite				3.23			
Brain tumor							
Meningioma					5		0.73
Glioblastoma					5		0.38
Metastatic					5		0.50
Kidney		1.04	1560	1.62	2		0.27
Eye							
Aqueous humor		1.00	1500	1.50			
Vitreous humor		1.00	1530	1.53			
Lens		1.14	1630	1.85			
Benzene		0.88	1320	1.17			
Rubber							
Rhoc		1.00	1560	1.56			
Soft		0.95	1050	1.00			
Spleen		1.05	1570	1.65			

From W. Welkowitz and S. Deutsch, *Biomedical Instruments: Theory and Design,* Academic Press, New York, 1976; by permission.

The velocity of sound propagation through a medium varies with the density of the medium and its elastic properties. It also varies with temperature. As shown in Table 9.1, the velocity through most body fluids and soft tissues is in a fairly narrow range around 1550 m/sec. The velocity in water is just slightly lower (1529 m/sec). Note that the velocity of sound through fat is significantly lower (1440 m/sec) and through bone is much higher (3360 m/sec).

Every material has an *acoustic impedance,* which is a ratio of the acoustic pressure of the applied ultrasound to the resulting particle velocity in the material. Since acoustic impedance is a complex value, consisting of both resistive and reactive components, a simpler term, called *characteristic impedance,* is more often used. The characteristic impedance of a material is the product of its density and the velocity of sound through it. Table 9.1 gives the characteristic impedances of several materials.

Also given in Table 9.1 is an *attenuation constant,* α, for each material. As ultrasound travels through the material, some of the energy is absorbed and the wave is attenuated a certain amount for each centimeter through which it travels. The amount of attenuation is a function of both the frequency of the ultrasound and the characteristics of the material. The attenuation constant, α, is defined by the equation.

$$\frac{\text{amplitude at point X}}{\text{amplitude at } X + 1 \text{ unit distance}} = \beta$$

As shown in Table 9.1,

$$\alpha \,(\text{per cm}) = cf_\beta$$

where c = proportionality constant
         f = ultrasound frequency
         β = exponential term determined by the properties of the material

This formula shows that attenuation increases with some power of the frequency, which means that the higher the frequency, the less distance it can penetrate into the body with a given amount of ultrasonic energy. For this reason, lower ultrasound frequencies are used for deeper penetration. However, lower frequencies are incapable of reflecting small objects. As a rule, a solid object surrounded by water or saline must be at least a quarter-wave thick in order to cause a usable reflection. Thus, for finer resolution, higher frequencies must be used. Ultrasound frequencies of 1 to 15 MHz are usually used for diagnostic purposes. At 2 MHz, distinct echoes can be recorded from interfaces 1 mm apart. Higher-frequency ultrasound is also more subject to scattering than ultrasound at lower frequencies. However, the high-frequency ultrasound beam can be focused for greater resolution at a given depth.

Another useful way of assessing the attenuation of ultrasound as it penetrates the body is the *half-value layer* of the medium given in Table 9.2. The half-value layer is the depth of penetration at which the ultrasound energy is attenuated to half the applied amount.

Table 9.2. ULTRASOUND ABSORPTION

Type of Tissue	Frequency (MHz)	Half-Value Layer (cm)
Blood	1.0	35.0
Bone	0.8	0.23
Fat	0.8	3.3
Muscle	0.8	2.1

(From Feigenbaum, Echocardiography, 2nd Edition, Lea and Febiger, 1976 by permission.)

A well-known characteristic of ultrasound frequently utilized in biomedical instrumentation is the *Doppler effect,* in which the frequency of the reflected ultrasonic energy is increased or decreased by a moving interface. The amount of frequency shift can be expressed in the formula:

$$\Delta f = \frac{2V}{\lambda}$$

where f =   shift in frequency of the reflected wave
  V = velocity of the interface
  λ = wavelength of the transmitted ultrasound

The frequency increases when the interface moves toward the transducer and decreases when it moves away. With an ultrasound frequency of 3 MHz, the shift is about 40 Hz for each cm/sec of interface velocity.

A useful way to understand this in general terms is to consider what happens if an automobile with its horn sounding passes by on the street. The pitch or perceived frequency of the sound seems higher as the car is approaching but seems lower as it goes away. This is an example of the Doppler frequency shift. When ultrasound is reflected from a moving object, the measured frequency shift is proportional to velocity.

9.2.2. Basic Modes of Transmission

Ultrasound can be transmitted in various forms. Following are the modes of transmission most commonly used in diagnostic medical applications:

1. *Pulsed ultrasound:* In this mode, ultrasound is transmitted in short bursts at a repetition rate ranging from 1 to 12 kHz.

Returning echoes are displayed as a function of time after transmission, which is proportional to the distance from the source to the interface. Movement of interfaces with respect to time can also be displayed. The burst duration is generally about 1 μsec. Pulsed ultrasound is used in most imaging applications.

2. *Continuous Doppler:* Here a continuous ultrasonic signal is transmitted while returning echoes are picked up by a separate receiving transducer. Frequency shifts due to moving interfaces are detected and recorded and the average velocity of the targets is usually determined as a function of time. This mode always requires two transducer crystals, one for transmission and one for receiving, whereas any of the pulsed modes can use either one or two crystals. Continuous Doppler ultrasound is used in blood flow measurements (see Chapter 6) and in certain other applications in which the average velocity is measured without regard to the distance of the sources.

3. *Pulsed Doppler:* As in pulsed ultrasound, short bursts of ultrasonic energy are transmitted and the returning echoes are received. However, in this mode frequency shifts due to movement of the reflected interfaces can be measured in order to determine their velocities. Thus, both the velocity and distance of a moving target can be measured. In a typical application, three cycles of 3-MHz ultrasound are transmitted per pulse at a pulse rate of 4 to 12 kHz.

4. *Range-gated pulsed Doppler:* This mode is a refinement of pulsed-Doppler ultrasound, in which a gating circuit permits measurement of the velocity of targets at a specific distance from the transducer. The velocity of these targets can be measured as a function of time. With range-gated pulsed Doppler ultrasound, the velocity of blood can be measured, not only as a function of time, but also as a function of the distance from the vessel wall.

In any of the above-described modes, the most effective frequency of the ultrasound depends upon the depth of penetration desired and the required resolution.

9.2.3. Ultrasonic Imaging

The most widely used applications of ultrasound in diagnostic medicine involve the noninvasive imaging of internal organs or structures of the body. Such imaging can provide valuable information regarding the size, location, displacement, or velocity of a given structure without the necessity of surgery or the use of potentially harmful radiation. Tumors and other regions of an organ that differ in density from surrounding tissues can be detected. In many instances, ultrasonic techniques have replaced more risky or more traumatic procedures in clinical diagnosis.

Imaging systems generally utilize the pulsed ultrasound or pulsed Doppler mode. Instrumentation must include an electrical signal source capable of driving the transmitter, which consists of a piezoelectric crystal. The same crystal can be used for receiving echoes or a second crystal may be used.

After amplification, the received information is displayed in one of several display modes. There is some confusion in the literature in the definition of some of these modes. For example, some authors consider the M-scan used in echocardiography (Section 9.3.2) as a form of the B-scan mode rather than a separate mode. While it is true that these two modes are very similar, differences in the information presented make it necessary to distinguish between the two in order to properly interpret the display. Also, some authors have adopted terminology from military sonar displays which is not appropriate to imaging in medical applications. The following definitions are those most generally found in the literature and are used consistently in this text:

1. *A-scan display [Figure 9.8(a)]:* This is the simplest form of display. Each transmitted pulse triggers the sweep of an oscilloscope. That pulse (often attenuated) and the returning echoes are displayed as vertical deflections on the trace. The sweep is calibrated in units of distance, and may provide several ranges in order to accurately determine the distance of the interfaces of interest. Often, the amplifier gain is varied with the sweep to compensate for the lower amplitude of more distant echoes. In most cases the transducer is kept stationary so that any movement of echoes along the trace will be the result of moving targets. An example of an A-scan display is that of the echoencephalogram.

2. *M-scan display:* As in the A-scan mode, each transmitted pulse triggers the oscilloscope sweep; however, the received pulses are used to brighten the trace rather than control the vertical deflection, as shown in Figure 9.8(b). The quiescent brightness level is set below the visibility threshold so that only the echoes, which appear as dots with brightness proportional to the intensity of each echo, can be seen. For the M-scan, the transducer is held stationary so that the movement of the dots along the sweep represent movement of received targets. If photographic paper is slowly moved past the face of the oscilloscope so that each

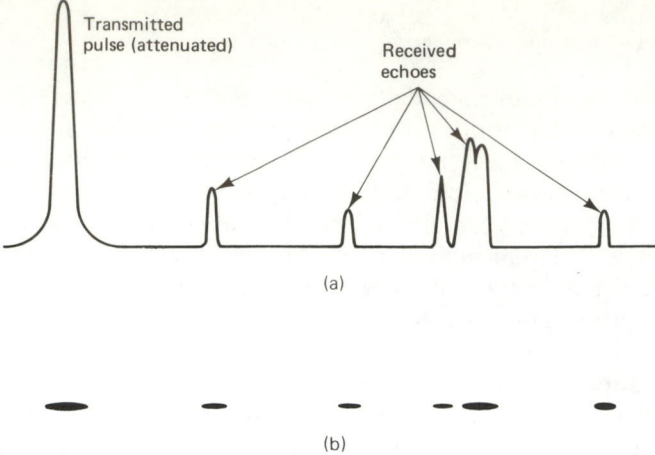

Figure 9.8. Ultrasound Display Principles. (a) Typical A-scan. Echoes cause vertical deflection of oscilloscope pattern. (b) Corresponding display in which echoes control brightness of oscilloscope beam. This principle is used in both B- and M- scan displays.

Figure 9.9 M-scan of moving and stationary target with corresponding A-scan.

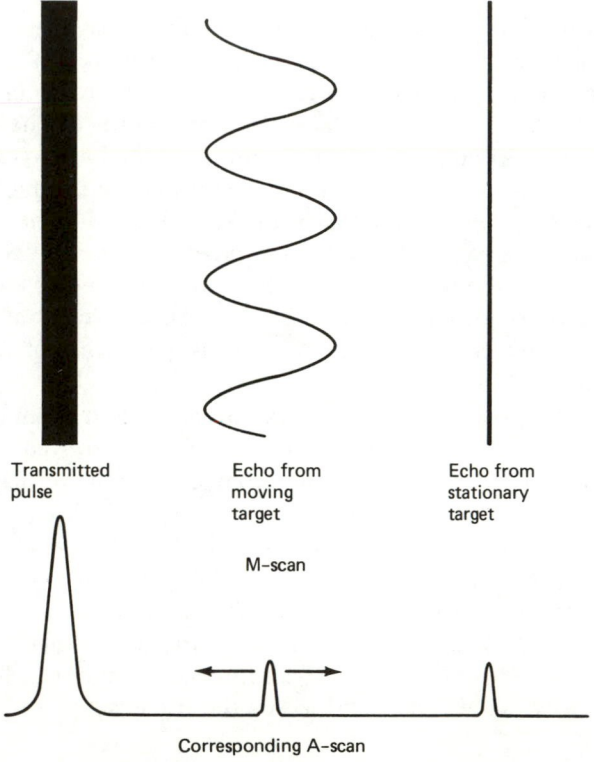

trace lies immediately adjacent to the one preceding it, the dot representing each target will trace a line on the paper as shown in Figure 9.9. A stationary target will trace a straight line, whereas a moving target will trace the pattern of its movement with respect to time. A light-pen recorder in which the intensity of the light source can be controlled may be used instead of an oscilloscope to produce a chart record of the movement of echoes with respect to time. An example of an M-scan recording is the echocardiogram shown in Figure 9.12.

3. *B-scan display:* While the M-scan is used to display the movement of targets with respect to time, the B-scan presents a two-dimensional image of a stationary organ or body structure. As in the M-scan, the brightness of the oscilloscope or light-pen beam is controlled by returning echoes; however, in the B-scan the transducer is moved with respect to the body while the vertical deflection of the oscilloscope or movement of the chart paper is made to correspond to the movement of the transducer. The movement may be linear, circular, or a combination of the two, but where it is anything other than linear, the sweep must be made to compensate for the variations in order to provide a true two-dimensional display of the segment being scanned. Examples of B-scan displays are shown in Figure 9.15.

9.3. ULTRASONIC DIAGNOSIS

The applications of ultrasonic methods to medicine are many and varied. The techniques are used in cardiology, for abdominal imaging, in brain studies, in eye analysis, and in obstetrics and gynecology. The records obtained have various names, which usually include the words "echo" or "sono." For example the *echocardiogram,* analogous to the electrocardiogram, is a record of ultrasonic measurements in the heart. The *echoencephalogram* is a record obtained from the brain. A general term used, especially with eye analysis, is the *ultrasonogram.* Examples of these will be presented later in this section. Multiple names have been coined for the instruments used, including many trade titles. Typical designations are *ultrasonograph, ultrasonoscope,* and *sonofluoroscope.* The latter is used essentially for moving structures, whereas some of the devices are used for static images.

Before discussing some of the components and specific medical fields, perhaps a general overview and illustration will give the reader a perspective. Figure 9.10 is a *phased-array ultrasonograph,* which represents a class of instruments that has recently appeared. It is virtually an entire medical ultrasound laboratory in one mobile instrument.

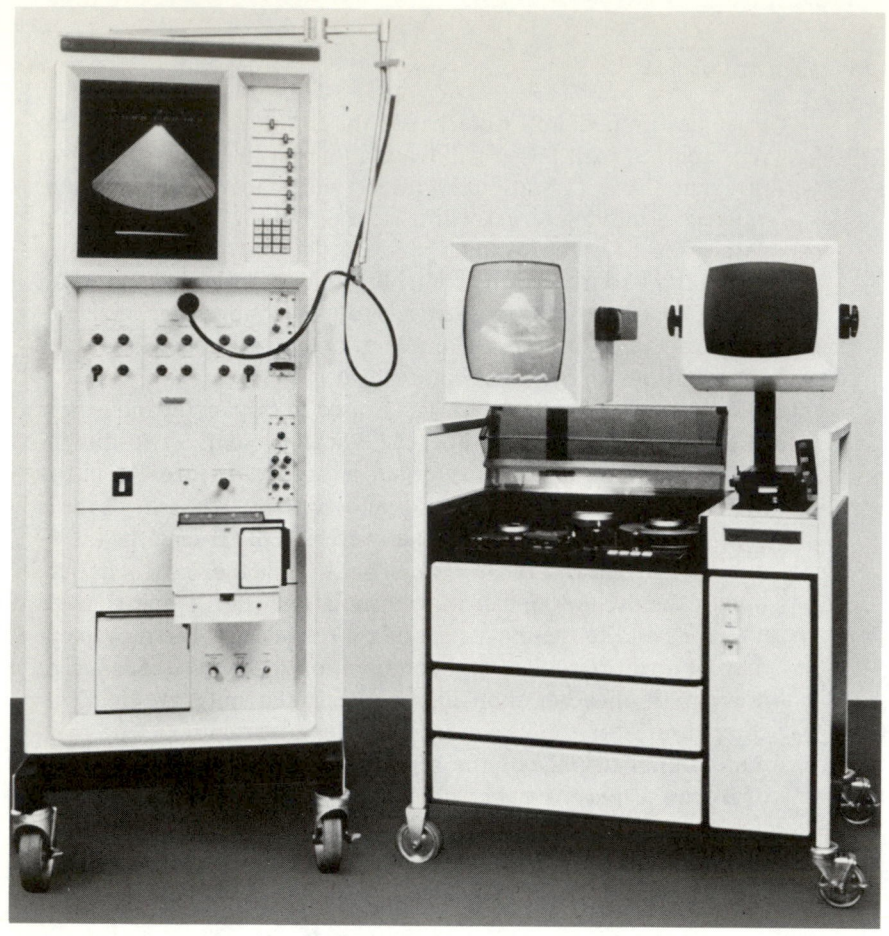

Figure 9.10. Phased array ultrasonograph. (Courtesy of Varian Associates, Medical Group, Palo Alto, CA.)

The V-3000 uses a large video display. The displayed scan image represents the area of the sector swept, up to the selected maximum depth, which may be 7, 15, or 21 cm. The image on the screen remains the same size regardless of the depth chosen. A calibrated scale appropriate to the chosen depth appears on the edges of the image and allows accurate measurement of the anatomy scanned.

The video monitor facilitates easy viewing and simultaneously presents ancillary information along with the diagnostic image. Orthogonal calibration marks appear on the edges of the image. Keyboard-entered patient identification, date, and time are displayed above the image along with the frame number and ECG trigger time relative to the R wave. A nonfade ECG trace is shown below the scan image. When the instrument is used in the A or M modes, a brightened ray indicates the spatial orientation chosen for the particular one-dimensional study being done.

Two separate sets of gain controls give maximum flexibility in optimizing the image for a specific examination. In addition to an overall gain control, a set of seven slide controls allows independent adjustment of gain for each depth interval, corresponding to one-seventh of the selected total penetration depth. Other controls are available which adjust gray scale and suppress noise.

M-mode and A-mode are available on the V-3000 in addition to the two-dimensional imaging capability. In a pure two-dimensional imaging application, the V-3000 operates at a scanning speed of 30 frames per second. For an M-mode or A-mode study, the frame rate is halved to 15 frames per second, with every other pulse used to accumulate data for the M- or A-mode. In M- or A-mode scanning, a brightened line is superimposed on the main display indicating the single ray along which the data are obtained. In the case of A-mode, these data are displayed directly along the brightened ray; for the M-mode, it appears on the optional slow oscilloscope. When a footswitch is depressed, the M-mode information is printed out on a strip-chart recorder.

For cardiac applications, an optional ECG amplifier is available. The trace appears on all displays, including the strip-chart recorder. ECG-triggered photographs may be made by positioning a cursor at the desired time position on the ECG trace.

The V-3000 includes computer-directed self-diagnosis, activated by a front-panel control. These diagnostic routines test the various circuit boards and show the results on the main display.

Phased-array sector scanning describes a technique in which the ultrasonic beam is electronically swept through an arc to produce sharp, high-resolution images in real time while the small transducer is held stationary in any desired position. Scanning is accomplished by pulsing individual crystals in the transducer at slightly different times under the control of a microcomputer. Because the emitted (and received) ultrasonic beam is swept electronically, a wide-angle image is instantly produced wherever the transducer is placed, allowing thorough and rapid imaging with no constraints on transducer positioning.

The video recorder cart is an essential element of the V-3000 system. It includes all the necessary components to do dynamic recording and allows off-line detailed study of these recordings. During an examination, video recordings can be activated by a foot switch. The cart contains a video tape recorder with slow- and stop-motion capabilities, a viewing monitor, and a camera directed at an internal live video monitor optimized for photography.

The video recorder cart is a complete video playback system, and is easily disconnected from the diagnostic module. It can be moved to remote locations for video-tape viewing and photographic record making. The off-line photographic capabilities, combined with the stop action of the video tape recorder, give a physician the opportunity to reanalyze difficult cases, freeze motion, and produce records when it was inconvenient to do so during the actual examination.

This ultrasonograph can be used in a wide variety of medical applications. The wide image plane and ability to scan between the ribs allow cardiac imaging from the precordial, apical, subxiphoid, and suprasternal positions. This versatility permits visualization of all four chambers and all four valves of the heart, plus the great arteries and the great veins.

The high-speed, high-resolution gray-scale imaging capabilities allow the physician to make rapid and thorough "fluoroscopic" examinations, free of distortion caused by involuntary motion. The small and unconstrained transducer permits thorough examination despite patient position or movement and is especially useful when imaging anatomy under the ribs, in the pelvis, or in the presence of sutures.

The small transducer is easily positioned to obtain, without patient discomfort, high-resolution images through any plane in the female pelvis. The continuous change in the real-time image, as this plane is swept through the pelvis, permits the visualization of very small structures such as the ovaries and tubes, or, in the case of the gravid uterus, the fetus as early as the 4-week stage.

9.3.1. Ultrasonic Transducers

Whatever the application, the basic ultrasonic system consists of a generator for the electric signal, a transducer, the necessary amplifiers, and other electronic processing devices and the display unit. It is the transducer that converts the electric signals into the mechanical vibrations and thus the acoustic waves which form the basis of the technique. Transducers are produced in a variety of configurations for the various applications, and with varying frequency capabilities. The acoustic wave from the transducer enters the body through the skin surface and is then propagated in a predetermined beam pattern (wide or narrow depending on requirements) toward the structure to be examined. When the ultrasonic waves strike an acoustic interface such as the boundary of an organ, some energy is reflected. This reflected energy is picked up on the transducer and is amplified, processed, and finally displayed on an oscilloscope.

Figure 9.11 shows a Dapco echocardiology transducer. These are available in a variety of internal focal arrangements. In transducers of this type, the length of the focal zone is determined by the distance from the transducer to the cardiac structures of interest. The choice of frequency depends on the amount of tissue attenuation encountered. Transducers are available at frequencies of 1.6, 2.25, 3.5, and 5.0 MHz and with crystal diameters of 6, 13, and 20 mm for each frequency.

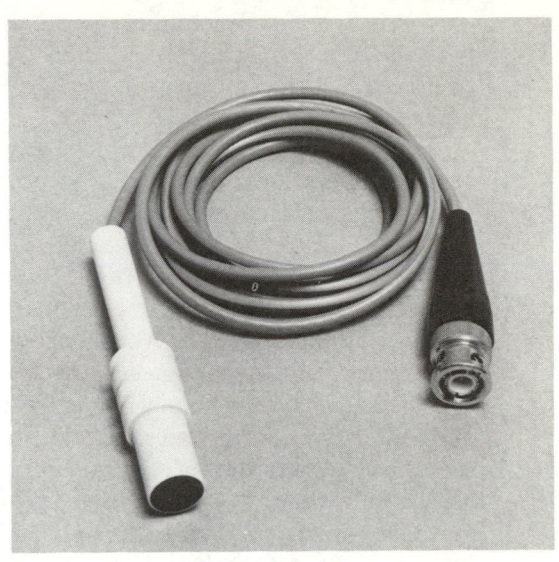

Figure 9.11. Echocardiology transducer. (Courtesy of Dapco Industries, Ridgefield, CT.)

The size of a transducer generally refers to the size of the active element (diameter in millimeters). The size is determined by the area to be examined. In the case of general abdominal B-scanning, there are few anatomical restrictions. However, a larger-diameter transducer provides ease of mechanical scanning and optimum focus at the required longer focal zones. Smaller transducers are best suited for investigations in anatomic areas of irregular shapes, reduced skin surface, and investigations necessitating improved resolution with minimal penetration (i.e., eye, neck, chest, and limbs). In echocardiography, the smaller active-element diameters are used to facilitate ease of placement and maneuverability within the intercostal space. The 13-mm active-element diameter transducer is most commonly used for investigations on the average adult. When investigating an extremely large and/or obese patient or when placement is not critical, a 20-mm diameter can be used. On the other hand, a 6-mm diameter is recommended for pediatric and neonatal echocardiography.

The type of tissue and the amount of tissue attenuation encountered during a diagnostic examination determines the frequency to be used. A frequency of 2.25 MHz is the most commonly used for general-purpose ultrasonography. Transducers of higher frequency, such as 3.5 or 5.0 MHz, are utilized when tissue penetration is easily accomplished and improved resolution at short distances is required. In view of the small size and shorter examination depth in children, a higher frequency is usually used to ensure good resolution. The 5.0-MHz transducer is recommended for the majority of pediatric echocardiography, reserving the 3.5 MHz for investigations of the older and larger children.

When tissue penetration becomes a concern, as with most obese patients, a low-frequency (i.e., 2.25 or 1.6 MHz) transducer should be employed, which can provide more information than can be obtained with a transducer of a higher frequency.

Echocardiology transducers are available nonfocused or in a choice of focal configurations. Generally, a focused transducer is preferred for echocardiography. The sound beams generated in a focused transducer have a reduced width to provide optimal lateral resolution and sensitivity at given depths. In selecting a focal length, two factors must be considered: the distance from the transducer to the structure to be investigated and the transducer's diameter. The focal zone of a transducer refers to the distance (in centimeters) the sound beam travels in water to a standard test object. The focal zones are commonly referred to as short (5.0 cm), medium (7.5 cm), and long (10.0 cm) in water. If echocardiography is to be performed on a small child, the distance from the transducer to the cardiac structure usually falls within 5.0 cm; therefore, a short focal length should be utilized.

Size of the transducer also affects the choice of focal lengths. For instance, a 20-mm active-element diameter can be focused best at the longer focal lengths, whereas a 13-mm or smaller diameter can accommodate all three with optimum results.

A variation of the standard echocardiology transducer is the cardiac suprasternal notch transducer. This type of transducer is designed with a small active-element diameter and is commonly utilized at a frequency of 3.5 or 5.0 MHz. When these transducers are positioned in the region of the suprasternal notch, simultaneous measurement of the aortic arch, right pulmonary artery, and left atrium can be easily obtained. The cardiac suprasternal notch transducer is extremely useful in diagnosing and monitoring a variety of heart diseases and defects, including mitral insufficiency, patent ductus arteriosus, ventricular septal defects, and hypertensive cardiovascular heart disease.

The transducers are available also as general scanners, standard medical transducers which can be utilized for A-mode or M-mode application, biopsy transducers for tissue biopsy under direct vision, plus special types for thyroid examination and ophthalmic ultrasonography. There are also microprobe needle transducers.

9.3.2. Echocardiography

One use of ultrasound in the cardiovascular system has already been discussed in Section 6.3. This is the Doppler technique in blood flow measurement. Pulsed Doppler ultrasound can be used to measure the velocity gradient across a blood vessel as well as the velocity of the heart wall or specific valves in the heart. Some additional applications will be discussed later in the chapter. The major application in cardiovascular diagnosis, however, is the *echocardiogram,* which utilizes an M-scan technique. In the echocardiogram movements of the valves and other structures of the heart are displayed as a function of time and usually in conjunction with an electrocardiogram.

Over the past several years, echocardiology has been extremely useful in diagnosing many cardiac abnormalities, among them are calcific aortic stenosis, pulmonary valve stenosis, mitral valve stenosis, left atrial myxoma, mitral valve prolapse, and rheumatic heart disease.

In selecting a transducer for an echocardiographic investigation, the following factors must be considered for optimum results:

1. The type of investigation to be performed.
2. Physical size of the patient.
3. The anatomic area involved.
4. The type of tissue to be encountered.
5. The depth of the structure(s) to be studied.

Figure 9.12 is a typical echocardiogram. Figure 9.13 is a sketch showing placement of the transducer. For this particular echocardiogram the transducer was placed so that the beam crossed the chest wall into the right ventricle, through the septum, into the left ventricle, and ultimately through the left atrium. The aorta and mitral valve are also imaged.

With ultrasound it is possible to distinguish between different soft tissues and to measure the motion of structures of the heart. This fact has made it a valuable method of analysis in cardiology. One important factor is that there is virtually no interference by echoes from other body structures, since the heart is surrounded by the lungs, which are literally air bags. This helps greatly in the interpretation. The heart has a number of acoustic interfaces, such as the atrial and ventricular walls, the septum, and the various valves. The position and movements of each interface can be measured by the reflected ultrasound. The echoes from these walls and valves are predictable since the components of the heart move in a known manner.

A good example of this technique is shown in Figure 9.12. This type of echocardiogram is useful in interpreting the movements of the mitral valve with respect to time. The mobility of the valve is measured by the displacement of the echo per unit time during diastole. If the mobility is reduced, as in the case of mitral stenosis (narrowing of the left ventricular orifice), its severity compared with other existing conditions can determine the appropriate action to take, including the possible use of surgery.

Another use of echocardiography is in the detection of fluids. When the pericardium (the sac surrounding the heart) is inflamed (a condition known as pericarditis), there is sometimes an escape of fluid. The presence of this fluid can be detected in the echocardiogram.

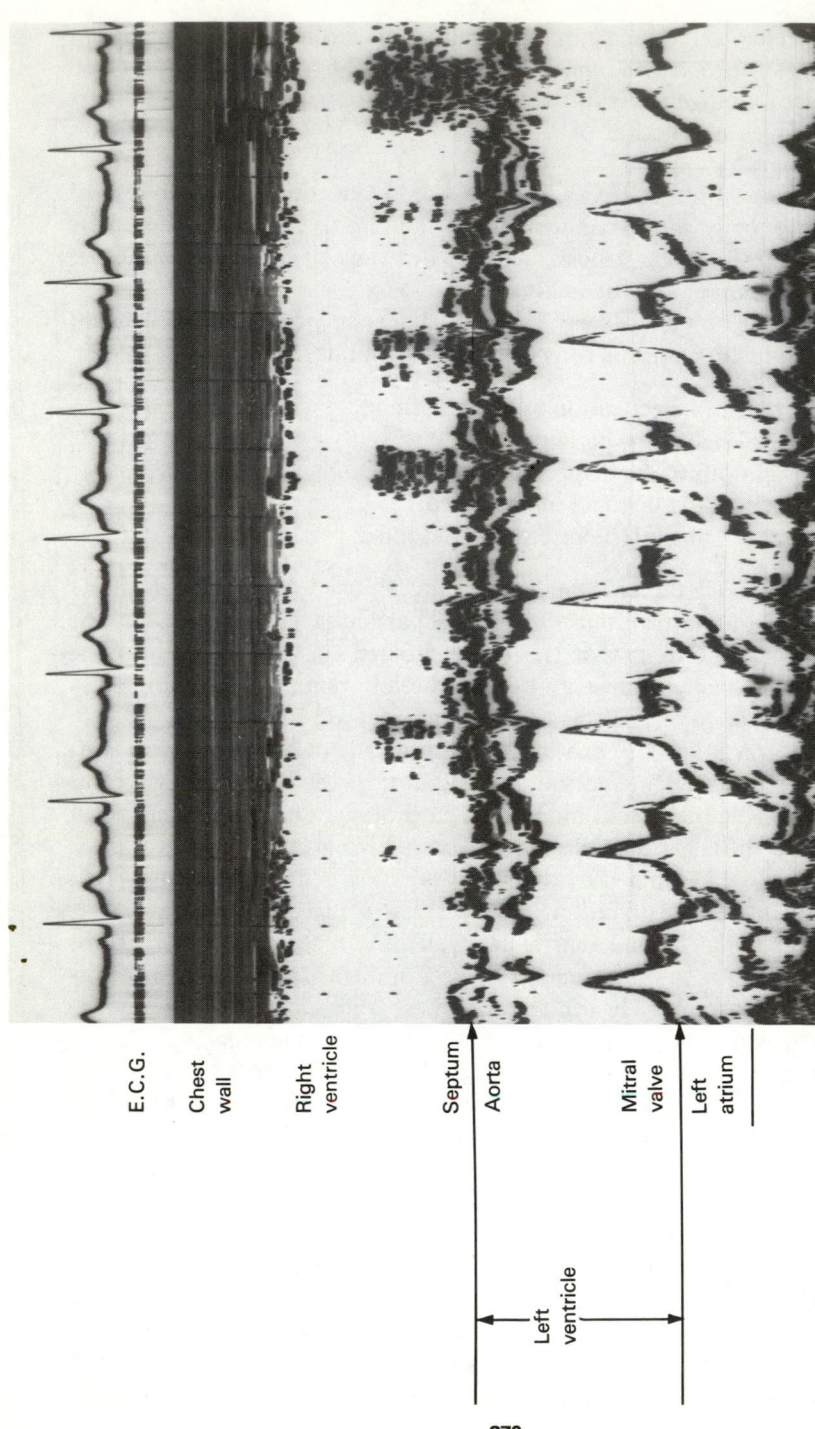

Figure 9.12. Echocardiogram. (Courtesy of Dept. of Cardiology, Cedars-Sinai Hospital, Los Angeles, CA.)

E.C.G.

Chest wall

Right ventricle

Septum

Aorta

Left ventricle

Mitral valve

Left atrium

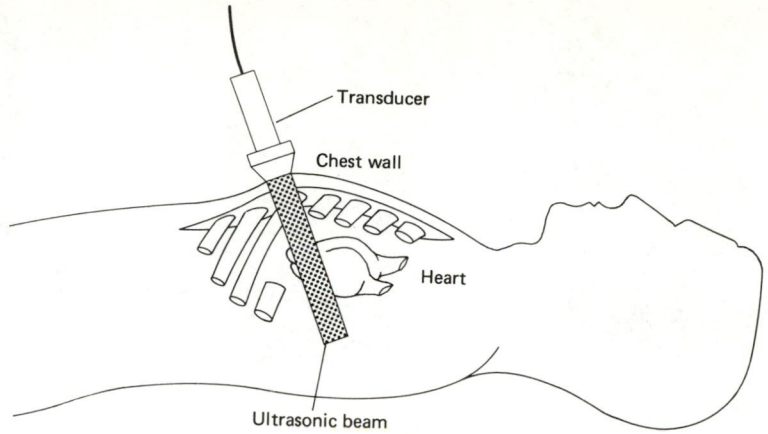

Figure 9.13. Typical transducer placement to obtain echocardiogram.

Fast scanning speeds are required to prevent blurring of the image due to heart movements. To give a thorough dynamic analysis, many simultaneous recordings can be taken in different positions so as to give cross-sectional images at various points along the length of the heart. As techniques and interpretations continue to advance, the medical profession believes that the potential of this diagnostic method will continue to increase.

9.3.3. Echoencephalography

A well-established clinical application of ultrasonic imaging using the A-scan mode of display is the echoencephalogram, which is used in determining the location of the midline of the brain. An ultrasonic transducer is held against the side of the head to measure the distance to the midline of the brain. The midline echoes from both sides of the head are simultaneously displayed on the oscilloscope, one side producing upward deflection of the beam and the other producing downward deflection.

In the normal brain these two deflections line up, indicating equal distance from the midline to each side of the head. Nonalignment of these deflections indicates the possibility of a tumor or some other disorder that might cause the midline of the brain to shift from its normal position. The instrument for this measurement produces ultrasound energy at a frequency of from 1 to 10 MHz. The pulse rate is 1000 per second.

9.3.4. Ophthalmic Scans

Another important application involves diagnostic scanning of the eye. Figure 9.14 shows an ophthalmic B-scan with camera. The transducer is shown at the bottom of the picture. The transducer is placed directly over

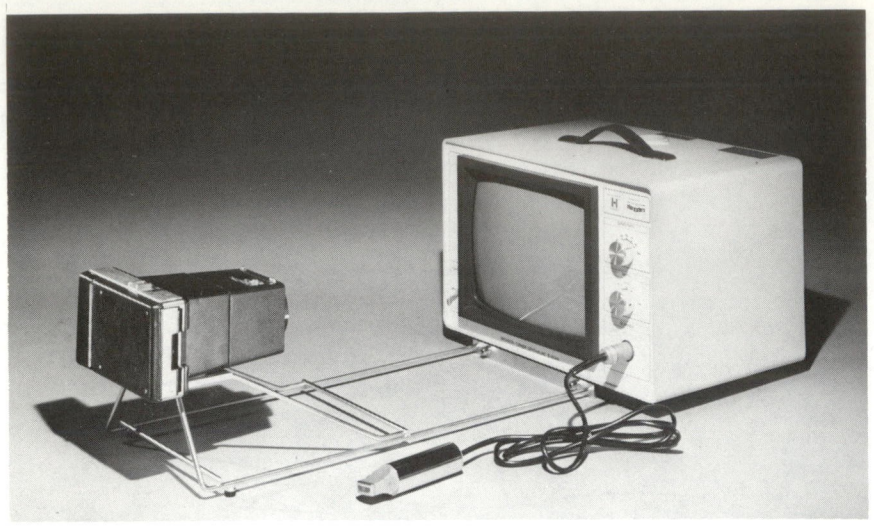

Figure 9.14. Ophthalmic B-scan with camera. (Courtesy of Storz Instrument Co., St. Louis, MO.).

Figure 9.15. Typical ultrasonograms of the eye. (a) Phthisis bulbi (wasting away of the eye) in a 13-year-old child. (b) Papilledema. (c) Retinitis proliferans with detachment of the retina associated with this condition. (d) Intraocular foreign body. (Courtesy of Storz Instrument Co., St. Louis, MO.)

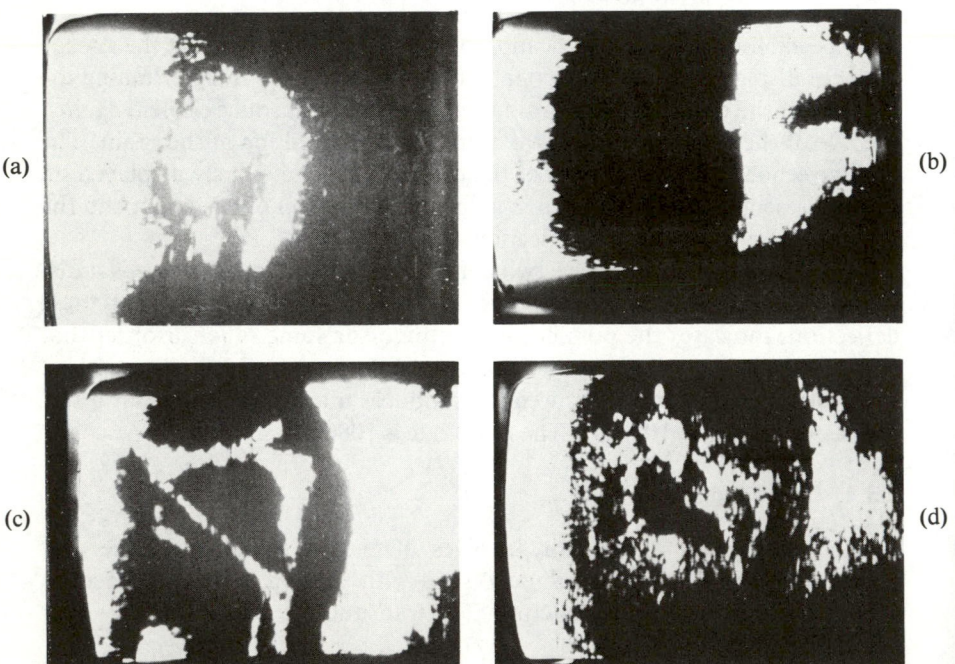

the eye of the patient. The result is a series of ultrasonograms of the type shown in Figure 9.15. Ultrasonic techniques are the only means of identifying intraocular pathology in the presence of opaque media.

9.3.5. Some Other Types of Ultrasonic Imaging

Ultrasonic techniques utilizing the B-scan mode are used for visualizing various organs and structures of the body, including the breasts, kidneys, and the organs and soft tissues of the abdomen. In obstetrics and gynecology, ultrasonic imaging permits visualization of very small structures, such as the ovaries and tubes, and permits examination of a fetus as early as the 4-week stage. Diagnostic ultrasonic equipment used for intracranial and abdominal visualization generally utilizes frequencies from 1 to 2 MHz, whereas for examination of the eye, breast, and body surfaces, frequencies in the range 4 or 5 to 15 MHz are used to obtain better resolution.

9.3.6. Other Applications

Ultrasonic tomography techniques, in which information from scans taken from many different vantage points are combined mathematically, provide increased detail in the visualization of certain parts of the body. The principle involved is very similar to that of computerized axial tomography utilizing X-ray information, which is described in detail in Chapter 15. However, because of the limited penetrating range, particularly at higher frequencies, and the requirement that no gaseous regions or bone lie in the ultrasonic path, the uses in which ultrasonic tomography is practical are limited. Where these techniques can be used, however, they provide a way of obtaining detailed cross sections without radiation exposure.

9.3.7. The Noninvasive Vascular Laboratory

In Section 9.3.2 the diagnosis of the heart and the use of ultrasound in cardiology were discussed. In Section 6.3.2, measurement of blood flow by ultrasonic methods was described. Also, in the introductory paragraphs of Section 9.3, the phased-array ultrasonograph was presented as an example of a medical ultrasound laboratory. To expand on these ideas and to help complete the picture on the status of the use of ultrasound in cardiology it is appropriate to introduce the *noninvasive vascular laboratory*. Although defined in many ways and by various manufacturers of the necessary equipment, this type of laboratory, in general, involves the scanning of the vascular portion of the cardiovascular system, and, as in inferred by the name, all methods are noninvasive. Also, the procedures can be performed with outpatients as well as those in the hospital. Some major hospitals have set up

Figure 9.16. Dopscan Ultrasonic Doppler arterial scanning system. (Courtesy of Carolina Medical Electronics, King, NC.)

subdepartments within their cardiology departments for many of these relatively new procedures.

A typical noninvasive vascular system is illustrated in Figure 9.16. The ultrasonic Doppler arterial scanning system utilizes the Doppler effect described in Sections 9.2.1 and 9.2.2. It is primarily used for the diagnosing of potential stroke conditions using Doppler ultrasound scanning of the carotid and other arteries in the neck area, and can be used in conjunction with X-ray angiography or as an alternative to it. The system is used for patients who are suspected of reduced cerebral circulation or who have arterial bruits (noises) caused by turbulence in the flow of blood through a constriction. It enables blood vessel mappings to be made, including photographic records, chart recordings of pulsatile directional blood flow, and magnetic tape recordings of arterial flow sounds along with comments by the operator. It is extremely useful for patients who are considered to be risk candidates in X-ray angiography.

In a normal procedure, the patient relaxes in a reclining position on an examination table while the carotid arterial system is transcutaneously scanned with a focused ultrasonic beam. The small, smooth probe is attached to an X-Y position-sensing arm. As the technician moves the probe in a prescribed scanning pattern over the skin, the probe's position is plotted on the screen of a storage oscilloscope whenever it senses blood flow. Repeated passes over the carotid bifurcation and adjacent arteries gradually construct on the oscilloscope a representation of the vessels scanned. Simultaneously, blood flow velocity is displayed on a monitor scope and the sounds characteristic of ultrasonically detected blood flow are available by means of a loudspeaker or headphones. All information is permanently recorded.

When the screening is complete, the vascular map is photographed, and this, along with the pulsatile flow tracings, the sound recordings, and the operator's comments provide permanent graphic and aural records for subsequent clinical evaluation.

Although the image obtained does not have as much intimate detail as one obtained by angiography, it is quite adequate in most cases. For example, it is quite easy to detect the narrowing of a major artery (arterial stenosis).

This system is also useful in investigating blood flow in the ophthalmic region and some of the pathways around the vertebrae and the base of the skull. It can be used on other major superficial vessels such as femoral, brachial, and popliteal arteries.

Figure 9.17. Directional Doppler. (Courtesy of Parks Electronic Laboratory, Beaverton, OR.)

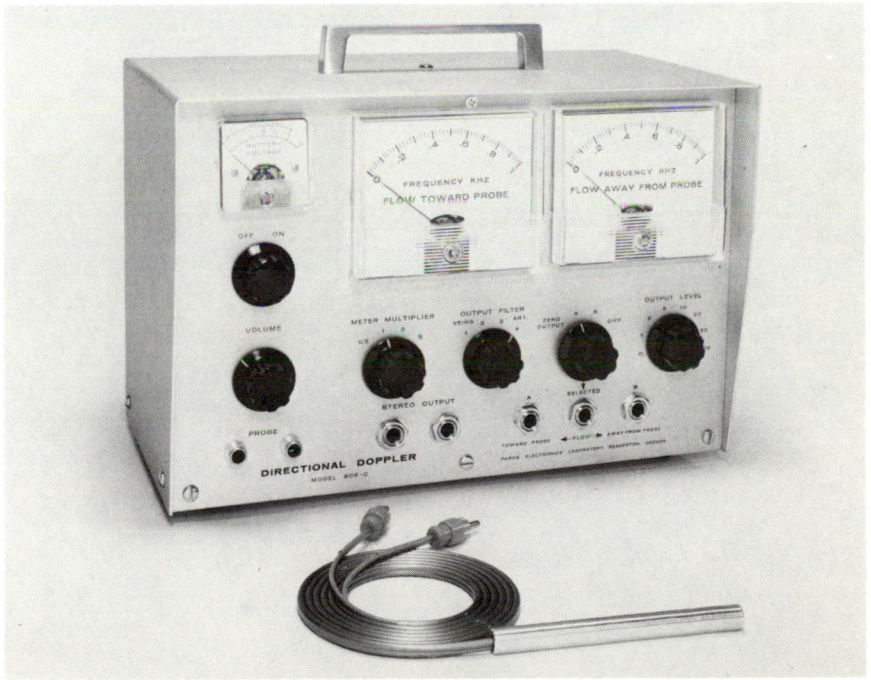

Another useful device for the noninvasive vascular laboratory is illustrated in Figure 9.17. This instrument is a single-frequency directional Doppler blood flow detector. A dual-frequency model is also available. The single-frequency model employs either 5 or 10 MHz, whereas the dual-frequency type has both frequencies available in the same machine. The 10-MHz ultrasound is better for the small arteries around the eye, whereas the lower frequency gives better results for the deep vessels of the thigh and the iliacs.

Examining the face of the instrument in the photograph, the directional aspects can be observed on the two meters showing blood flow away from and toward the probe. The probe is shown at the bottom of the picture and is plugged in when used. It should be noted that stereo headphones can be plugged into the stereo output. This procedure is useful in small-vessel studies because the stereo effect lets the operator know if the ultrasonic beam is intercepting more than one vessel. This device can be used for venous flow as well as arterial. An external speaker-amplifier is available if the instrument is to be used in teaching.

10

The Nervous System

The task of controlling the various functions of the body and coordinating them into an integrated living organism is not simple. Consequently, the nervous system, which is responsible for this task, is the most complex of all systems in the body. It is also one of the most interesting. Composed of the brain, numerous sensing devices, and a high-speed communication network that links all parts of the body, the nervous system not only influences all the other systems but is also responsible for the behavior of the organism. In this broad sense, *behavior* includes the ability to learn, remember, acquire a personality, and interact with its society and the environment. It is through the nervous system that the organism achieves autonomy and acquires the various traits that characterize it as an individual.

A complete study of the nervous system, with all its ramifications, would be far beyond the scope of this book. However, an overall view can be given that provides the reader with a physiological background for measurements within the nervous system, as well as some understanding of the effect of the nervous system on measurements from other systems of the body. To make this presentation more useful in the study of biomedical instrumentation, many of the concepts and theories are greatly simplified. This simplification is not intended to detract from the reader's understanding of the concepts and theories, but it should facilitate visualization of an extremely complex system and provide a better perspective for further detailed study, if required. The simplification requires, however, that caution must be used in attempting to extrapolate or generalize from the information presented.

10.1. THE ANATOMY OF THE NERVOUS SYSTEM

The basic unit of the nervous system is the *neuron.* A neuron is a single cell with a *cell body,* sometimes called the *soma,* one or more "input" fibers called *dendrites,* and a long transmitting fiber called the *axon.* Often the axon branches near its ending into two or more terminals. Examples of three different types of neurons are shown in Figure 10.1.

The portion of the axon immediately adjacent to the cell body is called the *axon hillock*. This is the point at which action potentials are usually generated. Branches that leave the main axon are often called *collaterals.* Certain types of neurons have axons or dendrites coated with a fatty insulating substance called myelin. The coating is called a *myelin sheath* and the fiber is said to be *myelinated.* In some cases, the myelin sheath is interrupted at rather regular intervals by the *nodes of Ranvier,* which help speed the transmission of information along the nerves. Outside of the central nervous system, the myelin sheath is surrounded by another insulating layer, sometimes called the *neurilemma*. This layer, thinner than the myelin sheath and continuous over the nodes of Ranvier, is made up of thin cells, called *Schwann cells.*

As can be seen from Figure 10.1, some neurons have long dendrites, whereas others have short ones. Axons of various lengths can also be found throughout the nervous system. In appearance, it is difficult to tell a dendrite from an axon. The main difference is in the function of the fiber and the direction in which it carries information with respect to the cell body.

Both axons and dendrites are called *nerve fibers,* and a bundle of individual nerve fibers is called a *nerve.* Nerves that carry sensory information from the various parts of the body *to the brain* are called *afferent nerves,* whereas those that carry signals *from the brain* to operate various muscles are called *efferent nerves.*

The *brain* is an enlarged collection of cell bodies and fibers located inside the skull, where it is well protected from light as well as from physical, chemical, or temperature shock. At its lower end, the brain connects with

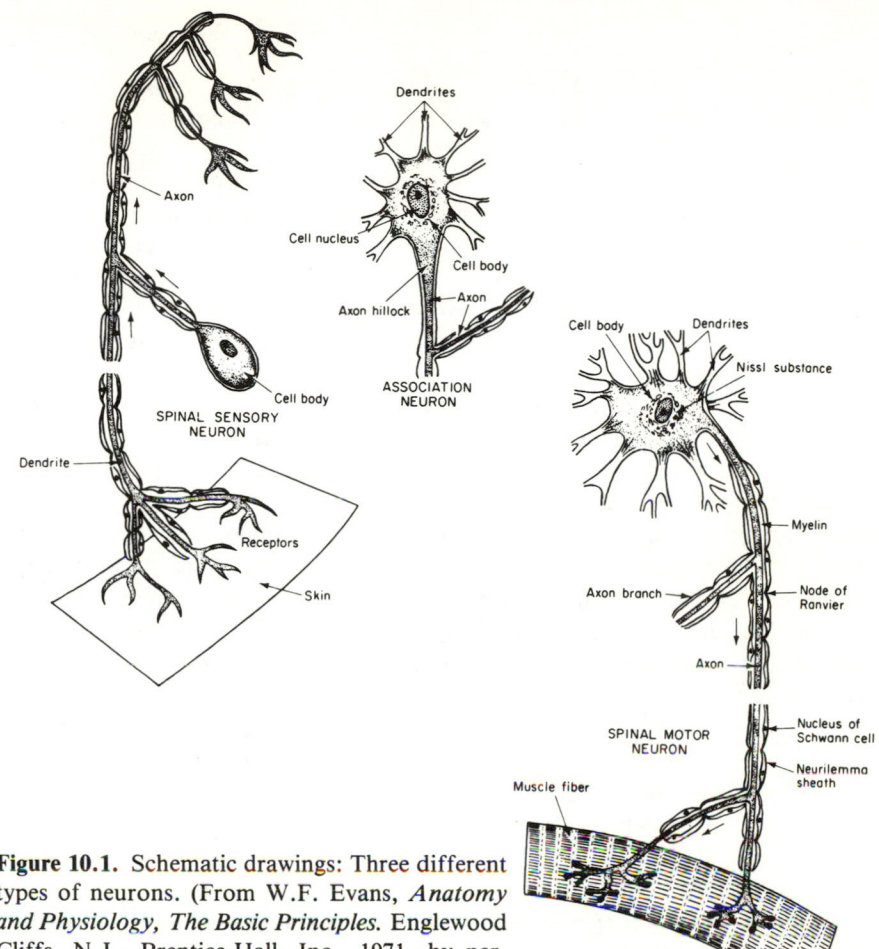

Figure 10.1. Schematic drawings: Three different types of neurons. (From W.F. Evans, *Anatomy and Physiology, The Basic Principles.* Englewood Cliffs, N.J., Prentice-Hall, Inc., 1971, by permission.)

the *spinal cord,* which also consists of many cell bodies and fiber bundles. Together the brain and spinal cord comprise one of the main divisions of the nervous system, the *central nervous system* (CNS). In addition to a large number of neurons of many varieties, the central nervous system also contains a number of large fatty cell bodies called *glial cells.* About half the brain is composed of glial cells. At one time it was believed that the main function of glial cells was structural and that they physically supported the neurons in the brain. Later it was postulated, however, that the glial cells play a vital role in ridding the brain of foreign substances and seem to have some function in connection with memory.

Cell bodies and small fibers in fresh brain are gray in color and are called *gray matter,* whereas the myelin coating of larger fibers has a white appearance, so that a collection of these fibers is referred to as *white matter.*

Collections of neuronal cell bodies within the central nervous system are called *nuclei,* while similar collections outside the central nervous system are called *ganglia.*

The central nervous system is generally considered to be *bilaterally symmetrical,* which means that most structures are anatomically duplicated on both sides. Even so, some functions of the central nervous system in humans seem to be located nonsymmetrically. Several of the functions of the central nervous system are *crossed over,* so that neural structures on the left side of the brain are functionally related to the right side of the body, and vice versa.

Nerve fibers outside the central nervous system are called *peripheral nerves.* This name applies even to fibers from neurons whose cell bodies are contained within the central nervous system. Throughout most of their length, many peripheral nerves are mixed, in that they contain both afferent and efferent fibers. Afferent peripheral nerves that bring sensory information into the central nervous system are called *sensory nerves,* whereas efferent nerves that control the motor functions of muscles are called *motor nerves.* Peripheral nerves leave the spinal cord at different levels, and the nerves that *innervate* a given level of body structures come from a given level of the spinal cord.

The interconnections between neurons are called *synapses.* The word "synapse" can be used as both a noun and a verb. Thus, the connection is called a *synapse,* and the act of connecting is called *synapsing.* All synapses occur at or near cell bodies. As explained in Section 10.2, mammalian neurons that synapse do not touch each other but do come into close proximity, so that the axon (output) of one nerve can activate the dendrite or cell body (input) of another by producing a chemical that stimulates the membrane of a dendrite or cell body. In some cases, the chemical is produced by one axon, near another axon, to inhibit the second axon from activating a neuron with which it can normally communicate. This action is explained more fully below. Because of the chemical method of transmission across a synapse from axon to dendrite or cell body, the communication can take place in one direction only.

The peripheral nervous system actually consists of several subsystems. The system of afferent nerves that carry sensory information from the sensors on the skin to the brain is called the *somatic sensory nervous system.* *Visual pathways* carry sensory information from the eyes to the brain, whereas the *auditory nervous system* carries information from the auditory sensors in the ears to the brain.

Another major division of the peripheral nervous system is the *autonomic nervous system,* which is involved with emotional responses and controls smooth muscle in various parts of the body, heart muscle, and the secretion of a number of glands. The autonomic nervous system is composed of two main subsystems that appear to be somewhat antagonistic to each other, although not completely. These are the *sympathetic nervous system,* which speeds up the heart, causes secretion of some glands, and inhibits other body functions, and the *parasympathetic nervous system,* which tends to slow the heart and controls contraction and secretion of the stomach. In general, the sympathetic nervous system tends to mobilize the body for emergencies, whereas the parasympathetic nervous system tends to conserve and store bodily resources.

A very general look at the anatomy of the brain should be helpful in understanding the functions of the nervous system. Figure 10.2 shows a side view of the brain and spinal cord, and Figure 10.3 is a cutaway showing some of the major structures.

The part of the brain that connects to the spinal cord and extends up into the center of the brain is called the *brainstem.* The essential parts of the brainstem are the *medulla* (sometimes called the *medulla oblongata*), which is the lowest section of the brainstem itself, the *pons* located just above the medulla and protruding somewhat in front of the brainstem, and the upper part of the brainstem called the *midbrain.* Above and slightly forward

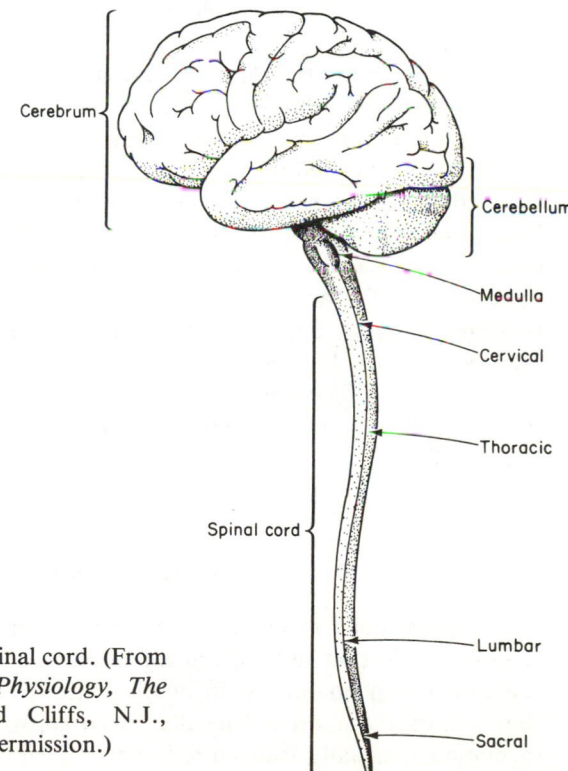

Figure 10.2. The brain and spinal cord. (From W.F. Evans, *Anatomy and Physiology, The Basic Principles,* Englewood Cliffs, N.J., Prentice-Hall, Inc., 1971, by permission.)

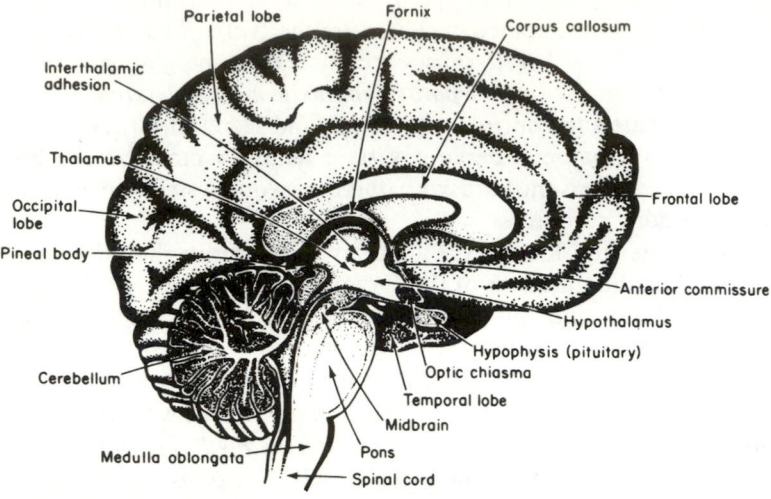

Figure 10.3. Cutaway section of the human brain. (From W.F. Evans, *Anatomy and Physiology, The Basic Principles,* Englewood Cliffs, N.J., Prentice-Hall, Inc., 1971, by permission.)

of the midbrain are the *thalamus* and *hypothalamus.* Behind the brainstem is the *cerebellum.* Almost completely surrounding the midbrain, thalamus, and hypothalamus are the structures of the *cerebrum.* The outer surface of the cerebrum is called the *cerebral cortex.* The *corpus callosum* is the interconnection between the left and right hemispheres of the brain. Structurally,the two hemispheres appear to be identical, but as indicated earlier, they seem to differ functionally in man. Just forward of the hypothalamus is the *hypophysis* or *pituitary gland,* which produces hormones that control a number of important hormonal functions of the body. Not shown in the figures, but surrounding the thalamus, is the *reticular activating system.* The specific functions of each of these major portions of the brain, as far as they have been discovered to date, are discussed in Section 10.3.

10.2. NEURONAL COMMUNICATION

As discussed in detail in Chapter 3, neurons are among the special group of cells that are capable of being excited and that, when excited, generate action potentials. In neurons, these action potentials are of very short duration and are often called *neuronal spikes* or *spike discharges.* Information is usually transmitted in the form of *spike discharge patterns.*

These patterns, which are simply the sequences of spikes that are transmitted down a particular neuronal pathway, are shown in Figures 10.4 and 10.5. The form of a given neuronal pattern depends on the firing patterns of other neurons that communicate with the neuron generating the pattern and the refractory period of that neuron (see Chapter 3). When an action potential is initiated in the neuron, usually at the cell body or axon hillock, it is propagated down the axon to the axon terminals where it can be transmitted to other neurons.

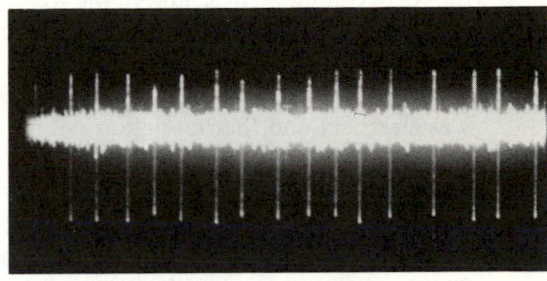

Figure 10.4. Spike discharge pattern from a single neuron in the red nucleus of a cat. The red nucleus is involved with motor functions. (Courtesy of Neuropsychology Research Laboratory, Veterans Administration Hospital, Sepulveda, CA.)

Figure 10.5. Spike discharge patterns from a single thalamic cell in a cat: (a) Random quiet pattern; (b) Burst pattern. (Courtesy of Neuropsychology Research Laboratory, Veterans Administration Hospital, Sepulveda, CA.)

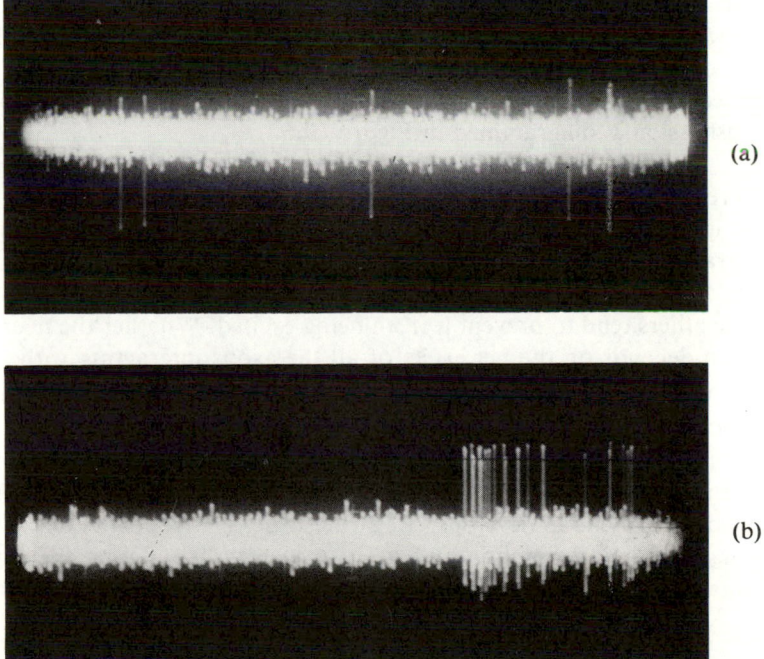

(a)

(b)

Given sufficient excitation energy, most neurons can be triggered at almost any point along the dendrites, cell body, or axon and generate action potentials that can move in both directions from the point of initiation. The process does not normally happen, however, because in their natural function, neurons synapse only in a certain way; that is, the axon of one neuron excites the dendrites or cell body of another. The result is a one-way communication path only. If an action potential should somehow be artificially generated in the axon and caused to travel up the neuron to the dendrites, the spike cannot be transmitted the wrong way across the gap to the axon of another neuron. Thus, the one-way transmission between neurons determines the direction of communication.

It was believed for many years that transmission through a synapse was electrical and that an action potential was generated at the input of a neuron due to ionic currents or fields set up by the action potentials in the adjacent axons of other neurons. More recent research, however, has disclosed that in mammals, and in most synapses of other organisms, the transmission times across synapses are too slow for electrical transmission. This has led to the presently accepted chemical theory, which states that the arrival of an action potential at an axon terminal releases a chemical— Probably *acetylcholine* in most cases—that excites the adjacent membrane of the receiving neuron. Because of the close proximity of the transmitting axon terminal to the receiving membrane, the time of transmission is still quite short. The possibility that some of the chemical may still be present after the refractory period is eliminated by the presence of *acetylcholine esterase,* another chemical that breaks down the acetylcholine as soon as it is produced, but not before it has been able to initiate its intended action potential in the nearby membrane. This chemical theory of transmission is diagrammed in Figure 10.6.

Actually, the situation is not quite as simple as has been described. There are really two kinds of communication across a synapse, excitatory and inhibitory. The same chemical appears to be used in both. In general, several axons from different neurons are in communication with the "input" of any given neuron. Some act to excite the membrane of the receiver, while others tend to prevent it from being excited. Whether the neuron fires or not depends on the net effect of all the axons interacting with it.

The effects of the various neurons acting on a receiving neuron are reflected in changes in the graded potentials of the receiving neuron. *Graded potentials* are variations around the average value of the resting potential. When this graded potential reaches a certain threshold, the neuron fires and an action potential develops. Regardless of the graded potential before firing, the action potentials of a given neuron are always the same and always travel at the same rate. An excitatory graded potential is called an

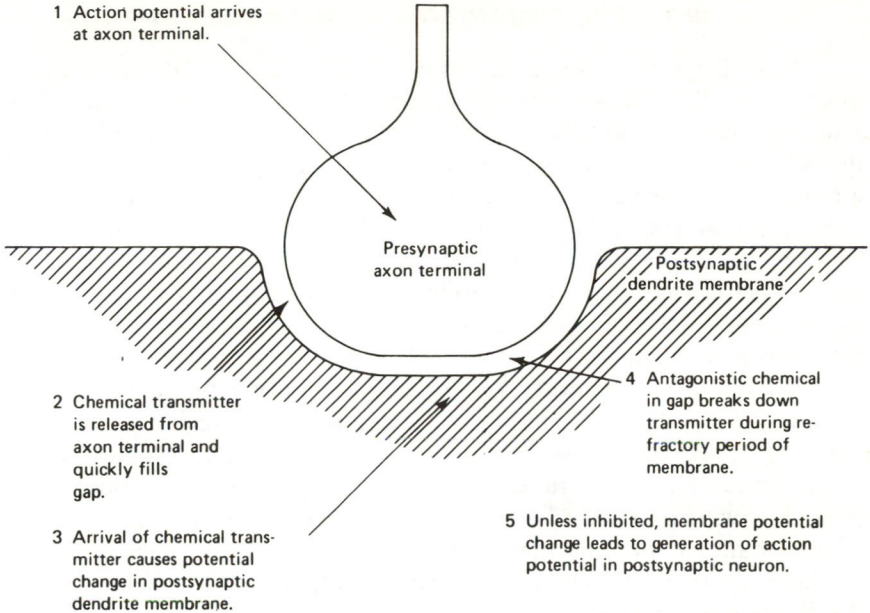

1 Action potential arrives at axon terminal.

Presynaptic axon terminal

Postsynaptic dendrite membrane

2 Chemical transmitter is released from axon terminal and quickly fills gap.

3 Arrival of chemical transmitter causes potential change in postsynaptic dendrite membrane.

4 Antagonistic chemical in gap breaks down transmitter during refractory period of membrane.

5 Unless inhibited, membrane potential change leads to generation of action potential in postsynaptic neuron.

Figure 10.6. Sequence of events during chemical transmission across a synapse.

excitatory postsynaptic potential (EPSP), and an inhibitory graded potential is called an *inhibitory postsynaptic potential* (IPSP).

There are several theories as to how inhibitory action takes place. One possibility is that the inhibitory axon somehow causes a graded potential (IPSP) in the receiving neuron which is more negative than the normal resting potential, thus requiring a greater amount of excitation to cause it to fire. Another possibility is that the inhibiting axon acts, not on the receiving neuron but on the excitatory transmitting axon. In this case, the inhibiting axon might set up a premature action potential in the transmitting axon, so that the necessary combination of chemical discharges cannot occur in synchronism as it would without the inhibition. Whatever method is actually used, the end result is that certain action potentials which would otherwise be transmitted through the synapse are prevented from doing so when inhibitory signals are present. Synapses, then, behave much like multiple-input AND and NOR logic gates and, by their widely varied patterns of excitatory and inhibitory "connections," provide a means of switching and interconnecting parts of the nervous system with a complexity far greater than anything yet conceived by man.

10.3. THE ORGANIZATION OF THE BRAIN

Knowledge of the actual function of various parts of the brain is still quite sparse. Experiments in which portions of an animal's brain have been removed (oblated) show that there is a tremendous amount of redundancy in the brain. There is also a great amount of adaptivity in that if a portion of a brain believed responsible for a given function is removed from an infant animal, the animal somehow still seems able to develop that function to some extent. This result has led to the idea of the "law of mass action," in which it is theorized that the impairment caused by damage to some portion of the brain is not so much a function of *what* portions have been damaged but, rather, *how much* was damaged. In other words, when one region of the brain is damaged, another region seems to take over the function of the damaged part. Also, while tests show that a particular region of the brain seems to be related to some specific function, there are also indications of some relationship of that function to other parts of the brain. Thus, when, in the following paragraphs, certain functions are indicated for certain parts of the brain, it must be realized that these parts only seem to play a predominant role in those functions and that other parts of the brain are undoubtedly also involved.

In the brainstem, the *medulla* seems to be associated with control of some of the basic functions responsible for life, such as breathing, heart rate, and kidney functions. For this purpose, the medulla seems to contain a number of timing mechanisms, as well as important neuronal connections.

The *pons* is primarily an interconnecting area. In it are a large number of both ascending and descending fiber tracts, as well as many nuclei. Some of these nuclei seem to play a role in salivation, feeding, and facial expression. In addition, the pons contains relays for the auditory system, spinal motor neurons, and some respiratory nuclei.

The cerebellum acts as a physiological microcomputer which intercepts various sensory and motor nerves to smooth out what would otherwise be "jerky" muscle motions. The cerebellum also plays a vital role in man's ability to maintain his balance.

The *thalamus* manipulates nearly all sensory information on its way to the cerebrum. It contains main relay points for the visual, auditory, and somatic sensory systems.

The *reticular activation system* (RAS), which surrounds the thalamus, is a nonspecific sensory portion of the brain. It receives excitation from all the sensory inputs and seems to be aroused by any one of them, but it does not seem to distinguish which type of sensory input is active. When aroused, the RAS alerts the cerebral cortex, making it sensitive to incoming information. It is the RAS that keeps a person awake and alert and causes him to pay attention to a sensory input. Most information reaching the RAS is relayed through the thalamus.

The *hypothalamus* is apparently the center for emotions in the brain. It controls the neural regulation of endocrine gland functions via the pituitary gland and contains nuclei responsible for eating, drinking, sexual behavior, sleeping, temperature regulation, and emotional behavior generally. The hypothalamus exercises primary control over the autonomic nervous system, particularly the sympathetic nervous system.

The *basal ganglia* seem to be involved in motor activity and have indirect connections with the motor neurons.

The main subdivision of the *cerebrum* is the *cerebral cortex,* which contains some 9 billion of the 12 billion neurons found in the human brain. The cortex is actually a rather thin layer of neurons at the periphery of the brain, which contains many fissures or inward folds to provide a greater amount of surface area. Some of the deeper fissures, also called *sulci,* are used as landmarks to divide the cortex into certain lobes. Several of the more prominent ones are shown in Figure 10.7, along with the location of the important lobes.

Figure 10.7. The cerebral cortex. (From W.F. Evans, *Anatomy and Physiology, The Basic Principles,* Englewood Cliffs, N.J., Prentice-Hall, Inc., 1971, by permission.)

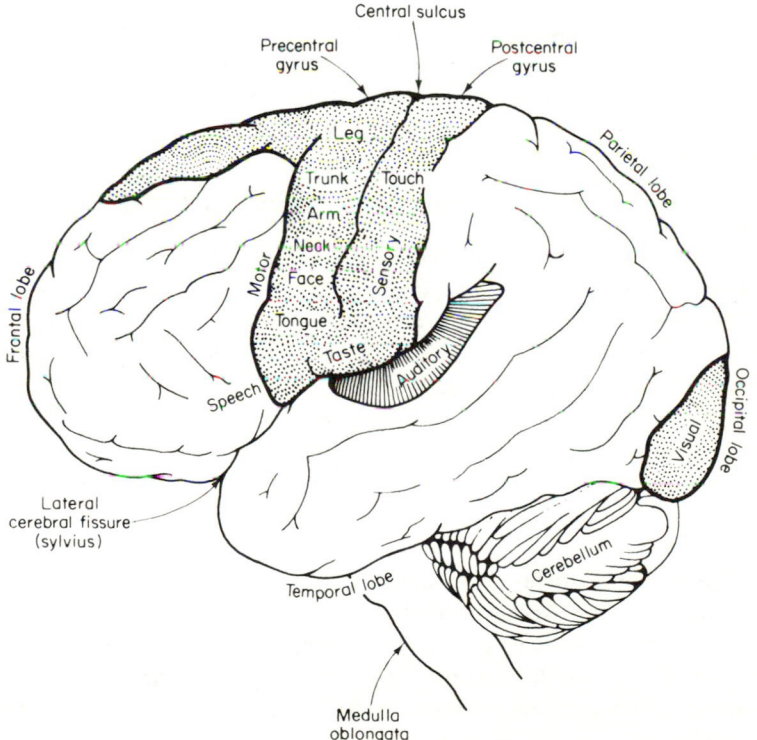

All sensory inputs eventually reach the cortex, where certain regions seem to relate specifically to certain modalities of sensory information. Other regions of the cortex seem to be specifically related to motor functions. For example, all somatic sensory (heat, cold, pressure, touch, etc.) inputs lead to a region of the cortical surface just behind the central sulcus, encompassing the forward part of the *parietal lobe.* Somatic sensory inputs from each part of the body lead to a specific part of this region, with the inputs from the legs and feet nearest the top, the torso next, followed by the arms, hands, fingers, face, tongue, pharynx, and, finally, the intra-abdominal regions at the bottom. The amount of surface allotted to each part of the body is in proportion to the number of sensory nerves it contains rather than its actual physical size. A pictorial representation of the layout of these areas, called a *homunculus,* is depicted as a rather grotesque human figure, upside down, with enlarged fingers, face, lips, and tongue.

Just forward of the central sulcus is the *frontal lobe,* in which are found the primary motor neurons that lead to the various muscles of the body. The motor neurons are also distributed on the surface of the cortex in a manner similar to the sensory neurons. The location of the various motor functions can also be represented by a homunculus, also upside down but proportioned according to the degree of muscular control provided for each part of the body.

Figure 10.8 shows both the sensory and motor homunculi, which represent the spatial distribution of the sensory and motor functions on the cortical surface. In each case, the figure shows only one-half of the brain in cross section through the indicated region.

The forward part of the brain, sometimes called the *prefrontal lobe,* contains neurons for some special motor control functions, including the control of eye movements.

The *occipital lobe* is at the very back of the head, over the cerebellum. The occipital lobe contains the visual cortex, in which the patterns obtained from the retina are mapped in a geographic representation.

Auditory sensory input can be traced to the *temporal lobes* of the cortex, located just above the ears. Neurons responding to different frequencies of sound input are spread across the region, with the higher frequencies located toward the front and low frequencies to the rear.

Smell and taste do not have specific locations in the cerebral cortex, although an olfactory bulb near the center of the brain is involved in the perception of smell.

The cerebral cortex has many areas that are neither sensory nor motor. In man, this accounts for the largest portion of the cortex. These *association areas* are believed by many scientists to be involved with integrating or associating the various inputs to produce the appropriate output responses and transmit them to the motor neurons for control of the body.

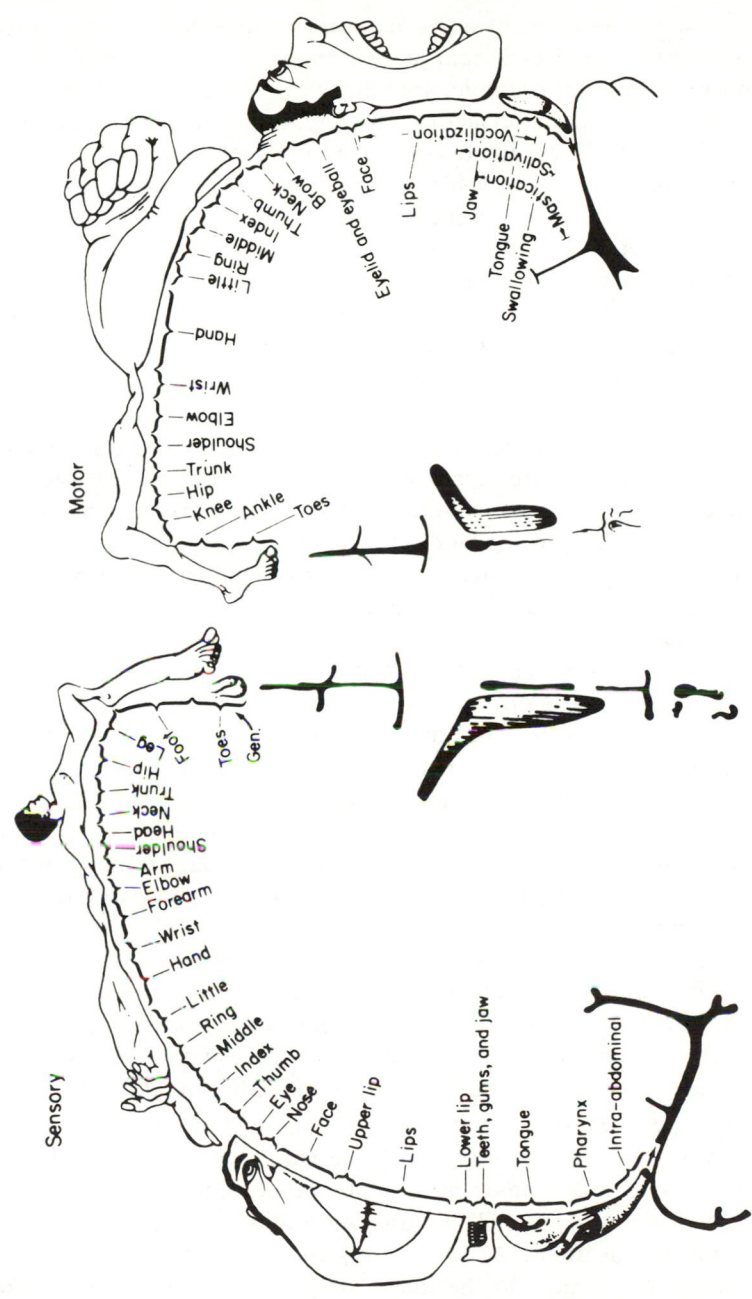

Figure 10.8. Human sensory and motor homunculi. (From W.F. Evans, *Anatomy and Physiology, The Basic Principles*, Englewood Cliffs, N.J., Prentice-Hall, Inc., 1971, by permission.)

10.4. NEURONAL RECEPTORS

Certain special types of neurons are sensitive to energy in some form other than the usual chemical discharge from axons of other neurons. Those in the retina of the eye, for example, are sensitive to light, while others, such as the pressure sensors at the surface of the body, are sensitive to pressure or touch. Actually, any of these sensors can respond to any type of stimulation if the energy level is sufficiently high, but the response is greatest to the form of energy for which the sensor is intended. In each case, the energy sensed from the environment produces patterns of action potentials that are transmitted to the appropriate region of the cerebral cortex. Coding is accomplished either by having a characteristic of the sensed energy determine which neurons are activated or by altering the pattern in which spikes are produced in a given neuron. In some cases, both of these methods are employed.

The *somatic sensory system* consists of receptors located on the skin that respond to pain, pressure, light, touch, heat, or coolness—each receptor responding to one of these modalities. The intensity of the sensory input is coded by the frequency of spike discharges on a given neuron, plus the number of neurons involved. There is considerable feedback to provide extremely accurate localization of the source of certain types of inputs, especially fine touch.

In the *visual system,* light sensors, both rods and cones, are located at the retina of the eye. Some processing of the sensed information occurs at the retina, from which it is transmitted via the optic nerve, through the thalamus, to the occipital cortex. The crossover in the visual system is interesting. Instead of all the information sensed by the left eye going to the right brain, as one might expect, information from the left half of each retina (which views the right visual field) is transmitted to the left side of the brain. Thus, the demarcation is according to the field of vision and not according to which eye generates the signals. The rods, which are more sensitive to dim light, are not sensitive to color, whereas the less-sensitive cones carry the color information. There is still a certain amount of speculation as to the exact manner in which color information is coded, but it is fairly well agreed that the color information is sensed as some combination of primary colors, each of which is carried by its own set of sensors and neurons.

In the *auditory system,* the frequency of sound seems to be coded in two different ways. After passing through the acoustical system into the inner ear, the sound excites the basilar membrane, a rather stiff membrane coiled in a fluid-filled chamber. The sound vibrations are actually carried to the membrane by the fluid. At lower frequencies, the entire membrane seems to vibrate as a unit, and the sound frequency is coded into a spike discharge frequency by the hair-cell sensory neurons located along the membrane. Above a certain crossover point (about 4000 Hz), however, the situation is different, and the frequencies seem to distribute themselves along the basilar membrane. Thus, for higher frequencies, the frequency is coded according to which sensors are activated, whereas for lower frequencies, the coding is by spike discharge frequency. Auditory information from both ears is transmitted to the temporal lobes on both sides of the cerebral cortex. Timing devices are provided so that if a sound strikes both ears a fraction of a millisecond apart, the ear receiving it first inhibits the response from the other ear. This gives the hearing a sense of direction.

This same directional characteristic also applies to smell and taste. That is, an odor reaching one nostril a fraction of a millisecond before it reaches the other causes inhibition that provides a sense of direction for the odor. Coding for taste and smell is not well understood, although intensity is somehow coded into firing rates of neurons as well as the number of neurons activated.

10.5. THE SOMATIC NERVOUS SYSTEM AND SPINAL REFLEXES

The somatic sensory nervous system carries sensory information from all parts of the body to corresponding sites in the cerebral cortex, whereas motor neurons carry control information to the muscles of the body. The sensory and motor neurons are not necessarily single uninterrupted channels that go all the way from the cortex to the big toe, for example. They may have a number of synapses along the way to permit inhibition as well as excitation. There are, of course, exceptions in which some of the motor control functions are carried out by extremely long axons. In this system countless feedback loops control the action of the muscles. The muscles themselves contain stretch and position receptors that permit precise control over their operation.

Many of the routine muscular movements of the body are not controlled by the brain at all but occur as reflexes of the spinal cord. The spinal cord has many nuclei of neurons that give almost automatic response to input stimuli. Actually, only the more complicated responses are controlled by the brain. In a simplified form, this process could be comparable to a large central computer (the brain) connected to a number of small satellite computers in the spinal cord. Each of the small computers handles the data processing and controls the functions of the system within which it operates. Whenever one of the small computers is faced with a situation beyond its limited capability, the data are sent to the central computer for processing. Thus, the spinal reflexes seem to handle all responses except those beyond their capability.

10.6. THE AUTONOMIC NERVOUS SYSTEM

The autonomic nervous system differs from the somatic and motor nervous systems in that its control is essentially involuntary. It was once thought that the autonomic system is completely involuntary, but recent experimentation indicates that it is possible for a person to learn to control portions of this system to some extent.

The major divisions of the autonomic nervous system are the *sympathetic* and *parasympathetic systems.* The sympathetic nervous system receives its primary control from the hypothalamus and is essentially a function of emotional response. It is the sympathetic nervous system that is responsible for the "fight-or-flight" reaction to danger and for such responses as fear and anger. When one or more of the sensory inputs to the brain indicate danger, the body is immediately mobilized for action. The heart rate, respiration, red blood cell production, and blood pressure all increase. Normal functions of the body, such as salivation, digestion, and sexual functions, are all inhibited to conserve energy to meet the situation. Blood flow patterns in the body are altered to favor those functions required for the emergency, and adrenalin, which is the chemical that apparently activates synapses in the sympathetic nervous system, is released throughout the body to maintain the emergency status. Other indications of activation of the sympathetic nervous system are dilation of the pupils of the eyes and perspiration at the palms of the hands, which lowers the skin resistance.

The sympathetic nervous system is designed for "global" action, with short neurons leaving the spine at all levels to innervate the motor systems affected by these nerves. In contrast, the parasympathetic system is responsible for more specific action. Although not completely antagonistic to the sympathetic system, the parasympathetic nervous system causes dilation of the arteries, inhibition or slowing of the heart, contractions and secretions of the stomach, constriction of the pupils of the eyes, and so on. Where the sympathetic system is primarily involved in mobilizing the body to meet emergencies, the parasympathetic system is concerned with the vegetating functions of the body, such as digestion, sexual activity, and waste elimination.

10.7. MEASUREMENTS FROM THE NERVOUS SYSTEM

Direct measurements of the electrical activity of the nervous system are few. However, the effects of the nervous system on other systems of the body are manifested in most physiological measurements. It is possible, in many cases to stimulate sensor neurons with their specific type of stimulus and measure the responses in various nerves or, in some cases, in individual neurons either in the peripheral or central nervous system. It is also possible to stimulate individual neurons or nerves electrically and to measure either the muscle movement that results from the stimulation or the neuronal spikes that occur in various parts of the system due to the stimulation.

When measuring responses to electrical stimulation, care must be taken to see that the stimulation does not create a wider response than that which would occur if the neuron were stimulated naturally. For example, if electrical stimulation is used, it is very easy to activate other neurons in the vicinity of the intended neuron inadvertently, thus causing responses that are not really related to the desired response.

10.7.1. Neuronal Firing Measurements

Several methods of measuring the neuronal spikes associated with nerve firings have been developed. They differ basically in the vantage point from which the measurement is taken. A *gross* nerve firing measurement is obtained when a relatively large (greater than 0.1 mm in diameter) electrode is placed in the vicinity of a nerve or a large number of neurons. The result is a summation of the action potentials from all the neurons in the vicinity of the electrode. For a more localized measurement, the action potentials of a single neuron can be observed either *extracellularly,* with a microelectrode located just outside the cell membrane, or *intracellularly,* with a microelectrode actually penetrating the cell. Figure 10.9 shows an example of a gross neuronal measurement; Figure 10.10 is an example of an extracellular measurement of a single neuron; and Figure 10.11 shows an intracellular measurement of a single neuron.

Figure 10.9. Gross measurement of multiple unit neuronal discharge. (Full width covers time span of 500 msec. Maximum peak-to-peak amplitude is approximately 145 microvolts.) (Courtesy of Neuropsychology Research Laboratory, Veterans Administration Hospital, Sepulveda, CA.)

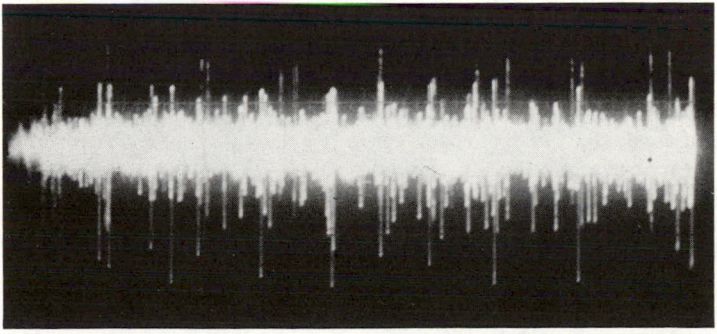

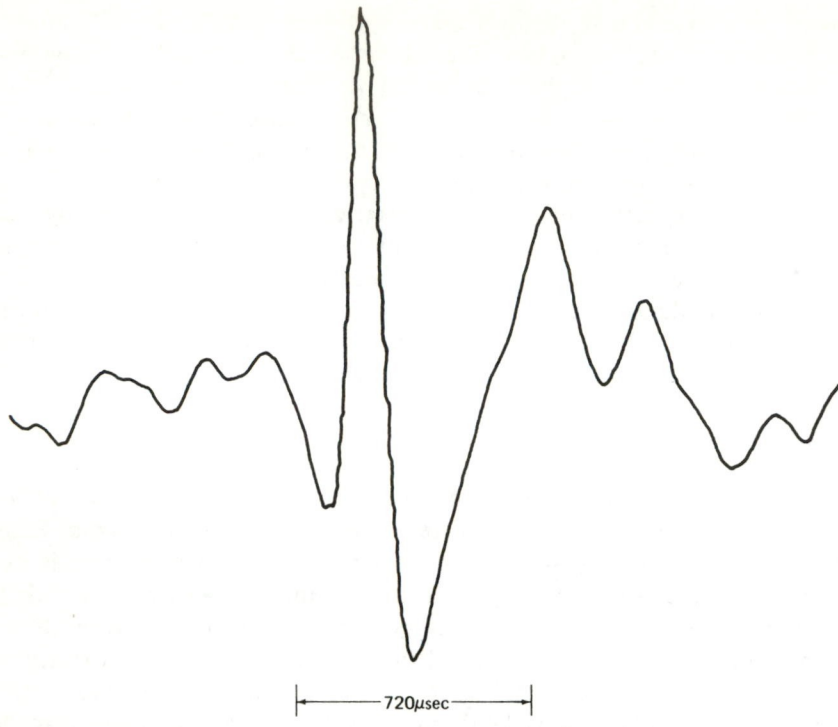

Figure 10.10. Extracellular measurement of unit discharge from red nucleus of a cat. Peak-to-peak height is approximately 180 microvolts.

Figure 10.11. Intracellular measurement of antidromic spike from abducens nucleus of a cat. (Part of motor control system for the eye.) Spike height is about 61 millivolts. Each horizontal division equals 0.5 millisecond. (Courtesy of Brain Research Institute, UCLA.)

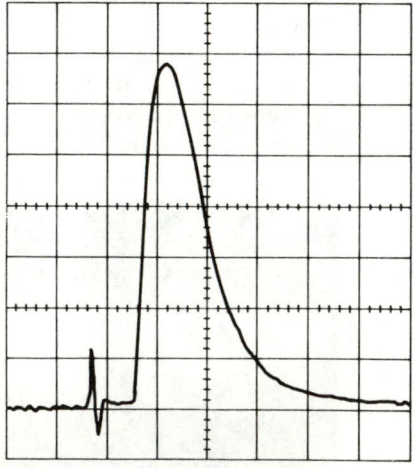

Because of the difficulty of penetrating an individual cell without damaging it and holding an electrode in that position for any length of time, the use of intracellular measurements is limited to certain specialized cell preparations, usually involving only the largest type of cells. Yet, although action potential spikes can be measured readily with extracellular electrodes, the actual value of resting and action potentials and the measurement of graded potentials require the use of intracellular techniques. Any form of single neuron measurement is much more difficult to obtain than gross measurements. In practice, the microelectrode is inserted into the general area and then moved about slightly until a firing pattern indicative of a single neuron can be observed. Even though this is done, identification of the neuron from which the measurement originates is difficult.

Electrodes and microelectrodes used in the measurement of gross and single neuronal firings are described in detail in Chapter 4. Single neuron measurements require microelectrodes with tips of about 10 μm in diameter for extracellular measurements and as small as 1 μm for intracellular measurements. A fine needle or wire electrode is used for gross neuronal measurements. When the measurement is made between a single electrode and a "distant" indifferent electrode, the measurement is defined as *unipolar*. When the measurement is obtained between two electrodes spaced close together along a single axon or a nerve, the measurement is called *bipolar*.

Neuronal firing measurements range from a few hundred microvolts for extracellular single-neuron measurements to around 100 mV for intracellular measurements. For most of these measurements, especially those less than 1 mV, differential amplication is required to reduce the effect of electrical interference. The amplifier must have a very high input impedance to avoid loading the high impedance of the microelectrodes and the electrode interface. Because of the short duration of neuronal spikes, the amplifier must have a frequency response from below 1 Hz to several thousand hertz.

Ordinary pen recorders are generally unsuitable for recording or display of neuronal firings because of the high upper-frequency requirement. As a rule, an oscilloscope with a camera for photographing the spike patterns or a high-speed light-galvanometer or an electrostatic recorder is used for these measurements.

Another measurement involving neuronal firings is that of nerve conduction time or velocity. Here a given nerve is stimulated while potentials are measured from another nerve or from a muscle actuated by the stimulated nerve. The time difference between the stimulus and the resultant firing is measured on an oscilloscope. Some commercial electromyograph (EMG) instruments, such as those described in Section 10.7.3., have provisions for performing nerve conduction velocity measurement.

10.7.2. Electroencephalogram (EEG) Measurements

Electroencephalography was introduced in Chapter 3 as the measurement of the electrical activity of the brain. Since clinical EEG measurements are obtained from electrodes placed on the surface of the scalp, these waveforms represent a very gross type of summation of potentials that originate from an extremely large number of neurons in the vicinity of the electrodes.

Originally it was thought that the EEG potentials represent a summation of the action potentials of the neurons in the brain. Later theories, however, indicate that the electrical patterns obtained from the scalp are actually the result of the graded potentials on the dendrites of neurons in the cerebral cortex and other parts of the brain, as they are influenced by the firing of other neurons that impinge on these dendrites. There are still many unanswered questions regarding the neurological source of the observed EEG patterns.

EEG potentials have random-appearing waveforms with peak-to-peak amplitudes ranging from less than 10 μV to over 100 μV. Required bandwidth for adequately handling the EEG signal is from below 1 Hz to over 100 Hz.

Electrodes for measurement of the EEG are described in Chapter 4. For clinical measurements, surface or subdermal needle electrodes are used. The ground reference electrode is often a metal clip on the earlobe. As discussed in Chapter 4, a suitable electrolyte paste or jelly is used in conjunction with the electrodes to enhance coupling of the ionic potentials to the input of the measuring device. To reduce interference and minimize the effect of electrode movement, the resistance of the path through the scalp between electrodes must be kept as low as possible. Generally, this resistance ranges from a few thousand ohms to nearly 100 kΩ depending on the type of electrodes used.

Placement of electrodes on the scalp is commonly dictated by the requirements of the measurement to be made. In clinical practice, a standard pattern, called the *10-20 electrode placement system,* is generally used. This system, devised by a committee of the International Federation of Societies for Electroencephalography, is so named because electrode spacing is based on intervals of 10 and 20 percent of the distance between specified points on the scalp. The 10-20 EEG electrode configuration is illustrated in Figure 10.12.

In addition to the electrodes, the measurement of the electroencephalogram requires a readout or recording device and sufficient amplification to drive the readout device from the microvolt-level signals obtained from the electrodes. Most clinical electroencephalographs provide the capability of simultaneously recording EEG signals from several regions of the brain.

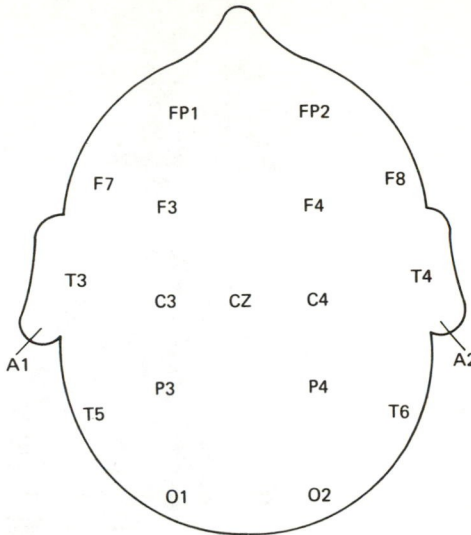

Figure 10.12. 10—20 EEG electrode configuration.

For each signal, a complete channel of instrumentation is required. Thus, Electroencephalographs having as many as 16 channels are available. A clinical instrument with eight channels and a portable unit are shown in Figure 10.13.

Because of the low-level input signals, the electroencephalograph must have high-quality differential amplifiers with good common-mode rejection. The differential preamplifier is generally followed by a power amplifier to drive the pen mechanism for each channel. In nearly all clinical instruments, the amplifiers are ac-coupled with low-frequency cutoff below 1 Hz and a bandwidth extending to somewhere between 50 and 100 Hz. Stable dc amplifiers can be used, but possible variations in the dc electrode potentials are often bothersome. Most modern electroencephalographs include adjustable upper- and lower-frequency limits to allow the operator to select a bandwidth suitable for the conditions of the measurement. In addition, some instruments include a fixed 60-Hz rejection filter to reduce powerline interference.

To reduce the effect of electrode resistance changes, the input impedance of the EEG amplifier should be as high as possible. For this reason, most modern electroencephalographs have input impedances greater than 10 MΩ.

Perhaps the most distinguishing feature of an electroencephalograph is the rather elaborate lead selector panel, which in most cases permits any two electrodes to be connected to any channel of the instrument. Either a bank of rotary switches or a panel of pushbuttons is used. The switch panel also permits one of several calibration signals to be applied to any desired channel for calibration of the entire instrument. The calibration

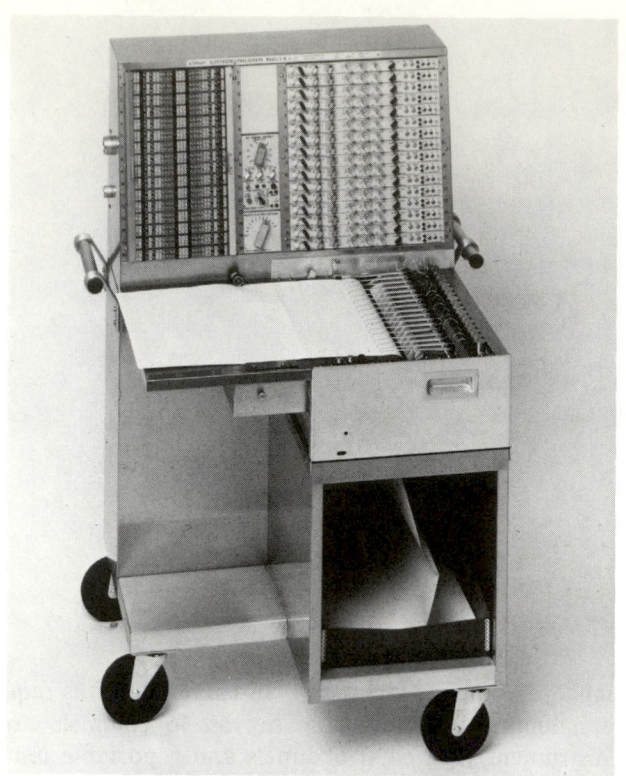

Figure 10.13. Electroencephalographs: (a) Grass model 16. (Courtesy of Grass Instrument Company, Quincy, Mass.); (b) Beckman portable model. (Courtesy of Beckman Instruments, Schiller, Park, IL.)

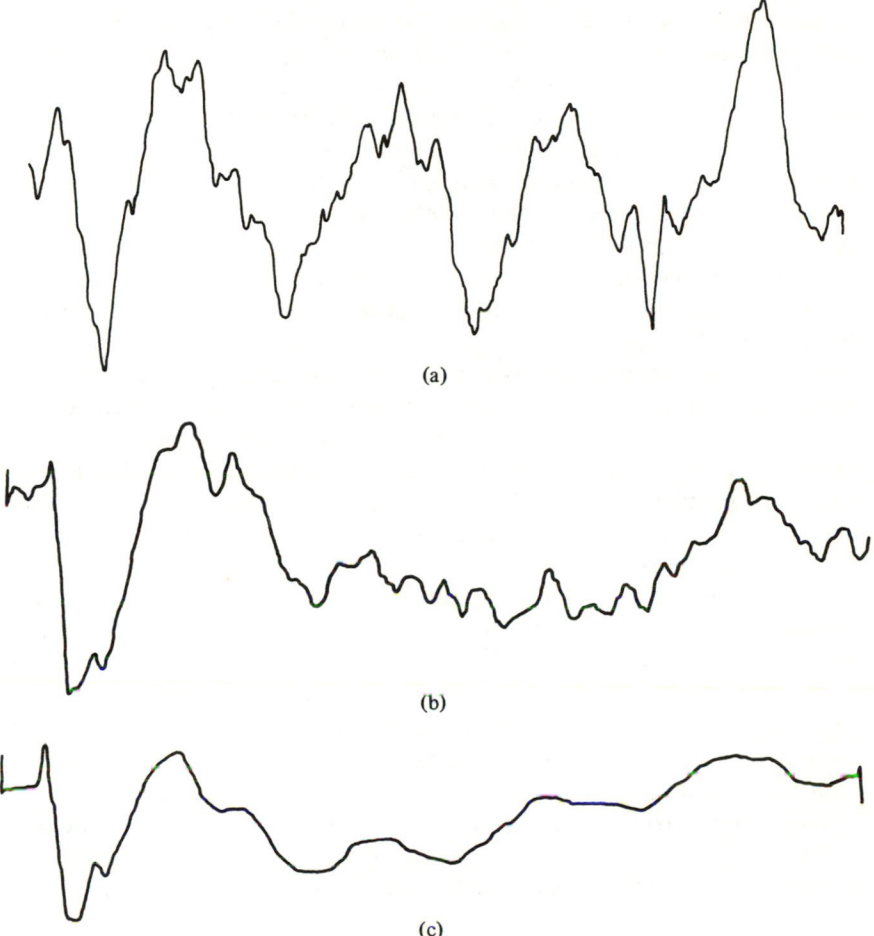

Figure 10.14. Averaging of EEG evoked potentials: (a) raw EEG of single response; (b) average of 8 responses; (c) average of 64 responses. (Courtesy of Dr. Norman S. Namerow, The Center for Health Sciences, Department of Neurology, UCLA, whose research was supported by M.S. Grant #516-C-3.)

signal is usually an offset of a known number of microvolts, which, because of capacitive coupling, results in a step followed by an exponential return to baseline.

The readout in a clinical electroencephalograph is a multichannel pen recorder with a pen for each channel. The standard chart speed is 30 mm/sec, but most electroencephalographs also provide a speed of 60 mm/sec for improved detail of higher-frequency signals. Some have a third speed of 15 mm/sec to conserve paper during setup time. An oscilloscope readout for the EEG is also possible, but it does not provide a permanent record. In some cases, particularly in research applications, the oscilloscope is used in conjunction with the pen recorder to edit the signal until a particular feature or characteristic of the waveform is observed. In this way, only the portions of interest are recorded. Many electroencephalographs also have provisions for interfacing with an analog tape recorder to permit recording and play-back of the EEG signal.

In some research applications, the EEG signals are separated into their conventional frequency bands by means of bandpass filters and the output signals of the individual filters are recorded separately (see Chapter 3). In some cases, they are displayed as biofeedback to the subject whose EEG is being measured (see Chapter 11). In other situations, the entire EEG signal is digitized for computer analysis and through Fourier analysis is converted into a frequency spectrum.

A special form of electroencephalography is the recording of *evoked potentials* from various parts of the nervous system. In this technique the EEG response to some form of sensory stimulus, such as a flash of a light or an audible click, is measured. To distinguish the response to the stimulus from ongoing EEG activity, the EEG signals are time-locked to the stimulus pulses and averaged, so that the evoked response is reinforced with each presentation of the stimulus, while any activity not synchronized to the stimulus is averaged out. Figure 10.14 shows a raw EEG record containing an evoked response from a single presentation of the stimulus and the effect of averaging 8 and 64 presentations, respectively.

10.7.3. Electromyographic (EMG) Measurements

Like neurons, skeletal muscle fibers generate action potentials when excited by motor neurons via the motor end plates. They do not, however, transmit the action potentials to any other muscle fibers or to any neurons. The action potential of an individual muscle fiber is of about the same magnitude as that of a neuron (see Chapter 3) and is not necessarily related to the strength of contraction of the fiber. The measurement of these action potentials, either directly from the muscle or from the surface of the body, constitutes the electromyogram, as discussed in Chapter 3.

Although action potentials from individual muscle fibers can be recorded under special conditions, it is the electrical activity of the entire muscle that is of primary interest. In this case, the signal is a summation of all the action potentials within the range of the electrodes, each weighted by its distance from the electrodes. Since the overall strength of muscular contraction depends on the number of fibers energized and the time of contraction, there is a correlation between the overall amount of EMG activity for the whole muscle and the strength of muscular contraction. In fact, under certain conditions of isometric contraction, the voltage-time integral of the EMG signal has a linear relationship to the isometric voluntary tension in a muscle. There are also characteristic EMG patterns associated with special conditions, such as fatigue and tremor.

The EMG potentials from a muscle or group of muscles produce a noiselike waveform that varies in amplitude with the amount of muscular activity. Peak amplitudes vary from 50 μV to about 1 mV, depending on the location of the measuring electrodes with respect to the muscle and the activity of the muscle. A frequency response from about 10 Hz to well over 3000 Hz is required for faithful reproduction.

Surface, needle, and fine-wire electrodes are all used for different types of EMG measurement. Surface electrodes are generally used where gross indications are suitable, but where localized measurement of specific muscles is required, needle or wire electrodes that penetrate the skin and contact the muscle to be measured are needed. As in neuronal firing measurements, both unipolar and bipolar measurements of EMG are used.

The amplifier for EMG measurements, like that for ECG and EEG, must have high gain, high input impedance and a differential input with good common-mode rejection. However, the EMG amplifier must accommodate the higher frequency band. In many commercial electromyographs, the upper-frequency response can be varied by use of switchable lowpass filters.

Unlike ECG or EEG equipment, the typical electromyograph has an oscilloscope readout instead of a graphic pen recorder. The reason is the higher frequency response required. Sometimes a storage cathode-ray tube is provided for retention of data, or an oscilloscope camera is used to obtain a permanent visual record of data from the oscilloscope screen. A typical commercial electromyograph is shown in Figure 10.15.

Most electromyographs include an audio amplifier and loudspeaker in addition to the oscilloscope display to permit the operator to hear the "crackling" sounds of the EMG. This audio presentation is especially helpful in the placement of needle or wire electrodes into a muscle. A trained operator is able to tell from the sound not only that his electrodes are making good contact with a muscle but also which of several adjacent muscles he has contacted.

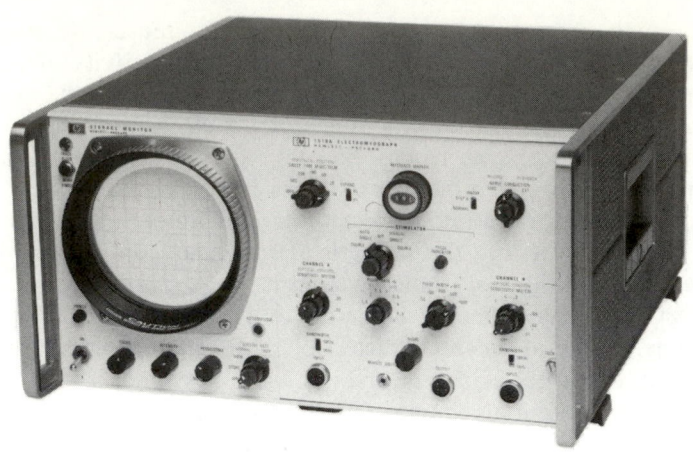

Figure 10.15. Electromyograph. (Courtesy of Hewlett-Packard Company, Waltham, MA.)

Another feature often found in modern electromyographs is a built-in stimulator for nerve conduction time or nerve velocity measurements. By stimulating a given nerve location and measuring the EMG downstream, a latency can be determined from the time difference displayed on the oscilloscope.

The EMG signal can be quantified in several ways. The simplest method is measurement of the amplitude alone. In this case, the maximum amplitude achieved for a given type of muscle activity is recorded. Unfortunately, the amplitude is only a rough indication of the amount of muscle activity and is dependent on the location of the measuring electrodes with respect to the muscle.

Another method of quantifying EMG is a count of the number of spikes, or, in some cases, zero crossings, that occur over a given time interval. A modification of this method is a count of the number of times a given amplitude threshold is exceeded. Although these counts vary with the amount of muscle activity, they do not provide an accurate means of quantification, for the measured waveform is a summation of a large number of action potentials that cannot be distinguished individually.

The most meaningful method of quantifying the EMG utilizes the time integral of the EMG waveform. With this technique, the integrated value of the EMG over a given time interval, such as 0.1 second, is measured and recorded or plotted. As indicated above, this time integral has a linear relationship to the tension of a muscle under certain conditions of isometric contraction, as well as a relationship to the activity of a muscle under isotonic contraction. As with the amplitude measurement, the integrated EMG is greatly affected by electrode placement, but with a given electrode location, these values provide a good indication of muscle activity.

In another technique that is sometimes used in research, the EMG signal is rectified and filtered to produce a voltage that follows the envelope or contour of the EMG. This envelope, which is related to the activity of the muscle, has a much lower frequency content and can be recorded on a pen recorder, frequently in conjunction with some measurement of the movement of a limb or the force of the muscle activity.

11

Instrumentation for Sensory Measurements and the Study of Behavior

The most obvious difference between inanimate and animate objects is that the latter move, respond to their environment, and show changes in their body functions. These properties of animate objects, in a general sense, are called *behavior*. In animals and men the behavior is controlled by the nervous system. The specialized field of medicine in which the nervous system is studied and its diseases are treated is called *neurology*. The *behavior* of organisms, on the other hand, is studied within the various fields of *psychology*. The *experimental psychologist* studies the behavior of animals and men by observing them in experimental situations. The way in which physical stimuli are perceived by men is studied in a specialty called *psychophysics*. The interaction between environmental stimuli and physiological functions of the body is studied in the field of *psychophysiology*. *Clinical psychologists*, as well as *psychiatrists* (who have medical training), deal with the study and treatment of abnormal (pathological) behavior. Behavior is considered abnormal if it interferes substantially with the well-being of the individual and with his interaction with society.

For the treatment of disorders involving the various senses, especially those related to communication, a number of specialized fields have evolved. The *audiologist* determines deficiencies in the acuity of hearing, which often can be improved by the prescription of hearing aids. The *speech pathologist* treats disorders of speech, which may be due to damage to the structures involved in the formation of sounds or may have a neurological cause. The *ophthalmologist* is a physician who specializes in disorders of the eye, whereas the *optometrist* has no medical training and treats only those visual disorders that can be corrected by the prescription of eyeglasses. For the measurement of the acuity of the senses, as well as for the study of behavior, large numbers of instruments have been developed, which can be highly specialized.

The results of behavioral studies seldom show a simple cause-and-effect relationship but are usually in the form of statistical evidence. This peculiarity requires large numbers of experiments in order to obtain results that are statistically significant. As a result, especially in animal experiments, automated systems are frequently used to control the experiment automatically and record the results. The diversity of the field, on the other hand, has resulted in commercially available instruments that are often in the form of modules and building blocks which can be assembled by the experimenter into specialized systems to suit the requirments of a particular experiment.

One obvious way to study behavior is to measure the electrical signals in the brain and the nervous system that control the behavior, as discussed in Chapter 10. However, because the voltages recorded on an electroencephalograph are the result of many processes that occur simultaneously in the brain, only events that involve larger areas of the brain, such as epileptic seizures, can be readily identified on the EEG recording. For this reason, mental disorders generally cannot be diagnosed from the electroencephalogram, although the EEG is usually used to rule out certain organic disorders of the brain (e.g., tumors), which can show symptoms similar to those of nonorganic types of mental illness. The instrumentation used to measure the EEG is described in Chapter 10.

11.1. PSYCHOPHYSIOLOGICAL MEASUREMENTS

As stated in Chapter 10, many body functions, including blood pressure, heart rate, perspiration, and salivation, are controlled by the autonomic nervous system. This part of the nervous system normally cannot be controlled voluntarily but is influenced by external stimuli and emotional states of the individual. By observing and recording these body functions, insight into emotional changes that cannot be measured directly can be obtained. A practical application of this principle is the *polygraph* (colloquially called the "lie detector"), a device for simultaneously recording several body functions that are likely to show changes when questions asked by the interrogator cause anxiety in the tested person.

For the measurement of blood pressure, heart rate, and respiration rate in psychophysiological studies, the same instruments are used as are utilized for medical applications (see Sections 6.1 and 6.2, and Chapter 8, respectively). For measuring variations in perspiration, a special technique has been developed. In response to an external stimulus, such as touching a sharp point, the resistance of the skin shows a characteristic decrease, called the *galvanic skin response* (GSR). The baseline value of the skin resistance, in this context, is called the *basal skin resistance* (BSR). The GSR is believed to be caused by the activity of the sweat glands. It does not depend on the overt appearance of perspiration, however, and the actual mechanism of the response is not completely understood. The GSR is measured most readily at the palms of the hands, where the body has the highest concentration of sweat glands. An active electrode, positioned at the center of the palm, can be used together with a neutral electrode, either at the wrist or at the back of the hand. In some devices clips are simply attached to two fingers. Frequently, in order to increase the stability of the measurement, nonpolarizing electrodes, such as silver–silver chloride surface electrodes (see Chapter 4), are used with an electrode jelly that has about the same salinity as the perspiration. In order to minimize the polarization at the electrodes, the current density is kept below 10 μA/cm^2.

Figure 11.1 shows a block diagram of a device that allows the simultaneous measurement, or recording, of both the BSR and the GSR. Here a current generator sends a constant dc current through the electrodes. The voltage drop across the basal skin resistance, typically on the order of several kilohms to several hundred kilohms, is measured with an amplifier and a meter that can be calibrated directly in BSR values. A second meter, coupled through an *RC* network with a time constant of about 3 to 5 seconds, measures the GSR as a change of the skin resistance of from several hundred ohms to several kilohms. The output of this amplifier can be recorded on a suitable graphic recorder. A measurement of the absolute magnitude of the GSR is not very meaningful. The change of the magnitude of the GSR, depending on the experimental conditions and its *latency* (the time delay between stimulus and response), can be used to study emotional changes. A polygraph for recording physiological functions, including GSR is shown in Figure 11.2.

Instead of the change of the skin resistance, the change of the *skin potential* has been used occasionally. This is actually a potential difference of between 50 and 70 mV that can be measured between nonpolarizing electrodes on the palm and the forearm and that also shows a response to emotional changes.

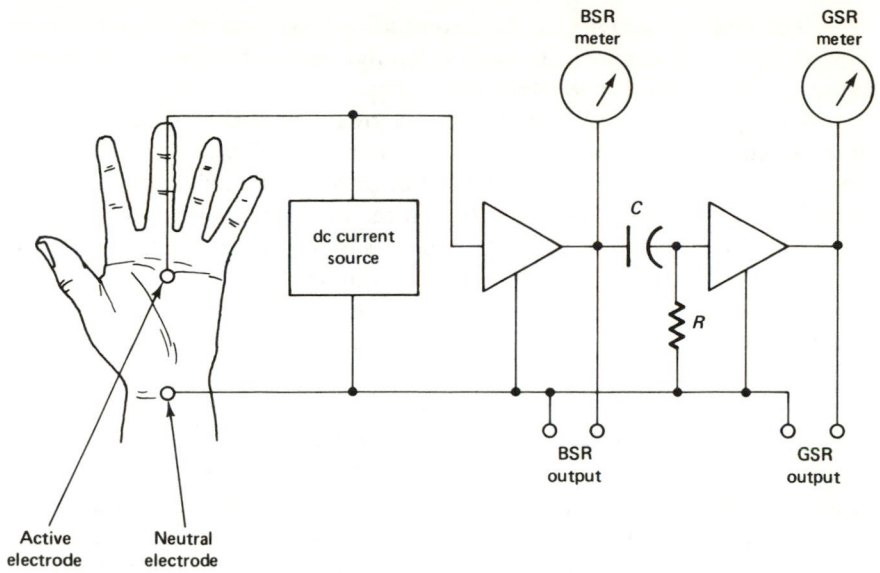

Figure 11.1. Block diagram of a device to measure and record basal skin resistance (BSR) and the galvanic skin response (GSR.)

Figure 11.2. Polygraph for the recording of four body functions. The sensors (right and bottom, clockwise) are for respiration (2 channels) galvanic skin response (GSR) and blood pressure changes. (Courtesy of Stoelting Co., Chicago, IL.)

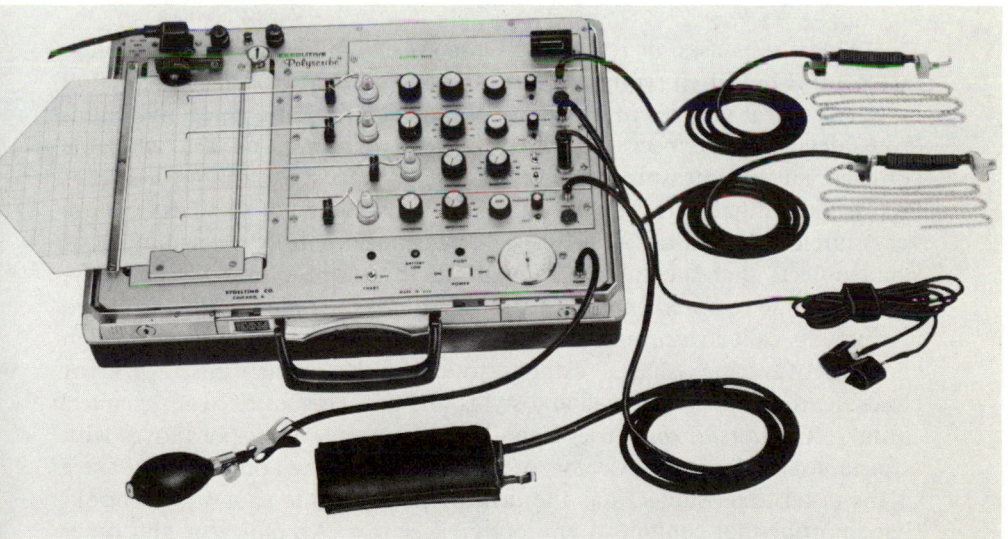

Although the activity of the autonomic nervous system cannot be controlled directly, it can be influenced in an indirect way by two mechanisms known as *conditioning* and *feedback*.

Certain physiological responses are normally elicited by certain external stimuli. The view of food, for instance, stimulates the production of saliva and causes "one's mouth to water." As discovered by Pavlov in his famous experiments with dogs, a previously neutral stimulus can be made to elicit the same response as the view of food if it is presented several times just before the natural stimulus. This process of making the autonomic nervous system respond to previously neutral stimuli is called *Pavlovian* (or *classical*) *conditioning*.

Experiments of this type require the continuous recording of one or more of the autonomic responses. Pavlov, for example, measured the flow rate of saliva. Sometimes the autonomic responses can be influenced by simply informing the subject when a change in the response occurs. This, again, requires that the response be measured and that certain characteristics of it be signaled to the subject in a suitable way. This principle is called *biological feedback* or *biofeedback*. Although this technique had been known for some time, it received renewed interest during the early 1970s for possible therapeutic uses in controlling variables like heart rate, blood pressure, and the occurrence of certain patterns in the electroencephalogram. Biofeedback is described in more detail in Section 11.5.

11.2. INSTRUMENTS FOR TESTING MOTOR RESPONSES

Motor responses, or responses of the skeletal muscles, are under voluntary control but often require a learning process for the proper interaction between several muscles in order to perform the response correctly. Numerous devices have been described in the literature, or are available commercially, to measure motor responses and to study the influence of factors like fatigue, stress or the effects of drugs. Some of these devices are very simple. Manual dexterity tests, for instance, consist of a number of small objects that the subject is required to assemble in a certain way, while the time required for completion of the task is measured. In related instruments called *steadiness testers* a metal stylus must be moved through channels of various shapes without touching the metal walls. An error closes the contact between wall and stylus and advances an electromechanical counter. The *pursuit rotor* uses a similar principle. A light spot moves with adjustable speed along a circular, or star-shaped, pattern on the top surface of the tester. The subject has the task of pursuing the spot with a hook-shaped probe that contains a photoelectric sensor. An indicator and timer automatically measure the percentage of time during which the subject is "on target" during a certain test interval.

The performance of certain muscles or muscle groups can be measured with various *dynamometers,* which measure the force that is exerted either mechanically or with an electric transducer.

11.3. INSTRUMENTATION FOR SENSORY MEASUREMENTS

The human senses provide the information inputs required by man to orient himself in his environment and to protect himself from danger. Many methods and instruments have been developed to measure the performance of the sense organs, study their functioning, and detect impairments. Some of the senses do not require ver sophisticated equipment. The temperature senses, for instance, can be studied with several metal objects, or water containers, which are maintained at certain temperatures. Some of the original work on touch perception, early in this century, was performed by stimulating the skin with bristles of horsehair that had been calibrated to exert a known pressure. The same method is still in use today except that nylon has replaced the horsehair. More complicated devices are necessary for studying optical perception. An example would be a measurement in which a spot of controllable brightness and size is viewed against a background whose brightness can also be varied. Variations in the size and brightness of the spot and the brightness of the background are all independently controlled. Another special device for studies of visual perception is the *tachistoscope.* Here a display of an illuminated card is presented to the viewer by means of a semitransparent mirror or by a slide projector. A second display is then presented for an adjustable short time interval, which may be followed by either a repeat of the original card or by a third display. The change of displays is achieved by switching the illumination or by means of electromechanical shutters. By varying the presentation time for the second display and by using displays of various complexity, the perception and recognition of objects can be studied. The purpose of the presentation of the third display is to mask optical afterimages, which might prolong the actual presentation time of the second display.

Acuity of hearing can be measured with the help of an instrument called an *audiometer.* Here the sound intensity in an earphone is gradually increased until the sound is perceived by the subject. The hearing in the other ear during this measurement is often masked by presenting a neutral stimulus (white noise) to this ear. Normally, the threshold of hearing is determined at a number of frequencies. This process is automated in the *Békésy audiometer* (named after George von Békésy, its inventor), shown in Figure

11.3. In order to perform a measurement, the subject first presses a control button, thus starting a reversible motor, which drives a volume control potentiometer and increases the amplitude of the stimulus signal until it is perceived by the subject. The subject then releases the button, opening the switch, and the motor reverses. By alternately closing and opening the switch, the subject maintains the volume at a level at which the tone can just be heard. A pen, connected to the volume-control mechanism, draws a line on a moving paper. At the same time the paper-drive mechanism, which is linked to the instrument's frequency control, slowly changes the frequency of the tone. Within about 15 minutes a recording, called an *audiogram,* is obtained. The audiogram is often calibrated, not in absolute values of the perception threshold but in relative values referred to the acuity of normal subjects (which is stored in the instrument in a mechanical cam, not shown in Figure 11.3). The resultant curve corresponds directly to the hearing loss as a function of frequency. Figure 11.4 shows a somewhat simplified version of the original Békésy audiometer, which changes the frequency of the stimulus in steps instead of continuously.

Figure 11.3. Békésy audiometer diagram.

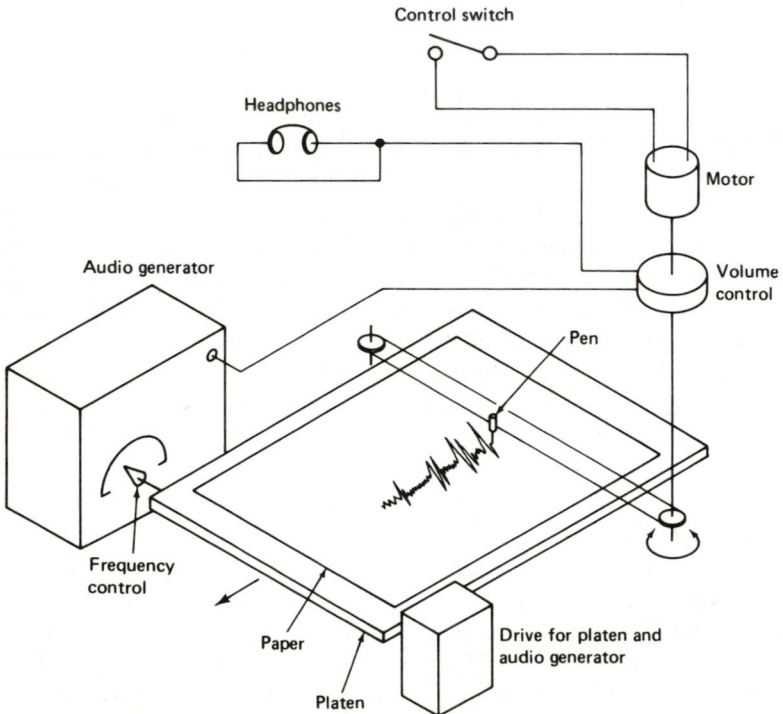

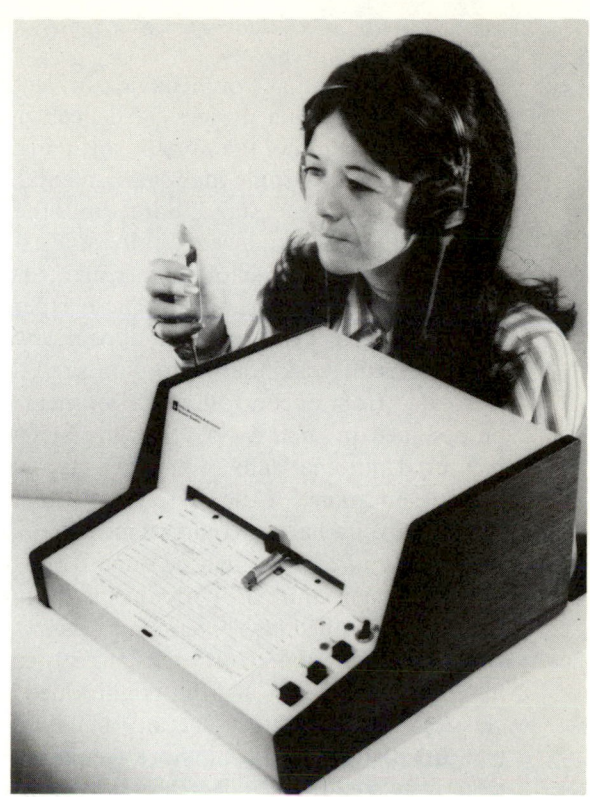

Figure 11.4. Bekesy audiometer. (Courtesy of Grason-Stadler. Subsidiary of General Radio Company, Concord, MA.)

Hearing acuity in infants or uncooperative subjects can be tested with the help of a conditioning method. A light electrical shock can cause a change in the galvanic skin resistance. An audible tone, when paired with the shock, can be made to elicit the same response. Once this conditioning has been completed, the skin reflex can be used to determine whether the subject can hear the same tone presented at a lower volume. This technique, however, is not always completely reliable. A better method is to measure the evoked EEG response when a tone with a certain intensity is presented. This requires the repeated presentation of the tone and an averaging technique to extract the evoked response from the ongoing activity (see Section 10.7.2).

11.4. INSTRUMENTATION FOR THE EXPERIMENTAL ANALYSIS OF BEHAVIOR

In order to describe and analyze behavior accurately, data must be recorded in terms other than the subjective report of an observer. Especially for a mathematical analysis, numerical values must be assigned to some aspects of behavior. For behavior involving motor responses and motor skills, special testing devices have been developed to obtain a numerical rating—for example, the pursuit rotor just described. Other tests required the completion of some manual or mental task in which the time required for completion is measured. Sometimes the number of errors is also used to compare the performance of individuals.

Many basic behavioral experiments are performed with animals (rats, pigeons, monkeys) as subjects. These experiments are made in a neutral environment provided by a soundproof enclosure, often called a "Skinner box" (after B.F. Skinner, who pioneered the method), in which the animal is isolated from uncontrolled environmental stimuli. Each experiment must be designed in such a way that the behavior is well defined and can be measured automatically. For example, such events as pressing a bar or pecking on a key, or the presence of an animal in one part of the cage or jumping over a barrier could be measured. In specially instrumented cages, the activity of animals can be quantified.

Behavior emitted by organisms to interact with and modify their environment is called *instrumental* or *operant behavior.* Such behavior, which is controlled by the central nervous system rather than by the autonomic nervous system, can also be conditioned but in a way that differs from classical conditioning. Operant behavior that is positively reinforced (rewarded) tends to occur more frequently in the future; behavior that is negatively reinforced decreases in frequency. In animal experiments, positive reinforcement is usually administered in the form of food or water given to animals that had been deprived of these commodities. This reinforcement can be administered easily by automatic dispensing devices. Negative reinforcement is in the form of harmless, but painful, electric shocks administered through isolated grid bars that serve as the floor of the cage. With suitable reinforcement, the animal can be conditioned to "emit certain behavior," such as the pressing of a bar, in response to a certain stimulus. From changes in the behavior that can occur under the influences of drugs, or when the stimulus is modified, valuable insight into the mechanisms of behavior can be obtained.

Figure 11.5 shows a setup as it might be used for the simpler types of such experiments, using rats as subjects. The Skinner box is equipped with a response bar and a stimulus light. Positive reinforcement is administered by an automatic dispenser for food pellets. An electric-shock generator is connected to the grid floor of the cage through a scrambler switch that makes it impossible for the animal to escape the shock by clinging to bars that are of the same electrical potential. An automatic programmer turns on the stimulus at certain time intervals and controls the reinforcements according to the animal's response, following a prescribed schedule (called the *contingency*). In many experiments, these schedules can be very complex.

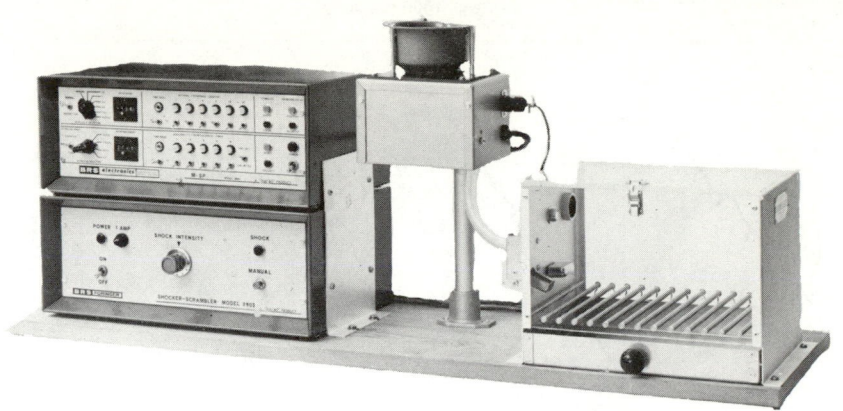

Figure 11.5. Skinner Box. (Courtesy of BRS-Foringer, Beltsville, MD.)

Elaborate modular control systems, either with relays or based on solid-state logic, are therefore available for programming stimulus contingencies and measuring response parameters. Simple behavior is often recorded on a *cumulative-event recorder*. In this device a paper strip is moved with a constant speed (4 in./hour). Each time the bar is pressed by the animal, a solenoid or stepping motor is energized and moves a pen a small distance over the paper perpendicular to the direction of paper movement. The pen is reset, either when it has traveled the full width of the paper or by a timing motor after a certain time interval (for example, every 10 minutes). The position of the pen at any time represents the total number of events (bar presses) that have occurred since the last resetting of the pen. Reinforcement is indicated by a diagonal movement of the pen. This recording method, despite its simplicity, is very informative. The slope of the curve corresponds to the response rate. When reset after fixed time intervals, the pen excursion directly represents a form of time histogram (see Figure 11.6).

Insight into behavior mechanisms obtained in animal experiments has been extrapolated to human behavior. Part of human behavior can be explained as having been conditioned by reinforcements administered by society and the environment. In a form of treatment called *behavior therapy,* behavioral and emotional problems are treated according to the principles of operant conditioning, sometimes using special equipment.

Perhaps the best-known example of a behavior-therapy method using electronic equipment is the treatment of bed wetting with the *Mowrer sheet* (named after the psychologists who first used it). This method uses a moisture sensor placed beneath the bed sheet, which activates an acoustical alarm and turns on a light to awaken the subject when the presence of moisture is first detected.

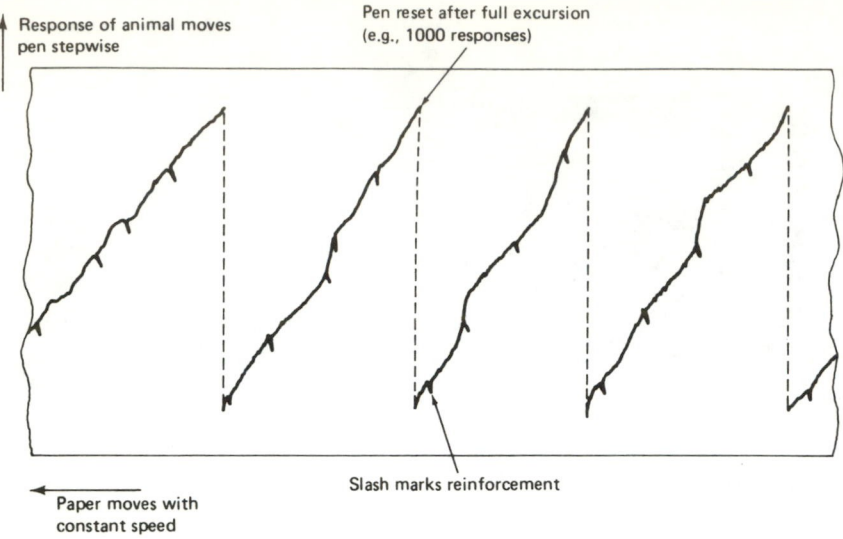

Figure 11.6. Graph from a cumulative event recorder.

11.5. BIOFEEDBACK INSTRUMENTATION

In general engineering terms, feedback is used to control a process. If this concept is applied to biological processes within the body, it is known as *biological feedback* or *biofeedback*. A variable produced by the process is measured and compared with a reference value and, based on the difference, action is taken to bring the variable to the reference value.

As stated in Chapter 10, the body functions that are controlled by the autonomic nervous system are not normally subject to voluntary control. In fact, most of these body functions are not consciously perceived. However, it has been found that if these functions are measured by some suitable method, and, if information pertaining to their magnitude can be conveyed to the subject, a certain degree of voluntary control can be exercised over some of the body functions hitherto believed uncontrollable. Biofeedback is not completely understood and there appears to be a certain overlap with Pavlovian and operant conditioning, but it is presently being used in clinical treatments.

Many different physiological processes have been evaluated for possible control by biofeedback methods, including EEG, EMG, heart rate, and blood pressure. For example, it had been observed that the duration or prevalence of certain brainwave patterns in the EEG, especially the alpha waves (see Chapter 3), could be influenced by biofeedback methods. It had also been observed that the alpha pattern is more prevalent in the EEGs of subjects when they are meditating or simply if their eyes are kept closed.

For a while "alpha feedback" was promoted in counterculture circles as a way of achieving a "drugless high," and a certain cult developed around the method. Among serious researchers this method is now very controversial. More promising are attempts to control the onset of seizures in certain forms of epilepsy by making the subject aware of certain EEG patterns that precede such seizures.

EMG voltages can be measured relatively easily and their presence or magnitude can be signaled to the subject. EMG feedback is used in two different ways. In *relaxation training* the patient is taught to maintain a low EMG-activity level, corresponding to relaxation of the muscles. In the rehabilitation of paralytic patients after traumatic injury or other nerve damage, on the other hand, EMG signals can be measured before muscle activity is detected by other means and can be used to train such patients in the use of paralyzed muscles. It might be mentioned that EMG feedback has also been used in the treatment of *bruxism*, the nocturnal grinding of the teeth.

Heart rate can be measured fairly easily. Blood pressure, on the other hand, is a fairly elusive variable. While it has been shown that both of these variables can be controlled to a certain degree by biofeedback methods, clinical applications for the treatment of hypertension have had disappointing results. There have been a number of experiments in the use of biofeedback for secondary effects. For example, by observing bioelectric data some patients have been able to control glandular secretions, such as insulin in the case of diabetics.

Biofeedback instrumentation includes a transducer and amplifiers to measure the body variable that is to be controlled by the biofeedback process. The magnitude of the measured variable, or, more commonly, changes in the magnitude, are converted into some suitable visual or auditory cue that is presented to the subject. Sometimes it is necessary to provide additional signal processing between the measurement and feedback part of the instrumentation. This is especially true when the variable to be controlled is subject to substantial fluctuations and only a statistical characteristic (e.g., the mean over a certain trial time) is to be controlled.

Some applications of biofeedback that have been demonstrated successfully include a group of medical students who were able to slow their heart rates by an average of 9 beats per minute, a group who were able to equate their own EEGs to their relaxation habits and some patients who have been able to control migraine headaches. Biofeedback has been represented by some to be the purest form of "self-control." The instrumentation is really an adaptation of many instruments discussed throughout this book. The success of biofeedback depends on interpretation of data and the training of the subjects so that they can use the results effectively.

12

Biotelemetry

There are many instances in which it is necessary to monitor physiological events from a distance. Typical applications include the following:

1. Radio-frequency transmissions for monitoring astronauts in space.
2. Patient monitoring where freedom of movement is desired, such as in obtaining an exercise electrocardiogram. In this instance, the requirement of trailing wires is both cumbersome and dangerous.
3. Patient monitoring in an ambulance and in other locations away from the hospital.
4. Collection of medical data from a home or office.
5. Research on unrestrained, unanesthetized animals in their natural habitat.

6. Use of telephone links for transmission of electrocardiograms or other medical data.
7. Special internal techniques, such as tracing acidity or pressure through the gastrointestinal tract.
8. Isolation of an electrically susceptible patient (see Chapter 16) from power-line-operated ECG equipment to protect him from accidental shock.

These applications have indicated the need for systems that can adapt existing methods of measuring physiological variables to a method of transmission of resulting data. This is the branch of biomedical instrumentation known as biomedical telemetry or biotelemetry.

12.1. INTRODUCTION TO BIOTELEMETRY

Literally, *biotelemetry* is the measurement of biological parameters over a distance. The means of transmitting the data from the point of generation to the point of reception can take many forms. Perhaps the simplest application of the principle of biotelemetry is the stethoscope, whereby heartbeats are amplified acoustically and transmitted through a hollow tube system to be picked up by the ear of the physician for interpretation (see Chapter 6).

Historically, Einthoven, the originator of the electrocardiogram, as a means of analysis of the electrical activity of the heart, transmitted electrocardiograms from a hospital to his laboratory many miles away as early as 1903. The rather crude immersion electrodes (see Figure 4.4), were connected to a remote galvanometer directly by telephone lines. The telephone lines in this instance were merely used as conductors for the current produced by the biopotentials.

The use of wires in the transmission of the biodata by Einthoven suited his purpose; however, a major advantage of modern telemetry is the elimination of the use of wires. Certain applications of biotelemetry utilize telephone systems, but essentially these are situations in which "hard-wire" connections are extended by the telephone lines. However, this chapter is concerned primarily with the use of telemetry by which the biological data are put in suitable form to be radiated by an electromagnetic field (radio transmission). This involves some type of modulation of a radio-frequency carrier and is often referred to as *radio telemetry*.

The purpose of this chapter is merely to outline the elements of the subject and to present an example of its application. For a comprehensive treatment, the reader is referred to the Bibliography.

12.2. PHYSIOLOGICAL PARAMETERS ADAPTABLE TO BIOTELEMETRY

Although there had been examples of biotelemetry in the 1940s, they did not receive much attention until the advent of the NASA space programs. For example, in the 1963 report of the Mercury program, the following types of data were obtained by telemetry:

1. Temperature by rectal or oral thermistor.
2. Respiration by impedance pneumograph.
3. Electrocardiograms by surface electrodes.
4. Indirect blood pressure by contact microphone and cuff.

As the field progressed, it became apparent that literally any quantity that could be measured was adaptable to biotelemetry. Just as with hardwire systems, measurements can be applied to two categories:

1. Bioelectrical variables, such as ECG, EMG, and EEG.
2. Physiological variables that require transducers, such as blood pressure, gastrointestinal pressure, blood flow, and temperatures.

With the first category, a signal is obtained directly in electrical form, whereas the second category requires a type of excitation, for the physiological parameters are eventually measured as variations of resistance, inductance, or capacitance. The differential signals obtained from these variations can be calibrated to represent pressure, flow, temperature, and so on, since some physical relationships exist.

In a typical system, the appropriate analog signal (voltage, current, etc.) is converted into a form or code capable of being transmitted. After being transmitted, the signal is decoded at the receiving end and converted back into its original form. The necessary amount of amplification must also be included. Sometimes it is desirable to store the data for future use. Before discussing these aspects, however, a discussion of the applications for these systems is necessary.

Currently, the most widespread use of biotelemetry for bioelectric potentials is in the transmission of the electrocardiogram. Instrumentation at the transmitting end is simple because only electrodes and amplification are needed to prepare the signal for transmission.

One example of ECG telemetry is the transmission of electrocardiograms from an ambulance or site of an emergency to a hospital, where a cardiologist can immediately interpret the ECG, instruct the trained rescue team in their emergency resuscitation procedures, and arrange for any special treatment that may be necessary upon arrival of the patient at the hospital. In this application, the telemetry to the hospital is supplemented by two-way voice communication. (See Section 12.5.3 for further details.)

The use of telemetry for ECG signals is not confined to emergency applications. It is used for exercise electrocardiograms in the hospitals so that the patient can run up and down steps, unencumbered by wires. Also, there have been cases in which individuals with heart conditions wear ECG telemetry units at home and on the job and relay ECG data periodically to the hospital for checking. Other applications include the monitoring of athletes running a race in an effort to improve their performance. ECG telemetry units are also common in human performance laboratories on some college campuses.

The actual equipment worn by the subject is quite comfortable and usually does not impede movement. In addition to the electrodes that are taped into place, the patient or subject wears a belt around the waist with a pocket for the transmitter. A typical transmitter is about the size of a package of king-size cigarettes. The wire antenna can be either incorporated into the belt or hung loosely. Clothing generally has convenient openings to allow for lead wires from the electrodes to come through to the transmitter. Power for the transmitter is from a battery, usually a mercury cell, with a useful life of about 30 hours.

Cardiovascular research performed with experimental animals necessitates some changes in technique. First, the electrodes used are often of the needle type, especially for long-term studies. Second, the animal is likely to interfere with the equipment. For this reason, miniature transmitters have been designed that can be surgically implanted subcutaneously. However, doing so is not always necessary. Many researchers have designed special jackets or harnesses for animals that have been quite successful. Some of the aspects of the particular problem are discussed later.

Telemetry is also being used for transmission of the electroencephalogram. Most applications have been involved with experimental animals for research purposes. One example is in the space biology program in the Brain Research Institute at the University of California, Los Angeles, where chimpanzees have had the necessary EEG electrodes implanted in the brain. The leads from these electrodes are brought to a small transmitter installed on the animal's head, and the EEG is transmitted. Other groups have developed special helmets with surface electrodes for this application. Similar helmets have been used for the collection of EEGs of football players during a game.

Telemetry of EEG signals has also been used in studies of mentally disturbed children. The child wears a specially designed "football helmet" or "spaceman's helmet" with built-in electrodes so that the EEG can be monitored without traumatic difficulties during play. In one clinic the children are left to play with other children in a normal nursery school environment. They are monitored continuously while data are recorded.

One advantage of monitoring by telemetry is to circumvent a problem that often hampers medical diagnosis. Patients frequently experience pains, aches, or other symptoms that give trouble for days, only to have them disappear just before or during a medical examination. Many insidious symptoms behave in this way. With telemetry and long-term monitoring, the cause of these symptoms may be detected when they occur or, if recorded on magnetic tape, can be analyzed later.

One problem often encountered in long-term monitoring by telemetry is that of handling the large amount of data generated. If the time to detect symptoms is very long, it becomes quite a task to record all the information. In many applications, data can be recorded on tape for later playback. A number of types of tape recorders can play back information at a higher speed than that at which data are recorded. Thus, an hour's worth of data can be played back in ½ minute. These rapid-playback techniques can be used effectively only if the observer is looking for something specific. That is, a certain voltage amplitude or a certain frequency can be sensed by a discriminator circuit and used to activate a signal, either a light or sound. The observer can then stop the machine and record the vital segment of the data on paper. He does not have to record the whole sequence, only that part of most interest.

The third type of bioelectric signal that can be telemetered is the electromyogram. This device is particularly useful for studies of muscle damage and partial paralysis problems and also in human performance studies.

Telemetry can also be used in transmitting stimulus signals to a patient or subject. For example, it is well known that an electrical impulse can trigger the firing of nerves (see Chapter 10). It has been demonstrated that if an electrode is surgically implanted and connected to dead nerve endings, an electrical impulse can sometimes cause the nerves to function as they once did. If a miniature receiver is implanted subcutaneously, the electrical signal can be generated remotely. This point brings up the possibility of using telemetry techniques therapeutically. One example is the use of telemetry in the treatment of "dropfoot," which is one of the most common disabilities resulting from stroke. This condition is essentially an inability of the patient to lift his foot, which results in a shuffling, toe-dragging gait.

A method for correcting "dropfoot" by transmitting a signal to an implanted electronic stimulator has been used successfully at Rancho Los Amigos Hospital in Los Angeles. An external transmitter worn by the patient delivers a pulse-modulated carrier signal of 450 kHz to an implanted receiver that demodulates the signal and delivers the resulting signal (a pulse train with a pulse duration of 300 μsec and a frequency that can be varied between 20 and 50 pulses per second) to the peroneal nerve. This nerve, when stimulated, causes muscles in the lower forepart of the leg to contract, thus raising the foot. Stimulation is automatically cycled during gait by a heel switch that turns the transmitter on and off so as to approximate the normal phasic activity of these muscles during gait.

By using suitable transducers, telemetry can be employed for the measurement of a wide variety of physiological variables. In some cases, the transducer circuit is designed as a separate "plug-in" module to fit into the transmitter, thus allowing one transmitter design to be used for different types of measurements. Also, many variables can be measured and transmitted simultaneously by multiplexing techniques.

The transducers and associated circuits are essentially the same as those discussed in earlier chapters. Sometimes they must be modified as to shape, size, and electrical characteristics, but the basic principles of transduction are identical with their hard-wire system counterparts. Not all types of transducers lend themselves to telemetry, however, and usually, in a typical application, a study of adaptable types is necessary.

One important application of telemetry is in the field of blood pressure and heart rate research in unanesthetized animals. The transducers are surgically implanted with leads brought out through the animal's skin. A male plug is attached postoperatively and later connected to the female socket contained in the transmitter unit.

Blood flow has also been studied extensively by telemetry. Both Doppler-type and electromagnetic-type transducers can be employed.

The use of thermistors to measure temperature is also easily adaptable to telemetry. In addition to constant monitoring of skin temperature or systemic body temperature, the thermistor system has found use in obstetrics and gynecology. Long-term studies of natural birth control by monitoring vaginal temperature have incorporated telemetry units.

A final application, discussed below in more detail, is the use of "radio pills" to monitor stomach pressure or pH. In this application, a pill that contains a sensor plus a miniature transmitter is swallowed and the data are picked up by a receiver and recorded.

It is interesting to note that biotelemetry studies have been performed on dogs, cats, rabbits, monkeys, baboons, chimpanzees, deer, turtles, snakes, alligators, caimans, giraffes, dolphins, llamas, horses, seals, and elks, as well as on humans.

12.3.   THE COMPONENTS OF A BIOTELEMETRY SYSTEM

With the many commercial biotelemetry systems available today, it would be impossible to discuss all the ramifications of each. This section is designed to give the reader an insight into the typical simple system. More complicated systems can be built on this base. In putting together a telemetry system, it should be realized that although parts of it are unique for medical purposes, most of the electronic circuits for oscillators, amplifiers, power supplies, and so on are usually adaptions of circuits in regular use in radio communications.

One of the earliest biotelemetry units was the *endoradiosonde,* developed by Mackay and Jacobson and described in various papers by these two investigators since 1957. The pressure-sensing endoradiosonde is a "radio pill" less than 1 cm³ in volume so that it can be swallowed by the patient. As it travels through the gastrointestinal tract, it measures the various pressures it encounters. Similar devices have also been built to sense temperature, pH, enzyme activity, and oxygen tension values by the use of different sensors or transducers. Pressure is sensed by a variable inductance, whereas temperature is sensed by a temperature-sensitive transducer.

One version of the circuit is shown in Figure 12.1. Basically, it is a transistorized Hartley oscillator having a constant amplitude of oscillation and a variable frequency to communicate information. The ferrite core of the coil is attached to a diaphragm, which causes it to move in and out as a function of pressure and, therefore, varies the value of inductance in the coil. This change in inductance produces a corresponding change in the frequency of oscillations. Inward motion of the ferrite core produces a decrease in frequency. Thus, changes in pressure modulate the frequency. An emitter resistor was used in earlier models, and the radio-frequency voltage across it was transmitted by a combined shield and antenna. In later models the oscillator resonator coil also acts as an antenna. The transmitted frequencies, ranging from about 100 kHz to about 100 MHz, can be picked up on any simple receiver.

Figure 12.1. Circuit of pressure-sensitive endoradiosonde. (From R. S. Mackay, *Biomedical Telemetry.* New York, John Wiley & Sons, Inc., 1968, by permission.)

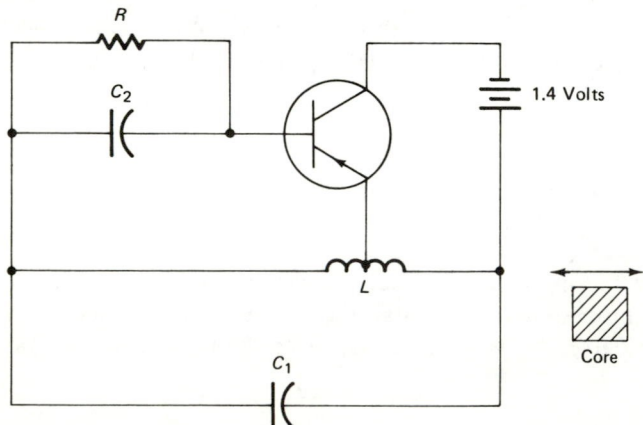

To illustrate the basic principles involved in telemetry, a simple system will be described. Most applications involve more circuitry. The stages of a typical biotelemetry system can be broken down into functional blocks, as shown in Figure 12.2 for the transmitter and in Figure 12.3 for the receiver. Physiological signals are obtained from the subject by means of appropriate transducers. The signal is then passed through a stage of amplification and processing circuits that include generation of a subcarrier and a modulation stage for transmission.

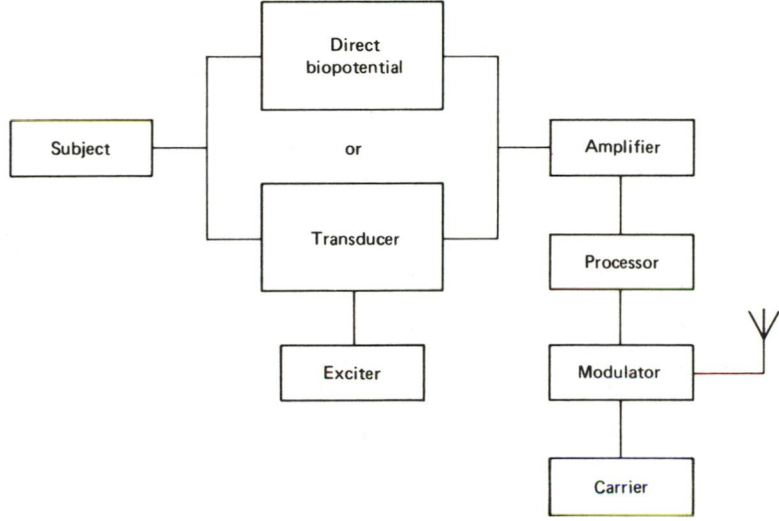

Figure 12.2. Block diagram of a biotelemetry transmitter.

The receiver (Figure 12.3) consists of a tuner to select the transmitting frequency, a demodulator to separate the signal from the carrier wave, and a means of displaying or recording the signal. The signal can also be stored in the modulated state by the use of a tape recorder, as shown in the block diagram. Some comments on these various stages are provided later.

Since most biotelemetry systems involve the use of radio transmission, a brief discussion of some basic concepts of radio should be helpful to the reader with limited background in this field. A *radio-frequency* (RF) *carrier* is a high-frequency sinusoidal signal which, when applied to an appropriate transmitting antenna, is propagated in the form of electromagnetic waves. The distance the transmitted signal can be received is called the *range* of the system. Information to be transmitted is impressed upon the carrier by a process known as *modulation*. Various methods of modulation are desribed below. The circuitry which generates the carrier and modulates it constitutes the *transmitter*. Equipment capable of receiving the transmitted signal and

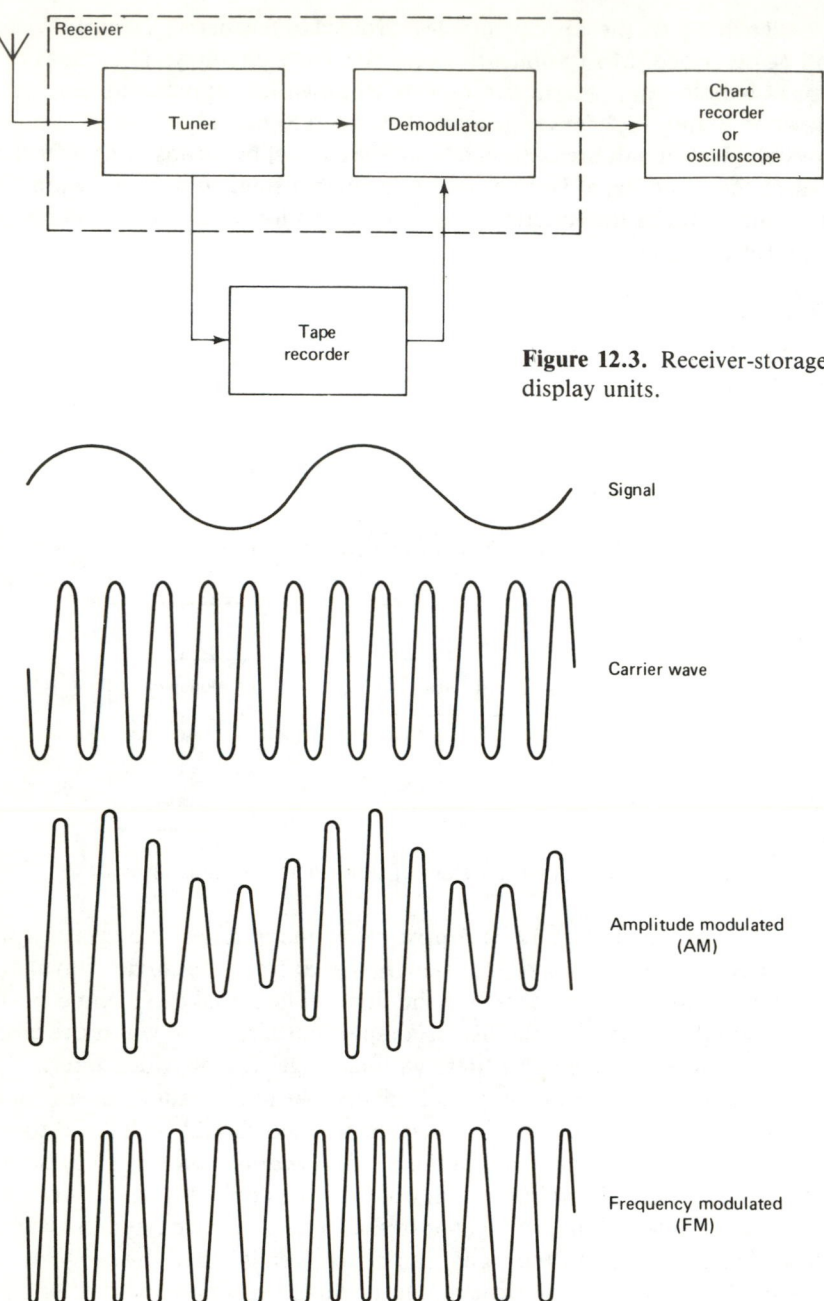

Figure 12.3. Receiver-storage-display units.

Signal

Carrier wave

Amplitude modulated (AM)

Frequency modulated (FM)

Figure 12.4. Types of modulation.

demodulating it to recover the information comprise the *receiver*. By tuning the receiver to the frequency of the desired RF carrier, that signal can be selected while others are rejected. The range of the system depends upon a number of factors, including the power and frequency of the transmitter, relative locations of the transmitting and receiving antennas, and the sensitivity of the receiver.

The simplest form of using a transmitter is to simply turn it on and off to correspond to some code. Such a system does not lend itself to the transmission of physiological data, but is useful for remote control applications. This is called *continuous wave* (CW) transmission, and does not involve modulation.

The two basic systems of modulation are *amplitude modulation* (AM) and *frequency modulation* (FM). These two methods are illustrated in Figure 12.4.

In an amplitude-modulated system, the *amplitude* of the carrier is caused to vary with the information being transmitted. Standard radio broadcast (AM) stations utilize this method of modulation, as does the video (picture) signal for television. Amplitude-modulated systems are susceptible to natural and man-made electrical interference, since the interference generally appears as variations in the amplitude of the received signal.

In a frequency modulation (FM) system, the *frequency* of the carrier is caused to vary with the modulated signal. An FM system is much less susceptible to interference, because variations in the amplitude of the received signal caused by interference can be removed at the receiver before demodulation takes place. Because of this reduced interference, FM transmission is often used for telemetry. FM broadcast stations and television sound also utilize this method of modulation.

In biotelemetry systems, the physiological signal is sometimes used to modulate a low-frequency carrier, called a *subcarrier,* often in the audio-frequency range. The RF carrier of the transmitter is then modulated by the subcarrier. If several physiological signals are to be transmitted simultaneously, each signal is placed on a subcarrier of a different frequency and all of the subcarriers are combined to simultaneously modulate the RF carrier. This process of transmitting many *channels* of data on a single RF carrier is called *frequency multiplexing,* and is much more efficient and less expensive than employing a separate transmitter for each channel. At the receiver, a multiplexed RF carrier is first demodulated to recover each of the separate subcarriers, which must then be demodulated to retrieve the original physiological signals. Either frequency or amplitude modulation can be used for impressing data on the subcarriers, and this may or may not be the same modulation method that is used to place the subcarriers on the

RF carrier. In describing this type of system, a designation is given in which the method of modulating the subcarriers is followed by the method of modulating the RF carrier. For example, a system in which the subcarriers are frequency-modulated and the RF carrier is amplitude-modulated is designated as FM/AM. An FM/FM designation means that both the subcarriers and the RF carrier are frequency modulated. Both FM/AM and FM/FM systems have been used in biotelemetry, the latter more extensively.

In addition to the basic modulation schemes already described, there are many other techniques. Factors that affect the choice of a modulation system may include size, as in an implantable unit (to be described later), or complexity, as in multichannel units, and also considerations of noise, transmission, and other operational problems.

The common denominator for most of the other approaches is a technique known as *pulse modulation,* in which the transmission carrier is generated in a series of short bursts or pulses. If the *amplitude* of the pulses is used to represent the transmitted information, the method is called *pulse amplitude modulation* (PAM), whereas if the *width* (duration) of each pulse is varied according to the information, a *pulse width modulation* (PWM) system results. In a related method called *pulse position modulation* (PPM), the timing of a very narrow pulse is varied with respect to a reference pulse. All three of these pulse modulation methods have certain advantages and are used under certain circumstances. Sometimes other designations have been used to describe the same process; for example, *pulse duration modulation* (PDM) is the same as pulse width modulation. Other designations are *pulse code modulation* (PCM) and *pulse interval modulation* (PIM). Most pulse-modulated telemetry systems use a subcarrier as well as the RF carrier to achieve better stability and greater accuracy. Direct modulation of the RF as in an AM or FM system makes the transmitter more sensitive to nearby electrical equipment and other transmitters. The double designation defined above can be used with all these systems such as PIM/FM, PWM/FM, and so on.

Pulse interval modulation (PIM) and a related modulation system, *pulse interval ratio modulation* (PIRM), are illustrated in Figure 12.5. Either coding system can use direct *pulsatile transmission* (PULSE) or frequency-modulated transmission (FM) (Figure 12.4), in which the frequency is shifted during the time duration defining each pulse. The FM method consumes much more energy than pulse modulation but has greater range. The pulsatile method radiates RF for only 3 to 5 percent of the time. High-quality RF tuners with special modifications are required to capture pulsatile signals. The pulse mode is shown in Figure 12.5(a).

Both systems use the principle of time-duration encoding. The leading edge of a pulse of radio frequency energy with pulsatile transmission (PULSE) or radio frequency shift (FM) defines the beginning and end of time duration.

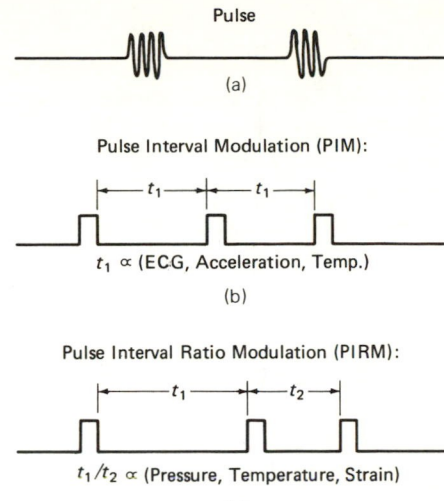

Pulse

(a)

Pulse Interval Modulation (PIM):

$t_1 \propto$ (ECG, Acceleration, Temp.)

(b)

Pulse Interval Ratio Modulation (PIRM):

$t_1/t_2 \propto$ (Pressure, Temperature, Strain)

Figure 12.5. Pulse modulation

(c)

Such denoted times, or ratios of times, are designed to be proportional to voltage and hence to the magnitude of the parameters to be measured (ECG, pressure, etc.). These time durations are very short, usually in the tens of microseconds. Thus, sampling frequency, and hence frequency response, can be very high.

In PIM, Figure 12.5(b), the length of the interval between successive pulses, t_1, is proportional to the signal input. This system is best suited to biopotential or accelerometer usage, but may be used for temperature as well. In PIRM, Figure 12.5(c), the ratio of two successive intervals in each sequential pair of pulses is proportional to a function of the signal input. This system is more complex than PIM but is less dependent on battery voltage. It is best suited for applications such as temperature and pressure.

As in amplitude and frequency modulation systems, multiplexing of several channels of physiological data can be accomplished in a pulse modulation system. However, instead of frequency multiplexing, *time multiplexing* is used. In a time-multiplexing scheme, each of the physiological signals is sampled briefly and used to control either the amplitude, width, or position of one pulse, depending on the type of pulse modulation used. The pulses representing the various channels of data are transmitted sequentially. Thus, in a six-channel system, every sixth data pulse represents a given channel. In order to identify the data pulses, an identifiable reference pulse is included in each set. If the sampling rate is several times the highest frequency component of each data signal, no loss of information results from the sampling process.

A full discussion of all possible methods is beyond the scope of this book, but the reader is referred to the Bibliography for further information. However, some typical examples are presented below.

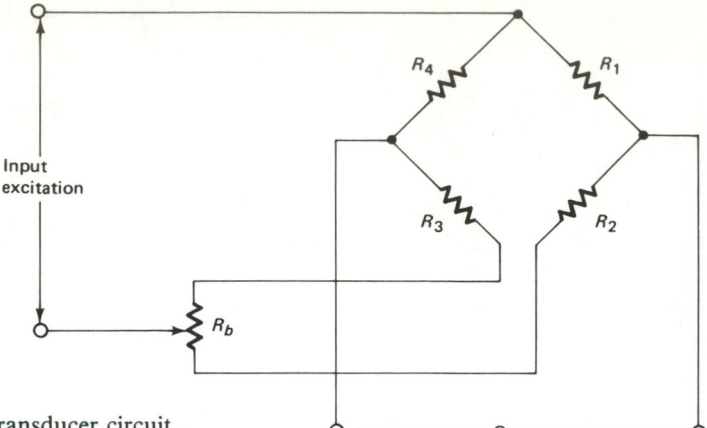

Figure 12.6. Transducer circuit.

A system for monitoring blood pressure is used to illustrate the FM/FM method of transmission. The transducer used in this case is the flush-diaphragm type of strain-gage transducer. Electrically, it can be represented by the bridge circuit of Figure 12.6. Resistors R_1 and R_3 decrease, whereas R_2 and R_4 increase in value as blood pressure increases. Resistor R_b is simply for balancing or zeroing. The transducer is connected in the transmitter circuit as shown in Figure 12.7.

Either direct current or alternating current can be used as excitation for strain-gage bridges. When dc is used, the amplifier following the bridge must be a dc amplifier, with its associated problems of stability and drift. When ac is used, the bridge acts as a modulator. A demodulator and filter are required in order to recover the signal.

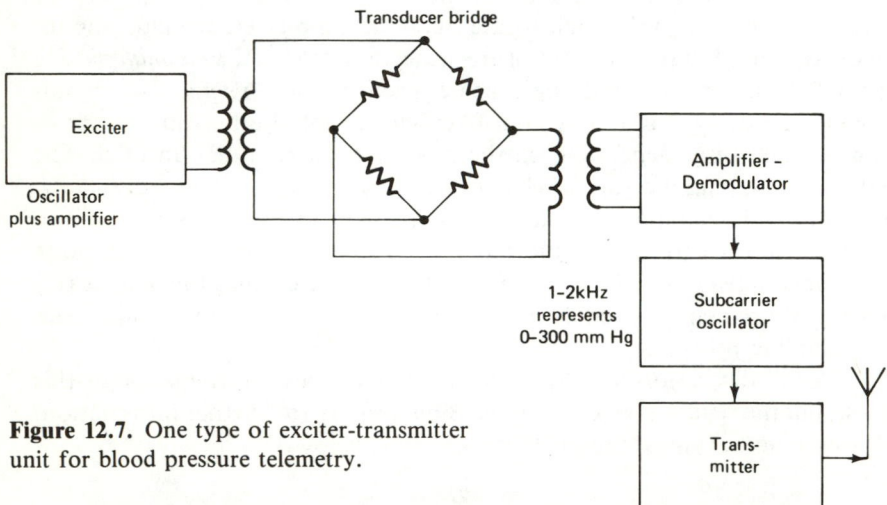

Figure 12.7. One type of exciter-transmitter unit for blood pressure telemetry.

The exciter unit, which in this example consists of a Colpitts transistor oscillator plus an *RC*-coupled common-emitter amplifier stage, excites the bridge with a constant ac voltage at a frequency of approximately 5 kHz. The exciter unit is coupled to the bridge inductively. The bridge is initially balanced both resistively and capacitively so that any changes in the resistance of the arms of the bridge due to changes in pressure on the transducer will result in changes of the output voltage.

This output voltage is inductively coupled to another common-emitter amplifier stage and *RC*-coupled to a further stage of amplification. However, whereas the previous stages are class A amplifiers and do not change the waveshape of the input voltage, the latter stage is a class C amplifier, which means that the transistor is biased beyond cutoff and the resulting output wave is rectified to obtain a signal representative of the pressure variation.

This rectified wave is put through a resistance–capacitance filter, and the resulting voltage controls the frequency of a unijunction (double-base) transistor oscillator. This is the FM subcarrier oscillator that is used to modulate the main carrier.

The system can be arranged so that there is a fairly linear relationship between the subcarrier oscillator frequency and the physiological parameter to be measured. For example, in the system for blood pressure illustrated in Figure 12.7, a frequency range of 1 to 2 kHz represents the range of 0 to 300 mm Hg (0 to 40 kPa) pressure. The transducer action can be traced very easily. The subcarrier is used to frequency-modulate the main transmitter carrier. This carrier is transmitted at low power on a frequency band specially designated for biotelemetry.

The same exciter-transmitter circuit could be used with small modifications if the blood pressure transducer were replaced by another type or by a thermistor or any other electrical resistance device. Also, the exciter-bridge combination could be replaced by a direct biopotential signal input, such as an electrocardiogram signal.

It should be noted that, with the transmission of radio-frequency energy, legal problems might be encountered. Many systems use very low power and the signals can be picked up only a few feet away. Such systems are not likely to present problems. However, systems that transmit over longer distances are subject to licensing procedures and the use of certain allocated frequencies or frequency bands. Regulations vary from country to country, and in some European countries they are more strict than in the United States.

The regulations that are of concern to persons operating in the United States are contained in the Federal Communications Commission (FCC) regulations for low-power transmission. In case of doubt, this material should be referred to in order to ensure compliance (see the Bibliography).

Returning to the system under discussion, the signal transmitted at low power on the FM transmitter is picked up by the receiver, which must be tuned to the correct frequency. The audio subcarrier is removed from the RF carrier and then demodulated to reproduce a signal that can be transformed back to the amplitude and frequency of the original data waveform. This signal can then be displayed or recorded on a chart. If it is desirable to store the data on tape for later use, the original data waveform or the modulated subcarrier signal is put on the tape. In the latter case, when playback is desired, the subcarrier signal is passed through the FM subcarrier demodulator.

There are systems that convert an analog signal, such as ECG, into digital form prior to modulation. The digital form is useful when used in conjunction with computers, a topic covered in Chapter 15.

An example of another type of telemetry system is shown in Figures 12.8 and 12.9. This is a pulse-width modulation (PWM) system capable of simultaneously transmitting four channels of physiological data. The transmitted signal is a composite of a positive synchronizing pulse and a series of negative signal pulses. The data to be telemetered cause the signal pulses to move back and forth in time with respect to the synchronizing pulse X and result in four varying time intervals (t_1, t_2, t_3, t_4). The position of each pulse with respect to its neighbors carries the data.

Figure 12.8. Biolink PWM transmitting system. (Courtesy of BIOCOM, Inc., Culver City, CA.)

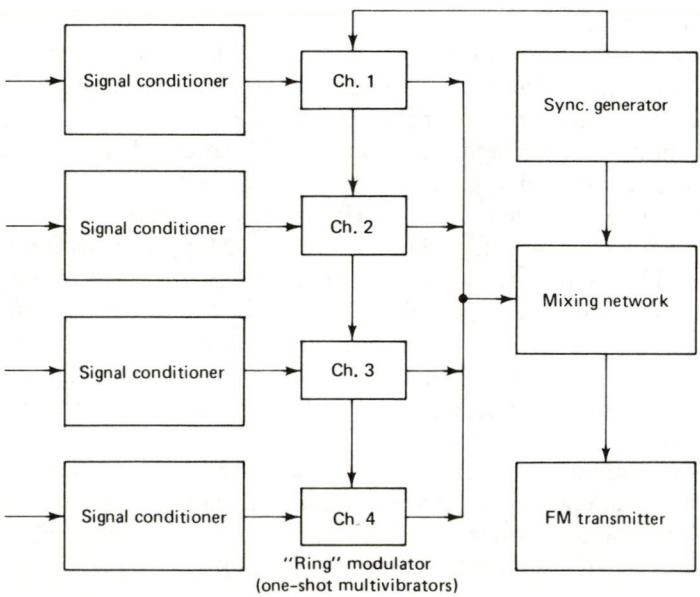

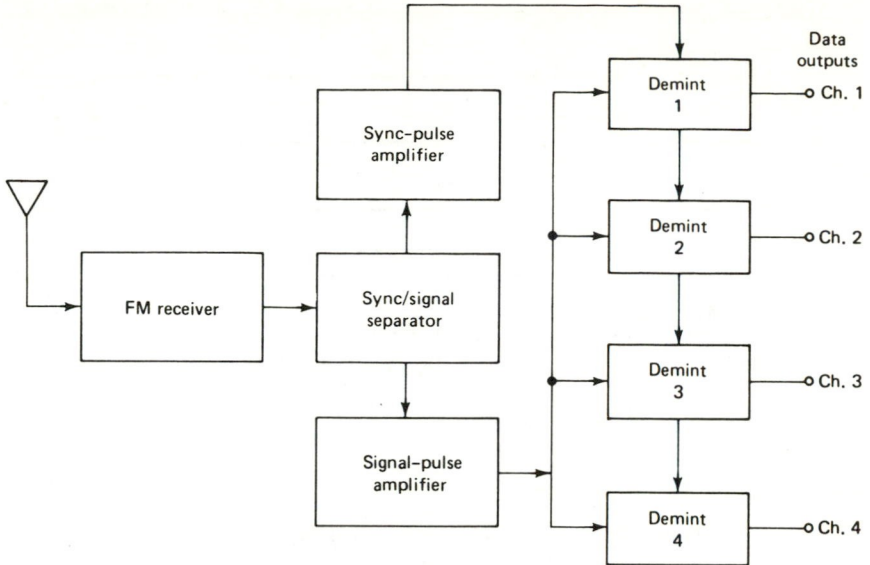

Figure 12.9. Biolink PWM receiving system. (Courtesy of BIOCOM, Inc., Culver City, CA.)

Referring to the block diagram of the transmitting system in Figure 12.8, it can be seen that the sync generator begins the action. Its pulse turns the first channel one-shot multivibrator *ON*. How long it remains on depends on the level of the data being fed into it at that instant of time. Its return to the *OFF* position triggers the next channel, and so on down the line. The resultant square waves are thus width-modulated by the data.

After reception, the composite signal must be separated and re-formed to be properly demodulated. The sync-signal separator and amplifiers perform this function, as shown in Figure 12.9. Each channel consists of a flip-flop and an integrating network. The signal pulses are fed through a suitable diode network to all channels. The sync pulse is fed to the first channel only.

In operation, the sync pulse turns the first flip-flop *ON*. The first signal pulse comes in and would turn any *ON* flip-flops *OFF*. Since channel 1 is the only unit that is *ON*, it is turned *OFF*. When it returns to the *OFF* position, it automatically triggers the channel 2 flip-flop *ON*. Subsequent signal pulses are used to turn off (or gate) each corresponding flip-flop after it has been turned on. This situation is shown in Figure 12.10. The resulting square wave out of each flip-flop varies in width corresponding to the original square wave in the transmitter. Simple integration yields the original data.

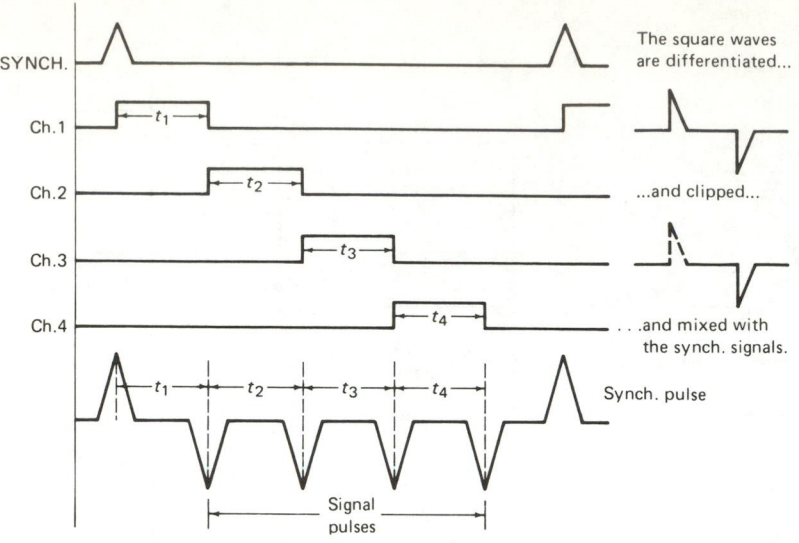

Figure 12.10. Forming the PWM composite signal. (Courtesy of BIOCOM, Inc., Culver City, CA.)

12.4. IMPLANTABLE UNITS

It was mentioned previously that sometimes it is desirable to implant the telemetry transmitter or receiver subcutaneously. The implanted transmitter is especially useful in animal studies, where the equipment must be protected from the animal. The implanted receiver has been used with patients for stimulation of nerves, as described in Section 12.1.

Although the protective aspect is an advantage, many disadvantages often outweigh this factor, and careful thought should be given before embarking on an implantation. The surgery involved is not too complicated, but there is always risk whenever surgical techniques are used. Also, once a unit is implanted, it is no longer available for servicing, and the life of the unit depends on how long the battery can supply the necessary current.

This section is primarily concerned with completely implanted systems, but there are occasions when a partial implant is feasible. A good example is a system used for the monitoring of the electroencephalogram where the electrodes have been implanted into the brain and the telemetry unit is mounted within and on top of the skull. This type of unit needs a protective helmet.

The use of implantable units also restricts the distance of transmission of the signal. Because the body fluids and the skin greatly attenuate the signal and because the unit must be small to be implanted, and therefore has little power, the range of signal is quite restricted, often to just a few feet. This disadvantage has been overcome by picking up the signal with a nearby antenna and retransmitting it. However, most applications involve monitoring over relatively short distances, and retransmission is not necessary.

Another problem has been the encapsulation of the unit. The outer case and any wiring must be impervious to body fluids and moisture. However, with the plastic potting compounds and plastic materials available today, this condition is easily satisfied. Silicon encapsulation is commonly used.

The power source is of great importance. Mercury and silver-oxide primary batteries have been used extensively and, more recently, lithium batteries have found many applications. Implantable telemetry batteries vary in physical size and electrical capacity, depending on the application. For field work with free-roaming animals, the power requirements are quite different from those needed in a closed laboratory cage. The size of the animal is also a factor. Requirements range from an electrical capacity of 20 mA-hr with a weight of 0.28 gram and a volume of 0.05 cm^3 to 1000 mA-hr, with weights of the order of 12 grams and 3.2 cm^3. Also, if power is not needed continuously, radio-frequency switches can be used to turn the system on and off on command.

In simple terms the complete implantable telemetry transmitter system consists of the transducer(s), the leads from the transducer(s) to the transmitter, the transmitter unit itself, and the power source. The latter can be encapsulated within the transmitter or may be a separate unit connected by suitable leads to the transmitter. The transducers are implanted surgically in the position required for a particular measurement, such as in the aorta or other artery for blood pressure. Figure 6.28 shows a typical pressure transducer implantation in a dog. The transmitters and power units have to be placed in a suitable body cavity close to the under surface of the skin and situated so that they give no physical or psychological disturbance to the animal. It is extremely important that all units and wires are adequately sealed since leakage of body fluids into equipment or the chemical effects of man-made materials on body tissues can cause malfunction or infection, respectively. Discussion of these materials is beyond the scope of this chapter, but this is an important topic in the general field of biomedical engineering. An antenna loop is also part of the transmitter.

Some implantable units presently in use are used as illustrations of the foregoing discussions of implantable systems. A basic unit is shown in Figure 12.11. This is a single-channel blood pressure transmitter. The module at the top contains the signal conditioning circuitry and RF transmitter. The second module contains a 200-mA-hour lithium power source and a 1.7-MHz RF switch for turning the system on and off remotely. The pressure transducer, shown in the lower left corner, is the same type described in Section 6.2.4.5.

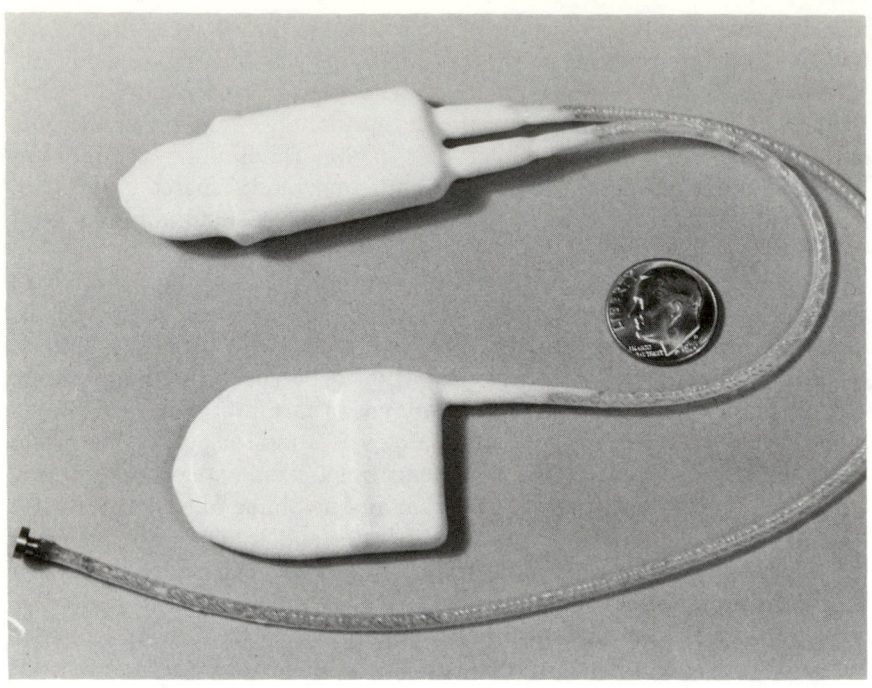

Figure 12.11 Single channel implantable transmitter for blood pressure. (Courtesy of Konigsberg Instruments Inc., Pasadena, CA.)

Figure 12.12. Cut-away single channel temperature transmitter. (Courtesy of Konigsberg Instruments Inc., Pasadena, CA.)

Figure 12.12 is a cutaway view of a single-channel temperature transmitter. A 1.35-V battery is contained inside the antenna loop at the top. Inside, one of the three hybrid packages is shown open. Figure 12.13 shows an array of all parts of the complete system. The top unit in the figure is the TD6 Telemetry Demodulator. It has six main channels and is designed to work with the 88 to 108-MHz receiver shown immediately below it. The receiver is modified to accept both continuous FM and pulsed-RF-mode telemetry signals. An inductive power control wand for turning the implant on and off is shown on the bottom right side. Below the wand there is an external

Figure 12.13. Complete implantable telemetry system. (Courtesy of Konigsberg Instruments Inc., Pasadena, CA.)

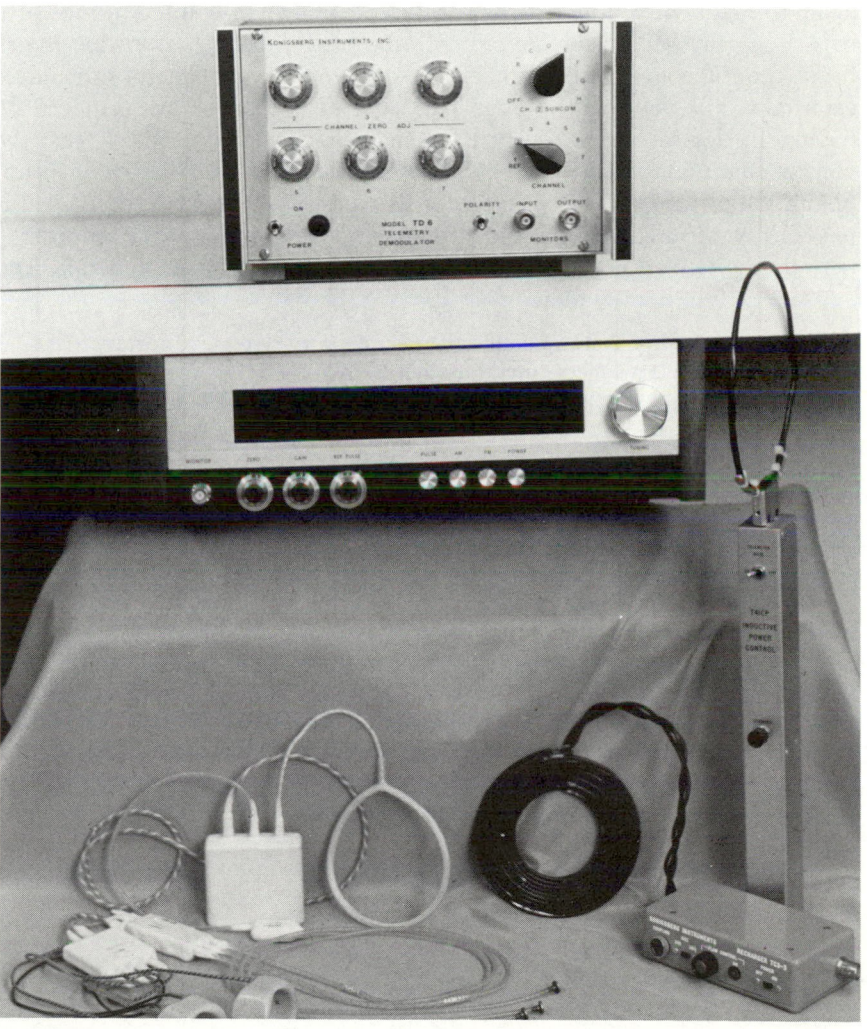

recharging transmitter. In use, the coil of the recharging unit must be placed from 1 to 2.5 cm axially from the implanted pickup coil, which is part of the battery system. This pickup coil is shown in the figure adjacent to the re-charger coil, and the battery unit to which it is connected is suitably encap-sulated. The implantable transmitter and transducer systems that are im-planted are shown on the bottom left side of the picture.

A cutaway view of an inductively-powered multichannel telemetry system is shown in Figure 12.14. Sensor leads and compensation components are shown on the right. Power and antenna leads are shown on the left. In the center, one of three hybrid packets is shown open. It contains six sensor-input amplifiers, an eight-channel multiplexer, an analog-to-PWM converter, and a 10-kHz clock and binary counter.

Finally, there are systems with only partial implantations. Referring again to Figure 6.28, a pressure transducer is shown implanted in the aorta of a dog. In that particular system, the lead from the transducer was brought out through the dog's back and connected to a telemetry transducer external to the body of the dog. This type of preparation is achieved by having the dog wear a jacket. Prior to surgery, dogs are trained to wear the jackets continuously so that they get used to them. After the surgical im-plantation of the transducer and after the chest wall is healed, the jacket is put back on the dog. It is made of strong nylon mesh so that it is comfortable, permits air circulation, but cannot easily be bitten into by the dog. The lead

Figure 12.14. Cut-away multi-channel telemetry system. (Courtesy of Konigsberg Instruments Inc., Pasadena, CA.)

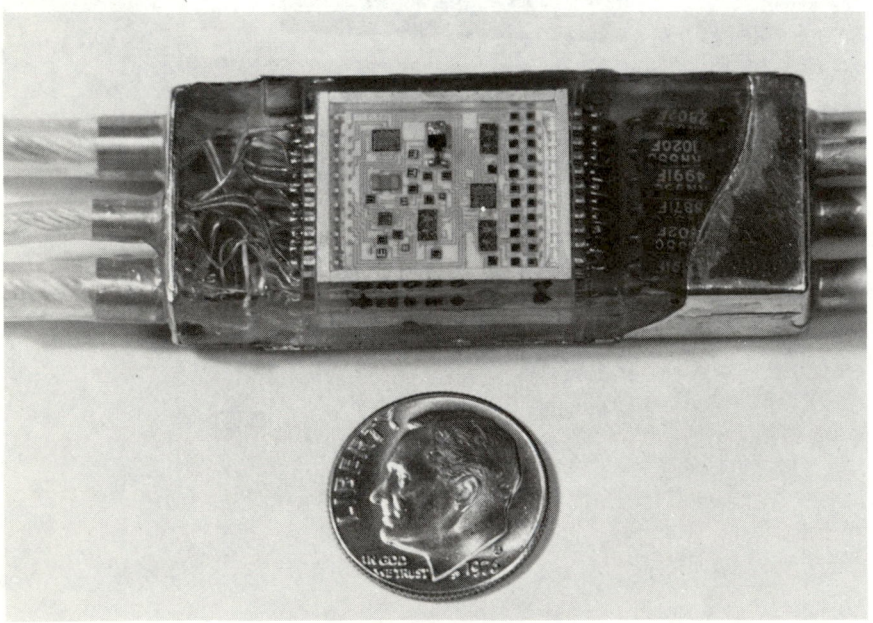

that comes out of the dog's back from the transducer is plugged into an external telemetry transmitter which is kept in a pocket of the jacket. The transmitter can be removed when not in use. Another pocket, on the opposite side of the jacket, is available for other equipment. For example, in an experiment concerned with the effect of the hormone, norepinephrine, on blood pressure, a small chemical pump was placed in the other pocket to inject norepinephrine into the blood stream at various rates. The effect on the blood pressure of the dog was observed and recorded by the use of the telemetry system. By using telemetry the dog is isolated, so that outside effects, such as fear of people, will not be present during the experiment. The telemetry transmitter is about the size of a pack of cigarettes. This type of semi-implantation, with implanted transducer and external transmitter, was used extensively during the early development stages of biotelemetry in animal research. The system is the same as that shown in Figure 12.2. A photograph of a dog wearing a jacket with the telemetry transmitter in the pocket is shown in Figure 12.15.

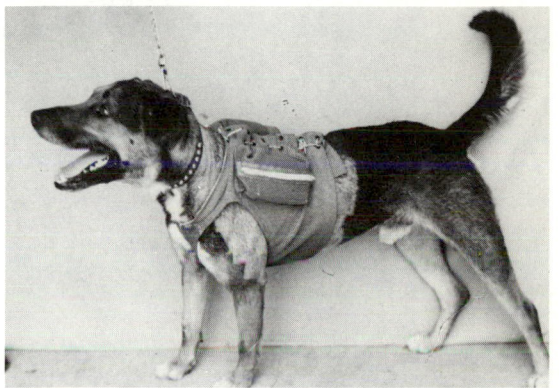

Figure 12.15. Jacket for partially implanted telemetry system.

12.5 APPLICATIONS OF TELEMETRY IN PATIENT CARE

There are a limited number of situations in which telemetry is practical in the diagnosis and treatment of hospital patients. Most involve measurement of the electrocardiogram. Some common applications are described below.

12.5.1. Telemetry of ECGs from Extended Coronary Care patients

Cardiac patients must often be observed for rhythm disturbances for a period of time following intensive coronary care. Such patients are generally allowed a certain amount of mobility. To make monitoring possible, some

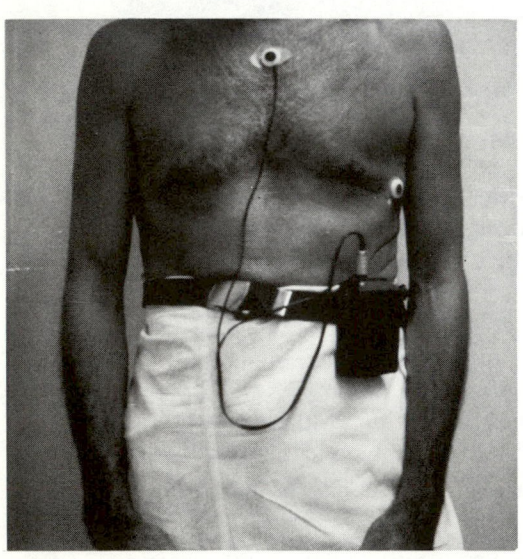

Figure 12.16. ECG telemetry transmitter: (a) Hewlett-Packard Type 78100A in hospital use (Courtesy of Hewlett-Packard Company, Waltham, MA). (b) Electrode placement for telemetered ECG.

hospitals have extended coronary-care units equipped with patient-monitoring systems that include telemetry. In this arrangement, each patient has ECG electrodes taped securely to his chest. The electrodes are connected to a small transmitter unit that also contains the signal-conditioning equipment. The transmitter unit is fastened to a special belt worn around the patient's waist. Figure 12.16 shows typical units. Batteries for powering the signal-conditioning equipment and transmitter are also included in the transmitter package. These batteries must be replaced periodically. Some systems include provisions for easy testing of the transmitter batteries. In other cases, the batteries must be replaced at some predetermined interval.

A telemetry receiver for each monitored patient is usually included as part of the monitoring system. The output of each receiver is connected to one of the ECG channels of the patient monitor. A potential problem in the use of telemetry with free-roaming patients concerns being able to locate a patient in case his alarm should sound. Telemetry equipment has no provision for indicating the location of a transmitter. The area in which the patients are allowed to move must be limited. There may also be a problem if patients are able to venture beyond the range of the telemetry transmitter. Most modern hospitals are constructed in such a way that radio waves cannot pass through the walls. Thus, unless special antennas are provided in hallways, reception may be confined to the ward itself or to a very small portion of the hospital. If a patient wanders beyond the range of the system, his ECG can no longer be monitored and the purpose of the telemetry is defeated.

12.5.2 Telemetry for ECG Measurements During Exercise

For certain cardiac abnormalities, such as ischemic coronary artery disease, diagnostic procedures require measurement of the electrocardiogram while the patient is exercising, usually on a treadmill or a set of steps. Although such measurements can be made with direct-wire connections from the patient to nearby instrumentation, the connecting cables are frequently in the way and may interfere with the performance of the patient. For this reason, telemetry is often used in conjunction with exercise ECG measurements. The transmitter unit used for this purpose is similar to that described earlier for extended coronary care and is normally worn on the belt. Care must be taken to ensure that the electrodes and all wires are securely fastened to the patient, to prevent their swinging during the movement of the patient. In most ECG telemetry systems movement of the wiring with respect to the body results in artifacts on the ECG tracing. However, with proper equipment and with the wiring skillfully tied down, excellent results can be obtained. If other physiological variables in addition to the ECG are to be measured from the exercising patient, suitable transducers and signal-conditioning equipment must be included in the transmitter unit, along with the provision for multiplexing to accommodate the additional channels. In general, the receiver and other equipment used for conditioning and displaying the signals received from the exercising patient are located very near the patient. Thus, the transmitter can operate with low power. The receiver must be able to retrieve the ECG and any other information transmitted, in addition to providing appropriate signals to the remainder of the instrumentation system.

12.5.3. Telemetry for Emergency Patient Monitoring

In many areas ambulances and emergency rescue teams are equipped with telemetry equipment to allow electrocardiograms and other physiological data to be transmitted to a nearby hospital for interpretation. Two-way voice transmission is normally used in conjunction with the telemetry to facilitate identification of the telemetered information and to provide instructions for treatment. Through the use of such equipment, ECGs can be interpreted and treatment begun before the patient arrives at the hospital.

Telemetry of this type requires a much more powerful transmitter than the two applications previously described. Often the data must be transmitted many miles and sometimes from a moving vehicle. To be effective, the system must be capable of providing reliable reception and reproduction of the transmitted signals regardless of conditions. In some cases, an emergency rescue squad can transmit physiological information from a portable transmitter to a receiver in their vehicle. The vehicle, which contains a more powerful transmitter and better antenna system, is able to retransmit the data to the hospital. This process of *retransmission* is necessary in cases where the emergency team might be working in some location from which they are unable to maintain direct communication with the hospital.

One type of system in use is illustrated in Figure 12.17. Figure 12.17(a) shows the portable telemetry unit itself, and 12.17(b) is an action photograph of the unit being used by a paramedic team. The coronary observation display console on the receiving end of the system in the hospital is illustrated in Figure 12.17(c).

The portable unit carried in the ambulance or paramedic vehicle has a nominal output of 12 W RF. It weighs less than 8.6 kg (19 lb) and can be carried by a handle or using a shoulder strap. It can transmit on any of 10 different channels. These are the eight approved MED frequencies and two EMS or public safety dispatch channels. The Federal Communications Commission (FCC) has set up rules and regulations concerning the use of "Special Emergency Radio Service" (see the Bibliography) in which the *MED frequencies* are defined. Table 12.1 shows these frequencies. To cover the band, the mobile telemetry transmitters are usually capable of operating in the range 450 to 470 MHz.

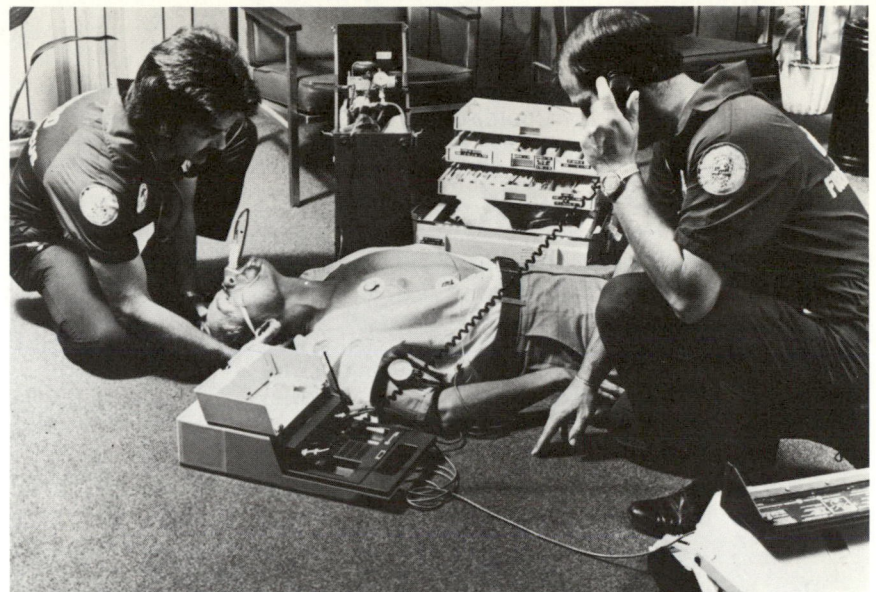

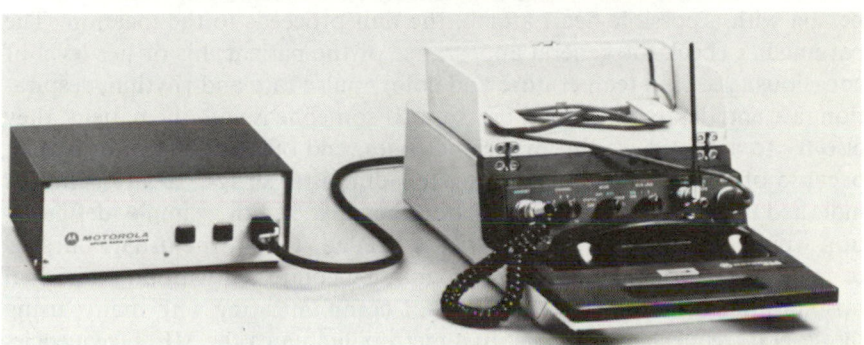

Figure 12.17 Emergency medical care system. (a) Portable transmitter unit. (b) Transmitter unit in use. (c) Hospital console. (Courtesy of Motorola Communications and Electronics Inc., Schaumburg, IL.)

Table 12.1. EMERGENCY MEDICAL SYSTEMS UHF FREQUENCIES (MHz)[a]

Channel Name	Primary Use	Base and Mobile	Mobile Only
Dispatch 1	Dispatch only	462.950	467.950
Dispatch 2	Dispatch only	462.975	467.975
Med 1	Medical voice and telemetry	463.000	468.000
Med 2	Medical voice and telemetry	463.025	468.025
Med 3	Medical voice and telemetry	463.050	468.050
Med 4	Medical voice and telemetry	463.075	468.075
Med 5	Medical voice and telemetry	463.100	468.100
Med 6	Medical voice and telemetry	463.125	468.125
Med 7	Medical voice and telemetry	463.150	468.150
Med 8	Medical voice and telemetry	463.175	468.175

[a]From FCC Rules and Regulations.

In a typical paramedic operation, after a call is received concerning a person with a possible heart attack, the unit proceeds to the location. The paramedics check the general appearance of the patient, his or her level of consciousness, skin temperature and color, pulse rate and rhythm, respiration rate and depth, and blood pressure. If someone is with the patient, they also try to ascertain weight, medical allergies, and other patient information, because of the possibility of having to administer drugs. If any action is indicated that is within their capabilities, they take it. For example, defibrillation would be performed on the spot if needed. If not, the usual course is to relay the ECG to a hospital. They may be connected with an individual hospital, but they have the capability of communicating with many, using the several frequencies available. In a metropolitan area the MED frequencies are designated to hospitals or groups of hospitals. Channels are 25 kHz apart. The paramedic operator has no need to tune since each frequency is independent and the control is by a single switch with the 10 channels marked on it. Since voice communication is also available, the ECG is usually relayed to the hospital, an interpretation made by a cardiologist, and action taken within minutes. The regulations are such that the receiving unit in the hospital must have the capability of operating on at least four of the eight MED channels.

Emergency medical care has become an important part of the overall health delivery system. Its importance cannot be overemphasized. In November 1976 the IEEE issued a special volume of its transactions (see the Bibliography) on emergency medical services communications, which is an excellent reference that not only gives the history and development of the field, but presents a view of the problem nationally, regionally, urban and rural. It covers equipment and philosophies and even some of the political aspects.

12.5.4. Telephone Links

Although it cannot be considered to be radio telemetry, the use of the telephone system to transmit biological data is becoming quite common. One application involves the transmission of ECGs from heart patients and (particularly) pacemaker recipients. In this case the patient has a transmitter unit that can be coupled to an ordinary telephone. The transmitted signal is received by telephone in the doctor's office or in the hospital. Tests can be scheduled at regular intervals for diagnosing the status and potential problems indicated by the ECGs.

13

Instrumentation for the Clinical Laboratory

Every living organism has within itself a complete and very complicated chemical factory. In higher animals, food and water enter the system through the mouth, which is the beginning of the digestive tract. In the stomach the food is chemically broken down into basic components by the digestive juices. From there it is transported into the intestine, where the nutrients and the excess water are extracted. The extracted nutrients are then further broken down in numerous steps. Some are stored for later use, whereas others are used for the building of new body cells or are metabolized to obtain energy. All life functions, such as the contraction of muscles or the transmission of information through the nervous system, require energy for their operation. This energy is obtained from the nutrients by a series of oxidation processes which consume oxygen and leave carbon dioxide as a waste product. The exchange of oxygen and carbon dioxide with the air takes place in the lungs (see Chapter 8). Many of the chemical processes are performed in the liver, which in an organ specialized for this purpose. Certain soluble waste products are eliminated through the kidneys and the urinary tract. To make all this activity possible, the organism requires an efficient mechanism to transport the various chemical substances between the locations where they are introduced into the organism, are modified, or are excreted.

13.1. THE BLOOD

In very primitive animals, especially in those living in an ocean environment, like the sea anemone, the exchange of nutrients and metabolic wastes between cells and the environment takes place directly through the cell membrane. This simple method is insufficient, however, for larger animals, particularly those that live on land. For these animals, including man, nature has provided a special transport system to exchange chemical products between the specialized cells of the various organs—namely, the blood circulation. The circulatory system of an adult male human contains about 5 liters of blood. Blood consists of a fluid, called the *plasma,* in which are suspended three different types of *formed elements* or *blood cells.* One cubic millimeter of blood (about ⅙ drop) contains approximately the following numbers of cells:

Red blood cells (RBC) or erythrocytes	4.5–5.5 million
White blood cells (WBC) or leucocytes	6000–10,000
Blood platelets or thrombocytes	200,000–800,000

Red blood cells are round disks, indented in the center, with a diameter of about 8 μm. A red blood cell has no cell nucleus, but it has a membrane and is filled with a solution containing an iron-containing protein, *hemoglobin.* Red blood cells transport oxygen by chemically binding the oxygen molecules to the hemoglobin. Depending on the oxygen content, the hemoglobin changes its color, which accounts for the difference in color between oxygen-rich arterial blood (bright red) and oxygen-depleted venous blood (dark red).

White blood cells are of several different types, with an average diameter of about 10 μm. Each contains a nucleus and, like the amoeba, has the ability to change its shape. White blood cells attack intruding bacteria, incorporate them, and then digest them.

Blood platelets are masses of protoplasm 2 to 4 μm in diameter. They are colorless and have no nucleus. Blood platelets are involved in the mechanism of blood clotting.

By spinning blood in a centrifuge, the blood cells can be sedimented. The blood plasma with the blood cells removed is a slightly viscous, yellowish liquid that contains large amounts of dissolved protein. One of the proteins, *fibrinogen,* participates in the process of blood clotting and forms thin fibers called *fibrin*. The plasma from which the fibrinogen has been removed by precipitation is called *blood serum*.

The mechanism of blood clotting serves the purpose of preventing blood loss in case of injury. This mechanism can, on the other hand, cause undesirable or even dangerous blood clots if foreign bodies, like catheters or extracorporeal devices, are introduced into the bloodstream. Blood clotting can be inhibited by the injection of *heparin,* a natural anticoagulant extracted from the liver and lungs of cattle.

Many diseases cause characteristic variations in the composition of blood. These variations can be a characteristic change in the number, size, or shape of certain blood cells (in anemia, for instance, the RBC count is reduced). Other diseases cause changes in the chemical composition of the blood serum (or some other body fluid, like the urine). In diabetes mellitus, for instance, the glucose concentration in the blood (and the urine) is characteristically elevated. A count of the blood cells, an inspection of their size and shape, or a chemical analysis of the blood serum can, therefore, provide important information for the diagnosis of such diseases. Similarly, other body fluids, smears, and small samples of live tissue, obtained by a *biopsy,* are studied through the techniques of *bacteriology, serology,* and *histology* to obtain clues for the diagnosis of diseases.

The purpose of *bacteriological tests* is to determine the type of bacteria that have invaded the body, in order to diagnose a disease and prescribe the proper treatment. For such a test, a sample containing the bacteria (e.g., a smear from a strep throat) is innoculated to the surface of various growth media (nutrients) in test tubes or flat petri dishes. These cultures are then incubated at body temperature to accelerate the growth of the bacteria. When the bacteria have grown into colonies, they can be identified by the color and shape of the colony, by their preference for certain growth media, or by a microscopic inspection, which may make use of the fact that certain stains show a selectivity for certain bacteria groups.

Serological tests serve the same purpose as bacteriological tests but are based on the fact that the organism, when invaded by an infectious disease, develops antibodies in the blood, which defend the body against the infection. These antibodies are selective to certain strains of organisms, and their action can be observed in vitro by various methods. In some methods, for example, agglutination (collecting in clumps) becomes visible under a microscope when a test serum containing the antigen of the organism is added. Because the tests are based not on the organism itself but on the antigen developed by the organism, serological tests are not limited to bacteria but can be used for virus infections and infections by other microorganisms.

Histological tests involve the microscopical study of tissue samples, which are sliced into very thin sections by means of a precision slicer called a *microtome.* The tissue slices are often stained with certain chemicals to enchance the features of interest.

Blood counts and chemical blood tests are often ordered routinely on admission of a patient to a hospital and may be repeated daily to monitor the process of an illness. These tests, therefore, must be performed in very large numbers, even in the smaller hospital. The physician in private practice often has samples analyzed by commercial laboratories specializing in this service. Automated methods of performing the tests have found widespread acceptance, and special instruments have been developed for this purpose.

13.2. TESTS ON BLOOD CELLS

When whole blood is centrifuged, the blood cells sediment and form a packed column at the bottom of the test tube. Most of this column consists of the red blood cells, with the other cells forming a thin, *buffy layer* on top of the red cells. The volume of the packed red cells is called the *hematocrit*. It is expressed as a percentage of the total blood volume. If the number of (red) blood cells per cubic millimeter of blood is known, this number and the hematocrit can be used to calculate the *mean cell volume* (MCV). As stated above, the active component in the red blood cells is the hemoglobin, the concentration of which is expressed in grams/100 ml. From the hemoglobin, the hematocrit and the blood cell count, the *mean cell hemoglobin* (MCH) (in *picograms*) and the *mean cell hemoglobin concentration* (MCHC) (in percent) can be calculated.

The hematocrit can be determined by aspirating a blood sample into a capillary tube and closing one end of the tube with a plastic sealing material. The tube is then spun for 3 to 5 minutes in a special high-speed centrifuge to separate the blood cells from the plasma. Because the capillary tube has a uniform diameter, the blood and cell volumes can be compared by measuring the lengths of the columns. This is usually done with a simple nomogram, as shown in Figure 13.1. When lined up with the length of the blood column, the nomogram allows the direct reading of the hematocrit.

The red blood cells have a much higher electrical resistivity than the blood plasma in which they are suspended, and so the resistivity of the blood shows a high correlation with the hematocrit. This factor provides an alternative method of determining the hematocrit that is obviously more adaptable to automation than the centrifugal sedimentation method.

The *hemoglobin concentration* can be determined by lysing the red blood cells (destroying their membranes) to release the hemoglobin and chemically converting the hemoglobin into another colored compound (acid hematin or cyanmethemoglobin). Unlike that of the hemoglobin, the color concentration of these components does not depend on the oxygenation of the blood. Following the reaction, the concentration of the new component can be determined by colorimetry, as described in Section 13.3.

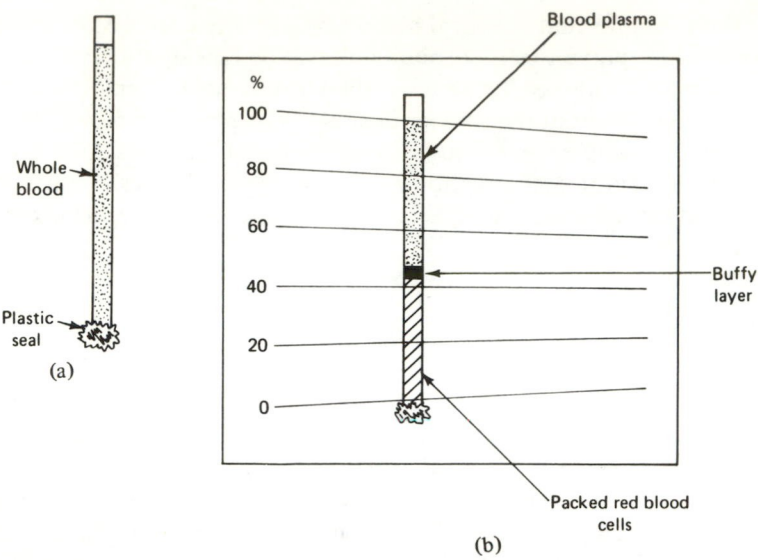

Figure 13.1. Hematocrit determination: (a) blood sample drawn in capillary and sealed with plastic putty; (b) capillary after centrifuging, placed on nomogram to read hematocrit (reading 43%).

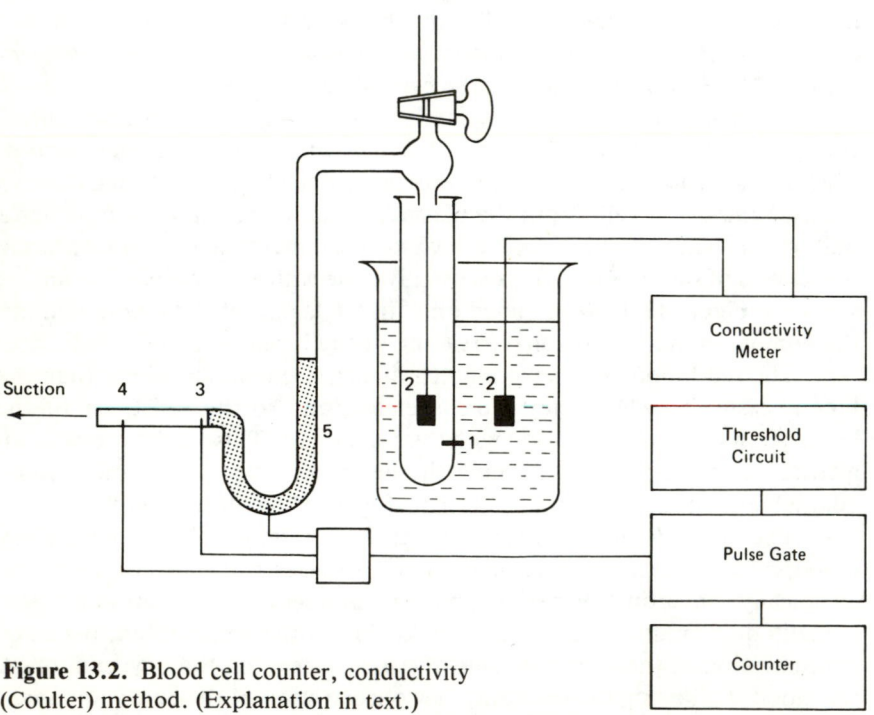

Figure 13.2. Blood cell counter, conductivity (Coulter) method. (Explanation in text.)

Manual blood cell counts are performed by using a microscope. Here the blood is first diluted 1:100 or 1:200 for counting red blood cells (RBC) and 1:10 or 1:20 for white blood cell count (WBC). For counting WBC, a diluent is used that dissolves the RBCs, whereas for counting RBCs, an isotonic diluent preserves these cells. The diluted blood is then brought into a counting chamber 0.1 mm deep, which is divided by marking lines into a number of squares. When magnified about 500 times, the cells in a certain number of squares can be counted. This rather time-consuming method is still used quite frequently when a *differential count* is required for which the WBCs are counted, according to their distribution, into a number of different subgroups. An automated differential blood cell analyzer uses differential staining methods to discriminate between the various types of white blood cells.

Today simple RBC and WBC counts are normally performed by automatic or semiautomatic blood cell counters. The most commonly used devices of this kind are based on the conductivity (Coulter) method, which makes use of the fact that blood cells have a much lower electrical conductivity than the solution in which they are suspended. Such a counter (Figure 13.2) contains a beaker with the diluted blood into which a closed glass tube with a very small orifice (1) is placed. The conductance between the solution in the glass tube and the solution in the beaker is measured with two electrodes (2). This conductance is mainly determined by the diameter of the orifice, in which the current density reaches its maximum. The glass tube is connected to a suction pump through a U-tube filled with mercury (5). The negative pressure generated by the pump causes a flow of the solution from the beaker through the orifice into the glass tube. Each time a blood cell is swept through the orifice, it temporarily blocks part of the electrical current path and causes a drop in the conductance measured between the electrodes (2). The result is a pulse at the output of the conductance meter, the amplitude of which is proportional to the volume of the cell. A threshold circuit lets only those pulses pass that exceed a certain amplitude. The pulses that pass this circuit are fed to a pulse counter through a pulse gate. The gate opens when the mercury column reaches a first contact (3) and closes when it reaches the second contact (4), thus counting the number of cells contained in a given volume of the solution passing through the orifice. A count is completed in less than 20 seconds. With counts of up to 100,000, the result is statistically accurate. Great care must be taken, however, to keep the aperture from clogging. Counters based on this principle are available with varying degrees of automation. The most advanced device of this type (shown in Figure 13.3) accepts a new blood sample every 20 seconds, performs the dilutions automatically, and determines not only the WBC and RBC counts but also the hematocrit and the hemoglobin concentration. From these measurements, the mean cell volume, the mean cell

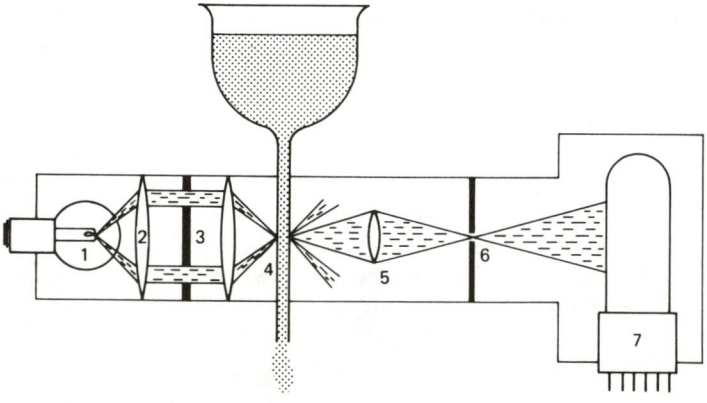

Figure 13.3. Coulter Counter® Model S. Sr. (Courtesy of Coulter Electronics Hialeah, FL.)

Figure 13.4. Blood cell counter, dark field method. (Explanation in text.)

hematocrit, and the mean cell hematocrit concentration are calculated and all results are printed out on a preprinted report form.

A second type of blood cell counter uses the principle of the dark-field microscope (Figure 13.4). The diluted blood flows through a thin cuvette (4). The cuvette is illuminated by a cone-shaped light beam obtained from a lamp (1) through a ring aperture (3) and an optical system (2). The cuvette is imaged on the cathode of a phototube (7) by means of a lens (5) and an aperture (6). Normally no light reaches the phototube until a blood cell passes through the cuvette and reflects a flash of light on the phototube.

13.3. CHEMICAL TESTS

Blood serum is a complex fluid that contains numerous substances in solution. The determination of the concentration of these substances is performed by specialized chemical techniques. Although there are usually several different methods by which any particular analysis can be performed, most tests used are based on a chemical color reaction followed by a colorimetric determination of the concentration. This principle makes use of the fact that many chemical compounds in solution appear colored, with the saturation of the color depending on the concentration of the compound. For instance, a solution that appears yellow when being held against a white background actually absorbs the blue component of the white light and lets only the remainder—namely, yellow light—through. The way in which this light absorption can be used to determine the concentration of the substance is shown in Figure 13.5.

In Figure 13.5(a) it is assumed that a solution of concentration C is placed in a cuvette with a length of the light path, L. Light of an appropriate color or wavelength is obtained from a lamp through filter F. The light that enters the cuvette has a certain intensity, I_0. With part of the light being absorbed in the solution, the light leaving the cuvette has a lower intensity I_1. One way of expressing this relation is to give the *transmittance, T,* of the solution in the cuvette as the percentage of light that is transmitted:

$$T = \frac{I_1}{I_0} \times 100\%$$

If a second cuvette with the same solution were brought into the light path behind the first cuvette, only a similar portion of the light entering this cuvette would be transmitted. The light intensity I_2 behind the second cuvette is

$$I_2 = TI_1$$

or

$$I_2 = T^2 I_0$$

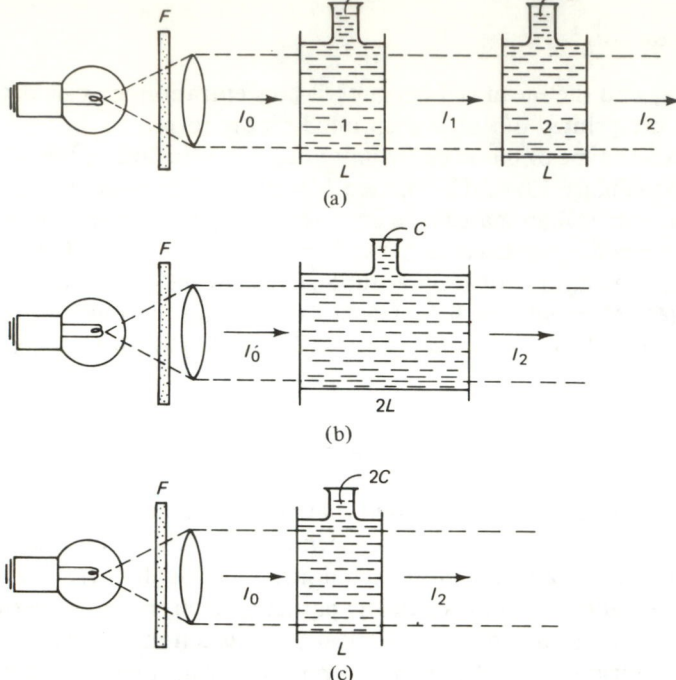

Figure 13.5. Principle of colorimeter analysis. (Explanation in text.)

The light transmitted through successive cuvettes decreases in the same manner (multiplicatively). For this reason, it is advantageous to express transmittance as a logarithmic measure (in the same way as expressing electronic gains and losses in decibels). This measure is the *absorbance* or optical *density, A*.

$$A = -\log\frac{I_1}{I_0}$$

or

$$A = \log\frac{1}{T}$$

The total absorbance of the two cuvettes in Figure 13.5(a) is, therefore, the sum of the individual absorbances.

The amount of the light absorbed depends only on the number of molecules of the absorbing substance that can interact with the light. If, instead of two cuvetttes, each with path length L, one cuvette with path length $2L$, were used [Figure 13.5(b)], the absorbance would be the same. The absorbance is also the same if the cuvette has a path length L, but the concentration of the solution were doubled [Figure 13.5(c)]. This relation can be expressed by the equation:

$$A = aCL \quad \text{(Beer's law)}$$

where L = path length of the cuvette

C = concentration of the absorbing substance

a = *absorbtivity,* a factor that depends on the absorbing substance and the optical wavelength at which the measurement is performed.

The absorbtivity can be obtained by measuring the absorption of a solution with known concentration, called a *standard.* If A_s is the absorption of the standard, A_u the absorption of an unknown solution, and C_s the concentration of the standard, then the concentration of the unknown is

$$C_u = C_s \frac{A_u}{A_s}$$

Corrections may have to be applied for light losses due to reflections at the cuvette or absorption by the solvent. Figure 13.6 shows the principle of a *colorimeter* or *filter-photometer* used for measuring transmittance and absorbance of solutions. A filter F selects a suitable wavelength range from the light of a lamp. This light falls on two photoelectric (selenium) cells: a reference cell C_R and a sample cell C_S. Without a sample, the output of both cells is the same. When a sample is placed in the light path for the sample cell, its output is reduced and the output of C_R has to be divided by a potentiometer P until a galvanometer (G) shows a balance. The potentiometer can be calibrated in transmittance or absorbance units over a range of 1 to 100 percent transmittance, corresponding to 2 to 0 absorbance units.

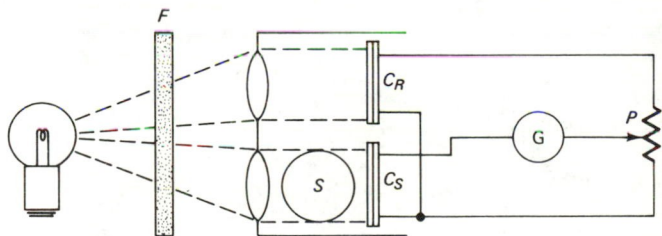

Figure 13.6. Colorimeter (filter-photometer).

Other colorimeters, instead of using the potentiometric method, use a meter calibrated directly in transmittance units (a linear scale) and in absorbance. Figure 13.7 shows such a device; the instrument allows measurement at different colors with a built-in filter wheel. If a standard with a known concentration of a certain substance is used as a reference, the scale can be calibrated directly in concentration units for this substance.

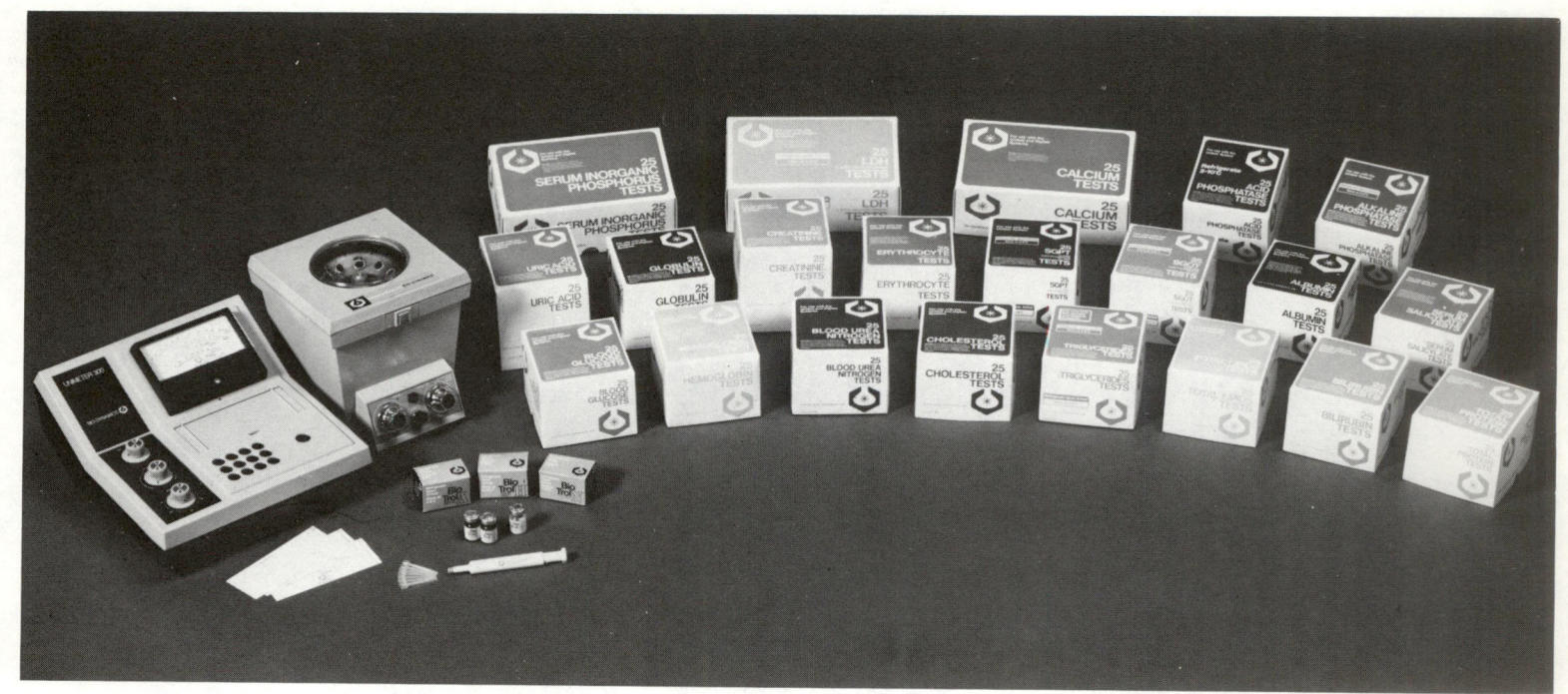

Figure 13.7. Unimeter—modern colorimeter system permits the physician to perform blood analysis in the office. The colorimeter with built-in incubator is at the far left, with a centrifuge by its side and packages with pre-measured reagent kits. Test values are read on exchangeable scales shown lying in front of the colorimeter. (Courtesy of Biodynamics/bmc, Indianapolis, IN.)

In order to use the colorimeter to determine the concentration of a substance in a sample, a suitable method for obtaining a colored derivative from the substance is necessary. Thus, a chemical reaction that is unique for the substance to be tested and that does not cause interference by other substances which may be present in the sample must be found. The reaction may require several steps of adding reagents and incubating the sample at elevated temperatures until the reaction is completed. Most reactions require that the protein first be removed from the plasma by adding a precipitating reagent and filtering the sample.

In most tests, an excess of the reagents is added and the incubation is continued until the *end point* of the reaction is reached (i.e., until all of the substance has been converted into its colored derivative). In *kinetic analysis* methods, the transmittance is measured several times at fixed time intervals while the chemical reaction continues. The concentration of the substance of interest then can be calculated from the rate of change of the absorbance.

Table 13.1. THE MOST COMMONLY USED CHEMICAL BLOOD TESTS

Test	Normal Ranges	Unit
1. Blood urea nitrogen (BUN)	8–16	mg N/100 ml
2. Glucose	70–90	mg/100 ml
3. Phosphate (inorganic)	3–4.5	mg/100 ml
4. Sodium	135–145	mEq/liter
5. Potassium	3.5–5	mEq/liter
6. Chloride	95–105	mEq/liter
7. CO_2 (total)	24–32	mEq/liter
8. Calcium	9–11.5	mg/100 ml
9. Creatinine	0.6–1.1	mg/100 ml
10. Uric acid	3–6	mg/100 ml
11. Protein (total)	6–8	g/100 ml
12. Albumin	4–6	g/100 ml
13. Cholesterol	160–200	mg/100 ml
14. Bilirubin (total)	0.2–1	mg/100 ml

The most commonly required tests for blood samples are listed in Table 13.1. This table also shows the units in which the test results are expressed* and the normal range of concentration for each test. Most of these tests can be performed by color reaction even though, in most cases, several different methods have been described that can often be used alternately.

For the measurement of sodium and potassium, however, a different property is utilized, one that causes a normally colorless flame to appear

*Depending on the test, the concentration is expressed in either grams or milligrams per 100 milliliters (0.1 liter) or in milliequivalents per liter, which is obtained by dividing the concentration in milligrams per liter by the molecular weight of the substance.

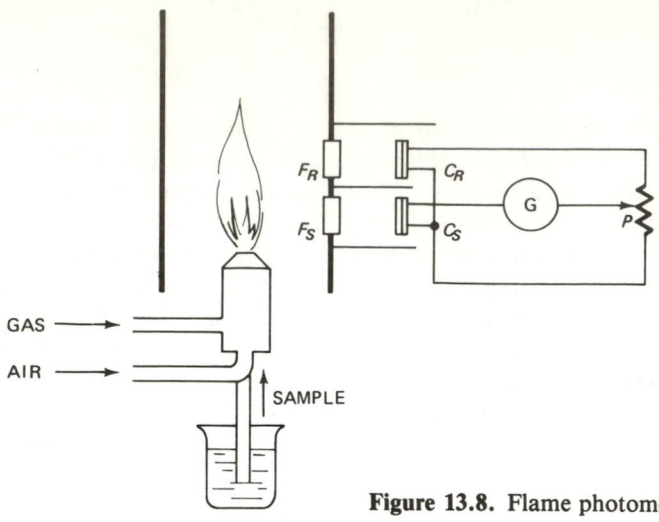

Figure 13.8. Flame photometer.

yellow (sodium) or violet (potassium) when their solutions are aspirated into the flame. This characteristic is used in the *flame photometer* (Figure 13.8) to measure the sodium or potassium concentration in samples. The sample is aspirated into a gas flame that burns in a chimney. As a reference, a known amount of a lithium salt is added to the sample, thus causing a red flame. Filters are used to separate the red light produced by the lithium from the yellow or violet light emitted by the sodium or potassium. As in the colorimeter, the output from the sample cell C_S is compared with a fraction of the output from a reference cell C_R. The balance potentiometer P is calibrated directly in units of sodium or potassium concentration.

For the determination of chlorides, a special instrument (*chloridimeter*) is sometimes used that is based on an electrochemical (*coulometric*) method. For this test, the chloride is converted into silver chloride with the help of an electrode made of silver wire. By an electroplating process with a constant current, the silver chloride is percipitated.

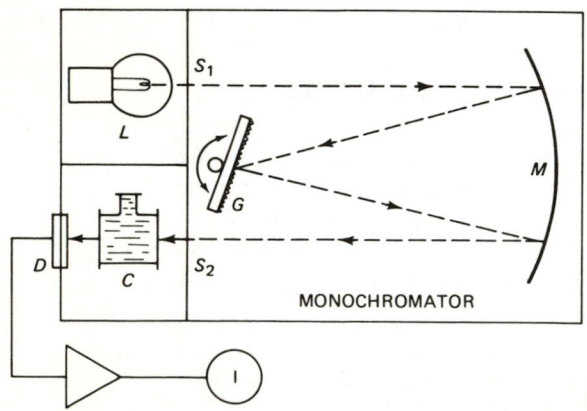

Figure 13.9. Spectrophotometer.

When all the chloride has been used up, the potential across the cell changes abruptly and the change is used to stop an electric timer, which is calibrated directly in chloride concentration.

The simple colorimeter (or filter-photometer) shown in Figures 13.6 and 13.7 has a sophisticated relative, the *spectrophotometer* shown in Figure 13.9. In this device the simple selection filter of the colorimeter is replaced by a *monochromator*. A monochromator uses a diffraction grating G (or a prism) to disperse light from a lamp that falls through an entrance slit S_1 into its spectral components. An exit slit S_2 selects a narrow band of the spectrum, which is used to measure the absorption of a sample in cuvette C. The narrower the exit slit, the narrower the bandwidth of the light, but also the smaller its intensity. A sensitive photodetector D (often a photomultiplier) is therefore required, together with an amplifier and a meter I, which is calibrated in units of transmittance or absorbance. The wavelength of the light can be changed by rotating the grating. A mirror M folds the light path to reduce the size of the instrument.

The spectrophotometer allows the determination of the absorption of samples at various wavelengths. The light output of the lamp, however, as well as the sensitivity of the photodetector and the light absorption of the cuvette and solvent, varies when the wavelength is changed. This situation requires that, for each wavelength setting, the density reading be set to zero, with the sample being replaced by a blank cuvette, usually filled with the same solvent as used for the sample. In *double-beam* spectrophotometers this procedure is done automatically by switching the beam between a sample light path and a reference light path, generally with a mechanical shutter or rotating mirror. By using a computing circuit, the readings from both paths are compared and only the ratio of the absorbances (or the difference of the densities) is indicated.

Certain chemicals, when illuminated by light with a short wavelength in the ultraviolet (UV) range, emit light with a longer wavelength. This phenomenon is called *fluorescence.* Fluorescence can be used to determine the concentration of such chemicals using a *fluorometer,* which, like the photometer, can be either a *filter-fluorometer* or a *spectrofluorometer,* depending on whether filters or monochromators are used to select the excitation and emission wavelengths.

13.4. AUTOMATION OF CHEMICAL TESTS

Even though most chemical tests basically consist of simple steps like pipetting, diluting, and incubating, they are rather time-consuming and require skilled and conscientious technicians if errors are to be avoided. Attempts to replace the technicians by an automatic device, however, were

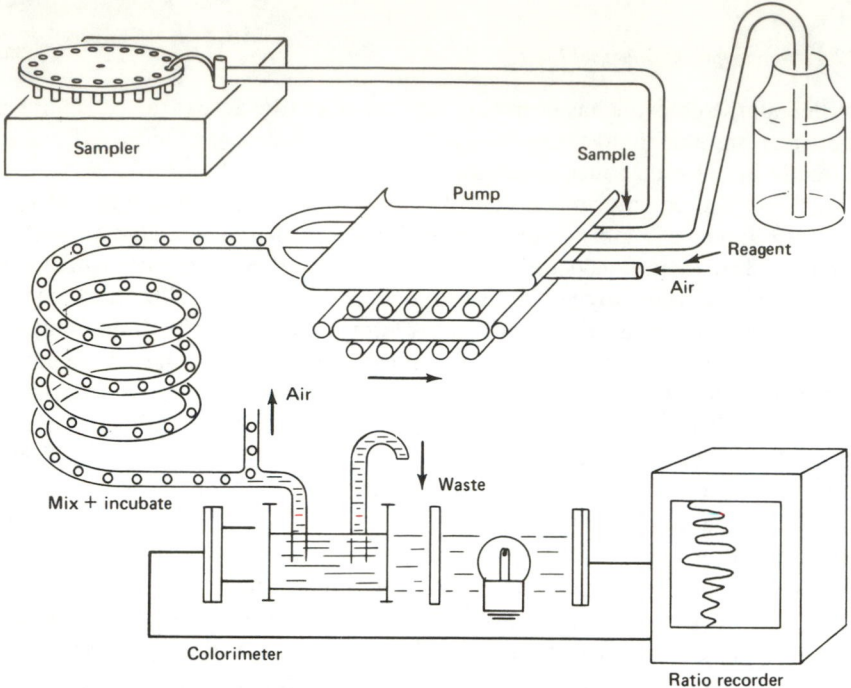

Figure 13.10. Continuous flow analyzer (simplified).

not very successful at first. The first automatic analyzer that found wide acceptance and that is still used at most hospitals in the *Autoanalyzer,* the principle of which is shown in Figure 13.10.

The basic method used in the Autoanalyzer departs in several respects from that of standard manual methods. The mixing, reaction, and colorimetric determination take place, not in an individual test tube for each sample but sequentially in a continuous stream. The sampler feeds the samples into the analyzer in time sequence. A proportioning pump, which is basically a simple peristaltic pump working simultaneously on a number of tubes with certain ratios of diameters, is used to meter the sample and the reagent. Mixing is achieved by injecting air bubbles. The mixture is incubated while flowing through heated coils. The air bubbles are removed, and the solution finally flows through the cuvette of a colorimeter or is aspirated into the flame of a flame photometer. An electronic ratio recorder compares the output of the reference and sample photocells. The recording shows the individual samples as peaks of a continuous transmittance or absorbance recording. The samples of a "run" are preceded by a number of standards that cover the useful concentration range of the test. The concentration of the samples is determined from the recording by comparing the peaks of the samples with the peaks of the standards. In this way the effects of errors (e.g., incomplete reaction in the incubator) are eliminated because they affect standards and samples in the same way.

Suitable adaptations of almost all standard tests have been developed for the Autoanalyzer system. The removal of protein from the plasma is achieved in the continuous-flow method with a *dialyzer* (not shown in Figure 13.10), which consists of two flow channels separated by a cellophane membrane that is impermeable to the large protein molecules, but not to the smaller molecules. The smallest model of the Autoanalyzer performs a single test at a rate up to 120 samples per hour. Large later models (one of which is shown in Figure 13.11) perform up to 12 different tests on each of 90 samples per hour. The results of these tests are directly provided in the form of a "chemical profile," drawn by a recorder on a preprinted chart. By the use of additional equipment, the results may also be provided as a digital output signal for recording on a storage medium, like punch cards or paper tape, or may be usable for direct computer processing.

A major problem with the continuous-flow process is the "carryover" that can occur when a sample with an excessively high concentration is followed by a sample with normal or low concentration. Methods of "carryover" correction are available.

Although the continuous-flow analyzer was the first to find wide acceptance, numerous other analyzers that use discrete samples are now available. Some of these analyzers perform all tests in test tubes mounted on a carousel-type carrier, or a chain belt, with the test tubes being rinsed after the completion of the analysis.

Figure 13.11. Technicon Autoanalyzer SMA II System. (Technicon Autoanalyzer and SMA II are registered trademarks of Technicon Instruments, Tarrytown, NY by permission.)

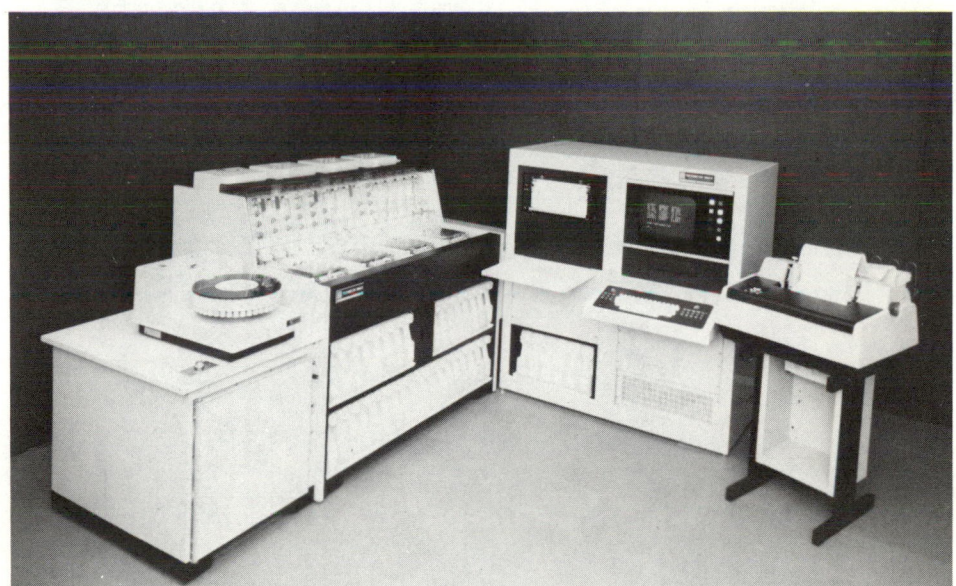

Figure 13.12. Automatic Clinical Analyzer. (a) The unit itself-test packs are entered in the U-shaped tray at the left, while the packages with large numbers contain the diluents. (b) Test pack containing liquid and solid reagents in the arrow-shaped compartments. (Courtesy of E.I. Du Pont de Nemours and Company Inc. Automatic Clinical Analysis Division, Wilmington, DE.)

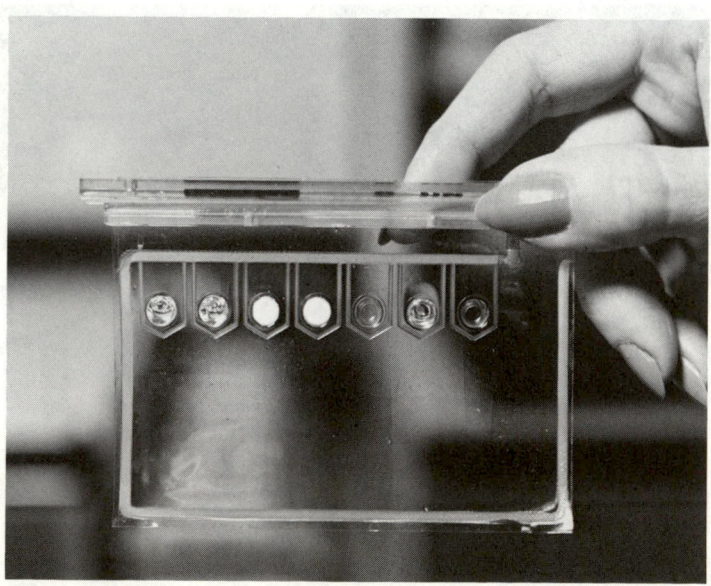

All automatic analyzers of this type use syringe-type pumps to dispense the sample and to add the reagents. After incubation the sample is aspirated into a colorimeter cuvette, where its absorbance is measured.

Discrete sample analyzers as well as continuous-flow analyzers require that all reagents be available in the proper dilution. One automatic analyzer uses a principle that always assures the correct quantities. In this analyzer (Figure 13.12) all reagents for a given test are sealed in premeasured quantities in dry form in compartments of a plastic pouch. The package also carries a machine-readable code which identifies the particular test. The patient sample in a carrier and pouchpacks for the tests to be run on that sample are inserted together. The analyzer identifies the test from the machine-readable code and injects the necessary amount of sample together with a suitable diluent into each test pack. The reagents are released by breaking the walls of their compartments, and are mixed with sample and diluent. After incubation the absorbance of the solution is measured directly in the transparent plastic pouch using a special colorimeter which forms the pouch into an optical cell with a defined path length.

Another type of analyzer, illustrated in Figure 13.13, processes a number of samples simultaneously by means of a fast-spinning disk which

Figure 13.13. Rotochem II Parallel Fast Analyzer (Courtesy of American Instrument Company, Silver Springs, MD.)

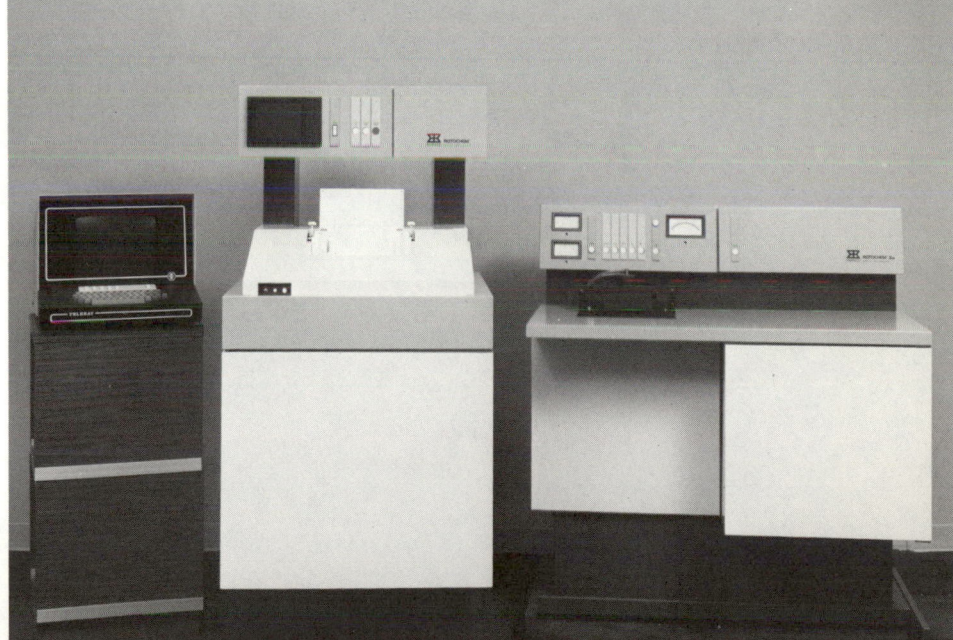

contains reagent and sample chambers and cuvettes. The transfer of sample and reagent to the cuvette and the mixing of both is accomplished by centrifugal force. The absorbance of the solutions is measured by one colorimeter which measures all samples in sequence while the rotation of the disk carries them through the colorimeter lightbeam. This arrangement makes the centrifugal analyzer especially useful for the kinetic analysis methods mentioned in Section 13.3.

A basic problem with automatic analyzers is the positive identification of samples. In early devices the small sample cups were identified only by their position in the sample tray and the technician loading the samples had to prepare a "load list" for this purpose. Machine-readable methods of sample identification are now available and greatly reduce the likelihood of mixups.

Many modern automatic analyzers utilize electronic data processing by built-in mini- or microcomputers to calibrate the system. They also convert absorbance measurements into concentration values and print out the results of the tests. The role of the computer in the clinical chemistry laboratory is described in some detail in Chapter 15.

14

X-Ray
and Radioisotope
Instrumentation

In 1895 Conrad Röntgen, a German physicist, discovered a previously unknown type of radiation while experimenting with gas-discharge tubes. He found that this type of radiation could actually penetrate opaque objects and provide an image of their inner structures. Because of these mysterious properties, he called his discovery *X rays*. In many countries X rays are referred to as *Röntgen rays* in honor of their discoverer, who received a Nobel prize in 1901 for his work.

Soon after the discovery of X rays, their importance as a tool for medical diagnosis was recognized. Later it was found that X rays could also be used for therapeutic purposes. Both applications of X rays are the domain of the medical specialty known as *radiology*. X-ray machines were the first widely used electrical instruments in medicine. In fact, hospitals still spend more money for the purchase of X-ray equipment than for any other type of medical instrumentation.

One year after Röntgen's discovery, Henry Becquerel, the French physicist, found a similar type of radiation emanating from samples of uranium ore. Two of his students, Pierre and Marie Curie, traced this radiation to a previously unknown element in the ore, to which they gave the name *radium*, from the Latin word *radius*, the ray. The process by which radium and certain other elements emit radiation is called *radioactive decay,* whereas the property of an element to emit radiation is called *radioactivity.*

14.1 BASIC DEFINITIONS

One of the characteristics of the radiation originating in the X-ray tube or in radioactive materials is that it ionizes the gases through which it travels. Therefore, the term *ionizing radiation* is used to differentiate between this type of radiation and other, *nonionizing* types of radiation, such as radio waves, light, and infrared radiation.

Many man-made radioisotopes are now available along with the X-ray tube and radium as sources of radiation. The ability of this radiation to penetrate materials that are opaque to visible light is utilized in numerous techniques in medical diagnosis and research. The ionizing effects of radiation are also used for the treatment of certain diseases, such as cancer. The use of radiation for treatment of diseases has become an important subfield of medicine, called *radiation therapy,* which is discussed briefly in Section 14.5.

Another related topic is computerized axial tomography. While this technique involves X rays, its principles are primarily computer-related. For this reason it is discussed in detail in Chapter 15 as a computer application.

There are three different types of radiation, each with its own distinct properties. More than one type of radiation can emanate from a given sample of radioactive material. The properties of the three types of radiation are defined below.

Alpha rays are positively-charged particles that consist of helium nuclei and that travel at the moderate velocity of 5 to 7 percent of the velocity of light. They have a very small penetration depth, which in air is only about 2 in.

Beta rays are negatively-charged electrons. Their velocity can vary over a wide range and can almost reach the velocity of light. Their ability to penetrate the surrounding medium depends on their velocity, but generally it is not very great. Both alpha and beta rays, when traveling through a gaseous atmosphere, interact with the gas molecules, thereby causing ionizing of the gas.

Gamma rays and *X rays* are both electromagnetic waves that have a much shorter wavelength than radio waves or visible light. Their wavelengths can vary between approximately 10^{-6} and 10^{-10} cm, corresponding to a

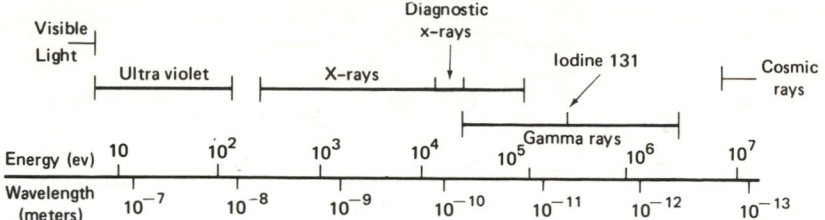

Figure 14.1. Part of the electromagnetic spectrum showing the location of the X rays and gamma rays.

frequency range of between 10^{10} and 10^{14} MHz, with the X rays at the lower and the gamma rays at the higher end of this range. The ability of these rays to penetrate matter depends on their wavelengths, but it is much greater than that of the alpha and beta rays. Gamma rays do not interact with gases directly but can cause ionization of the gas molecules via photoelectrons released when the rays interact with solid matter.

Gamma rays are usually not characterized by their frequency but by their energy, which is proportional to the frequency. This relationship is expressed in the Planck equation:

$$E = hf$$

where E = energy, ergs
h = Planck's constant = 6.624×10^{-27} erg sec
f = frequency, hertz

The energy of radiation is usually expressed in electron volts (eV), with 1 eV = 1.602×10^{-12} erg.

Figure 14.1 shows the position of gamma rays and X rays within the spectrum of electromagnetic waves.

14.1.1. Generation of Ionizing Radiation

X rays are generated when fast-moving electrons are suddenly decelerated by impinging on a target. An X-ray tube is basically a high-vacuum diode with a heated cathode located opposite a target anode (Figure 14.2). This diode is operated in the saturated mode with a fairly low cathode temperature so that the current through the tube does not depend on the applied anode voltage.

The intensity of X rays depends on the current through the tube. This current can be varied by varying the heater current, which in turn controls the cathode temperature. The wavelength of the X rays depends on the target material and the velocity of the electrons hitting the target. It can be varied by varying the target voltage of the tube. X-ray equipment for diag-

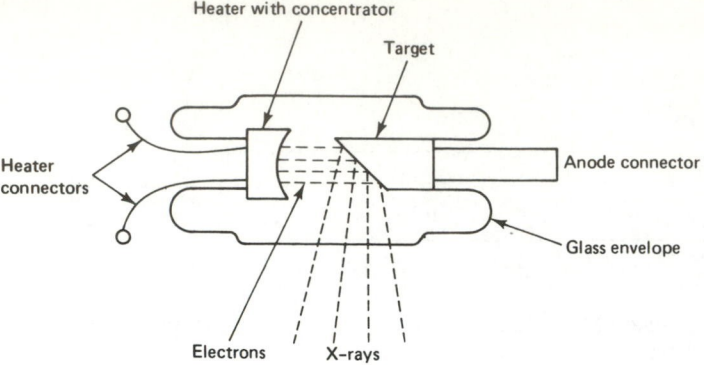

Figure 14.2. X-ray tube, principle of operation.

nostic purposes uses target voltages in the range of 30 to 100 kV, while the current is in the range of several hundred milliamperes. These voltages are obtained from high-voltage transformers that are often mounted in oil-filled tanks to provide electrical insulation. When ac voltage is used, the X-ray tube conducts only during one half-wave and acts as its own rectifier. Otherwise high-voltage diodes (often in voltage-doubler or multiplier configurations) are used as rectifiers. For therapeutic X-ray equipment, where even higher radiation energies are required, linear or circular particle accelerators have been used to obtain electrons with sufficiently high energy. When the electrons strike the target, only a small part of their energy is converted into X rays; most of it is dissipated as heat. The target, therefore, is usually made of tungsten, which has a high melting point. It may also be water- or air-cooled, or it may be in the form of a motor-driven rotating cone to improve the dissipation of heat. The electron beam is concentrated to form a small spot on the target. The X rays emerge in all directions from this spot, which therefore can be considered a point source for the radiation.

Radioactive decay is the other source of nuclear radiation, but only a very small number of chemical elements exhibit natural radioactivity. Artificial radioactivity can be induced in other elements by exposing them to neutrons generated with a cyclotron or in an atomic reactor. By introducing an extraneous neutron into the nucleus of the atom, an unstable form of the element is generated that is chemically equivalent to the original form (*isotope*). The unstable atom disintegrates after some time, often through several intermediate forms, until it has assumed the form of another, stable element. At the moment of the disintegration, radiation is emitted, the type and energy of which are characteristic of a particular decay step in the process. The time after which half of the original number of radioisotope atoms have decayed is called the *half-life*. Each radioisotope has a characteristic half-life that can be between a few seconds and thousands of years.

Radioisotopes are chemically identical to their mother element. Chemical compounds in which a radioisotope has been substituted for its mother element are thus treated by the body exactly like the nonradioactive form. With the help of the emitted radiation, however, the path of the substance can be traced and its concentration in various parts of the organism determined. If this procedure is to be done in vivo, the isotope must emit gamma radiation that penetrates the surrounding tissue and that can be measured with an extracorporeal detector. When radioactive material is introduced into the human body for diagnostic purposes, great care must be taken to ensure that the radiation dose that the body receives is at a safe level. For reasons explained below, it is desirable that the radioactivity be as great as possible during the actual measurement. For safety reasons, however, the activity should be reduced as fast as possible as soon as the measurement is completed. In certain measurements, the radioactive material is excreted from the body at a rapid rate and the activity in the body decreases quickly. In most measurements, this "biological decay" of the introduced radioactivity occurs much too slowly. In order to remove the source of radiation after the measurement, isotopes with a short half-life must be used. However, there is a dearth of gamma-emitting isotopes of elements naturally occurring in biological substances that have a half-life of suitable length. The radioisotopes most frequently used for medical purposes are listed in Table 14.1. Iodine 131 is the only gamma-emitting isotope of an element that occurs in substantial quantities in the body. H-3 (tritium) and carbon 14 are beta emitters; hence their concentration in biological samples can be measured only in vitro because the radiation does not penetrate the surrounding tissue.

Table 14.1. RADIOISOTOPES

Isotope	Radiation	Half-Life
^3H	Beta	12.3 days
^{14}C	Beta	5570 years
^{51}Cr	Gamma	27.8 days
^{99m}Tc	Gamma	6 hours
^{131}I	Gamma	8.07 days
^{198}Au	Gamma	2.7 days

14.1.2. Detection of Radiation

Pierre and Marie Curie discovered that radioactivity can be detected by three different physical effects: (1) the activation it causes in photographic emulsions, (2) the ionization of gases, and (3) the light flashes the radiation causes when striking certain minerals. Most techniques used today are still based on the same principles. Photographic films are the most commonly used method of visualizing the distribution of X rays for diagnostic purposes. For the visualization of radioisotope concentrations in biological samples, a photographic method called *autoradiography* is used. In this technique thin slices of tissue are laid on a photographic plate and left in contact (in a freezer) for extended time periods, sometimes for months. After processing, the film shows an image of the distribution of the isotope in the tissue.

When the gas ions caused by radiation are subjected to the forces of an electric field between two charged capacitor plates, they move toward these plates and cause a current flow. Above a certain voltage, all ion pairs generated reach the plates, and further increases of the voltage cause no additional increase of the current (saturation). The current flow (normally very small) can be used to measure the intensity of the radiation. This device is called an *ionization chamber*.

The number of ion pairs generated depends on the type of radiation. The number is greatest for alpha and lowest for gamma radiation. If the voltage is increased beyond a certain value, the ions are accelerated enough to ionize additional gas molecules (gas amplification, *proportional counter*). If the voltage is increased even further, a point can be reached at which any initial ion pair causes complete ionization of the tube (*Geiger counter*). Further increase of the voltage, therefore, does not increase the current (plateau). The ion generation, however, is self-sustaining and must be terminated, usually by reducing the voltage briefly. The Geiger counter cannot discriminate between the different types of radiation, but it has the advantage of providing large output pulses.

The physical configuration of the various detectors based on the principle of gas ionization can actually be the same. The mode of operation, as shown in Figure 14.3, is determined solely by the operating voltage applied to the device.

Another type of device related to the Geiger counter is the *spark chamber,* which consists of an array of opposed electrodes that have a voltage applied between them that by itself is not high enough to cause a discharge. The ionization caused by the passage of radiation, however, triggers a spark that momentarily discharges the circuits of the two electrodes between which it occurs. The spark can be detected either by photographic methods or by the sound waves that it produces.

Certain metal salts (e.g., zinc sulfide) show fluorescence when irradiated with X rays or radiation from radioisotopes. When observed under a microscope under favorable circumstances, the minute light flashes (scintillations) caused by individual radiation events can actually be seen. In earlier days these scintillations were used to measure radioactivity by simply counting them. Both scintillation and fluorescence, however, are light events of such low intensity that they can be seen only with eyes that are well adapted to the dark. Only through use of electronic devices for the detection and visualization of low-level light has their usefulness been increased to such an extent that today most isotope instrumentation is based on this principle.

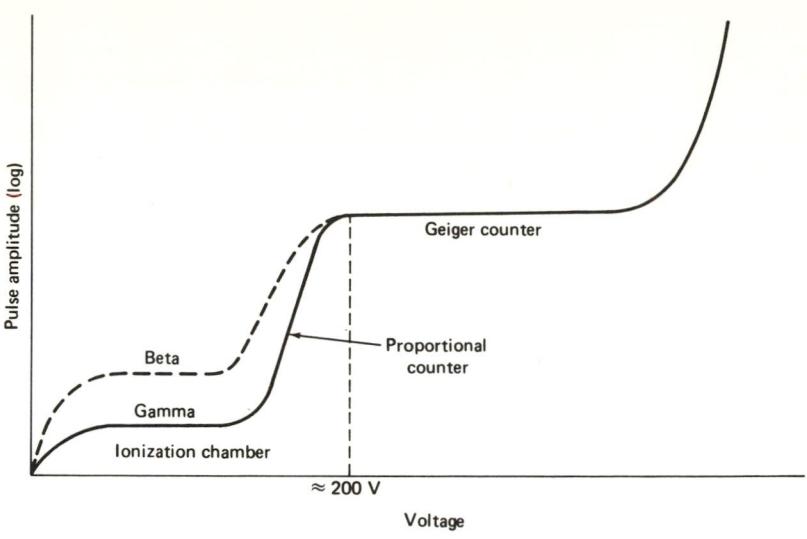

Figure 14.3. Detection of nuclear radiation by the ionization of gas between two capacitor plates. The curve shows the logarithm of the current pulse amplitude as a function of the applied voltage for a constant rate of nuclear disintegrations generating either beta or gamma radiation.

14.2. INSTRUMENTATION FOR DIAGNOSTIC X RAYS

The use of X rays as a diagnostic tool is based on the fact that various components of the body have different densities for the rays. When X rays from a point source penetrate a body section, the internal structure of the body absorbs varying amounts of the radiation. The radiation that leaves the body, therefore, has a spatial intensity variation that is an image of the internal structure of the body. When, as shown in Figure 14.4, this intensity distribution is visualized by a suitable device, a shadow image is generated that corresponds to the X-ray density of the organs in the body section.

Figure 14.4. Use of X rays to visualize the inner structure of the body.

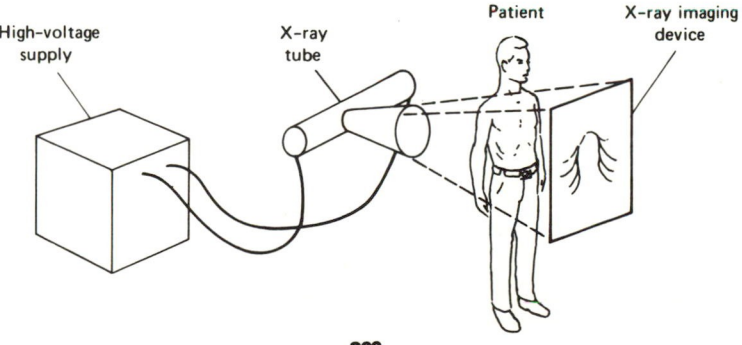

Bones and foreign bodies, especially metallic ones, and air-filled cavities show up well on these images because they have a much higher or a much lower density than the surrounding tissue. Most body organs, however, differ very little in density and do not show up well on the X-ray image, unless one of the special techniques described later is used.

14.2.1. Visualization of X Rays

X rays normally cannot be detected directly by the human senses; thus, indirect methods of visualization must be used to give an image of the intensity distribution of X rays that have passed through the body of a patient. Three different techniques are in common use.

14.2.1.1. Fluoroscopy. Rontgen actually discovered X rays when he noticed that certain metal salts glowed in the dark when struck by the radiation. The brightness of this *fluorescence* is a function of the radiation intensity, and cardboard pieces coated with such metal salts were first used exclusively to visualize X-ray images. Early *fluoroscopes* were simply cardboard funnels, open at the narrow end for the eyes of the observer, while the wide end was closed with a thin cardboard piece that had been coated on the inside with a layer of fluorescent metal salt. The fluoroscopic image obtained in this way is rather faint, however, and the X-ray intensity necessary to obtain a reasonably bright image is of such a magnitude that it can be harmful to both the patient and the observer. If the radiation intensity is reduced to a safer level, the fluoroscopic image becomes so faint that it must be observed in a completely darkened room and after the eyes of the observer have adapted to the dark for 10 to 20 minutes. Because of these inconveniences, direct fluoroscopy now has only limited use.

14.2.1.2. X-ray films. Although X rays have a much shorter wavelength than visible light, they react with photographic emulsions in a similar fashion. After processing in a developing solution, therefore, a film that has been exposed to X rays shows an image of the X-ray intensity. The sensitivity of this effect can be increased by the use of *intensifying screens,* which are similar to the fluoroscopic screens described above. The screen is brought into close contact with the film surface so that the film is exposed to the X rays as well as to the light from the fluorescence of the screen. X-ray films, with or without intensifying screens, are packaged in light-tight *cassettes* in which one side is made of thin plastic that can easily be penetrated by the X rays.

14.2.1.3. Image intensifiers. The faint image of a fluoroscopic screen can be made brighter with the help of an electronic *image intensifier,* as shown in Figure 14.5. The intensifier tube contains a fluorescent screen, the

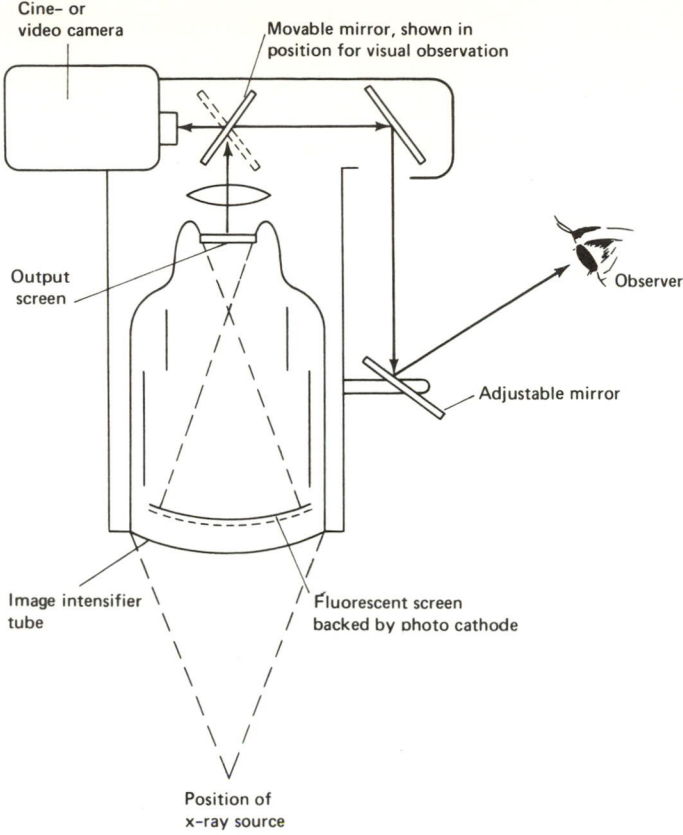

Cine- or
video camera

Movable mirror, shown in
position for visual observation

Output
screen

Observer

Adjustable mirror

Image intensifier
tube

Fluorescent screen
backed by photo cathode

Position of
x-ray source

Figure 14.5. X-ray image intensifier for visual observation and recording of the picture with a cine (movie) camera or with a video tape recorder—diagram simplified.

surface of which is coated with a suitable material to act as a photocathode. The electron image thus obtained is projected onto a phosphor screen at the other end of the tube by means of an electrostatic lens system. The resulting brightness gain is due to the acceleration of the electrons in the lens system and the fact that the output image is smaller than the primary fluorescent image. The gain can reach an overall value of several hundred, and not only allows the X-ray intensity to be decreased but makes it possible to observe the image in a normally illuminated room. The intensifying tube, however, is rather heavy and requires a special suspension. For chest or pelvic examinations with the patient in a supine position, the screen on which the intensified image appears is high above the patient and requires a system of lenses and mirrors to present the image to the radiologist, who normally stands right next to the patient. For this reason, a TV camera is now used frequently to pick up the intensified image, which can then be observed on conveniently placed TV monitors. This TV picture can also be recorded on a TV tape recorder. Similarly, a movie camera can be used to record directly the intensified X-ray image during an examination.

14.2.2. X-ray Machines

In order to obtain an X-ray image from a certain part of the body, the region to be examined must be positioned between the X-ray tube and the imaging device, as shown in Figure 14.4 for a chest X ray. Similar to a light that throws the shadow of an object on a wall, the X-ray tube projects the "shadow" of the structures inside the body on the imaging device. In order for the X-ray image to be a sharp and well-defined replica of these structures, the part of the body being X rayed must be as close as possible to the imaging device. The X-ray tube, on the other hand, should be as far away as possible.

With mobile X-ray machines, such as the one shown in Figure 14.6(a), the cassette with the X-ray film is usually placed directly beneath the patient. The X-ray tube is mounted on an arm and can be adjusted to the desired height. "Aiming" of the tube is simplified by a small light that projects the shadow of cross hairs along the axis of the X-ray beam. Mechanical shutters can be adjusted to limit the size of the beam to the area over which an X ray is to be taken.

In stationary X-ray machines the support arm for the X-ray tube is mounted on the wall or ceiling of the examination room in such a way that its height can be adjusted and the direction of the beam can be changed. For chest X rays, a holder for the film cassette is mounted adjustably on a wall of the room. For most other X rays, the cassette is inserted in the top of an adjustable table while the X-ray tube is placed above the patient, who lies on the table in a suitable position. With image intensifiers, however, the tube is normally below the table while the intensifier is positioned above the patient. In some X-ray machines the X-ray tube and image intensifier are mounted at either end of a C-shaped structure in such a way that they face each other. Figure 14.6(b) shows this arrangement in a mobile X-ray machine.

The high voltage for the operation of the X-ray tube is provided from a transformer, often mounted in an oil-filled enclosure, which is connected to the tube housing by a pair of heavy cables. The control panel of an X-ray machine normally provides for three different controls. The tube voltage, expressed in kilovolts-peak (kVP), determines the hardness or penetration power of the X-ray beam. The beam current, expressed in milliamperes, determines the intensity of the X-ray beam. The third control simply determines the time (expressed in seconds or fractions of a second) that the beam is turned on for X-ray photos. Battery-powered mobile X-ray machines,

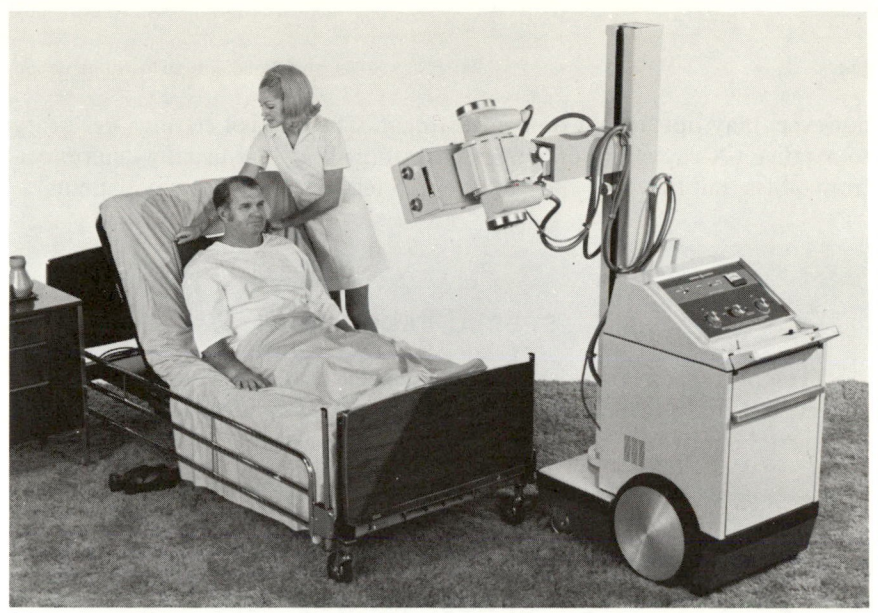

(a)

Figure 14.6. (a) Mobile X-ray machine. (Courtesy of General Electric Company, Medical Systems Division, Milwaukee, WI.) (b) Mobile X-ray unit (with image intensifier and television monitor). The box below the video monitor is a video disc recorder which permits the recording and playback of X-ray images. (Courtesy of Picker Corporation, Cleveland, OH.)

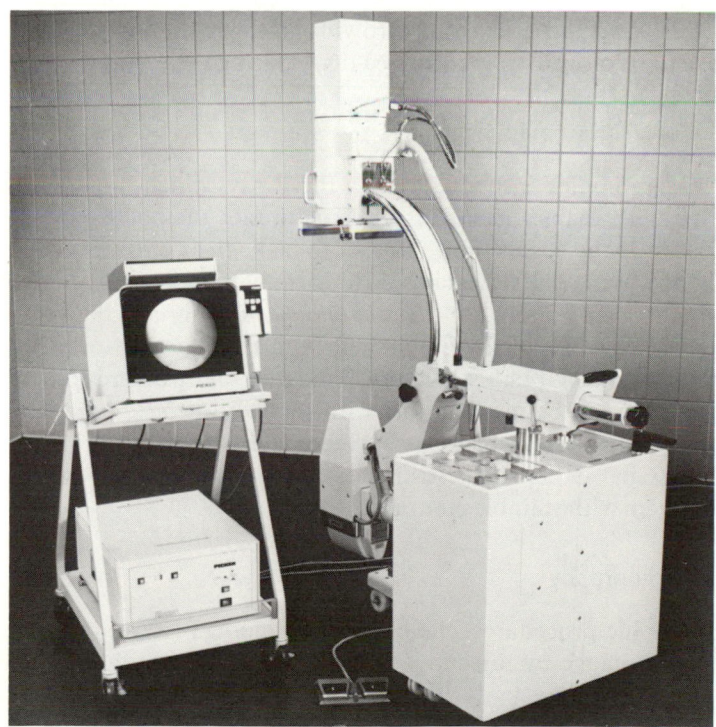

(b)

however, may not have a time adjustment. The control settings necessary to obtain an X-ray photo of a given part of the body are usually determined from tables, but they may have to be corrected for obese or bony patients.

14.3.   SPECIAL TECHNIQUES

The previous section described the general principle of obtaining X-ray images, but often special techniques must be used to obtain usable images from certain body structures.

14.3.1. Grids

As mentioned before, some of the X rays entering the body of a patient are actually scattered and no longer travel in a straight line. If the body section examined is very thick and if the X-rayed area is large, the scattered X rays can cause a blurring of the X-ray image. This effect can be reduced by the use of a *grid* or a *Bucky diaphragm* (named after Gustav Bucky, its inventor). This device consists of a grid-like structure made of thin lead strips that is placed directly in front of the X-ray film. Like a venetian blind that lets sun rays through only when they strike parallel to the slats of the blind, the grid absorbs the scattered X rays while those traveling in straight lines can pass. In order to keep the grid from throwing its own shadow on the film, it may have to be moved by a motorized drive during the exposure of the film.

14.3.2. Contrast Media

While foreign bodies and bone absorb the X rays much more readily than soft tissue, the organs and soft tissue structures of the body show very little difference in X-ray absorption. In order to make their outlines visible on the X-ray image, it may be necessary to fill them with a *contrast medium* prior to taking the X-ray photo. In the *pneumoencephalogram,* the ventricles of the brain are made visible by filling them with air, which absorbs X rays less than the surrounding brain structures. Similarly, the structures of the gastro-intestinal tract can be made visible with the help of *barium sulfate,* given orally or as an enema, which has a higher X-ray absorption than the surrounding tissue. Other body structures and organs can also be visualized by filling them with suitable contrast media.

14.3.3. Angiography

In angiographic procedures, the outlines of blood vessels are made visible on the X-ray image by injecting a bolus of contrast medium directly into the bloodstream in the region to be investigated. Because the contrast medium is rapidly diluted in the blood circulation, an X-ray photo or a series of such photos must be taken immediately after the injection. This procedure is often performed automatically with the help of a power-operated syringe and an electrical cassette changer.

14.3.4. Cardiac Catheterization

Cardiac catheterization is a technique used primarily to diagnose valve deficiencies, septal defects, and other conditions of the heart characterized by hemodynamic changes. For this purpose, a special catheter is inserted through an artery, vein, or occasionally, directly through the chest wall into the heart. Under fluoroscopic control (with an image intensifier), the catheter is manipulated until its tip is in the desired position within the heart. By means of the catheter, intracardiac pressures can be measured in various parts of the heart that show characteristic changes if the heart valves are either narrowed or do not close completely. Septal defects can be detected by withdrawing blood samples from various heart chambers and measuring the oxygen concentration of the samples. Similarly, pumping efficiency can be assessed by measuring pressures within the ventricles at various points of the cardiac cycle. By injection of an indicator through the catheter the cardiac output can be measured. By the injection of a contrast medium through a suitably placed cardiac catheter (*selective angiography*), the vascular structures of the heart, including the coronary arteries (*coronary arteriography*), can be visualized. Catheterization in general is discussed in more detail in Chapter 6.

14.3.5. Three-Dimensional Visualization

A basic limitation of X-ray images is the fact that they are two-dimensional presentations of three-dimensional structures. One organ located in front of or behind another organ therefore frequently obscures details in the image of the other organ. In *stereoradiography* two X-ray photos are taken from different angles, which, when viewed in a stereo viewer, give a three-dimensional X-ray image. In *tomography* (from the Greek word *tomos*, meaning slice or section) the X-ray photo shows the structure of only a thin slice or section of the body. Several photos representing slices taken at different levels permit three-dimensional visualization. Tomographic X-ray photos can be obtained by moving the X-ray tube and the film cassette in opposite directions during the exposure of the film. This procedure causes the image of the structures above and below a certain plane to be blurred by the motion, whereas structures in this plane are imaged without distortion. Special tomography machines that scan body sections with a thin X-ray beam and that determine the X-ray absorption with a radiation detector have been developed. The image of the section is reconstructed from a large number of such scans with the help of a digital computer (see Chapter 15).

14.4. INSTRUMENTATION FOR THE MEDICAL USE OF RADIOISOTOPES

The radiation exposure during X-ray examinations occurs only during a very short time interval. In diagnostic methods involving the introduction of radioisotopes into the body, on the other hand, the exposure time is much longer, and therefore the radiation intensity must be kept much smaller in order not to exceed a safe radiation dose. For this reason, the techniques used for radiation detection and visualization with radioisotopes differ greatly from those used for X rays. Radioiosotope techniques are all based on actually counting the number of nuclear disintegrations that occur in a radioactive sample during a certain time interval or on counting the radiation quanta that emerge in a certain direction during this time. Because of the random nature of radioactive decay, any measurement performed in this way is afflicted with an unavoidable statistical error. When the same sample is measured repeatedly, the observed counts are not the same each time but follow a gaussian (normal) distribution. If the mean number of counts observed is n, the standard deviation of this distribution curve will be the square root of n. The concentration of radioactive material in an unknown sample can be determined by comparing the count with that of a known standard. A much greater accuracy is obtained if the number of disintegrations counted for the measurement is high. Higher counts can be obtained either by counting over a longer time interval or by increasing the activity of the sample, both ways being limited in medical applications in which the radioactivity is measured inside the body.

Almost all nuclear radiation detectors used for medical applications utilize the light flashes caused by radiation in a suitable medium. Such *scintillation detectors* (also called *scintillation counters*) for gamma rays use a crystal made from thallium-activated sodium iodide, which is in close contact with the active surface of a photomultiplier tube. Each radiation quantum passing the crystal causes an output pulse at the photomultiplier, the amplitude of which is proportional to the energy of the radiation. This property of the scintillation detector is used to reduce the *background*, (counts due to natural radioactivity) by means of a *pulse-height analyzer*. This is an electronic circuit that passes only pulses within a certain amplitude range. The limits of this circuit are adjusted in such a way that only pulses from the radioisotope used can pass, whereas pulses with other energy levels are rejected. Figure 14.7 shows two types of scintillation detectors used for the determination of the concentration of gamma-emitting radio-isotopes in medical applications. In the *well counter,* the scintillation crystal has a hole into which a test tube with the sample is inserted. In this con-figuration almost all radiation from the sample passes the crystal and is counted while a lead shield reduces the background count.

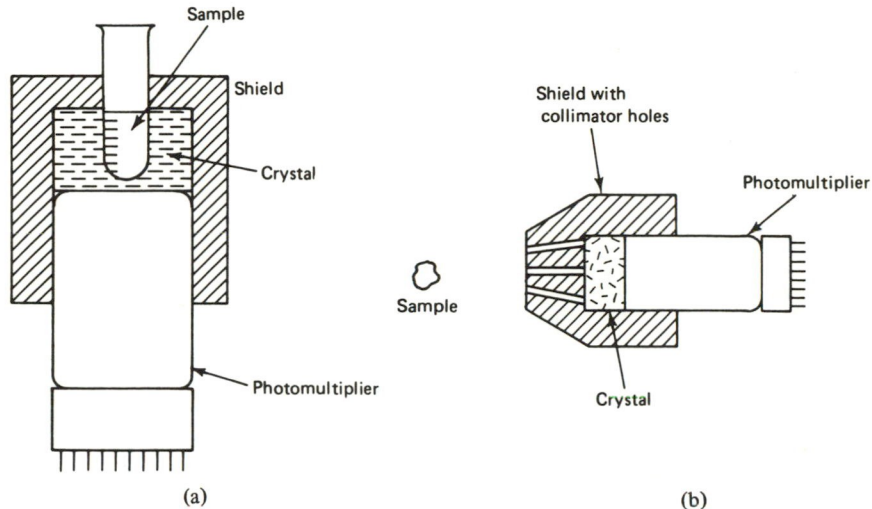

(a)   (b)

Figure 14.7. Scintillation detectors for gamma radiation. (a) Well counter for in vitro determinations; (b) Detector with lead collimator for in vivo determinations.

For activity determinations inside the body, a *collimated detector,* also shown in Figure 14.7, is used. In this detector, a lead shield around the scintillation crystal has holes arranged in such a way that only radiation from a source located at one particular point in front of the detector can reach the crystal. Only a very small part of the radiation coming from this source, however, passes the crystal. This detector, therefore, is much less sensitive than the well counter type.

Figure 14.8 shows the other building blocks that constitute a typical instrumentation system for medical radioisotope measurements. The pulses from the photomultiplier tube are amplified and shortened before they pass through the pulse-height analyzer. A timer and gate allow the pulses that occur in a set time interval to be counted by means of a *scaler* (decimal counter with readout). A *rate meter* (frequency meter) shows the rate of the pulses. Its reading can be used in aiming the detector toward the location of maximal radioactivity and to set the pulse-height analyzer to where it passes all pulses from the particular isotope used.

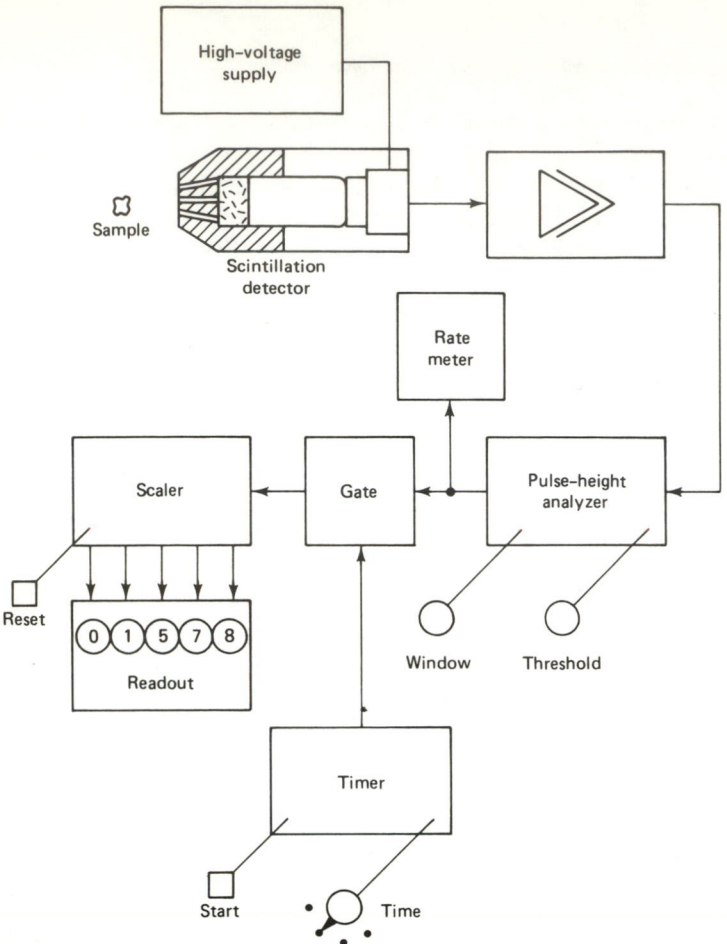

Figure 14.8. Block diagram of an instrumentation system for radioisotope procedures.

An automatic system for the measurement of radioactivity in "in vitro" samples is shown in Figure 14.9. The automatic sample changer arm (right) obtains test tubes containing the samples from a carousel and drops them into a counting well. The number of radioactive disintegrations measured over a preselected time interval is printed out on the printer shown on the left side. A background correction can be made if desired.

The principle of the collimated scintillation detector can be used to visualize the spatial distribution of radioisotopes in a body organ. In a *radioisotope scanner* the detector is slowly moved over the area to be examined in a zigzag fashion. Attached to the mounting arm of the detector is a recording mechanism that essentially produces a plot of the distribution of the radioactivity. In early scanners this recorder was a solenoid-operated printing mechanism that was connected to the output of a binary divider

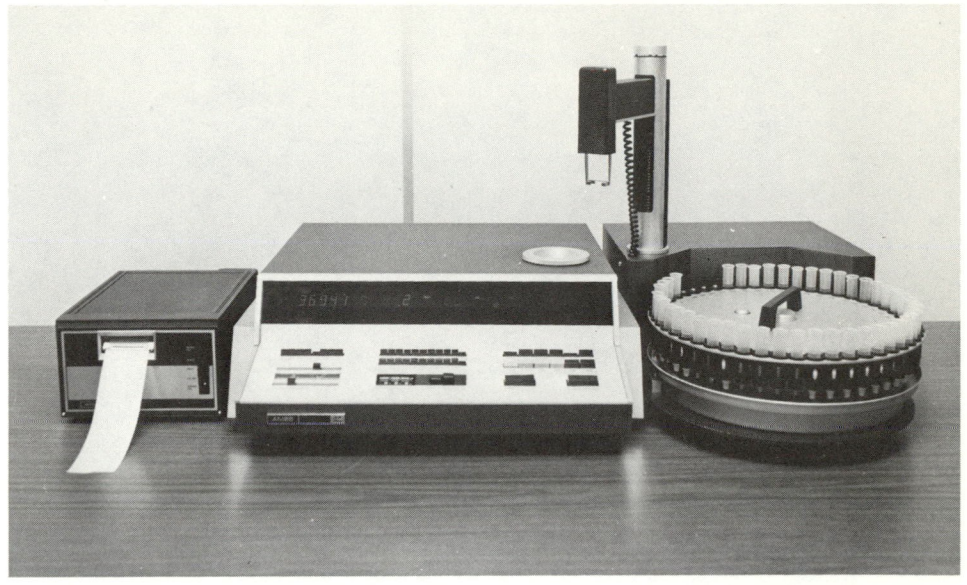

Figure 14.9. Automatic system for measurement of radioactivity in in-vitro samples. (Courtesy of Ames Co., Division of Miles Laboratory Inc., Elkhart, IN.)

and that produced a dot after a certain number of detector pulses had occurred. The density of dots along a scanning line reflected the amount of radioactivity, and when observed from a distance, the completed scan resembled a halftone picture. Interesting medical details are often manifested in rather small differences of the activity, which are not readily visible in this simple kind of scan presentation.

Such variations can be made more easily visible by use of contrast-enhancement methods, which usually employ a photographic recorder. In this recorder, a flashing light leaves a dot on an X-ray film. While the light source is triggered from the output of a digital divider, its intensity is also modulated by a rate meter circuit and, therefore, also depends on the radio-activity. The rate-meter signal is manipulated by amplification and zero suppression so that a small range of variation in radioactivity occupies the entire available density range of the X-ray film. A similar contrast enhancement can be achieved with the mechanical dot printer by an attachment. This device moves a multicolored ink ribbon under the printer head in accordance with the output from a rate meter and thus reflects small changes in radioactivity by changes of the color of the dots. A basic problem with radioisotope scanners is that the detector must travel very slowly in order to give a high-enough count rate for detecting small variations in activity.

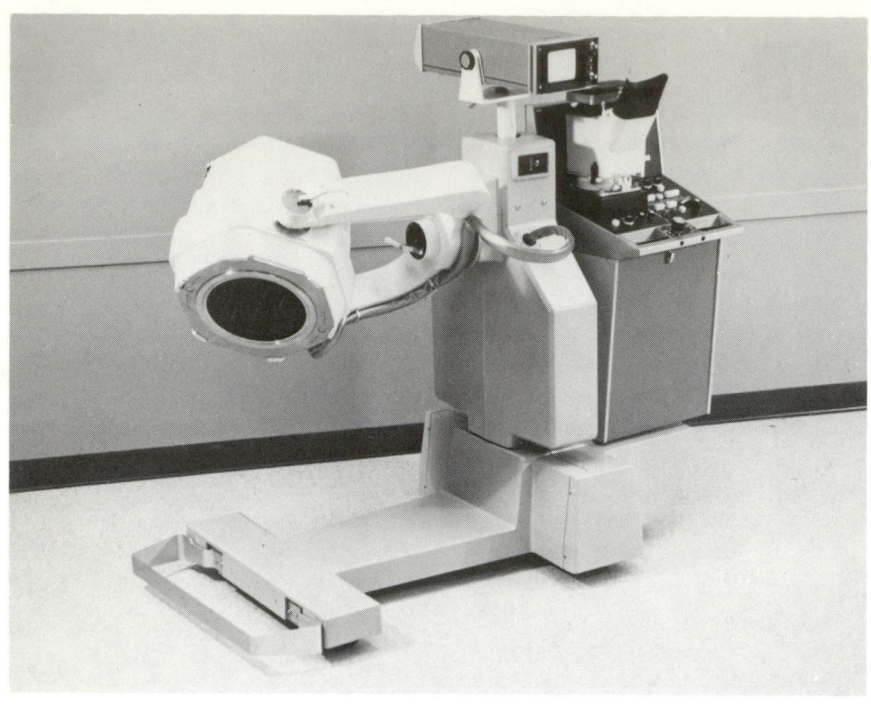

Figure 14.10. (a) Radioisotope camera. (b) Radioisotope camera in use. (Courtesy of Searle Radiographics Division of Searle Diagnostics, Inc., Des Plaines, IL.)

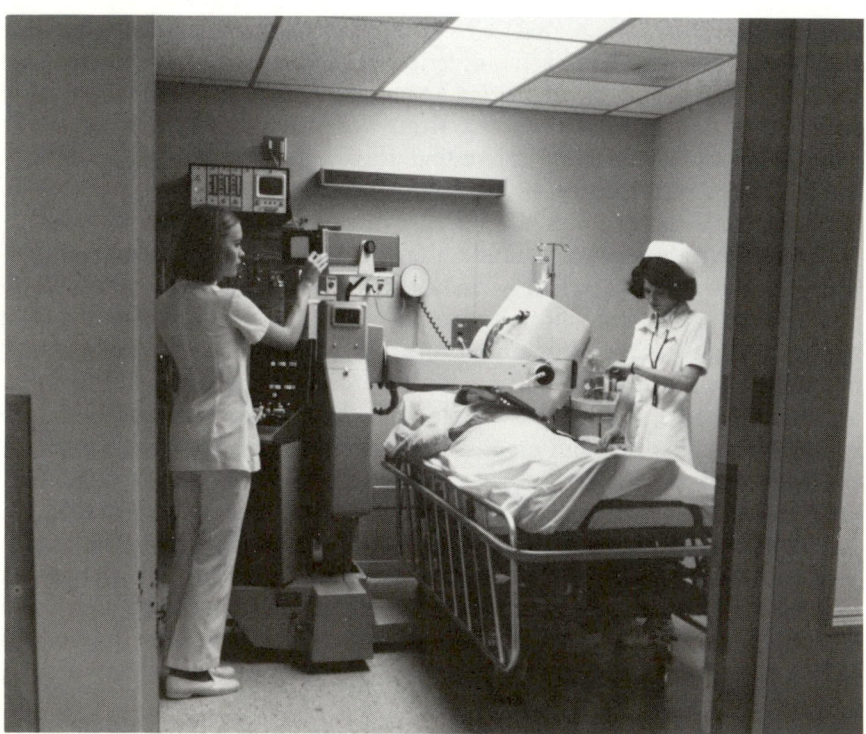

Therefore, the scan of a larger organ can take a long time. For this and other reasons, scanners of this type are being replaced by *radioisotope cameras,* a portable model of which is shown in Figure 14.10. Figure 14.10(a) is a close-up view of the machine and Figure 14.10(b) shows the machine in use in a hospital. Instead of the smaller moving crystal of the scanner, this type of device has one large, stationary scintillation crystal. The position of a light flash in this crystal is determined by means of a resistor matrix from the output signals of an array of several photomultiplier tubes mounted in contact with the rear surface of the crystal. The detection of a nuclear event at a certain point in the crystal causes a light flash at the corresponding location on the screen of a cathode-ray tube, which is photographed with a Polaroid camera or with a special camera that uses X-ray film. Computers or computer techniques are being increasingly utilized to store the signals from the radioisotope camera and to process images in order to enhance the details. Different types of collimators are used in this camera, depending on the geometry of the organ to be examined.

Scans of the thyroid gland can be obtained fairly easily with iodine 131. They show cysts as areas of reduced activity and possible malignant tumors as "hot nodules" with increased activity compared to the rest of the gland. Other organs are less-easily visualized and require the use of contrast enhancement in the scanner or camera and the administration of large doses of short-lived radioisotopes. The logistics of obtaining such isotopes can be simplified by use of technetium 99m, which, although it has a half-life of only 6 hours, is the decay product of molybdenum 99, which has a half-life of 66 hours. The molybdenum 99 is contained in a special device aptly called a "cow" because the technetium 99m is "milked" from it by *eluting* it—letting a buffer solution trickle through the device. These short-lived radioisotopes do not occur naturally in the body, and, unlike iodine, the organs of the body do not have a natural selectivity to these elements. Physical effects, such as variations in blood flow, account for differences in the isotope distribution that outline the organs. The organs that can be visualized include the lungs, brain, and liver.

Despite the substantial technical effort involved in obtaining X-ray pictures or radioisotope scans, a very experienced physician is required to interpret the results. Techniques to apply computer image processing to this field are also finding increasing applications, especially with cameras.

Hydrogen and carbon, the two elements that constitute the largest percentage of all organic substances, have useful radioisotopes that are only beta emitters. With these radioisotopes, many natural and synthetic substances, including chemicals, nutrients, and drugs, can be made radioactive and their pathways in the organism can be traced. The radioactivity of these isotopes, however, can be measured only in vitro, and special detectors have to be used. For older methods, the sample is placed in a *planchet,* a round, flat dish made of aluminum or stainless steel, in which the solvent is evaporated. In a *planchet counter* [shown in Figure 14.11(a)] the planchet becomes part of a Geiger–Müller tube. The thin layer in which the sample is spread and its close contact with the collection electrodes result in a fairly high counting efficiency for beta radiation. The counting cell is continuously purged by a flow of gas that removes ionization products.

For the soft beta radiation from tritium, a radioactive isotope of hydrogen, however, the sensitivity of the planchet counter is marginal and *liquid scintillation counters* are now normally used instead. In these devices the sample is placed in a small counting vial, where it is mixed with a solvent containing chemicals that scintillate when struck by beta rays. The vial is then placed in a detector [Figure 14.11(b)], in which it is positioned between two photomultiplier tubes. The light signal picked up is very weak,

Figure 14.11. Detectors for beta radiation: (a) Planchet or gas flow counter; (b) Liquid scintillation counter.

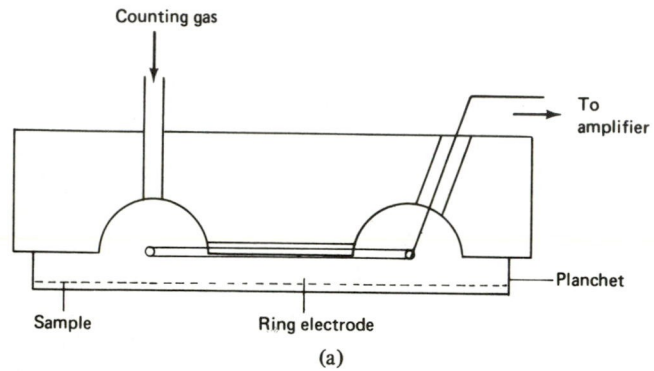

(a)

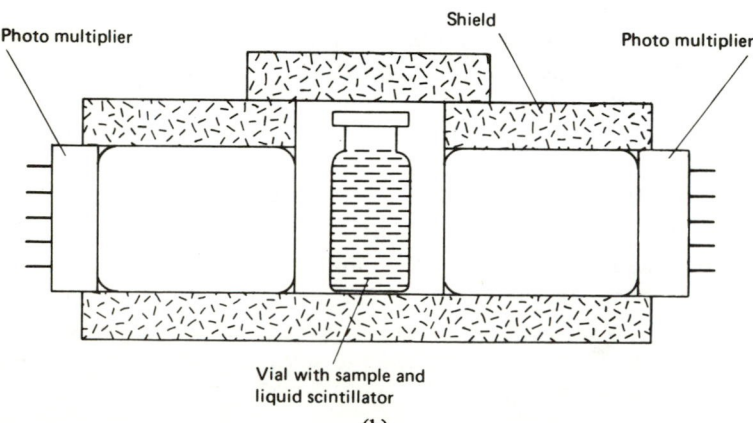

(b)

and erroneous counts from tube noise must be reduced by a coincidence circuit, which passes only pulses that occur at the outputs of both tubes simultaneously. The remainder of the circuit is similar to the gamma measurement system shown in Figure 14.7. The low activities often encountered in measurements of this type sometimes require very long counting times. This situation has lead to the development of systems that automatically change the samples and print out the results.

14.5. RADIATION THERAPY

The ionizing effect of X rays is utilized in the treatment of certain diseases, especially of certain tumors. In dermatology very soft X rays that do not have enough penetration power to enter more deeply into the body are used for treatment of the skin. They are called *Grenz rays* (from the German word *Grenze,* meaning border) because in the spectrum they are actually at the border between the normally used X rays and the ultraviolet range (see Figure 14.1). In the therapy of deep-seated tumors, on the other hand, very hard X rays that are generated with voltages much higher than those for diagnostic X rays are used. Sometimes *linear accelerators* or *betatrons* are used to obtain electrons with a very high voltage for this purpose. Changing the direction of entry of the beam in successive therapy sessions or rotating the patient during a session reduces the radiation damage to unafflicted body parts while concentrating the radiation at the site of the tumor.

15

The Computer
In
Biomedical Instrumentation

In the relatively short time since its development, the digital computer has had a pronounced effect on almost every aspect of modern-day life. Its presence is evident in the bank, the supermarket, and at the airline ticket counter. Computerized TV games, automobiles, and microwave ovens are fast becoming commonplace. Pocket-sized calculators with enormous computational capability are now obtainable within the budget of the average student. Even so, all evidence indicates that the full impact of computer technology is yet to be realized.

Historically, the digital computer has its roots in the work of four pioneers. The first of these was Charles Babbage, a mathematics professor at Cambridge University, who devised a machine in 1812 to perform certain simple computations and originated ideas that led to the stored-program concept of automatic computation. The second was George Boole, an English mathematician who developed the logic system used in digital circuit design. The next major contribution was that of Herman Hollerith, who originated the machine-readable punched card, which was first used in the 1890 census, and became the standard form of data entry for many years. The fourth pioneer was Howard Aiken of Harvard University, who developed the first automatic-sequence-controlled calculator, proposed in 1937 and completed in 1944. Although essentially a hugh mechanical calculator, Aiken's machine led to development of several early electronic computers in the late 1940s and early 1950s which used numerous banks of vacuum tubes with extensive power and air-conditioning requirements. In the late 1950s, transistorized computers began to appear, bringing with them smaller size, lower power requirements, fewer heat problems, and more important, greater reliability and lower cost. Integrated-circuit technology continued the trend toward smaller and less expensive computers through the 1960s and led to the low-cost calculators and microprocessors which made their appearance in the late 1960s and early 1970s.

The earliest computer applications in the medical field were those related to billing and the other business aspects of running a hospital, where techniques already in use in other parts of the business world could be adopted. In the latter 1950s and early 1960s, computerized ECG and EEG analysis, pulmonary function analysis, multiphasic screening, and automated clinical laboratories began to emerge, in some cases on an experimental basis. The introduction of lower-cost minicomputers and on-line, real-time (these terms are defined in Section 15.1) computer systems in the mid-1960s made many of these applications both economically and technically feasible for clinical use. The 1960s also brought about the first computerized patient-monitoring systems, initially using large computer systems, and later, incorporating minicomputers. Experimental work with totally computerized hospital systems dates back to the late 1950s and early 1960s. This idea, in which all information generated in the hospital is handled through an integrated computer system, has yet to find widespread application among hospitals, although some systems that approach the total-hospital concept are presently in operation.

The advent of the microprocessor has markedly affected medical instrumentation, as it has most disciplines involving measurement or control. Microprocessors are now incorporated in many commercially available clinical instruments to enhance their capabilities or automate their operation. In some systems, such as certain patient monitors, microprocessors have replaced minicomputers, substantially reducing their cost.

Most medical applications of computers and microprocessors involve specific instrumentation systems; in fact, the computer often becomes an integral part of an instrument. It is therefore essential that anyone involved in the field of medical instrumentation be familiar with the basic concepts of digital computation and some of the more important medical applications. Furthermore, it is important that the biomedical engineer or technician be given an understanding of the techniques involved in interfacing a computer or microprocessor with the rest of the instrumentation system.

This chapter is intended to bring to the reader a brief background of the basic concepts of digital computation, a look into some of the more important applications of computers in medicine, and a discussion of interfacing techniques. The chapter also includes a presentation on microprocessors and their role in medical instrumentation.

15.1 THE DIGITAL COMPUTER

The modern digital computer is a special type of calculating machine capable of automatically performing a long and complicated sequence of operations as directed by a set of instructions stored within the machine. In addition to its computational ability, the computer can store and retrieve large quantities of information and can automatically alter its sequence of instructions on the basis of calculated results. The sequence of instructions required for the computer to perform a given task is called a *program.*

Digital computer technology is generally divided into two main areas of interest, the electronic circuitry and other physical equipment involved, called the computer *hardware,* and the programs with which the computer operates, called the *software.* Both hardware and software must be considered in discussing basic computer concepts.

15.1.1 Computer Hardware

Although a wide variety of digital computers can be found in biomedical applications, ranging from a one-chip microprocessor to a large multimillion-dollar computer complex, they all contain four basic elements: an *arithmetic unit* to perform the mathematical and decision-making functions, a *memory* to store data and instructions, one or more *input–output* (I/O) devices to permit communication between the computer and the outside world, and a *control unit* to control the operation of the computer. A block diagram showing the relationship of these elements is presented in Figure 15.1 where the dashed lines indicate the control functions and the solid lines show the data flow.

Under the direction of the control unit, data from the instrumentation system or from some other source enter the computer via one of the input devices. The data may be transferred directly to memory, where it is stored until needed, or through the arithmetic unit. After processing, results are either stored in the memory for future recall, or they may be presented to the outside world via one or more of the output devices.

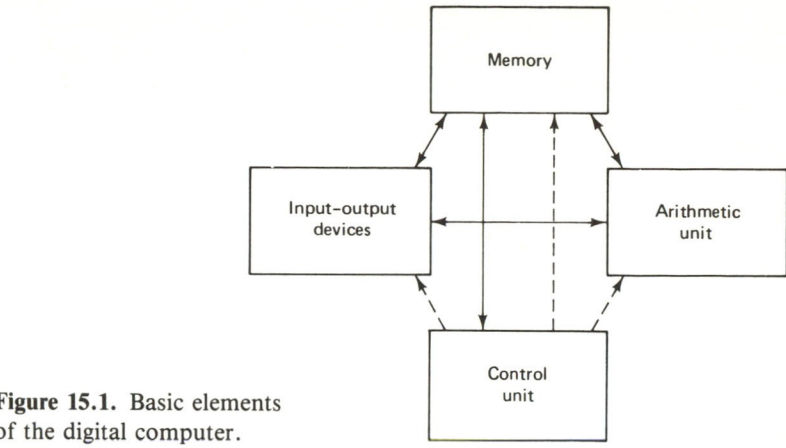

Figure 15.1. Basic elements of the digital computer.

As its name implies, the digital computer accepts, manipulates, and presents data in digital form. Although various digital codes are used, each has as its base the binary number system in which all values are represented by a set of 1s and 0s. Bistable elements are used in the computer to represent these values. Each binary digit is known as a *bit,* and the number of bits that are normally stored or manipulated together in the computer constitute the computer *word.* Computers are often designated by their word lengths. For example, a computer that handles information in 16-bit words is referred to as a 16-bit computer. Some computers work with a variable-length word, dividing each value into 8-bit segments called *bytes.* In these machines, the word length may be any number of bytes up to some limit. Also, in some machines, both alphabetic characters and decimal numbers can be coded into 6- or 8-bit groups called *characters.*

The arithmetic unit includes the circuitry that performs the computation and logic functions of the computer as well as some registers for temporarily storing the data being manipulated by that circuitry. A *register* is a set of bistable circuits capable of storing one computer word or a part of a word. The exact number and type of registers depends on the architecture of the computer. In some computers, the arithmetic unit utilizes registers physically located in the memory. One of the registers in the arithmetic unit, called the *accumulator,* is generally used to hold the results of the computation.

The circuitry that actually performs the computational and logic operations within the arithmetic unit is called the *arithmetic/logic unit* (ALU). Its functions include addition, subtraction, and the logic functions of AND, OR, and exclusive OR (XOR). Hardware multiplication and division are also provided in the ALUs of some computers, whereas others perform these operations through addition, subtraction, and shifting operations, using short sequences of instructions called *microprograms.*

Together, the arithmetic and control units constitute the *central processing unit* or (CPU). The *control unit* consists of registers and decoders which sequentially access instructions from the memory, interpret each instruction, and send appropriate control signals to all parts of the computer to carry out the program being executed.

The *memory of a digital computer* is used to store data (numbers or information in alphabetical form) with which the computer may be required to operate at some future time. A certain portion of the memory which must be readily accessible for storage or retrieval of information at any instant of time is called *random-access memory* (RAM) and generally serves as the computer's *primary memory.* This memory might be functionally visualized as a hugh array of mailboxes, each just large enough to store exactly one word of information and each identified by a number called an *address,* which is unique to the storage location. As the term "random access" implies, any address in the RAM is accessible with equal speed, regardless of the order in which information is called for. Data are *written into* storage and *read out.*

In many computers, particularly the older ones, tiny magnetic cores serve as the storage elements in the RAM. More modern computers use integrated circuits for this purpose in order to reduce size and attain higher operating speeds. Since each core or integrated-circuit element is able to hold one bit of information, each word of memory requires a number of elements equal to the word length of the computer. Thus, a 16-bit computer with 32,768 words of RAM must have 524,288 storage elements in that part of its memory.

Magnetic core memories are inherently *nonvolatile;* that is, they retain their contents without electrical power. In contrast, integrated-circuit RAMs are inherently *volatile* and lose all stored information whenever the power is removed. Such memories often have a backup battery supply to protect the memory in case of power failure. RAMs that consist of flip-flop circuits are called *static memories,* since stored data, once entered, remain intact unless power is lost until replaced by new data. Some integrated-circuit RAMs, however, use metal oxide silicon transistors that store data in the form of capacitive charges. These devices are called *dynamic* memories because the charges must be continually refreshed in order to retain the stored information.

A basic characteristic of the modern digital computer is that the programs required for each job to be done are internally stored within the memory of the computer. Generally, a portion of RAM is used for this purpose, permitting the programs to be changed as required. However, in some computers, and nearly all microprocessors, all or a portion of the programs for fixed operations or control functions may be stored in a *read-only memory* (ROM). A ROM also provides random access, but its contents cannot be changed during the normal operation of the computer. ROMs, which are also integrated-circuit memories, are generally lower in cost than RAMs; and like RAMs, they can be fabricated by *large-scale integration* (LSI) techniques in which a large number of elements are packed on a single chip. They also permit very fast access times. Most ROMs are programmed at the time of their manufacture, which requires that the ROM be replaced in order to change the program. A special type of ROM, called a *programmable read-only memory* (PROM) is available for applications in which it may be necessary to occasionally alter the program. Like ROMs, PROMs are fixed-program devices; however, with special equipment, their contents can be changed by their users. Special versions of the PROM are the erasable *programmable read-only memory* (EPROM), in which the memory can be erased by exposure to ultraviolet light and reprogrammed electrically, and the *electrically alterable read-only memory* (EAROM), which permits changes by means of electrical inputs. All these devices require special equipment, called *PROM programmers* or *burners,* to alter the program content.

The amount of random access storage (RAM and ROM) a computer can have is limited by the cost of these memories. The size of a computer's primary memory is generally dictated by the amount of information that must be readily accessible at a given time, and it varies with the particular application for which it is used. Additional information, often in vast quantities, may be stored in a *secondary memory,* which is less costly but requires longer access time. Both magnetic tapes and disks are used for this purpose.

A *disk memory* consists of one or more disks mounted in a *disk drive.* Each disk looks very much like a metal phonograph record coated on both sides with a magnetic material. Some types of disks are removable, whereas others remain a permanent part of the disk drive and cannot be removed. A set of read/write heads is provided in the drive for each disk surface. Data are arranged in circular tracks around the disk, which is constantly rotated at high speed by the drive. To access a given address on the disk, the heads must move radially to the designated track and the disk must rotate to the point at which the address is beneath the heads. The actual time required to reach a given location, usually a number of milliseconds, depends on how far the heads and disk must move. Once a location has been reached, however, a block of data, consisting of a large number of words, can be transferred very rapidly to or from the primary memory, where it can be accessed randomly. Fixed-disk systems generally have shorter access times than those with removable disks.

A less expensive form of disk storage which has achieved considerable popularity, particularly in small computer systems, is the *diskette* or *floppy disk,* a very thin oxide-coated Mylar disk, slightly under 8 in. diameter, with a hole in the center like a phonograph record. The term "floppy" is used because these disks are flexible in contrast to the rigid construction of conventional disks. Each diskette is enclosed in a protective envelope, within which it rotates during use. Because they are so thin, the diskettes require very little storage space. Floppy disks are slower than conventional disks and can store less data per disk, but their small size and low cost tend to compensate for these limitations.

Another form of secondary memory is one or more *digital magnetic tape drives* connected to the computer. Data are stored on an oxide-coated plastic tape similar to that used in analog instrumentation or home stereophonic tape recorders. Nine parallel read/write heads record a finely packed sequence of 8-bit characters along the tape. A ninth bit, called a *parity* bit, is added to each character for error-detection purposes. In this manner extremely large quantities of data can be stored on a single reel of tape. To gain access to a particular set of data on the tape, however, the tape must be wound to the location of the data. This process may take several seconds or even minutes, but once the location of the data has been reached, information can be transferred at a very high rate. This characteristic makes the use of tape most practical in applications where large, continuous blocks of data can be written into or read out of the tape and transferred to or from the primary memory.

Cassette tapes can also be used for data storage, but are slower and hold much less data than the conventional (reel-to-reel) tapes described above. Even so, they are often used with microprocessors or small computer systems because of their compact size and lower cost. A slightly larger (and also more expensive) *tape-cartridge system* is also sometimes used with small computers.

Two newer forms of digital memory, *magnetic bubble memory* (MBM) and *charge-coupled devices* (CCD), are beginning to appear on the scene. Both invented at Bell Laboratories at about the same time, these memories fill a gap in speed and cost between integrated-circuit RAMs and magnetic disks. They are also similar in principle in that they are both *sequential-access memories,* in which strings of bits circulate through one or more designated pathways. A given word can be accessed only when the beginning of that word circulates past a readout point. These memories are much faster than disks, however, because the circulation of data does not involve mechanical movement of the storage medium.

In magnetic bubble memories, microscopic domains of magnetic polarization, called *bubbles,* are generated sequentially in a thin magnetic film on the surface of a garnet chip and circulated by a rotating magnetic field. Patterns of Permalloy metal are deposited on the film to define the pathways through which the bubble domains move. Bubbles are generated by pulsing current through a microscopic one-turn loop just above the magnetic film. In a typical arrangement, one bit of data is introduced every

10 μsec. The presence of a bubble during a 10-μsec period constitutes a logic 1, whereas the absence of a bubble during that period constitutes a 0. Data are read out by means of an array of detector elements that change their resistance when bubbles pass under them. By organizing data into blocks and utilizing a combination of major and minor loops, maximum access time for any specified block of data is 1 msec, which is 8 to 100 times the speed of a disk. Since they have no moving parts, MBAs have much higher reliability and lower error rate than disks. MBMs are non-volatile and thus require no special provision to preserve data in case of power failure. Their cost is about the same per bit as that of floppy disk or movable-head rigid disk storage. Because of their higher speed, MBMs are likely to replace disks as secondary storage in some future computer systems.

Charge-coupled devices (CCDs) are considerably faster than bubble memories, but are also more expensive. The elements of a CCD memory are somewhat similar to those of a dynamic integrated-circuit random-access memory in which data are stored in the form of small capacitive charges. However, in the CCD memory, these charges are shifted from one element to the next, along a designated path, through a silicon chip, upon receipt of each clock pulse. The output of each path is recirculated back to the input to provide continuous circulation of a string of bits. CCD memories may have a number of such paths, each typically containing 64 or more bits. The charges are refreshed as they pass through a special circuit for this purpose in each path. Although a given bit can be accessed only when it circulates past the readout, entire words or groups of words can be carried in parallel paths and accessed in microseconds (or less). Thus, their access times are short enough to make CCD memories useful as primary memories in certain applications as well as for more-rapid-access mass storage. Their lower cost, which is only about one-fourth that of an integrated-circuit RAM, makes them attractive contenders in both areas. Unlike bubble memories, CCD memories are volatile and can lose their contents when power is removed, unless a protective battery supply is provided.

Technologies associated with both MBMs and CCDs are constantly improving, with the promise that the near future will bring greater packing densities and lower costs in both areas. Another new technology, *electron-beam addressable memories* (EBAM), is also emerging. This type of memory, in which large quantities of data are stored as electrostatic charges on elements of a target of silicon dioxide or some similar material, will permit faster access than CCDs at a relatively low cost per bit. Data are written onto the target and read out by means of a high-resolution electron beam that can be directed to any portion of the array.

The versatility of a computer is largely determined by the various *input-output* (I/O) devices attached thereto. This portion of the system, which is the computer's only means of communicating with its users, is also its interface with the medical instrumentation system with which it works. Some of the more commonly used I/O systems include equipment to read and punch cards and/or paper tape, record and play back digital magnetic tapes or disks, accept input directly from a keyboard, and provide a typewriter or line-printer output. In its application with biomedical instrumentation, the I/O equipment might also include an analog-to-digital converter to convert data from analog form into the digital (usually binary) form required for computer input, a digital-to-analog converter to provide an analog representation of the output for display or control purposes, or a cathode-ray-tube display. Analog-to-digital and digital-to-analog conversion, plus other aspects of interfacing the digital computer with biomedical instrumentation, are discussed in detail in later sections of this chapter.

Figure 15.2. Optical character recognition (OCR) reader for computer entry (Courtesy of ECRM Inc., Bedford, MA.)

A new, emerging form of input device is *optical character recognition* (OCR) equipment, capable of reading information directly from a typewritten page. The OCR unit shown in Figure 15.2 can read text in single-, double-, or triple-space type from any of three character fonts that can be used on any Selectric typewriter.

Input–output equipment can either be *on-line* (connected to a computer) or *off-line* (not connected, but used in preparation of data for later computer input). It can either be *local* (at the same location as the computer) or *remote* (at some other location and either directly wired to the computer or connected through telephone lines). Remote equipment might be located in the same building as the computer or may be several thousand miles away.

Although the above-described components are essentially common to all computers, their implementation can assume a wide variety of forms, ranging from a large-scale computer of the type shown in Figure 15.3 to a microcomputer of the type shown in Figure 15.4. *Microcomputers* are small low-cost computers generally built around microprocessors (see Section 15.2).

Large-scale computers of the type shown in Figure 15.3, often costing millions of dollars, are designed to process large amounts of data at high speeds, usually for a sizable number of users, either in a batch-processing

Figure 15.3. Large-scale digital computer installation (Courtesy of IBM Corp.)

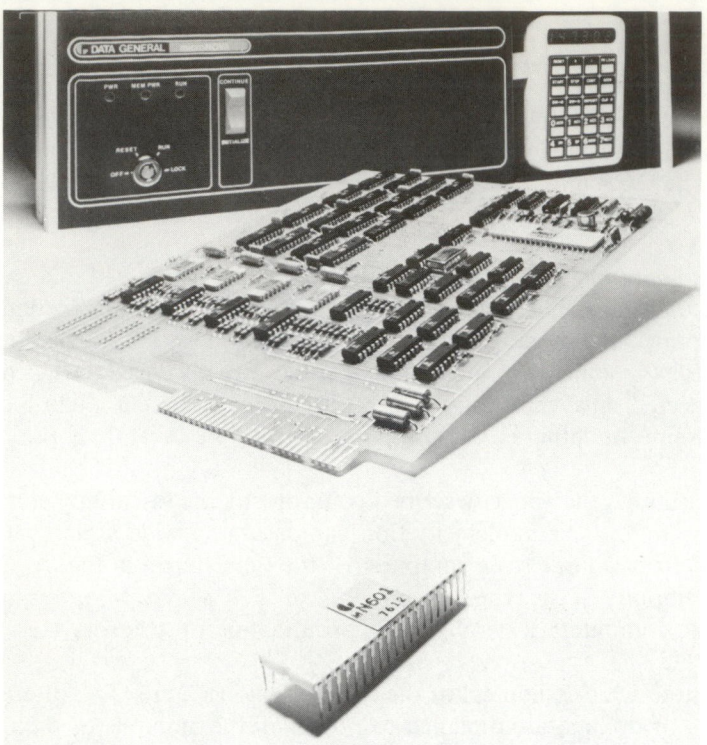

Figure 15.4. Microcomputer (top), Microcomputer board (center), and Microprocessor (bottom). (Courtesy of Data General Corporation, Westboro, MA.)

or time-sharing mode. *Batch processing* is a term used to define a method of operation in which all data for a given problem must be entered into the computer before processing begins. Once the data have been entered, the entire computational resources of the computer are devoted to that problem. When available, the results are printed out or otherwise presented to the user, and the computer begins work on the next problem. In most systems of this type, the results from a previous problem may be printed out while the current problem is being processed. At the same time, data for the next job may be entering the computer.

In contrast, many of the larger computers utilize some form of time sharing. *Time sharing* is a method of computer operation in which a number of users at various locations can use a computer simultaneously. Each user submits data and receives results via his own terminal connected to the computer either directly or via a telephone line. Although it may appear that the computer is working on a large number of jobs and processing many sets of data simultaneously, it is really sharing its time among the users, sequentially alloting a certain amount of time to each. The division of time depends upon the problems being solved and a previously determined priority schedule. Provided that the number of users is not excessive, the high operating speed of the computer allows it to service each user as rapidly as if he alone were using the machine.

The user's *terminal* is his interface with the computer. It can range from a simple teletypewriter to a very elaborate input–output system, perhaps including an analog-to-digital converter for interfacing with an instrumentation system. A typical terminal with keyboard entry and cathode-ray-tube (CRT) display is shown in Figure 15.5.

Computers are often programmed to communicate with their users in an *interactive* or *conversational mode,* allowing the users to exchange messages with the computer as though they were communicating with a person operating a keyboard at the other end of a line. Interactive programs are able to guide the user through the various steps involved in requesting information and obtaining results, and thus are suitable for situations where access is provided to physicians, nurses, or other hospital personnel unfamiliar with computer languages or conventional methods of computer operation.

Communication between the computer and a remote terminal is generally by telephone line. For this purpose data are placed on an audio-frequency *carrier* within the voice-frequency range. The modulator-demodu-

Figure 15.5. Computer remote terminal with keyboard and cathode-ray tube (CRT) display. (Courtesy of IBM Corporation.)

lator device by which the data are encoded on the carrier and by which received data are decoded is called a *modem* (a derivation combining the terms MODulator and DEModulator). The modem may be of a type that connects directly to the telephone line and applies the modulated carrier signal electrically, or it may be equipped with an acoustical coupler in which a conventional telephone receiver–transmitter cradle may be placed. The telephone line may be leased specifically for transmission of data, or it may be an ordinary telephone line normally used for voice conversation.

As an alternative to using a remote terminal on a large time-shared computer, a hospital or other medical facility may have one or more smaller computers of its own. These smaller computers are generally known as *minicomputers,* and the very smallest, *microcomputers.* Actually, the definition is usually based on cost and physical size rather than the amount of storage or complexity of the CPU. Minicomputers generally range in price from about $1000 to $25,000 for the basic unit. Most minicomputer systems include perpherals, which may add considerably to the cost. Although they are small and relatively inexpensive, modern minicomputers can be extremely fast and powerful and can provide large storage capability. Minicomputers can also have time-sharing capability. Thus, a minicomputer may be able to service a number of remote terminals around the hospital.

Although there may be some overlap, a *microcomputer* generally costs less than $1000, and incorporates a microprocessor for its CPU. Microprocessors are discussed in Section 15.2. Microcomputers are generally smaller and have less capability than minicomputers, but with ever-changing technology, some microcomputers compare favorably in many ways with some of the smaller minis.

Generally, minicomputers and microcomputers are used *on-line* and often become a part of the instrumentation system with which they serve. In many applications, they operate in *real time,* an arrangement in which the computer is able to process data as rapidly as it is received.

15.1.2. Computer Software

In a general sense, the term *software* is defined to include all the programs used by a computer system, as well as documentation and other nonhardware items supplied by manufacturers to facilitate the purchaser's efficient operation of the equipment. The software cost for a given system is usually much greater than that of the hardware involved.

There are two basic types of software: (1) *system software,* supplied by the computer manufacturer for managing the operation of the system, translating programs, performing diagnostic checks, and so on, and (2) *application software,* for carrying out the specific functions involved in the user's application. The system programs are usually specific to the computer involved, whereas application programs are most often written in a form that can be used on different kinds of computers.

The set of basic operations a computer is able to perform is called its *repertoire* or *instruction set.* The set of symbolic instructions and rules for formatting and combining these instructions, called *syntax,* constitutes a *programming language.* The language used internally by the computer itself is called *machine language,* and consists of a numeric code for each operation in the computer's repertoire. Although application programs could be written in machine language, long lists of operation codes would have to be memorized in order to write them. Instead, computers generally have system programs that accept mnemonic instructions, such as "ADD" or "SUB," and convert each to its machine language equivalent. These programs are called *assemblers,* and the mnemonic language is called an *assembly language.* Assembly language programming is much easier for programmers to use than machine language, but it is still specific to a given type of computer. That is, a program written in assembly language for one computer cannot be expected to be used on a different kind of computer. Assembly language has a one-to-one relationship with machine language in that the program must include a mnemonic statement for every step the computer is to perform. One exception is a *macroassembler,* which permits a symbolic macroinstruction to be substituted for a sequence of instructions. Another advantage of assembly language is that it permits the use of *symbolic addressing* of memory locations rather than absolute addressing, which is required in machine language. With *symbolic addressing,* the programmer assigns a name to a specified memory location rather than its absolute numerical address and allows the computer to determine the actual address to be used.

To further aid the programmer, most computers have additional software, called *compilers* and *interpreters,* which accept instructions in languages that are more problem-oriented than assembly language, and convert them into machine language. In most cases, a single statement in one of these high-level languages initiates a sequence of machine-language instructions, sometimes rather lengthy, thus reducing the length and complexity of the necessary programs. In addition, such languages involve terminology, symbols, and operations with which the user is already familiar. For example, instructions to carry out mathematical operations are written in the form of equations.

Although a compiler and an interpreter both translate high-level languages into machine languages, there is a basic difference. A *compiler* goes through an entire program after it has been entered into the computer and translates every instruction before execution is begun. On the other hand, an interpreter translates the high-level program a step at a time and executes each step as it proceeds.

There are a number of high-level languages, some suited to specific types of applications. Among the more important of these are FORTRAN (an abbreviation of FORmula TRANslation), COBOL (COmmon Business-Oriented Language), and BASIC (Beginners' All-purpose Instruction Code). Compilers and/or interpreters for these and many other languages are available for most computers, especially the larger ones.

The system software that manages the operation of the computer includes programs that control the flow of data into and out of the computer and between primary and secondary memory, and assure that all the necessary operations are carried out as efficiently as possible. These programs are called by such names as *supervisor, monitor, executive,* and *operating system.* In a time-sharing system, these programs also control the interaction of the computer with the various terminals it services and determine the priorities with which different functions are handled.

Application software is necessary to adapt a computer to each specific job it is to do. Some computers involved with medical instrumentation are used for many purposes and consequently require a variety of application programs, while others, particularly minicomputers and microcomputers, are *dedicated* to one specific task. If the task for which a dedicated computer is to be changed, a new set of application software must usually be entered, and often the computer must by physically disconnected from one set of instrumentation and connected to another. In many applications, particularly those related to research, application programs require frequent modification or rewriting. In contrast, dedicated computers in clinical instrumentation systems are often provided with software that remains unchanged and requires no programming on the part of the user. A computer system of this kind is called a *turnkey* system, since the user must do no more than turn it on in order to use it.

15.2. MICROPROCESSORS

The first all-electronic computer (ENIAC, completed in 1945) contained 18,000 vacuum tubes. The poor reliability of such early devices and the need to shut the computer down to replace defective tubes would have made much larger computers impractical. The invention of the transistor in 1947 removed this limitation and made possible the development of the first generation of computers which employed large numbers of (discrete) transistors and semiconductor diodes. In the mid-1950s, semiconductor technology had developed photolithographic and diffusion methods, which led to the planar transistor in 1958, followed shortly by the first integrated circuits in 1959. Since then the number of circuit components that can be integrated into a circuit chip has approximately doubled every year. The first step, in which up to about 16 gate functions (64 components) are contained in one integrated circuit, was called *small-scale integration* (SSI). SSI circuits range in complexity up to dual-flip-flops and one-bit binary adders. By about 1965, *medium-scale integration* (MSI) had evolved, making it possible to include up to 200 gate functions (1000 components) on one chip. The more complex MSI circuits include a complete 4-bit ALU (see Section 15.1.1). By about 1969 the number of components per circuit exceeded the 1000 limit. *Large-scale integration* (LSI) technology had come into existence. *Very large scale integration* (VLSI) technologies presently under development promise even greater concentrations of components on a single chip.

The logical continuation of the development that had begun with placing an ALU on a chip has now made it possible to put a complete computer central processing unit on a chip. The first device of this kind was announced in 1969. Because it was a complete CPU, albeit with somewhat limited performance, its developers coined the term *microprocessor* (sometimes abbreviated MPU or μP). Progress has since continued to the point where the performance of microprocessors now equals that of the minicomputer CPUs of a few years ago. The number of components that can be placed on one integrated-circuit microprocessor chip has greatly exceeded 18,000, the number of vacuum tubes used by the first, room-sized electronic computer in 1945.

A computer, however, contains more than just the CPU. Thus, in conjunction with microprocessors, large-scale integration has been applied to the other computer components, such as RAMs and ROMs, introduced in Section 15.1.1. Output ports for parallel as well as serial interfaces and controllers for disk drives are also now available as LSI circuit chips.

While the complexity of integrated circuits has increased dramatically over the years, their price has actually decreased. As a result, complete microcomputers are now available which are comparable not only in physical size, but also in price, to electronic controllers implemented with discrete components or small-scale-integration ICs. Their performance, however, is more nearly comparable to rack-size minicomputers, originally costing several tens of thousands of dollars. These developments have made it possible to incorporate a microcomputer as an integral part of many electronic instruments. The designers of biomedical instruments were among the first to utilize this possibility. As a result, biomedical devices have greatly benefited from this technology.

15.2.1. Types of Microprocessors

The first microprocessor introduced in 1969 was a 4-bit device with a rather limited instruction set. From this beginning, development evolved in several directions. Even when utilizing LSI chips for memories and input–output ports, a complete microcomputer normally includes at least a dozen integrated circuits in addition to the microprocessor. In a *single-chip computer*, all these components have been integrated into one circuit package! To achieve this feat, however, the size of the program ROM and the data RAM must be limited. Also, the allowable number of pins in the large IC packages (usually 40) limits the number of I/O ports.

The 4-bit design as it was used in the first microprocessor made it necessary to perform mathematical operations one decimal digit at a time. A word length of 8 bits is more common in modern microprocessors, which can operate in the (binary-coded) decimal system, two digits at a time, or with signed or unsigned binary numbers. Because the resolution of an 8-bit word is frequently insufficient, multiple-precision arithmetic may be employed. Sixteen-bit microprocessors are also available, some of which are compatible with the instruction sets of certain minicomputers. Another type of microprocessor, called a *bit-slice processor,* requires several chips to form a complete central processing unit. Each chip, called a *bit-slice unit,* contains circuitry for 2 or 4 bits. By combininng chips, words of any desired length can be used. Because bit-slice microprocessors are available in fast bipolar Schottky and ECL technologies, the central processing units of all but the largest mainframe computers can actually be implemented by such microprocessors. At the other end of the scale is the 1-bit microprocessor, which is intended to replace digital logic for control applications.

15.2.2. Microprocessors in Biomedical Instrumentation

The first biomedical instruments incorporating microprocessors began to appear on the market around 1975. While the first devices were mainly laboratory-type instruments, microprocessors are now used in all areas of biomedical instrumentation. Although microprocessors were originally advocated mainly as replacements for controllers using digital logic, it was soon found that the new technology could be extended much further. Following are some examples of the ways in which microprocessors are employed in contemporary medical instruments.

15.2.2.1. Calibration. Many instruments require zeroing and recalibration at certain time intervals, sometimes every few hours. A software or hardware timer in a microprocessor system can initiate a calibration cycle. As with manual calibration, this cycle requires the introduction of a blank and standard, each of which might be in the form of a voltage, gas or liquid. In manual calibration methods, zero and gain-control potentiometers are normally adjusted until the readout indicates the proper values. Microprocessor-equipped devices usually perform the calibration in digital form. During the calibration, offset and gain correction factors are determined and stored in memory to be applied to the measured data during the measurement.

15.2.2.2. Table lookup. In analog systems, nonlinear functions (e.g., those required for the correction of a transducer characteristic) are usually implemented by straight-line approximations. In microprocessor-equipped systems, table lookup with interpolation can be used. This procedure is less limited and more accurate and also permits the determination of parameters that are dependent on more than one variable.

15.2.2.3. Averaging. Microprocessors can easily average data over time or over successive measurements and can thus decrease statistical variations.

15.2.2.4. Formatting and printout. Because medical equipment using microprocessors usually processes data in digital form, the microprocessor can be utilized to format the data, convert the raw data into physical units, and print out the results in a form that does not require further transcribing or processing.

15.3. INTERFACING THE COMPUTER WITH MEDICAL INSTRUMENTATION AND OTHER EQUIPMENT

To operate effectively with or as part of a medical instrumentation system, a computer or microprocessor must properly interface with the various devices comprising the rest of the system. Input data must be requested and received in an acceptable form and output signals must be provided wherever control functions are required or where data must be transmitted to other equipment. Several important factors must be considered in interfacing, including the type of output data produced by each instrument, the logic and formatting requirements of the computer, the input requirements of any devices that are to receive signals from the computer, the method by which these signals are to be transmitted, and the commands required to control input–output traffic.

Many biomedical instruments with which a computer may be interfaced generate analog data in the form of voltages proportional to the variables represented. For computer entry, these analog signals must be converted into digital form. On the other hand, where the computer is required to provide analog output signals for display or control purposes, digital output data must be converted into analog form. Following a discussion of digital interfacing requirements, a brief introduction to analog-to-digital and digital-to-analog conversion is presented.

15.3.1. Digital Interfacing Requirements

Interfacing a computer with other devices that handle data in digital form involves both software and hardware. The software is usually a part of the computer's system software and is often an extension of the input–output package that controls the flow of information to and from such peripheral devices as disks and magnetic tape drives. Programs are included to monitor input lines, generate commands, identify the various sources of input data, accept each word of data as it arrives, and route it to the arithmetic unit or memory as appropriate.

Interfacing hardware is required to format the data, provide buffer registers to temporarily hold each word until it can be dealt with, and where necessary, convert input or output signals from one system of logic to another.

Formatting is the arranging of data into a form that can be accepted and recognized by the computer or device receiving computer output. It involves such factors as the number of bits to be received or sent out at a time and the way in which the bits of a word are arranged among the input or output lines. Data may be received or sent out in either *serial* or *parallel* form. In serial form, the bits of each word or character are received or sent one at a time over a single line, whereas in parallel transmission, a separate line is provided for each bit. Serial transmission is generally used where data are sent over long distances via telephone lines or for connection to Teletypewriter keyboard-printers or CRT terminals. On the other hand, most computer input–output (I/O) ports accept and produce data in parallel form, requiring a *parallel-to-serial* or *serial-to-parallel converter*. In such a converter the serial data are shifted into or out of a shift register, which is parallel-interfaced with the computer I/O port via a buffer register. A parallel-to-serial converter also generates and frames each word or character with *start* and *stop bits* which can be recognized by the receiving device. A serial-to-parallel converter uses these bits to control formatting of the parallel data.

Where the interface includes more than one digital device, a separate I/O port can be provided for each, or all the devices can be interconnected via a common set of data lines called an *an input/output data bus* or *party-line bus*. When this type of bus arrangement is used, additional interconnecting lines must be provided to address each individual device so that data are transferred to or from only one device at a time and to assure that each communicating device is properly identified to the computer.

When the digital devices with which the computer must interface can be controlled so that transfer of data always occurs in time correspondence with the computer's internal clock, the I/O operation is said to be *synchronous*. Most situations, however, require *asynchronous* input/output in which bidirectional control of the transfer may be accomplished through a process called *handshaking*. In this procedure, the computer and the I/O device exchange signals, indicating first that a valid character is on the line, ready to be received, and then that the transfer has been successfully accomplished.

Transmission of data in serial form via a telephone line requires not only that the data be converted into serial form and framed with appropriate start and stop bits, but also that the string of bits be placed on a carrier signal by a modem. The rate of data transmission is given in *baud*, the total number of bits transmitted per second, including start and stop bits. The rate for a 10-character-per-second teletype is 110 baud (a total of 11 bits is required for each 8-bit data character).

Devices with which a computer must interface may produce digital data in pure binary form, binary-coded decimal, or in some type of alphanumeric code, such as ASCII (American Standard Code for Information Interchange) or EBCDIC (Extended Binary-Coded Decimal Interchange Code). Both of these codes are used extensively in digital communication. In each code, an 8-bit character is defined for each numeral, letter of the alphabet, both upper- and lower-case, and punctuation mark. In addition, each code contains a number of special characters for control of a printing device, such as a Teletypewriter, or for identification of the beginning of a block of data. Limited-character 6- and 7-bit ASCII codes are also available and are used in some applications.

15.3.2. Analog-to-Digital and Digital-to-Analog Conversion

Whenever a digital computer must communicate with an instrumentation system that generates or requires data in analog form, the interface must include equipment to convert analog signals into digital data or numerical information in digital form into analog voltages. In the process of digitizing data, most analog-to-digital (A/D) converters incorporate digital-to-analog (D/A) conversion circuitry, as indicated below. For this reason D/A converters are discussed first.

15.3.2.1. Digital-to-analog conversion. In order to obtain a continuous analog signal from a sequence of values in digital form, a voltage must be generated proportional to the value of each digital word as it appears in the sequence. The circuitry by which this is accomplished is called a *digital-to-analog converter*.

Generation of a voltage proportional to a digital word can be accomplished in various ways. One method is illustrated in Figure 15.6, which shows the weighted resistor (summing amplifier) digital-to-analog

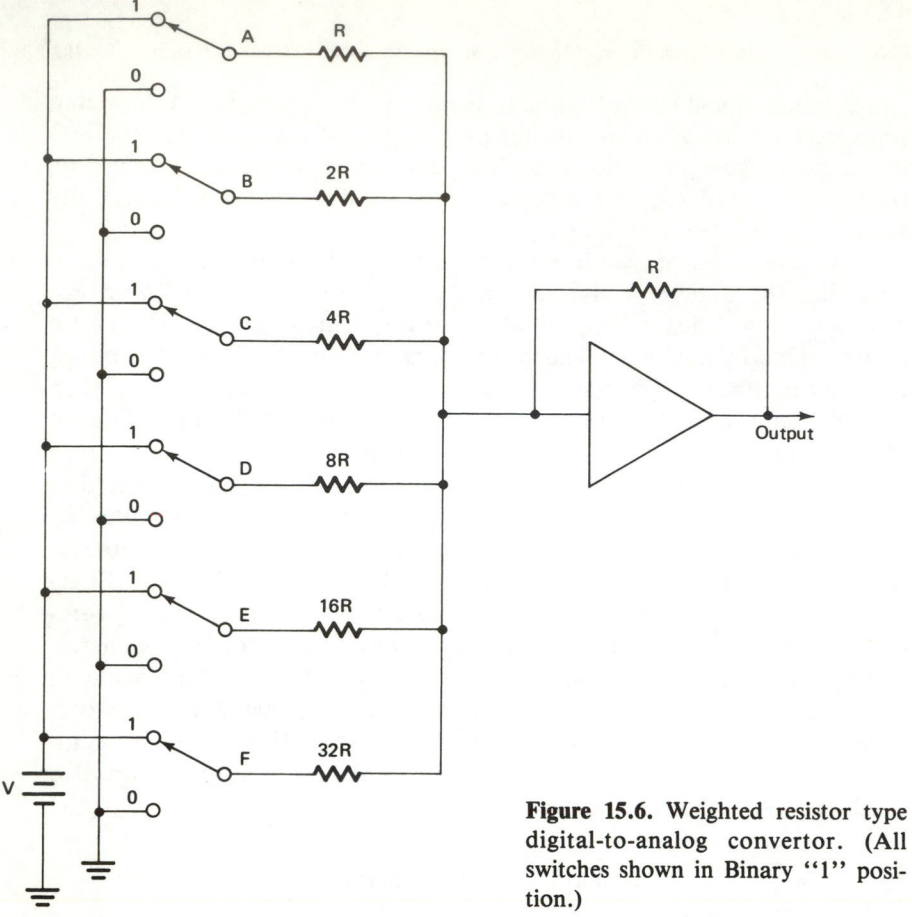

Figure 15.6. Weighted resistor type digital-to-analog convertor. (All switches shown in Binary "1" position.)

converter. This circuit is an operational amplifier connected as an *analog adder*. The output is the sum of the contributions of the various inputs. At each input the common input voltage is weighted or multiplied by the ratio of the feedback resistor to the associated input resistor.

For example, in the circuit shown in Figure 15.6, each bit of a 6-bit binary word controls the switch to one input. If a given bit has a value of 1, its corresponding switch places the appropriate input at a reference voltage V. If that bit has a value of 0, however, the input is set to ground (0 V). The most significant bit (labeled A in the figure) then contributes a voltage equal to V to the output of the circuit when that bit is a 1, but when that bit has a value of 0, it contributes nothing. Because the input resistor for bit B has twice the value of that for bit A, a 1 in bit B contributes exactly half the voltage of V to the output. Similarly, bit C contributes one-fourth the voltage of $V,$ and so on down to the least significant bit, F, which, when given a value of 1, contributes only $\frac{1}{32}V$. These contributions correspond exactly to the relative values of the bits in the binary word. Thus, the output of the operational amplifier is proportional to the sum of the value of all bits that have the value 1, and consequently is proportional to the value represented by the digital word. For a binary word of greater length (a greater number of bits), an additional input resistor and switch are required for each additional bit. For an *n*-bit word, the input resistor for the least significant bit would have a value of $2^{n-1} R$.

Figure 15.7 shows a *binary ladder circuit*. The output of the ladder circuit is connected to the input of an operational amplifier. As in the case of the analog adder, the ladder has an input corresponding to each bit of the binary word. Again, each input has a switch controlled by the value of its corresponding bit. As before, when a bit has a value of 1, its input is switched to ground. The ladder network is so arranged that each input switched to voltage V contributes a voltage to the input of the amplifier proportional to the value of the corresponding binary bit, while the output voltage of the circuit is proportional to the sum of all bits with a value of 1. All resistors are either of value R or $2R$. The accuracy of this circuit is not dependent upon the absolute value of resistors, but upon their relative values. Also, the ladder is so arranged that, regardless of the combination of switch positions, the input impedance seen by the amplifier is constant and equal to R.

Figure 15.7. Binary-ladder type digital-to-analog converter. (All switches shown in Binary "1" position.)

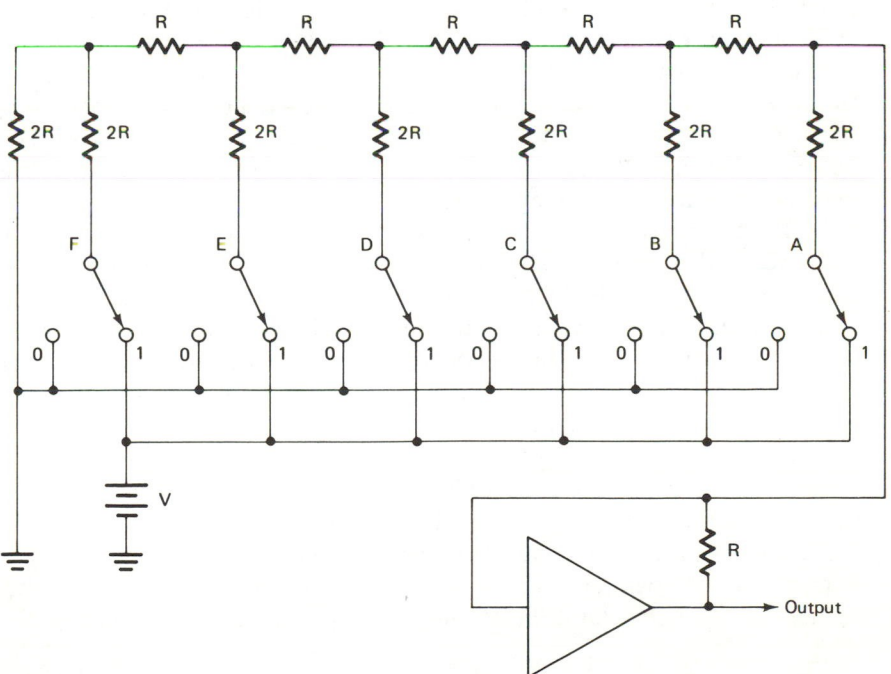

In the circuit shown in Figure 15.7, switch A is controlled by the most significant bit and switch F is controlled by the least significant bit. To accommodate digital words of greater length, the network can be extended to provide an input for each additional bit which contributes the correct voltage for that bit.

In both types of digital-to-analog converters, the switching is usually done by solid-state switching circuits. Although many circuit configurations of this type are in use, they all essentially accomplish the same purpose of providing the reference voltage with a digital input of 1 and ground with an input of 0.

There are several ways of estimating the value of the analog signal at the output of the converter between the occurrence of digital data points, all of which involve analog filters. The simplest, called *zero-order hold,* assumes that the signal remains constant at the level of each digital value until the next one occurs. Then it jumps immediately to the level of the new value, where it again remains until another value is received. Unless abrupt changes in the data can be expected which could result in excessive error, this method is usually used. More complex (and more expensive) methods are also available, such as *first-order hold,* in which the signal at any time is caused to change at the same rate as it did between the two previous digital data points.

15.3.2.2. Analog-to-digital conversion.

15.3.2.2. Analog-to-digital conversion. An analog-to-digital converter is a device that accepts a continuous analog voltage signal as input and from that signal generates a sequence of digital words that represent the analog voltage as it varies with time. There are actually two processes involved in the *digitizing* of analog data. The first is *sampling*—the process of measuring the analog voltage at discrete points in time. The sampled voltage must then be *quantized.* Quantizing is the selection of a digital word of specified length to represent the analog voltage.

The simplest form of A/D converter involves a *voltage-to-frequency converter* and a counter. The voltage-to-frequency converter produces a sequence of output pulses at a frequency proportional to the voltage of the analog signal. The counter counts the number of pulses in a specified unit of time. The frequency range of the converter and the time period for counting are selected to provide an output count that corresponds numerically to the voltage of the analog signal.

Another simple A/D converter is called a *ramp* or *pulse-width converter.* At the beginning of each reading a capacitor is discharged and allowed to begin charging at a fixed rate, until it has reached a voltage equal to the voltage of the analog signal, as determined by an analog comparator. The output of the comparator is a pulse whose width is proportional to the analog voltage. During the duration of the pulse, a digital counter counts the output of a fixed-rate digital clock so that the count at the end of each pulse is proportional to the analog voltage at that time.

A slightly more complex but inherently more accurate type of A/D converter is a *dual-slope* or *up-down integrator converter*. In this device, the input of an analog integrator is alternately switched between the analog voltage being digitized and a constant reference voltage. As in the pulse-width converter, a capacitor is charged at a rate proportional to the analog voltage for a fixed time period, so that the height of the ramp at the end of the period is proportional to that voltage. The integrator is then switched to a reference voltage, and the capacitor discharges at a constant rate until the ramp reaches a predetermined level. The counter counts the clock output during this discharge interval, which is proportional to the analog input voltage.

All three of the A/D converters described so far are relatively inexpensive, but are too slow for any application in which the analog voltage varies at a rapid rate. Thus, for most A/D converters, faster and more accurate but also more expensive techniques are employed. In these techniques, the heart of the A/D converter is a D/A converter of a type described above. The basic arrangement is shown in block diagram form in Figure 15.8. In this figure the divider network is a binary ladder which, in conjunction with the reference supply, constitutes a D/A converter of the type shown in Figure 15.7. The flip-flop register is a set of bistable (flip-flop) circuits, each of which can represent a value of binary 0 or 1 and can thus store one bit of a binary digital word. The entire set of flip-flops that constitutes the register represents each digital word to be generated by the converter. Through the

Figure 15.8. Analog-to-digital converter incorporating digital-to-analog converter. (Copyright 1964, Digital Equipment Corporation, Maynard, MA. All rights reserved.)

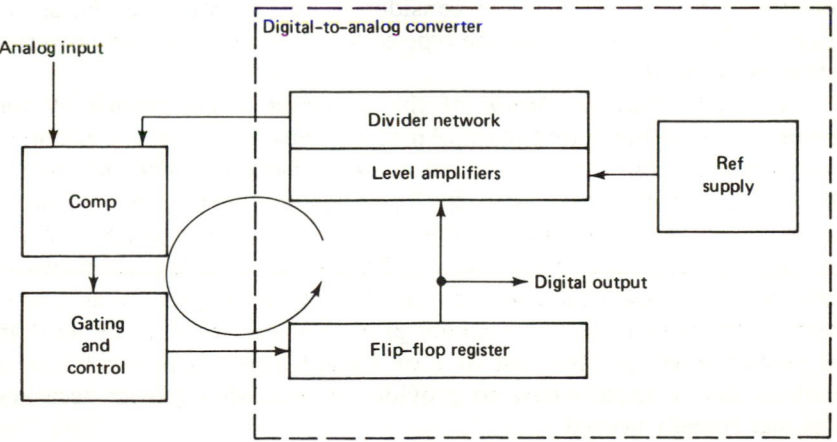

level amplifiers, each flip-flop controls a corresponding input to the ladder network, and together they produce an analog output with the same voltage as that represented by the flip-flop register. At the time of sampling, this voltage is compared with the analog input voltage in an analog comparator circuit. When these two voltages differ, the bits in the flip-flop register are adjusted through appropriate gating and control circuitry until agreement is reached. At that time, the value represented by the flip-flop register is the nearest digital equivalent to the analog input voltage and is caused to appear at the output of the converter.

Although nearly all analog-to-digital converters use this comparison method of matching the value of the register with the input voltage, the methods by which the digital value of the register is adjusted to match the input signal can differ widely. The most common method is called the *successive approximation method,* in which each bit of each digital word is successively tested to determine whether its addition to the value of the register would cause the input signal to be exceeded. If not, that particular bit is set to 1. If the bit would have caused the value of the register to be greater than the input signal, then the bit is left at 0. The process begins at the bit representing the largest value (most significant bit) and continues from "left to right" down the register. The advantage of this type of system is that the conversion time is fixed and does not depend on the input signal. Furthermore, this type of converter gives a good response to large, rapid changes in input, such as might be expected with a multiplexer. To avoid changes in the input signal during the time the converter is in the process of checking each bit, a sample-and-hold circuit is often used to read the voltage at the beginning of each conversion period and to maintain that voltage during conversion period. The result is a closer approximation of the analog signal.

Important factors in selecting an analog-to-digital converter are the resolution of the quantizing process, the conversion rate, and the conversion aperture time. Also to be considered are the computer input requirements for formatting and the type of logic circuitry that the converter output must match.

The quantizing *resolution* of the converter is determined by the number of bits in the output word. An 11-bit-plus-sign word, for example, is capable of dividing the full range of the input signal into 4095 increments of level. This number includes 2047 positive increments and a similar number of negative increments, plus zero. The accuracy of any voltage reading, then, cannot exceed about 1 in 2000, or about 0.05 percent of full scale. Most physiological data do not require that degree of accuracy, however, for many transducers cannot provide accuracies much better than 1.0 percent. But since the cost of 1 or 2 additional bits of resolution is relatively low, it usually pays to provide for somewhat greater accuracy than that actually needed.

The *conversion rate* of an analog-to-digital converter depends on the conversion method used and the speed of the control circuitry. Extremely high rates of conversion are available. Shannon's sampling theorem requires that, to reproduce a periodic signal without severe distortion, the sampling rate be at least twice the highest frequency component that the system is able to pass. For nonperiodic waveforms, it is generally good practice to use a sampling rate of at least five times the highest frequency component. Obviously, the higher the sampling rate, the more accurate will be the representation of the analog signal; but higher digitizing rates mean that more data must be stored and handled by the computer. This usually results in a greater computation cost.

The *aperture time* is the period of time during which the analog signal is actually being sampled for conversion. A long aperture time might result in a change of data during the sampling interval. Most modern analog-to-digital converters have sufficiently short aperture times for the conversion rates at which they operate.

The process of sequentially taking readings from two or more analog data channels with a single analog-to-digital converter is called *time multiplexing*. If N data channels are multiplexed into a converter which operates at R conversions per second, and all channels are converted at the same rate, the conversion rate for any given channel is R/N conversions per second. This means that with multiplexing, the conversion rate of the converter must be the required conversion rate for each channel multiplied by the number of channels. If it is important that all of the channels be converted in exact time correspondence, sample-and-hold circuitry must be incorporated into the multiplexer to "hold" values until they can be digitized.

15.4. BIOMEDICAL COMPUTER APPLICATIONS

Applications of the digital computer in medicine and related fields are so numerous that even listing all of them is beyond the scope of this textbook. Most of these applications, however, utilize a few basic capabilities of the computer which provide an insight to ways in which computers can be used in conjunction with biomedical instrumentation. These basic capabilities include:

1. *Data acquisition:* The reading of instruments and transcribing of data can be done automatically under control of the computer. This not only results in a substantial saving of time and effort, but also reduces the number of errors in the data. When data are expected at irregular intervals, the computer can continuously scan all input sources and accept data when-

ever they are actually produced. If the data originate in analog form, the computer usually controls the sampling and digitizing process as well as identification and formatting of the data. In some cases, the computer can be programmed to reject unacceptable readings and provide an indication of possible trouble in the associated instrumentation. Sometimes the computer provides automatic calibration of each input source.

2. *Storage and retrieval:* The ability of the digital computer to store and retrieve large quantities of data is well known. The biomedical field provides ample opportunities to make use of this capability. In a modern hospital, large amounts of data are accumulated from many sources. These include admission and discharge information, physicians' reports, laboratory test results, and several other kinds of information associated with each patient. In addition, the hospital also generates a considerable amount of non-patient-oriented data, such as pharmacy records, inventories of all types, and accounting records. Without a computer, the storage of this vast amount of information is both space- and time-consuming. Manual retrievel of the data is tedious, and for some types of information, almost impossible. The digital computer, however, can serve as an automated filing system in which information can be automatically entered as it is generated. These files can be stored as long as necessary and updated whenever appropriate. Any or all of the information can be retrieved on command whenever desired and can be manipulated to provide output reports in tabular or graphic form to meet the needs of the hospital staff or other users.

3. *Data reduction and transformation:* The sequence of numbers resulting from digitizing an analog physiological signal such as the ECG or EEG would be quite useless if retrieved from the computer in raw form. To obtain meaningful information from such data, some form of data reduction or transformation is necessary to represent the data as a set of specific parameters. These parameters can then be analyzed, compared with other parameters, or otherwise manipulated. For example, the electroencephalogram (EEG) signal can be subjected to *Fourier transformation* to obtain a frequency spectrum of the signal. Further analysis can then be performed using the frequency-related parameters rather than the raw EEG data. The electrocardiogram (ECG) signal can also be subjected to data reduction methods, as shown in section 15.4.1, or heart rate information can be extracted for patient-monitoring purposes. Special trans-

formations are also required to reconstruct images in computerized axial tomography (see Section 15.4.4). The size and complexity of many of these transformation and data reduction problems are such that manual methods would be completely impractical.

4. *Mathematical operations:* Many important physiological variables cannot be measured directly, but must be calculated from other variables that are accessible. For example, many of the respiratory parameters described in Chapter 8 can be calculated from the results of a few simple breathing tests and gas concentration measurements. Also, the calculation of cardiac output by a dye or thermal dilution method as described in Chapter 6 can easily be done by computer. If a digital computer is connected on-line with the measuring instruments, the calculated results can often be obtained while the patient is still connected to the instruments. This not only enables the physician to conduct further tests if the results so indicate, but can also inform him immediately if any measurements were not properly made and require repetition.

5. *Pattern recognition:* To reduce certain types of physiological data into useful parameters, it is often necessary that important features of a physiological waveform or an image be identified. For example, analysis of the ECG waveform requires that the important amplitudes and intervals of the electrocardiogram be recognized and identified. Digital computer programs are available to search the data representing the ECG signal for certain predetermined characteristics that identify each of the important peaks. In Section 15.4.1 the technique by which this is accomplished is described. Somewhat different techniques are used in other pattern recognition problems, such as the identification and labelling of chromosomes, but since each type of pattern has unique features that must be identified, programming for pattern recognition is a highly specialized process.

6. *Limit detection:* In applications involving monitoring and screening, it is often necessary to determine when a measured variable exceeds certain limits. For example, in the analysis of the electrocardiogram, each important parameter of the ECG can be checked to determine whether it falls within a preestablished "normal" range. By comparison of the measured parameter with each limit of the range, the computer can indicate which parameters exceed the limit and the amount by which they deviate from normal. Using this technique, patients can be

screened to select those with ECG irregularities that should receive further attention. In most cases, the "normal" range is defined in advance, but sometimes the computer is programmed to establish normal ranges for each patient based upon the averages of repeated measures taken under specified conditions.

7. *Statistical analysis of data:* In the diagnosis of disease, it is often necessary to select one most likely cause out of a set of possible causes associated with a given set of observed symptoms, measurements, and test results. Similarly, medical research investigators must decide at times whether an observed change or condition in a person or animal is due to some treatment imposed by the researcher, or whether the result could be attributed to some other cause or just to chance alone. Both of these situations require the use of inferential statistical procedures, some of which are quite complex. Fortunately, most statistical methods lend themselves well to computer techniques, especially when large numbers of variables must be analyzed together or where data from a large number of patients are used. Even simple descriptive statistics, such as means, standard deviations, and frequency distributions can be computerized, resulting in significant savings of time and effort.

8. *Data presentation:* An important characteristic of any instrumentation and data-processing system is its ability to present the results of measurements and analyses to its users in the most meaningful way possible. By virtue of appropriate output devices, a digital computer can provide information in a number of useful forms. Table printouts, graphs, and charts can be produced automatically, with features clearly labeled using both alphabetic and numeric symbols. If the necessary computer peripherals are available, plots and cathode-ray-tube displays can also be generated. In addition to controlling the output devices, the computer can be programmed to organize the data for presentation in the most meaningful form possible, thus providing the user with a clear and accurate report of his results.

9. *Control functions:* Digital computers are capable of providing output signals that can be used to control other devices. In such applications, the computer is programmed to influence or control physiological, chemical, or other measurements from which its input data are being generated. The computer can also be used to provide feedback to the source of its data.

For example, while reading and analyzing the results of a chemical process, the computer can be made to control the rate, quantity, or concentration of reagents added to the process, or it could control the heating element of a temperature bath. By controlling these and other possible inputs, the process can be regulated to achieve desired results. In addition, the computer can be programmed to recognize certain characteristics of the measured results that would indicate possible sources of error. Sometimes other parameters are monitored in addition to the actual results to increase the sensitivity of the computer to conditions that could result in erroneous measurements. The computer can automatically compensate for some sources of error, such as a gradual drift in the baseline, by either altering the process itself or by mathematically adjusting the results before printing them out. When more serious types of error occur, the computer can alert the operator to the condition or, if necessary, can automatically stop the process.

The extent to which each of the described capabilities above can actually be utilized in a given situation depends on the available hardware and software. Obviously, some of these capabilities require greater resources than others.

Following are some specific examples of computer applications in clinical medicine and research. Although they represent only a few of the many possible ways in which computers can be used in medicine and biology, they serve to illustrate the role of each of the above-described capabilities. In each example, the computer techniques are described in conjunction with their associated biomedical instrumentation.

15.4.1. Computer Analysis of the Electrocardiogram

The use of computers for the clinical analysis of the electrocardiogram (ECG) has developed over the span of many years. There are several reasons for this. First, ECG potentials are relatively easy to measure. Second, the ECG is an extremely useful indicator for both screening and diagnosis of cardiac abnormalities. In addition, certain abnormalities of the ECG are quite well defined and can be readily identified.

Measurement of the electrocardiogram for computer analysis is essentially the same as is used for manual ECG interpretation. Most computerized systems use the 12 standard leads described in Chapter 6. There are more elaborate systems, however, that simultaneously measure three orthogonal components of the ECG vector. For some of these systems, a special orthogonal lead configuration is used.

Entry of the ECG into a digital computer requires that the analog ECG signals be converted into digital form. Although some attempts have been made to partially reduce the ECG data in analog form, nearly all presently used systems incorporate an analog-to-digital converter operating at a constant rate. The actual sampling rate depends upon the desired bandwidth of the signal to be analyzed. Sampling rates ranging from 100 readings per second up to 1000 readings per second are in current use. Analog filtering is often used ahead of the converter to eliminate noise and interference above the upper limit of the desired frequency band.

Once inside the computer, the ECG signal can be subjected to additional smoothing by means of digital filtering methods. This smoothing process eliminates high-frequency variations in the signal that might otherwise be mistaken for features of the ECG.

Pattern recognition techniques are next employed to identify the various features of the ECG. These features are shown in Figure 3.6. The most stable reference point of the ECG pattern, and one of the most reliably identified, is the downward slope between the R and S waves of the QRS complex. This slope can be characterized as the most negative peak that occurs in the first derivative of the ECG waveform. To recognize this point, the ECG signal must be differentiated to obtain a signal representing the first derivative, and the first derivative signal must be scanned to locate its most negative peaks. Other tests are then applied to both the ECG and its derivative to verify that a true RS slope has been located.

From this reference point, the computer scans the ECG data in a backward direction with respect to time to locate the positive peak just preceding the reference. This peak is identified as the R wave. The negative peak of the ECG just subsequent to the reference slope is the S wave, and the negative peak just ahead of the R wave is the Q wave.

A predetermined interval of the ECG signal prior to the QRS complex is scanned for a positive peak to locate the P wave. Actually, the P wave is often identified on the basis of both the ECG waveform and its first derivative. The T wave is identified as a peak within a predetermined interval of the ECG signal following the QRS complex. In most ECG analysis programs, identification of the various waves is based on at least two leads.

The baseline of the ECG waveform is usually defined as a straight line from the onset of the P wave in one ECG cycle to the onset of the P wave in the next cycle. The amplitude of each of the waves (P, Q, R, S, and T) is measured with respect to that baseline. Also, a few points along the S-T segment are measured to determine their deviation from the baseline. Deviations from the baseline of the ECG signal as well as characteristics of the first derivative waveform are used to locate the onset and ending times of all waves. From this information the duration of each wave and the intervals between waves are measured. The duration of the QRS complex, the P-R interval, and the S-T interval are especially significant.

Each of the measured amplitudes, durations, and intervals is a characteristic parameter of the ECG signal. Another important parameter is the heart rate (determined by measuring the time intervals between successive R waves). Each of these parameters can be averaged over several cycles with the means and standard deviations being printed out for each of the leads measured.

For screening purposes, each of the parameters can also be checked to see if it falls within a normal range for that parameter. Any parameters that lie outside the normal range are indicated on the computer-generated report. A report of this type is shown in Table 15.1. This is the result of a test run on a 36-year-old male who was presumably normal, but was found by this screening analysis to have bradycardia (slow heart rate).

Identification and other patient information is printed at the top. The mean values for the various parameters are then presented in a matrix form. The columns represent the 12 standard leads while the rows indicate the parameters. Data from lead V3 were purposely omitted to show the response of the system to missing data. Below this matrix, values for the P-R, QRS, and Q-T intervals and the heart rate for each of the leads are printed out. The heart rate varies from lead to lead because in this system each lead is measured at a different time. Calibration information for each lead and the calculated angle of the axis of the heart (see Chapter 6) for each portion of the ECG cycle are also given. At the bottom of the printout are indications of any noted abnormalities. In the example, the condition of bradycardia (heart rate below 60 beats per minute) is noted as well as the absence of data from one lead.

In more sophisticated systems for computer analysis of the ECG, additional ways of representing the ECG are derived to further aid in distinguishing an abnormal ECG from a normal one. One such representation is a three-dimensional time-variant vector derived from the simultaneous measurement of three orthogonal leads. The behavior of this vector tells much more about the electrical activity of the heart than does the instantaneous calculation of the axis angle for a given portion of the ECG cycle.

Another parameter is the *time integral* of the ECG waveform. To obtain this integral, the areas of each wave above and below the baseline are determined and the sum of the areas below the baseline (negative) is subtracted from the sum of the areas above the baseline (positive). This integral can be determined for any portion of the ECG cycle. The sum of the time integral of the QRS complex and that of the T wave is sometimes called the *ventricular gradient,* and is believed to indicate the difference in the time course of depolarization and repolarization of the ventricles. The time integrals of the three orthogonal leads can be added vectorially to obtain three-dimensional time integrals.

Some systems for computer analysis of the ECG use statistical methods in an attempt to classify ECG patterns as various types of abnormalities

Table 15.1. ECG COMPUTER ANALYSIS DATA

```
RUN

H456789A      13:54     11/5/70

U.S.P.H.S. CERTIFIED E.C.G. PROGRAM PROCESSED BY THE BECKMAN HEARTLINE
FOR    BECKMAN INSTRUMENTS, INCORPORATED              LOC 10
 STAT
PAT  123456789          DATE 11- 5-70        SERIAL 126         OPERATOR 5
36 YR          MALE     5 FT 11 IN      190 LBS
BP  NORMAL              MEDS  NONE
```

	I	II	III	AVR	AVL	AVF	V1	V2	V3	V4	V5	V6	
PA	.08	.13	.00	-.07	.05	.08	.05	.12		.12	.08	.07	PA
PD	.13	.12	.00	.08	.09	.10	.05	.10		.10	.11	.08	PD
Q/SA	-.07	.00	.00	-.91	-.11	.00	.00	.00		.00	.00	.00	Q/SA
Q/SD	.02	.00	.00	.06	.02	.00	.00	.00		.00	.00	.00	Q/SD
RA	.86	.97	.13	.00	.61	.58	.16	.41		1.67	1.72	1.33	RA
RD	.05	.09	.05	.00	.05	.09	.02	.03		.08	.10	.09	RD
SA	-.10	.00	-.21	.00	-.08	.00	-.95	-2.64		.00	.00	.00	SA
SD	.01	.00	.02	.00	.02	.00	.05	.07		.00	.00	.00	SD
RPA	.00	.00	.07	.00	.00	.00	.00	.00		.00	.00	.00	RPA
RPD	.00	.00	.02	.00	.00	.00	.00	.00		.00	.00	.00	RPD
STO	.03	.00	.00	-.03	.03	.02	-.03	.09		.08	.01	.00	STO
STM	.03	-.01	-.02	-.01	.04	.02	.04	.29		.04	.01	.00	STM
STE	.04	.00	-.04	-.02	.06	.00	.03	.38		.08	.04	.02	STE
TA	.28	.27	.07	-.30	.24	.19	-.15	1.15		.61	.43	.34	TA

	I	II	III	AVR	AVL	AVF	V1	V2	V3	V4	V5	V6	
PR	.16	.18	.00	.21	.15	.19	.17	.19		.19	.18	.18	PR
QRS	.08	.09	.09	.06	.09	.09	.07	.10		.08	.10	.09	QRS
QT	.38	.39	.43	.37	.39	.39	.38	.39		.39	.40	.41	QT
RATE	60	71	61	55	59	58	56	54		62	53	56	RATE

	I	II	III	AVR	AVL	AVF	V1	V2	V3	V4	V5	V6	
CODE	3	2	2	3	2	2	3	3	A	3	2	2	CODE
CAL	99	99	99	99	99	99	99	99		99	99	99	CAL

```
    AXIS IN     P   QRS  T    Q    R    S   STO                ST-T  QRS-T
    DEGREES    53    47  28       37  253   23                  05     19
```

```
      MSDL APPROVED VERSION        .
      D 41-42-25-11               .
1131 RATE UNDER 60                . BRADYCARDIA
      1 LEAD NOT MEASURED         .
                                  . ATYPICAL ECG
                                  . ------------ M.D.


TIME 1 SECS.
```

or as being normal. Obviously, the more information available about the ECG, the better will be the discriminating ability of the computer programs. Multivariate statistical analysis techniques are sometimes employed, both for one-dimensional and three-dimensional data. Because of the wide inter-personal variation even among normals, accurate computer classification is difficult.

15.4.2 The Digital Computer in the Clinical Chemistry Laboratory

The modern clinical laboratory includes various types of automated instruments for the routine analysis of blood, urine, and other body fluids and tissues. Some of these devices are described in Chapter 13. While automated equipment can be used for most laboratory tests, there are still many determinations which are performed manually, either because of insufficient volume for certain tests or because satisfactory automated tests have not yet been devised. As a result, data from the clinical laboratory are generated in many forms, many of which require manual transcription of the test results.

In the chemistry laboratory, Autoanalyzers and other types of automated clinical chemistry equipment produce charts on which the test results are recorded. To produce laboratory reports which eventually become a part of the patients' records, data must be transcribed from these charts and combined with results from manually performed tests. Care must be taken to assure that data are accurately transcribed and that each test result is associated with the correct patient information.

To accommodate the large output of test results from the automated clinical chemistry equipment and to assimilate those data with patient information and the results of manually performed tests, a number of clinical chemistry laboratories have installed computer systems for data acquisition and processing. Computers of various sizes including micro processors, can be used in such systems, depending upon the extent to which the computer participates in the operation of the laboratory. In a highly automated system, the computer accepts test requisitions, prepares lists for blood drawing, schedules the loading of sample trays, reads test results, provides on-line quality control of the process, assimilates data, performs calculations, prepares reports, and stores data for possible comparison with future test results.

In a typical computerized system such as those discussed in Section 13.4, the medical staff may order tests directly via a remote terminal on the hospital ward or by use of machine-readable requisition forms which are automatically read by computer input equipment in the laboratory. From this requisition information, the computer schedules the drawing of blood by printing out blood drawing lists and preprinted specimen labels. These labels, which may be machine-readable, contain identification information to be used for all tests, automated and manual, from a given patient during that day. As the specimens arrive in the laboratory, the computer prepares a loading list which assigns a specific sample position in the analyzer loading tray for each test.

Patient information is entered into the computer either at the time the patient is admitted to the hospital or when the medical staff orders tests. This information is usually entered by keyboard, either from a remote terminal or in the laboratory.

Once a test run is begun, the output readings of all automated instruments are automatically entered into by the computer. Entry is usually accomplished by means of retransmitting slide wires attached to the recorder pens which produce analog voltages proportional to the output of each instrument. These analog voltages are sampled and converted to digital form by means of a time multiplexer and an analog-to-digital converter. The computer is programmed to recognize legitimate peaks as they arrive and to reject questionable or improperly shaped peaks. The computer also performs the necessary calculations to convert the value of each measured peak into medically useful units. By virtue of its position in the sequence of measured peaks or machine-readable ID labels, each test result is identified and associated with the correct patient. Control samples, placed randomly (by computer assignment) throughout the run, are used to periodically check the calibration of the system. By monitoring these control samples and the measured values from patient samples, the computer is able to perform "on-line quality control." In some cases, the computer can automatically correct the output values for drift and certain other types of error. In case of severe error, the computer may provide a warning to the operator, who may then choose to stop the test because of equipment malfunction.

The computer, after assimilating data from all automatically performed tests, may also receive results from manually performed tests. These manual test results would be entered by keyboard or via machine-readable data sheets specially prepared for each type of test. Once all test results have been received, credibility checks can be run to search for any impossible or unlikely combinations of results or any impossible changes in a given patient's test results from one day to the next. After the data have been checked and verified, the computer provides a physician's report, either in printed form or on a cathode-ray-tube terminal. This terminal can either be located in the laboratory, on the patient's ward, or in the physician's office. In addition, the computer might incorporate the test results into a patient filing system, so that whenever desired, the physician can request a profile of test results for a given patient over a specified number of days. Such a profile allows the physician to note changes in a patient's condition over time.

Another feature of most clinical laboratory computer systems is the capability of handling emergency requests. Such emergencies often require that a specimen of blood or urine be entered into the system ahead of routine samples. When patient identification is controlled by the position of a sample in the sample tray of the automated instrument, changes in sample positions to accommodate emergency needs must also be made known to the computer, either by keyboard notification of each change or by some automatic means of reading sample cup labels.

Provision must also be made for a physician to obtain results of a specific test before other tests on that patient have been completed and prior to the normal reporting of results. The inquiry is usually made by keyboard, either at the computer or from a remote terminal. Results of that specific test, if available, are given at the same terminal. If the test has not been completed at the time of the inquiry, the physician is so notified.

15.4.3. The Digital Computer in Patient Monitoring

Instrumentation systems for monitoring patients in intensive- and coronary-care units are described in Chapter 7. In recent years, especially since the advent of the microprocessor, an increasing number of patient-monitoring systems include some form of digital computer.

The type of computer involved and the extent of its role in the overall patient monitoring system may vary widely. In some systems, a small computer, usually a microprocessor, is used to store a limited amount of data and control a nonfade display of the ECG and other variables in an analog system. The waveforms either move across the screen with uniform brightness or remain stationary until replaced by new information, which appears to sweep across the screen and replace the old trace. Computer-controlled displays of this type usually include on-screen digital readouts of such parameters as systolic and diastolic blood pressures and heart rate.

In another type of computerized patient-monitoring system, the computer is simply attached to a conventional analog patient monitor to store and analyze information. Except for the interface through which the computer receives its data, the two systems are completely independent. A computer failure would have no effect whatever on the monitoring of patients. Waveform and trend plots are displayed on cathode-ray screens which are separate from the basic patient-monitoring system.

More often, the computer is an integral part of the patient-monitoring system and, in addition to storing and analyzing data, takes over many of the functions otherwise performed by analog circuitry, such as the filtering of signals to remove noise and artifacts and the controlling of alarms in case of an emergency. Some of the more recent systems utilize microprocessors for this purpose. The PDS 3000 shown in Figures 7.6 and 7.7 (Chapter 7) is a system of this type.

In a few very large hospitals, the patient monitoring system is integrated into a more extensive computer system in which patient records, laboratory test results, pharmacy records, and related information are combined with the ongoing data obtained from the patient monitor. Such systems may also tie in with the operating suite, cardiac catheterization laboratory, and other special diagnostic laboratories. By bringing together data from many sources, the computer can provide more complete information to assist the medical staff in their diagnoses and in monitoring the treatment of patients.

As stated in Chapter 7, the physiological variables typically measured by a patient-monitoring system include the ECG, temperature, a means of obtaining respiration rate, and often arterial and central venous blood pressures. Blood gas and pH measurements are also sometimes included. In a computerized system, the computer generally controls the collection and logging of data from their various sources to assure that readings are taken at the required intervals and properly recorded. Even where the computer is merely an adjunct to a conventional analog monitoring system, this data-acquisition function is required. Since most of the measured variables occur in analog form, control of an A/D converter is also involved. Digital filtering techniques are usually employed to smooth the data for display.

Computerized patient-monitoring systems generally involve most of the basic functions listed and described at the beginning of Section 15.4. Data acquisition and logging and the basic storage and retrieval functions have already been discussed. Data reduction and transformation techniques and mathematical operations are employed extensively in the calculation of a number of parameters, many of them indirect. The derived parameters usually include heart rate, respiration rate, systolic and diastolic blood pressures, and mean arterial and venous pressures. Other parameters, such as cardiac output, stroke volume, blood gas values, urine output, and various lung volumes and capacities are also sometimes calculated. Pattern-recognition techniques are utilized in the detection of arrhythmias and combinations of conditions that may require special attention. Limit detection and statistical analysis are used in checking the validity of data, monitoring for alarm conditions, and comparing results with normal values. The computer is also very much involved in the presentation and display of data. In addition to providing nonfade display of ECG and other raw data, the system may also produce many forms of graphical display, including histograms, trend plots, and plots showing the relationship of two or more variables. In some cases, the computer can also be used to control the infusion of blood or medication, based on the measured values of affected variables. For example, it can monitor a patient's urine output and actuate a pump to infuse a diuretic agent whenever the output falls below a predetermined quantity.

15.4.4. Computerized Axial Tomography (CAT) Scanners

A highly acclaimed application of the digital computer to clinical medicine is *computerized axial tomography (CAT)*. This procedure, which combines X-ray imaging (see Chapter 14) with computer techniques, permits visualization of internal organs and body structures with greater definition and clarity than could ever be attained by conventional methods. Although X rays have been in use since their discovery in 1895 and the reconstruction methods used in axial tomography date back to 1917, a practical combination of these techniques could not be achieved until the availability of the modern computer.

The basic principles involved in conventional X-ray imaging are discussed in Chapter 14, in which it is pointed out that the X-ray photograph is literally a shadow of all organs and structures in the path of the rays. If two radiopaque objects lie, one behind the other, in the X-ray path, as shown in Figure 15.9, the smaller of the two may be completely hidden by the larger. To partially circumvent this problem, a method of *linear tomography* was developed in which the X-ray source and film are simultaneously moved in opposite directions, as shown in Figure 15.10. For any given combination of source and film velocities, there will be one single plane perpendicular to the path of the rays in which objects will appear to remain stationary with respect to the film during the movement. In contrast, the shadows of objects at all other distances from the source will move on the film and produce a blur. In Figure 15.10, the sphere lies in the plane that appears stationary, whereas the cube does not. The shadow of the sphere is therefore reinforced as the X-ray vantage point is changed.

The principle of obtaining X-ray images from a number of vantage points is also used in computerized axial tomography, but in a different way. As the name implies, the vantage points for *axial tomography* are taken around the axis of the body. Instead of sending X rays through the entire portion of the body to be visualized, a very narrow pencil-like X-ray beam scans a single slice perpendicular to the body's axis. By scanning two or more such slices, a three-dimensional representation can be produced. Rather than obtaining an image on an X-ray film, the intensity of the X rays, after penetrating the body, is measured by means of one or more sodium iodide, xenon, or calcium chloride crystal detectors, which scintillate in proportion to the intensity (see Chapter 14). The scintillation light is measured by photomultiplier tubes. In the original computerized axial tomography (CAT) scanners, the source of the pencil-like beam was mechanically moved across the region of the slice, as shown in Figure 15.11. At the same time, the detector moved linearly in parallel with the source to receive a signal whose variations with respect to time represented the density pattern across the slice from one vantage point. The mechanism containing the source and

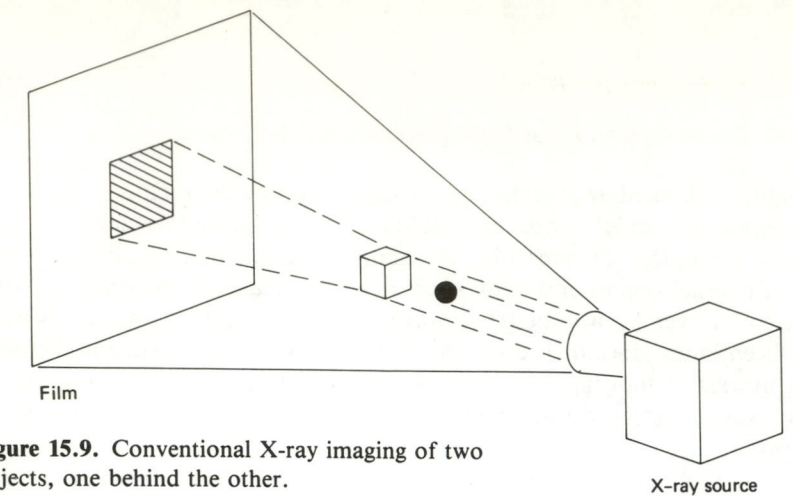

Figure 15.9. Conventional X-ray imaging of two objects, one behind the other.

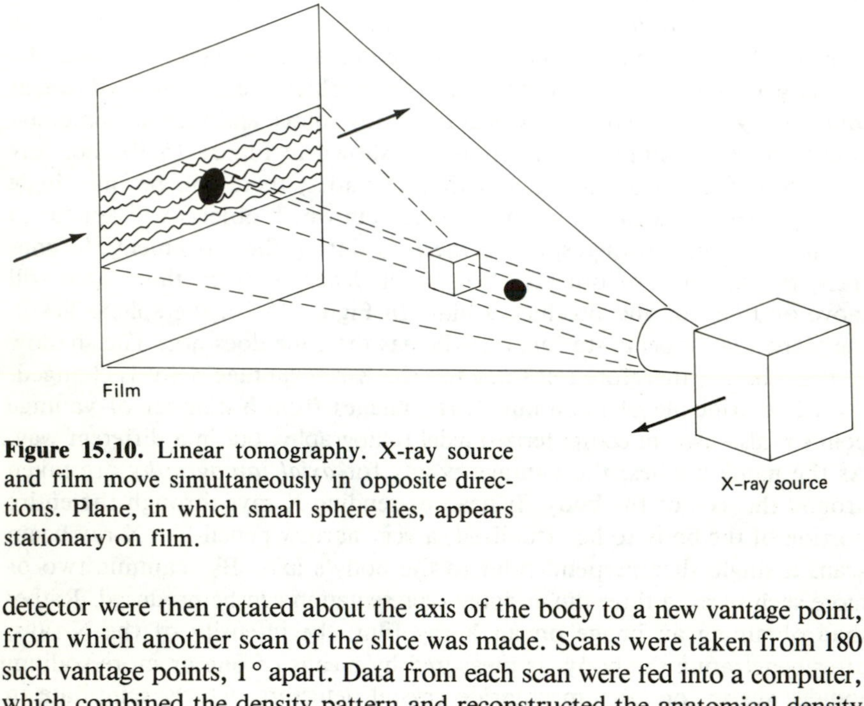

X-ray source

Figure 15.10. Linear tomography. X-ray source and film move simultaneously in opposite directions. Plane, in which small sphere lies, appears stationary on film.

Film

X-ray source

detector were then rotated about the axis of the body to a new vantage point, from which another scan of the slice was made. Scans were taken from 180 such vantage points, 1° apart. Data from each scan were fed into a computer, which combined the density pattern and reconstructed the anatomical density of the two-dimensional slice. By repeating this process for several slices, a detailed three-dimensional representation could be obtained. The early instruments usually scanned two slices at a time, this process requiring about 5 minutes. Because the region to be scanned had to remain stationary for this length of time, such scans were limited to the brain and other structures of the head, which could be kept immobilized in the necessary position by water bags.

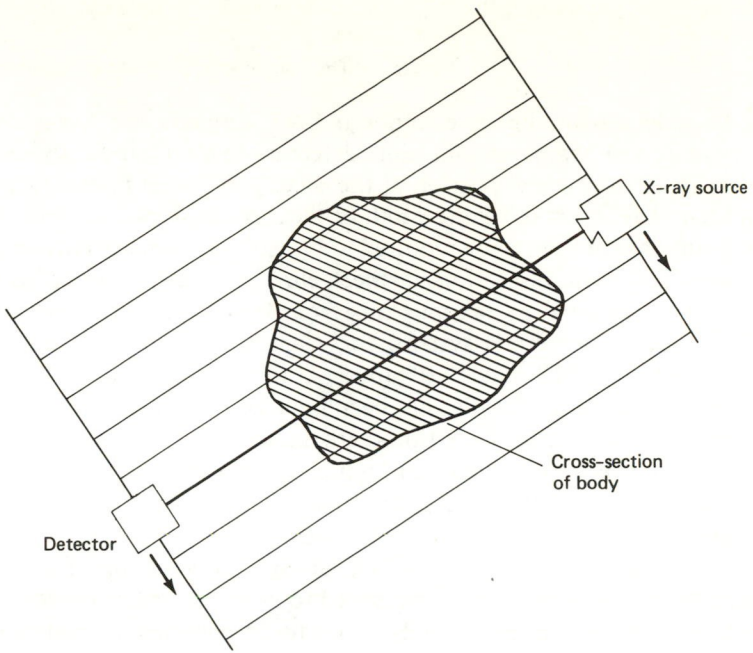

Figure 15.11. Scanning pattern of early computerized axial tomography (CAT) scanners. X-ray source and detector move simultaneously in linear parallel paths to measure density through slice. Entire unit rotates about body to obtain scans from 180 vantage points, 1° apart.

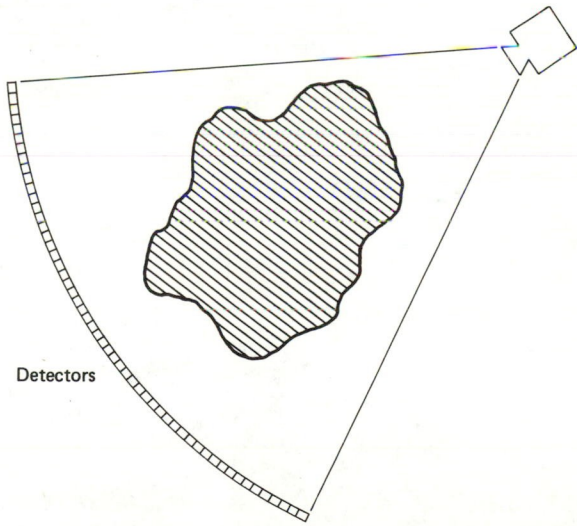

Figure 15.12 Fan beam covering entire cross-section of body with large array of detectors. Elimination of need for linear motion of source and detectors reduces scanning time.

To reduce scanning time, modern CAT scanners use X-ray sources that produce fan beams and multiple detectors to simultaneously measure the density across a wider portion of the slice. The fastest instruments have a fan beam that covers the entire width of the slice, as shown in Figure 15.12. Several hundred detectors are required to measure the density pattern of the slice with sufficient resolution to meet clinical needs. Greater scanning speed is also obtained by taking scans from fewer vantage points around the body. One commercial system, for example, uses only 15 scans, 12° apart; another uses 18 scans, 10° apart. Using these techniques, the time for complete scanning of a slice has been reduced to as little as 2½ seconds. Scanners with 100-msec scan times are under development. A modern instrument of this type is shown in Figure 15.13. Some instruments offer a choice of two scanning rates, permitting a trade-off between speed and resolution.

The higher scanning rates now available permit scanning of all sections of the body, since a patient can be asked to hold his or her breath and lie completely still for the few seconds necessary to complete a procedure. By synchronizing scans with the ECG, it is even possible to reconstruct slices of the heart in various phases of the cardiac cycle.

Figure 15.13. Modern computerized axial tomography (CAT) scanner (Courtesy of EMI Medical Inc., Northbrook, IL.)

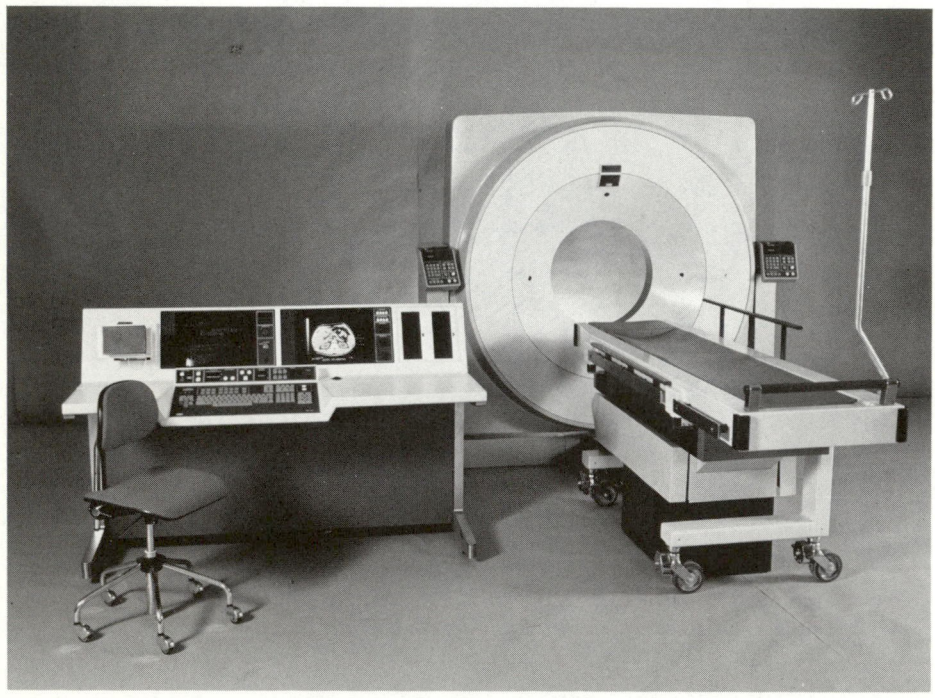

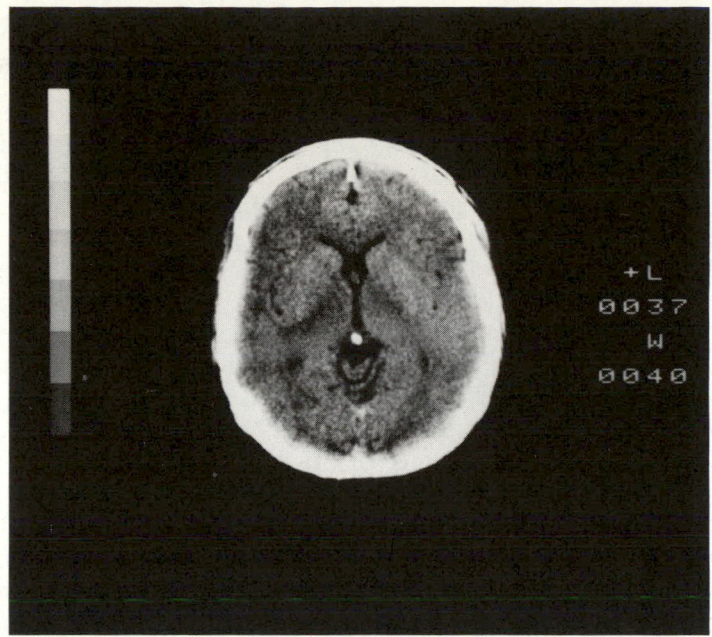

Figure 15.14. Reconstructed image of slice through brain. (a) Non-contrast CT scan of the mid-brain demonstrating the third ventricle, frontal horns of the lateral ventricles, and quadrageminal cistern. (b) Quadrant magnification of scan in (a). (Courtesy of EMI Medical Inc., Northbrook, IL.)

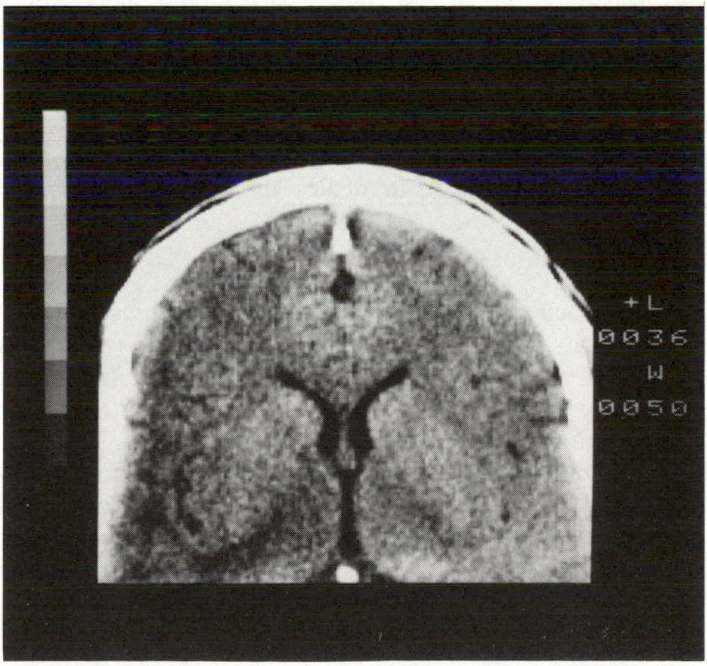

In the computer the cross section to be reconstructed is divided into tiny picture elements called *pixels.* The greater the number of pixels, the greater the resolution. An image of 180 × 180, or a total of 32,400 pixels, is typical. Each pixel is given a value proportional to the X-ray density of that element.

Several different mathematical techniques can be used to construct an image from the set of density patterns obtained during the individual scans. Most involve Fourier transformations and some require iterative operations, both of which are well suited to computer techniques. Digital spatial filters are usually employed to remove the blurring effects of the shadows created by more dense regions. In the final result, each pixel of the computer-generated image is given a degree of brightness proportional to its X-ray density. Figure 15.14 is an example of a reconstructed image of a slice through the brain. Figure 15.15 shows an image of the abdominal region. In some systems, the contrast between regions of different density can be enhanced by assigning each level of brightness a different color on a color TV monitor. This process, called *color enhancement,* provides a further aid in the detection of tumors and other abnormalities that might go unnoticed in a black-and-white display.

Because the CAT scanner can provide information about internal organs and body structures unobtainable by any other available means, and with radiation exposure to the patient no greater than that of conventional X-ray photographs, this instrument brought about a revolution in diagnostic radiology. Its popularity has resulted in scanners being installed in numerous hospitals throughout the United States, Europe, and many other parts of the world. The number of these installations and their high cost (ranging from $250,000 to nearly $1,000,000) have drawn criticism from those who fear technology as a contributor to increasing medical costs. Attempts to regulate the number of scanners on the basis of population have received considerable support. Nonetheless, this instrument is widely regarded as one of the major developments in medical instrumentation in recent years.

15.4.5. Other Computer Applications

The examples discussed in the previous sections represent only a small sample of the many ways in which computers are used in medical instrumentation. Although the proliferation of computers and microprocessors has extended into almost all types of medical instrumentation, a few more specific applications should be mentioned.

In the pulmonary function laboratory, pulmonary function tests and arterial blood gas analysis are often computerized. Measured values of lung volumes, vital capacity, flow rates, FEVs, blood gas levels, and related variables are compared with predicted normal values, based on the height,

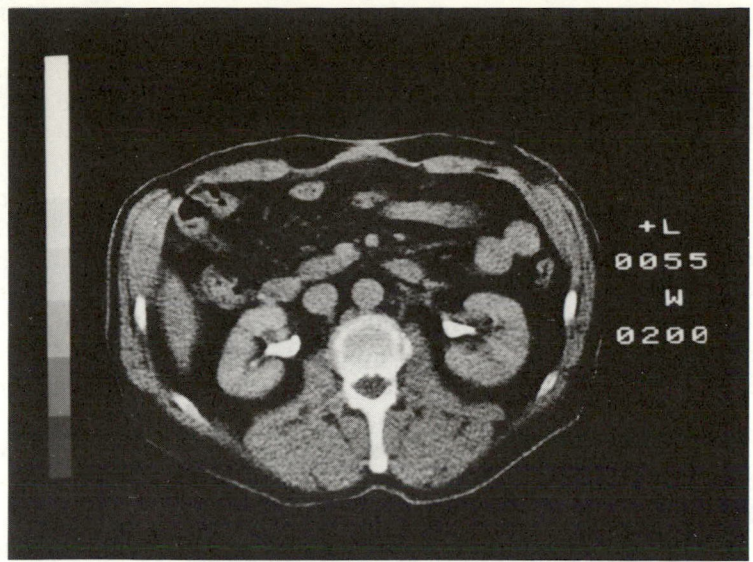

Figure 15.15. Reconstructed image of abdominal slice. (a) CT scan at the mid-renal level; Normal study; Contrast filled renal pelvis is well demonstrated; The infundibula are clearly visualized bilaterally. The vena-cava and aorta are also visible. (b) CT scan of the same patient at a slightly higher level; Shows the lower aspect of the gall bladder and tip of the spleen; Both kidneys are well demonstrated with contrast noted in the collecting system; The left renal vein can be seen in its entirety extending anterior to the aorta and entering the inferior vena-cava. (Courtesy of EMI Medical Inc., Northbrook, IL.)

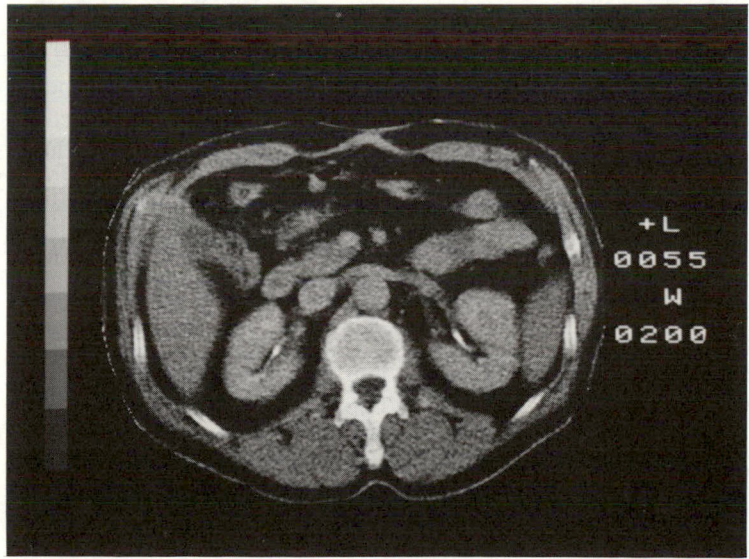

weight, and age of the patient. Variables not directly measurable are calculated and results may be interpreted for the physician. In some systems, each set of measurements is compared with data from previous analyses for determination of trends.

An extension of computerized ECG analysis are various computer-assisted systems for exercise. In such systems preliminary data are gathered to establish a preexercise cardiac template and to search for any contraindications to exercise for the patient. During the exercise, the ECG is monitored to determine the changes in a number of specific features of the waveform and to detect various exercise end-point indicators, such as attainment of a target heart rate, supraventricular tachycardia, a predetermined amount of S-T depression, and certain PVC patterns.

The cardiac catheterization laboratory provides another area in which the computer is able to make a significant contribution. Intracardiac blood pressures and pressure gradients across heart valves, vascular resistance values, and other parameters of importance to the physician in locating and defining cardiovascular abnormalities are measured or calculated using data from one or more catheters within the chambers of the heart. With an on-line computer, results can be obtained almost immediately, giving the physician the assurance that the catheter is in the desired location and often eliminating the need for the patient to return for a repeat of the test.

The success of computerized axial tomography to obtain detailed X-ray images of slices of the body (Section 15.4.4) has led to the development of similar techniques for other forms of imaging. A promising example is *emission computerized tomography,* an application of computerized tomographic techniques to nuclear medicine, which permits detailed visualization of the distribution of radioisotopes throughout the body. As explained in Chapter 14, radioactive isotopes of certain elements can be used to trace the metabolism, pathways, and concentrations of these elements. Through emission computerized tomography, the physician can be provided a detailed three-dimensional distribution map of an isotope which has been injected into the body and allowed to distribute itself. The three-dimensional image is created by taking a number of slice scans, similar to the X-ray slice images obtained by CAT scanner. The instrumentation for emission computerized tomography is more complicated, however. In one configuration the body or section of the body to be imaged is surrounded by 66 sodium iodide detectors, 11 on each side of a hexagonal array. The detectors are scanned sequentially and coincident pulses on opposite sides of the hexagon are detected and counted. The entire array is rotated through 60° during the course of a normal scan. The count of coincident events for each pair of detectors is fed into a computer which, using techniques similar to those employed in CAT scanners, produces a radioactivity map of each slice scanned.

Computerized tomographic methods are being developed for ultrasonic imaging of the heart and abdominal organs. Computer techniques are also involved in *zeugmatography,* a new noninvasive imaging method utilizing the measurement of *nuclear magnetic resonance* (NMR). The benefits to be obtained from these and other new computer applications in medical technology must yet be assessed in light of their costs before their clinical significance can be determined.

16

Electrical Safety
of
Medical Equipment

Each year in the United States about 100,000 people are killed in accidents. About half the fatal accidents occur in motor vehicles, about 20 percent involve falls, and only about 1 percent of the fatalities are caused by electric current, including lightning. The majority of accidental electrocutions occur in industry or on farms. The statistics, which consider medical facilities to be industries, do not specifically show how many of these accidents occur in hospitals, but the number is probably not large. Most electrical accidents, however, are not fatal, but incidents in which staff members or patients receive nonfatal electrical shocks are much more common than the fatality statistics show.

Over the years electrical and electronic equipment has found increasing use in the hospital. Little attention was paid at first to the hazards that this proliferation might create. Some sensational reports published around 1970 on *microshock hazard,* which supposedly had killed a large number of patients in intensive-care units, suddenly drew attention to this subject. While the reports on microshock accidents were frequently anecdotal and no concise statistical analysis ever seems to have been published, growing concern about electrical hazards nevertheless resulted in numerous regulations and standards which attempted to improve electrical safety in the hospital. While some of the requirements have come under attack for unnecessarily increasing the cost of health care, this development has definitely contributed to improved design of electrical and electronic equipment for hospital use.

16.1 PHYSIOLOGICAL EFFECTS OF ELECTRICAL CURRENT

Electrical accidents are caused by the interaction of electric current with the tissues of the body. For an accident to occur, current of sufficient magnitude must flow through the body of the victim in such a way that it impairs the functioning of vital organs. Three conditions have to be met simultaneously [see Figure 16.1(a)]: two contacts must be provided to the body (arbitrarily called first and second contacts), together with a voltage source to drive current through these contacts. The physiological effects of the current depend not only on their magnitude but also on the current pathway through the body, which in turn depends on the location of the

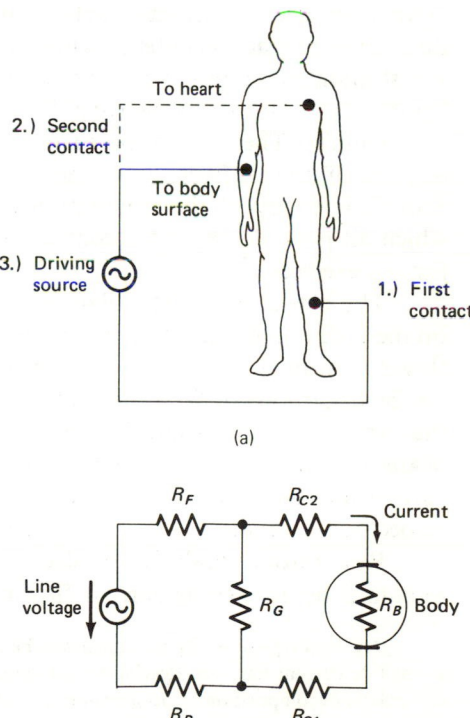

Figure 16.1. The electrical accident. (a) The three necessary conditions. (b) The generalized model where R_F is the fault or leakage resistance, R_{C1} and R_{C2} are first and second contact resistance, R_B is body resistance and R_R is the ground return resistance.

first and second contacts. Two particular situations have to be considered separately: when both contacts are applied to the surface of the body and when one contact is applied directly to the heart. Because the current sensitivity of the heart is much higher in the second case, the effect of current applied directly to the heart is often referred to as *microshock*, while in this context the effect of current applied through surface contacts is called *macroshock*.

Figure 16.1(b) is a generalized model of an electrical accident and will be referred to later in the chapter in various appropriate sections.

Basically, electric current can affect the tissue in two different ways.* First, the electrical energy dissipated in the tissue resistance can cause a temperature increase. If a high enough temperature is reached, tissue damage (burns) can occur. With household current, electrical burns are usually limited to localized damage at or near the contact points, where the density of the current is the greatest. In industrial accidents with high voltage, as well as in lightning accidents, the dissipated electrical energy can be sufficient to cause burns involving larger parts of the body. In *electrosurgery*, the concentrated current from a radio-frequency generator with a frequency of 2.5 or 4 MHz is used to cut tissue or coagulate small blood vessels.

Second, as shown in Chapter 10, the transmission of impulses through sensory and motor nerves involves electrochemical action potentials. An extraneous electric current of sufficient magnitude can cause local voltages that can trigger action potentials and stimulate nerves. When sensory nerves are stimulated in this way, the electric current causes a "tingling" or "prickling" sensation, which at sufficient intensity becomes unpleasant and even painful. The stimulation of motor nerves or muscles causes the contraction of muscle fibers in the muscles or muscle groups affected. A high-enough intensity of the stimulation can cause tetanus of the muscle, in which all possible fibers are contracted, and the maximal possible muscle force is exerted.

The extent of the stimulation of a certain nerve or muscle depends on the potential difference across its cells and the local density of the current flowing through the tissue. An electric current flowing through the body can be hazardous or fatal if it causes local current densities in vital organs that are sufficient to interfere with the functioning of the organs. The degree to which any given organ is affected depends on the magnitude of the current and the location of the electrical contact points on the body with respect to the organ.

Respiratory paralysis can also occur if the muscles of the thorax are tetanized by an electric current flowing through the chest or through the

---

*A third type of injury can sometimes be observed under skin electrodes through which a small dc current has been flowing for an extended time interval. These injuries are due to electrolytic decomposition of perspiration into corrosive substances and are, therefore, actual chemical burns.

respiratory control center of the brain. Such a current is likely to affect the heart also, because of its location.

The organ most susceptible to electric current is the heart. The peculiar characteristics of its muscle fibers cause it to react to electric current differently than other muscles. When the current density within the heart exceeds a certain value, extra systolic contractions first occur. If the current density is increased further, the heart activity stops completely but resumes if the current is removed within a short time. This type of response, however, appears to be limited to a fairly narrow range of current density. An even further increase in current density causes the heart muscle to go into fibrillation. In this state the muscle fibers contract independently and without synchronism, a situation that fails to provide the necessary gross contraction. When the fibrillation occurs in the ventricles (ventricular fibrillation) the heart is unable to pump blood. In human beings (and other large mammals) ventricular fibrillation does not normally revert spontaneously to a normal heart rhythm. Ventricular fibrillation and resulting cessation of blood circulation is the cause of death in the majority of fatal electrical accidents. It can be converted to a regular heart rhythm, however, by the application of a defibrillating current pulse of sufficient magnitude. Such a pulse, applied from a defibrillator (see Section 7.6), causes a momentary contraction of many or all muscle fibers of the heart, which effects a synchronization of their activity. If, in an accidental situation, the heart receives enough current to tetanize the entire myocardium and assuming the current is removed in time, the heart will revert to normal rhythm after cessation of the current.

The magnitude of electric current required to produce a certain physiological effect in a person is influenced by many factors. Figure 16.2 shows the approximate current ranges and the resulting effects for 1-second exposures to various levels of 60-Hz alternating current applied externally to the body. For those physiological effects that involve the heart or respiration, it is assumed that the current is introduced into the body by electrical contact with the extremities in such a way that the current path includes the chest region (arm-to-arm or arm-to-diagonal leg).

For most people, the perception threshold of the skin for light finger contact is approximately 500 μA, although much lower current intensities can be detected with the tongue. With a firm grasp of the hand, the threshold is about 1 mA. A current with an intensity not exceeding 5 mA is generally not considered harmful, although the sensation at this level can be rather unpleasant and painful. When at least one of the contacts with the source of electricity is made by grasping an electrical conductor with the hand, currents in excess of about 10 or 20 mA can tetanize the arm muscles and make it impossible to "let go" of the conductor. The maximum current level a person can tolerate and still voluntarily let go of the conductor is called his *let-go current level*. Ventricular fibrillation can occur at currents

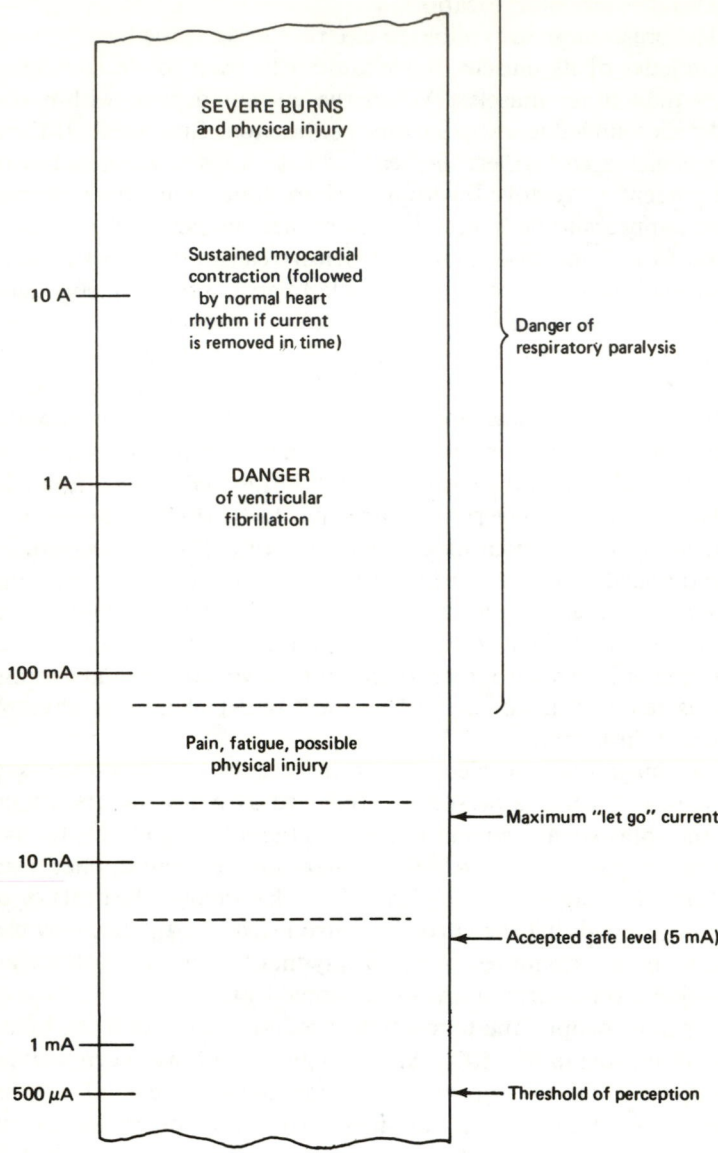

Figure 16.2. Physiological effects of electrical current from 1-second external contact with the body (60 Hz ac).

above about 75 mA, while currents in excess of about 1 or 2 A can cause contraction of the heart, which may revert to normal rhythm if current is discontinued in time. This condition may also be accompanied by respiratory paralysis.

Data on these effects are rare for obvious reasons and are generally limited to accidents in which the magnitude of the current could be reconstructed, or to experimentation with animals. From the data available it appears that the current required to cause ventricular fibrillation increases with the body weight and that a higher current is required if the current is applied for a very short duration. From experiments in the current range of the perception threshold and let-go current, it is known that the effects of the current are almost independent of frequency up to about 1000 Hz. Above that limit, the current must be increased proportionally with the frequency in order to have the same effect. It can be assumed that, at higher current levels, a similar relationship exists between current effects and frequency.

In the foregoing considerations, the electrical intensity is always described in terms of electric current. The voltage required to cause the current flow depends solely on the electrical resistance that the body offers to the current. This resistance is affected by numerous factors and can vary from a few ohms to several megohms. The largest part of the body resistance is normally represented by the resistance of the skin. The inverse of this resistance, the skin conductance, is proportional to the contact area and also depends on the condition of the skin. Intact, dry skin has a conductivity of as low as 2.5 $\mu \upsilon$ cm^2. This low conductivity is caused mainly by the horny, outermost layer of the skin, the epithelium, which provides a natural protection against electrical danger. When this layer is permeated by a conductive fluid, however, the skin conductivity can increase by two orders of magnitude. If the skin is cut, or if conductive objects like hypodermic needles are introduced through the skin, the skin resistance is effectively bypassed. When this situation occurs, the resistance measured between the contacts is determined only by the tissue in the current path, which can be as low as 500 Ω. Electrode paste used in the measurement of bioelectric potentials (see Chapters 4, 6, and 10) reduces the skin resistivity by electrolyte action and mechanical abrasion. Many medical procedures require the introduction of conductive objects into the body, either through natural openings or through incisions in the skin. In many instances, therefore, the hospital patient is deprived of the natural protection against electrical dangers that the skin normally provides. Because of the resulting low resistance, dangerously high currents can be caused by voltages of a magnitude that normally would be rendered safe by the high skin resistance.

In certain medical procedures, a direct contact to the heart may even be established. This contact can occur in three different ways:

1. Electrically conductive catheters are inserted through a vein into the heart to apply stimulating signals from an externally worn pacemaker. Such pacing catheters provide a connection with a resistance of only a few ohms. Patients with such catheters are normally located in the coronary-care or intensive-care unit of the hospital.

2. Fluid-filled catheters provide a conductive pathway only incidentally because the insulating catheter wall retains the current in the conductive fluid that fills the catheter lumen. These catheters provide a current path with a much higher resistance than that of a pacing catheter (0.1 to 2 M Ω, depending on the size and length of the catheter). Fluid-filled catheters are used for a number of medical procedures. For cardiac catheterization—normally performed in a specially equipped X-ray suite—pressures in the heart are measured and blood samples are withdrawn through similar catheters. Similarly, dyes or saline solution are injected and blood samples are withdrawn to determine the cardiac output (see Chapter 6), a procedure that is sometimes even performed at the bedside of patients. In (selective) angiocardiography, catheters are used to inject a radiopaque dye into the heart or the surrounding blood vessels to facilitate their visualization on a series of X-ray photos, often taken in rapid succession (see Chapter 14). This procedure is often performed in the regular X-ray suite.

3. While in the procedures described, a conductive path is created either intentionally or incidentally, a contact to the heart can also be established accidentally without the physician being aware of that fact. This situation can occur when an electrical device (e.g., a thermistor catheter, which is supposed to be insulated, see Chapter 6) has an insulation failure, or when a fluid-filled catheter is inadvertently positioned inside the heart rather than in one of the major veins.

Information on the current necessary to cause ventricular fibrillation when applied directly to the heart was obtained mainly from experiments with dogs, since human data are very limited. While fibrillation has occasionally been observed at currents as low as 20 μA, in most cases the necessary current is much higher.

16.2. SHOCK HAZARDS FROM ELECTRICAL EQUIPMENT

An example of a typical hospital electric-power-distribution system, is shown in somewhat simplified form in Figure 16.3. From the main hospital substation, the power is distributed to individual buildings at 4800 V, usually through underground cables. A stepdown transformer in each building has a secondary winding for 230 V that is center-tapped and thus can provide two circuits of 115 V each. This center tap is grounded to the earth by a connection to a ground rod or water pipe near the building's substation. Heavy electrical devices, such as large air conditioners, ovens, and X-ray machines, operate on 230 V from the two ungrounded terminals of the transformer secondary. Lights and normal wall receptacles receive 115 V through a black "hot" wire from one of the ungrounded terminals of the transformer secondary and a white "neutral" wire that is connected to the grounded center tap, as shown in Figure 16.3.

In order to be exposed to an electrical macroshock hazard, a person must come in contact with both the hot and the neutral conductors simultaneously, or with both hot conductors of a 230-V circuit. However, because

Figure 16.3. Electric power distribution system (simplified).

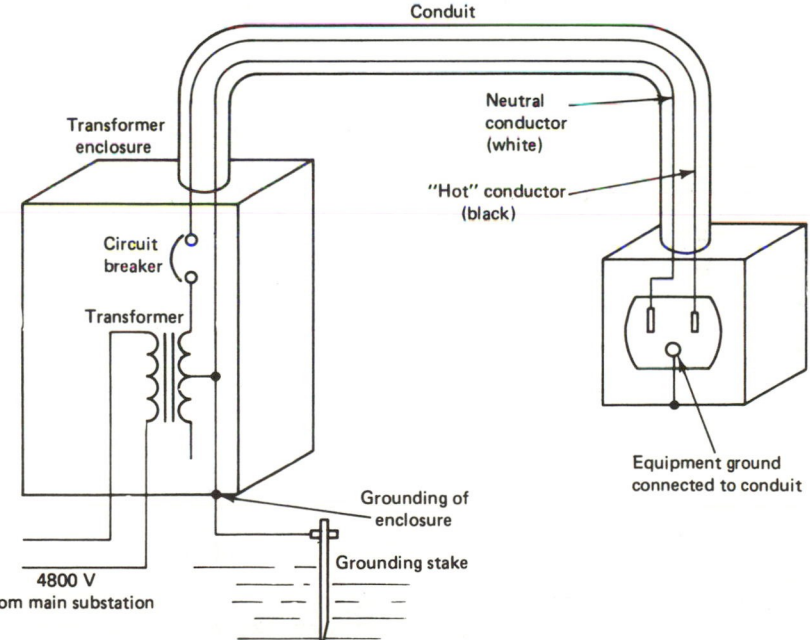

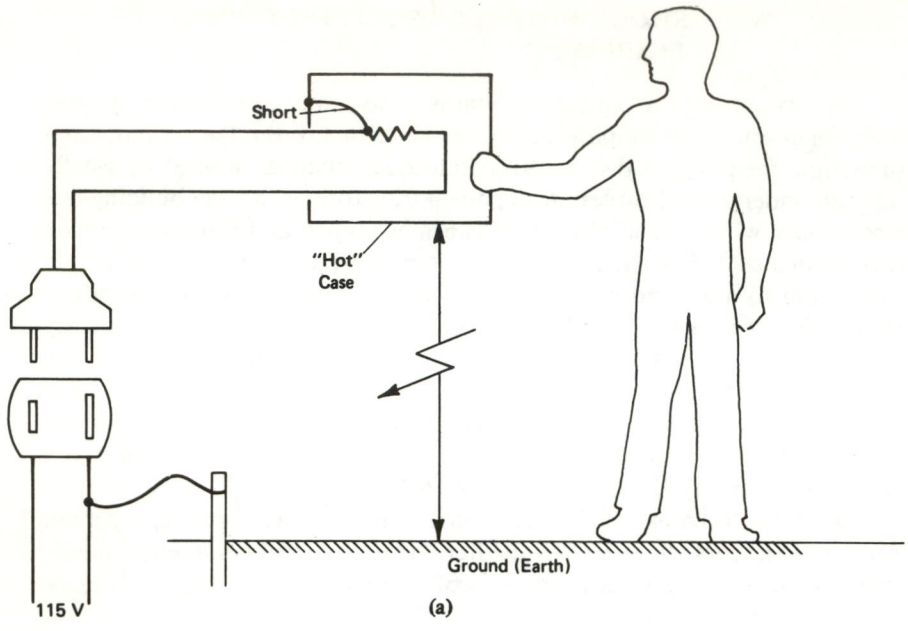

Figure 16.4. Ground shock hazards.

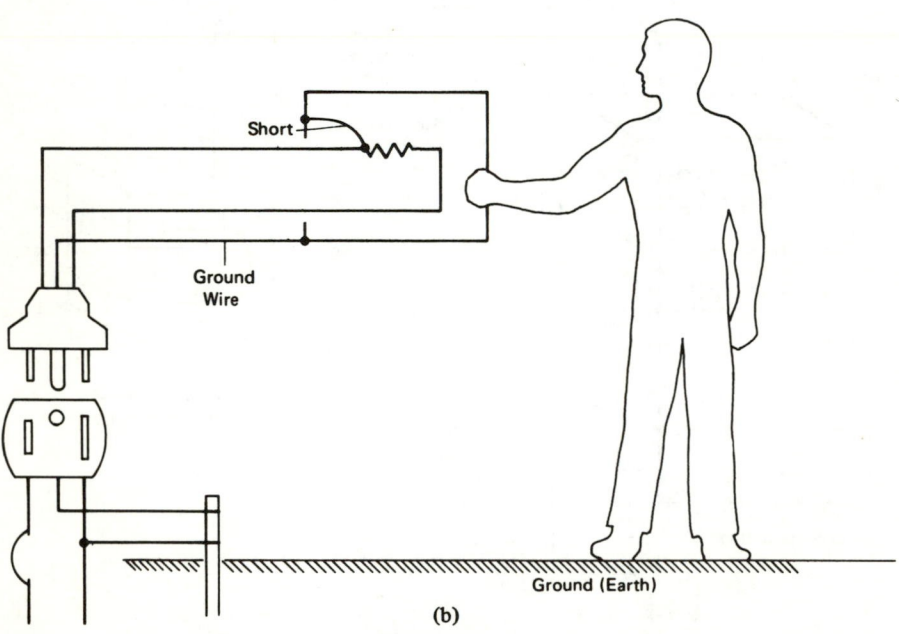

the neutral wire is connected to ground, the same shock hazard exists between the hot wire and any conductive object that is in any way connected to ground. Included would be such items as a room radiator, water pipes, or metallic building structures. In the design of electrical equipment, great care is taken to prevent personnel from accidentally contacting the hot wire by the use of suitable insulating materials and the observation of safe distances between conductors and equipment cases. Through insulation breakdown, wear, and mechanical damage, however, contact between a hot wire and an equipment case can accidentally occur.

Figure 16.4(a) shows the scenario of such an accident. A defect in the equipment has caused a short between the hot wire of the line cord and the (conductive) equipment case, placing the case at a potential of 115 V ac with respect to ground. A user whose body is in contact with ground (the first contact of Figure 16.1) will be placed in jeopardy when a (second) contact between his body and the case of the faulty equipment is established. The generalized model for electrical accidents, shown in Figure 16.1(b), permits a more detailed analysis of the situation. The model represents a network consisting of a voltage source and six resistances. The *fault resistance* (or leakage resistance), R_F, represents the short between the hot conductor and the case of the equipment. The *first and second contact resistance,* R_{C1} and R_{C2}, represent, respectively, the resistances of the first and second contacts to the body of the accident victim. Together with the *body resistance,* R_B, they form the resistance of the current path through the victim's body. The *grounding resistance, R_G* (which in Figure 16.4 is infinitely large), is connected in parallel with the current path through the body. The *ground return resistance, R_R,* is essentially the resistance between ground and the center tap of the transformer shown in Figure 16.3. This resistance is normally very small.

An electrical accident can occur when the six resistances shown in the figure assume any combination of values such that the resulting current through the body of the victim reaches a dangerous magnitude. All measures taken to reduce the probability of electrical accidents are, in effect, attempts to manipulate the value of one or more of the resistances.

16.3. METHODS OF ACCIDENT PREVENTION

In order to reduce the likelihood of electrical accidents, a number of protective methods have evolved. Some are used universally, some are required in areas that are generally considered especially hazardous, and still others have been developed essentially for use in hospitals.

16.3.1. Grounding

The protection method used most frequently is proper *grounding* of equipment. The principle of this method is to make the grounding resistance R_G in Figure 16.1(b) small enough that for all possible values of the fault resistance R_F, the majority of the fault current bypasses the body of the victim and the body current remains at a safe level even if contact and body resistances are small. The practical implementation of this method is shown in Figure 16.4(b), where the metal case of the equipment is connected to ground by a separate wire. In cord-connected electrical equipment this ground connection is established by the third, round, or U-shaped contact in the plug. If a short occurs in a device whose case has been grounded in this way, the electric current flows through the short to the case and returns to the substation through the ground wire. Ideally, the short circuit will result in sufficient current to cause the circuit breaker to trip immediately. This action would remove the power from the faulty piece of equipment and thus limit the hazard.

Protection by grounding, however, has several shortcomings. Obviously, it is effective only as long as a good ground connection exists. Experience has shown that many receptacles, plugs, and line cords of the conventional type do not hold up under the conditions of hospital use. Many manufacturers now make available *Hospital Grade* receptacles and plugs which are designed to pass a strict test required by the Underwriters Laboratory for devices to qualify for this specification. Hospital Grade plugs and receptacles are marked by a green dot.

A second disadvantage is that in the case of a short, protection is provided by removing the power from the defective device by tripping the circuit breaker. This action, however, also removes the power from all other devices connected to the same branch circuit. In a hospital setting, one defective device could disable a number of other devices, which might include life-saving instruments.

16.3.2. Double Insulation

In double-insulated equipment the case is made of nonconductive material, usually a suitable plastic. If accessible metal parts are used, they are attached to the conductive main body of the equipment through a separate (protective) layer of insulation in addition to the (functional) insulation that separates this body from the electrical parts.

The intention of this method is to assure that the fault resistance R_F is always very large. Double-insulated equipment need not be grounded, and therefore it is usually equipped with a plug that does not have a ground pin. Equipment of this type must be labeled "Double Insulated." Double insulation is now widely used as a method of protection in hand-held power tools and electric-powered garden equipment such as lawn mowers. However, double insulation is of only limited value for equipment found in a hospital environment. Unless the equipment is also designed to be waterproof, the double insulation can easily be rendered ineffective if a conductive fluid such as saline or urine is spilled over the equipment or if the equipment is submerged in such a fluid.

16.3.3. Protection by Low Voltage

In the generalized accident model of Figure 16.1(b) it was assumed that the voltage source was the line voltage (115 or 230 V ac). If, instead, another voltage source were used, and if the voltage of this source could be made small enough, the body resistance R_B would be sufficient to limit the body current to a safe value, even if the fault and contact resistances become very small. One way of creating this situation is to operate the equipment from batteries. Aside from its lower voltage there is the additional advantage that battery-operated equipment does not have to be grounded. Normally, battery operation is limited to small devices such as flashlights and razors, but occasionally equipment as large as portable X-ray machines may use this method of protection. A low operating voltage can also be obtained by means of a step-down transformer. In addition to lowering the voltage the transformer provides isolation of the supply voltage from ground. Where power requirements are small, the transformer can be made an integral part of the line plug, a design now frequently employed in small electronic equipment as well as in such medical devices as ophthalmoscopes and endoscopes.

16.3.4. Ground-Fault Circuit Interrupter

Statistical evidence indicates that most electrical accidents are of the type in which the body of the victim provides a conductive path to ground, as shown in Figure 16.4. Normally all current that enters a device through the hot wire returns through the neutral wire. However, in the case of such an accident, part of the current actually returns through the body of the victim and through ground. In the *ground fault circuit interrupter,* the difference between the currents in the hot and neutral wires of the power line is monitored by a differential transformer and an electronic amplifier. If this difference exceeds a certain value, usually 5 mA, the power is interrupted by a circuit breaker. This interruption occurs so rapidly that, even in the case of a large current flow through the body of a victim, no harmful effects are encountered.

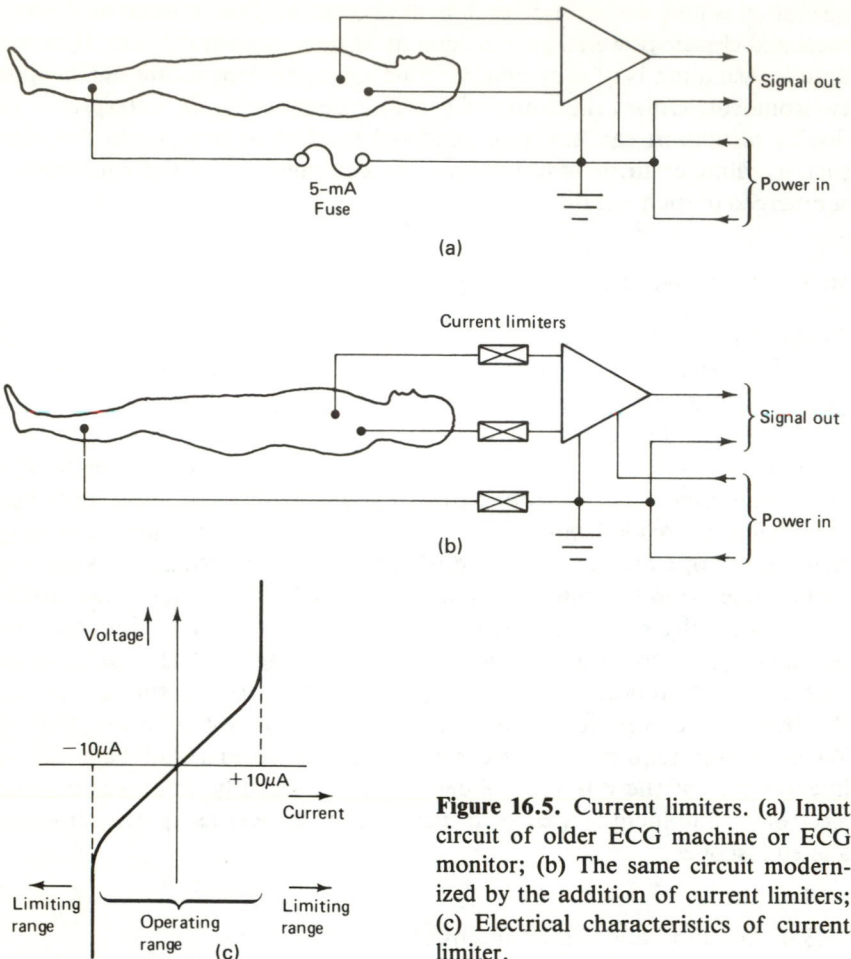

Figure 16.5. Current limiters. (a) Input circuit of older ECG machine or ECG monitor; (b) The same circuit modernized by the addition of current limiters; (c) Electrical characteristics of current limiter.

16.3.5. Isolation of Patient-Connected Parts

Many types of medical equipment require that an electrical connection be established to the body of the patient, either to measure electrical potentials, such as in ECG machines, or to apply electrical signals, such as in electrical pacemakers. These electrical connections, however, could also serve as a path for dangerous electrical currents should the equipment malfunction. For example, in older ECG machines and patient monitors, it was common practice to connect one of the patient leads (the RL lead) to a power-line ground. This effectively grounded the patient and established one of the two connections necessary for an electrical accident. Modern technology makes it possible to design circuits that isolate the patient leads from ground. For patient leads that connect to an amplifier, this isolation is most commonly achieved by the use of an isolated input amplifier, as shown in Figure 16.5. This type of amplifier is completely isolated from the rest of the equipment, with the power provided through a low-capacitance transformer. A second transformer is used to couple the amplified signal to the rest of the equipment. Because signal transformers are difficult to design for the frequency range of biological signals, a modulation scheme is normally employed. The amplifier shown in the figure uses amplitude modulation of the carrier signal used to provide power for the isolated amplifier. Other designs use frequency modulation.

Figure 16.6. Input circuit of modern ECG machine or ECG monitor with isolated patient leads achieved by the use of a carrier amplifier.

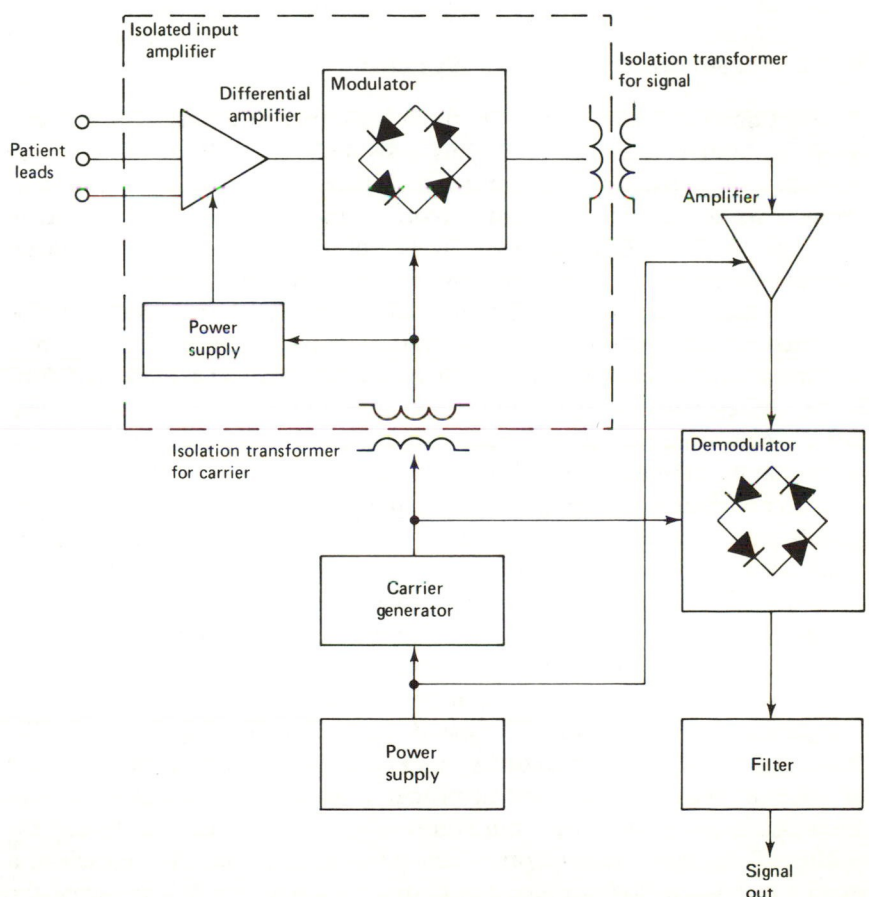

Occasionally, isolation protection is provided by connecting a *current limiter* into each patient lead. The characteristics of these devices are shown in Figure 16.6. For low currents these devices act as resistors, but when a certain current level is approached they change their characteristics and prevent the current from exceeding a predetermined limit. Although current limiters are less desirable than isolated amplifiers, they are nevertheless used where many patient leads have to be protected, such as in EEG machines.

In biomedical devices that provide electrical energy to the body of a patient, such as pacemakers or electrosurgical devices, protection is achieved by isolating the patient leads from ground. In pacemakers, this is now normally accomplished by using only battery-operated types. Modern electrosurgical devices use output transformers to isolate patient leads.

Every one of the methods described in this section is concerned with making the contact resistances R_{C1} and R_{C2} in Figure 16.1(b) very large.

16.3.6 Isolated Power Distribution Systems

As mentioned earlier, the ground return resistance of a normal power distribution system is very low. If this resistance could be made large by operating the substation transformer of Figure 16.3 without grounding its center tap, all electrical accidents involving ground contact of the victim could be avoided. Unfortunately, it is not possible to operate general purpose electrical distribution systems in this way. Special power distribution systems which serve a limited number of devices and receptacles can be operated through transformers with ungrounded secondaries, however, and an increased safety margin can result from their use. As a matter of fact, in the United States, safety standards require that all "anesthetizing locations" (operating rooms and other rooms in which gaseous anesthetizing agents are used) be equipped with such power distribution systems.

In an isolated distribution system, the power is not supplied from the transformer substation directly, but is obtained from a separate isolation transformer for each operating room. This transformer, together with the associated circuit breaker and the line isolation monitor described below, is mounted in a separate enclosure, either in the operating room or adjacent to it. The panel of such an installation is shown in Figure 16.7.

If a short between the case and one of the two wires occurs in a piece of equipment powered from an isolated system, the result will be quite different from that of the grounded system described earlier. Even if the case of the equipment is not grounded properly, someone touching the equipment and a grounded object simultaneously will not receive a shock, for neither of the power conductors is connected to a ground. Nevertheless, a small current can flow through the body of such a person because of the

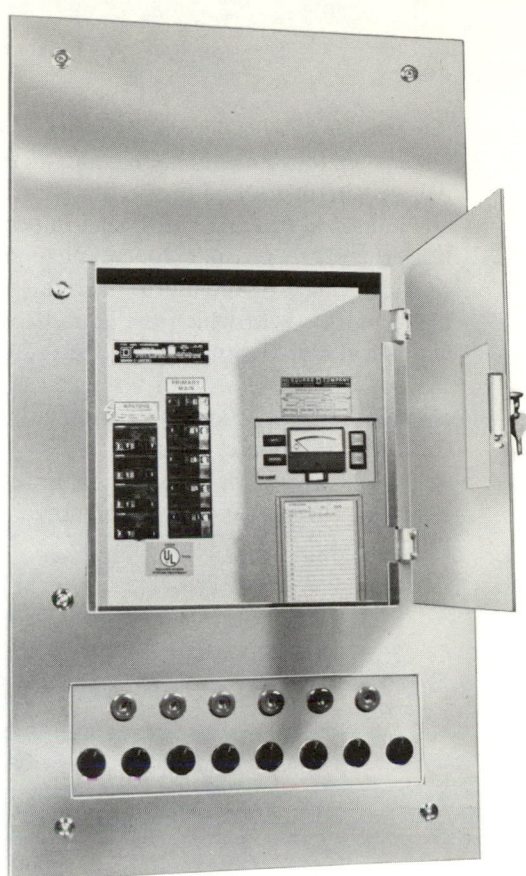

Figure 16.7. Panel of isolated power distribution system. (Courtesy of Sorgel Electric Corporation, subsidiary of Square D Company, Oshkosh, WI.)

capacity between the conductors of the system and ground. This current, however, will be of a magnitude of at most 1 or 2 mA, which may be perceived without being harmful. If the equipment in which the short occurs is properly grounded, this *leakage current* will return through the ground connection. In this case, however, the short in the faulty equipment effectively grounds one of the conductors of the isolated distribution system. As a result, the isolated system is changed back to a grounded distribution system and all the protection provided by the isolated system is obviated. In order to provide a warning in the event that this situation occurs, isolated power systems employ *line isolation monitors* (LIM). This device alternately checks the two wires of the distribution system for isolation from ground. The degree of isolation, expressed as the *risk current* or *fault hazard current,* is indicated on an electric meter.

In addition to the meter, two warning lamps are provided. When the system is adequately isolated, a green lamp (sometimes labeled "SAFE") will be on. If the isolation begins to deteriorate or if a short occurs between one of the wires and ground anywhere in the system, a red lamp (sometimes labeled "HAZARD") will light up. At the same time, an acoustical alarm will begin to sound. This hazard alarm merely indicates that the system has lost its protective properties. It still requires a second fault before an actual hazard can arise. If the alarm occurs while an operation is in progress, it is therefore possible to complete the procedure before attempting to find the cause of the alarm. For this situation, the line isolation monitor has a button with which the acoustical alarm can be silenced. Even if the acoustical alarm is turned off, the red warning lamp remains on, indicating the continued presence of an alarm condition. The line isolation monitor also has a button that allows it to be tested for proper functioning. Pressing this button simulates a short.

Receptacles powered from an isolated system are not always the common three-prong type but may be a special locking type, shown in Figure 16.8.

Figure 16.8. Locking plug for isolated power distribution system. (Courtesy of Veterans Administration Biomedical Engineering and Computing Center, Sepulveda, CA.)

In addition to the isolated distribution system, a special high-quality grounding system is also required for all anesthetizing locations. This system not only protects the patient and staff by shunting all leakage currents to the ground but is also necessary for the proper functioning of the line isolation monitor. The special grounding system is called an *equipotential grounding system* because it keeps all metallic objects in the area that could possibly come in contact with staff or patients at the same electrical potential. For this purpose, not only the enclosures of electrical equipment but also all other metal objects—operating tables, anesthesia machines, and instrument tables—that might come in contact with electrical equipment must be interconnected by the grounding system. Portable items of this kind may require the use of separate ground wires connected to a common grounding point near the head of the operating table. Special bayonet-type plugs are used on this ground wire. Similar equipotential grounding systems are also required in intensive-care units. Such a system is shown in Figure 16.9.

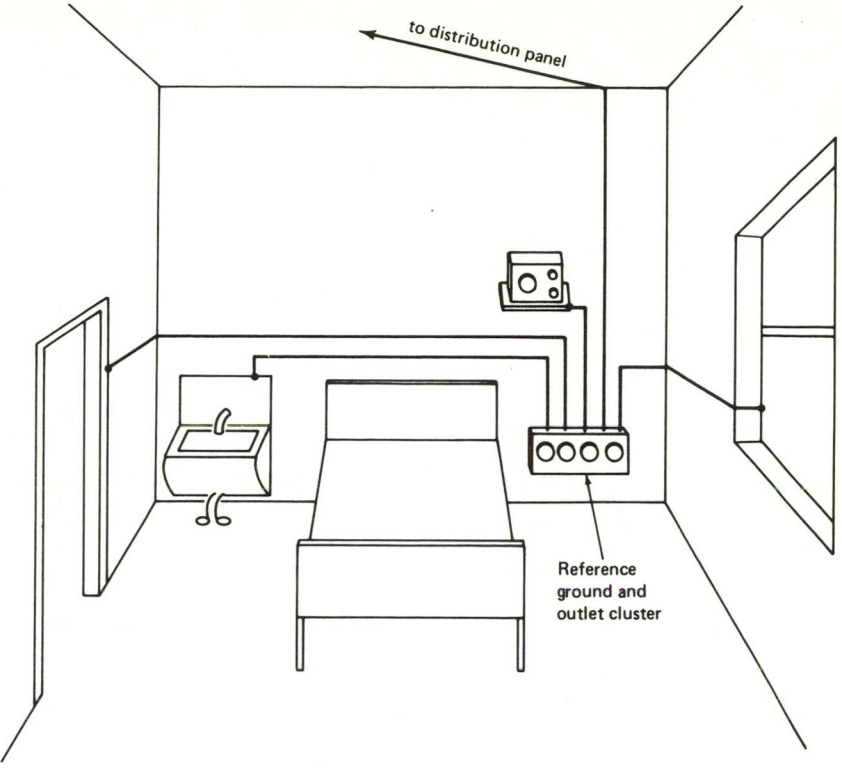

Figure 16.9. Principle of an equipotential grounding system in one room or cubicle of an intensive care or cardiac care unit.

Appendices

A

Medical Terminology and Glossary

A.1. MEDICAL TERMINOLOGY

One of the problems of an interdisciplinary field is that of communication between the disciplinary components that make up the field. Engineers and technicians have to learn enough physiology, anatomy, and medical terminology to be able to discuss problems intelligently with members of the medical profession.

The typical technical person faces enough difficulty with language, but when confronted with medical terminology, his or her problems are compounded. However, with a few simple rules, medical terminology can be understood more easily. Most medical words have either a Latin or Greek origin, or, as in engineering, chemistry, and physics, the surnames of prominent researchers are used.

Most words consist of a root or base which is modified by a prefix or suffix or both. The root is often abbreviated when the prefix or suffix is added.

The following list gives some of the more common roots, prefixes, and suffixes.

PREFIXES

a	without or not	*mal*	bad
ab	away from	*medio*	middle
ad	to, toward	*mes*	middle
an	absence of	*meta*	beyond, over
ante	before	*micro*	small
antero	in front	*ortho*	straight, correct
anti	against	*para*	beside
bi	two	*patho*	disease
brady	slow	*peri*	outside, around
dia	through	*poly*	many
dys	difficult, painful	*pseudo*	false
endo	within	*quadri*	fourfold
epi	upon	*retro*	backward
eu	well, good	*sub*	beneath
ex	away from	*supra*	above
exo	outside	*tachy*	fast
hyper	over	*trans*	across
hypo	under or less	*tri*	three
infra	below	*ultra*	beyond
intra	within	*uni*	single, one

ROOTS

aden	gland	*gaster*	stomach
arteria	artery	*haemo* or *hemo*	blood
arthros	joint	*hepar*	liver
auris	ear	*hydro*	water
brachion	arm	*hystera*	womb
bronchus	windpipe	*kystis (cysto)*	bladder
cardium	heart	*larynx*	throat
cephalos	brain	*myelos*	marrow
cholecyst	gallbladder	*nasus*	nose
colon	intestine	*nephros*	kidney
costa	rib	*neuron*	neuron
cranium	head	*odons*	tooth
derma	skin	*odynia*	pain
enteron	intestine	*optikas*	eye
epithelium	skin	*os*	bone
esophagus	gullet	*osteon*	bone
ostium	mouth, orifice	*pyretos*	fever
otis	ear	*ren*	kidney
pes	foot	*rhin*	nose

pharynx	throat	*rhythmos*	rhythm
phlebos	vein	*spondylos*	vertebra
pleura	chest	*stoma*	mouth
pneumones	lungs	*thorax*	chest
psyche	mind	*trachea*	windpipe
pulmones	lungs	*trophe*	nutrition
pyelos	pelvis	*vene*	vein
pyon	pus	*vesica*	bladder

SUFFIXES

algia	pain	*emia*	blood
centeses	puncture	*iasis*	a process
clasia	remedy	*itis*	inflammation
ectasis	dilatation	*oma*	swelling, tumor
ectomy	cut	*sclerosis*	hardening
edema	swelling		

ia is also used as a suffix in many combinations and it indicates a state or condition.

Examples of how the words are formed are easily illustrated. Arteriosclerosis means hardening of the arteries. The heart is the *cardium* and the loose sac in which it is contained is called the *pericardium* (outside the heart). If the pericardium is diseased, it is called *pericarditis.* Note that some letters are dropped or changed for the new word, but the construction is easily recognizable.

Another example would be the root *trophe,* literally meaning nutrition. The prefix *a* means absence of, and *hyper* means over. Therefore, *atrophy* is to waste away, and *hypertrophy* is to enlarge.

English usage is sometimes peculiar and utilizes the Greek and Latin words together. An example is the kidney—*ren* in Latin and *nephros* in Greek. We talk about kidney function as *renal* function, but inflammation of the kidney is *nephritis.*

Descriptions for relative position are frequently used in medical usage. These are:

anterior	situated in front of; forward part of.
distal	away from the center of the body.
dorsal	a position more toward the back of an object of reference.
frontal	situated at the front.
inferior	situated or directed below.
lateral	a position more toward the side of flank.
proximal	toward the center of the body.
sagital	relating to the median plane of the body or any plane parallel to it.
superior	situated or directed above.

A.2. MEDICAL GLOSSARY

Throughout this book many biomedical words have been used which are possibly unfamiliar to the reader. To help achieve a better understanding, the following glossary of medical terms is presented in alphabetical order for easy reference. There are many sources for these definitions, including the authors' own interpretation, but among well-known reference books used are various Webster's dictionaries (G. & C. Merriam Co., Springfield, Mass.) and *Dorland's Illustrated Medical Dictionary,* 25th ed. (W. B. Saunders Company, Philadelphia, 1974).

A

acetylcholine- a reversible acetic acid ester of choline having important physiological functions, such as the transmission of a nerve impulse across a synapse.

acidosis- a condition of lowered blood bicarbonate (decreased pH).

afferent- conveying toward the center or toward the brain.

alkalosis- a condition of increased blood bicarbonate (increased pH).

alveoli- air sacs in the lungs formed at the terminals of a bronchiole. It is through a thin membrane (0.001 mm thick) in the alveoli that the oxygen enters the bloodstream.

anaerobic- growing only in the absence of molecular oxygen.

anoxic- oxygen insufficient to support life.

aorta- the great trunk artery that carries blood from the heart to be distributed by branch arteries through the body.

aortic valve- outlet valve from left ventricle to the aorta.

apnea- absence of breathing.

arrhythmia- an alteration in rhythm of the heartbeat either in time or force.

arteriole- one of the small terminal twigs of an artery that ends in capillaries.

artery- a vessel through which the blood is pumped away from the heart.

atrioventricular- located between an atrium and ventricle of the heart.

atrium- an anatomical cavity or passage; especially a main chamber of the heart into which blood returns from circulation.

auscultation- the act of listening for sounds in the body.

autonomic- acting independently of volition; relating to, affecting, or controlled by the autonomic nervous system.

axon- a usually long and single nerve-cell process that, as a rule, conducts impulses away from the cell body of a neuron.

B

baroreceptors- nerve receptors in the blood vessels, especially the carotid sinus, sensitive to blood pressure.

bifurcation- branching, as in blood vessels.

bioelectricity- the electrical phenomena that appear in living tissues.

brachial- relating to the arm or a comparable process.

bradycardia-   a slow heart rate.

bronchus-   either of two primary divisions of the trachea that lead, respectively, into the right and the left lung; broadly, bronchial tube.

bundle of His-   a small band of cardiac muscle fibers transmitting the waves of depolarization from the atria to the ventricles during cardiac contraction.

C

cannula-   a small tube for insertion into a body cavity or blood vessel.

capacity, functional residual-   the volume of gas remaining in the lungs at the resting expiratory level. The resting end-expiratory level is used as the baseline because it varies less than the end-inspiratory position.

capacity, inspiratory-   the maximal volume of gas that can be inspired from the resting expiratory level.

capillaries-   any of the smallest vessels of the blood-vascular system connecting arterioles with venules and forming networks throughout.

cardiac-   pertaining to the heart.

cardiac arrest-   standstill of normal heartbeat.

cardiology-   the study of the heart, its action and diseases.

cardiovascular-   relating to the heart and blood vessels.

catheter-   a tubular medical device inserted into canals, vessels, passageways, or body cavities, usually to permit injection or withdrawal of fluids or to keep a passage open.

cell-   a small, usually microscopic, mass of protoplasm bounded externally by a semipermeable membrane, usually including one or more nuclei and various nonliving products, capable (alone or interacting with other cells) of performing all the fundamental functions of life, and of forming the least structural aggregate of living matter capable of functioning as an independent unit.

cerebellum-   a large, dorsally projecting part of the brain especially concerned with the coordination of muscles and the maintenance of bodily equilibrium.

cerebrum-   the enlarged anterior or upper part of the brain.

computerized axial tomography-   a technique combining x-ray and computer technology for visualization of internal organs and body structures.

coronary artery and sinus-   vessels carrying blood to and from the walls of the heart itself.

cortex-   the outer or superficial part of an organ or body structure; especially the outer layer of gray matter of the cerebrum and cerebellum.

cortical-   of, relating to, or consisting of the cortex.

cranium-   the part of the head that encloses the brain.

cytoplasm-   the protoplasm of a cell exclusive of that of the nucleus.

D

defibrillation-   the correction of fibrillation of the heart.

defibrillator-   an apparatus used to counteract fibrillation (very rapid irregular contractions of the muscle fibers of the heart) by application of electric impulses to the heart.

dendrite- any of the usual branching protoplasmic processes that conduct impulses toward the body of a nerve cell.

depolarize- to cause to become partially or wholly unpolarized.

diastole- a rhythmically recurrent expansion, especially the dilatation of the cavities of the heart as they fill with blood.

diastolic- of or pertaining to the diastole (e.g., diastolic blood pressure).

dicrotic- having a double beat; being or relating to the second expansion of the artery that occurs during the diastole of the heart (hence dicrotic notch in the blood pressure wave).

dyspnea- difficulty in breathing.

E

ECG- abbreviation for electrocardiogram.

echocardiogram- an ultrasonic record of the dimension and movement of the heart and its valves.

ectopic- located away from the normal position.

EEG- abbreviation for electroencephalogram.

efferent- conveying away from a center.

electrocardiogram- a record of the electrical activity of the heart.

electrocardiograph- an instrument used for the measurement of the electrical activity of the heart.

electrode- a device used to interface ionic potentials and currents.

electroencephalogram- the tracing of brain waves made by an electroencephalograph.

electroencephalograph- an instrument for measuring and recording electrical activity from the brain (brain waves).

electrolyte- a nonmetallic electric conductor in which current is carried by the movement of ions.

electromyogram- the tracing of muscular action potentials by an electromyograph.

electromyograph- an instrument for measurement of muscle potentials.

electromyography- the recording of the changes in electric potential of muscle.

electrophysiology- the science of physiology in its relations to electricity; the study of the electric reactions of the body in health.

embolus- an abnormal particle (air, clot, or fat) circulating in the blood.

embryo- a human or animal offspring prior to emergence from the womb or egg; hence, a beginning or undeveloped stage of anything.

EMG- abbreviation for electromyography.

epilepsy- any of a variety of disorders marked by disturbed electrical rhythms of the central nervous system, typically manifested by convulsive attacks, usually with clouding of consciousness.

expiratory reserve volume- that volume capable of being expired at the end-expiratory level of a quiet expiration.

external respiration- movement of gases in and out of lungs.

extracellular- situated or occurring outside a cell or the cells of the body.

extracorporeal- situated or occurring outside the body.

extrasystole- premature contraction of the heart independent of normal rhythm.

F

fibrillation- spontaneous contraction of individual muscle fibers; specifically, nonsynchronized activity of the heart.

fluoroscopy- process of using an instrument to observe the internal structure of an opaque object (as the living body) by means of X rays.

forced expiratory flow- ($FEF_{200-1200}$) the average rate of flow for a specified portion of the forced expiratory volume, usually between 200 and 1200 ml (formerly called maximum expiratory flow rate).

forced expiratory volume- (qualified by the subscript indicating time interval in seconds, FEV_T—e.g., FEV_{10})- the volume of gas exhaled over a given time interval during the performance of a forced vital capacity. FEV can be expressed as a percentage of the forced vital capacity ($FEV_{T\%}$).

forced midexpiratory flow ($FEF_{25-75\%}$)- the average rate of flow during the middle half of the forced expiratory volume.

forced vital capacity- (FVC)- the maximum volume of gas that can be expelled as forcefully and rapidly as possible after maximum inspiration.

functional residual capcity- *see* Capacity, functional residual.

G

galvanic- uninterrupted current derived from a chemical battery.

ganglion- any collection or mass of nerve cells outside the central nervous system that serves as a center of nervous influence.

H

heart block- a delay or interference of the conduction mechanism whereby impulses do not go through all or a major part of the myocardium.

heparin- an acid occurring in tissues, mostly in the liver. It can be produced chemically and can make the blood incoaguable if injected into the bloodstream intravenously.

hyperventilation- abnormally prolonged, rapid deep breathing or overbreathing.

hypoventilation- decrease of air in the lungs below the normal amount.

hypoxia- lack of oxygen.

I

infarct- an area of necrosis in a tissue or organ resulting from obstruction of the local circulation by a thrombus or embolus.

inferior vena cava- main vein feeding back to the heart from systemic circulation below the heart.

inspiratory capacity- *see* Capacity, inspiratory.

inspiratory reserve volume- maximal volume of gas that can be inspired from the end-inspiratory position.

ion- an atom or group of atoms that carries a positive or negative electric charge as a result of having lost or gained one or more electrons.

ischemic- a localized anemia due to an obstructed circulation.

isoelectric- uniformly electric throughout; having the same electric potential, and hence giving off no current.

isometric- having the same length: a muscle acts isometrically when it applies a force without changing its length.

isotonic- having the same tone: a muscle acts isotonically when it changes length without appreciably changing the force it exerts.

isotropic- exhibiting properties with the same values when measured along axes in all directions.

K

Korotkoff sounds- sounds produced by sudden pulsation of blood being forced through a partially occluded artery and heard during auscultatory blood pressure determination.

L

latency- time delay between stimulus and response.

lobe- a somewhat rounded projection or division of a body organ or part.

lumen- the cavity of a tubular organ or instrument.

lung capacity, total- the amount of gas contained in the lung at the end of maximal inspiration.

M

maximal breathing capacity- same as maximal voluntary ventilation.

maximal voluntary ventilation- the volume of air that a subject can breathe with maximal effort over a given time interval.

membrane- a thin layer of tissue that covers a surface or divides a space or organ.

metabolism- the sum of all the physical and chemical processes by which the living organized substance is produced and maintained.

mitral stenosis- a narrowing of the left atrioventricular orifice.

mitral valve- valve between the left atrium and ventricle of the heart.

motor- a muscle, nerve, or center that effects or produces movement.

myelin- the fat-like substance forming a sheath around certain nerve fibers.

myocardium- the walls of the chamber of the heart which contain the musculature that acts during the pumping of blood.

myograph- an apparatus for recording the effects of a muscular contraction.

N

necrosis- death of tissue, usually as individual cells, groups of cells, or in small localized areas.

nerve- a cord-like structure that conveys impulses from one part of the body to another. A nerve consists of a bundle of nerve fibers either efferent or afferent or both.

neuron- a nerve cell with its processes, collaterals, and terminations—regarded as a structural unit of the nervous system.

nodes of Ranvier- nodes produced by constrictions of the myelin sheath of a nerve fiber at intervals at about 1 mm.

O

oxyhemoglobin- a compound of oxygen and hemoglobin formed in the lungs—the means whereby oxygen is carried through the arteries to the body tissues.

P

partial pressure of oxygen in air- the pressure of the oxygen contained in air. Since air is about 21 percent oxygen, partial pressure is 21 percent of 760 mm of mercury, or 159 mm Hg. That is, oxygen needs can be supplied by pure oxygen at 159 mm Hg, which is equivalent to breathing air at 760 mm Hg P_{O_2} (at sea level).

perfuse- to pour over or through.

permeate- to pass through the pores or interstices.

plethysmography- the recording of the changes in the volume of a body part as modified by the circulation of the blood in it.

pneumograph- an instrument for recording the thoracic movements or volume change during respiration.

prosthesis- an artificial substitute for a missing or diseased part.

pulmonary- relating to, functioning like, or associated with the lungs.

pulmonary atelectasis- lung collapse.

pulmonary minute volume (pulmonary ventilation)- volume of air respired per minute = tidal volume × breaths/min.

pulse pressure- the difference between systolic and diastolic blood pressure (usually about 40 mm Hg).

R

radioisotope- an isotope that is radioactive, produced artifically from the element by the action of neutrons, protons, deuterons, or alpha particles in the chain-reacting pile or in the cyclotron. Radioisotopes are used as tracers or indicators by being added to the stable compound under observation, so that the course of the latter in the body (human or animal) can be detected and followed by the radioactivity thus added to it. The stable element so treated is said to be "labeled" or "tagged."

residual capacity- *see* Capacity, residual functional.

residual volume- air left in the lungs after deep exhale (about 1.2 liters).

respiratory center- the center in the medulla oblongata that controls breathing.

respiratory quotient- ratio of volume of exhaled CO_2 to the volume of consumed O_2 (0.85).

S

semilunar pulmonary valve- outlet valve from the right ventricle into the pulmonary artery.

sinoatrial node- the pacemaker of the heart—a microscopic collection of atypical cardiac muscle fibers which is responsible for initiating each cycle or cardiac contraction.

sphygmomanometer- an instrument for measuring blood pressure, especially arterial blood pressure.

spirometer- an instrument for measuring the air entering and leaving the lungs.

stenosis- narrowing of a duct or canal.

stroke volume- amount of blood pumped during each heartbeat (diastolic volume of the ventricle minus the volume of blood in the ventricle at the end of systole).

superior vena cava- main vein feeding back to the heart from systemic circulation above the heart.

synapse- the point at which a nervous impulse passes from one neuron to another.

systemic- pertaining to or affecting the body as a whole.

systole- the contraction, or period of contraction, of the heart, especially that of the ventricles. It coincides with the interval between the first and second heart sound, during which blood is forced into the aorta and the pulmonary trunk.

systolic- of or pertaining to systole (e.g., systolic blood pressure).

T

tachycardia- relatively rapid heart action.

thorax- the part of the body of man and other mammals between the neck and the abdomen.

thrombus- a clot of blood formed within a blood vessel and remaining attached to its place of origin.

tidal volume- volume of gas inspired or expired during each quiet respiration cycle.

tissue- an aggregation of similarly specialized cells united in the performance of a particular function.

trachea- the main trunk of the system of tubes by which air passes to and from the lungs.

tricuspid valve- the valve connecting the right atrium to the right ventricle.

V

vasoconstriction- narrowing of the lumen of blood vessels, especially as a result of vasomotor action.

vasodilation- dilation or opening of blood vessel by vasomotor action.

vasomotor- having to do with the musculature that affects the caliber of a blood vessel.

ventricle- a chamber of the heart which receives blood from a corresponding atrium and from which blood is forced into the arteries.

ventricular fibrillation- convulsive nonsynchronized activity of the ventricles of the heart.

venule- a small vein; especially one of the minute veins connecting the capillary bed with the larger systemic veins.

vital capacity- volume of air that can be exhaled after the deepest possible inhalation.

B

Physiological Measurements Summary

B.1. BIOELECTRIC POTENTIALS

Electrocardiogram (ECG or EKG). A record of the electrical activity of the heart. Electrical potentials: 0.1 to 4 mV peak amplitude. Frequency response requirement: dc to 100 Hz. Used to measure heart rate, arrhythmia, and abnormalities in the heart. Also serves as timing reference for many cardiovascular measurements. Measured with electrodes at the surface of the body.

Electroencephalogram (EEG). A record of the electrical activity of the brain. Electrical potentials: 10 to 100 μV peak amplitude. Frequency requirement: dc to 100 Hz. Used for recognition of certain patterns, frequency analysis, evoked potentials, and so on. Measured with surface electrodes on the scalp and with needle electrodes just beneath the surface or driven into specific locations within the brain.

Electromyogram (EMG). A record of muscle potentials, usually from skeletal muscle. Electrical potentials: 50 μV to 1 mV peak amplitude. Required frequency response: 10 to 3000 Hz. Used as indicator of muscle action, for measuring fatigue, and so on. Measured with surface electrodes or needle electrodes penetrating the muscle fibers.

Other bioelectric potentials

1. Electroretinogram—a record of potentials from the retina.
2. Electrooculogram—a record of corneal-retinal potentials associated with eye movements.
3. Electrogastrogram—a record of muscle potentials associated with motility of the GI tract.
4. Individual nerve action potentials—potentials generated by information being transmitted by the nervous system.

B.2. SKIN RESISTANCE MEASUREMENTS

Galvanic skin response (GSR). Measurements of the electrical resistance of the skin and tissue path between two electrodes. A variation of resistance from 1000 to over 500,000 Ω. Variations are associated with activity of the autonomic nervous system. Used to measure autonomic responses. Principle behind "lie detection" equipment. Variations occur with bandwidth from 0.1 to 5 Hz. Measured with surface electrodes.

Basal skin resistance (BSR). Same as GSR, except that the BSR is a measure of the slow baseline changes instead of the variations caused by the autonomic system. Frequency-response requirements: dc to 0.5 Hz.

B.3. CARDIOVASCULAR MEASUREMENTS

Blood pressure measurements

1. *Arterial:* Pressure variations from 30 to 400 mm Hg. Pulsating pressure with each heart beat. Frequency-response requirements: dc to 30 Hz. Measured at various points in the arterial circulatory system. Measured directly by implanted pressure transducer; transducer connected to catheter in bloodstream, or manometer; indirectly by sphygmomanometer, and so on.
2. *Venous:* Pressure variations from 0 to 15 mm Hg. An almost static pressure with some variations with each heart beat. Frequency-response requirements: 0 to 30 Hz. Measured at various points in the venous circulatory system. Measured

by manometer, implanted pressure transducer, or external transducer connected to catheter.

Blood volume measurements

1. *Systemic volume:* Measure of total blood volume in the system. Measured by injection of an indicator such as a dye and subsequent measurement of indicator concentration.
2. *Plethysmograph measurement:* A measure of local blood volume changes in limbs or digits. This is an actual change in volume measured as a displacement change in a closed cup or tube. Volume pulsations occur at rate of heart beat. Required frequency response: dc to 40 Hz. Can also be measured indirectly with photoelectric device or tissue impedance measurement. Used to measure effectiveness of circulation, and in pulse-wave velocity measurements.

Blood flow measurements. A measure of the velocity of blood in a major vessel. In a vessel of a known diameter, this can be calibrated as flow and is most successfully accomplished in arterial vessels. Range is from -0.5 to $+1650$ ml/sec. Required frequency response: dc to 50 Hz. Used to estimate heart output and circulation. Requires exposure of the vessel. Flow transducer surrounds vessel. Methods of measurement include electromagnetic and ultrasonic principles.

Ballistocardiogram. Slight movement of body due to forces exerted by beating of the heart and pumping of blood. Patient placed on special platform. Movement measured by accelerometer. Required frequency response: dc to 40 Hz. Used to detect certain heart abnormalities.

Pulse and cardiovascular sound measurements

1. *Pulse pressure measurements:* Pressure variations at surface of the body due to arterial blood pulsations. Used for timing of pulse waves, pulse-wave velocity measurements, and as an indirect indicator of arterial blood pressure variations. Required frequency response: 0.1 to 40 Hz. Measured by low-frequency microphone or crystal pressure pickup.
2. *Heart sounds:* An electrically amplified version of the sounds normally picked up by the conventional stethoscope. Frequency response: 30 to 150 Hz. Picked up by microphone.

3. *Phonocardiogram:* A graphic display of the sounds generated by the heart and picked up by a microphone at the surface of the body. Frequency response required is 5 to 2000 Hz. Measured by special crystal transducer or microphone.
4. *Vibrocardiogram:* A measure of the movement of the chest due to the heart beat. Frequency response required: 0.1 to 50 Hz. Special pressure or displacement transducer placed on the appropriate point on the chest.
5. *Apex cardiogram:* A measurement of the pressure variations at the point where the apex of the heart beats against the rib cage. Frequency response required: 0.1 to 50 Hz. Measured with special pressure sensitive-microphone or crystal transducer.

B.4. RESPIRATION MEASUREMENTS

Respiration flow measurements. A measurement of the rate at which air is inspired or expired. Range: 250 to 3000 ml/sec, peak. Frequency response: 0 to 20 Hz. Used to determine breathing rate, minute volume, depth of repiration. Measured by pneumotachometer or as the derivative of volume measurement.

Respiration volume. Measurement of quantity of air breathed in or out during a single breathing cycle or over a given period of time. Frequency response required: 0 to 10 Hz. Used for determination of various respiration functions. Measure by integration of respiration flow-rate measurements or by collection of expired air over a given period. Indirect measurement by belt transducer, impedance pneumograph, or whole-body plethysmograph.

B.5. TEMPERATURE MEASUREMENTS

Systemic temperature. A measure of the basic temperature of the complete organism. Measured by thermometer, rectal or oral, or by rectal or oral thermistor probe.

Local skin temperature. Measurement of the skin temperature at a specific part of the body surface. Measured by thermistors placed at the surface of the skin, infrared thermometer or thermograph.

B.6. PHYSICAL MOVEMENTS

Various measurements of *displacement, velocity, force,* or *acceleration.* Measured by transducers sensitive to the parameter desired or derived indirectly from related parameters. Special measurement of movement by ultrasound techniques.

B.7. BEHAVIORAL CHARACTERISTICS

Measurement of response of organism to various stimuli. Responses measured may be any of the above, or may be subjective. Includes such measures as speech, visual and sound perception, tactile perception, smell, and taste. Measuring devices include generation of the appropriate stimulus as well as transducers for the various responses.

C

SI Metric Units and Equivalencies

SI Unit	Quantity	Equivalency
degree Celsius	temperature	degree centigrade
		⅝ (degree Fahrenheit − 32°)
gram	mass	0.03527 ounce
hertz	frequency	cycles per second
kilogram	mass	2.2 pounds
kilopascal	pressure	7.5 mm Hg
liter	fluid volume	1.06 quarts
meter	length	3.28 feet
meter per second	velocity	39.37 inches per second
newton	force	0.2247 pound force

Problems and Exercises

D.1. INTRODUCTION

This book is both a reference and a textbook. In the latter function, problems and exercises are needed to aid the student. In a book of this nature, which is primarily descriptive, quantitative problems are not as necessary as in the usual technical book. A few have been provided, but most of these exercises are designed to test the student's knowledge of the key portions of the text, and provide an opportunity to expand on it. The problems are relatively short and do not include long essay-type questions. Such questions are left to the instructor to pose.

Chapter 1

1.1. There are many factors to consider in the design and application of a medical instrumentation system. Discuss what you think are the 10 most important and state why.

1.2. The book lists a number of qualities important to a medical instrumentation system. Suggest one additional quality not listed and state your reasons.

1.3. How would you state the sensitivity characteristics of the following instruments?
(a) An electrocardiograph to give a 2-in. deflection on a recorder for a 2-mV peak reading.
(b) An electroencephalograph to give a 1.5-cm deflection for a 50-μV peak reading.
(c) A thermistor-temperature measuring system to record body temperature at a normal value, plus or minus 5 percent, on a 3-in. scale.

1.4. Check elsewhere in this book or in other references for the required frequency response of:
(a) An electromyogram.
(b) Blood flow measurements.
(c) Phonocardiogram.
(d) Plethysmogram.

1.5. Discuss the possibility of other errors not listed in Section 1.3.

1.6. By using the table of roots, prefixes, and suffixes in Appendix A, determine what the following medical names mean:
(a) Periodontitis.
(b) Bradyrhythmia.
(c) Tachycardia.
(d) Endoesophagus.
(e) Exostosectomy.
(f) Hepatitis.
(g) Dysentery.
(h) Epidermitis.

1.7. Name six body functions and relate them to a field or topic normally studied as an engineering-type subject—for example, cardiovascular system and fluid mechanics.

1.8. The text lists three basic differences that contribute to communication problems between the physician and the engineer. What are they? How can they be overcome? Can you think of any others?

1.9. Discuss the major differences encountered between measurements in a physiological system as distinct from a physical system.

1.10. What are the objectives of a biomedical instrumentation system?

1.11. Explain the difference between in vivo and in vitro measurements.

1.12. Name the major physiological systems of the body.

1.13. What specific features might be incorporated into an instrument designed for clinical use as opposed to one designed for research purposes?

1.14. In designing an instrumentation system for measurement of physiological variables, which of the components shown in Figure 2.1 should be determined first? Why? Which would you next determine?

1.15. Draw a diagram showing the hydraulic (cardiovascular) system of the body, using the terminology common to the engineering analogy given in this chapter.

Chapter 2

2.1. Discuss four different types of transducers, explaining what they measure and the principles involved.

2.2. What do you understand by the term "gage factor"?

2.3. Discuss the relationship among displacement, velocity, acceleration, and force.

2.4. Explain the difference between isometric and isotonic transducers.

2.5. What is a mercury strain gage? Describe its operation and list as many biomedical applications as you can.

2.6. Which of the following types of physical transducers, in basic form, are capable of a direct measurement of displacement and which are primarily velocity transducers?
(a) Potentiometer transducer.
(b) Piezoelectric crystal.
(c) Differential transformer.
(d) Bonded strain gage.
(e) Unbonded strain gage.
(f) Capacitance transducer.
(g) Induction-type transducer.

2.7. You have invented a device that changes its resistance linearly as a function of ozone in a sample of air (smog). Draw a transducer-type circuit excited by an audio oscillator and explain how it would operate and how you would use it.

2.8. What is the difference between an active and a passive transducer?

2.9. When a student takes courses in physics, the topics usually include the concepts of mechanics, heat, light, sound, electricity, and magnetism. For each of these topics, specify a transducer which belongs in that field and discuss the basic energy form and how transduction is effected.

2.10. Invent a new transducer. Explain the energy form you wish to use and what you would transduce to. Later you may wish to adapt your new transducer to a biomedical application, but you should read a few more chapters first.

Chapter 3

3.1. Draw an action potential waveform and label the amplitude and time values.

3.2. Explain polarization, depolarization, and repolarization.

3.3. What is a biopotential? Name six types of biopotential sources.

3.4. Explain the electrical action of the sinoatrial node.

3.5. Do you think the electroencephalogram is subject to frequency discrimination? Explain.

3.6. How are the potentials in muscle fibers measured, and what is the record called that is obtained therefrom?

3.7. How does an evoked EEG response differ from a conventional electroencephalogram?

Chapter 4

4.1. Name the three basic types of electrodes for measurement of bioelectric potentials.

4.2. For a patient, which type of electrode would be the least traumatic?

4.3. Why are microelectrodes sometimes needed?

4.4. What are the problems involved in using flat electrodes in terms of interference or high impedance between electrode and skin? How could you help eliminate this problem?

4.5. What do you understand by the term "reference electrode"?

4.6. What is a glass electrode used for?

4.7. What is an ear-clip electrode used for?

4.8. Why are the partial pressure of oxygen and the partial pressure of carbon dioxide useful physical parameters? Explain briefly how each can be measured.

4.9. Calculate the potential difference across a membrane separating two very dilute solutions of a monovalent ion, one concentration being 100 times as great as the other. Assume a body temperature of 37 °C.

4.10. What is the major advantage of floating-type skin surface electrodes?

4.11. What is the hydrogen ion concentration of blood with a pH of 7.4?

Chapter 5

5.1. A patient has a cardiac output of 4 liters/min, a heart rate of 86 beats per minute, and a blood volume of 5 liters. Calculate the stroke volume and the mean circulation time. What is the mean blood velocity in the aorta (in feet per second) when the vessel has a diameter of 30 mm?

5.2. Explain the operation of the heart and the cardiovascular system briefly. Draw an analogous electric circuit and show how Ohm's law and Kirchoff's laws could apply in the analog.

5.3. Develop a time-phase diagram showing the correlation of the mechanical pumping of the heart, including the opening of the valves, with the electrical-excitation events.

5.4. Draw the waveshape of blood pressure on a time base and explain it. What is the dicrotic notch?

5.5. What is the difference in the information contained in a phonocardiogram and an electrocardiogram?

5.6. In a harmonic analysis of the following waveforms, what range of frequencies could be expected in the human being?
(a) The ECG.
(b) The phonocardiogram.
(c) The blood pressure wave.
(d) The blood flow wave.

5.7. Would you expect blood flow to obey Bernoulli's equation, even with reservations? Explain why.

5.8. If a person stands up, does his blood pressure increase? Why?

5.9. If a person eats a large meal, does his heart rate increase? Why?

5.10. What part of the cardiovascular system normally contains the greatest volume of blood?

5.11. Define systole and diastole.

Chapter 6

6.1. Draw an electrocardiogram (in lead II), labeling the critical features. Include typical amplitudes and time intervals for a normal person.

6.2. A differential amplifier has a positive input terminal, a negative input terminal, and a ground connection. ECG electrodes from a patient are connected to the positive and negative terminals, and a reference electrode is connected to ground. A disturbance signal develops on the patient's body. This will appear as a voltage from the positive terminal to ground and a similar voltage from the negative terminal to ground. How does the differential amplifier amplify the ECG signal while not essentially amplifying the disturbance signals? Draw a sketch showing the patient connected to the amplifier.

6.3. Why are the vector sums of the projections on the frontal-plane cardiac vector at any instant onto the three axes of the Einthoven triangle zero?

6.4. Explain the difference between indirect and direct measurement of blood pressure.

6.5. The "thermostromuhr" and the indicator-dilution method with cool saline as an indicator both use thermistors for detectors. What is the difference between the two methods?

6.6. For a cardiac-output determination, 5 mg of Cardiogreen was injected into a patient and a calibration mixture with a concentration of 5 mg/liter was prepared from a previously withdrawn blood sample. The calibration mixture gave a deflection of $\frac{3}{4}$ cm on the recorder used, which had a paper speed of 1 cm/sec. The area under the extrapolated curve (obtained by the Hamilton method) was 86 cm². What is the cardiac output in liters per minute? (*Answer:* 0.872 liter/min)

6.7. Explain the basic operation of the following blood pressure transducers:
(a) A resistance-bridge type.
(b) A linear variable differential transformer type.

6.8. Explain what is meant by "plethysmography"? Discuss one way to make measurements and their clinical implications.

6.9. You are to measure the blood pressure of a dog during heavy exercise on a treadmill by using a catheter-type resistance strain-gage transducer. What is the desirable frequency response for your whole system? Explain.

6.10. Laplace's law can be used for cylindrical blood vessels. Simply stated in this context, the tension in the wall of a vessel is the product of the radius and internal pressure. Given that 1 mm Hg is equivalent to approximately 1300 dynes/cm^2, find the tension in the wall of:

(a) An aorta with a mean pressure of 100 mm Hg and a diameter of 2.4 cm. (*Answer:* 1.56×10^5 dynes/cm^2)

(b) A capillary with a mean pressure of 25 mm Hg and a diameter of 8 μm. (*Answer:* 13 dynes/cm^2)

(c) The superior vena cava with a mean pressure of 10 mm Hg and 3 cm diameter. (*Answer:* 19.5×10^3 dynes/cm^2)

6.11. Assume that blood flow obeys Bernoulli's equation:

$$\frac{p}{w} + \frac{v^2}{2g} + z = \text{constant}$$

where p = pressure
$\quad\quad\; w$ = specific weight
$\quad\quad\;\; v$ = velocity
$\quad\quad\;\; g$ = gravitational constant
$\quad\quad\;\; z$ = elevation head

The three terms are often referred to as the pressure head, the velocity head, and the potential or elevation head, respectively. In measurements on a patient, the elevation is a constant, so the equation can be expressed as

$$p + \frac{wv^2}{2g} = \text{constant}$$

A certain type of blood pressure transducer positioned in the aorta will measure this value, but since the lateral blood pressure is simply the p term, $wv^2/2g$ represents an error. If the density of the blood, w/g, is estimated to be 1.03 grams/cm^3, and the blood is flowing at a velocity of 100 cm/sec, calculate the error in blood pressure measurement. Given 1 mm Hg is equivalent to 1330 dynes/cm^2. (*Answer:* 3.88 mm) Do you consider this to be a significant error in the aorta?

6.12. You are employed by a hospital research unit on a certain project to measure the blood pressure and blood flow in the femoral artery of an anesthetized dog lying on an operating table.

(a) Design a system to do this by (1) describing the transducers, if any, you would use; (2) specifying all necessary instrumentation; (3) discussing surgical or medical methods used to ensure that your physiological measurements are taken correctly—for example, catheterization, implantation, and so on. Draw block diagrams to illustrate.

(b) How would you zero and calibrate your blood pressure measurements?

6.13. Blood shows certain conductive properties. Discuss an instrument that uses this property.

6.14. Discuss the advantages and disadvantages of four types of blood pressure transducers that can either be implanted or placed in the bloodstream through a catheter.

6.15. Draw a simplified model using block diagrams to show how the brain, pressoreceptors, and hormonal secretion could control the heart rate. Use the brain and the heart as your feed-forward loop and other parameters as feedback loops.

6.16. Why is the impedance plethysmograph sometimes called a pseudo-plethysmograph?

6.17. Explain the difference between a phonocardiogram and a vibrocardiogram. How do transducer requirements differ for these two measurements?

6.18. What is meant by mean arterial pressure? How do you measure it?

6.19. Discuss the automatic and semiautomatic methods of measuring blood pressure. Can you suggest any modifications?

6.20. What is the difference between a single-lumen catheter and a multiple-lumen "floatation" catheter?

6.21. Discuss measurement of blood pressure and possible errors due to trauma or other psychological effects on the patient.

6.22. What are the relative merits of dyes and cold saline methods in cardiac output measurements?

Chapter 7

7.1. Design a coronary-care hospital suite. Show all rooms in a layout plan. Illustrate all your instrumentation systems by block diagrams.

7.2. Discuss warning devices to be used in intensive-care units.

7.3. Explain the operation of a pacemaker and why it is needed.

7.4. What do you understand by fibrillation? How do you correct for it? Draw a circuit of a direct-current defibrillator.

7.5. What part of the electrocardiogram is the most useful for determining heart rate? Explain.

7.6. A certain patient-monitoring unit has an input amplifier with a common-mode rejection ratio of 100,000:1 at 60 Hz. At other frequencies, the common-mode rejection ratio is 1000:1. Do you consider these ratios adequate for monitoring the ECG? Explain.

7.7. Discuss possible causes of a patient-monitoring system falsely indicating an excessive high heart rate.

7.8 What is a "demand" pacemaker and when is it used?

7.9. What is the difference between a "bouncing ball" and nonfade display? Discuss their relative merits.

7.10. Discuss instrumentation and methods for rapid diagnosis and repair of instrumentation in an intensive-care unit.

7.11. What equipment would you need in a diagnostic catheterization laboratory?

7.12. Design the cardiology department of a small hospital to include facilities for intensive-care monitoring, surgery, and diagnostics. Specify all the equipment and instrumentation necessary, including the possibility of emergencies.

Chapter 8

8.1. Using the correct anatomical and physical terms, explain the process of respiration, tracing the taking of a breath of air through the mouth to the using of the oxygenated blood in the muscle of an athlete's leg.

8.2. How many lobes are there in the lungs? Explain.

8.3. Boyle's law is an important law in physics. How does it relate to the breathing process? (*Hint: PV* is a constant at a constant temperature.)

8.4. What is the difference between death by carbon monoxide poisoning and death by strangulation? Explain.

8.5. Define the important lung capacities and explain them.

8.6. A person has a total lung capacity of 5.95 liters. If the volume of air left in the lungs at the end of maximal expiration is 1.19 liters, what is his vital capacity? (*Answer:* 4.76 liters)

8.7. If the volume of air expired and inspired during each respiratory cycle varies from 0.5 to 3.9 liters during exercise, what is this value called and what does it mean?

8.8. During a typical day, a person works for 8 hours, rests for 4 hours, walks for 1 hour, eats for 2 hours, and sleeps for 9 hours. How many pounds of oxygen would he consume during the whole day? (During sleep and rest he can be assumed to consume 0.05 pound/hour; during eating, this figure will double; during walking, consumption will triple; and during work it will quadruple.) (*Answer:* 2.6 pounds)

8.9. Explain the operation of a pulmonary measurement indicator.

8.10. Since the lungs contain no musculature, what causes them to expand and contract in breathing?

8.11. For what measurements can a spirometer be used? What basic lung volumes and capacities cannot be measured with a spirometer? Why?

Chapter 9

9.1. What do you understand by the term "noninvasive methods"?

9.2. Explain the difference between a thermistor and a thermocouple in temperature measurement.

9.3. Discuss how temperature is controlled in the body by the brain.

9.4. Explain the technique of thermography. Comment on its usefulness.

9.5. Why is skin surface temperature lower than systemic temperature measured orally?

9.6. What are the important characteristics to be considered in selecting a thermistor probe for a specific medical application?

9.7. Discuss the properties of ultrasound and how ultrasound can be used for diagnostics.

9.8. What do you understand about the following terms?
(a) Doppler effect.
(b) Half-value layer.
(c) Acoustic impedance.
(d) Attenuation constant.

9.9. What are the four basic modes of transmission of ultrasound? Describe each briefly.

9.10. What is meant by "ultrasonic imaging"?

9.11. What is echoencephalography?

9.12. Discuss the applications of ultrasound in medicine. Can you suggest some possible applications that are not discussed in the chapter?

9.13. A patient has a heart problem that seems to suggest mitral valve stenosis. Discuss the transducer you would specify to perform a diagnosis.

9.14. Compare ultrasonic diagnosis with X-ray diagnosis discussed in Chapter 14.

9.15. Discuss the relative differences between high- and low-frequency ultrasound.

9.16. An ultrasonic imaging system is capable of operating at both 5 and 12.5 MHz. What is the advantage of being able to select between two frequencies? Under what circumstances would you use each?

9.17. What is range-gated pulsed Doppler ultrasound? Describe at least two possible applications.

Chapter 10

10.1. What is the difference between afferent and efferent nerves?

10.2. Explain the difference between a motor nerve and a sensory nerve.

10.3. How does the action of the sympathetic nervous system differ from that of the parasympathetic system? Quote an example from a body system.

10.4. What is a neuronal spike? Draw a typical spike showing amplitude and duration.

10.5. What is a 10–20 electrode placement system and with what bioelectric instrument is it used?

10.6. Discuss some possible uses of electromyography.

10.7. Draw a sketch of a neuron and label the cell body, dendrite, axon, and axon hillock.

10.8. What are the nodes of Ranvier and what useful purpose do they serve?

10.9. Explain the way in which a neuronal spike is transmitted from one neuron to another.

10.10. What are graded potentials?

10.11. Explain the function of:
(a) The cerebral cortex.
(b) The cerebellum.
(c) The reticular activation system.
(d) The hypothalamus.

10.12. What is a spinal reflex, and how is it related to the functions of the brain?

10.13. If the same neuronal spike were measured intracellularly and extracellularly, what would be the difference between the two measurements?

10.14. What are the differences in amplification and bandwidth requirement of amplifiers for ECG, EMG, and EEG?

Chapter 11

11.1. You want to determine what concentration of salt in water can be detected by the human taste sense. How would you set up the experiment?

11.2. For a "differential response" experiment, an animal box contains two lamps and one bar. The positive reinforcement is food, dispensed by a magnetic feeder, the negative reinforcement is electric shock. Devise a simple programming circuit, using relays, that causes positive reinforcement when the animal presses the bar while either light is on and negative reinforcement if the animal presses the bar while both lights are on. Bar pressing while no light is on shall have no effect.

11.3. List some of the possible difficulties that might be encountered in using GSR measurements as a lie detector test.

11.4. Explain the principle of the Békésy audiometer.

11.5. What is a cumulative recorder and how it is used?

11.6. You have been assigned the task of measuring all possible responses of the autonomic nervous system. Design a system for providing various forms of stimuli that would be expected to actuate the autonomic system for measurement of each response. Describe the type of instrumentation you would use.

Chapter 12

12.1. List some advantages and disadvantages of biotelemetry.

12.2. Draw a block diagram of a system to send an electrocardiogram from an ambulance to a hospital by telemetry.

12.3. Why do you think measurements of physiological parameters on an unanesthetized animal may be more useful than those on an anesthetized one?

12.4. What do you see as some of the problems of telemetrized systems in the future?

12.5. It is desirable to monitor the temperature of a man very accurately while he is climbing a mountain and then record the data on tape for later computer analysis. You are to remain in a cabin at the foot of the mountain. Explain how you would do this accurately, and draw a block diagram of any equipment stages used in your system.

12.6. Explain how four physiological parameters can be monitored and telemetered simultaneously.

12.7. If subdermal needles connected to a telemetry transmitter are implanted into a muscle, explain how a trained physician might recognize different effects from another room by using a sense other than vision to monitor.

12.8. Design a hospital with a telemetry system, explaining why you would telemetrize the functions you have selected.

12.9. Discuss telemetry of electrocardiograms and advantages or disadvantages over a wired system for:
(a) A hospital bed patient.
(b) A convalescing patient.
(c) An athlete being measured on a treadmill.

12.10. Discuss telemetry as an emergency care tool.

12.11. What are medical transmitting frequencies? Why is it necessary to specify them?

12.12. Design a system that is capable of transmitting the ECG of a patient at home by radio to a hospital, then by telephone line to a computer center, and finally sending the data to a cardiologist to diagnose.

Chapter 13

13.1. List the most important components of the blood.

13.2. List the main types of bood tests and explain each briefly.

13.3. What do you understand by the term "blood count"?

13.4. Describe the operation of a blood counter.

13.5. Describe the colorimetric method of determining chemical concentration.

13.6. When counting red blood cells with one of the automatic counting methods described in Section 13.1, you will, by necessity, also count the white blood cells in the process. Why is the error introduced by this negligible? Why must the blood be diluted for all the automatic blood cell counters? How do the automatic cell counters avoid counting the platelets?

13.7. For a glucose determination, a standard with a known glucose concentration of 80 mg/100 ml is used. After the color reaction has taken place, this standard shows a transmittance of 38 percent. A patient sample shows a transmittance of 46 percent. What is the glucose concentration in this patient sample? Another patient sample shows a transmittance below 10 percent, which is hard to read accurately. What can be done to this sample to bring the transmittance into a more suitable range, and what correction has to be applied in the calculation?

13.8. Explain the difference between the continuous-flow method and the discrete sample method of automated clinical chemistry equipment. What are some of the shortcomings of each?

Chapter 14

14.1. In both X-ray and radioisotope procedures, potentially harmful ionizing radiation is used for diagnostic purposes. Why is the safe radiation intensity for X rays much higher than that for isotope methods?

14.2. X-ray and radioisotope methods for diagnostic purposes both make use of the tissue-penetrating properties of radiation. What is the principal difference between the two methods?

14.3. Why is the use of radioisotopes for in vivo methods limited to those isotopes that emit gamma radiation?

14.4. Describe the principle of visualizing body organs by radioisotope methods.

Chapter 15

15.1. Define each of the following terms as related to digital computation:
(a) Word length.
(b) Register.
(c) Memory.
(d) Character.
(e) Address.
(f) Byte.
(g) Time sharing.
(h) Modem.
(i) Real time.
(j) On line.
(k) Software.

15.2. Describe the processes required to enter each of the following types of data into a digital computer:
(a) numerical data written in tabular form on sheets of paper

(b) the output of a pneumotachograph transducer

(c) the output of a digital electronic counter

(d) an electrocardiogram signal

15.3. A microcomputer has three types of memory: integrated-circuit RAM, integrated-circuit ROM, and floppy disk. How might each be used in biomedical instrumentation?

15.4. What role does a digital-to-analog converter play in an analog-to-digital converter?

15.5. What is the purpose of a parallel-to-serial converter? When would such a device be used?

15.6. Several applications of digital computers to medicine are given in Chapter 15. Can you suggest other possible applications?

15.7. You have been assigned the task of designing a computerized patient-monitoring system for an intensive-care unit in a medium-size community hospital. What parameters would you monitor? What role would you have the computer play? Draw a block diagram of a typical system and explain the purpose of each block.

15.8. Explain the principle of computerized axial tomography and compare its method of visualization with conventional X-ray methods.

15.9. In a high-speed CAT scanner, 20 scans are taken using a fanbeam X-ray source and an array of 350 detectors. What is the maximum number of pixels that can result from this arrangement?

Chapter 16

16.1. Name two different ways in which electricity can harm the body.

16.2. List the various effects of electrical current that occur with increasing current intensity.

16.3. What is the difference between electrical macroshock and microshock? In what parts of the hospital are microshock hazards likely to exist?

16.4. What is the basic purpose of the safety measures used with electrically susceptible patients?

16.5. Why is it so important to maintain the integrity of the grounding system for protection against microshock?

16.6. A fluid-filled catheter is used to measure blood pressure in the right atrium of the heart. Resistance of the fluid path is $1\text{M}\Omega$. The external end of the catheter is grounded to the equipment ground of a receptacle at the left side of the patient's bed. The patient's right leg is grounded via a patient monitor to another receptacle at the right side of the patient's bed. Because of a malfunction in a vacuum cleaner, a fault current of 10 A flows through the ground wire connecting the two receptacles. What is the maximum allowable resistance for the ground wire connecting the receptacles to prevent exceeding the 10 μA safe current limit for microshock in the patient?

Bibliography

General

The following works may be considered as general references throughout the book. Other references are cited for particular chapters.

BEST, C. M. AND N. B. TAYLOR, *The Living Body, a Text in Human Physiology.* New York: Holt, Rinehart and Winston, Inc., 1971.

BEST, C. M. AND N. B. TAYLOR, *Physiological Basis of Medical Practice.* 9th ed., John K. Brobeck (ed.). Baltimore, Md.: The Williams & Wilkins Company, 1973.

BILOON, F., *Medical Equipment Service Manual.* Englewood Cliffs, N.J.: Prentice-Hall, Inc., 1978.

BRONZINI, J. D., *Technology for Patient Care.* St. Louis, Mo.; The C. V. Mosby Company, 1977.

BROWN, J. H. U., J. E. JACOBS, AND L. STARK, *Biomedical Engineering.* Philadelphia: F. A. Davis Company, 1971.

CACERES, C. A. (ed.), *The Practice of Clinical Engineering.* New York: Academic Press, 1977.

CROMWELL, L., M. ARDITTI, F. J. WEIBELL, E. A. PFEIFFER, B. STEELE, AND J. LABOK, *Medical Instrumentation for Health Care.* Englewood Cliffs, N.J.: Prentice-Hall, Inc., 1976.

Dorland's Illustrated Medical Dictionary. 25th ed. Philadelphia: W. B. Saunders Company, 1974.

EVANS, W. F., *Anatomy and Physiology, the Basic Principles.* Englewood Cliffs, N.J.: Prentice-Hall, Inc., 1971.

FLEMING, D. G., AND B. N. FEINBERG, *CRC Handbook of Engineering in Medicine and Biology.* Cleveland, Ohio: CRC Press, Inc., 1976.

GEDDES, L. A., AND L. E. BAKER, *Principles of Applied Biomedical Instrumentation,* 2nd ed. New York: John Wiley & Sons, Inc., 1975.

GRAF, R. F. AND G. J. WHALEN, *The Reston Encyclopedia of Biomedical Engineering Terms.* Reston, Va.: Reston Publishing Company, Inc., 1977.

GUYTON, A. C. *Textbook of Medical Physiology.* Philadelphia: W. B. Saunders Company, 1971.

JACOBSON, B., AND J. G. WEBSTER, *Medicine and Clinical Engineering.* Englewood Cliffs, N.J.: Prentice-Hall, Inc., 1977.

KLINE, J. (ed.), *Biological Foundations of Biomedical Engineering.* Boston: Little, Brown and Company, 1976.

MOUNTCASTLE, V. B. (ed.), *Medical Physiology,* 12th ed., Vols. I and II. St. Louis, Mo.: The C. V. Mosby Company, 1968.

RAY, C. D. (ed.), *Medical Engineering.* Chicago: Year Book Medical Publishers, Inc., 1974.

SCHMIDT-NIELSEN, K., *Animal Physiology*. Englewood Cliffs, N.J.: Prentice-Hall, Inc., 1974.

STRONG, P., *Biophysical Measurements*. Beaverton, Ore.: Tektronix, Inc., 1970.

WEBSTER, J. G. (ed.), *Medical Instrumentation, Application and Design*. Boston: Houghton Mifflin Company, 1978.

WEISS, M. D., *Biomedical Instrumentation*. Philadelphia: Chilton Book Company, 1973.

WELKOWITZ, W., AND S. DEUTSCH, *Biomedical Instruments, Theory and Design*. New York: Academic Press, 1976.

YANOF, H. M., *Biomedical Electronics,* 2nd ed. Philadelphia: F. A. Davis Company, 1972.

Chapter 2

BARTHOLOMEW, D., *Electrical Measurements and Instruments*. Boston: Allyn and Bacon, Inc., 1963.

COBBOLD, R. S. C., *Transducers for Biomedical Instruments: Principles and Applications*. John Wiley & Sons, Inc., 1974.

NORTON, H. N., *Handbook of Transducers for Electronic Measuring Systems*. Englewood Cliffs, N.J.: Prentice-Hall, Inc., 1969.

Chapter 3

BRAZIER, M. A. B., *The Electrical Activity of the Nervous System*. Baltimore, Md.: The Williams & Wilkins Company, 1968.

BURCH, G. E., AND T. WINSOR, *A Primer of Electrocardiography*. Philadelphia: Lea & Febiger, 1960.

KEELE, C. A., AND E. NEIL, *Samson Wright's Applied Physiology*. New York: Oxford University Press, inc., 1961.

SCHER, A. M., "The Electrocardiogram," *Scientific American,* 205, No. 5 (November, 1961), 132–141.

THOMPSON, R. F., *Foundations of Physiological Psychology*. New York: Harper & Row, Publishers, 1967.

Chapter 4

BAIR, E. J., *Introduction to Chemical Instrumentation*. New York: McGraw-Hill Book Company, 1962.

BATES, R. G., *Determination of pH—Theory and Practice*. New York: John Wiley & Sons, Inc., 1954.

DURST, R. A. (ed.), *Ion-Selective Electrodes*. Washington, D.C.: National Bureau of Standards Publication No. 314, 1969.

GEDDES, L. A., *Electrodes and the Measurement of Bioelectric Events*. New York: John Wiley & Sons, Inc., 1972.

HILL, O. W., AND R. S. KHANDPUR, The Performance of Transistor ECG Amplifier," *World Medical Instruments,* 7, No. 5 (May, 1969), 12–22.

MILLER, H. A., AND D. C. HARRISON (eds.), *Biomedical Electrode Technology: Theory and Practice*. New York: Academic Press, 1974.

PFEIFFER, E. A., "Electrical Stimulation of Sensory Nerves with Skin Electrodes for Research, Diagnosis, Communication and Behavioral Conditioning: A Survey," *Medical and Biological Engineering,* 6 (1968), 637–651.

Chapters 5 and 6

BURCH, G. E., AND N. P. DE PASQUALE, *Primer of Clinical Measurement of Blood Pressure*. St. Louis. Mo.: The C. V. Mosby Company, 1962.

BURCH, G. E., AND T. WINSOR, *A Primer of Electrocardiography,* 4th ed. Philadelphia: Lea & Febiger, 1960.

CAPPELEN, C. (ed.), *New Findings in Blood Flowmetry*. Olso, Norway: Universitetsforlaget, 1968.

Cardiac Output Computer. Promotional material available from Columbus Instruments, Columbus, Ohio, 1970.

ECG Measurements, Application Note AN711. Waltham, Mass.: Hewlett-Packard, Inc., Medical Electronics Division, 1970.

FLEMING, D. G., W. H. KO, AND M. R. NEUMAN, *Indwelling and Implantable Pressure Transducers*. Cleveland, Ohio: CRC Press, Inc., 1977.

FRANKLIN, D. L., "Techniques for Measurement of Blood Flow Through Intact Vessels," *Medical Electronics and Biological Engineering,* 3, (1965), 27–37.

GEDDES, L. A., *The Direct and Indirect Measurement of Blood Pressure*. Chicago: Year Book Medical Publishers, Inc., 1970.

GEE, W., et al., "Ocular Pneumoplethysmography in Carotid Artery Disease," *Medical Instrumentation,* 8, No. 4 (July–August, 1974).

The Measurement of Cardiac Output by the Dye Dilution Method. Monograph available from Lexington Instruments Corporation, Waltham, Mass., 1963.

Pressure Measurements, Application Note AN710. Waltham, Mass.: Hewlett-Packard, Inc., Medical Electronics Division, 1969.

Recommendations for Human Blood Pressure Determination by Sphygmomanometers. New York: American Heart Association, 1967.

"Recommendations for Standardization of Instruments in Electrocardiography and Vectorcardiography," AHA Subcommittee Report, *IEEE Transactions on Biomedical Engineering,* January, 1967, 60–68.

Sphygmomanometers—Principles and Precepts. Copiague, N.Y.: W. A. Baum Co., 1965.

WARBASSE, R. J., et at., "Physiologic Evaluation of a Catheter Tip Electromagnetic Velocity Probe," *American Journal of Cardiology,* 23 (March, 1969), 424–433.

ZIERLER, K. L., "Circulation Times and the Theory of Indicator Dilution Methods for Determining Blood Flow and Volume," in *Handbook of Physiology,* Sec. 2, Vol. 1, J. Field (ed). American Physiological Society, 1959, 585–615.

Chapter 7

Cardiac Pacemaking. Somerville, N.J.: Hoechst Pharmaceuticals, Inc., 1974.

"Coronary Arteriography," *Directions in Cardiovascular Medicine,* Vol. 1. Somerville, N.J.: Hoechst Pharmaceuticals, Inc., 1970.

"The Coronary Care Unit," *Directions in Cardiovascular Medicine,* Vol. 3. Somerville, N.J.: Hoechst Pharmaceuticals, Inc., 1970.

A Guide to Fetal Monitoring. Waltham, Mass.: Hewlett–Packard, Inc., 1973.

HENDRIE, W. A., "Patient Monitoring," *World Medical Instrumentation,* 7, No. 7 (July, 1969).

HILL, D. W., AND A. M. DOLAN, *Intensive Care Instrumentation,* New York: Academic Press, 1976.

KARSELIS, T., *Descriptive Medical Electronics and Instrumentation.* Thurofare, N.J.: Charles B. Slack, Inc., 1973.

KLEIN, B., *Introduction to Medical Electronics.* Blue Ridge Summit, Pa.: Tab Books, 1973.

LOWN, B., "Intensive Heart Care," *Scientific American,* 219, No. 1 (July, 1968), 19–27.

An Overview of Pacing. Minneapolis, Minn.: Medtronic Inc., 1973.

Planning a Patient Monitoring System. Waltham, Mass.: Hewlett–Packard, Inc., Medical Electronics Division, 1973.

Note: Many instrument manufacturers have published description booklets on patient-monitoring systems.

SWAN, H. J. C., W. GANZ, et al., "Catheterization of the Heart in Man with Use of a Flow-Directed Balloon-Tipped Catheter," *New England Journal of Medicine,* 283 (August 27, 1970), 447.

Chapter 8

BASS, B. H., *Pulmonary Function in Clinical Medicine.* Springfield, Ill.: Charles C Thomas, Publisher, 1964.

BELINKOFF, S., *Introduction to Inhalation Therapy.* Boston: Little, Brown and Company, 1969.

BUIST, A. S., "Clinical Significance of Pulmonary Function Tests: Early Detection of Airways Obstruction by the Closing Volume Technique," *Chest,* 64, No. 4 (1973), 495–499.

CHERNIAK, R. M., AND L. CHERNIAK, *Respiration in Health and Disease.* Philadelphia: W. B. Saunders Company, 1965.

COMROE, J. H., et al., *The Lung Clinical Physiology and Pulmonary Function Tests.* Chicago: Year Book Medical Publishers, Inc., 1963.

EGAN, D. F., *Fundamentals of Respiratory Therapy,* 2nd ed. St. Louis, Mo.: The C. V. Mosby Company, 1973.

FILLEY, G. F., *Pulmonary Insufficiency and Respiratory Failure.* Philadelphia: Lea & Febiger, 1968.

GAENSLER, E. A. AND G. W. WRIGHT, "Evaluation of Respiratory Impairment," *Archives of Environmental Health,* 12 (February, 1966).

The Riker Pulmonitor, Vol. 1 in a series of monographs on spirometry. Northridge, Calif.: Riker Laboratories.

SCHOLANDER, P. F., "Analyzer for Accurate Estimation of Respiratory Gases in One-half Cubic Centimeter Samples," *Journal of Biological Chemistry, 167 (1947).*

WOOLMER, R. F., *A Symposium of pH and Blood Gas Measurements.* Boston: Little, Brown and Company, 1959.

Chapter 9

BARNES, R. B., "Thermography of the Human Body," *Science,* 140, No. 3569 (May 24, 1963), 870–877.

BARTHOLOMEW, D., *Electrical Measurements and Instruments.* Boston: Allyn and Bacon, Inc., 1963.

BENCHIMOL, A., *Non-invasive Diagnosing Techniques in Cardiology.* Baltimore, Md.: The Williams & Wilkins Company, 1977.

FEIGENBAUM, H., *Echocardiography,* Philadelphia: Lea & Febiger, 1976.

KELLY, E., (ed.), *Ultrasonic Energy.* Urbana, Ill.: University of Illinois Press, 1965.

LAWSON, R. N., AND L. L. ALT., "Skin Temperature Recording with Phosphors: A New Technique," *Canadian Medical Association Journal,* 92 (February 6, 1965), 255–260.

LINZER, M. (ed.), *Ultrasonic Imaging.* New York: Academic Press, 1979.

LION, K. S., *Instrumentation in Scientific Research.* New York: McGraw-Hill Book Company, 1959.

SEGAL, B. L., "Echocardiography, Clinical Application in Mitral Stenosis,"

Journal of the American Medical Association, 1951 (January 17, 1966), 161–166.

THOMPSON, R. F., *Foundations of Physiological Psychology.* New York: Harper & Row, Publishers, 1967.

WELLS, P. N. R., *Biomedical Ultrasonics.* New York: Academic Press, 1977.

WELLS, P. N. R., *Ultrasonics in Clinical Diagnosis.* Edinburgh: Churchill Livingstone, 1972.

WHITE, D. (ed.), *Ultrasound in Medicine.* New York: Plenum Publishing Corporation, Vol. III, 1977; also Vol. I, 1975; Vol. II, 1976.

WHITE, D. N., *Recent Advances in Ultrasound in Biomedicine.* Forest Grove, Ore.: Research Studies Press, 1977.

Chapter 10

BRAZIER, M. A. B., *The Electrical Activity of the Nervous System,* 3rd ed. Baltimore, Md.: The Williams & Wilkins Company, 1968.

THOMPSON, R. F., *Foundations of Physiological Psychology.* New York: Harper & Row, Publishers, 1967.

VENABLES, P. H., AND MARTIN, I. (eds.), *A Manual of Psycho-Physiological Methods.* New York: John Wiley & Sons, Inc., 1967.

Chapter 11

BÉKÉSY, G. V., "A New Audiometer," *Acta Oto-Laryngologica,* 34 (1947), 411–422.

GRINGS, W. W., *Laboratory Instrumentation in Psychology.* Palo Alto, Calif.: The National Press, 1954.

SCHWITZGEBEL, R. L., "Survey of Electro-mechanical Devices for Behavior Modification," *Psychological Bulletin,* 70, No. 6 (1968), 444–459.

SCHWITZGEBEL, R. L., AND R. K. (eds.), *Psychotechnology-Electronic Control of the Mind and Behavior.* New York: Holt, Rinehart and Winston, Inc., 1973.

SIDOWSKY, J. R., "Buyers Guide for the Behavioral Scientist," *American Psychologist,* 24 (1969), 309–384.

STEBBINS, W. C., "Behavioral Techniques," in *Methods in Medical Research,* Vol. 11, B. F. Rushmer (ed.). Chicago: Year Book Medical Publishers, Inc., 1966.

VENABLES, P. H., AND I. MARTIN, *A Manual of Psychophysiological Methods.* New York: John Wiley & Sons, Inc., 1967.

Note: See also the following company manuals on the programming of behavioral experiments:

Bits of Digi—An Introductory Manual to Digital Logic Packages, 4th ed. Beltsville, Md.: BRS—Foringer Division of Technical Services, Inc., 1970.

Lafayette Data Systems Operation and Program Manual. Lafayette, Ind.: Lafayette Instrument Company, 1970.

Solid State Control—A Handbook for the Behavioral Laboratory. Fogelsville, Pa.: Lehigh Valley Electronics, 1970.

Chapter 12

CROMWELL, L., "Use of Telemetry in Cardiovascular Research," in *Proceedings of the 1968 National Telemetry Conference IEEE,* Houston, Texas, April, 1968.

CROMWELL, L., "Some Advances in Techniques for Remote Monitoring of Blood Pressure, in *IEEE Conference Record, Fifth Annual Rocky Mountain Bioengineering Symposium,* Denver, Colo., May, 1968.

CROMWELL, L., "Investigation of Cardiovascular Phenomena by the Use of Biotelemetry," in *Journal of Physiology (London),* 198, No. 2 (September, 1968), 114.

CROMWELL, L., "Biotelemetry Applied to the Measurement of Blood Pressure." Doctoral dissertation, U.C.L.A., December, 1967.

FRANKLIN, D. L., R. L. VAN CITTERS, AND N. W. WATSON, "Applications of Telemetry to Measurement of Blood Flow and the Pressure in Unrestrained Animals," in L. Winner (ed.), *Proceedings of the National Telemetry Conference,* New York, 1965, 233–234.

FRYER, T. B., *Implantable Biotelemetry Systems,* Washington, D.C.: NASA Publication SP-5094, 1970.

HANISH, H. M., *Biolink Telemetry Systems Application Notes.* Culver City, Calif.: Biocom, Inc., 1971.

KIMMICH, H. P. (ed.), *Biotelemetry,* Vol. 1. Basel: S. Karger, 1974 Fryer, T. B. "Power Sources for Implanted Telemetry Systems," (especially pp. 31–40; Fryer; T. B., and H. Sandler, "A Review of Implant Telemetry Systems," pp. 351–374.).

KONIGSBERG, E., "A Pressure Transducer for Chronic Intravascular Implantation," in *Fourth National Biomedical Sciences Instrumentation Symposium,* Anaheim, Calif., May, 1966.

MACKAY, R. S., *Bio-Medical Telemetry.* New York: John Wiley & Sons, Inc., 1968.

"Special Issue on Emergency Medical Services Communications," *IEEE Transactions on Vehicular Technology,* VT-29, No. 4 (November, 1976).

Chapter 13

ANNINO, J. S., *Clinical Chemistry.* Boston: Little, Brown and Company, 1964.

LEE, L. W., *Elementary Principles of Laboratory Instruments,* St. Louis, Mo.: The C. V. Mosby Company, 1970.

WHITE, W. L., M. M. ERICKSON, AND S. C. STEVENS, *Practical Automation for the Clinical Laboratory.* St. Louis, Mo.: The C. V. Mosby Company, 1965.

WILLARD, H. H., L. L. MERRITT, AND J. A. DEAN, *Instrumental Methods of Analysis,* 2nd ed. New York: Van Nostrand Reinhold, 1974.

Chapter 14

BLAHD, W. H., *Nuclear Medicine.* New York: McGraw-Hill Book Company, 1965.

"Cardiac Catheterization," *Directions in Cardiovascular Medicine,* Vol. III. Somerville, N.J.: Hoechst Pharmaceuticals, Inc., 1973.

CHASE, G. D., AND J. L. RABINOWITZ, *Principles of Radioisotope Methodology,* 3rd ed. Minneapolis, Minn.: Burgess Publishing Company, 1967.

HINE, G. H. (ed.), *Instrumentation in Nuclear Medicine.* New York: Academic Press, Vol. 1, 1967, Vol. 2, 1974.

JAUNDRELL-THOMPSON, F., AND W. J. ASHWORTH, *X-Ray Physics and Equipment.* Philadelphia: F. A. Davis Company, 1970.

JOHNS, W. E., *The Physics of Radiology.* Springfield, Ill.: Charles C Thomas, Publisher, 1961.

QUIMBY, E. H., S. FEITELBERG, AND W. GROSS, *Radioactive Nuclides in Medicine and Biology: Basic Physics and Instrumentation.* Philadelphia: Lea & Febiger, 1970.

SELMAN, J., *The Fundamentals of X-Ray and Radium Physics.* Springfield, Ill.: Charles C Thomas, Publisher, 1965.

SILVER, S., *Radioactive Nuclides in Medicine and Biology: Medicine.* Philadelphia: Lea & Febiger, 1968.

Chapter 15

BARNA, A., AND D. I. PORAT, *Introduction to Microcomputers and Microprocessors.* New York: John Wiley & Sons, Inc., 1976.

DAVID, D. O., AND B. D. PRESSMAN, "Computerized Tomography of the Brain," *Radiologic Clinics of North America,* 12, No. 2 (August, 1974), 297–313.

DAVIS, G. B., *An Introduction to Electronic Computers.* New York: McGraw-Hill Book Company, 1965.

GORDON, R., G. T. HERMAN, AND S. A. JOHNSON, "Image Reconstruction from Projections," *Scientific American,* 233, No. 4 (October, 1975), 56–68.

HAGA, E. (ed.), *Computer Techniques in Biomedicine and Medicine.* Philadelphia: Querbach Publishers, Inc., 1973.

HOESCHELE, D. F., JR., *Analog-to-Digital/Digital-to-Analog Conversion Techniques,* New York: John Wiley & Sons, Inc., 1968.

JULIUSSEN, J. E., "Magnetic Bubble Systems Approach Practical Use," *Computer Design,* October, 1976, 81–92.

JURGEN, R. K., "Electronics in Medicine," *IEEE Spectrum,* January, 1978, 68–72.

KNIGHTS, E. M., JR., (ed.), *Mini-Computers in the Clinical Laboratory.* Springfield, Ill.: Charles C Thomas Publisher, 1970.

KORN, G. A., *Microprocessors and Small Digital Computer Systems for Engineers and Scientists.* New York: McGraw-Hill Book Company, 1977.

MILHORN, H. T., JR., *The Application of Control Theory to Physiological Systems.* Philadelphia: W. B. Saunders Company, 1966.

PEATMAN, J. B., *Microprocessor-based Design.* New York: McGraw-Hill Book Company, 1977.

SIPPL, C. J., AND D. A. KIDD, *Microcomputer Dictionary and Guide.* Champaign, Ill.: Matrix Publishers, Inc., 1976.

SOUCEK, B., *Microprocessors and Microcomputers.* New York: John Wiley & Sons, Inc., 1976.

STACY, R. W., AND B. D. WAXMAN (eds.), *Computers in Biomedical Research,* Vols. I, II, III, and IV. New York: Academic Press, 1965–1974.

WELLS, P. N. T., AND J. P. WOODCOCK, *Computers in Ultrasonic Diagnosis.* Forest Grove, Ore.: Research Studies Press, 1977.

ZAKLAD, H., "Computerized Multiple X-Rays Give a New View of the Body's Interior," *Electronics,* October 14, 1976.

Chapter 16

BRUNER, J. M. R., "Hazards of Electrical Apparatus," *Anesthesiology,* 28 (1967), 396–424.

Electricity in Patient Care Facilities. Boston: National Fire Protection Association, 1973.

KEESEY, J. C., AND F. S. LETCHER, *Minimum Threshold for Physiological Responses to Flow of Alternating Electric Current Through the Human Body at Power-Transmission Frequencies.* Bethesda, Md.: Naval Medical Research Institute Research Report MR 005.08-00300B #1, September, 1969.

Manual for the Safe Use of Electricity in Hospitals. Boston: National Fire Protection Association, 1971.

National Electrical Code 1971. Boston: National FIre Protection Association, 1971.

Patient Safety, Application Note AN718. Waltham, Mass.: Hewlett–Packard, Inc., Medical Electronics Division, 1970.

PFEIFFER, E. A., "Electrical Stimulation of Sensory Nerves with Skin Electrodes for Research, Diagnosis, Communication and Behavioral Conditioning: A Survey," *Medical and Biological Engineering,* 6 (1968), 637–651.

Pfeiffer, E. A., "Plugs and Receptacles in the Hospital: A Compendium for the Clinical Engineer," *Clinical Engineering,* 2, No. 1 (1976), 46–55.

Pfeiffer, E. A., and F. J. Weibell, "Safe Current Limits: Are They Too Low?" *Biomedical Safety and Standards,* 2, No. 8 (August 10, 1972), 92–94.

Weibell, F. J., "Electrical Safety in the Hospital—1974," *Annals of Biomedical Engineering,* 2 (1974), 126–148.

Index

(*Medical terms are only indexed for reference in the body of the text. For further definition, please see the glossary commencing on page 491.*)

A-to-D Conversion (*see* Analog-to-digital conversion)
Absolute refractory period, 53
Absorbance (optical), 301
Absorbance or Optical density, 352
Absorbtivity, 353
Accumulator, 387
Accuracy, 8
AC defibrillation, 207
Acetylcholine, 284
Acetylcholine esterase, 284
Acoustic impedance, 258
Action potentials, 20, 50-54, 284, 300
  generation of, 50-52
  propagation of, 53

Action electrode (*see also* Indicator electrode), 76
Active transducers, 27
Activity (of ions), 66
Activity coefficient, 64-66,
Address, 386-88
Afferent nerves, 278
After potentials, 50
Aging of electrodes, 81
Aikin, Howard, 385
Air flow measurements, 221, 224, 227
Airway resistance, 217, 222
Airway resistance measurements, 230
Alarm system, in patient monitoring, 175-80
All-or-nothing law, 53
Alpha rays, 364
Alpha waves (EEG), 59
Alveoli, 86, 216
Amplifier:
  buffer, 119
  operational, 405

Amplitude modulation (AM), 325
Analog-to-digital conversion, 406-9
Analyzer:
  carbon monoxide (*see* Carbon monoxide analyzer)
  gas analyzer (*see* Gas analyser)
Anatomy of nervous system, 278-82
AND logic function, 387
Angina pectoris, 100
Angiography, 135, 374
Angular displacement transducers, 36
Anterior-anterior, 211
Anterior-posterior, 211
Aorta, 88
Aperture time, 409
Apex (of the heart), 55
Apex cardiogram, 103, 414
Apex cardiograph, 171
Application software, 396
Arithmetic logic (ALU), 387
Arithmetic unit, 386
Arrhythmia, 107, 196
Arterial blood pressure, 88-89, 94-95, 102, 135-50, 463
Arterial diagnostic unit, 190
Arterial system, 96
Arteries, coronary, 85
Artifacts:
  movements, 23, 179
A-scan display, 261
ASCII Code, 403
Aspiration, 242
Assemblers, 397
Assembly language, 397
Assist, 238
Assist control, 238
Assist controller, 239
Association areas (of the brain), 288
Astrup technique (for gas analysis), 236
Asynchronous, 402
Asynchronous 200 (*see* Fixed rate)
Atria (singular: atrium), 18, 86-88
Atrial fibrillation, 207
Atrial heart sound, 101
Atrial-programmed pacers, 200
Atrioventricular node, 54, 89
Atrioventricular ring, 91
Atrioventricular valves, 101
Attenuation constant, 258
Audiodisplay of EMG, 301
Audiogram, 258
Audiologist, 305

Audiology, 305
Audiometer, 309
Auditory nervous system, 280
Auditory system, 290
Augmented unipolar limb-leads, 115
Ausculation, 100
Ausculatory method, 128
Autoanalyzer R, 358, 417-19
Automatic-blood pressure, 128
Automatic control, 16
Automatic nervous system, 280
Automatic storage or processing, 15
Automatic three-channel records, 123
Auxiliary input, 119
Averaging, 407
AV node, (*see* Atrioventricular node)
Axial tomography, 421
Axis, electrical, 107
Axon, 278

Babbage, Charles, 384
Background, 376
Background count, 377
Bacteriological tests, 346
Bacteriology, 346
Balancing, 8
Ballistocardiogram, 103, 464
Barium sulfate, 374
Baroreceptors, 19, 92
Basal ganglia, 287
Basal skin resistance (BSR), 306
Baseline, 56
Baseline stability, 9
Basilar membrane, 290
Batch processing, 394
Becquerel, Henry, 364
Bedside patient monitor, 174-80
Beers law, 353
Behavior, 277, 304
Behavior therapy, 313
Békésy audiometer, 309
Békésy, George Von, 309
Bell (spirometer), 223
Beta rays, 364
Betatrons (*see* Linear accelerators)
Beta waves (EEG), 59
Bicuspid valve, 88
Bifurcates, 86
Bilaterally symmetrical, 280
Bimorph, 30
Binary coded decimal, 403
Binary computer input, 403

Binary ladder circuit, 405
Binary numbering system, 387
Biochemical system, 18
Biochemical transducers, 76
Bioelectric potential electrodes, 66
Bioelectric potentials, 54-62
Bioengineering (*see also* Biomedical engineering), 2-3
Biofeedback, 308, 314
Bioinstrumentation (*see also* Biomedical instrumentation), 2
Biomedical engineer, 3
Biomedical engineering, 1-9
Biomedical equipment tech (BMET), 3
Biomedical instrumentation
  classification, 12
  definition of, 2, 4-9
Biometrics, 2, 4-9
Bionics, 3
Biophysicist, 3
Biopotential electrodes (*see* Bioelectric potential electrodes)
Biopsy, 346
Biotachometer, in operating room monitoring system, 187, 189
Biotelemetry, definition, 317
Bipolar, 75, 113, 295
  electrodes, 203
  measurements of neuronal firing, 293-94
Bipolar limit-lead, 113
Bistable elements, 387
Bit, 387
Bit-slice processor, 400
Black box, 10-12
Blood, 64, 345
Blood cell counter, 347-49
Blood cells, 345
Blood clotting, 345
Blood drawing lists, 417-18
Blood flow:
  angiography, 135
  bolus, 160
  characteristics, 98-100
  values, 98
Blood flow measurements, 150, 464
Blood flow meters:
  Doppler, 155
  magnetic, 151-54
  radiographic, 157
  thermal, 157
  ultrasonic, 154-56
Blood flow (transducers) probes, 151

Blood gas electrodes, 79-82
Blood oxygenation (*see* Oxygenation of blood)
Blood plasma, 345
Blood platelets, 345
Blood pressure:
  computerized analysis, 370
  diastolic, 89, 94-95
  direct, 135
  indirect, 126
  measurement, 98, 126-50
  systolic, 89, 94, 95
Blood pressure kymograph, 135
Blood pressure transducers, 137-50
Blood serum, 346
Blood vessels, 18, 64-68, 75
Blood volume (*see also* Plethysmography), 84, 132-38
Body plethysmograph, 229
Body surface electrodes, 42, 45-50.
Body temperature (*see* Temperature)
Bolus method, 232
Bonded, 138
Bonded silicon element, 139
Bonded strain gage (*see also* Strain gage, bonded), 29
Boole, George, 384
Bourbon tube, 47
Bradycardia, 106, 196
Brain, 278
Brain midline location by ultrasound, 219-20
Brainstem, 281
Brain waves, (*see* Electroencephalogram)
Breast lesion detection by ultrasound, 219-20
Breathing, mechanics of (*see* Mechanics of breathing)
Bridge circuit (*see* Wheatstone bridge)
Bronchi, 213
Bronchioles, 215, 216
Brocho-spirometer, 225
Bruxism, 315
B-scan, 263
BTPS, 178
Bubbles, 390
Bucky diaphragm (or a grid), 374
Buffer amplifier, 119
Buffer register, 402
Buffy layer, 347
Bundle of His, 55, 91
Burns, chemical, 372
Burns, electrical, 371
Bus, ground, 383
Bytes, 387

Calibration, 9, 400
Calomel electrode, 77
Cannula-type transducer, 152
Capacitance manometer, 137
Capacitance plethysmograph, 164
Capacitance transducers, 41
Capacity:
  diffusion, 189
  forced vital, 270
  function residual, 270
  inspiratory, 220
  maximal breathing, 221
  timed vita, 220
  total lung, 220
  vital, 220
Capillaries, pulmonary, 218
Capillary electrometer, 108
Capture, 200
Carbon dioxide analyzers (*see* P_{CO_2} measurements)
Carbon monoxide analyzer, 233-35
Cardiac care unit, 194
Cardiac catheterization, 125, 428
Cardiac catheterization laboratory, 428
Cardiac monitoring (*see* Coronary care monitoring)
Cardiac observation unit, 194
Cardiac output, 98
Cardiac output computer, 160-62
Cardiac output measurement, (*see also* Blood flow measurement), 157-62
Cardiotachometer, 176
Cardiovascular system, the, 18-19, 84
  typical values (*table*), 98
Cardioversion, 212
Carrier, 395
Carriers, radio-frequency (RF), 272
Cassettes, 370
Cassette tapes, 390
Catheter, 135
Catheter, pacing, 197
Catheterization, 135
Catheterization (cardiac) laboratory, 428
Catheter-tip blood pressure transducer, 136-37
Cath. lab, 192
Cathode-ray-tube display, 181, 395
Caudal plethysmograph, 169
Cell body (neuronal), 278
Cell hemoglobin concentration, mean, 296
Cells, blood, 345
Cells, glial, 279
Cell volume, mean, 347

Central nervous system, 279
Central nurses' station, 279
Central processing unit (CPU), 388
Central sulcus (of the brain), 287
Central terminal, 115
Cerebellum, 282, 286
Cerebral cortex, 282, 287
Cerebrovascular accident (CVA), 99
Cerebrum, 282, 287
Channels, 325
Characteristic impendance, 258
Characters (computer), 387
Charge-coupled devices (CCD), 390, 391
Chemical analysis methods, 233
Chemical systems, 18
Chemical transmission (of action potentials), 286
Chemoreceptors, 19
Chest electrodes, 113
Chloridimeter, 356
Cine (or video) angiography, 371
Circulation:
  pulmonary, 85
  systemic, 85
Clark electrode, 80
Clinical engineer, 3
Clinical instrumentation, 12
Clinical laboratory, 417-19
Clinical psychologist, 304
Closed-circuit technique (for residual volume), 228
Closing volume, 221, 232
CNS (*see* Central nervous system)
CO analyzer (*see* Carbon monoxide analyzer)
CO_2 electrodes, 81
CO_2 elimination, 218
Coding of neuronal information, 290-91
Collaterals, 278
Collimated detector, 377
Color enhancement, 426
Colorimeter (*see also* filter photometer), 353
Color vision, 291
Common-mode gain, 110
Common-mode rejection ratio, 111
Competitive, 200
Compilers, 397
Compliance, lung, 217, 222
Computer, 384
  application to medicine, 409-29
  capabilities, 409-13
  control of process, 412, 413
  digital, 386-98

Computer *(continued):*
  elements of, 386-96
  hardware, 386
  history, 384-85
  in patient monitoring, 184, 419, 420
  interface to instrumentation, 392
  interfacing, 401-9
  program, 387-88
  software, 386, 396-98
  terminal, 393
  word, 387
Computerized axial tomography (CAT), 244, 421
Concentration, hemoglobin, 347
Conditioning (classical or Pavlovian), 308
Conductor, inhomogeneous volume, 432
Conduit, electrical, 437
Cones of the retina, 290
Configuration of thermistors, 248
Contact potential (of thermocouple), 248
Contingency, 312
Continuous Doppler, 260
Continuous wave (CW) transmission, 325
Contrast medium (X-ray), 374
Control functions, 412
Control of experiment or process by computer, 412
Control unit, 386, 388
Conventional or bouncing-ball display, 181
Conversational mode *(see* Interactive)
Conversion rate, 409
Converter:
  analog-to-digital, 406
  digital-to-analog, 403
Cord, spinal, 281
Core memory, 388
Coronary arterial system, 85
Coronary arteries, 85
Coronary arteriography, 375
Coronary care monitoring, 174-85
Coronary care unit, 194
Coronary sinus, 86
Corpus callosum, 282
Cortex, cerebral, 282
Coulometric, 356
Coulter counter (blood cells), 248-51
Counters:
  Geiger, 368
  liquid scintillation, 382
  planchet, 381
  proportional, 367
  scintillation, 376

Countershock, 207
Cow (radioisotopes), 381
Crossed over, 280
Cumulative-event recorder, 313
Curie, Pierre and Marie, 367
Current, let-go, 433
Current limiter, 444
Curvilinear, 109

D-to-A conversion *(see* Digital-to-analog conversion)
Data acquisition, 409
Data acquisition by computer, 409
Data presentation, 412
Data processing equipment, 16
Data reduction and transformation, 410
DC defibrillation, 207
Dead space, 221
Dead space air, 221
Dead space ventilation per minute, 221
Decay, radioactive, 364
Dedicated computer, 398
Defibrillation, *(see* Defibrillators)
Defibrillators, 206-12
Delta waves (EEG), 50
Demand, 200
Demand pacemaker, 199
Demodulating, 325
Dendrites, 278
Density of materials, 256-57
Density, packing, 347
Depolarization of cell, 51
Depolarized cell, 51
Detector, collimated, 377
Detectors, scintillation, 377
Diagnosis, 12
Dialyzer, 359
Diaphragm, 217
Diastole, 88
Dicrotic notch, 94
Differential amplifier, 109
Differential count, 349
Differential gain, 111
Differential pressure, 143
Differential pressure transducers, 47
Differential transformer, 137
Differential transformer transducers, 41, 145
Diffusing capacity *(see also* Transfer factor), 188, 189-91
  for the whole lung, 233
  using CO infrared analyzer, 233

Diffusion, gas, 233
Digital computer, 386
Digital data, 387
Digital magnetic tape drives, 390
Digital-to-analog converter, 403
Digitizing, 406
Discharges, spike, 282
Disk drive, 389
Diskette or floppy disk, 389
Disk, magnetic (for computer), 383
Disk memory, 389
Displacement, measurement of, 46
Displacement transducer, 46
Display, 15
Display equipment, 14-16
Disposal electrodes, 73
Distribution systems, electrical, 437
Divalent ion electrodes, 82
Doppler effect, 259
Doppler shift, 134
Double-beam, 357
Dual insulation, 440
Dual-slope or up-down integrator converter, 407
Dye dilution methods, 157
Dynamometers, 309

Ear-clip electrodes, 73
Ear oximeter, 237
EBCDIC Code, 403
Echocardiogram, 263, 269
Echoencephalogram, 263
ECG (see Electrocardiogram; Electrocardiograph)
ECG electrodes, 69-73, 112-13
  placement patterns, 112-17
ECG lead configurations, 114, 415
EEG (see Electroencephalogram)
EEG electrode placement system, 296
EEG electrodes, 73-75, 296-97
EEPSP (see Excitatory postsynaptic potential)
Efferent nerves, 278
EGG (see Electrogastrogram)
Einthoven triangle, 113
Einthoven, Willem, 49
EKG (see Electrocardiogram; Electrocardiograph)
Elastic strain gage (see Strain gage, elastic)
Electret, 29
Electrical axis, 107
Electrical accidents, 430-36

Electrical isolation, 9
Electrically alterable read-only memory (EAROM), 389
Electrical transmission (of action potentials), 284
Electric induction, 29
Electrocardiogram, 54, 106-15, 181, 413
  analysis by computer, 413-16
  monitoring, 181-87
Electrocardiograph, 55, 107, 117-21, 176
  in automated system, 419
Electrochemical activity of calls, 50-53
Electrode, 112
Electrode impedance, 66-68
Electrode offset voltage, 67
Electrode placement system, 296
Electrode potential, 64
Electrodes:
  active, 76
  biochemical, 76
  bioelectric potential, 66
  bipolar, 75
  blood gas, 79
  body surface, 69
  calomel, 77-78
  Clark, 80
  CO_2, 80-81
  disposable, 73
  divalent ion, 82
  ear-clip, 74
  ECG, 70-73
  EEG, 74-75, 296-300
  electro-chemical, 64
  EMG, 61, 75, 300
  exploring, 115
  floating skin surface, 71-73
  flow-through, 82
  glass, 78-79
  hydrogen, 76
  immersion, 70
  indicator, 76
  indifferent, 115
  mercurous chloride, 77-78
  micro, 68
  needle, 74
  oxygen, 81
  P_{CO_2}, 80-81
  pH, 78
  plate, 70
  platinum, 79
  P_{O_2}, 81
  polarographic, 80

Electrodes *(continued):*
  potentials of, 65
  reference, 76, 77
  scalp, 74
  Severinghaus, 81
  silver-silver chloride, 77
  skin surface, 70, 72, 74
  sodium, 82
  specific ion, 82-83
  spray-on, 73
  stomach pH, 78
  suction cup, 71
  theory of, 64-66
  unipolar, 75, 115
Electroencephalogram, 57, 243-47, 411
Electroencephalograph, 297
Electrogastrogram, 62, 463
Electrolyte bridge, 71
Electrolyte jelly *(see* Electrolyte paste)
Electrolyte paste (for electrodes), 70
Electromagnetic spectrum, 365
Electrometer, capillary, 108
Electromyogram (EMG), 61, 75, 300
Electromyograph, 119, 300
Electronic spirometers, 225
Electronic stethoscope, 170
Electronic thermometer *(see* Thermometer)
Electron-beam addressable memories
      EBAM), 391
Electrooculogram, 62, 463
Electroretinogram, 62, 463
Electrosphygmomanometer, 130
Electro surgery, 432
Eluting, 381
Embolism, 100
Emergency room, 194
Emission computerized tomography, 428
EMAC, 398
EMG *(see* Electromygram)
EMG, electrodes, 61, 75, 300
End expiratory level, 219
End inspiratory level, 219
Endorradiosonde, 322
End point, 355
Energy conversion, 27
Engineering biomedical *(see* Biomedical
      engineering)
Environmental health engineering, 3
Environment internal, 213
EOG *(see* Electrooculogram)
Epiglottis, 215
Equipment ground, 437-40

Equipotential grounding system, 446
ERG *(see* Electroretinogram)
Errors in biomedical measurements, 7
Erythrocytes *(see* Red blood cells)
Evaluation, 12
Evoked potentials, 300
Exercise stress test, 124
Excitation (neuronal), 293-95
Excitation voltage, 27
Excitatory postsynaptic potential, 285
Executive (system), 398
Exercise ECE, 428
Experimental pyschologist, 304
Expiratory flow *(see* Air flow measurements)
Expiratory reserve volume (ERV), 219
Exploratory electrode, 115
External environment, 171
External pacemaker *(see* also Pacemaker,
      external), 198
External respiration *(see also* Respiration,
      external), 213
Extracellularly, 293

Fault hazard current, 445
Fault resistance (or leakage resistance), 439
Feedback, 308
  biological, 308
  in physiological systems, 23, 292
  of EEG, 246
Fetal pulse detector, 221
FEV *(see* Forced expiratory volume)
Fever, 246
Fibers, nerve, 278
Fibrillation:
  artrial, 206
  ventricular, 207
Fibrin, 346
Fibrinogen, 345
Fick method, 158
Fight or flight reaction, 292
Filter-fluorometer, 357
Filter-photometer *(see also* Colorimeter), 353
First and second contact resitance, 439
First-degree block, 196
First heart sound, 101
First order hold, 407
Fixed-rate or asynchronous *(see also* Asyn-
      chronous), 200
Flame photometer, 356
Floatation catheter, 146
Floating skin surface electrodes, 71
Floppy disk (or diskette), 389

Flow air (*see* Air flow):
  blood (*see* Blood flow)
Flow-through electrodes, 82
Fluorescence, 357, 370
Fluorometer, 357
Fluoroscopes, 370
Foil strain gage, 39
Force, measurement of, 43
Forced expiratory volume (FEV), 220
Forced transducer, 43
Forced vital capacity (FVC), 220
Force-summing member, 43
Formatting and printout, 401, 402
Formed elements (blood), 345
Formed elements or blood cells, 345
Fortran, 398
Fourier transformation, 410
Frank lead system, 123
FRC (*see* Functional residual capacity)
Frequency modulation (FM), 325
Frequency multiplexing, 325
Frequency response, 8
  of heart sounds, 171
Frontal lobe (of the brain), 288
Functional residual capacity (FRC), 220

Gage factor, 37
Gain, 119
Galvani, Luigi, 49
Galvanic skin response (GSR), 306
Gamma rays, 364
Ganglia, 280
Gas analyzers, 232-37
Gas chromatograph, 235
Gas diffusion (*see* Diffusion, gas)
Gas exchange in lungs, 217-18, 232
Geiger counter, 368
Glass electrode, 78, 79
Glial cells, 279
Graded potentials, 284
Gray matter, 279
Grenze, 383
Grenz rays, 383
Grid (or a Bucky diaphragm), 374
Ground (*see* Equipment ground; Equipotential
      ground)
Ground fault circuit interrupter, 441
Grounding, 440
Grounding resistance, 439
Ground return resistance, 439

Haldane gas analyzer, 233
Half-cell, 78
Half-cell potential, 78
Half life, 366
Half-value layer, 259
Hall generator, 42
Hamilton method, 160
Handshaking, 403
Hardware, 386
Heart, 18, 89-92
  attack, 100
  electrical axis, 107
  murmurs, 101
  pump analogy, 86-88
  rate, 19, 92
Heart block, 196
Heart sounds, 100-4, 169-72
Heat, stylus, 119
Helium, use of in residual volume measure-
      ments, 228
Hematocrit, 347
Homoglobin, 345
Hemoglobin concentration, 347
Hemoglobin, mean cell (MCH), 347
Heparin, 346
Hierarchy of organization (man), 16-18
Histological tests, 346
Histology, 346
Hollerith, Herman, 384-85
Homunculus, 288
Hospital engineer, 3
Hospital grade, 440
Human factors engineering, 3
Hydraulic system in body, 18-19
Hydrogen electrode, 76
Hydrogen ion (*see also* pH), 78-79
Hypercapnia, 218
Hypertension, 126
Hyperventilation, 218
Hypophysis (*see also* Pituitary gland), 282
Hypothalamus, 282
Hypoventilation, 218
Hypoxia, 218
Hysteresis, 7

Idioventricular focus, 196
Image intensifier, 370
Immersion electrodes, 69
Impedance:
  of electrodes, 66

Impedance plethysmograph, 167
Impedance pneumograph, 232
Implantable transducers, 147
Implantation techniques, 137
Implanted pacemaker (*see* Pacemaker, implanted)
Inaccessibility of variables, 21
Indicator electrode, 76
Indicator or dye dilution methods, 157
Inductance transducers (*see* Variable inductance transducers; Induction type transducers)
Induction type transducers, 40
Infarct myocardial (or coronary), 100
Inferior vena cava, 85
Information:
    assimilation and organization, 410
    gathering, 11, 409
    storage and retrieval, 410
Infrared thermometer, 252
Inhalation therapy, 237
Inhalator, 237
Inhibition (neuronal), 284-85
Inhibitory postsynaptic potential, 285
Innervate, 280
Input-output data bus or party-line bus, 402
Input-output devices, 386, 391
Insertion, percutaneous, 135
Inspiratory capacity (IC), 220
Inspiratory flow (*see* Air flow measurements)
Inspiratory reserve volume (IRV), 219
Instrumental behavior (or operant), 312
Instrumentation, biomedical (*see* Biomedical instrumentation)
Instrumentation system objectives, 12
Insulated power distribution system, 444
Integrated circuit memory elements, 399
Integration of EMG waveform, 302
Intensifying screens, 370
Intensive care monitoring, 174-84
Intensive care unit, 194
Interaction:
    among physiological systems, 22
    of transducer with measurement, 22-23
Interactive (or conventional mode), 395
Interfacing the digital computer to instrumentation system, 401
Interference, 8
Intermediate coronary care unit, 194
Internal environment, 213
Internal pacemakers (*see also* Pacemaker, internal), 198

Internal respiration (*see* Respiration, internal)
Interpreters, 397
Intra-alveolar pressure, 217
Intracellularly, 293
Intrathoracic pressure, 217
In vitro measurements, 13
In vivo measurements, 13
Ionic potentials (*see* Bioelectric potentials)
Ionization chamber, 368
Ionizing radiation, 364
Ions, 50
IPSP (*see* Inhibitory postsynaptic potential)
Isolation, 8
Isolation transformer, 443
Isometric, 43
Isometric force transducer, 46
Isopotential line, 57
Isotonic, 43
Isotonic displacement transducer, 46
Isotope, 366

Junction potential (of thermocouple) (*see* Contact potential)

KCl (in electrodes), 77
Kinetic analysis, 355
Korotkoff sounds, 103, 127
Kymograph, 135, 224

Large-scale integration (LSI), 389, 399
Larynx, 215
Latency, 306
Leads, augmented unipolar limb, 114
Leads, limb, 113
Lead selector panel (in EEG), 297
Lead selector switch, 117
Leakage current, 445
Leakage current, electrical, 380
Left ventricle, 91
Let-go-current, 433
Leukocytes (*see* White blood cells)
"Lie detector" (*see* Polygraph)
Limit detection, 411
Limiters, current, 444
Linear accelerators (or) betatrons, 383
Linearity, 7
Linear tomography, 421
Linear variable differential transformer, 40, 143

Liquid scintillation counters, 382
Loading list (for autoanalyzer), 362
Lobes:
　of brain, 286-87
　of lungs, 215, 216
Local (at the same location as the computer),
　393
Lown, Bernard, 207
Lown, method:
　of defibrillation, 207
　waveform (for defibrillation), 209
Lung:
　capacities, 220
　volumes, 219
Lung compliance (*see* Compliance, lung)
Lungs, 213

Machine language, 397
Macroassembler, 397
Macroshock, 432
Magnetic blood flow meters, 115, 151
Magnetic bubble memory (MBM), 390
Magnetic disk, 389
Magnetic drum, 389
Magnetic induction, 28
Man-instrument system, the, 11
　block diagram, 15
　components of, 13-16
Marker, 119
Marriott lead, 117
Mass spectrometer, 235
Mathematical operations by computer, 411
Maximal breathing capacity (MBC), 221
Maximal expiration flow (MEF), 221
Maximal voluntary ventilation (MVV), 221
Maximum midexpiratory flow, 221
MBC (*see* Maximal breathing capacity)
Mean arterial pressure, 131
Mean cell hemoglobin concentration
　(MCHC), 347
Mean circulation time, 98
Mean velocity, 98
Measurement of closing volume, 232
Measurement of residual volume, 228
Mechanics of breathing, 218
MED frequencies, 340
Medical engineer, 3
Medical engineering (*see also* Biomedical
　engineering), 2
Medical technologist, 3
Medium-scale integration (MSI), 399

Medulla, 281, 286
Medulla oblongata, 281
MEF (*see* Maximal expiration flow)
Membrane potentials:
　of cells, 50-53
　in electrodes, 64-65
Memory of a digital computer, 386-90
　core, 388
　integrated circuit, 388
　random access, 388
　volatile, 388
Mercurous chloride electrode (*see* Calomel
　electrode)
Mercury strain gage, 165
Mercury strain gage plethysmograph (*see*
　Plethysmograph):
　mercury strain gage, 38
Mercury thermometer (*see also* Thermometer),
　246
Metabolism, 212
Microcomputers, 393, 396
Microelectrodes, 66-69
Microphones, 170-71
Micropipette electrode, 69
Microprocessor, 399
Microprograms, 385, 387
Microshock, 432
Microshock, electrical, 432
Microshock hazard, 430
Microtome, 346
Midbrain, 281
Minicomputers, 393, 396
Mitral valve, 88
Mobile emergency care units, 194
Modem, 396
Modified chest lead I, 117
Modulation, 323
　amplitude, 325
　frequency, 325
　pulse, 326
　pulse amplitude, 326
　pulse position, 326
　pulse width, 326
Modulation of a carrier signal, 27
Mole, 64
Monitor (system), 398
Monitoring, patient (*see* Patient monitoring)
Monochromator, 357
Monophasic, 2
Motor nerves, 280
Motor response, 308
Mowrer sheet, 313

M-scan display, 261
Multiplier (analog), 403
Murmurs, 101
Muscle potentials (*see* Electromyogram)
Muscle tone (*see* Tone, muscle)
Muscles, intercostal (*see* Intercostal muscles)
MVV (*see* Maximal voluntary ventilation)
Myelin, 278
Myelinated nerve fiber, 278
Myelin sheath, 278
Myocardial (or coronary) infarct, 100

Nasal cavities, 215
National Aeronautics and Space Agency
    (NASA), 4-5
National Electrical Code (NEC), 491
Nebulizer, 241
Needle electrodes, 66, 296
Nernst equation, 64
Nerve, 278
Nerve conduction rate, 53
Nerve conduction time measurements, 295
Nerve conduction velocity, 53
Nerve fibers, 278
   afferent, 278
   efferent, 278
Nerves, 20-21, 277-86
   afferent, 278
   efferent, 278
   motor, 278, 280
   peripheral, 278
   sensory, 278, 280
Nerve velocity (*see* Nerve conduction time)
Nervous system, 20-21, 277-92
Net height, of action potential, 53
Neurilemma, 278
Neurology, 304
Neuron, 278
Neuronal communication, 282-86
Neuronal firing measurements, 293-95
   extracellular, 293
   gross, 293
   intracellular, 293
Neuronal receptors, 290
Neuronal spikes, 282
Neutral wire, 437
Nitrogen washout method (for residual volume
    measurements), 228
Nodes of Ranvier, 278
Noise, 8

Noncompetitive, 200
Nonfade displays, 182
Noninvasive, 243
Noninvasive techniques, 106
Noninvasive vascular laboratory, 273
Nonionizing, 364
Nonvolatile, 388
NOR logic gates, 233
Nuclear, 364
Nuclear magnetic resonance (NMR), 429
Nuclei, 280

Occipital lobe (of the brain), 288
Oculo pneumo plethysmograph, 168
OFF, 331
Off-line, 393
Olfactory bulb, 290
ON, 331
On-line (connected to a computer), 393
Open-circuit (or nitrogen washout method),
    228
Operant behavior (or instrumental), 312
Operant conditioning, 312
Operating (system), 398
Operating room, patient monitoring system,
    194
Operational amplifier, 404-05
Ophthalmodynamometer, 169
Ophthalmologist, 305
Optical character recognition (OCR), 393
Optical density, 353
Optometry, 305
Oral temperature, 245
OR logic function, 381
Oscilloscope:
   for EEG evoked responsed, 300
   for EMG display, 301
   in operating room monitor, 194
   in patient monitoring display, 176
Oximeter, ear (*see also* Ear oximeter), 237
Oxygen, intake of, 213-20
Oxygen analyzers (*see* P_{O_2} measurements)
Oxygenation of blood, 217-18
Oxygen electrode, 81
Oxygen-reduction electrode, 81
Oxygen tension, 79
Oxyhemaglobin, 218

P wave (ECG), 57, 414
P_{CO_2} electrodes (*see* CO_2 electrodes)

P_{CO_2} measurements, 233
P_{O_2} electrode (*see* Oxygen electrode)
P_{O_2} measurements, 233
P_{O_2} sensors, 19
Pacemaker, 55, 91, 198
Pacemaker system (*see also* Sinoatrial node), 196
  demand, 199-203
  external, 198-99
  implanted, 198-203
  internal, 198
Pacing artifacts or spikes, 201
Pacing catheter, 191
Paddles, defibrillator, 207-8
Palpatory method, 128
Paradoxical sleep, 59
Parallel-to-serial converter, 402
Paramedic service, 194
Parasympathetic nervous system, 281
Parasympathetic (and sympathetic systems), 292
Parietal lobe of the brain, 288
Parietal pleura (*see* Pleura, parietal)
Parity bit, 390
Passive transducers, 35
Pathologist, speech, 305
Pathways, visual, 380
Patient cable, 117
Patient data file, 358
Patient monitoring, 174
Patient safety (*see* Safety, patient)
Pattern recognition, 411
  of ECG, 414
Pavlov, 308
Pavlovian conditioning (or classical), 308
Peak flow (in respiration), 221
Pediatric paddles, 211
Peltier effect, 34, 248
Pen amplifier, 119
Pen motor, 119
Perception threshold, electrical, 434
Percutaneous insertion, 147
Percutaneous transducers, 147
Pericardium, 89
Period absolute refractory, 28
  relative refractory, 29
Peripheral devices (computer), 391-92
Peripheral nerves, 280
Persistence, 181
Perspiration, 196
pH, 78, 236-38
pH electrode, 78

pH meter, 78
Pharynx, 215
Phased-array sector scanning, 263
Phased-array ultrasonograph, 263
Phonocardiogram, 103, 170, 414
Phonocardiograph, 170
Photocell transducers, 42-46
Photo Darlington, 42
Photodiode, 37
Photoelectric displacement transducers, 45
Photoelectric plethysmograph, 166
Photoemissive cell, 37
Photometer, filter, 351-53
  flame, 356
Photomultiplier, 42
Photoresistive cells, 37
Photo transistor, 42
Physiological resistance transducer, 138
Physiological temperature (*see* Temperature)
Physiological variable, 26
Piezoelectric effect, 30,
Piezoelectric transducers, 31
Pituitary gland, 282
Pixels, 426
Planchet counter, 381
Plasma, 345
Plate electrodes, 70
Platelets, 345
Platinum electrode, 79
Plethysmograph, 150, 163
  body, 229-30
  capacitance, 165
  impedance, 167
  mercury strain gage, 165
  photoelectric, 166
Plethysmography, 163
Pleura, 216
  parietal, 217
  pulmonary, 217
  visceral, 217
Pleural cavities, 216
  pressures, 186
Pleural pressure, 216
Pneumatic system in body, 19-20
Pneumoencephalogram, 374
Pneumo-encephalography, 316
Pneumotachograph (*see also* Pneumotacho-meter), 231
Pneumotachometer, 231
Polarization:
  of cell, 50
  of electrodes, 68

Polarized cell, 50
Polarographic electrode, 80
Polygraph, 305
Pons, 286
Posistor, 249
Potassium ions in producing bioelectric
potentials, 51-53
Potential:
action, 51
electrode, 65
graded, 284
half-cell, 78
resting, 50
skin, 306
Potentials, bioelectric (*see* Bioelectric potentials)
Potentiometer transducers, 35
Power source for pacemaker, 200-1
Power switch, 119
P-Q interval, 57
Preamplifier, 117
Precordial electrodes, 113
Prefrontal lobe (of the brain), 288
Pressorecepters, 92
Pressure:
blood (*see* Blood pressure)
intra-alveolar (*see* Intra-alveolar pressure)
pleural (*see* Pleural pressure)
Pressure-cycled, 239
Primary memory, 388
Probes for blood flow measurements, 119
Problems:
of engineers working with physicians, 3-4
of obtaining measurements from living
organism, 11-12, 21-24
Program (computer), 386
Programmable read-only memory (PROM), 389
Programmed electrosphygmomanometer
PE300, 125
Programming language, 397
Propagation counter, 368
Propagation of action potentials, 53
Propagation rate, 53
Proportional counter, 368
Pseudoplethysmograph, 134-38
Psychiatrists, 304
Psychology, 304
Psychophysics, 304
Psychophysiology, 304
Pulmonary alveoli (*see also* Alveoli), 216
Pulmonary capillaries (*see* Capillaries,
pulmonary)

Pulmonary circulation, 85
Pulmonary function, 215
Pulmonary function indicator, 183-85, 343
Pulmonary function measurements, 426-28
Pulmonary pleura (*see also* Pleura,
pulmonary), 217
Pulmonary valve, 86
Pulsatile transmission (PULSE), 326
Pulse, 103
Pulse amplitude modulation (PAM), 326
Pulse code modulation (PCM), 326
Pulsed Doppler, 260
Pulsed ultrasound, 259
Pulse duration modulation (PDM), 326
Pulse generator, 198
Pulse height analyzer, 376
Pulse interval modulation (PIM), 326
Pulse interval ratio modulation (PIRM), 326
Pulse modulation, 326
Pulse position modulation (PPM),
Pulse width modulation (PWM),
"Purkinje fibers," 55, 92
Pursuit rotor, 308

Q wave (ECG), 57
QRS complex (ECG), 57
Quantizing analog data for computer entry,
406

R wave (ECG), 57, 106, 414
R wave control of pacemakers, 201
Radiation, ionization, 364
Radiation therapy, 364
Radioactive decay, 364, 366
Radioactivity, 364
Radio-frequency (RF) carrier, 323
Radioisotope camera, 381
Radioisotopes, 366
Radioisotope scanner, 378
Radiology, 363
Radio telemetry, 317
Radium, 364
Radium therapy, 364
Radius, 364
Ramp (or pulse-width converter), 406
Random access memory (RAM), 388
Range, 7, 323
Range-gated pulsed Doppler, 260
Ranvier, nodes of, 278
Rapid eye movement (REM) sleep, 59

RAS (see Reticular activating system)
Rate:
  heart, 88
  respiration, 232
Rate meter (radioisotope), 377
Rate meter for ECG monitoring, 176
Read-only memory (ROM), 388
Readout, 388
Real time, 396
Receptors, neuronal (see Neuronal receptors)
Rectilinear, 109
Recognition, pattern, 411
Recorder, 15
Recorder, chart in operating room monitoring
    system, 187
Recording equipment, 16, 187
  for neuronal firing measurement, 295
Recording spirometer (see also Spirometer),
    223
Rectal temperature, 243
Red blood cells, 345
Reduction of date, 411
Reference electrode, 76
Reflexes, spinal, 21, 291
Refractory period, 52
Register, 387
Reinforcement (behavior), 312
Relative refractory period, 53
Remote, 393
REM sleep (see Rapid eye movement sleep)
Repertoire or instruction set, 397
Repolarization, 52
  of cells, 52
Research instrumentation, 12
Residual volume (RV), 219
Resistance, airway (see Airway resistance)
Resolution, 7, 408
  of A-to-D converter, 407
Respiration, 213-18
  external, 213
  internal, 213
  rate, 232
Respirator, 238
Respiratory bronchioles, 216
Respiratory center, 19-20
Respiratory flow (see Air flow measurements)
Respiratory minute volume, 220
Respiratory system, 19-20, 170-94
Respiratory therapy, 237
Response, evoked, 37
Response, galvanic skin (see Galvanic skin
    response)
Resting potential, 50

Resuscitation, 165-69
Reticular activating system, 282, 286
Retina, 237-38
Retransmission, 340
Rheoencephalography (see also Impedance
    plethysmography), 168
Rhythm strips, 122
Rib cage (role in respiration), 174
Right atrium, 86, 89
Right heart, 85
Right ventricle, 86, 89
RIHSA, 321
Risk current, 445
Rods (in the retina), 290
Röntgen, Conrad, 363
Röntgen rays, 363
Rotational potentiometer, 36
Rubber resistor (see Strain gage, elastic)

S wave (ECG), 57, 414
Safety, patient, 24, 430
Sampling (for computer), 406
SA node (see Sinoatrial node)
Scaler, 377
Scalp surface electrodes (for EEG), 73
Scanner, radioisotope (see Radioisotope
    scanner)
Scanning modes, ultrasonic, 261-62
Scholander gas analyzer, 233
Schwann cells, 278
Scintillation detectors (see also Scintillation
    counters), 376
Secondary memory, 389
Second-degree block, 196
Second heart sound, 101
Seebeck effect, 34
Selenium cell, 35
Selective angiography, 375
Semiconductor strain gage (see also Strain
    gage, semiconductor), 39
Semilunar valves, 101
Sensitivity, 7, 119, 203
Sensors, chemoreceptors, 19
Sensory nerves, 280
Serial or parallel form, 402
Serial-to-parallel converter, 402
Septum, 89
Sequential-access memories, 390
Serological tests, 346
Serology, 346
Severinghaus electrode, 81

Shock, 100
Signal conditioning equipment, 15
Signal processing, 15
Signal-to-noise ratio, 8
Silver-silver chloride electrode, 67, 77
Single-chip computer, 400
Sinoatrial node, 55, 91
Sinus, coronary, 85
Skinner, B. F., 312
Skinner-box, 312
Skin potential, 306
Skin resistance, electrical, 435
Skin surface electrodes (*see also* Body surface
    electrodes), 66
Skin surface temperature (*see* Temperature,
    skin surface)
Skopein, 169
Sleep, patterns in the electroencephalogram,
    59-60
Slip-on, 151
Slowly, 232
Small-scale integration (SSI), 399
Smell, 237-38
Sodium ion electrode, 82
Sodium ions in producing bioelectric
    potentials, 51
Sodium pump, 52
Software, 386, 396
Solarcell, 35
Soma (*see also* Cell body), 278
Somatic sensory nervous system, 280
Sonic gas analyzer, 188, 191-92
Sonofluoroscope, 263
Spark chamber, 368
Specific ion electrodes, 82
Spectrofluorometer, 357
Spectrophotometer, 357
Speech pathologist, 305
Sphygmomanometer, 126
Spike discharge patterns, 282
Spike discharges, 282
Spinal cord, 279
Spinal reflexes (*see* Reflexes, spinal)
Spirogram (*see also* Spirometer), 225
Spirometer, 223-27
    broncho, 225
    electronic, 225
    recording, 224
    waterless, 224
    wedge, 224
Spirometer factor, 227
Spray-on electrodes, 73
Stability, 9

Standard, 353
Standardization, 117
Standardization adjustment, 119
Standby, 200
Start bits, 402
Static compliance, 222
Static memories, 388
Statistical analysis of data, 412
Steadiness tester, 308
Stereoradiography, 375
Stereotaxic instrument, 75
Stethoscope, 169
    electronic, 170
Stimulation of nervous system, 293
Stimulus, 13
Stochastic process, 21
Stomach pH electrode, 78
Stop bits, 402
Storage and retrieval, 410
Strain gage, 37
    bonded, 39
    bridge, 38
    foil, 39
    mercury, 38
    semiconductor, 39
    transducers, 39
    unbonded, 38
Stroke, 99
Stroke volume, 98
Stylus heat, 119
Subcarrier, 325
Subject, the, 13
Successive approximation method, 408
Suction apparatus, 242
Suction cup electrodes, 71
Suffix gram, 54
Suffix graph, 54
Sulci (of the brain), 287
Superior vena cava, 85
Supervisor (system), 398
Surface electrodes, 73
Surface or skin temperature, 246
Swan-Ganz catheter, 160
Symbolic addressing, 397
Sympathetic nervous system, 281
Sympathetic (and parasympathetic system),
    292
Synapses, 280
Synchronous, 402
Syntax, 397
Systemic circulation, 85
Systemic temperature (*see also* Temperature,
    systemic), 245

System software, 396
Systole, 88

T wave (ECG), 57, 414
Table lookup, 401
Tachistoscope, 309
Tachycardia, 106, 212
Tape-cartridge system, 390
Tape, magnetic (for computers), 390
Taste, 291
Technetium 99m, 381
Telemetry:
  blood pressure, 328-30
  ECG, 337-39
  EEG, 332
  implantable, 332-37
  in emergency care, 340
  multichannel, 330
  pH, 322
Telemetry system, 321
Temperature:
  body, 241, 244
  control center, 245
  measurement of, 244-55
  oral, 245
  rectal, 245
  skin surface, 246
  systemic, 245-46
  underarm, 245
Temperature coefficient (of thermistor), 248
Temporal lobes (of the brain), 288
Ten-twenty (10-20) EEG electrode placement
    system, 297
Terminal bronchioles, 216
Terminal, computer, 395
Thalamus, 286
Therapy:
  behavior, 313
  radium, 364
  X-ray, 364
Thermal convection measurement of blood
    flow, 160
Thermistor, 247
  self heating, 248
Thermocouple, 33, 247
Thermogram, 254
Thermograph, 252
Thermometer:
  electronic, 246
  infrared, 252
  mercury, 246
Thermostat of body, 245

Thermostromuhr, 157
Thermovision, 254
Theta waves (EEG), 59
Third-degree block, 196
Third heart sound, 101
Thoracic cavity, 215
Thorax, 215
Threshold, 200
Thrombocytes (*see* Platelets)
Thrombosis, 100
Thrombus, 99
Thyroid, 321
Tidal volume, 219
Time-cycled ventilors, 239
Timed vital capacity, 220
Time integral of EDG, 415
Time multiplexing (*see also* Multiplexing, time)
    327, 409
Time sharing, 394
Tomography, computerized axial (CAT), 244
Tone, 92
Tomography, 375
Tomos, 375
Total lung capacity (TLC), 220
Trachea, 215
Transducers, 14, 27
  acceleration, 46
  bloodflow, 152
  blood pressure, 135
  capacitance, 45
  catheter tip, 139
  digital, 48
  displacement-force, 43, 46
  implantable, 139, 147, 149
  inductive, 29
  linear variable differential transformer, 40,
    145
  passive, 35
  photoelectric, 34, 45
  piezoelectric, 30
  pressure, 47
  resistance, 35, 138
  thermoelectric, 33
  variable inductance, 40
  velocity, 46
Transducers for telemetry system, 328
Transfer factor, 233
Transformer, linear variable differential, 40,
    145
Transit time ultrasonic flow meter, 155
Transmission of action potentials (*see*
    Neuronal communication)
Transmit, 15

BETWEEN ACTOR
AND CRITIC

Selected Letters of

EDWIN BOOTH

and

WILLIAM WINTER

BETWEEN ACTOR AND CRITIC

Selected Letters of

EDWIN BOOTH

and

WILLIAM WINTER

Edited with an Introduction and Commentary by

DANIEL J. WATERMEIER

Princeton University Press
Princeton, New Jersey
1971

Publication of this book has been aided by
a grant from the Moorhead State College Foundation,
Moorhead, Minnesota

This book has been composed in Linotype Monticello

Printed in the United States of America

by Princeton University Press, Princeton, New Jersey

Contents

	List of Illustrations	vi
	Preface	vii
	Introduction	3
I.	Booth's Theatre: 1869-1875	18
II.	American Tours: 1875-1877	51
III.	The Prompt Books: 1877-1879	84
IV.	American Tours: 1879-1880	132
V.	England: 1880-1881	155
VI.	The American Tour: 1881-1882	190
VII.	England: 1882	208
VIII.	Germany and Austria: 1883	226
IX.	American Tours: 1883-1886	250
X.	The Booth-Barrett Tours: 1886-1890	279
	Biographical Note	307
	Selected Bibliography	310
	Index	317

Illustrations*

(following page 150)

* All illustrations are courtesy of the Walter Hampden Memorial Library at The Players, New York, except numbers 1, 7, and 11 which are courtesy of the Folger Shakespeare Library, and numbers 18 and 19 which are courtesy of Robert Young.

1. Manuscript page of Letter 2. Booth to Winter, February 7, 1869.

2. Edwin Booth, his daughter Edwina, and his second wife Mary McVicker Booth about 1875.

3. Edwin Booth as Richelieu.

4. Edwin Booth as Iago.

5. Edwin Booth about 1880.

6. Manuscript pages of Letter 48. Winter to Booth, May 3, 1879.

7. Manuscript pages of Letter 60. Booth to Winter, August 17, 1880.

8. Edwin Booth in 1888.

9. William Winter about 1890.

10. Booth's Theatre, 1869.

11. Manuscript page of Letter 123. Booth to Winter, March 17, 1890.

12. Edwin Booth as Hamlet.

13. Edwin Booth as Benedick in *Much Ado About Nothing*.

14. Edwin Booth as Pescara in *The Apostate*.

15. The Players, 16 Gramercy Park, New York City.

16. Edwin Booth's bedroom at The Players.

17. Edwin Booth's living room at The Players.

18. William Winter in February, 1916.

19. William Winter's home. New Brighton, Staten Island, New York.

Preface

DURING the course of a friendship which lasted over twenty years, Edwin Booth wrote the drama critic William Winter, by Winter's own estimation, between three and four hundred letters. This collection, the largest to any single correspondent, was for many years presumed to have been sold at auction after Winter's death and, according to Ernest Sutherland Bates in the *Dictionary of American Biography*, "hopelessly scattered." They were sold at auction by Winter's youngest son Jefferson between 1923 and 1927, but they were not "hopelessly scattered." Most were purchased by agents of Henry Clay Folger and are preserved in the collection of Booth letters at the Folger Shakespeare Library in Washington, D.C. A few others found their way into the collections of the New York Public Library, the Walter Hampden Memorial Library at The Players in New York, and into private collections such as that of Winter's great-grandson Robert Young of Montpelier, Vermont. In all I have examined about three hundred and fifteen of these letters.

With but a few exceptions, none of these letters have ever appeared in print, nor, to my knowledge, were they examined by any of Booth's modern biographers, who have consequently neglected his long, intimate friendship with Winter. The letters represent then a new and significant source of information about Booth. Indeed, they constitute one of the fullest day-by-day records of his career for nearly twenty years.

Of the hundreds of letters that Winter must have written Booth, I have been able to find only six. In all probability, these six are all that remain. Booth had a passion for privacy. He regarded his letters and those of his correspondents, especially when they disclosed "indelicate" matters, as confidences. Once in a moment of agonizing regret over one of his "unlucky slips" he directed Winter to destroy his letters "as I shall your letters that relate to either's folly." Fortunately, Winter did not on this occasion follow Booth's directive, but Booth undoubtedly destroyed many of Winter's letters because he thought their contents "private." Booth, unlike Winter, moreover, does not seem generally to have sensed the historical importance of letters from his correspondents or, for that matter, of his own letters. He regarded most of them as mere communications to be read, answered if necessary, and discarded. Of the thousands of letters that he must have received during his lifetime, one is surprised to find in the collection of the Walter Hampden Memorial Library how few he preserved. He was, furthermore, constantly on the move. Only during the sum-

mers did he remain stationary for more than a month at a time. Rarely did he have a permanent residence. Much of his correspondence was probably abandoned in hotel rooms or dressing rooms or was lost or discarded during moves from one residence to another. In 1888, for example, when he moved his belongings from storage in Boston to his new quarters at The Players, he told Winter that he destroyed "lots of letters & old papers that have accumulated for many years."

The loss of most of Winter's letters to Booth is frustrating, but it does not diminish the value of Booth's letters. They contribute to a richer and fuller understanding of Edwin Booth, unquestionably one of the most significant artists in the history of the American theatre.

From the collection of Booth-Winter letters (many of which are in fact brief notes of relatively insignificant content) I have selected those letters which I think most reveal Booth's personality as a man and artist and best chronicle his professional activities during the years of his correspondence with Winter.

One hundred eighteen of Booth's letters are presented in their entirety (and relevant excerpts from twenty more are quoted in the footnotes and the narrative augmentation). Because there are so few and because I would like to think they are representative, I have included Winter's six letters. I have also included one lengthy letter from Booth's daughter Edwina to Winter describing in detail her father's reception in Germany in 1883.

The letters are arranged chronologically and divided into ten chapters corresponding to periods of Booth's professional activity. The letters are not, however, presented in a vacuum. Through a series of headnotes, I have tried to orient the reader towards the content of each letter, put it in context and provide a narrative continuity from one letter to the next.

I have tried to let Booth speak with his own voice and have thus preserved his idiosyncrasies of spelling, punctuation, and capitalization, for the most part without editorial comment. Booth frequently used abbreviations, often with the final letter raised and underscored. "Which," "could," and "February," for example, are written "w^h," "c^d," and "Feb^r." I have retained the abbreviations, but in accordance with modern usage I have lowered the final letter and added a period. To avoid needless repetition and typographical clutter most of the salutations and complimentary closes, which are consistently similar, have been omitted, and matters of dating have been reduced to a regularized heading. Winter and his son Jefferson frequently added notes to the letters, as did subsequent owners, but I have retained only those notes which add information to the letter. Postscripts have been normalized flush to the left margin.

Each of the letters has been annotated as completely and as comprehensively as appears necessary for proper understanding. I have, for the most part, determined the dates of New York engagements from George C. D. Odell's *Annals of the New York Stage* (New York, 1927-1949). The dates of engagements outside of New York I have determined from various newspaper reviews and playbills, often collected in scrapbooks of Boothiana.

I am most grateful to Charles H. Shattuck of the University of Illinois who first suggested that I investigate the Booth-Winter correspondence and who magnanimously put at my disposal numerous reviews, articles, and letters which he had collected for his work with Booth's Hamlet.

Robert Young not only gave me permission to print letters and photographs in his collection, but he also made available his own yet unpublished biography of Winter and also Winter's journals, preserved over the years from the auctioneer's block by his late grandmother Mrs. Cornelius Cole Brown, née Viola Rosamund Winter (1881-1970).

The Folger Shakespeare Library, the New York Public Library, Astor, Lennox and Tilden Foundations and the Walter Hampden Memorial Library at The Players have permitted me to reproduce letters in their collections. For countless courtesies, I am indebted to many librarians, but especially to Paul Myers, Curator of the Theatre Collection of the New York Public Library at Lincoln Center, Dorothy Mason of the Folger Shakespeare Library, Alma de Jordy of the University of Illinois Library, and, above all, Louis Rachow, Librarian of the Walter Hampden Memorial Library.

My friends Klaus Hanson of Schiller College at Bönnigheim and Ronald Engle of the University of North Dakota assisted me in the translation of German materials.

Tice L. Miller of the University of West Florida, William Morris of Glassboro State College, and Lisbeth J. Roman of the State University of New York at Binghamton provided me with bits of information gleaned from their own studies in American theatre history.

I am grateful to Barbara Volker for assistance in the preparation of the manuscript; to my sister-in-law Marsha Kane Camera for help in the selection of illustrative material; and to my colleague Timothy Y. C. Choy for critical advice.

To my wife Roberta for her patience and encouragement, I am forever indebted.

DANIEL J. WATERMEIER

Moorhead, Minnesota
May 10, 1970

BETWEEN ACTOR
AND CRITIC

Selected Letters of

EDWIN BOOTH
and
WILLIAM WINTER

Introduction

In *The Rambler* (No. 60. Saturday, October 13, 1750) Samuel Johnson wrote that "the business of the biographer is often to pass slightly over those performances and incidents, which produce vulgar greatness, to lead thoughts into domestick privacies, and display the minute details of daily life, where exterior appendages are cast aside, and men excel each other only by prudence, and by virtue." Booth has been the subject of several biographies and the "performances and incidents" which led to his anything but "vulgar greatness" are familiar to most students of American theatre history.[1] However, none of the biographies adhere as closely to the spirit of Johnson's dictum, none capture as subtly and delicately their subject, as did the subject himself in his letters. During his life, Booth wrote thousands of letters to his many friends, members of his family, professional colleagues, and chance acquaintances. Unlike Joseph Jefferson, he did not write an autobiography, nor like William Charles Macready did he keep a diary. While he certainly never intended them as such, his letters are both his autobiography and diary. Like his acting, they are expressions of his own gentle, unaffected personality, the mirror of his life and soul.

His letters to Winter are familiar, straightforward, unpretentious letters, impulsively written in the mood of the moment. Winter himself characterized them "as sometimes pensive, sometimes gay, sometimes humorous, sometimes satirical, often as fluently pungent as those of Byron, of whose epistolary manner I have often been reminded when reading them." Booth, he continued, "probably never saw Sterne's precept 'write naturally and then you will write well,' but he certainly obeyed it." The letters vary in length. Some are fifteen page epistles, neatly and carefully written on a high quality stationery. Booth often wrote these in the small hours of the morning, or the "raven hours" as he called them, when the excitement and excess energy of a just finished performance were not yet expelled and sleep prevented. Others are terse business notes, scrawled across whatever writing paper was available between acts or at brief whistle-stops.

All the letters are written in the context of sincere and unselfish

---

[1] Winter published the "standard" biography, *Life and Art of Edwin Booth* (New York, 1893; revised 1894), shortly after Booth's death. He has also been the subject of three twentieth-century popular biographies: Richard Lockridge, *The Darling of Misfortune* (New York, 1932), Stanley Kimmel, *The Mad Booths of Maryland* (Indianapolis, 1940) and Eleanor Ruggles, *Prince of Players* (New York, 1953).

friendship. They are not the letters of an eminent actor self-consciously writing to an equally eminent drama critic, but the letters of a man who happened to be an actor corresponding with his close friend who happened to be a critic. Long before their correspondence and friendship began, Winter had indicated in numerous reviews his admiration for Booth's acting. Booth, who initiated the correspondence, did not, however, set out to take advantage of Winter's admiration, to "use" him as an instrument to communicate his ideas about acting or particular interpretations to the general public, nor did he ever use him in such a way. Booth, moreover, never asked Winter to reply for him to adverse or disparaging critics, nor is it likely that Winter would have accommodated himself to such a task. Occasionally, he did ask Winter to "set the record straight" as when he asked him to write an "authorized" account of the failure of Booth's Theatre (which, incidentally, Winter did not do) or observed that "a word in print" might help a particular theatrical cause, as when Booth and others were trying to raise money to rebuild McVicker's Theatre, destroyed in the Chicago Fire of 1871. Such relatively innocent requests were rare and were always made candidly. Booth could act Iago, but he was not by nature a devious man, nor, for that matter, was Winter. Theirs, like all true friendships, was a relationship based not on their professional positions, not on mutual benefit and advantage, but on similarities of temperament, mutual admiration and trust, and a common devotion to the art of the theatre. Their friendship began and endured because, as Booth once wrote, they "felt in unison."

Booth's letters to Winter cover a wide range of subjects, but most often they are concerned with incidents in his private and professional life, his reactions to various theatrical events and gossip, and his reflections on his triumphs, ambitions and defeated hopes. The correspondence begins in earnest shortly after the opening of Booth's Theatre in 1869. The letters touch upon his years of management at Booth's Theatre and his ultimate failure, as injudicious direction of the theatre's financial affairs, the duplicity of his associates, poor legal advice, and the Panic of 1873, combined to force him from his theatre into bankruptcy in 1874. The letters then chronicle all but a few months of his remaining fifteen years as a travelling star, reporting his movements from one engagement to the next, his squabbles with petty critics, his difficulties with penurious managers and doltish actors, or "dogans" as he called them, and his relationships with dozens of nineteenth-century theatrical persons, some as notable as Joseph Jefferson, Augustin Daly, and Lawrence Barrett, others as obscure as Samuel Piercy, J. Newton Gotthold, and Louis Vider.

One of the largest groups of letters—almost twenty percent of the

total—and unquestionably the most interesting, report Booth's travels through the British Isles and the Continent during the summer of 1880, his engagement at London's Princess's Theatre that fall and winter, his association with Henry Irving at the Lyceum Theatre in the spring, and his tour of the English provinces, Germany, and Austria in 1882-83. The letters record not only his activities, but also his reactions to the various landscapes and peoples, his attitude towards his foreign accomplishments and failures, and his opinion of the English, French, and German theatre of the time and of such stars as Irving, Ellen Terry, Genevieve Ward, Constant Coquelin, Sophie Croizette, Pauline Lucca, and Adolph Sonnenthal. For example, the following observation on English attitudes towards Americans: "I see no American papers—impossible to get them & the English ones give very little notice of our affairs. We are so unimportant. Queer folks these English! A young lady at dinner today told Edwina she was in New York in '67 (probably 17-67) & remembered the houses being mostly of *wood*! An English officer one day asked me if our N.Y. houses were not *wooden*! Jolly bright aren't they?"

Generally, the letters reveal that America's foremost tragedian was not impressed with European acting. He thought Coquelin, for example, "really superb," but the supporting acting in Paris was "pretty good—no more." Of the English theatre he complained: "There seems to be a dearth of good stuff here—we have far better in America, if we'd only think so. Actresses particularly are scarce." He was, however, "charmed" by Irving's production of *Much Ado About Nothing*: "I have returned from the Lyceum. 'Much Ado' is superbly 'staged' and very finely acted. Irving's conception & treatment of the hero is excellent & were it not for his unfortunate voice & manner it wd. be as good as Miss Terry's Beatrice—which is perfect save for a little lack of power in the great scene with Benedict: Edwina and I were charmed with all we saw and heard tonight, despite the little exceptions I have noted. Terry will set 'em wild in America if she does this part. The scenery and sets are the finest I ever saw!" Yet he felt that such scenic splendor was detrimental to the actor's art. In the middle of his disappointing engagement at the Princess's Theatre, he observed bitterly: "What have I gained by acting here? I haven't knocked the dust out of old Drury's Cushions (I think you prophesied I could) nor scattered the old owls from their roosts—'or words to them effex.' It seems to me a loss of time & labor. I shall leave no impression—there seem to be few minds here worth impressing. The actor's art is judged by his costumes and the scenery, if they are not 'esthetic' (God save the mark!) he makes no stir. . . . Chas. Kean, Fechter, & Irving have feasted the Londoners so richly they cannot

relish undecorated dishes. I was at the same ill work at home but was fortunately checked, by fire first and afterwards by bankruptcy. I do not regret my losses now—since I have seen the evil result of grand revivals!"

An even larger group of the letters is concerned with the preparation of the "Prompt Book," the series of acting versions of most of the plays in Booth's standard repertory which he and Winter arranged and published between 1877 and 1878. Many of these letters are filled with the myriad and frequently tedious details of publishing, but they also reveal how carefully and comprehensively Booth studied the plays, particularly the Shakespearian plays, he acted. One finds, for example, his ideas about the seldom-acted *Richard II*, notes on his restoration of the original text of *Richard III*, and his opinion of such minor pieces as *Ruy Blas*, which he characterized as "a monstrous mass of twaddle" but which he continued to perform because the public liked it. Booth was continually altering or cutting lines and rearranging scenes. Occasionally he shows a pedantic side, as in his concern over the phrase "mobled queen" in *Hamlet* or "base Judean" in *Othello*. But most often his changes were dictated by the exigencies of the theatre. For example, after he had acted *Richard III* in the original text for a few months—"enough to feel its effect on the public"—he suggested to Winter that they cut a few more lines and divide the first acts differently because much of it was "wearisome to the audience."

Booth was above all else a serious and dedicated artist, yet there is an almost disappointing lack of reference in these letters to his interpretation of characters. As he wrote Winter, he found it "very difficult in cold blood to make notes and comments on any part I play." His moments of insight came most often, he noted, only during actual performance or immediately following a performance when he was still "heated and alive" with the play. Occasionally, he would get in an "analytical mood" and "bore some unfortunate sleepy head after the play with wise reflections on the 'whys' and 'whats' of the part I have been sacrificing to the (Gallery) Gods!" But usually he was "too fatigued" after a performance to indulge in such "fancies." Sometimes, however, Booth presents a thought-provoking idea about his interpretation of a character. On Hamlet, he wrote: "I have always endeavored to make prominent the femininity of *Hamlet's* character and therein lies the secret of my success—I think. I doubt if ever a robust and masculine treatment of the character will be accepted so generally as the more womanly and refined interpretation. I know I frequently fall into effeminacy, but we can't always hit the proper keynote." And once in response to an unidentified friend's opinion of Othello,

he wrote: "A Shakespearian (!) friend—who gloats over my Iago—fails to discover in Othello—either read or acted (by me or anyone)—anything but a 'beast' as Shakespeare intended him to be! Did you ever?!!! I cannot possibly see the least animalism in him—to my mind he is poetically pure & noble; even in his rage, which wrings from him occassional [sic] *curse*'ry remarks, not over-refined, perhaps, I perceive no beastiality [sic]—such as my learned friend discovers." But he never presented a fully developed explanation of any of his interpretations. Winter, of course, saw him perform hundreds of times with varying degrees of effectiveness so that there was really no need to furnish Winter with his "ruminations," as Booth called them, although he undoubtedly did so when they communicated vis-à-vis.

The letters also reveal the private Booth, fretting over the frustrated love affairs of his daughter, on whom he doted; fraught with anxiety over his second wife's physical and mental deterioration; privately seething with rage but publicly silent in the face of a vicious scandal; vexed with frequent attacks of dyspepsia; busy with real estate ventures, and (rarely) gay with too much wine and social life. These and countless other "domestick privacies" and "minute details of daily life" contribute to one's understanding of the man behind the actor.

What do these letters reveal about the character of this man who was so many characters? They disclose the personality of the mature Booth from ages thirty-five to fifty-seven. At the very beginning of the correspondence, he is seen as a man whom personal tragedy and failure have sobered and made reflective, tolerant, and compassionate. He becomes even more so as he experiences new tragedies and failures. The Booth of these letters is decidedly not the "Fiery Star" of the 1850's who drank to excess and boasted of his philandering. Indeed, in these letters he is revealed as almost a teetotaler, limiting himself to an occasional glass of the "lightest lager," and rarely does he mention anything having to do with sex. There are, however, rare glimpses of a sharp temper. Once he becomes "half-wild" with irritation at a bit of theatrical puffery which recounted the assassination of Lincoln and the capture, death, and burial of John Wilkes, and threatens to "quit the town and let the d--ned theatre go to pot!" On another occasion he berates his "son-(of a b---h) in-law, who does nothing but spend money, smile, and live on his wife's father" and damns him to "*Hell* where he belongs!"

The letters reveal a mind which is sensitive, intelligent, and imaginative with an intuitive grasp of what was beautiful and right. Booth was not an intellectual, nor was he even particularly well-read. He

often reminded Winter that he was "not literary" and had "read very little of the novelists—Thackeray, whom I can't remember, Dickens, whom I've half forgotten, are all that I can recall just now." There are no indications that he was acquainted with such "modern" authors as Henry James, or Walt Whitman. The few books that he mentions in these letters are invariably on theatrical subjects or Shakespeare, although occasionally one finds him reading such a book as Darwin's *Descent of Man* or *The Journals of Caroline Fox*. He seems, however, to have had no real interest in general ideas and was oblivious to the social, political, and scientific forces which even then were reshaping the course of Western culture. When he was not "mentally acting," he wrote, his life "drifted away in regrets for the Past or building airy castles for the future."

Although his letters to Winter do not generally indicate such, Booth had a keen interest in and appreciation of contemporary painting and sculpture and had many artist friends and acquaintances, among whom were Jervis McEntee, Eastman Johnson, Sanford Gifford, Emanuel Leutze, Launt Thompson, and Frederick Church. He was, moreover, acquainted with some singers and musicians, among whom were Clara Louise Kellogg, the American soprano, and Ole Bull, the Norwegian violinist. While in Vienna in 1883, he mentions having heard Lucca sing "on three occasions . . . with great delight," and generally seems to have enjoyed attending the opera on his "off-nights." His artistic tastes had no spelled-out rationale, however. "I like what I like, & may be very far wrong in poetry, painting, & sculpture too (I know I am as regards acting)," he wrote Winter, "but I have these friends, whose works please me very much—not because they are my friends, but because they always seem to strike a harmonious chord in my own 'instrument' which responds at once, and this is as far as my judgement extends in matters of *Art*."

Possessing little formal education, with limited intellectual interests, Booth, nevertheless, utilized effectively what he did know. He was thoroughly familiar with the plays of Shakespeare, even some of those which he did not act, and expectably his letters are filled with Shakespearian quotations and allusions. He quotes also from such diverse sources as Longfellow, Martial, Goethe, and Rochefoucauld. His friend Adam Badeau, and both of his wives had tried in vain to teach him French, but they at least succeeded in making a few French words and phrases a part of his vocabulary. His style, although direct and plain, is fresh, fluent, and always interesting. Rarely, even in the haste of composition does he commit a grievous grammatical error and even his misspellings are almost always limited to proper names or to mistakes common even among the highly

educated and literate of the times such as the transposition of "ei" for "ie" or double for single "s."

One of his most signal character traits revealed in these letters was his modesty. It exhibited itself most obviously in the form of an intense shyness of public functions such as testimonials or dedications, especially if he was expected to give a speech, and he went to great lengths to avoid them. Often abused by critics, he rarely struck back. He occasionally became annoyed, especially if the attacks were more against his person than his acting, but late in his career his annoyance seemed to mellow. Once he admitted that perhaps the poor opinion of the press during one of his engagements was correct. On another occasion, after an especially abusive attack by two critics, he wrote that if he were "not so indifferent to praise or blame, so indolent & forgetful of injuries (not virtuously but cowardly so) I'd let out the truth about those cusses." But after he had cursed a little his temper was always soothed. "Tis not manly," he concluded, "& I despise myself for not hating better." The older he became, however, the more willing he was to "forgive and forget."

In a profession where vanity is the norm, Booth displayed neither a trace of vanity or pretentiousness, nor did he tolerate pretentiousness in other actors. In 1879, for example, Booth wrote to Henry Irving to propose that during Irving's planned visit to America, he should fill certain weeks at the Lyceum at whatever terms Irving should desire. In exchange, Booth would do whatever he could to assure the success of Irving's tour. It was a long and courteous letter and coming from Irving's American counterpart, should have commanded a prompt and courteous reply. Irving, however, never answered the letter and reportedly dismissed it with the excuse "I never answer letters." Such neglect piqued Booth considerably. To Winter he complained: "I have no sympathy with *mooney* actors who live in clouds. There is no excuse he can frame to palliate such a discourtesy." The entire incident was undoubtedly an unfortunate misunderstanding, a breakdown in communications, for after Booth had met and worked with Irving and had seen his "patience and untiring energy—his good taste and superior judgment in all pertaining to stage-craft," he was more than willing to "forgive and forget."

He scorned theatrical puffery which tended to "gild the lily," and actors who did not take their art seriously. "Actors," he once wrote about the majority of his profession, "are a set of mere vain, selfish, brainless idiots—seeking only their own personal glorification which consists in paper-puffery and large type on the play-bill." He never sought personal glory, although it came to him. He sought only the betterment of his art and profession. He strove to eliminate the prac-

tice of benefits because he thought they reduced the art of acting to little more than beggary. It was his standing policy neither to give nor to appear in benefits, but his generosity often compelled him to make exceptions. His generosity also caused him to hire actors whom he knew to be incompetent, simply because they needed a job or because they were old friends. "Save me from my friends!" he once cried, yet nothing seemed to please him more than to share his good fortune; his pocketbook, as well as his influence, was always available for the asking. He gave with complete magnanimity and never expected a favor or repayment in kind. In this spirit he founded The Players at a personal expenditure of over $200,000. He asked only that his friends be as truthful and as honest with him as he was with them, and he was deeply hurt whenever his trust was betrayed, or the duplicity of seeming friends was revealed. "I've felt so keenly & *so often* the falseness of flatterers," he wrote, "that—I'll swear—I've grown old before my time. I sincerely believe that I can count with the fingers of one foot the exact number of true men that I have known &, you know, in my capacity of 'mummer' I've known many & tried not a few." The loss of his theatre, mainly through the duplicity of men he thought were his friends, was a bitter pill the taste of which he never forgot.

Few lives have been so filled with personal catastrophe as Booth's. Prior to the beginning of his correspondence with Winter, he had experienced his father's eccentricities and alcoholism, his own rescue from drunkenness, the death of his beloved first wife, the assassination of Lincoln by his brother, and the loss of the Winter Garden. In the twenty-one years of his correspondence with Winter he suffered the loss of Booth's Theater, bankruptcy, the death of an infant son, a generally unhappy second marriage, a near fatal carriage accident, an assassination attempt, and the long illness, insanity, and death of his second wife. There can be no question that much of the sadness and melancholy introspection of his letters can be traced to these painful incidents. Eventually he became resigned to the fact that "all must suffer" and one had to try to endure this suffering with patience and grace. He adopted a stoic attitude by which he was almost able to transcend "this miserable little life with its ups and downs." Spiritualism was the closest he ever seems to have come to a formal religious philosophy and he firmly believed in an after-life where he would be reunited with his loved ones. Death he regarded as a "dear old doctor" who would give us "a life more heathful & enduring than all the physicians, temporal or spiritual can give." With this attitude he instilled a somewhat hollow optimism into his life. One of his favorite quotations was Edgar's "The worst is not, so long as we can say:

'This is the worst.'" It was an attitude which was in perfect harmony with the era in which he lived, an era in which as Lewis Mumford in *The Brown Decades* has written, life was viewed through "the brown spectacles . . . of renounced ambitions and defeated hopes," in which the prevalent mood "was sometimes less than tragic; but at bottom, it was not happy."

Booth was, for the most part, a somber, quiet, solitary man, fond of his privacy, his pipe, and his study, but he was not without an amiable sense of humor. His letters disclose that he was fond of puns, of inverting grammatical structures and twisting quotations for humorous effect, of hyperbole, understatement and mock heroics. On London's weather, for example, he commented: "Since I was here (17 years ago) they have introduced sun-light. A great improvement too, you would not recognize the old town." On "great actors" he observed: " 'Great actors' are very queer cusses to handle; besides, there are so many of 'em; nearly every company counts a dozen sich." He was, as perhaps the last passage indicates, addicted to writing in "southern," "western," "Negro" or "cockney" dialects. "Such" is frequently spelled "sich," "ass" is spelled "hass," "French" becomes "Frinch," and *je ne sais quoi*, "Genessee Squaw." Once he even exclaims in minstrelese "I have bin dar, Honey."

One of Winter's critical tenets was that an actor's stage characters were reflections of his own personal character. This judgment, at least in regard to Booth, seems substantiated. Booth's intelligent but never overly intellectual, sensitive, and imaginative interpretations were to a great extent a reflection of his own mental qualities. The quiet, introspective, refined quality of his acting was but an extension of his own personal modesty, pensiveness, and gentility. His personal integrity revealed itself in his careful preparation and long study of the roles he played. The profoundly human and spiritual quality of his acting was but a reflection of his own humanity and spirituality. Even his sense of humor displayed itself in his fondness for the playful roles of Petruchio and Don Caesar de Bazan. Perhaps even more than with other great actors of the past, Booth's stage life was a mirror of his own life. The similarity did not escape his notice. Late in his career he wrote Winter that in the "mimic world" of the stage "I almost nightly find the actual sufferings of my real life rehearsed."

Although a versatile actor, except in his portrayals of lovers and most comic characters, he was at his best in the portrayal of brooding, melancholy characters like Brutus and Hamlet and lively histrionic characters like Richelieu, Iago, and Bertuccio. Undoubtedly his greatest creation and the most complete expression of his own personality was his Hamlet. Dark, melancholy, lyrical, shadowy but not

vague, repeatedly revised into an external form apparently simple yet actually rich in nuance, spiritual but intensively alive, Booth's Hamlet, like some of the haunting paintings of Albert Pinkham Ryder, created an atmosphere, as Winter once wrote, of "dread sublimity and awe." It was a creation finely attuned to the mood of the times, an almost mythic reflection of the common experience of an entire generation. That it, as well as Booth himself, captured the imagination of such large and so many audiences is not strange.

The tone and sometimes the content of Booth's letters were inevitably colored by the life and personality of Winter. Yet the few surviving letters from Winter to Booth provide only the scantiest of clues to the former's personality, much less his life, nor is such information generally familiar or easily accessible.[2] To provide a complementary context for the reading of Booth's letters, a brief survey of Winter's life may, therefore, be useful.

Winter's background was totally unlike Booth's. Booth was a Southerner, the son of a famous actor, who seemed predestined for a life on the stage. Winter was a "down-East Yankee," born in the seaport of Gloucester, Massachusetts, on July 15, 1836, the scion of a family of merchant sea captains. About 1843 Winter's father, Charles Winter, a former sea captain, left the sea for a shore job as the assistant wharfinger of the Boston commercial wharf. There was nothing in Winter's family background which indicated that he would have a predilection for literature and the theatre. He received his early education in the Boston public school system and in September of 1854 he entered the Dane Law School of Harvard College. His three years at Harvard were active ones. In December of 1854 he published his first volume of poems, which he later repudiated as "the most absurd collection of verses ever put into print"; made the acquaintance of Henry Wadsworth Longfellow, who encouraged him in

---

[2] Winter has been the subject of a generally excellent doctoral dissertation: Richard M. Ludwig's "The Career of William Winter, American Drama Critic: 1836-1917" (Harvard University, 1950). His theatrical criticism has also been given detailed study in Charles J. McGaw's doctoral dissertation "An Analysis of the Theatrical Criticism of William Winter" (University of Michigan, 1940). His Shakespearean criticism has been given special study by Gilbert M. Rubenstein's doctoral dissertation "The Shakespearean Criticism of William Winter: An Analysis" (Indiana University, 1951). His great-grandson Robert Young has also written a biography, still unpublished, based on family records and Winter's own letters and journals. I am indebted to all of these sources, but especially to Mr. Young, for the following biographical survey.

his poetic efforts; contributed poems, book and theatre reviews to various papers, among them the Cambridge *Chronicle* and the Boston *Olive Branch*, *Evening Transcript*, and *Gazette*; worked as a fee collector for the owners of a Boston harbor tugboat; campaigned as a political speaker for General John C. Fremont in Charlestown and South Boston in 1856; and, not incidentally, attended classes in law, languages, and literature at Harvard.

In 1857 Winter was graduated from law school and was admitted to the Suffolk County Bar. For about two years he worked in the law offices of Lyman Mason and Aurelius D. Parker in Boston, continuing meanwhile to contribute poems and articles to Boston papers. In December of 1859 he abandoned law and went to New York to pursue a career in literature. He was hired by the founding editor of *The Saturday Press*, the celebrated "King of Bohemia," Henry Clapp, Jr., as that paper's drama critic. For two years Winter wrote critical articles and poems for *The Saturday Press* under the pseudonyms "Personne" and "Quelqu'un." From Clapp, Winter learned the fundamentals of his craft. Like Clapp himself, he became intolerant of pretense and affectation and he discovered how useful a critical instrument satire could be. To this association with Clapp, Winter owes the beginnings of his critical reputation.

Winter's association with Clapp, his second volume of poetry, *The Queen's Domain* (New York, 1859), and his close friendship with Thomas Bailey Aldrich brought him into early contact with that coterie of "Bohemians" which included Clapp, Aldrich, Walt Whitman, William Dean Howells, and Artemus Ward; and such literary lesser lights as the writers Fitz-James O'Brien, George Arnold, and Fitz-Hugh Ludlow; the poetess and actress (the beautiful "Queen of Bohemia"), Ada Clare; the sculptor Launt Thompson, and the painter Eastman Johnson. The Bohemians lived by night in a smoky *Bierstube* at 647 Broadway near Bleeker Street, known as Pfaff's "cave." Here they wrote, gossiped, and argued. During the day most of them worked as journalists or magazine engravers and lithographers. In the decade before the Civil War, the Bohemians were America's artistic "avant-garde," diametrically opposed to the growing materialism in American life, to middle-class prudery, and, especially, to the effete, conservative literary and artistic circles of New York and Boston. However, Winter's commitment to Bohemianism was not intense. Unlike Clapp, he did not repudiate his New England heritage, nor lose his admiration for the Boston Brahmins, especially Longfellow. He soon broke with the Bohemian circle and, like Aldrich, aligned himself more comfortably with the "genteel school" of Richard Henry Stoddard, Bayard Taylor, and Edmund Clarence Stedman.

The Civil War disrupted the Bohemian coterie and brought *The Saturday Press* to a halt. Winter, like many destined to become eminent in the arts, literature, finance, and science thirty or forty years later, who were of fighting age in the early 1860's, did not go to war, but continued to pursue his own particular interests. He worked briefly for the New York *Leader*; then from 1861 to 1865, worked principally as the drama critic for the New York *Albion* where his articles appeared under the pseudonym "Mercutio." During this period he was at various times the editor of the *Petroleum Gazette*, the managing editor and entire staff of the *Insurance Monitor*, and the managing editor of *Frank Leslie's Magazine*; and he contributed literary and theatrical critiques and poems to *Vanity Fair*, *The Galaxy*, *Harper's Weekly*, the *Round Table*, *Philobiblion*, and the *Spirit of the Times*.

On December 8, 1860, Winter married twenty-year-old Elizabeth "Lizzie" Campbell, whom he had met the previous year when she tried to sell her short story "The Oasis in the Desert" to *The Saturday Press*. Born in Loch Awe, Scotland in 1840, Lizzie was taken as a child by her parents, John and Janet Campbell, to Toronto, Canada. In 1859, reportedly at the urging of Ada Clare who had befriended her, she came to New York to pursue a career as an authoress. In 1861 the Winters' first child Percy Curtis was born. Three more sons and a daughter followed: Arthur Elliot (1872), Louis Victor (1873), William Jefferson (1878), and Viola Rosamund (1881). In 1864, Lizzie temporarily abandoned her literary pursuits for a stage career. She made her first appearance in New York at the Olympic Theatre on April 11, 1864. In 1866 she played Katherine to Booth's Petruchio at the Winter Garden. At various times during the late 1860's and 1870's she was associated with the companies of Booth, Augustin Daly, and Lester Wallack. In 1877 she even had a brief starring engagement in Toronto, but finally in 1878 she retired from the stage and returned to writing. Using such pen-names as Elizabeth Campbell, Isabella Castellar, E. C. Winter, and Elsie Snow, she wrote numerous short stories, several novels, and adapted plays, notably *Mary Stuart* for Helena Modjeska and *The First Violin* for Richard Mansfield. Attractive, undoubtedly independent and intellectual, probably with definite "feminist" ideas then popular, she was throughout their long marriage apparently devoted to Winter.

In 1865 Winter was hired at the salary of thirty-five dollars a week—twenty-five dollars more than he was getting at the *Albion*—as the drama critic of Horace Greeley's New York *Tribune*, replacing Edward Howard House. During his forty-four-year tenure with the *Tribune*, Winter rose to the top of his profession, becoming one of

the most prolific, respected, and influential of nineteenth-century American drama critics.

He was preeminently qualified for this position. He was thoroughly familiar with classical and contemporary drama and literature, was especially knowledgeable of Shakespeare's plays and the history of the English and American theatre, read Latin, German, and probably French, and wrote easily and fluently in a typically nineteenth-century elegant style. By the time he came to the *Tribune*, Winter's basic critical tenets and style of dramatic criticism were already well established and they changed but little during the rest of his career.

While he never explicitly expounded the esthetic principles upon which his criticism was based, an examination of his critical writings reveals that the foundation of his criticism was essentially Aristotelian and derived from such critics as Johnson, Hazlitt, and Coleridge. Like many of the prominent drama critics of his time such as Joseph Knight, Clement Scott, and John Ranken Towse, Winter concentrated more on acting than on the play. He did not consider the drama unimportant; on the contrary, he held it to be an ennobling, uplifting, and powerful influence for good in society. He considered the actor, however, to be the primary artist in the accomplishment of drama's high aims. Acting to him was the "soul of theatre," and throughout his life he championed the acceptance of the actor as an artist. He admired acting which attempted to conceive of characters in terms of universal truths, so that an audience would gain knowledge and understanding of the best in human nature. He preferred a style of acting which endeavored to effect a balance between emotion and technique. He generally disdained acting which was either extremely emotional or extremely intellectual. He especially disliked acting which aimed only at reproducing the external forms of behavior. He never hesitated to praise, even to flatter, an actor who he thought had effectively expressed the ideal of a character, nor did he hesitate to admonish, often in the most caustic terms, an actor who had not. He almost always suggested ways in which a performance could be improved. Winter was an intimate friend of many of the actors, actresses, and producers whom he criticized, among them Irving, Barrett, Mansfield, John McCullough, John Drew, Ada Rehan, Adelaide Neilson, Helena Modjeska, Augustin Daly, and David Belasco. There is no question that these friendships occasionally affected his judgments, yet Winter always considered himself to be completely fair and objective. He maintained that the actor's personality significantly influenced the personality of the character he impersonated. Thus, familiarity with an actor could only be an asset to a critic and not a handicap.

Throughout his career as a drama critic, Winter continued to write poetry and to contribute critical and biographical articles to leading periodicals of the time, most of which he later collected and published. An ardent Anglophile, he made at least eight trips to England, recording his experiences in several volumes of travelogue. He also maintained friendships with a number of literary men of his time, among them Longfellow, Dickens, and Wilkie Collins, and he later published reminiscences of these friendships. Ultimately his bibliography contained over fifty volumes of theatre criticism, history and biography, literary reminiscence, travelogue, and poetry.

Physically, he was a slight, small-boned man, about five feet seven inches tall and probably never weighing more than one hundred twenty pounds. His hair was thick and chestnut colored, and from his early days in New York he sported a large drooping mustache. In later life both hair and mustache turned snow white. His handsome face was distinguished by eyes which have been described as very light blue, or as almost violet, with a peculiar white rim surrounding the iris. He was by no means a prude, but, in the tradition of genteelism, he was a virtuous man, who admired nobility, forbearance, and purity. Like Booth, he seems to have been detached or at least uninterested in contemporary issues. The pursuit of Art, Beauty, the Ideal absorbed almost all of his attention. In his work he was industrious, diligent, and scholarly, with an uncommon sense of personal and artistic integrity. He had idiosyncrasies, of course. He reportedly hated elevators and later in life, telephones. He was inclined to hypochondria, sentimentalism, and to a certain display of world-weariness or romantic melancholy. He also had moments of testiness, but he seems seldom to have lost his temper. He was never satisfied with his accomplishments as a drama critic, but wanted to succeed as a poet, which he undoubtedly viewed, as did many in the "genteel school," as the highest accomplishment of man. The lack of recognition as a poet was a continued frustration to him.

In the 1890's, Winter was generally recognized as the "dean of American drama critics." His reputation, however, gradually began to decline because of his violent attacks against the realistic, socially and morally oriented plays of Pinero, Ibsen, Sudermann, Shaw, and Maeterlinck. While his attacks were not without a rational, esthetic basis, they were, nevertheless, totally blind to the merits of the "New Drama." His caustic reviews of some of these new plays and the editor's consequent excisions from his copy, forced Winter's resignation from the *Tribune* on August 8, 1908. From his resignation until his death, Winter lived as an independent man of letters. He added thirteen volumes to his already extensive list of publications, revised

some of his previous books, and continued to contribute occasional reviews and articles to *Harper's Weekly, The Saturday Evening Post* and other periodicals. After a short illness he died from uremic poisoning on June 30, 1917, just two weeks short of his eighty-first birthday.

That more of Winter's letters to Booth have not survived is unfortunate, for without them a complete understanding of his relationship with Booth remains impossible. Yet Winter's personal dedication to Booth cannot be questioned. Both as a man and as an artist, he regarded Booth as "an exceptional person, an honor to human nature, and a blessing to his time." No better support for this high praise could be found than the Booth letters in this volume. They cast aside the "exterior appendages," as Johnson called them, and reveal a gifted, sensitive, humane man and artist, a man and artist who commands not only one's attention but ultimately one's admiration and affection.

Chapter I

Booth's Theatre

1869-1875

WHEN the personal acquaintance of Edwin Booth and William Winter began is uncertain. Winter saw Booth act for the first time on April 20, 1857, at the Boston Theatre as Sir Giles Overreach in Philip Massinger's *A New Way to Pay Old Debts*.[1] They probably met when Booth returned to Boston the next year. Booth's earliest extant letter to Winter is the following hastily written cryptic note, possibly in reference to a personal dresser and a hotel.

1. Baltimore, July 29, 1859

Thanks for your advice—I have made the first offer—think it reasonable—hope he'll take it—as I must stay at the finest hotels his board would be heavy to have him with me.

Can't you give me some items about the 'Royal'?

Between this letter and the next a decade passed. But Booth and Winter probably met occasionally during this period if they did not correspond. They were undoubtedly acquainted. They shared mutual friends and Mrs. Winter had played Katherine to Booth's Petruchio at the Winter Garden in 1866 (see Letter 3, fn. 7). In a letter to Jervis McEntee, late in 1868, however, Booth mentions that Winter is among the critics "who do not know me personally."[2]

This same letter to McEntee suggests some of the circumstances which may have motivated Booth to write Winter once again. Booth asked McEntee to ask Launt Thompson,[3] who lived in the same studio building, to contact Winter and explain to him the "villainy of

[1] William Winter, *Vagrant Memories* (New York, 1915), p. 152.

[2] Jervis McEntee, American painter, was a close friend of Booth. In 1880 he completed a series of full-length portraits of Booth in various characters. The letter is quoted in Richard Lockridge, *The Darling of Misfortune* (New York, 1932), p. 194. The original is in the collection of the Walter Hampden Memorial Library at The Players in New York.

[3] Launt Thompson, American sculptor noted mainly for his monumental works, was a personal friend of both Booth and Winter. In 1864, he completed a life-size bust of Booth as Hamlet which is presently at The Players.

that d----d rascal Stuart." This Stuart was William Stuart, Booth's former friend and partner at the Winter Garden, who, after Booth terminated their partnership, had returned to his former career as a "yellow" journalist and was now attacking Booth in the press as he had once attacked Edwin Forrest.[4] To McEntee Booth wrote: "Stuart is black at the core and will kill himself in time, but he can spit venom yet and should be hanged." Booth was afraid that Stuart would damage his reputation with the more respectable New York critics, among whom, he wrote, "Winter ranks first." With his new theatre about to open Booth desperately desired to retain the good opinion of these critics not because, as he wrote, "I fear their ill—but my good will towards them demands it."

In this context of magnanimous good will, Booth wrote to Winter a few days after the opening of Booth's Theatre, February 3, 1869. At first it seems an astonishing letter from an actor to a critic. Booth is offering Winter financial assistance so that he can become a member of the Century Club! Booth was a member of the club and Winter had been elected in 1866, but had been unable to afford the $100

---

[4] Stuart, whose real name was Edmund O'Flaherty, was born in Ireland in 1821. In 1853 he changed his name and emigrated to America because of his involvement in a political scandal. Charles A. Dana hired him to write for the New York *Tribune* and in the spring of 1855, he contributed a series of reviews viciously ridiculing Edwin Forrest, then appearing at the Broadway Theatre. Between 1857 and 1859 Stuart was involved in various theatrical activities, mainly at Wallack's Theatre. For one season, 1859-60, he was the co-manager of the Winter Garden with the noted Irish playwright Dion Boucicault. After Boucicault retired from the management, Stuart was associated from 1860 to 1864 with the next manager of the Winter Garden, T. B. "Black" Jackson. In 1864, after Jackson withdrew, he joined Booth and John Sleeper Clarke in the management of the Winter Garden. Stuart's astuteness in the box-office and, especially, in publicity was responsible for much of the success that Booth enjoyed at the Winter Garden down to 1865. He was, however, a Janus-faced opportunist who, while profiting from Booth, publicly praised him but privately derided him. Booth eventually became aware of Stuart's duplicity. He terminated their partnership when the Winter Garden burned in 1867 and refused all further association. Stuart, who claimed he had "made" Booth, was piqued at his ingratitude and for the next twenty years seldom neglected an opportunity to malign him. According to Winter, Stuart originated "all the stock misrepresentations of Booth that have drifted through the American press." See *Life and Art of Edwin Booth* (New York, 1894), p. 119. Unless otherwise indicated, this and subsequent references are to the revised, 18mo edition of this work, hereafter cited as *Life and Art.*

initiation fee and the $36 annual dues, so that he had never been officially installed.

Winter's integrity as a critic and as a man was not to be taken lightly. Only a few weeks before the opening of Booth's Theatre, Winter had been offered by an agent of the powerful and notorious Jim Fisk an annual salary of $2500 if only he would write an occasional line about Fisk—"anything that might do him good." Winter, who valued his freedom above everything, replied that he had "never been carried in anybody's pocket" and that he didn't intend to begin. A few weeks later, Winter met Fisk in the lobby of the Grand Opera House. Fisk extended his hand and Winter promptly turned his back and cut his acquaintance.[5] If Winter had thought that Booth was offering a bribe or attempting to buy his services, he would undoubtedly have cut his acquaintance also. The sincere and unselfish tone of Booth's letter is, however, unmistakable. Nevertheless Winter refused Booth's generosity. At the end of the letter he scribbled: "I did not accept the offer."

2.   [New York] February 7, 1869

I am sorry I have not seen you since the 'first night'—I know not how to explain to you my unseemly and crazy excitement of that evening—when we meet I may be able to do so. To the subject I referred to in my last. What I am about to do is in obedience to a heart impulse—if I offend blame the head. You know perhaps, that I am a member of the 'Century,' and what is done and said in that mutual admiration society is generally well known by all its members. I ascertained that you were desirous of joining the Club and were unanimously elected—but that you were obliged to decline on account of pecuniary considerations. Now, bearing a grateful memory of your past services and wishing to refute what Stuart says of actors in general & of myself in particular, I venture even at the risk of wounding your delicacy to offer you assistance knowing you to to be a thorough gentleman, and possessing a keen sense of dignity myself—I would not suggest this if I had any object in view other than that of aiding a deserving fellow being, and testifying in some *useful* way my obligation to him. "What a damned inequality in the lot of mankind," says [illegible name]. You toil, and wear your life out in vain—while I (a toiler too)

[5] William Winter, *Other Days* (New York, 1908), pp. 115-116.

reap the reward of your labor as well as my own. True—you do not share the abuse I get with the dollars, but I think you are entitled to at least the thanks of the better paid fellow. Now are you 'out of the dark'? If I have offended you, my dear boy, don't retain any ill feeling toward me beyond the moment, but believe that my action is prompted by the purest motives. I often do foolish, hasty things—without duly considering the result. But for some time past, in reviewing the successful career I have had throughout this country I have felt that much was due to this New York press (not to Stuart—for I paid him well for all *he* did)—and to no one individual member thereof was I more beholden than to yourself. At such times I have wished for an opportunity to acknowledge the debt I owe you, and a few weeks after the story of the Century reached my ears—Now, said I, if Will Winter will accept this trifling favor in return for the many he has given me, and will acquit me of any designs upon his 'quill,' I will prove to him that I am not unmindful of the past. I touched the matter gently the other day—for I know you are sensitive, and I am clumsy in these affairs of the heart, and, as I have said, have often caused pain where I intended to give pleasure. I hoped you would have seen me at the theatre and demanded an explanation of my 'mystery'—when I would have blustered out my meaning awkwardly. It is better as it is. And I enclose a check which you may light your segar with in disgust, or add to the pleasure you have already so often given me by using for whatever purpose may please you best. I have nothing to gain by this—for you have already said all that you can conscientiously say of the very bad performance on my first night (which, I think, you would like better if seen again);[6] but, I candidly admit, if I should need your aid in order to carry out the

[6] Winter's review of *Romeo and Juliet* appeared in the *Tribune* (February 4). He wrote that Booth's Romeo "showed the utmost study." The characterization was "ardent," "poetic and picturesque," and "full of art" but it "lacked in wholehearted passion and a certain dashing manliness of personality." Other critics were not as kind as Winter. The critic of the *Spirit of the Times* (February 6) wrote that in the balcony scene Booth "hopped about like a clog-dancer or a somnambulic-velocipedist, who fancied himself astride a bicycle." And the critic of the *Daily Star* (February 5) wrote that Booth made Romeo "look like a sheep, cavort like a monkey, and chatter like a parrot." After this revival Booth dropped Romeo from his repertory and never appeared in the role again.

intention I have regarding the drama I will not hesitate to ask it, and if you knew my motives I am sure you would not refuse it. Of that hereafter. At present I have reference only to the past; and I have faith enough in man yet to believe that you appreciate me as a man as well as an actor. No one is more conscious of his faults (in both relations) than I, and undeserved praise is distasteful to me, while that I feel I merit affords but little gratification, yet I do not slight it—though I may accept it silently; nor do I forget those who have cheered and encouraged me in my toilsome journey to the goal that I have reached. As for adverse criticism—I think you would be surprised if you could hear how frequently I endorse it—for my sense of justice runs away with me sometimes to the very extreme, and those who know me best have often checked me in doing what my conscience dictates as 'fair play.' I give myself a very good 'Karakter,' you see & know that I deserve it—and in time you will endorse it, by accepting me as a friend.

I have on other occasions offered my 'mite' to those who had no claim beyond our hearts' sympathy without giving offense— and I do not believe that in this instance you will be very angry with me, knowing that you have served me far, very far beyond the trifling acknowledgement I make. There may be others to whom I am likewise indebted, but they are, perhaps, more fortunate in a wordly sense than you—at all events, I do not know them personally and have no sympathy with them. You have a gentle, sensitive, sympathetic nature, and (albeit my tongue wags rudely sometimes) I—from the first—far back—felt in unison with you. Prompted as I am [words cut out] and a sense [words cut out] fellow—now it will pain me beyond expression if I have again permitted my heart to bewilder my head. 'Abuse me like a pickpocket' if you will (as an actor), but don't let this little act of mine lessen your esteem for me as a man. Sensitive souls are apt, you know, to turn from those who bungling offer kindness. As for the future—I shall never ask or expect more than you can in justice do, and the records of the past assure me that you never shrink from that. I have wearied you, I know, but attribute all this verbosity to the delicacy I feel in approaching a subject that may be painful. Again—if I have hurt you, say so gently and forgive; if you have faith in me

you will be gentle too, and not wring my hand off when next we meet. In my little 'sanctum' I have a box of cigars, and liesure [sic] during the play to smoke and chat with friends—come in and 'pipe with me.'

During Booth's management of Booth's Theatre he usually made starring tours in the fall and spring while other stars were engaged at his theatre. These tours were designed mainly to raise money, for only thus could Booth hope to present the most attractively mounted legitimate drama in New York and still remain financially solvent. During his tour in the fall of 1869, Booth wrote Winter a casual and affectionate letter on the theme of "friendship." His theme was motivated by an article by Harriet Beecher Stowe, noted author of *Uncle Tom's Cabin*, entitled "The True Story of Lady Byron's Life." In the article Mrs. Stowe charged that the real cause of Lord Byron's separation from his wife in 1816 after a marriage of only a year was her discovery of his incestuous affair with his half-sister Augusta. The article, based on the testimony of Lady Byron herself, was published simultaneously in the September, 1869 issues of *Atlantic Monthly* and its British counterpart, *Macmillan's Magazine*. It created a sensation on both sides of the Atlantic and for months the columns of leading American and British journals screamed with repudiations and defenses of Mrs. Stowe's claim. Mrs. Stowe later expanded her essay to book-length and published it as *Lady Byron Vindicated* (New York, 1869). Mrs. Stowe's article reminded Booth of Edward John Trelawney's less sensational, but no less uncomplimentary portrait of his supposed friend Byron entitled *Recollections of the Last Days of Byron and Shelley* (London, 1858). Trelawney was much abused for his sketch of Byron. He was variously accused of duplicity, of deserting Byron in his hour of need at Missolonghi, even of perversion in examining Byron's deformity after his death. To both Winter and Booth such disclosures made by Mrs. Stowe and Trelawney displayed an unforgivable lack of taste and judgment and, in Trelawney's case, a breach of friendship. As Winter, an admirer of Byron, later wrote in his review of Mrs. Stowe's book, they should be better left to "medical hospitals or lunatic asylums" or best to "the mercy of God."

3. Philadelphia, October 24, 1869

Many thanks for yr. kind remembrances—The subject possesses great interest to me, and you treat it *feelingly* and—with all yr. might. Poor Harriet will be "safely *stowed*," with that festive beast Trelawney—(who some years ago did little less than

she)—among the filthy things in memory's locker to be used as
unclean similes when we feel more than ordinarily vindictive. I
remember I used the word 'Trelawney' for a long while after I
read his lies—as an epithet expressing my profoundest disgust of
one who had *slobbered* me [with] love & adultation & [turned]
cold & callous from me [when] I most needed help.—

I am pleased to learn from good authority that your wife made
a decided *character* of the trifle I was obliged to give her in
Leah.[7] My compliments to her, and assure her of my sincere
regards and good wishes. I've said enough to you in this vein, my
boy, and only hope that I may some day have an opportunity to
be of real service to you. As a critic—you must know—I do not
always agree with you & wd. like to chat with you occassionally
on the subject of "play-acting" & its requirements, &c—if ever the
time does come for a quiet little "dine", en famille, at my 'flat' in
Booth's building. I'm glad you occassionally think of me—
for—I tell you, Will—there are so d----d few *hearts* in the world
that I like to cherish 'em when found. I've felt so keenly &
so often the falseness of flatterers that—I'll swear—I've grown old
before my time. I sincerely believe that I can count with the
fingers on one foot the exact number of *true* men that I have
known &, you know, in my capacity of "mummer" I've known
many & tried not a few. Eh bien! I'm prematurely grey-bearded
about the heart, mayhap,—but I don't think that organ is
entirely rusted out—it yet vibrates with a kind & tender yearning
for the good old fashioned friendships that you poets sing of.
Apropos: can't I get a copy of yr. "heart-throbs"[8]—I like what
I've read exceedingly—but I'm not literary, you know, and have no
right to say so—I think you as true in poesy as in the *principle*
which activates you in criticism—albeit in the latter you may
chance to err, the *spirit* is "plumb" and "level"—that is always

[7] From September 30 to October 16, 1869, Kate Bateman, the famous
child actress now adult, appeared at Booth's Theatre in the title role of
Augustin Daly's *Leah the Forsaken*, an adaptation of S. H. Mosenthal's
Deborah. Mrs. Winter supported her as Hannah, the servant girl. During
this season at Booth's, Mrs. Winter also appeared as Mrs. Page in
Falstaff and Julia Mannering in *Guy Mannering*.

[8] "Heart-throbs" is a jocular reference to Winter's poems, evidently to
some that had appeared in a current magazine. Winter had not published a
volume of poems since *The Queen's Domain* in 1859.

evident. Never let it become warped by prejudice nor yet loosened by indifference.

In 3 weeks I shall—*D[eo] V[olente]* return to York—there to abide months many. Adieu!

After his fall tour of 1869, Booth reappeared at his theatre in a spectacular revival of *Hamlet* on January 5, 1870. This production, reportedly fifteen months in preparation, ran continuously through March 19. New York's drama critics greeted the production ecstatically. The critic for the *Herald* (January 6) reported that "it was a genuine feast of reason, of beauty, of fashion, and of historical intelligence and splendors both as regards actors, scenery and audience." Other critics were less florid but no less enthusiastic.[9]

About the time of the *Hamlet* revival, Winter was trying with the help of his friends Joseph Jefferson and Lester Wallack, the actor, dramatist, and manager of Wallack's Theatre, to arrange a benefit for the aging and ill comic actor George Holland. He had asked Booth's advice and assistance. As indicated by the following letter, Booth doubted that such a benefit would be productive but was not unwilling to help make it so.

4. Booth's Theatre, New York, Wednesday [February], 1870

Don't trouble yourself about my "phelinks" all is well. I'm trying to hook that queer fish Joe & if he don't take my bait now he's a "Jack" (John Dory).[10] Hope Clarke[11] will *do*—do all you can to work him up—for I've hopes to run him through the Summer. Drop Joe a line at Balto. & urge him to play a long engagement here before he goes abroad & advise him to begin early in the fall. I don't know whether he listens much to advice in such matters—but it's for his good. For my part—my only (or rather my chief) object is to get away that I may let my

[9] For a superb account of this production see Charles H. Shattuck, *The Hamlet of Edwin Booth* (Urbana, 1969).

[10] "John Dory" is a type of common fish found in American and European sea waters and contemptuously referred to as "jack fish."

[11] In 1867 John Sleeper Clarke had emigrated to England where, except for occasional starring tours of America, he remained for the rest of his life. At various times he successfully managed several London theatres, among them the Haymarket and the Strand. He also maintained his interest in the management of the Walnut Street Theatre in Philadelphia. Clarke made a starring engagement at Booth's Theatre from April 18 to May 28, 1870.

ground lie fallow & gather *i'the West* the harvest waiting there for me. If he don't come here (of course, I know, he won't go to any other theatre in the city) I must pitch in myself or set Fetcher [sic] or Janaushek [sic][12] at work—to let me off.

My hesitation in regard to the benefit arises from the knowledge of these charity affairs always being such infernal failures. I wish it could transpire on Shakespeare's *23d.* (a day that should be always set apart as a holiday & for some good purpose), and be *dead sure* that *all* who promise to act will do so—as they *invariably always* don't. Selah! Jefferson & Clarke, Wallack & your uncle[13] et al &c.

In fact all of any note in the city should 'buckle up' & be doing. Don't be too hasty—but slowly get the thing in trim—then blow away.

If you suffer as I do after these rosy bouts I fear Seltzer won't save you, but try it. Get at least six syphons of Schultz & Warker's Seltzer Water[14] drink at leisure till you're chock full & then— fizzle, bang, pop! Your brains will clear, your nerves be stilled & hell be quiet. I've been there, Will, & know it well. Keep your bowels well relaxed & read "Paradise Lost." On the "whole"— I think you'll be relieved. Finis.

Shakespeare's birthday passed and Winter was still trying to organize the benefit for Holland. He asked Booth for the use of his theatre, but Booth refused because he had refused other such requests and because he opposed, in principle, *all* charity benefits. He was willing, however, to act in the benefit and suggested Wallack's as an

---

[12] On January 10, the Anglo-French actor Charles Albert Fechter made his first American appearance at Niblo's Garden in Hugo's *Ruy Blas*. Fechter's Hamlet had been creating a sensation in London since 1861 and for a time there were rumors that Booth wanted to engage Fechter to play Hamlet at Booth's Theatre. But on February 14, Fechter presented *Hamlet* at Niblo's Garden for the last week of his engagement. Most of the critics preferred Booth's interpretation, but Booth still hoped to engage Fechter for his theatre in the fall. He was also trying to engage the famous Polish-German actress Fanny Janauschek with whom he had acted in a bilingual production of *Macbeth* in 1868. At this time Madame Janauschek was studying English in hope of an English-speaking engagement in the fall.

[13] "Your uncle" is Booth's way of referring to himself.

[14] Carl H. Schultz and Thomas G. H. Warker were New York manufacturers of mineral water.

appropriate theatre since Holland had been associated with that theatre for almost fifteen years.

Winter was eventually forced to abandon the benefit project mainly because the noted manager Augustin Daly, at whose theatre Holland was then engaged, complained that such a benefit implied "invidious reflections on him by other theatrical managers."[15] But Daly himself arranged that a performance of his then current production, *Frou-Frou*, be given for the benefit of Holland on May 16, 1870.

5. [New York], Monday, April 25 [1870]

I would suggest that you fix on June the 30th (Thursday) for Holland[16] benefit. At that time Jefferson & the others will then be idle. To do it earlier will, I am sure, interfere materially with several of their projects. I cannot well give this house after what I've done—refused several such applications, from my Masonic brothers & others, on the ground that I wish to abolish (as far as lies within my power) these charity benefits, in fact benefits of every description. I gave 2 or 3 the first season & then by mere lack of thought[17]—having originally resolved that the words benefit & testimonial should not appear on my bills. I think it tends to bring contempt upon the actor—it places him in the light of a pauper asking alms. I am willing to play *Petruchio*—if we can get a full complement of names & I do not doubt but at the time I mention all can be had. Associated for years with Wallack's the benefit should take place there by all means & properly worked the tickets could be sold at double price for Matinée & evening. For the day get in Clarke (as he will doubtless be acting here at night)—his days he will act—in some short farce; myself

---

[15] William Winter, *The Wallet of Time*, I (New York, 1913), 53.

[16] After Holland's death on December 20 Winter did succeed in organizing a benefit which occurred on January 19 and 21, 1871. The performances provided Holland's widow with a fund of over $13,000. See William Winter, *Life and Art of Joseph Jefferson* (New York, 1894), p. 266.

[17] George C. D. Odell in his *Annals of the New York Stage* (New York, 1927-1949)—hereafter cited as *Annals*—IX, 427-429, cites three benefits at Booth's Theatre in 1869: a performance of *Othello* on May 28 for Mary McVicker; a performance of *The Lady of Lyons* with Edwin Adams as Claude Melnotte for his own benefit on May 31; and a performance of *Rip Van Winkle* with Jefferson in the title role for the benefit of the family of the late J. G. Hanley (1828-1869), Booth's stage manager at both the Winter Garden and Booth's Theatre, on September 8.

and some others one or two act performances. At night—Joe in a comedy wherein one or both of the Wallack's and as strong a cast as you can—two pieces. At this date I [illegible word] all—but Clarke will be idle, and Joe may not be—for Clarke may want him after his Cincinnati engagement for 2 weeks. Joe is really uncertain about his movements—I see he has suddenly determined to lengthen his engagement in Boston & his Cincinnati trip must be again deferred.

It will be just one month later than the date you fix. Think of it. What was it about the autograph? You shall have it any day.

Booth's most notable production of the 1870-71 season was a scenically splendid revival of Bulwer-Lytton's *Richelieu*, which opened on January 9 and played continuously through March 4, 1871. While most critics praised *Richelieu* with the same enthusiasm that had greeted last season's *Hamlet*, the critic of *The Season*, John S. Moray—a pseudonym for Augustus R. Cazauran[18]—excoriated Booth and his production in a four page "*Richelieu* Supplement" published on January 14. In his review, Moray recalled seeing Booth as Richelieu for the first time in Cincinnati in 1860. Booth's performance on this occasion, he noted, was limited to throwing up his dinner in the wings and stumbling across the footlights in a drunken stupor. He then attacked the current portrayal, parodying Booth's articulation of liquid consonants and sibilants, accusing him of excessive angularity of gesture, and ridiculing his stature. Returning to the theme of drunkenness, he ended his "review" with a warning to the aspiring actor to maintain his decency, taste, scholarship, and, especially, his sobriety or "when manhood comes he will find his voice gone, his brain unable to healthily analyze his work, and his heart, dulled by dissipation, unable to throb with that fervor which is the germ of the beautiful in his art, as in all others."

Usually, Booth gave little attention to such "yellow" journalism, but on this occasion he by chance remarked to his business manager, John Henry "Harry" Magonigle,[19] that such critics should be barred

[18] The career of Augustus R. Cazauran is largely obscure. He was the house dramatist and adapter of French plays for Palmer's Union Square Theatre in the late 1870's and 1880's. He is also credited with the reconstruction and remarkable success of Bronson Howard's *The Banker's Daughter* (1879).

[19] Magonigle was married to Catherine Devlin, a sister of Booth's first wife Mary Devlin and thus Booth's brother-in-law. In *Life and Art* Winter writes, p. 82, that Magonigle "was a devoted friend of Edwin Booth, and

from his theatre. Magonigle took Booth at his word and barred not only Cazauran but also the editor of *The Season*, Paul F. Nicholson. On April 13, 1871, Cazauran and Nicholson sued before Justice John R. Brady for an alternative mandamus against Booth. Justice Brady upheld their claim and issued a writ against Booth to be returnable no later than April 24, 1871.

Booth wrote Winter to plead for him to "put the case in a fairer light." Winter forwarded the letter to Whitelaw Reid, then the managing editor, later the editor, of the *Tribune*. On April 18, an editorial appeared in the *Tribune* under the headline "Managers and Critics." It sympathized with, but did not justify, Booth's action. It placed the blame for such unfortunate incidents not on the theatre managers nor even on such critics "who disgrace their calling, parasites of art who poison and deform the profession to which they cling," but rather on the demoralizing practice of granting "privileges" to the press. The editorial ended with a call for reform among newspapers for "a new and sounder rule in their dealings with theatres."

6. Launt's Studio [New York] Friday A.M. [ca. April 14, 1871][20]

In consequence of Harry's lack of judgement in excluding Cauzaran [sic] & Nicholson I find myself posted in all the papers as an ass. God knows I have no objection to fair & honest criticism, and that I can stand & smile at any quantity of abuse—I would never have played so blindly into my adversary's hands by doing what he so frequently sought, but Harry supposed from some careless remark I one day made concerning such fellows as Nicholson & Cauzaran—[that] I did not wish [them] in the theater at all & so he gave them a trump card you see.

Can't you put the case in a fairer light? I don't want the newspaper folks or the public either to think that I am such an ass as to be hurt by any adverse criticism. From all I've heard of the two fellows I consider them unfit for decent society & think they

---

stood by him with tireless devotion and affectionate fidelity, to the last moment of his life." Magonigle's career is largely obscure. He was Booth's business manager at Booth's Theatre, and during Booth's national tour in 1886-87 he acted as treasurer and unofficial "steward." His theatrical activities seem to have been limited to the business area, but he occasionally acted, excelling in comedy (see letter 56, fn. 28). He was one of the executors of Booth's will and Booth left him and his wife $10,000 each.

[20] The letter is dated from a notation on it by Whitelaw Reid concerning the editorial mentioned above. The notation is dated April 14, 1871.

should be kept on the outside—but not from any opinion they might express concerning me.

I thought you might perhaps put the sensible part of the Tribune readers at least in the light to see me clearer than I am judged by its article of today.

[P.S.] I have expressed myself badly—but you know what I mean.

In the fall of 1871, Winter was working on a biographical sketch of Booth to be included as part of a handsome pictorial work entitled *Edwin Booth in Twelve Dramatic Characters*, scheduled for publication before Christmas by James R. Osgood of Boston. Apparently, the dates which Booth had supplied of his first appearance in New York did not agree with the dates in Joseph Norton Ireland's *Records of the New York Stage, From 1760 to 1860* (New York, 1866-67) and Winter had sent an inquiry.

Booth found a book of letters, now preserved at the Walter Hampden Memorial Library, with praise of his early work by various people in Boston, New Orleans, Baltimore, and other cities. One of the letters from Thomas Barry, the manager of the Boston Theatre from 1854 to 1859, to Benjamin A. Baker, the manager of Booth's first starring tour of the East in 1857, confirmed that Booth's engagement at the Boston Theatre began on April 20, 1857. Booth concluded that "Ireland is right about May in New York." He made his first New York appearance on May 4, 1857, at Burton's Metropolitan Theatre.

7.   September 1, 1871

By accident I have this morning stumbled over a book of preserved letters I did not dream was in this locality. I find by referring to one from Mr. Barry to my agent Ben Baker that my first engt. in Boston was April /57—so Ireland is right about May in New York. I remember only that it was intensely cold & a snowy, sleety night nearly froze the scanty audience & I attributed their boisterous applause to the desire of keeping awake & warm rather than to the result of any effort of mine. I wish I had seen this book while you were here—it contains only a very few letters of any worth whatever; but it is curious to me to find my *Hamlet* spoken so well of by men whose opinions are considered valuable—but who now I suspect have been Fechterized & unBoothed. As my Forgettery is so good and my memory so

bad—you had better take Ireland's report in preference to mine.
To speak truth the past is like a dream to me—I kept no record—
thought little & drank a heap—hence my muddle. I hope "My
witness"[21] will prove a solid success—I am anxious to get at it—
the two or three poems of yours that I have read went right to
the core & made me thirst for more—but I like many things in
Art that other folks condemn & so I may be out. I suppose, however, like acting "it is a matter of opinion—some like an apple,
some an inion," but as my *Ma* hankered for apples when she
carried I—methinks I have a sweet tooth & cling to my likes in
spite of criticism. I have loved pictures that painters have
pooh-poohed, and liked actors others have condemned, and it may
be, my opinion in all such matters is worthless—

Hope you'll have a chance to come here again before I quit
October 6th. Adams[22] has four watercolors by Joe[23]—by Jove!
They are (one of them particularly) superb! he has improved
wonderfully since he promised to paint one for me—some 3 or 4
years ago—I'll be after him for it now.

Lotta[24] does not help me much—I never did and will not believe

[21] *My Witness* is the title of Winter's third volume of poems, published
by James R. Osgood in 1871.

[22] Edwin Adams, American actor, appeared as Mercutio in Booth's
production of *Romeo and Juliet*. He subsequently starred at Booth's
Theatre, most notably in the title role of Arthur Matthison's dramatization
of Tennyson's *Enoch Arden*.

[23] Jefferson was an exceptionally talented amateur painter. His paintings,
mostly landscapes, were exhibited at least four times: at the Pennsylvania
Academy in 1868, the National Academy in 1890, and the Fischer gallery,
Washington, D.C., in 1899 and 1900. He was also a collector and possessed paintings by Corot, Reynolds, Gainsborough, Rembrandt, Israels,
and Neuhuyses, among others. See Francis Wilson, *Joseph Jefferson:
Reminiscences of a Fellow Player* (New York, 1906), pp. 46-96.

[24] Charlotte "Lotta" Crabtree, popular "specialty" actress for whom
plays were adapted in order for her to display her versatility as a banjo
player, dancer, singer, and quick change artist, was not considered to be
"legitimate" theatre by Booth, but he hired her at the insistence of his
partner Richard A. Robertson. She appeared at Booth's Theatre from
August 14 to September 25, 1871, in *Little Nell*, a dramatization by John
Brougham of Dickens', *The Old Curiosity Shop*. Despite her popularity, as
Eleanor Ruggles reports in *The Prince of Players* (New York, 1953), p.
233, the management of Booth's Theatre lost twelve hundred dollars
during her six-week engagement.

that style of performance can be popular in my theatre—but que voulez vous?

Booth began his theatrical season of 1871-72 with an engagement of three weeks at the Boston Theatre, from October 23 to November 13, 1871. During the second week of his engagement, Booth wrote Winter in regard to a banquet celebrating the fourth anniversary of the opening of the Globe Theatre, Boston, which was held at the Parker House on October 30. One of the testimonial speeches was given by his father-in-law, the actor and Chicago theatre manager, James Hubert McVicker. McVicker had recently lost his theatre in the great Chicago fire of October 7 to 9, and Booth reminded Winter that every effort should be made to help McVicker rebuild it.[25]

McVicker was not the only theatre manager with difficulties. Booth's relationship with his partner Robertson was becoming increasingly strained since Lotta's engagement. Robertson wanted to convert the theatre into a variety house like Niblo's Garden, but Booth would have no part of such a plan. As long as he was co-manager, Booth's Theatre would remain a "temple to the dramatic art." Robertson offered to purchase Booth's interest. Booth refused and offered instead to purchase Robertson's interest for $250,000, to be conveyed mainly in real estate. Robertson accepted and Booth became the sole lessee and manager of Booth's Theatre in October, 1871.[26]

8. Tremont House, Boston, November 1, 1871

Expecting to see you here I did not acknowledge the receipt of your last—in which you speak of the *Aldine*;[27] I sent for it several times & have not yet obtained it—if you are coming hither bring it with you. When will it be—if you *do* come? I was not at the *Globe* dinner the other day—but McVicker was &

[25] The banquet was reported in the Boston *Evening Transcript*, October 31, 1871. McVicker's Theatre was rebuilt and opened again on August 15, 1872.

[26] On February 18, 1876, from Montgomery, Alabama, Booth wrote McVicker a lengthy letter describing his relationship with Robertson. The letter was later forwarded to Winter who used parts of it in writing *Life and Art*. It is preserved in box T.b.5 at the Folger Shakespeare Library.

[27] *The Aldine, A Typographical Art Journal* was a monthly illustrated magazine published and printed by "The Aldine Press" of James Sutton and Company in New York. Sutton printed *Edwin Booth in Twelve Dramatic Characters*.

speechified quite successfully—your letter was read & all things passed off well.

I wish some attention could be called to one fact: it is to the interest of every 'star' in America that Mac's theatre should go up, & several friends of his—among whom are Jefferson, Floyd, my brother June[28] & self—are trying to interest others in the work—a word in print to this effect would give weight to it, I think. He is the oldest & only manager who has struggled to maintain the legitimate drama in Chicago & it is one of the strongest points in the country for *stars* of every class.

I suppose you know that I have purchased Robertson's interest in the theatre—I have now a fearful responsibility—but [what] would life be without an aim? I feel the weight & groan, sometimes beneath it, but I'll carry it through though it break my back—and feel the while my life is not altogether useless. Robertson has been a good friend to me & I have paid him well in dollars—but in friendship I must ever be his debtor. It is better so—I will be freer to act, without a fear of marring his interests. My business has been very fine—the Richard 3d. house perfectly tremendous—I've rec'd more applause than New York has given me in years.[29] Hope you will come before I leave—next Monday week—for I would like to see [the] skitch and chat with you of other things. With best wishes.

On Christmas night, 1871, Booth presented a new production of *Julius Caesar*. After a month of continuous performance, it was apparent that the play was to have a long run. Desiring to share his good fortune, Booth, according to a note appended to this letter by Jefferson Winter, sent Winter, who was always in financial straits, a check for a "considerable sum of money." Winter, as with Booth's earlier offer (Letter 2), refused this gift and returned the check, accompanied by a "rather 'touchy' letter," which prompted the following apologetic reply.

[28] William A. "Billy" Floyd, popular American comic actor specializing in Irish character roles, was at this time the manager of the Globe Theatre in Boston.

Junius Brutus Booth, Jr. was at this time one of the managers of the Boston Theatre, a position which he held from 1867 to 1873.

[29] Booth presented *Richard III* at the Boston Theatre on October 28 and on November 6 and 11. The Boston *Evening Transcript* (October 30) reported that Booth's audiences "were among the largest and most intelligent that have gathered since the summer vacation."

9. Booth's Theatre, New York, January 25, 1872

I hardly know what to say to you. I fear I have been guilty of
an indelicacy from wh. my soul shrinks—I do the most infernally
stupid things sometimes—the results of an impulsive desire to
do some little good in return for the great blessings God has
heaped on me, and somehow I always bungle it. Believe me, my
dear Will, I respect & esteem you too dearly to offer even a hint
of offense towards you.

I have felt for a long time past that it would make me entirely
happy if I could share with several dear friends so much of my
worldly goods that do not of right belong to others & if ever the
day shall come when I can freely say "my debts are paid"—my
happiness would not be complete until I had set up a chosen few
of my less fortunate friends.

Pray don't dream that I had any other thought than that of
pure brotherly sympathy—I was really enjoying the belief that
I had done some little good & that you were happy.

On one condition only will I forgive you, & that is when you *do*
need me you will frankly, without the slightest reserve, tell me
so & let me enjoy so much of the Christ that may be in my
wicked composition.

God bless you, my boy; don't bother about that *Caesar*
business[30]—I know how you are taxed & am more than grateful
for what you have done for

Your friend Edwin

While Winter was reluctant to accept any financial assistance from
Booth, occasional necessity compromised his principles. Such an
occasion arose during Booth's fall tour in 1872. At Winter's request,
Booth sent his father, Captain Charles Winter of Cambridgeport,
Massachusetts, a money order for sixty dollars.

Unlike his previous tours which were restricted to engagements of
one or more weeks at major metropolitan theatres, Booth's tour this
season was comprised of one-night stands. Traveling with his own
company and a minimum of scenery and costumes, the experiences
recalled his younger days in California and Hawaii.

[30] Possibly a reference to Winter's review of Booth's Brutus in the
Tribune (December 26) which, while generally praising Booth's portrayal,
implied that "the part is rendered more impressive upon the stage when
played by a man of massive size and great physical stateliness." Booth
being slight of figure and not more than five-feet seven-inches tall was
hardly such a man.

10. Portland [Maine], October 29, 1872

I this day recd. yr. letter & lost no time in sending the amt. required—I wish I had got your letter sooner, but the Ex: man promised to send it through without delay & would give his personal attention to it.[31]

I enclose the receipt for it—in my haste I neglected to have it made out in *your* name, wh. I might have done—but let us hope there'll be no need to use or refer to it.

I wish my dear boy, that I could have made it more—but I did the best I could at this particular time, when J. Henry [Magonigle] is drawing heavy on me to meet 1st of Nov. calls. Don't think for a moment that this has inconvenienced me—I would say so frankly if it had—Grieved as I am that you should be so annoyed I am yet pleased to be allowed to serve you. Did not old Rochfaucault [sic], or some old 'buffee' say something about the thingamajig of a friend's whatdyecallit—the distresses of our best friends affording us something akin to pleasure—?[32] I don't know *what* he said, but I appreciate it all the same. And I also appreciate your feeling in not forgetting me when I could be of even so slight a service; may it never happen that I will be unable to come at the call of friendship. Enough of that. I am doing well—tho' not better than if I had gone on in the old way of *starring* at regular theatres without my company & I don't know as I like this hurry from place to place; it is rather novel—for tho' I had lots of it in my California & Sandwich Island days— the edge was rubbed off & it does not afford such keen enjoyment as in my *salad* days[33] it did. Besides, its too confoundedly civilized for romance; I find good halls & regular theatres in many places & comfortable hotels, with respectable, well-dressed audiences—all of which is disgusting when compared with my

[31] The cancelled money order is in the collection of Robert Young. It is dated the same day as the letter, October 29, 1872. On the order, Booth wrote: "Of great importance. Deliver *tomorrow* sure."

[32] François Duc de La Rochefoucauld (1613-1680), author of a book of moral maxims first published in 1665. Booth attempted to paraphrase the following maxim: *Dans l'adversité de nos meilleurs amis, nous trouvons quelque chose que ne nous deplaît pas.* In the adversity of our best friends, we often find something that is not displeasing to us.

[33] *Antony and Cleopatra*, I, v, 73-74: "My salad days/ When I was green in judgement." This and subsequent line citations from Shakespeare's plays are to the *Globe Edition*.

Kanaka & digger-injiun experience! I sigh for the Sierras & pine for *lone-gulch* & *shirt-tail-bend* to say nothing of cock-tail cañon & the rough & tumble patrons of callow California. But it's a change from the monotony of Booth's Theatre at all events & teaches me to know when I am well off. I wish you could see me *do* 'Richard' tonight—it'll be good as a play I warrant—we do not bother our brains here about such trifles as Joyce & Witham[34] rack their wits upon. No sir! *Petruchio* lends his doublet & trunks to *Gloster*, and *Baradas* 'goes back' on *King Louis XIII* to help piece out the Crookback's regal robes; *Claudius* of Denmark abdicates for the nonce & yields both crown & throne to his *tributary* England. All this is *State secret*—between us Kings & Courtiers, you know, these little tricks of 'trade' are winked at & are supposed to go no further than behind the scenes.[35]

Still on tour, Booth wrote the following letter from Winter's birthplace.

11. Gloucester, Mass., November 10, 1872

Since this is your birth-place I *must* give you a line—if *but* a line—by way of welcome into this harsh world, tho' I fear I have writ myself out; I've been at it all day & my back aches terribly.— Do you know the Pavillion hotel? A Summer resort on the verge of the little bay—Mary, I and our *yaller gal* are the only guests, & with the *proprietoress*, her old mother & a servant girl, are the sole inmates of this vast *corral*. Closed since Septr.—the landlady could not accommodate my company, but has made *us* very comfortable in a plain & homelike way. The superb view & the

---

[34] Thomas Joyce was Booth's costumer at both the Winter Garden and Booth's Theatre.

Charles Witham was Booth's principle scenic designer at the Winter Garden and Booth's Theatre. His more notable designs included the "Hundred Night's" *Hamlet* in 1864 at the Winter Garden and *Romeo and Juliet, Hamlet, Richelieu*, and *The Winter's Tale* at Booth's Theatre. Later he designed scenery for Augustin Daly and Edward Harrigan. A collection of his designs and sketches is preserved at the Museum of the City of New York.

[35] The remainder of this letter is missing. What might have happened to it is indicated by a notation in Jefferson Winter's holograph: "Copy the clear—then burn the marked passages."

delightful weather have made our three days sojourn here very pleasant. The Hall in which I 'lecture' is a fine one, & the audiences are quiet, appreciative (tho' they be your fellow town-folk) and "applaud in the proper places"—which cannot be remarked of numerous other localities I have meandered thro'—. A few weeks (or less) ago I was in Portsmouth, the cradle of another sweet song-bird who claims my dear soul's love—yclept Aldrich, Baily, Thomas,[36]—as the alphabetical Directory hath it; and here I am at yours—I like the place exceeding much; I think it very quaint & interesting, & have obtained some excellent views for Edwina's edification—as, indeed, I have done at many of the points I touched—*en route* as J. Henry would put it.

Say no more, my dear boy, about the trifling favor. I was only so happy to render you and your dear old Father—I return his letter rather than destroy it, for I think such things are precious, and I perceive so much tenderness in this little bit of paper—and something so homely and gentle even in the description of his occupation at that time, that I feel an interest even in his chickens; you'll hardly believe it—but my mind really began tearing down that shed and to build the coop, while the old cock (not your respectable Papa) and his hens stood cackling at me—consequently I couldn't destroy the note.

I know you must be happy in your house, and wish you had a roof & a few sods you could call your own.[37] I am sure Harry would be happy to assist you in any purchase—by advice, I mean—that you may make; in all my important business he is *cute* in getting things into shape for easy payments, &c. If you—or rather —*when* you begin seriously to consider the matter have a talk with him & he may be able to place it in a clearer light than it may appear at present—you see I take it for granted that your business-bump is like my own—deficient in bulk and so old Papa G. is not to be our "Great Father"—poor old man! his defeat

[36] Thomas Bailey Aldrich, American poet and man-of-letters, was an intimate friend of Booth and Winter. At this time, he was the editor of the Boston journal, *Every Saturday.* For further information on his friendship with Winter see William Winter, *Old Friends* (New York, 1909), pp. 132-152.

[37] About this time, the Winter family moved from 230 West 43rd Street, New York to a rented house on Staten Island.

following so close upon a much heavier blow cannot affect him much, however.[38]

Do you *really* think Shakespeare predicted Grant's re-election when he said a *Tanner* would last you some eight or nine year? and because he *keeps out water* a long while?[39] I hardly credit it—and yet it forced itself on my mind four years ago—while I was playing Hamlet (as I was on this occasion) on the ascension of *Ulyses* [sic], and every time—that old Yorick has grinned at me since then—I have had my doubts of Horace.—Good night! that last effort has strained me so that "my back fairly opens & shuts," as the women say, and I had better unlace my stays, so "ta-ta" with my love & blessing—in which my wife joins me—for you and your dear ones.

About this time Booth had apparently written Winter asking his advice on possible stars and revivals for the season at Booth's Theatre. While Booth's letter has been lost, Winter's reply has fortunately survived.

12. Staten Island, New York [ca. fall, 1872]

Dear Edwin:

I received your kind and thoughtful letter a week ago, but I have been compelled to delay answering it because I have been driven almost to death by work for the paper, and because I did not wish to write you until I could reply properly to what you say.

I still think Mrs. Lander would be a good card at Booth's Theatre—but not in *Mesalliance*. I saw that piece when it was first brought out at the Winter Garden, in the days of old Black Jackson, under the name of *Lucie D'Arville*; and I saw it again, in Mrs. Lander's version, when she revived it at Niblo's. It has, dramatically speaking, one good scene—and only one. It is not

---

[38] Horace Greeley, founding editor of the New York *Tribune* and social reformer, ran against Ulysses S. Grant for the presidency in 1872. In one of the most vicious political campaigns in American history, Greeley was overwhelmingly defeated. The "heavier blow" was the death of Greeley's wife on October 20, 1872.

[39] *Hamlet*, V,i,179-199. According to the First Gravedigger, a tanner's corpse will last nine years because "his hide is so tanned with his trade that he will keep out water a great while." Booth is alluding to rumors of Grant's excessive drinking.

to my way of thinking, a fit play to be produced anywhere—least of all under your management and at Booth's. Mrs. Lander's version is a better one than the other—but it is beyond me to understand why such a woman should wish to appear in such a play. Its utter improbability is just as glaring in her version as in the older one. Mary Provost, by the way, played the principal part under Jackson's management; its coarseness is just as offensive; its subject—an innocent wife overwhelmed by a vile and absolutely impossible plot—is just as hackneyed.[40]

The spectacle of a lecherous scoundrel in silk tights getting surreptitiously into a decent woman's apartment, there to wait until the action of a narcotic shall have rendered her a helpless prey to his villainous lust, is one that ought not to be presented on the stage. It can delight only the vicious—and of necessity inspire only the deepest disgust in the minds of all persons of refinement and taste—that is, in the minds of all persons of the very class of community which looks up to you and follows your banner. Read the piece for yourself, by all means, if you have the time to waste; but, having read it, you surely will agree with me about it. And do not engage the lady—excellent actress though she is—for your theatre unless she will agree to present herself in other plays—and not in that one.

I must concede the truth and force of what you say about Shylock. But nobody who sees you play the part could ever suspect how little you like it. However cruel and terrible he is, you must admit the terrible injustice and the brutal cruelty which have made him so. I cannot get up any sympathy with a man who spits in another's face and kicks him about the market place because he is of a different religious persuasion. And I surmise that old Shylock's prosperity in business was as obnoxious to Antonio as his religion.

I wish you would restore the last act of the comedy; the play has a most lame and impotent conclusion, ending it with Shylock's exit.[41]

[40] Jean Margaret Davenport Lander, American actress, appeared as Leonie Arnould in *Mesalliance; or, Faith and Falsehood* for two weeks beginning February 6, 1865, at Niblo's Garden. Mary Provost (or Prevost), American actress and theatre manager, appeared in the title role of *Lucie d'Arville* from March 2 to March 5, 1863, at the Winter Garden.

[41] Booth did restore the last act during his 1889 American tour with the

As to Hamlet: I have possessed your grace of what I think—long ago, and have little if anything new to say. I somewhat doubt whether it be possible—practicable rather—to play Hamlet fully and exactly according to the poet's ideal, because upon the stage he would then become a very tiresome person.

The true appeal to the character, as I see it, is to the highest mood of spirituality. It heats the imagination to a white heat—my imagination, anyway; and I know that it is the same with you. It involves no sensual excitements, no sensuous delights, no gorgeousness of colors, no celerity of movement. Its passion—if that be the right word—is that of intense intellectuality. Its atmosphere is that of dread sublimity and awe. How many audiences, collected within theatres, are attuned to influences fluent from such sources?

In New England, where the pervasive dish for the Sunday dinner is baked beans and pork, I have heard of a clergyman who complained that he was compelled to preach the religion of Christ not to men and women reverently thoughtful of divine things, but to so many bushels of beans. To what, I wonder, does the actor usually have to play Hamlet?

I think we differ only as to the word, and not at all as to the fact, when we come to the insanity of Hamlet. I do not regard him as a lunatic—in the sense in which that word would commonly be taken. But he is disordered—shocked—deranged—unsettled, by the effects of grief, shame, disappointment, supernatural solicitings; and, above all, by incessant brooding on the pathos of human life and the dread mystery of death—and the something or the nothing, after death. No man who has distempered his mind by ceaseless philosophic speculations as to the purpose and destiny of the universe; who is meditating both suicide and murder; and who does instant, impulsive homicide upon a person unseen and unknown, can, it seems to me, be regarded as a man entirely sane and normal. That is all I mean.

One of the most effective, perhaps, the most effective of all the attributes of your Hamlet is its direct applicability to general

---

Polish-American actress Helena Modjeska, who played Portia. For Booth's opinion of Shylock see the notes he contributed to H. H. Furness' New Variorum Edition of Shakespeare's *The Merchant of Venice* (Philadelphia, 1888), especially Booth's letter to Furness, pp. 383-384. See also letter 99, fn. 12.

experience. All human creatures, save those cursed with an indurated insensibility to suffering, perceive and feel the misery which makes a burden of so much of life. That is a matter of emotion, not intellect; therefore it is very general. We all feel at times that there should be—almost that there must be—something after death, something so much better and greater than our little life here that we cannot even picture it in imagination. Yet there comes to most a revulsion to doubt and despair. If we only knew! But whatever falls upon us; however heart-rendering and mind-wrecking our experience of affliction may be; however much we may wish "that the fever called living were over at last,"[42] still we must endure and strive and go on living out our lives until the natural end and doing our duty as well as we can until that end, even though the rest be silence. It is your perception of this common experience; the clarity with which this common experience is felt by your audience to be perceived and understood by you; the exquisite sympathy and beauty with which you interpret it, more perhaps than anything else, that endears your performance to the people. That is what I was trying to convey in my recent notice.

As to the Marble Heart: It is well enough in its way; but I counsel you urgently not to revive it. I remember you especially in the dream scene: but that is only one scene, and in the first act. Anything you do is interesting, of course, because you do it. But not even you can greatly commend to the respect and sympathy of serious persons the silly rhapsodies of a youth foolishly enamoured of a selfish, pleasure-seeking, money-loving coquette.[43] Do you ever read Bon Gaultier? [sic] I do—and get much amusement from so doing.

[42] "And the fever called 'Living'/ Is conquered at last." Edgar Allan Poe, "For Annie" (1849).

[43] Charles Selby's *The Marble Heart; or, The Sculptor's Dream* (1854) is a vapid, romantic melodrama. Booth first played Raphael Duchatlet, the love-sick sculptor, to Catherine Sinclair's Mademoiselle Marco, "the most fickle, fascinating, marble-hearted coquette in Paris," in the original American production at the Metropolitan Theatre in San Francisco in 1855. Although this production was a failure in San Francisco, Booth added Raphael to his repertory. He seems to have played this role for the last time in the spring of 1864 at Niblo's Garden. It was, however, twice revived at Booth's Theatre—once on June 7, 1869, with Edwin Adams and again on April 12, 1871, with Lawrence Barrett.

Anent Raphael, I commend to you the following:

> *It is the most infernal bore,*
> *Of all the bores I know,*
> *To have a friend who's lost his heart*
> *A short time ago!*[44]

Daphnis himself is no doubt an excellent good fellow; but Daphnis deserted by Chloe becomes a blight on social existence! Don't do it, Ned. Let us meet soon.

> Ever Yours,
> William Winter

The net profits of Booth's Theatre during its first year were $102,000; during its second, $85,000; during its third, $70,000.[45] Although the profits steadily declined from season to season, they were still considerable. The liabilities, however, were correspondingly great and almost all of the profit went back into the amortization of the mortgages on the building. Booth thought that if he could free himself from the worries of management, he could devote himself full time to starring tours, thus making enough to reduce the debt. In January, 1873, he persuaded his brother Junius to take over the management of the theatre. According to the terms of the contract, Junius was to lease the theatre for five years and pay his brother an annual rent of $73,000. At this time, the total debt of the theatre, secured by several short term mortgages, was $350,000. For the remainder of the season, Junius was moderately successful as a manager and he paid the rent. On September 18, however, the powerful Philadelphia banking firm of Jay Cooke failed, precipitating a sharp decline in security prices. By October, bankruptcies among banks and brokers had multiplied, businesses were failing, and unemployment was widespread. An overexpanded American economy found itself in the grip of a wholesale depression, from which it would not be completely released for over a decade. The public began to stay away from the theatre. The receipts of Booth's Theatre dwindled; at the end of 1873, it showed a net loss of $40,000. Junius could not possibly pay the rent. Booth's creditors, many of them doubtless facing bankruptcy, were anxious. They refused to renew the short-term loans by which much of Booth's Theatre had been financed. The

[44] Theophile Gautier, French poet, novelist, and critic. The verses seem, however, to be Winter's, not Gautier's.

[45] The finances of Booth's Theatre are complicated and to a great extent obscure. Lockridge, pp. 182-227, gives an efficient and seemingly accurate account of its "decline and fall."

short term mortgages came due. On the advice of his attorney, T. J. "Judge" Barnett, Booth conveyed the entire property of Booth's Theatre to Clarke Bell[46] on November 12 "for no consideration, being led to believe that it would be protected and carried until a favorable sale could be made and the property pay the debts."[47]

At the height of the Panic, William Stuart returned to New York from his seaside home near New London, Connecticut, to announce that in association with Charles Albert Fechter he was assuming management of the new Park Theatre on the southeast corner of Broadway and Twenty-second Street. To Booth, this news was like pouring salt on an open wound.

13.   Providence, December 14, 1873

Many thanks for your kindly remembrance of me & the splendid notice you sent.[48]

I presume you have some suspicion of what has happened in the Booth locality; the papers, I believe, have more than once referred to the transfer of the property to Clark [sic] Bell.

I cannot tell you more than this—I believed myself to be worth at least what I have put into that concern—all I have earned since '63—some *$500,000* & lo! I awake from the dream without a penny & a debt of some *$100,000* hanging over me! Is not this a cheerful look out? Truly a check to young ambition! I had a hope—a foolish delusive dream—that in a few more years I would be free of debt, when I wd. be able to place my family and a few good friends of mine out of reach of pecuniary cares, and establish a sort of "Charitable Institution" in connection with the Theatre;

[46] The T. J. Barnett who advised Booth to convey the theatre to Bell was, according to Lockridge, p. 219, a former dramatic critic of the *National Intelligencer*. He introduced himself to Booth in this capacity in 1866. A number of his letters of advice are preserved at the Walter Hampden Memorial Library. Clarke Bell was a prominent New York attorney, specializing in medical jurisprudence and later the president of the Medico-Legal Society of New York. Prior to his medico-legal activities, Bell had been an attorney for several large companies, notably the Union Pacific Railway. In this latter capacity, he perhaps became acquainted with the financiers and industrialists, Oakes A. and Oliver Ames, whose father had built the Union Pacific, and who were among Booth's principal creditors.

[47] James H. McVicker in a letter to Winter as quoted in *Life and Art*, p. 77.

[48] Booth celebrated his fortieth birthday on November 13.

I had a glorious project—one that wd. do more real good than any I've ever heard of—and it only seemed five years ahead of me, but it's gone up now' & after all these years of hard work—and enormous gains (none of which did I enjoy—for every dollar was sunk in the theatre) I came out—figuratively—"all tattered & torn." Here we go up-up-up; there we go down-down-down —As our "dear old boy," Stuart, emerges from his quiet nest at New London to grasp again the helm of-of—what? the dramatic *Monitor*,— presto! "Your uncle" turns up tails & is lost! A few years more of wrestling with *Hamlet* & perhaps my creditors will cease to howl, & I may be permitted to rest beneath my wig & pine tree. *Mtges* come due in the midst of all the crash— dollars were not, & Booth's Theatre is but a monument to the subscriber's d----d Jackassicality! Peace to his *manes*.

Sauntering 'round the Classic groves of Cambridge last Sunday eve—with Aldrich—"son of song" & old Mr. & Mrs. A—I chatted much with him of you—also while acting in Boston lately I met (for the first time I am sure) a brother of yours[49]—you see how much you have lately been near me—and while thinking that I would write & tell you of my sudden fall of fortune—lo! your letter came.

I hope, my dear boy, the sky will brighten for you, & that health & comfort may be brought round to you by that capricious jade who has played me such a scurvy trick. Adieu!

My wife joins me in good wishes to you & your family—I hope to be with Edwina after next week.

The conveyance of Booth's Theatre to Clarke Bell did not solve Booth's financial dilemma. The theatre could not be sold, nor could an agreement with Booth's creditors be obtained. On January 26, 1874, again at Barnett's advice, Booth filed a voluntary petition of bankruptcy in the United States District Court in New York. The newspapers estimated his liabilities at about $200,000 and his assets, consisting almost entirely of personal property, at under $10,000. After an engagement at the Brooklyn Theatre the week of January 26, Booth quit New York and began a tour of the Midwest.

From Toledo, he wrote Winter the following letter expressing his disappointment at being unable to establish the "true drama" in New York, his bitterness at the pettiness of some actors, and his concern

[49] Probably Charles Winter, Jr., Winter's elder and only full brother.

with newspaper articles ostentatiously sympathizing with his difficulties. But despite these setbacks and annoyances, Booth was essentially optimistic, writing that he would "weather the storm & ride safely into port yet."

14. Toledo, February 17, 1874

Your article and one in the *Mail* pleased me very much—what else the N.Y. papers have said I know not, but here—in the West—(particularly in Chicago) there has been such a flood of *bosh* & *twaddle* (all done with the best of feeling) that not only I—but, if I mistake not, nearly all the community are nauseated & it will do me more harm than good.[50] I know—from long experience—just how apt people are to suspect this sort of thing to be paid for & worked up as theatrical puffery, & I shrink from being so misjudged. I mention this that you may defend me— should all this well meant silliness be so construed by sensible people who do not know me. Once before—in sadder times than these, my friend, I was censured for permitting my friends to disgust the public with a perpetual parade of my misfortune which I should have endeavored to hide, &c; God knows I was the first to be disgusted & grieved by it—but I could not stop it. So in this case—I have read columns of "N.Y. correspondence" in the Western papers that have not only made me blush but swear in solid Saxon. It is not unlikely that some one will begin

[50] Winter's article on Booth's bankruptcy appeared in the *Tribune* (February 7). It is reprinted in *Life and Art*, p. 79. The Chicago *Inter-Ocean* (February 12) wrote that "the day of the legitimate drama appears to be over—at least in New York." "Edwin Booth," it continued, "in his earnest and praiseworthy endeavors to elevate American taste to an appreciation of sterling dramatic worth, surrounded by all the appointments and draperies which artistic taste could suggest, or arduous study and lavish taste supply, has filed a voluntary petition of bankruptcy. . . . The journals of that city lament the departure of the legitimate, and see nothing in the dim horizon, or rather setting sun, but spectacular monstrosities, emotional extravaganzas, society inanities, variety performances and the permanent hippodrome of P. T. Barnum." The Chicago *Evening Journal* (February 12), after reviewing Booth's deficiencies as a businessman, noted that it was "sad to contemplate that while Booth's magnificent performances end with financial ruin to himself, such a licentious play as *Led Astray* should bring a mint of money to the owner of the Union Square Theatre." It reminded the reader however that "anything which 'hangs over the precipice of immorality' appears to swell the pockets of the managers."

to regard it as coming indirectly from the dilapidated victim of this grand "fizzle"—if you see an occassion to set me right, I know you will do it; and I write this merely to prepare you for such an emergency.

Its a terrible blow indeed—but [not] the *worst* that I have felt; the [mere] loss of money (so long as God grants me health to work) does not disturb me much; the fear of being misjudged by my creditors and the disappointment in not being able to establish the true Drama in New York—are very painful reflections. With all this my faith in the "legitimate" is in no degree lessened—it was only *that* which paid in Booth's Theatre, and could I have possibly got about me a company of sufficient strength to do it justice (which is simply an impossibility—either here or in England), and with the real love of art to steady them there could have been very little loss from the *business*. As it is actors are a set of mere vain, selfish, brainless idiots— seeking only their own personal glorification which consists in paper-puffery and large types on the play-bill.

Let it go: my heart is firm & my conscience clear; I *tried* to do my duty—I'm not the first that has failed in a good work— there are far more deserving men than I who struggle thro' a saddened life & go out in darkness without a flicker.

The sympathy I have received is of far more value than that whose loss has called it forth, and I deeply appreciate it—but that little "inquisitor" within will ask—"are there not others more unfortunate, who are far more deserving of it?"

This, and the "stalwart" little wife beside me keep me up to "high water mark," and I look cheerily out to sea—bracing myself for another long pull o'er the waves. She's a "tight little craft" &, tho' I've proved myself to be a pretty bad helmsman thus far, we'll weather the storm & ride safely into port yet—it may be some years hence, but we'll get there—*sure*, if the Great Captain doesn't pipe "all hands below" too soon.—There! with a hitch at my trousers and a pipe full o' baccy you have as breezy a (stage) sailor as ever signed himself, dear Will, thine ever

Edwin

Two and a half months later, Booth's mood was, indeed, much more "breezy." The ultimate fate of Booth's Theatre was still unde-

termined, but Booth was less depressed about it and could even joke
—although not without a hint of bitterness—about its current status.

15. Burnet House, Cincinnati, April 26, 1874

After a most delightful "chat" (of several hours duration)
with "J. J. Rip Esqre,"[51] who starts tonight for "Akron"—
wherever that may be, I turn to my correspondents who have
been lying in wait for me these many days. Foremost & topmost
o' the pile your "queer fist"[52] appears—doubled & shaking—ready
to "pummel" me for not acknowledging its kindly contents.

I did not permit Mr. Shanks[53] to remain so long neglected,
however, but wrote him soon after the receipt of his kindly
information; as much thanks to you therefor. I desire to set your
mind at rest concerning McVicker & Cauzaron;[54] he, *Mac*,
gave *C.* a sound & stinging bit of his mind, & the fellow (in the
presence of Billy Floyd, I think) confessed his villainy & expressed
himself sorry &c, &c, for what he had done to E.B. & all that
sort of thing. *Mac's* dealing with the *dog* was as with the agent
for some actress whom Mac had engaged to play pieces written
for her by a *M. Louis Videre* [sic],[55] and beyond that & his
thorough scouring of the villain, had nothing to say or do with
him. As *Mac* says—"the beastly cur is not worth the ink it takes
to write his name in full." So let him go. Joe & I have been acting
at C.P. [?] houses here & in Louisville, both drawing tremendous
houses, & we've had two or three charming "seances"—the one
just ended was full of genial "spirits," who seemed to revel in
our society—they led us from "grave to gaye" & back again,

[51] Joseph Jefferson.

[52] "Queer fist" refers to Winter's peculiar, tight, spidery scrawl which
is often impossible to decipher.

[53] William F. G. Shanks, American journalist, was at this time on the
staff of the New York *Tribune.* He later became the city editor of the New
York *Daily Star.* He is credited by *The National Cyclopedia of American
Biography*, III, 459, with writing much dramatic criticism.

[54] Presumably the same Cazauran of letter 6.

[55] Possibly Louis Vider who wrote a number of plays in the 1870's and
1880's several of which were adapted from French plays or novels. The
titles of plays which he copyrighted are *The Amazon Queen* (1882),
Coney Island (1879), *Gotham* (1878), *Mexico* (1872) and *The Miracle
of the Roses* (1872).

'till the "brazen tongue," or rather the tiny ticker warned that "Time & Trains wait for 'no uns' "[56]—he'll soon be at "Ho-hokus-pokus" beneath his roof-tree,[57] while I go plodding on & yet on-er, grubbing for the where-withal to pay my debts; by June, however, I hope to take a deep, lung-full breath of *fresh—salt* air at *Cos Cob*, there to repose for a *long*, *long* vacation—for I am determined to take the next year easily, so far as acting is concerned. "At Peace" & "Homage"[58] were both cut from papers for my daughter's "scrap-book," which I am preparing as a surprise for her. Poor Ada![59] my acquaintance with her was brief, but that little convinced me that her heart was golden; & that suffering had mellowed not hardened the *true woman*. I should like to hear of Stuart's success—but I doubt it, for I fear Fechter is *unlucky*, & Stuart really possesses very little, if any theatrical business capacity. F.'s remarkable talents both as actor & stage manager shd. insure the success of any theatre—could *he* be managed; otherwise I doubt the safety of any enterprise he has to do with.[60]—In looking back I find that I dismissed your poems rather abruptly—I did not mean it so; they are charming—as I think are all that I have read of yours (this is but trifling

[56] Both Jefferson and Booth were actively interested in Spiritualism and they occasionally attended seances. Whether or not on this occasion these were real seances or merely "chats" I have been unable to determine.
Poems Grave and Gay (Boston, 1867) was the title of a book of George Arnold's poems edited by Winter. "Time and tide stayeth for no man." Richard Brathwaite, *The English Gentlemen* (1620).

[57] Jefferson established a home at Hohokus, New Jersey in 1868. He also had homes at Orange Island near New Iberia, Louisiana, Buzzard's Bay, Massachusetts and in later life at Palm Beach, Florida.

[58] Two poems by Winter, subsequently reprinted in *The Poems of William Winter* (Boston, 1881), pp. 45-46, 56-57.

[59] A reference undoubtedly to the death of Ada Clare from hydrophobia on March 4, 1874. See Winter's eulogy in his *Brief Chronicles* (New York, 1889), pp. 48-49.

[60] The New Park Theatre opened under the management of Stuart and Fechter on April 13, 1874, starring Fechter in a new play entitled *Love's Penance* which he had adapted from *Le médecin des enfants* by Count d'Avrigny (?Charles-Joseph Loeillard d'Avrigny 1760-1823). Shortly after the close of this play on May 6, Fechter withdrew from the management. Stuart relinquished the theatre to Henry E. Abbey in the fall of 1876.

praise to you, however); I cannot criticize, & always make bungles of compliments; I like what I like, & may be very far wrong in poetry, painting, & sculpture too (I know I am as regards acting), but I have these friends whose works please me very much—not because they are my friends, but because they always seem to strike a harmonious chord in my own "instrument" which responds at once, and this is just as far as my judgement extends in matters of *Art*.

I won't bore you with my own "worries"—you have enough of your own; "Booth's theatre" & its "affairs" are "statty-quoish" —I know nothing, but live in hope that ere long something definite may be determined by the creditors. Surely, I have done my part well in establishing the "leg (if not the *leg*itimate) drama" in New York; I have provided for it a superior *home*, where J. & P. may revel in all the gorgeousness of nudity & nastiness.[61]

I see poor Brougham[62] is very ill—he has had some severe turns these past two years, & I doubt if the poor old man will remain long with us.

My wife & I have both been remarkably well this season— she joins me in all kinds of good wishes for you & yours.

[61] Junius Brutus also went bankrupt in April, 1874, and conveyed the lease of Booth's Theatre to Henry C. Jarrett and Henry David Palmer. Jarrett and Palmer had achieved success through their promotion of the musical fantasy *The Black Crook* during their management of Niblo's Garden from 1868 to 1874. The chief attraction of *The Black Crook* was the female *corps de ballet* who daringly exposed an unprecedented amount of leg. According to Robert Young, Winter had once chided the public for their preference for *The Black Crook* to Jefferson's *Rip Van Winkle* with the following doggerel:

> For our taste in amusement
> Three syllables less we would beg:
> We want not the drama Legitimate
> The drama we want is Leg!

Despite the censure of critics such as Winter, and the disgust of actors such as Booth, the "leg drama" retained its popularity.

[62] John Brougham, popular American actor, playwright, adapter, and theatre manager, recovered from this illness and was active on the New York stage for the next four years. His last appearance was as Felix O'Reilly in Boucicault's melodrama *Rescued* at Booth's Theatre on October 25, 1879.

As late as September 30, 1874, Booth indicated in a letter to Winter his hope that a settlement might be reached with his creditors. Booth's Theatre might yet "weather the storm & escape the breakers (i.e. the jews wh. could gobble it up)" (Folger Shakespeare Library, Y.c. 215 [224]). All hope vanished, however, in February when the theatre went on foreclosure to the estate of Oakes Ames. McVicker bought up all his son-in-law's debts in March, becoming his sole creditor and releasing him from bankruptcy. After completing a successful engagement at the Boston Theatre from March 15 to April 3, Booth retired to his summer home at Cos Cob, Connecticut.

On August 17, he suffered a broken left arm, several broken ribs, and multiple bruises in a carriage accident. The day after the accident, Winter hurried to Cos Cob, but Booth was asleep, "drugged with morphine." Several weeks later, on September 8, Winter again took the train to Cos Cob. In his journal, he recorded that Booth looked "pale and worn," but was "mending rapidly." They chatted for a couple of hours in the library and strolled about looking at the horses, and finally Booth brought up the subject which must have been foremost in his mind—the loss of his theatre. He complained that "Robertson, Clarke Bell, and Judge Barnett were scoundrels and swindlers," that the theatre "paid for itself thrice or four times over" and that they had "stripped him of the property." He even accused Harry Magonigle of having "no head for business." A year later, the duplicity of Robertson, Bell, and Barnett, and his own folly, still rankled in him. The Booth's Theatre "romance," as he would later refer to it, was ended, however. Never again would he attempt to manage a theatre. The remainder of his career was spent making starring tours. For the next two years his primary concern was to recover his former financial security and to repay the enormous debt that he now owed to McVicker.

Chapter II

American Tours

1875-1877

By the fall of 1875, Booth had been absent from the New York stage for two seasons. On June 2 of that year he contracted with Augustin Daly for a six-week engagement at the Fifth Avenue Theatre beginning October 4. His terms were for one half the gross receipts of every performance up to $1500 and two thirds of all above $1500. The injuries he sustained in the carriage accident forced him to postpone his opening for three weeks and the engagement was reduced from six weeks to four. On October 25 he opened in *Hamlet*. He subsequently played nine other roles, including for the first time the seldom acted *Richard II*. The gross receipts for the run of thirty performances was $47,909, of which Booth's share was $24,646.[1] This engagement was, artistically as well as financially, one of the most successful of Booth's career. With it he began to recover from the disastrous setbacks of the previous year.

His reception by the opening night audience was wildly enthusiastic. John Ranken Towse, the newly appointed drama critic of the *Evening Post*, reported (October 26) that the "applause began before he had fairly entered upon the stage and was renewed again and again with a vigor that was the result of genuine feeling and no organized combination." No one, however, was more enthusiastic than William Winter. In the *Tribune* (October 26) he lauded Booth's Hamlet as "poetic," "spiritual," and "entirely Shakespearean." Booth ideally personified for Winter every facet of Hamlet's character— "the spiritualized character, the masculine strength, the feminine softness, the over-imaginative reason, the lassitude of thought, the autumnal gloom, the piteous, tear-freighted humor, the princely

[1] In *The Life of Augustin Daly* (New York, 1917), pp. 198-206, Judge Joseph Francis Daly gives a brief account of his brother's negotiations with Booth together with several of Booth's letters and the figure for the gross receipts. From this figure I would calculate Booth's share at $24,440. The figure of $24,646 is recorded on an interleaf at p. 172 of a copy of Asia Booth Clarke's *The Elder and the Younger Booth* (New York, 1882) which Mrs. Clarke gave to her son Wilfred. This copy is in the special collections division of the University of Illinois Library.

grace of condition, the brooding melancholy, the philosophic mind, and the deep heart." Booth seemed "to live Hamlet rather than act it."

On the Saturday before the opening of *Richard II*, Booth wrote Winter to thank him for this "glorious treatment."

16. Gilsey House [New York], November 6, 1875

I have been as closely boxed as yourself since I arrived in town—never in my life before have I been so over-run by visitors; as early as 10 A.M., & several times at nine they begin & keep it up 'till play-time.

With this super-flux of callers, long rehearsals & consequent fatigue I have been obliged to let you & several other friends rest quietly in my table drawer until this moment. At nine this morning some ladies called on Mrs. B., at half past, 2 doctors came to stretch my arm[2]—an hour later another lady dropped in, & so we go from morn 'till eve. I suppose its because we are at a first class hotel, for while we were *poor* & lodged in rooms at the theatre we were not so eagerly sought.

I have been wanting all this while to thank you, my dear boy, for your glorious treatment on my return, & have silently been doing it in the "innermost depths of my soul" but this is the only chance I've had to let you know it. You deserve more from me, however, than I can e'er repay—& if there should ever be any means whereby I can do more than *say* so, you must add one more to your many favors & let me know it. I hope the new venture on Monday will be favorably received & that I may not do what Macready says all others have done—i.e. fail to make the play attractive.[3] Mother tells me that, in his early days, Father played

---

[2] Booth's left arm never healed properly and it remained shorter than his right. He never completely recovered its full use. The English actress Mrs. Charles Calvert, née Adelaide Helen Biddles or Bedells, who later toured with Booth reports, however, that "his fearfully shrivelled and distorted arm" resulted in a new and original piece of stage business. In *Richard III* he drew up his sleeve and displayed his arm as an example of Jane Shore's witchcraft. Mrs. Calvert writes that "a shudder used often to pass around the house" and Booth said to her that he was "often complimented on the way in which it is made up!" See Mrs. Charles Calvert, *Sixty-Eight Years on the Stage* (London, 1911), p. 207.

[3] Macready played *Richard II* for the first time at Newcastle in the spring of 1813 and used it for his benefit in Dublin in 1815. The play was "applauded, but did not attract." He dropped it from his repertory for the

the character very frequently & very successfully, & never knew why he dropped it from his repertory.

If I break no more bones I shall continue my work now 'till June next—by wh. time I expect to be clear of debts & have a little to *set* on for future hatching.

I hope I shall see you before I leave the city[4]—I have 2 weeks longer & if I can't get to you, I hope you'll catch me some day.
I shall make a desperate effort to do *Lear* (original text) during my last week.

Another caller bids me *let you up*—so does my paper.

Booth's engagement at the Fifth Avenue Theatre ended on November 20. Three days later he began a four-week engagement at the Walnut Street Theatre in Philadelphia. Competing for his audience were the evangelists Dwight Lyman Moody and Ira D. Sankey and the opera company of the American soprano Clara Louise Kellogg. With such competition, Booth was skeptical of his financial "prospects." In an earlier letter, Winter had apparently told him of Joe Jefferson's successful engagement at the Princess's Theatre in London, where he had been playing *Rip Van Winkle* since November 1. Jefferson had painted a picture for Booth and had sent it in care of Winter.

17.   Continental Hotel, Philadelphia, November 23, 1875

Rehearsals, Matinées, doctors & lawyers ate up all my time in New York else had your last recd. a thankful recognition ere I left that city.

Beautiful as are both your poems I like the Poe 'un best[5] (don't try to twist that into a pun)—"After my death I wish no

---

next forty years, reviving it on December 1, 1850, when again it did not attract. See *Macready's Reminiscences and Selections from His Diaries and Letters*, ed. Sir Frederick Pollock (New York, 1875), pp. 48, 53, 73.

[4] Winter records at the end of the letter that he talked with Booth on November 16 after the fourth act of *King Lear*.

[5] "Dirge: In Memory of Poe" was read by Winter at the dedication of the Edgar Allan Poe Memorial in Baltimore on November 19, 1875. The other poem is probably "Comrades" read at a banquet in honor of the actor and dramatist, George Fawcett Rowe, at the Lotos Club in New York on August 29, 1875. See *The Poems of William Winter* (New York, 1909), pp. 97-101. Unless otherwise indicated all subsequent citations of *Poems* are to this edition.

other herald to keep mine honor from corruption, but such an honest chronicler as Winter"—I suppose many have before said this, but let it not therefore be less welcome.

I cannot just now scratch more than a few thankful lines for what you say of *Lear* & *Richard*; I will stick to 'em so long as the people will endure my agonies therein, & I am vain enough to believe they will, by & by, sympathize with & encourage me as you have done.[6]

I am delighted to know that Joe J. has grasped the Lion firmly by the jaw with ungloved hands & hope he will shake lots of "shekels" from old Bull's money-bags, but poor fellow! he has his sorrow heavy on him while he struggles to lighten the cares of others; in a letter today from my sister, Mrs. Clarke, the death of his youngest child is mentioned—after a brief illness & about the fourth or fifth day of his engagement.[7]

If you can send his picture to my *Cos Cob* residence direct to care of H.A. Rose Esq—*C.O.D.*, if not retain it 'till my return next Spring. With Moody & Sankey & the Kellogg troupe against me I can form no notion of my prospects here—I remain four weeks, however, & may gain a hearing on some of their off-nights—if it don't rain.

If I can serve you, Will, pray let me know it without hesitation—it will be a pleasure for me to go as far as in me lies to do so. With sincere wishes for yr. health, success, & happiness.

At Baltimore on January 3, 1876, under the management of John T. Ford,[8] Booth began a starring tour through the South, his first

[6] Booth's *Richard II* opened on November 8. Winter published reviews of it on November 9 and November 12. The first is principally a source study and stage history of the play; the second is an evaluation of Booth's performance. Winter admitted that *Richard II* was a play only for the "scholar and student of human nature" and would probably never "thrill and fascinate the public." He praised Booth's performance, however, as "a fresh and novel dramatic identity. . . . intellectually worthy to stand beside his *Hamlet*." His review of *King Lear* on November 17 called it the "greatest of Mr. Booth's achievements." In a slightly expanded form this review is printed in *Life and Art*, pp. 272-283.

[7] Jefferson was at the Princess's from November 1, 1875, to April 29, 1876. His son Henry died on November 5, 1875.

[8] John Thompson Ford, noted American theatre manager, at various times during his career managed theatres in Baltimore, Washington, D.C.,

since 1859. After two weeks in Baltimore and one week in Richmond, Booth visited Charlotte, Augusta, Charleston, Savannah, Macon, Columbus, Montgomery, Atlanta, Chattanooga, and Bowling Green. The tour consisted of 52 performances and it ended on March 3. The tour was highly profitable to both Booth and Ford. Booth was annoyed, however, that Ford published false financial figures in order to justify increased ticket prices. Except for this one "indelicacy" Booth thought the tour "delightful." Booth and Ford parted at Mammoth Cave where Booth rested for a week before making a further starring tour to Louisville, Cincinnati, and Chicago.

18.   Mammoth Cave, March 5, 1876

I have never recd. an answer to my last letter (from Phila., concerning the disposition of the picture you told me J. J. had sent me) but I daresay a few lines more will not bore you much. I received two beautiful poems of yours, also the one by Linton[9]— sent by you, for all of wh. (particularly the former) accept my thanks. My trip with Ford terminated on Friday last at Bowling Green, a place I should not have visited (but Joe acted there— so I'm not alone). Everything has been very pleasant and agreeable except the fact that the great bulk of the money made was by Ford for his trifling expense & *very mild* company. He has surely cleared $50,000 by me, while, after deducting *my* expenses, about $3,000, I recd. 27,000. I mention these *private* matters for this reason: F. is fond of advertising & actually published my terms as so large (much smaller with him than with any other except Clarke) that he was compelled to raise his prices on this acct.; expecting to realize *3* or *4000* by my engagement. It certainly was the most profitable engagement for the manager that was ever made. I was fool enough to give him *2* nights free because I was led to believe the theatres were very small & the people very poor; all of wh. was *not so*.[10] Why I bore

---

Alexandria and Richmond, Virginia, and Philadelphia. An excellent account of his life and career is given by his grandson, John Ford Sollers, "The Theatrical Career of John T. Ford," an unpublished doctoral dissertation, Stanford University, 1962.

[9] William James Linton, American wood engraver, illustrator, printer, and poet, engraved the drawings of William John Hennessy for *Edwin Booth in Twelve Dramatic Characters*.

[10] Booth agreed to tour with Ford for a flat fee of $30,000. Booth would have earned much more if he had instead contracted for a share of the

you with this is to let you know the truth of what appeared in the "N.Y. Times," & what other statement you may receive—in the *advertising* way—from others. Jno. T. is a good fellow & I like him in all but the indelicate manner he makes known what should be private & thus laying the blame on my high terms for his excessive prices for the "Show"—I strongly suspect he is "after" sending you a history of the trip[11] & he will doubtless try to impress you with the idea that he has made his 3 or 4000 dollars on this "unprecedented" tour; this is only for your own quiet amusement over the tricks of speculative managers. I have named to you the size of my "nest egg"—all claims against me being satisfied—around wh. I hope, in a few years, to build (not a *theatre*, not yet a church—although the latter might carry my memory perhaps for a year; about five minutes & a half longer than the former did)—I hope to build up a *competency* for my antiquity.

My wife & I have explored all that can be risked at this season down in the "bowels of the Earth" & am completely awe-struck! Shakespeare might have given a fair description of it, but no "busted" actor, nor literary "cuss" of this degenerate age need attempt it. I won't even except another "Sweet Will" I wot of. J.J. is still *on deck* in London, & is about to have his lost baby replaced by a younger one (Mum!)[12]

Richard 2d. has had a splendid effect everywhere I've played it—so has *Lear* & all agree with what you so kindly & beautifully said of them: I wish they had each had a better theatre, a better company & a run of several weeks in N.Y. It has been better

---

house. Sollers writes, p. 335, that the first night in Atlanta grossed $2,700 and that Booth's four performances in Atlanta grossed $9,000. He estimates the total gross of the tour at close to $90,000.

Booth and his critics continually complained about his supporting companies. Frederick Barkham Warde, American actor and lecturer on Shakespeare and the drama, was Booth's principal supporting actor on this tour. In his reminiscences, *Fifty Years of Make-Believe* (New York, 1920), p. 116, he characterized the company as "a thoroughly competent one."

[11] A letter from Ford to Winter dated March 6, 1876, concerning the tour is preserved in the collection of Robert Young. It contains no allusions to the finances of the tour. Winter used the material in the letter for an account of the tour which appeared in the *Tribune*, March 9, 1876. This article was later printed in *Life and Art*, pp. 98-100.

[12] William Winter Jefferson was born in London on April 29, 1876.

acted & better mounted in every respect elsewhere than at Daly's, where the piece had to struggle against his "duffers" & cheap, tawdry costumes & scenery.[13] He is after me for a return next fall, but I am waiting to see how very anxious Wallack may be for me at his house—which he professed to be last fall, after I had closed with Daly.

My tour has been delightful & all the stages easy; with the exception of murderous cooking at several hotels—all has been *digestable*; F. tried his best to please me & at the close his company (started by his property-man) presented him with a gold masonic cross; the humblest best, appreciated his kindness to them. Altho' as an actor I stand bolt upright for my fellows, as a delapidated manager—who tried to help them up the slippery hill to respectability—I despise them. Three cheers for "ye propertie-manne!"

After a week's rest I open with Macauley[14] in Louisville (2 weeks) thence with him in Cincinnati for one—thence to Chicago. I do not look for home 'till June. I presume McCullough[15] has

[13] Most of the New York critics agreed that *Richard II* might have been more spectacularly presented, but none described the costumes or scenery as "tawdry." The critic of the *Herald* (November 9) on the contrary, wrote that Daly clothed the play with a "wealth of scenery and decoration" and that Booth's first appearance in the robes of a king was "a masterpiece of imperial adornment." The critic for the *Spirit of the Times* (November 13) while admitting that Daly's production was not the equal of Charles Kean's production at the Princess's Theatre, wrote that "the scenery was very good and picturesque" and "the costumes generally rich and correct." All the critics agreed that the supporting company was generally "mediocre." A complete cast list is given by Odell, *Annals*, x, 14, who writes, "Where in the world did Daly get them?" Warde, p. 126, describes playing *The Merchant of Venice* against "a background of a modern American street with advertisements painted on it," Hamlet interviewing the Ghost in "a dense wood" and other scenic anachronisms. One wonders if Booth was not exaggerating the quality of production on his Southern tour.

[14] Bernard "Barney" Macauley was the manager of theatres in Louisville, Cincinnati, and Detroit in the 1870's and 1880's.

[15] John McCullough, American actor born in Ireland, was a close friend of both Booth and Winter. From 1866 to about 1877, he managed the California Theatre, for the first four years in partnership with Lawrence Barrett. Because of a heavy financial loss, he retired from management and, like Booth, spent the rest of his career making starring tours. At the time of this letter, McCullough was playing at the Arch Street Theatre in Philadelphia but was scheduled to appear at the California Theatre on March 27.

told you that I go to his place in Septr. I shall start in August &
take Edwina with us, we three with McVicker & wife will make
a family party & John has promised to have a special car at
Omaha, with him in it, to take us royalty to Frisky. It will be just
20 years since I left there—*Sept. 1856*. Now I'll ease up on you.
My wife joins me in kind regards to yours & you. Goodnight.

During the Southern tour Booth was annoyed by frequent allusions
to his father's excesses and to the assassination. Many articles even
implied that Booth's success in the South was due not to his abilities
as an actor but to the fact that he was the brother of John Wilkes
Booth. From Louisville he wrote Winter to complain about these
"disgraceful anecdotes."

19. Louisville, March 14, 1876

Forgive my boring you again so soon after the receipt of your
good, long letter—that shd. suffice for awhile I know, & so it shall
—don't answer this until you feel like doing so.

This is merely to direct your attention to a paper I send with
it; it contains a very kind welcome (by Watterson, I think, who
is very sick) and in the "over-page" a most brutal ghoul-feast.[16] I
want you to see what I have to endure in the midst of my "glory,"
&c.

Except in Phila. I never find in the North or East any reference
made to these miserable affairs—wh. shd. in decency & charity
have been long ago forgotten. But all through the South—from

[16] Henry Watterson, American statesman, humorist, and editor of the
Louisville *Courier-Journal*, was a personal friend of both Booth and
Winter. The "kind welcome" appearing in the *Courier-Journal* (March 12)
was a virtual paean of praise. It proclaimed that Booth came from a
"noble line" and his inheritance was "more royal than that to which kingly
nothings are heir, for whom the populace shout." Booth's artistic integrity
and perseverance had "fixed for him a name which glows like an independent light under the luster of his birth." Booth was "the very oak's
heart" and at the "zenith of his powers." The "over-page" contained an
article entitled the "Graves of the Booths" which was reprinted from the
Toledo *Blade*. It was a description of the Booth monument in Greenmount
Cemetery in Baltimore. It was generally complimentary to Junius Brutus
Booth, but it did note that he was "a slave to the demon alcohol." It described the new grave "where lie mouldering the bones" of John Wilkes,
"the recreant, renegade son of a great actor." Then it briefly reiterated the
assassination and the capture and burial of John Wilkes.

Balto. to this place I have been greeted with disgraceful anecdotes about my Father (all in the main false or exaggerated), and the flaunting in my face of buried cerements—raked up by these hyenas. Every little piddling village has stabbed me through & through, wherefore—I know not; it certainly is the most heartless, uncalled for brutality! I am sorry I had not called yr. attention to this before your article appeared that you might have expressed my feelings on this point; tho' I have recd. much to be proud of & grateful for at the hands of the Southern people this one shameless, devilish act of the press has destroyed all pleasant remembrances I might have borne on this trip through life. I sincerely hope I shall not be invited South again. If ever you have a chance to shame them on this subject I wish you wd.—

Thanks for the fine account you gave of the trip. Apropos of Ford & the 'ducats'—I alone am to blame for the disparity in shares—he offered me a better chance; I mentioned it solely because he had been publishing my large terms & the *small* profit he received; his reference to the erroneous statements &c, in his letters to you, was the result of a good "talking-to" on the subject.

Lent will effect me here—I lose the run of *holy* days, & always find myself just where I should not be at such times.

I scribble as a hasty "howdye"—not as answer to yr. letter— that will come hereafter.

On March 27 Booth began a one-week engagement in Cincinnati. At the end of the week, he wrote Winter to express regret at Winter's illness, to correct rumors about his own health, and to sketch his future theatrical plans.

20. Cincinnati, April 2, 1876

I shall bore you with but a few lines tonight—merely to express my regret at your illness & the hope that both yourself & family are out of doctors hands. I hold them to be next of kin to lawyers, and the least we know of them—the better. Your wife & little ones were ill when last you wrote & now this brings news of your own "upset"—pray God all are well by this time. I am reminded here of a very frequent allusion to my own health as being poor—now, as I do more work than any other actor I know or have ever heard or read of, I consider myself in pretty good condition; indeed I verily believe my tumble did me good—I have

enjoyed better health since than I can remember for some years previous to it. This *twaddy* sort of theatrical chat does no good—it will injure me by-&-by, & as I have no ambition to be thought an "interesting invalid" I wish in yr. next mention of theatrical matters you would *cure* me. I did not reply to the notice you sent (& which Osgood also sent me) concerning the plates, &c—for I could see no use they would be to me, and really I find my dollars go quick enough without indulging in such luxuries.[17] Thanks for your timely warning—all the same. I send you something about *Richd. 2d.*—wh. seems to be much liked everywhere. What a pity I had not a chance to "run" it in N.Y. on its first production! By-the-by, you speak of my not going to Daly's—I surely do not desire it, but Wallack shows no evidence of any desire for me at his house, while my stomach turns at *Booth's* under its present management. I fear Wallack will not want tragedy at his house—unless it were on the "off" [season?] terms; it may be many months ere I again appear in N.Y.—tho' I would like to do so next winter. I am to give a benefit for the Stratford "Memorial Theatre" fund on Shakespeare's next birthday—at Chicago. I am also to go to Balto. when I get through my season's work to aid the ladies of my native state ("My Maryland," you know) in doing their centennial glories.[18] I hear Barrett & a party of gentlemen wish to buy Booth's Theatre—Larry would then be "high-cock-a-lorum."[19]

Business in Louisville & here much better than could be expected from the times (Lent) & weather (rain, *toujours*) I don't pretend to write letters—merely scratchy sort of *how-dye-*

[17] The "plates" are Linton's engravings for *Edwin Booth in Twelve Dramatic Characters.*

[18] Booth gave a benefit performance in Baltimore in the latter part of May to help a ladies' club in that city celebrate the American centennial.

"Maryland! My Maryland!" the state anthem was written by James Ryder Randall in 1861.

[19] Lawrence Barrett, American actor and manager, was a close friend of both Booth and Winter, although at this time Booth and he were estranged because of a quarrel they had had in 1873. In the winter of 1880, however, they were reconciled and their relationship continued to be cordial for the rest of their lives. Much of their correspondence is preserved in the Harvard Theatre Collection and some of these letters are printed in Otis Skinner's *The Last Tragedian* (New York, 1939), pp. 129-209. Despite his apparent interest, Barrett did not buy or even lease Booth's Theatre.

do's? with sincere good wishes for you all from both of us. Edwina, you know, is at school in Pha.—well & rapidly improving. When you can't work scribble a few to yours ever

Edwin.

One week later, Winter seems to have recovered from his illness and was contemplating a biography of the actress Mary Ann Duff. He had asked Booth for information about her, and also for a biography of the elder Booth. From Chicago, where he had opened at McVicker's Theatre on April 10, Booth answered these queries, reported his own activities and his wife's health, and commented on various theatrical news of the day.

21. Chicago, April 12, 1876

I am very glad you are so much better. Why don't you stick to quinine & iron? Whenever I am run down, which is of frequent occurrence, I go for my pills—about 6 grains a day, & up I come again. You know its a splendid tonic & builds up the whole man quicker than ought else that I have ever tried.

I'm sorry I can give you no *news* of Mrs. Duff—I remember hearing my Father speak very highly of her as an actress, very beautiful &c, &—if I mistake not, was in some way *connected* with "Tom Moore"—all of wh. you doubtless know more of than I can tell you.[20] Nor do I know where a book of Mrs. Clarke's can be had[21]—I gave mine away long ago—to whom I know not; she has none, & was quite annoyed when I once asked her for a copy— she seemed to feel ashamed of so lame a work. Carleton published it, but I doubt if he has any copies. Can those plates be of any future service to *you?* let me know; I mean of course the Hen-

[20] Mary Ann Duff, née Dyke, Irish-American actress, was considered to be the leading American tragedienne in the 1820's and 1830's when she frequently acted with Junius Brutus Booth. Thomas Moore, the Irish poet, once made her a marriage proposal, but she rejected him in favor of the actor John Duff. Several of Moore's early poems lament his thwarted love. Later Moore married her sister Elizabeth "Bessie". Winter did not write Mrs. Duff's biography.

[21] The reference is to Asia Booth Clarke's *Passage's, Incidents, And Anecdotes in the Life of Junius Brutus Booth, the Elder* published in 1866 by G. W. Carleton in New York. In 1882 this book, slightly altered and expanded to include the career of Edwin Booth, was re-issued by James R. Osgood, Boston, as *The Elder and the Younger Booth*.

nessey [sic] cuts. Do the N.Y. theatres give "memorial benefits?"[22]

Apropos of what you say regarding *Richd. 2d.* having his day—see what Chicago thinks of him; it looks very much as tho' it is now having its day—the audiences, however, are large & enthusiastic, notwithstanding there is a terrible excitement over the local politics here; last night some 40 or 50,000 people were jammed into the Exposition building while I entertained over $1000 worth of them at McVicker's with "poor Richard"—the two writers requested seats together, & as you see they went hand in hand for me.[23]

Poor old beat Stuart![24] I feel just as you do towards him—I wish him well, but well out of my way also. I hear sad things of Barney;[25] my wife has just recd. a letter from a neighbor of his & a friend of ours, one of those folks who are not usually sympathetic with actors, & she speaks of Mrs. Williams & poor

[22] The reference is to the Shakespeare Memorial benefit in Chicago mentioned in the preceding letter. No such "memorial benefits" in New York City are cited in *Annals* for 1876. See Letter 23, fn. 38.

[23] On April 11, Chicago's Democrats and Republicans met jointly in the Exposition Building to contest what they considered to be an illegal extension of the mayor's term of office and to nominate and elect a new mayor.

The "two writers" to whom Booth refers were undoubtedly critics from two of Chicago's newspapers. For example, the critic of the *Tribune* (April 11) was enthusiastic in his praise of Booth's portrayal of Richard II, but could not endorse the play. He asserted that "there could not be any better arrangement of the play for stage representation" but still thought that it did not prove to be "good acting drama" because it was "too refined and spiritual in its construction" and lacked the "essential dramatic element of action." Booth performed Richard II in Chicago on April 10, 11, 12, and 22.

[24] Presumably the reference is to the failure of William Stuart's production of *The Crucible*, a melodrama written by and starring New York's ex-mayor A. Oakey Hall. It was presented at the Park Theatre from December 18, 1875, to January 8, 1876. After this production Stuart's success as a manager began to wane. In the fall of 1876 he relinquished his theatre to Henry E. Abbey.

[25] Barney Williams and his wife Maria Pray Mestayer, popular American starring team, achieved considerable success, both in America and Great Britain, in romantic Irish comedies. His last appearance was at Booth's Theatre on Christmas night, 1875 in Charles Gayler's *The Connie Soogah*. According to Winter, Williams died of "disease of the brain" on April 25, 1876. See *Brief Chronicles*, pp. 329-334.

Barney very feelingly—he seems to be going the way of Fox.[26] I prevailed on Wheatley[27] not to risk management again, & now he has done the next desperate thing; he evidently determined to suicide some how. McVicker has pencilled a sketch of my affairs for you—*I* think just the thing, but he doubts if it is worth publishing anything now about it; it gives Bell & Barnett such gentle h-ll, however, in their own words & under their own hands, that I am anxious to have it in print; he thinks it best to wait till we are in N.Y. in June & then he will see you—I shall urge him to send it sooner if I can.[28]

I've scribbled in such haste that I fear you will not be able to decipher what I've written; by-the-by, *your* peculiar calligraphy is quite clear to me now—for a long while it set my teeth on edge & made my eyes water.

My wife is not well—I wish I could go at once to Colorado & keep her there for the next six months; physic does her no good, & only the natural influences of air & water can set her up. I can't go, however, bound as I am by engagements & other shackles— such as my wife holds very dear—namely, the home & household gods (2 darkey servants particularly). Remember us both most kindly to your wife & the little ones.

With his production of *Hamlet* at the Winter Garden during the 1864-65 season—the so-called "Hundred Nights *Hamlet*"—Booth had hoped to begin a series of scenically spectacular revivals of Shakespearean and other classical English plays. He was undoubtedly

[26] George Washington Lafayette Fox, American actor, manager, and pantomimist, was principally known for his portrayal of the Clown in his pantomime *Humpty Dumpty* (1868). In 1870 he presented a highly successful travesty of Booth's Hamlet which ran continuously for 10 weeks, only one week less than the original. His final appearance was at Booth's Theatre on November 27, 1875. His last years were spent in an insane asylum where he died on October 24, 1877.

[27] William Wheatley, American theatre manager, was principally known for his successful management of Niblo's Garden, where in 1866 he presented the musical extravaganza *The Black Crook*. He made so much money from this production that he retired in 1868. The "next desperate thing" is possibly Wheatley's remarriage about this time.

[28] McVicker's "sketch" is preserved in box T.b.5 at the Folger Shakespeare Library. Winter printed portions of the sketch in *Life and Art*, pp. 122, 126-127.

motivated in this enterprise by Charles Kean's highly successful revivals at the Princess's Theatre from 1850 to 1859. Like Kean, Booth also had in mind to issue in conjunction with his productions a series of "acting versions" of the plays as they were arranged for presentation at the Winter Garden. Each of these acting versions was to be an "exact copy" in book form of the production it represented. It was to contain the text as it was cut and arranged for performance, all of the stage business, historical and explanatory notes, and selected illustrations of the actual scenery and costumes. In 1865 Booth himself had arranged the texts of *Hamlet* and *Richelieu* and had Charles Witham make sketches of various scenes from these plays in preparation for future publication. Booth was motivated in this project, so he later wrote to Henry L. Hinton, by Stuart, who wanted an acting version of the Winter Garden *Hamlet* prepared undoubtedly for publicity purposes. About the time Booth was arranging *Hamlet* and *Richelieu*, he was approached by Hinton, "a young actor in his company" according to Winter,[29] who wished to edit the plays and sell them in the theatre. The plans were disrupted by the assassination of Lincoln.

On January 15, 1866, two weeks after his return to the stage, Booth wrote Hinton to ask if he were still interested. "All the plays that I do at this theatre," he continued, "will be—(so far as scenery and costumes are concerned)—done with care and strict regard to truth, & the book should be correct in this respect as well as in the matter of 'business,' as it is technically termed." A few days after this letter Booth sent Hinton his promptbook of *Hamlet* which he had arranged a year earlier. Periodically during the next two years he supplied Hinton with the promptbooks of other plays. Between 1866 and 1868 Booth and Hinton published at least three plays "as produced at the Winter Garden"—*Hamlet, Richelieu,* and *The Merchant of Venice.*

Booth seems generally to have been satisfied with Hinton's work. He occasionally suggested "very slight" alterations and corrected "one or two trivial mistakes." He said that *The Merchant of Venice* was as he wished "in all respects"; and as late as April, 1868, he admitted that he was "perfectly satisfied" with Hinton's "revised version" of *Othello*. "You are fully capable, I am sure," he wrote, "of doing this work with but a little aid from me concerning the *cuts* &c., and I trust you will be successful in the undertaking." In 1868-69, "Booth's Series of Acting Plays" were issued under Hinton's editorship. Included in this series, published by Hurd and Houghton in New York, were *Richard III, Macbeth, The Merchant of Venice,*

[29] *Life and Art*, p. 168.

*The Fool's Revenge, Romeo and Juliet, Brutus; or, The Fall of Tar-
quin, The Lady of Lyons* and *Othello.*[30]

Despite his earlier commendations, after 1870 Booth became dis-
satisfied with Hinton's work. He complained that the books were not
correctly cut or marked and that the text was clumsily printed in
double columns on both sides of the page so that they were difficult
to use as prompters' copies or actors' rehearsal copies. His annoyance
increased as Hinton continued between 1869 and 1875 to issue other
plays "as played by E. Booth." Thus in 1876 Booth suggested that
Winter undertake a "correct" edition to be called the "Prompt Book."
Winter agreed and in the next two years published fifteen of Booth's
acting versions (see letter 24, fn. 39). Gradually the subject of
the Prompt Book began to dominate Booth's letters to Winter.

22. Chicago, April 18, 1876

Would it be of any service to you to edit an acting edition of
all the plays I perform?[31] The stupid mess Hilton [sic] made of
his books did him (not me) no good—they are useless & not at
all correctly cut or marked.

Thinking of the Hennessey cuts, the idea suddenly occurred
that a correct edition of the "Prompt Book" (a good title for it)
with remarks by *W. W.*—footnotes &c, if necessary—would sell
like "hot cakes" wherever I perform & you could thus derive a
pretty good return for your labor. I can *mark* the books precisely
as we do the plays, & wherever I go they alone should be permitted
to be sold at the box-office & in the theatre. I know nothing of
the profits of such a thing—but believe it pays pretty well.
McVicker tells me that many ask for *my* version of the plays &
will not accept Hinton's book. The title would attract the curious
(& *all* are so regarding theatrical affairs)—I suggested it to *H.*
but he couldn't see it. It would become the standard book for
actors also. The portrait in H's. books represents me as a drunken

---

[30] Sixteen of Booth's letters to Hinton are preserved in the Folger
Shakespeare Library, Y.c. 215 (118-133). See Jaggard's *Shakespeare
Bibliography* for a listing of Hinton's editions. I find evidence of other edi-
tions than those in Jaggard at the Walter Hampden Memorial Library,
the Folger Shakespeare Library, the Library of Congress, and the New
York Public Library.

[31] Winter in *Life and Art*, writes, p. 104, that he proposed the Prompt
Book project. I have found no evidence to support his statement. This
letter indicates that the idea seems to have originated from Booth.

Dumas[32]—(or some other nigger) & excites the indignation of my wife; eh bien! if requisite a better *mug* might be used for your book. I have the Mss of *Lear* & *Richd. 2d.* (not yet printed in acting form—from the original) and the arrangements of all my plays (hem!) are mostly new & my own.

I don't know how to go about this sort of thing, 'tis but a fancy that took me t'other day & I thought I'd give it you.

Business here is good—in spite of almost universal press opposition—they are down on *Mac* for his independence & dislike me because I decline to meet them—perhaps however[33] their opinions of my acting are correct—who knows?[34]

A letter from Tom McDonough[35] to McVicker says Joe has a picture for the Academy, but fear it will be crowded out—there are so many more offered than can be accommodated; Joe was urged by some high up party to send his picture (I forget who), it is said to be excellent & McD. says we'll some day find J. J. ranked among our great painters, I hope so—that is, if he sends us those pictures.

On Saturday (22d.) we give a Shakespearean day & night to the Stratford-on-Avon Memorial Fund, & thought all the theatres in the country were to do likewise, but I see no mention of it. Have you any knowledge of the affair?

Let me know—when you have leisure—what you think of the prompt-book & give me whatever suggestions you can regarding

[32] The "portrait" shows Booth as a long-haired, lugubrious Hamlet seated in a chair, presumably speaking "To be or not to be." The allusion is to Alexandre Dumas père whose grandmother, Marie Cesette Dumas, was a Negress from San Domingo.

[33] The following part of this letter was found in the collection of Robert Young.

[34] The critic of the Chicago *Tribune* (April 14) praised Booth's Shylock as "nearly perfect . . . as a man could be." He was more reserved in his appraisal of Othello writing on April 15, that Booth portrayed "with all the power and deftness of a master all the passion of the character" but that he failed "in representing the massive simplicity of Othello's nature." On April 18 he condemned Booth's Lear as "the least credible thing that has been done at McVicker's during the engagement." On April 21, after Booth's appearance as Claude Melnotte in *The Lady of Lyons*, he wrote that "Mr. Booth is not a bad *Claude*; he is only less than himself by being *Claude* at all."

[35] Presumably Thomas B. McDonough, American theatrical agent about 1870 to 1890.

it, also let me know if the Hennessey plates can be of service to you in any way hereafter.

My wife joins me in kindest regard to you & yours.
[P.S.] can't find yr. address

More than a year had passed since Booth filed a petition of bankruptcy. His case, however, was still being processed in the courts and rumors were still afloat concerning his failure. Booth wanted Winter to write an authorized history of the fiasco which for once and all would clear the air of "ridiculous stories." For this purpose he sent Winter a number of letters and statements from himself, McVicker, and Magonigle concerning Booth's Theatre from which Winter was to select whatever information he needed.[36]

23. Chicago, April 30, 1876

I send you by this mail a batch of papers concerning the Booth's Theatre romance, as Mac terms it, mine were written solely for his eye—supposing that he would use whatever he could for his statement. Of course, it would not do to publish the incoherent, rambling stuff that I have written hurriedly, nor would it be kind towards poor Harry to let the world know how far his inability served to upset me—You can, *if* you can, pick from the chaff whatever grains may be of service in the matter. You will also find *Mac's* account & letters to him & me from Magonigle. I looked it all over and attempted to erase that which I thought unkind or unnecessary to the purpose, but found it would require so much careful revision in order to preserve all the truth that I gave it up as a performance beyond my ability & send it in the rough. Don't bother about answering my letters—I know how busy you must be & how ill you are—I wish I could ease both disorders. Our prompt-book business cannot interfere at all with Hinton—his books are a failure & of no earthly value. The prompt-book might embrace all the standard plays that Wallack does also—Such a play as performed by Wallack, such &c as performed by Booth (if the names would add any value to them) —Now tell me first how to get *Lear* & *Richard* in shape—I have

[36] Several of these letters and statements are preserved at the Folger Shakespeare Library, box T.b.5. Winter did not publish anything about Booth's Theatre until his sketch of Booth for *Harper's Magazine*, LXII (June, 1881), 61-68.

the *Mss.* of both &, I presume, they could be printed therefrom, but as for footnotes, explanatory remarks, &c, &c, what am I to do? —in what shape should the work be when first submitted to you for these details? Would it be well to get up the book (as a regular promptbook) one side of the page *blank* for stage-directions &c? I have at *Cos Cob* also a beautifully marked prompt-book of Hamlet—three to begin with.[37] Could the books be gotten up in such a style at a low enough cost to pay you? McVicker thinks so. What can I get the Henessey plates for do you think? I have destroyed the letter I rec'd (I think) from Osgood on the subject; did he mean only the Henessey plates—or those of the text likewise? the former are very bad, I think—but if they could be of any service to you in the future & I can get them I'll do so. Disliking the portraits & not knowing if the text-plates could be used, I gave the matter no consideration. This "prompt-book" affair looks to me as a good thing, & I will cheerfully do all I can to further it. Can the type-writing process be of service in getting the thing in shape? The benefit from the Memorial Theatre yielded very little—the word *benefit* freezes the public heart I think, but I have just sent a draft for *£100* to Gaston Murray, Esq. the secretary—as my contribution; we played *Much Ado* for the benefit-Matinée, & in the evening Richard 2d.[38] The papers have abused me pretty well particularly my Lear & Richard, but the houses have been good & the applause unusually copious.

I have been negociating for an exchange of Cos Cob for a valuable plot [of] ground here on the grand boulevard, which in a few years will pay me two fold. I hope to effect the "swap" during this week—it may come however to *nix*. I must get my wife away from sea-air, & if I can rid myself of encumbrances I shall spend my leisure henceforth in a more genial climate for her sake—Denver is highly recommended by many who have derived great benefit from a few months' residence there. Mac astounded

[37] Presumably the "Hamlet at Cos Cob" is the *Hamlet* promptbook now preserved in the Harvard Theatre Collection. For a description of this promptbook see Charles H. Shattuck, "Edwin Booth's Hamlet: A New Promptbook," *Harvard Library Bulletin*, xv (January, 1967), 20-48.

[38] Booth was subsequently elected one of the governors of the Stratford Memorial Theatre. His reception in Stratford, when he visited there in 1880, was for this reason especially generous (see Letter 60).

me when he related what you had heard regarding him relative to my failure! He never had a dollar there, nor ever loaned me one for such a purpose; all sorts of ridiculous stories are afloat concerning the causes of my disaster—all of which, I hope, will be explained by *Mac's* statement & my higgeldy-piggeldy letters. I fell among thieves & came away plucked of every feather.

I'm scratching in the dark and must brush up to receive callers —will write again soon—don't bother about replying. Expect to be in N.Y. (or Cos Cob) about June 7th. Wife joins me in best wishes for you and yours.

Following his Chicago engagement, which ended on May 6, Booth, under the management of McVicker, made a starring engagement in Detroit, then went into Canada for appearances at London, Hamilton, Toronto, and St. Catharines, and recrossed the border for a final engagement in Buffalo. In the latter part of May he played a benefit engagement in Baltimore to aid a ladies' club celebration of the American centennial. He spent the month of June in settling his business affairs in Greenwich, Connecticut, and New York and in visiting his mother at Long Branch, New Jersey. About July 1, Booth left the East for Chicago enroute to San Francisco for his scheduled engagement there in September.

Shortly after his arrival in Chicago, Booth sent Jarrett and Palmer a letter suggesting that they change the name of his former theatre because the use of his name caused him some injury whenever he played in New York. What suddenly motivated Booth to this action is unclear. Jarrett and Palmer, at any rate, replied that as the lessees of the theatre under the name which it was dedicated, leased, and sold, they had no authority to change the name and referred the matter to the owner, Oakes Ames. Booth, in turn, referred the matter to his attorney, William Bruorton, with instructions, presumably, to seek an injunction legally restraining the use of his name. Booth also seems to have written directly to Ames requesting that he change the name of the theatre. By the end of the month the problem had not been resolved and Booth was resigned to a costly and lengthy suit.

On July 20, after two and a half years of proceedings, John H. Platt, the assignee, filed his final report on Booth's bankruptcy. According to the New York *Tribune* (July 21, 1876), the proved claims against Booth amounted to $147,057.50 and the unproved claims amounted to $38,787.06. At this time the assignee had a total balance on hand of $406.33. Since McVicker had purchased—at 5¼%—the claims of almost all of Booth's creditors, the report had

little financial significance to him, but he did see it as an excellent "cue" for the publication of a history of his failure which for several months he had been urging Winter to undertake.

Booth's progress with the "Prompt Book," meanwhile, had been postponed but not forgotten. As the following letter indicates, Booth had left five of his promptbooks with Winter and planned to send him an additional thirteen.[39] Faced with the editing of eighteen plays and a history of Booth's Theatre, Winter had some qualms about the practical value to him of the enterprise. But Booth assured him that he knew the "value of brains as well as time," and that he did not expect him to do this work for the "pure love of God."

24. Chicago, July 28, 1876

The enclosed—from this evenings paper—seems to be an excellent "cue" for the "Romance" we have talked of; how does it strike you? I am sorry I had no chance to see more of you while I was East—to go over this matter & the "Prompt Book" likewise. The latter I have not forgotten—altho' I've done nothing further towards it. I think I left 5 plays for you—do you need any more just now? Be sure & avoid double columns for the Book, it is not good for theatrical purposes. Unless you require more books & the headings for critical remarks &c (of wh. we

[39] Booth and Winter eventually published a total of fifteen Prompt Books. Between December 1877, and November 1878, they brought out *Hamlet, Katherine and Petruchio* (an abridged version of *The Taming of the Shrew*), *King Lear, Macbeth, The Merchant of Venice, Much Ado About Nothing, Othello, Richard II, Richard III*, John Howard Payne's *Brutus; or, The Fall of Tarquin* (1818), Gilbert A. Beckett and Mark Lemon's adaptation of P.F.P. Dumanois and A. P. Dennery's *Don Caesar de Bazan* (1844), Tom Taylor's *The Fool's Revenge* (1859) based on Victor Hugo's *Le Roi s'amuse*, Edward Bulwer-Lytton's *Richelieu* (1839) and Hugo's *Ruy Blas* (1838), as adapted by Edmund Falconer (1861). Each of these plays was first issued by Francis Hart and Co. in New York in uniform gray paperbound octavo volumes. The text was printed on recto only, so that the left hand pages would be free for prompters' or actors' annotations. (See letter 36.) In 1887 for the Booth-Barrett tours *Julius Caesar* was added to the series. Booth had virtually dropped from his repertory such plays as August von Kotzebue's *The Stranger* (*Menschenhass und Reue*), as adapted by Benjamin Thompson (1798), George Colman the Younger's *The Iron Chest* (1796), Richard Lalor Sheil's *The Apostate* (1816), Philip Massinger's *A New Way to Pay Old Debts* (1625), and Bulwer-Lytton's *The Lady of Lyons* (1838). These were not published in the Prompt Book series.

spoke) at an early date I will defer doing anything further 'till I get to California; let me know—The "Julius Caesar" of *Booth's Theatre* is my arrangement—so is the "Winter's Tale," which I'm told they will produce next season.[40] These 2 could be added —if you think it well; we will, however, get at the ones I perform regularly & add to the edition afterwards if necessary.

Write me & let me know all about the *Book* & what we shall do in the other matter. I am anxiously awaiting some word from Bruorton regarding Ames & the effect of my letter to him— requesting the withdrawal of my name; I suppose there will be more fight and law expenses for me; Bruorton will *post* you on any point in his profession on this subject—he is 206 B'dway— Room 24, if you desire to see him. I do not expect you, my dear boy, to do this *Romance* business for the "pure love o' God" for I know the value of brains as well as time. As to the "Prompt Book" I hope to do something towards making it a source of some little profit to you—desiring none for myself beyond the gratification of having such a book in use. I think I left you "Lear"—"Richd. 2d."—"Hamlet"—"Macbeth" & "Othello" I'll get at "Richelieu" & "Shylock" next. I presume I had better not

[40] Booth is referring to the spectacular production of *Julius Caesar* presented at Booth's Theatre by Jarrett and Palmer from December 27, 1875, to April 1, 1876, and returned for the week of May 22. It starred Lawrence Barrett as Cassius, Edward Loomis Davenport as Brutus and Frank Bangs as Antony. The production became part of Barrett's repertory for the next ten years. It was to a great extent based on Booth's 1871 production. See Richard Lee Benson, "Jarrett and Palmer's 1875 Production of *Julius Caesar*: A Reconstruction," an unpublished doctoral dissertation, University of Illinois, 1968.

The Winter's Tale was not produced at Booth's Theatre during the 1877-78 season. Booth had presented a spectacular revival of *The Winter's Tale*, starring Lawrence Barrett as Leontes, on April 25, 1872, at Booth's Theatre. A letter from Booth to Hinton, April 26, 1866 (Folger Shakespeare Library, Y.c. 215 [119]), reveals that as early as that date Booth was planning a revival and an acting version of *The Winter's Tale*. He admitted to Hinton that he had, for the most part, followed Charles Kean's arrangement of the play, "but not in every instance." He noted moreover that his scenic designer Henry Hillyard substituted several scenes for Kean's designs. He advised Hinton to "select such of Kean's notes &c. as you deem necessary—but be careful in whatever differences there may be in scenes or costume to omit or change anything that our version may demand."

do anything to Shakespeare's *Richard* until after I've acted it, but get Cibber's version in shape for you; what say you? There will remain *Fool's Revenge—Much Ado—Stranger—New Way—Iron Chest—Apostate—Lady of Lyons—Don Caesar—Brutus* —in all about *18* plays, and most of them differently arranged (by me) for the stage from other versions. I hope you have been at Wallack on the subject & that he will do something toward it.

It is yet undetermined where I shall act in N.Y.—I don't want to go again to Daly (nor to Booth's—under existing circumstances) and Wallack seems to decidely dislike the idea of my acting at his place, tho' he pretends otherwise; he's had two good chances to engage me—last season & this, yet he can't arrange time etc, etc., & so forth! Ah, ye silly players! Artistic cusses!

I start on Augt. 10th for San Francisco, I hope and expect to remain there six weeks, at least; playing at St. Louis & this city on my return—leaving after January & February for New York; I shall not be precipitate, but wait the possibilities of the next few months.

My wife & Edwina are having a delightful time here & look forward to the trip to California with great glee.

Remember us most kindly to your wife, & me especially to the two little *roosters* I met the day I called.

Believe me, dear Will, ever yours

Edwin

On September 4, 1876, Booth opened at the California Theatre in *Hamlet* under the management of John McCullough. It was his first trip to California in twenty years. The critic of the San Francisco *Chronicle* (September 5) reported that the public which regarded Booth as a home-town boy, gave him "a right royal welcome," applauding him at his initial entrance for a full five minutes and at every "point" thereafter. At the end of the performance, he was presented with a laurel wreath and further complimented with "thunders of applause" and "hearty, ringing cheers." At the close of his eight-week engagement on October 28, his share of the receipts was estimated to have totalled $50,000.[41]

Some of the critics, however, voiced reservations. The critic for

[41] Lockridge says, p. 234, that Booth's share averaged $6,000 a week. Winter records in *Life and Art*, p. 162, that the gross receipts were $96,000, but he does not report Booth's share.

the *Chronicle*, for instance, wrote that Booth's Hamlet was "an impersonation that will challenge comparison with the finest in the history of the stage," but that it was not an impersonation "absolutely free from faults." He complained of Booth's inability to make points where "vastly inferior actors make them tell unerringly," of the excessiveness of his "forehead thumping gesture," of his "faults of articulation," "*outre* pronunciation" and ambiguous readings. Booth's Hamlet did not have the "strongly marked broadness" of Charles Fechter's, nor the "scholarly attention" of Barry Sullivan's, but still it had "a grace which makes every gesture appropriate and invests every word with a fresh charm." The reception of his other roles was similar: Richelieu was "a good performance . . . but not by any means wonderful"; Othello was "an even but barely excellent performance," but yet preferable to John McCullough's; Lear was "totally out of his line"; Shylock was "smooth, even, generally natural, yet it did not appeal to the sympathies"; Cassius was "strikingly original" but not the equal of Lawrence Barrett's; Richard III was "a distinct and scholarly performance" but not the equal of Barry Sullivan's; and so on.

During this engagement Booth was "brooding" over the "Prompt Book" plans and was considering asking McCullough, as well as Lester Wallack, to join in the scheme. He was also annoyed by the legal entanglements of withdrawing his name from Booth's Theatre. Winter meanwhile had moved back to New York from Staten Island, but he still hoped he could locate in the country but close enough to the city so that he would not be too inconvenienced in his work. Booth suggested he investigate "a queer little place" at Greenwich.

25. Palace Hotel, San Francisco, September 17, 1876

Your letter to John was interpreted by me while he was supping with us after the play a few nights ago. It reminded me that I had a long letter yet unanswered somewhere in the dead letter box—which I have at last found, and—tho' half sick with bile & dyspepsia—will proceed at once to "give my tendance to"—Your entire plan of "The Prompt Book" exactly coincides with my view of it, and I will let you have the result of my occasional "broodings" over *Hamlet*, & other parts perhaps, in ample time for press. I do not see much choice between the two forms of title—only I would make the words—"The Prompt Book" head the page prominently; perhaps the words "edition of Standard acting plays" following the above in small print would read better & *feel* more solid without the "new," which might look experimental. Macbeth, J. Caesar & Winter's Tale had better

come later on the list and let Shylock & Brutus with Fool's Revenge come earlier; Hamlet & Richelieu are the plays usually called for first—I find it very difficult in cold blood to make notes and comments on any part I play—while when I am heated & alive with my book & my wits are all awake I frequently have an idea of what I am about & what the author means to convey—this is, perhaps, a lazy condition of the brain—a sluggishness resulting from a kind of surfeit of *devilled brains*; but sometimes I get with an analytical mood & bore some unfortunate sleepy head after the play with wise reflections on the "whys" & the "whats" of the part I have been sacrificing to the (Gallery) Gods!—I am anxious to know whether Wallack will aid the cause. McCullough will, I'm sure; I'll punch him up on it occasionally—tho' I hardly think he'll need that, for he is very thoughtful & altogether a big hearted fellow; I like him ever so much & my closer connection with him has strengthened the good opinion I ever had of him. He is a great favorite here & I hope he will come out of his managerial venture better than another chap I know who tried it once. My engagement here has thus far gone beyond anything they have yet had—the dollars and spectators are numerous & in every respect satisfactory to all parties interested, and the applause I nightly receive assures me that I am appreciated—for it is judicious—but the critics compare & cauterize with questionable judgement. It may be I shall act 8 weeks here—we cannot yet decide, & after—perhaps —a couple of nights at Sacramento, I shall wend my way New Yorkwards. It is rather odd that your letter suggests an engagement at the Lyceum under McVicker (had I spoken of such a possibility?) for it now looks very much that way—but of course, must not be mentioned officially until *Mac* deems fit to do it. I presume you have seen him—he has been in N.Y. during the past three weeks—I wish almost as earnestly as you do— that you had that little country home where your family could enjoy the "calm comforts" of green fields & healthful breezes, & where you could be untrammeled by the harness of ceaseless toil, but have you considered well the care & annoyances of the country gentleman, and the expense of "help" & potatoe-bugs— et al?

There was (*is*, I presume) a queer little place at Greenwich —some 90 years old, I believe; in the village & nearly opposite

"our" church (!) that always brought you to my mind whenever
I passed it; there is no more healthful or lovely place within so
short a distance from N.Y.—so easy of access &c, and certainly it
is as quaint and picturesque as any poet could desire for a *den*.
I never question of its price or whether it was in the market—
for I knew if I should hint even at a fancy for it its price would
go up among the millions. I wish you cd. see it—it belongs to a
Mr. Towes, who lives in a pretty place just opposite, but beyond
this I know nothing. I have a friend there named Benedict[42]
who will make all necessary enquries & it may be the owner will
be glad to get rid of it on easy terms—I believe there are *acres*
which can [be] whittled down to feet, at all events such is the case
when the house is rented.

I presume the J & P party will do all they can to annoy me
about the *title*—I've seen already some ugly things about it.
Your paper reached me duly & your *dig* at the managers went
to the right spot; so long as the theatre is in the hands of
cheap-John speculators there can be no hope for the acted drama.
John has about the best company in the country—far better
than any I know of in New York or the larger cities & the plays
are well mounted in every respect. I hear Barrett is to make a
run of *Lear* at Booth's in the fall or winter I wonder if Bangs will
lead the ballet? Surely they'll have the ballet introduced to
illustrate the "quick cross lightning" and other dreadful things
in the storm scene. It would not be safe to cast Bangs for "Tom"
—he wd. not, I fear, reserve even a blanket.[43] Goodnight. Love
to your loved ones.

[42] Elias Cornelius Benedict, prominent New York banker and stockbroker, was Booth's lifelong friend.

[43] Frank C. Bangs, American actor, who had played Mark Antony in
Booth's production of *Julius Caesar* in 1871, had recently appeared in the
title role of a spectacular production of Byron's *Sardanapalus* at Booth's
Theatre. The production and Bangs were generally excoriated by the New
York critics. The critic of the *Herald* (August 15) wrote, for example,
that Bangs' "principal office . . . is to fill up waits between the marching
and the dancing and form a central figure in the tableaux." His strongest
point, besides his beautiful costumes, was his legs, which the *Herald* considered "beauties in their blue silk casements." Undoubtedly Booth's
tongue-in-cheek references to the "ballet" and "blanket" allude to this
production. Barrett did produce *King Lear* at Booth's Theatre on December
4, 1876. Bangs, however, was not a member of the cast.

Following his return from California, Booth's activities flowed smoothly in their customary channel. On November 20, 1876, he appeared at the Lyceum Theatre in New York, formerly the Theatre Français, under the management of McVicker. This engagement continued for ten weeks through January 27, 1877, during which Booth appeared in 16 different roles. The critics were generally enthusiastic and the audiences appreciative and receptive. The engagement was marred, however, by rainy weather; and after the Brooklyn Theatre fire on December 5, which claimed 295 lives, New Yorkers were reluctant to go to the theatre. Booth, however, continued to be paid $4,200 a week and McVicker was the loser to the amount of $25,000.[44] Odell, nevertheless, has cited this engagement as "one of the finest dramatic festivals of 1876-77" (*Annals*, x, 205).

Booth apparently was to be given a banquet at the Lotos Club at the close of this engagement. Although he was pleased by such honors, he was reluctant to accept them for reasons which he disclosed in the following humorous letter.

26. Everett House, New York, January 14, 1877

I know you will think me perverse & "stoopid" in declining the banquet; you are perfectly right—I should accept—it would do me good, and all that; but I have declined so many similar honors tendered me here and elsewhere that I cannot consistently accept—at least 'tis thus I feel about it.

Although I have been for nearly 28 years before the public my ridiculous timidity still shrinks from *lionization*, and were I fit for any other pursuit I should not now be strutting & fretting my et cetera you know.

For one whose sole desire is to shrink into the homely shell of domestic privacy 'twould be a terrible ordeal to stand like a man at a mark with a whole army of mild-eyed McCouchy Lotus-eaters shooting sharp & critical eyes at him.[45] To be glorified

[44] According to an unidentified newspaper clipping, "Our Dramatic Stars," 1878, in the Harvard Theatre Collection.

[45] *Much Ado About Nothing*, II,i,254: "I stood like a man at a mark, with a whole army shooting at me."

"McCouchy" is possibly a reference to Alexander E. Macdonald who was the secretary of the Lotos Club from 1874 to 1877. The banquet was apparently to be given by the Lotos Club of which Booth and Winter were members. Booth continued to resist the banquet and it was never held. Indeed an historian of the Lotos Club says that Booth rarely visited the

and speechified and—scarified by fellows of infinite jest would place me at such a disadvantage that a pall of disappointment would eclipse the gaiety of the occasion.

Once at a Boston "tea"—I was a silent & unnoted guest; the rest—some ten or a dozen ladies & gentlemen—had said their little say and all were attentively listening to the delightful rattle of the Rt. Revd. Dr. O. Wendell H. M.D. Esqre.[46] (and when he talks you know he *talks* while all *must* listen); well, in the course of his charming converse he related an incident so very like my own experience that I rashly ventured to ope my lips for something else than *tea* & *wittles*—to speak upon the cue thus offered me to "say something." At once all eyes, ears & mouths were agape to catch the pearly eloquence about to flow from my mellifluous lips. Terrible to relate—I was stricken dumb! The sound of my dulcet voice quite stunned me & I nearly "swooned" with a vertigo, produced by my embarrassment. The gaping eyes & ears & mouths were so many yawning chasms! I could not articulate more than a *gulp.*

An hour after (it seemed this long at least) a generous knight, with true chivalric purpose & hair serenely separated centrally, rushed to my rescue & dashed between me & the oppressive silence which threatened to overwhelm me.

When'er I lay me down to sleep—I pray the Lord that man to keep!—for I have never recovered from the fearful shock of that terrible tournay!

Would you wish to see me—a 1st. class North American tragedian—crumpled up like a napkin 'mid the bread crumbs on the banquet table; where, surrounded by brilliant lights, I wd. be expected to shine—if only by the reflection of their lustre?

Spare me W.W.! and let me, when not forced by cruel Fate to face the footlights, let me enjoy the quiet seclusion of my wigwam with my peaceful pipe.

I must say *NO*—much to your & my own disappointment (for I do painfully regret my inability to support such honors), and

---

club. See John Elderkin, *A Brief History of the Lotos Club* (New York, 1895), p. 91.

[46] The reference is, of course, to the noted American poet and physician Oliver Wendell Holmes.

will rely on your good judgement & honest love to get me gracefully out of the scrape.

In return I will send you something (if you tell me where) that you will be pleased to have; after several enquiries & a few days search among some old boxes of rubbish—the debris of my earthquate at Booth's, the "Clarke & Jefferson" picture was unearthed & it is *yours!*[47]

There now! won't you "save me from my friends" for this?

Do, and win the eternal gratitude of

Yours ever

Edwin

After his engagement at the Lyceum, Booth appeared at the Brooklyn Academy of Music on January 29, 1877, for an engagement of two weeks through February 10.[48] After a week's rest, he went on tour, playing at Mrs. John Drew's Arch Street Theatre in Philadelphia from February 19 through March 11 and in Baltimore from about March 12 through March 24. On March 30, he was back in New York for another week's rest before beginning a seven-week tour of the New England towns of Stamford, Hartford, New Haven, Worcester, Providence, and Boston. From Philadelphia, New York, and Boston he wrote Winter brief notes about his activities, the Prompt Books, his future plays, and his reaction to various bits of theatrical gossip.

27.   Philadelphia, March 2, 1877

All my books—save a single copy of Shakespeare—are packed & stored in far Chicago, consequently my *Witness*[49] is absent. I fear my opinion in such matters would be of little value, for with poetry as with painting I like that which liketh me, without any pretense to a critical knowledge of either, and have frequently found my preference condemned by better judges. I remember

[47] The "picture" is a full-length portrait of Joseph Jefferson and John S. Clarke as Dr. Pangloss and Zekiel Homespun, respectively, in George Colman the Younger's *The Heir-at-Law*. According to Robert Young, the original painting has been lost. It is reproduced however in the illustrated, 12mo edition of *Life and Art*, p. 94.

[48] According to the newspaper article "Our Dramatic Stars" (see above, fn. 44), McVicker recouped his losses at the Lyceum with this engagement.

[49] Winter had apparently asked Booth's opinion of one or more of his poems in *My Witness*.

well there were several poems in the volume that pleased me very much, but I cannot now recall their titles. I wish I had the book with me & you should have my opinion for what 'tis worth.

I am really glad that you have got into the mood for hard work, & hope your interest will not *flag* until you have accomplished all you have laid out for the season.

I forget how many "prompt-books" you have, but if *Brutus* is among them I'd rather furnish another arrangement of that play—also the *Fool's Revenge.*

It is to be regretted that your original plan cannot be carried out, but after all it is a question if any matter beyond the mere text as arranged for representation would be appreciated—I mean in that form.[50] I hoped to furnish you with some "ruminations" of my own in reference to the various characters I personate, but it is a very difficult thing to do when one is fatigued after the play—and 'tis only at such times that my wits are alive to such fancies.

Perhaps in some future edition, or in some other shape you can accomplish it.

Philadelphia *has* treated me well, but the effect of "hard times" is still felt by "ye player-man."[51] After next week I go to Balto. for two; then a few days rest—prior to a brief tour through the Eastern towns (Hartford, New Haven, Worcester et al) — en route to Providence and Boston where my season terminates about the close of May.

[50] Winter implies in *Life and Art*, p. 166, that his intention was to publish all of Booth's stage business in the Prompt Books. This proved too complicated and was abandoned. The Prompt Books preserved much of the general stage movement and business, such as entrances and exits, music and sound cues, and directions for the handling of properties but very little of Booth's personal stage business.

[51] The depression resulting from the Panic of 1873 was still being felt throughout the country and was undoubtedly affecting theatre attendance.

Not all of the critics, however, "treated him well." The critic of the *Philadelphia Inquirer* (February 20), for example, praised Booth's Hamlet as "an artistic and scholarly performance," but complained that the role still lacked a certain "vital spark" which lifted an impersonation beyond mere technical skill. On February 26 he wrote that Richelieu was "artistic & picturesque" but not of "particular excellence." Richard III was characterized on March 6 as "not remarkable or even adequate" but effective in that it pleased an "indulgent audience."

I see by the Herald of today that *J&P* decline to continue their management of *Booth's* & that *Shook, Palmer*[52] and myself are after it. I wonder what will eventually be the fate of my *Folly*!

McCullough was here last week & told me how badly poor Adams' condition is, and yet how he laughs and jokes his flickering life away. It is a repetition of Harry Perry's case—"a fellow of infinite jest," who died as he had lived; full of rare but uncultivated talent.[53]

I see Ford has been bolstering up Mary Anderson[54] with my name—so worded as to deceive many with the idea that she & I are acting together in "Romeo & Juliet" &c—vive la humbug-a-telle!

28. Everett House, New York, March 30, 1877

I did hope to see you during this week of "loaf" & have a chat over the "Prompt Book" but I am really so tired—literally "played-out" that I've been unable to step abroad but once—and then perforce on business.

I send you with this a copy of Irving's arrangement of "Richard," sent by Clarke who had not yet seen the performance.[55] You know I intended, despite the adverse advice of actors, to produce the play in its original garb (curtailed, of course) last season, but my unfortunate fracture forbade my risking the rough & tumble plays, & this season I foolishly neglected to carry out that purpose through sheer laziness. In the meantime Irving—

[52] Sheridan Shook, New York capitalist, was the owner and with Albert Marshall Palmer the co-manager of the Union Square Theatre in New York. When Henry Jarrett and Henry Palmer relinquished the management of Booth's Theatre in April, 1877, the theatre was subsequently leased by a number of managers, including George Rignold, James C. Duff, and Augustin Daly.

[53] Edwin Adams had become seriously ill in 1876 during a tour of Australia. He returned to San Francisco where McCullough gave him a benefit at the California Theatre.

Harry Perry, American actor specializing in "light comedian" roles, was the first husband of Agnes Land Rookes, American actress born in Australia, who in 1867 married Junius Brutus Booth, Jr.

[54] Mary Anderson, a popular American actress, was a close friend of both Booth and Winter. She never, however, acted with Booth.

[55] Henry Irving produced *Richard III* in London at the Lyceum Theatre on January 29, 1877. His acting version was published in London the same year by E. S. Boot.

across the way—gets in his oar & leaves me in the wake; but I shall do it anyhow—some of these days—next fall, I think. I wish you could see my arrangement of the play—I am vain enough to believe that you will, when you do see it, endorse it as the better of the two for representation.

Baltimore, in spite of Lent & bad weather, did me good service. Mrs. Jennings,[56] as I presume you know, was obliged to leave us after the first week because of her child's illness—it died a few days ago, & she rejoins us on Monday at Stamford, where I renew my work for a seven weeks tour.

I hope all's well with you—all is with me, except my excessive fatigue.

I trust the next Autumn will find you fresh in the field, full of work, health & good luck.

29. The Brunswick, Boston, May 6, 1877

Glad am I to have your phiz:[57] I like it very much. The morning I arrived I met your brother at the stage-door; just as he asked me when I last heard from Will a package of letters was put into my hands by someone in the dark & among them I found yours. I shall send you a *cut* copy of Richelieu during the week & let you have my version of Richd. 3d. (which I hope to produce next fall) just so soon as I can get a copy made of the one I have—which is so cut up & interlined that you could make no use of it. I expect to hear again from you when you learn the cost of printing, &c.

I'm glad to see Clarke's contradiction of his reported failure; it startled me somewhat—for he has a large family, & works very rarely—loves his ease & as is usually the case with such folks does not know *how* to spend money.

I hope you will not be disappointed in your anticipated trip abroad, and that it will build you up—which I am sure it will, if you are careful. I hope you will be able to take your wife & babbies with you—it's a lonely ride & requires congenial company.

Shall see McCullough on Wednesday I think; wish he had

[56] The career of Clara Jennings is obscure. She seems to have been one of the leading supporting actresses of her time. Presumably she was married to John W. Jennings, minor American comic actor, whose career is equally obscure.

[57] Winter must have sent Booth a photograph of himself.

dined in better company than that rascal Clarke Bell's—I ought to have had that fellow disbarred long ago. I think John will take Magonigle to California to take charge of his accounts there; I hope so—for Harry needs a help, & I think he & his family would do well in Frisko. Have not heard from Jefferson—*since* he sent us *those pictures*; but Clarke says he has worked his plans splendidly. Daly & Duff[58] both want me next season—have not yet determined what I'll do. Sent you a notice of Richd. 2d. the other day.[59] I received the play you sent—*Commodus*; it is pretty D[amn] B[ad], read one called *Maternus* (some subject) by a Baltimorean for McCullough, it was much better.[60] Lear seemed to make a great hit last night. Business good. Health ditto—but very tired. Mrs. B. & Edwina send greeting. Had a pleasant call from Warren today. Adieu!

Booth ended his engagement at the Globe Theatre and his theatrical season of 1867-77 on May 20. According to one estimate, from the beginning of his Lyceum engagement in November to the close of his season in May, Booth made $72,492.75.[61] This would bring his total

[58] Augustin Daly was still the manager of the Fifth Avenue Theatre at this time. On September 10, 1877, however, Daly withdrew from the Fifth Avenue after the failure of his play *Dark Hours*. See Daly, pp. 238-243.

James C. Duff, American theatre manager, was Daly's brother-in-law and the son of the veteran manager John A. Duff. James assumed the management of the Broadway Theatre on April 30, 1877, and during the 1877-78 season made it a first-class theatre. See *Annals*, x, 400-405.

[59] Booth opened his engagement at Boston's Globe Theatre on April 30, with *Richard II*. On the following day the noted drama critic Henry Austin Clapp wrote in the *Advertiser* that Booth's impersonation was "exceedingly interesting and artistic," "a fine character study, drawn with lines delicate and clear, and illuminated by brilliant histrionic skill," and "worthy of Mr. Booth's best powers." Perhaps this was the notice Booth sent to Winter.

[60] *Commodus* was an historical play centered around the Roman emperor Lucius Aelius Aurelius Commodus who ruled from 177 to 193 A.D. It was written by Lew Wallace, the author of *Ben Hur*. In the *Catalog of the Sale of William Winter's Library by Jefferson Winter, Part 1, No. 230, The Walpole Galleries, April 28, 1922*, is listed a copy of the play, item 323, inscribed "Lew Wallace to W. Winter, esq."

A play entitled *Maternus: A Tragedy in Five Acts* was copyrighted in 1875 by Edward Spencer of Randallstown, Maryland.

[61] Sally MacDougall, "Edwin Booth Counted His Ducats," *Century*

seasonal earnings, including his San Francisco engagement, to over $120,000. His sister Asia reports that from his engagement at Daly's Fifth Avenue in October, 1875, to the end of this season Booth was able to completely discharge his debt of over $200,000 to McVicker.[62] For the first time in almost five years he was free from debt and financially solvent. He spent the summer of 1877 at the home of his friend Benedict at Greenwich, resting and making plans for his next season and for the publication of the Prompt Books.

Magazine, cvii (December, 1928), 198-204, purportedly based on McVicker's financial accounts.

[62] *The Elder and the Younger Booth*, p. 175.

Chapter III

The Prompt Books

1877-1879

About the middle of June, 1877, as Booth was beginning his vacation at Greenwich, Winter left New York for a ten-week tour of England and France. This was the first of almost annual visits to England. In the company of Jefferson, then completing a two and a half year "tour" of England, Winter visited the historical, theatrical, and literary shrines about which he had read and dreamed since he was a boy. His fresh, almost innocent impressions of "the equable English climate and the lovely English scenery," of Windsor and Eton, of Stratford-on-Avon and Warwickshire, and, especially, of "mighty" London were recorded in almost weekly "Letters" to the *Tribune*. In 1878 these were collected and published as *The Trip to England* by Lee and Shepard in Boston.

Winter returned to New York about August 25, a week before Booth was to leave for Chicago and an extended tour of the Midwest. Booth was anxious to resume work on the Prompt Books, wanting them ready for his January engagement at his former theatre. Winter, however, was unable to do much about the Prompt Books because of domestic and financial problems. His father was seriously ill and he was now responsible not only for his own growing family but also for several half brothers and sisters, one of whom was mentally retarded. Shortly after his return from England, he decided to move Lizzie and the children to Toronto to live with his mother-in-law, probably in order to reduce an almost impossible strain on his pocketbook.

About the middle of September during Booth's engagement at McVicker's Theatre in Chicago, Winter seems to have been sufficiently settled to resume his Prompt Book efforts. He apparently had written Booth about the cost of printing the Prompt Books and about the Booth's Theatre "romance," which Booth was still urging him to complete but which Winter thought was by now a dead issue.

30. Chicago, September 29, 1877

Fatigue and the oppressively hot weather have prevented my writing you; I have two of your letters before me now, and—despite the "grunt and sweat" which this heat induces, I will let you have an acknowledgement of some sort.

The last shall be first. Imprimis:

The estimate for printing is, as you wisely judge, far in excess of what would afford the least profit. While at "Booth's Theatre" I had a man in my employ (named Moore) as a door-keeper, who printed the play of "Ruy Blas" for me from a Ms. I had; he charged but $45 I think—it was less than $50 I am sure, and the work he did was in every respect excellent. The proof sheets were not corrected & consequently the book was a jumble of errors; but the paper & printing were first class. I forget what he told me the *plates* would cost, but even at Russell's[1] estimate therefor—at Moore's rate for printing &c—there would be some chance for profit. We must try again. I think Moore's address can be ascertained at *Booth's*.

Apropos of the aforesaid "Temple"—I do not understand why you consider your anonymous correspondent's plea "too late." *I* think it is too early yet; it will only be fully ripe when I reappear there in January next.

Some such efforts by the Press *then* would at least revive the dying memories of my exertions in a worthy cause, and serve as a requiem to the buried hopes I once had for my profession. Use your own judgement, however.

As often as my duties permit I visit poor Adams—nearly every day. The dear, good boy patiently endures the tortures of his terrible disease. The hearts of all who see him now, and know how free his life has been from ill toward others, must ache as mine does.

Like Mercutio he gasps forth an occasional jest, yet reverently realizes the awfulness of his position. His clear conscience, sparkling with harmless mirth breaks—every now and then—through the clouds which gather closer and closer daily about his poor frail body, but which does not, I am sure, cast even a shadow on the soul whose greatest sin was the neglect of its tenement. He is thoughtful & resigned, and oh, so patient in the midst of his great suffering!

He reads all the papers & is annoyed at their frequent reference to his condition. His chief regret seems to be for his wife, who is

---

[1] Possibly Benjamin F. Russell, a well-known theatrical attorney, who represented Jarrett and Palmer, John A. Duff, and other New York managers. Russell had originally been a printer and might have still been consulted regarding matters of printing.

most devoted to him and has nursed him for eighteen months as only a loving woman can do. God grant that the promised pecuniary relief, at least, may be afforded her.[2]

You have my warmest wishes, Will, for your dear ones in their new home, and my sympathy in your cruel separation from them. An adverse fate seems to pursue you, but remember—"the worst is not, so long as we can say—'this is the worst!' "[3]

I saw but one of your London letters (I so seldom read the papers) but I expected—from what my man, Henry,[4] told me, that I would receive others from you. What I did read was delightfully warm & glowing—yet withal quite Winter-ish (Peccavi!) Your experience, so different from mine in England, made me yearn to try a bout with the "Island mastiffs" once again, and it may be I shall do so, when I am old enough—before my teeth fall out. I hope your recent illness is one of the bye-gones now, and that you are on your pins & full of vim.

My wife and Edwina send kind remembrances. Business excellent.

After his Chicago engagement, which ended about October 6, Booth began a tour of several Midwestern cities. The publication of the Prompt Book and his opening at Booth's began more and more to dominate his letters to Winter. Winter in addition to his critical duties and the Prompt Book project, was involved in a feud with Dion Boucicault.

In the September 1877 issue of the *North American Review*, Boucicault had published an essay on "The Decline of the Drama." He cited the "mischievous influence" of newspaper critics as one of the principal reasons for the decline. The newspaper critic, said Boucicault, had never been able to function, as a true critic should, as an arbiter of taste. Although Winter and Boucicault had been friends for more than a decade, Winter could not let this charge go unanswered. For the next twelve years, they carried on a feud about the role of the newspaper critic, and their friendship was broken. The

[2] Edwin Adams died in Philadelphia, his birthplace, on October 28, 1877.

[3] *King Lear*, IV,i,29.

[4] William Henry Flohr, American actor and stage manager, was Booth's assistant stage manager and dresser at the Winter Garden and at Booth's Theatre. He seems to have continued as Booth's dresser into the 1880's.

issue was more or less settled in 1889 by the publication of Winter's *The Press and the Stage*, and it was ended with Boucicault's death the following year.[5]

31. Kennard House, Cleveland, November 25, 1877

I have two *Toronto* letters before me both of which shall now be answered as clearly as possible. First for *Richard*: I have concluded to begin my engagement with that character instead of *Hamlet*. These plays had better be ready & let the others follow—leisurely. I think they (the two named) will fill most of my time—six weeks. The *text* I have not attempted to correct or arrange in any way—I mean in the copies I have sent you, but merely cut & slashed them for stage action; the text I left for your better judgement. I follow no particular edition but cull from several such readings as best fit my understanding. I have some dozen or fourteen editions—in each of which I find much *bosh* & some sense—perhaps I sometimes mistake the former for the latter. If I remember rightly the *cut* copies of the plays I sent you are very 'ragged' & imperfect—I supposed you understood that & would put 'em in shape. Since acting *Richard* I have made other curtailments—which, I presume, can be regulated when the proof-sheets are printed; that's all I can do—the literary corrections you must make, of course. Yes, I have twice read Warwick[6]—years ago & recently, just before I produced the present version of *Richd*. Buckingham's name was *Henry Bohun*. Would it not be better to follow the cast as *Shak.* gives it—or rather as it stands in print or thus: [1]*Edward IV* (as he is the reigning King when the play begins)—[3]*Richard* of *Gloster* (as next in age); [2]*George* of *Clarence* (as youngest brother). *Henry* of *Richmond; Buckingham; Hastings; Stanley; Bishop of Ely; Dukes* of *Norfolk* & *Surrey; Rivers, Dorset, Oxford, Grey, Catesby, Ratcliff, Ld. Mayor, Tyrell, Murderers* (as they rank in importance). *Q. Margaret; Q. Elizabeth; Duchess of York*

[5] For an account of the Winter/Boucicault feud see Richard Ludwig, "The Career of William Winter, American Drama Critic: 1836-1917," an unpublished doctoral dissertation, Harvard University, 1950, pp. 194-205. Winter's account of his friendship with Boucicault is given in *Other Days*, pp. 124-151.

[6] The reference is to Bulwer-Lytton's novel *The Last of The Barons* (1843), an historical romance centered around Richard Neville, Earl of Warwick, and King Edward IV.

and *Lady Anne*. Ghosts of Clarence, Hastings, Anne, the 2 princes & Buckingham. But use your own judgement—if this like you not. I think "good" of *Osgood* if he'll "print 'em" at easy rates—*Richd.* & *Hamlet* anyway can be done, & if it doesn't pay—we can let some *cheaper* feller do the rest. Your suggestion of the sketch is good—but would it not be too bulky for such a book? Besides, how can we change, or rather—add thereto unless I were with you in a chatty mood?

An attempt was made yesterday to draw me into the Boucicault business; the city-editor of the *Sunday Voice* called to interview me relative to his letter about the critics &c, but I politely declined, & daresay have a rap for it today, but as I avoid looking at all abusive papers—I shall not suffer.

I will send you some autographs—if I do not forget to do so ere I close this letter. Edwina is very much pleased at the prospect of receiving "Thistle-down"[7]—her Daddy will be no less gratified thereat. I hope Joe will come out *whole* from Booth's—I mean pecuniarily;[8] I have my doubts of any more large successes there —or indeed at any theatre in N.Y. for some years to come.

I must spare you now. Excuse the paper—'tis all I have & the shops are closed.

Wife & daughter join me in good wishes for you & yours.
P.S. I made a slip in my princes' ages, you perceive.[9]

32.   Buffalo, December 4, 1877

I wrote you on Sunday week from Cleveland in reply to two letters relative to the Prompt-book. Yesterday yours of Novr. 29th

---

[7] *Thistle-down* (London, 1878) was the title of Winter's fourth volume of poems.

[8] Joseph Jefferson was engaged at Booth's Theatre, then under the management of Augustin Daly, from October 29 to December 1, 1877. Joseph Daly reports, p. 244, that because of his absence from New York for over two years Jefferson expected a highly profitable engagement with nightly receipts exceeding $1800. The nightly receipts, however, averaged only $1274. The loss was sustained by Daly; Jefferson still received $700 a performance. One of the principal causes of the poor theatrical business during this engagement, according to Daly, was a violent railroad strike during the summer of 1877, which had a nationwide effect on business of every sort.

[9] Booth corrected his "slip" with the superscript numbers preceding the names of Edward, Richard, and George above.

arrived & I judge by it that you had either written again to Cleveland or that my letter had not reached you. In it I mentioned the name of Bohun as being the family one of Buckingham. You now say it was Stafford, but was not that a *title*? "Poor Edward Bohun"—Duke of Buckingham, Earl of Hereford, *Stafford* and Northampton was beheaded by Henry 8th, and he—according to Shakespeare was son to the one we're after.[10] All my books on costume, arms &c, have been stored in a loft of McVicker's Theatre since I left Cos Cob, and I can furnish no detailed description thereof. Chas. Kean's edition of Cibber's play contains a correct account—from Planché.[11] Of course in my production of the play at Booth's I must trust entirely to a hired set of dresses, properties, &c, furnished by the costumer, Eaves,[12] who doubtless will follow Planché's suggestions, as he promises to supply *correct* dresses, &c.

I think I told you that since acting *Richard* I've found it necessary to omit & alter a few passages here & there; I presume that can be done when the proofs are struck off. I begin January 7th & have six weeks at B's, then I go to the Academy (Brooklyn) for one week only. I cannot afford to do much in the way of new scenery &c, for so short a time, but I am told that most of my Shakespearian scenery is still intact & but little worn. I look for no *run* of any play, but shall be ready to follow *Richard* with *Hamlet* on the 2d. week if necessary, I was in negociation with Rose Etyinge [sic][13] for some time & she was, at first, anxious

---

[10] Winter was correct. The family name of Henry, second Duke of Buckingham (1454?-1483), was Stafford not Bohun. He was, however, recognized by Richard III as the sole heir of the Bohun estates and was appointed Lord High Constable, an office traditionally held by the Bohun family. Henry revolted against Richard. He was captured and beheaded on November 2, 1483, and his lands were confiscated. Edward Stafford, Henry's eldest son and third Duke of Buckingham (1478-1521), was beheaded by Henry VIII on May 17, 1521.

[11] James Robinson Planché, English dramatist, herald, and antiquarian, wrote and illustrated *The History of British Costumes* (1834) which became the primary source for "historically accurate" productions of Shakespeare's history plays.

[12] Henry J. Eaves, the founder, in 1863, of Eaves Costume Company, New York.

[13] Rose Eytinge, popular and versatile American actress, made her first New York appearance with Booth in *The Fool's Revenge* at Niblo's Garden

to join me, but her agents badly advised her & she regrets her inability to assist, &c, &c. My company will be composed of favorite & reliable actors. I prefer the *Shake.* in full—but won't quarrel over an *e* or two. (You perceive I am answering your letter 'on the jump' as it were, skipping from one point of it to another). I wish we could have had all the details fully prepared long ago—I fear that now being hurried, we shall fail to make the book complete. Failing to obtain the services of Etynge *I* shall be obliged to fill the Matinées and Saturday nights; my wish was to let her have one of those occasions each week. I cannot act Richard—nor Hamlet twice on one day, therefore I must do some lighter part—say Shylock—for the first Matinée & second Sat: night &c—for which we can have no books ready. The *job* has been too long delayed—we should have had *all* ready; however, we must be content with what we have. As to your young friend McWilters[14]—I do not see anything I can give him, every position being filled; why could he not have the sale of books in the theatre? Some one must attend to it—let me know at once.

I am sincerely glad of your wife's success and hope it may prove a source of happiness & profit.[15] I hope she has forgiven me for advising her against the stage—I know she must have hated me "orfull!" for it, but, you know what *Romeo* says—"I thought all for the best."[16]—I like your idea of title-page &c— but the book must not be unwieldy, nor beyond the reach (in price) of the playgoer.

Does not "Warwick" furnish the ages of the folks we are dealing with? I think Bulwer states in that novel that Gloster was about 19 at the time Shakespeare first introduces him to us, and—if I remember rightly—he was but 33 at Bosworth.[17]

---

in 1862. She recorded her life and career in *The Memories of Rose Eytinge* (New York, 1905).

[14] The "young friend" is probably Winter's secretary and assistant whose name was Thomas McWatters. He was apparently given charge of the sale of the Prompt Books. See next letter.

[15] About this time Lizzie Winter returned to the stage in Toronto where, according to Young, she enjoyed a popular and critical success.

[16] *Romeo and Juliet*, III,i,109.

[17] In *Henry VI, Part 3*, Richard of Gloucester is represented as a mature warrior. In reality he was only 8 at the battle of Wakefield with which

Love to the Young Collegian & t'other chap—forget his name, and remember us three very kindly to your wife.[18] I have not written to J. J. since he sent us those pictures (!) but spiritually I have welcomed him home. I see that Jno. Clarke has just come over.

This has been desperate effort for me this morning, a raging headache has more than once tempted me to throw the scrawl into the fire, but I know you will wait for an answer & therefore let it go as it is. Wish I cd. give you the information you require —all I can give is the general Statement that the jonfon (or hauberk) was worn very short, with an upper garment (or sleeveless robe)—the hair long and *clubbed*, cut straight across the forehead; no hair on the face (although I have an engraved portrait of Richd. with a moustache!)[19]—Adieu!

After his engagement in Buffalo, which ended about December 9, Booth visited Lockport, Rochester, Syracuse, and Albany; he arrived in New York about Christmas. On January 7, 1878, he opened at Booth's Theatre as Richard III, using for the first time in New York Shakespeare's text instead of Cibber's. The Prompt Book of *Richard III* was published in time to be sold during performances. The critical reaction to Booth's restoration was generally favorable. The critic of the *Telegram* (January 8) thought Booth's acting version "worthy of all praise," adding that "the sublime tragedy is at last fitly redeemed from the coarse vulgarity and cheap fustian rhetoric of Colley Cibber." The critic of the *World* (January 8) wrote that it was "a great improvement." The *Herald* (January 8), however, thought Booth's version might be better reading but Cibber's was the better acting version. *Richard III* ran continuously for two weeks. During the next four weeks, Booth appeared in eleven different roles, closing his engagement on February 16 with King Lear. On Monday,

---

the play begins and only 19 when the play concludes. He was 33 at his death at Bosworth Field.

[18] The "young collegian" is Winter's eldest son Percy Curtis, who at this time was enrolled at a military academy in Toronto. The "other chap" is possibly Winter's second son Arthur, although it seems unlikely that Booth would forget his name.

[19] A description of the costumes and hair styles of fifteenth-century England was printed in the Appendix to the Prompt Book of *Richard III*.

Booth is probably referring to an engraving of his father, who wore a moustache and goatee as Richard III, as did Edmund Kean and George Frederick Cooke.

February 18, he opened at the Brooklyn Academy of Music for one week. During this engagement the Prompt Book of *Hamlet* was published.

33. Everett House, New York, February 24, 1878

For the past two weeks I have been 'run to earth'—I have positively had no time to do—or think of anything outside of the theatre. Thank God! I have a few days respite, though barely more than will allow me to 'catch breath.'

Brooklyn did well but less than any former engagement there. Horace[20] is obliged to go at once to the Arkansas Springs. His health is rapidly failing and he must start tonight. This will of course prevent the arrangement you propose regarding the books in Boston. He has, however, secured the services of him, who under McWatters had charge of them at Booth's, and the latter understands the case. Silverstone is the man's name; he will render his acct. to me. I will, as you wish, apply the proceeds toward expenses, but I want you to be paid at all hazards.

I like the *Hamlet* exceedingly; only I wish I had strained a point about the matter of costume for tho' I agree with what you say in reference thereto—I have made it a special point to mount the play in nearly all the leading cities, both with scenery & costume as correctly as Planché & other authorities have enlightened us on the subject of early Danish dress &c, &c. If just a line or two additional were given by way of information to the reader (especially to the actors) as to the costume used by me, it would have been of service, for many may suppose the dress I wear is of Shakespeare's time. So far as I am capable of judging I assuredly give your remarks the preference to the others you have selected. You certainly *hit* my idea of *Hamlet*, whether or not I succeed in so illustrating it is questionable. Sometimes I think I do, at others—no.[21] I wish

[20] Horace McVicker, the eldest son of James H. McVicker, was occasionally Booth's business manager.

[21] In the Appendix to the Prompt Book of *Hamlet*, Winter devoted only one sentence to Booth's costume. "Mr. Booth . . . has been accustomed to dress this piece in conformity with the usages of an ancient period in the history of Denmark, in order to invest its scenes with something of the character of the age to which its story relates."

The "remarks" are Winter's essay on "The Character of Hamlet" and excerpts from the Shakespearean criticism of François Guizot, Edward Dowden, Samuel Taylor Coleridge, and Hermann Ulrici.

Lear was to be the next instead of Richd. 2d., for I must do the former during my first week in Boston. The management there have blundered in bringing two tragedians in quick succession. McCullough plays several of my parts the week before I begin, which must, of course, weaken one or the other, if not both. Apropos of costume—it is safest to refer to Planché, Herbé and Vichellio [sic] (I'm not sure of its orthography) for information regarding my manner of dressing the characters.[22] Since my production of Richd. 2d. at Daly's I have made alterations, but I am pretty certain that I sent you a copy as now acted—when I was last in N.Y. No proofs have yet been sent to me. I shall start about Saturday next for Boston; after 3 weeks there I shall rest a few days & start for Pittsburgh, thence to Baltimore— where I close my season on May 4th. I am sorry I did not see Taylor's article on Richard;[23] he was in my wife's box and was very pleased. *Lear* & *Richelieu* would sell better in Boston than (except Hamlet) any of the others I think—if they were ready at the time of the performance.

I have not yet looked at the "Memoir"[24]—but will do so as soon as I can possibly get leisure; I hardly know what I can do for it, but perhaps when I read it over the *mood* will come. The *death's-head* & *crown* will be more suggestive than the mere cognizance of *Richard* which conveys no idea whatever.[25]

I have scratched off about all I know this morning—with

[22] Charles Auguste Herbé, a French painter and art historian, wrote a history of French costume entitled *Costumes Français civils, militaires, et religieux . . . 1834* (Paris, 1842).

Cesare Vecellio, an Italian painter, wrote a history of costume entitled *Habiti antichi et moderni di tutto il mundo* (Venice, 1598). In 1860 Vecellio's book was published in a new edition in both Italian and French by Fermin Didot frères fils Cie. in Paris.

Booth's personal copies of Herbé's and Vecellio's books are preserved at the Walter Hampden Memorial Library.

[23] This is possibly the Douglas Taylor who was an American drama critic and theatre historian about 1860 to 1900. I have been unable, however, to find an article about Booth's *Richard II* which I could identify as Taylor's.

[24] The reference is perhaps to Winter's biographical essay on Booth for *Harper's Magazine* in June, 1881.

[25] Each of the Prompt Books had a symbolic emblem printed on the front cover. The emblem for *Hamlet* was a raven; for *Richard III*, two intertwined roses; for *Macbeth* a bat over a crown; for *Richard II* a skull and crossbones over a crown.

my wife & Horace figuring up accounts at my free elbow, but I have managed to touch all the points of your two unanswered letters. I shall now drop a line to McWatters.

I hope all your dear ones are well and that their influence & your few days rest have lifted you up again. I am pleased to hear from Tompkins[26] that John [McCullough] is doing finely with *Coriolanus*. Scwhab [sic] I see, is going for the *Exiles*—The Boston folks have taken *Booth's* for its production in the Spring.[27] I hope to have a good long rest next season—acting occassionally, of course, but not continuously as heretofore.

Remember us both kindly to Mrs. Winter—not forgetting Edwina's good wishes, and believe me

Your friend ever
Edwin.

Late in February, Winter's father died of Bright's disease. Shortly before Booth opened at the Boston Theatre on March 4, he wrote Winter to express his sympathy and to answer several questions regarding the Prompt Books.

34. The Brunswick, Boston, March 2, 1878

I return the sad letter from your brother. Believe me, dear Will, I sympathize cordially with you, for I have experienced the most poignant filial sorrow; not only for the untimely death of a beloved father, but for the broken life and suffering of a

[26] Orlando Tompkins was one of the original stockholders of the Boston Theatre and its manager from 1862 to 1878 when his son Eugene took over.

[27] Frederick A. Schwab, American music and drama critic, adaptor, and theatre agent, was the drama critic of the New York *Times* from 1875 to about 1878 and the music critic of the *Times* from 1872 to 1890. In 1879 he seems to have negotiated for the management of Booth's English tour (see letter 50).

The Exiles was the American title of a spectacular melodrama by Victorien Sardou. An adaptation by Limington R. Shewell had played at the Boston Theatre from December 10, 1877, to February 9, 1878, and was scheduled for Booth's Theatre where it played from April 10 to May 11. An adaptation by George Fawcett Rowe played at the Broadway Theatre from March 2 to April 20. In the *Times* (February 24) Schwab wrote a "puff" for the Broadway Theatre production and suggested that this production would be so attractive that the production at Booth's Theatre would be "unnecessary."

dear mother—who is fading, as it were, before my eyes. God
alone can heal the wound—and my words are not only weak,
but useless.—I arrived here last night & found three letters from
you. The first thing I did this morning (after breakfast) was to
send the check & bill to De Vinne,[28] and requested him to send
the receipt to you. I sent also the design for *Richd.* and suggested
a different style of crown—the one selected is, I am sure, too
modern—it did not 'come in', I think, 'till *Richard 3d's* time (I
mean the bands & cap). I sent a *queer* sketch of my own. Horace
had made some arrangements with Silverstone about the books,
& as I thought an interested party would use more energy in
pushing the sale I told S. to bring the books & I also notified
McWatters to this effect. McGlenan is all right, but he has other
matters to attend to & might not give it that attention the subject
requires. I fear that 40 cents is a high price—since people can
buy playbooks for 25. I wonder if the books cd. not be sold on
the trains—with other books & pamphlets: it would not be
amiss to try it. I told De Vinne (or *Hart* rather) to let you see
my corrections—(?)—I fear there are fearful mistakes yet *extant.*
I followed the punctuation of the doubtful *Globe,*[29] but did not
agree in all points with it; my own lack of surety, however,
decided me to follow the book. I wrote to you the other day on
this business, and answered your questions concerning *Hamlet*
—yes, that *is* satisfactory. McW. has sent me the books as you
directed; more than I require—what I do not use shall be returned.
I hope I may be able to help out the *memoir*—I shall 'go' at it
now—during my three weeks here, for here we have fewer

[28] Theodore S. DeVinne, American publisher and printer, was at this
time the president of Francis Hart Co., who published the first edition of
the Prompt Books. Later DeVinne established his own press and printed
most of the publications of the Dunlap Society in the 1880's and 1890's.

[29] *The Works of William Shakespeare* edited by William G. Clark and
William A. Wright (London, 1864) is known as the *Globe Edition.* It
has the same text as the *Cambridge Edition* (London, 1863). The *Globe
Edition* had an enormous sale because of the purity of its text. Booth,
however, seems to have preferred Knight's edition (see letter 36). In a
brief letter to Winter dated New York, February 27, 1878 (Folger
Shakespeare Library, Y.c. 215 [262]), Booth wrote: "I told you I had no
other authority at hand, for reference, than this mischievous *Globe* stuff,
and by that have I been guided—save where my own sense refused its
twaddle. But in the punctuation business I followed it—often in spite of my
own notions though they are very vague on such matters."

callers & other interruptions; I must meet Modjeska after rehearsal on Tuesday, at Field's lunch,[30] however; something I usually decline.—The arms of Richelieu are as suggestive (perhaps more so) than the cross—or any other device we could use; *the cardinal's hat placed above the crown* it can be seen on my old play-book (claimed by Hinton)—drawn by Hennessey. An antique crown—broken in two parts—on a cracked shield might do for *Lear*; I use the dragon as his cognizance (when I can get *anything* made for the purpose) but does that suggest more than the "fiery quality" of the *critter* he likens himself to? What for Othello? The strawberry h'd'k'f pinned to the shield by a dagger; or a white & black hand clasped; or a heart with a serpent twined about it?[31] These come up as I write & may be weak—you can doubtless suggest better.

McCullough, I hear, has done well—he closes tonight; I *must* see Modjeska, or I'd see his *Cade*.[32] This is my last day of rest & I'm as tired as a jaded horse. One idle week but lets me down, one or two more would doubtless bring me up again.

I have three weeks here—then go to Pittsburgh (a long jump —taken solely to let Edwina see the Alleghennies & the great horse-shoe curve there)—after two weeks with Gotthold[33] in P'burgh I have a spare week, & then begin my ending with Ford

---

[30] Helena Modjeska filled an engagement of two weeks at the Boston Museum beginning on February 23. The Field who had a lunch for her was probably Richard Montgomery Field, the manager of the Boston Museum from 1864 to 1898. See *The Memories and Impressions of Helena Modjeska* (New York, 1910), pp. 358-362; also Marion Moore Coleman, *Fair Rosalind: The American Career of Helena Modjeska* (Cheshire, Connecticut, 1969), pp. 173-181.

[31] A cardinal's hat placed above Richelieu's coat-of-arms eventually became the emblem for *Richelieu*. The emblem for *King Lear* was a crown above which was a heart. Through the crown and heart a serpent was entwined. The emblem for *Othello* was a serpent coiled on a heart-shaped shield above which was a crescent enclosing a five pointed star.

[32] *Aylmere, or The Bondman of Kent*, also entitled *Jack Cade, the Captain of the Commons*, was an historical tragedy centered around Jack Cade's Rebellion (1450). It was written by Robert Taylor Conrad for Edwin Forrest, who first acted it on May 24, 1841, at the Park Theatre in New York. McCullough, whose style resembled Forrest's, adopted the play as part of his repertory.

[33] J. Newton Gotthold, an actor in Booth's Winter Garden company from 1864 to 1866, was the manager of the Pittsburgh Opera House.

in Balto. Next season I shall work & travel less. We have taken apartments in a house on Madison Ave: near the square, & shall 'house-keep' for awhile. Then I hope to have leisure for social enjoyment & to see more of you & other friends, who, for the most part of life, are "lost to sight."

I have gone over your letters while scribbling this, & have no more to tell you just now.—Do as your brother bids you; take all the rest you can & get strong. I met an old acquaintance in the cars from N.Y. (his name I could never remember five minutes after hearing it—is it Faxon?) who knew you as a boy in Gloster—we had a chat about you & your poetry. When does your book come out? have you heard from London?—Clarke plays at 5th Ave:[34] next week (or the following) & Fiske is after me again to go there. He is but agent for Gilsey who wants my name associated with his theatre, while Tweed, for Ames, wants me to take *Booth's*[35]—Goodbye!

Wife & daughter join me in good wishes for you all.

The Prompt Books did not sell in Boston as Booth had hoped they would. He nevertheless determined to publish the remainder of his standard repertory. His immediate concern was *King Lear*.

35. The Brunswick, Boston, March 7, 1878

The books are not such a 'go' as we anticipated they would be. The sale is very slow and all complain of the price: I told the boy to try 35 cts as in N.Y., but I fear the *gush* is more for Modjeska now—though my *trade* is good, yet not as rushing as formerly. Richelieu is played next week, & *Lear* will be done for the second & last time tomorrow, so these need not be so desperately dispatched—if you'd like more time. You cannot have *Richelieu* ready for Monday.—Richd. 2d. & 3d. week after next, my last.—I'm afraid the books will not yield an immediate fortune.—I like your preference of the title page and your choice of titles too. Let it be "Drama" by "Bulwer," of course. The *property-list* for *Lear* is not necessary; since we have none for

[34] John S. Clarke was engaged at the Fifth Avenue Theatre from March 9 to March 30, 1878.

[35] Stephen Ryder Fiske, American theatre manager and drama critic, assumed management of the Fifth Avenue Theatre on October 15, 1877. He continued managing the theatre, which was owned by the Gilsey estate, through 1879.

the other plays I would not give it in this one. I selected the earliest costumes, from several authorities, that succeeded skins & nudity; making those for *Lear* of the rudest stuffs; so also for Macbeth—but less coarse in the material, with but little variation for Hamlet—save in texture. The same styles of drapery, gartered leggings and trimmings are seen through several of the centuries preceeding, and subsequent to the ['fustian'?] era; at least this is my recollection of my researches in *W.G.* and *Booth Theatre* days. There are no necessary doubles in *Lear* and *Richelieu*. The first act occupies 2 weeks—"What fifty of my followers at a clap! Within a fort-night." Act 1st. Sc. 4th—Then as you say, the 2nd. act begins the 2d. month, at the close of which the *storm* begins & I presume it to be the same night on which *Lear*, Act 3d. meets with Edgar &c, as Gloster goes forth to seek him in the very next scene it would seem that but a few hours had elapsed since the King's departure, in Act 2d. S. 4th. In Act 4th, Gloster says speaking of 'Mad Tom'—"In *last night's* storm I such a fellow saw"—consequently but a few hours have passed 'twixt the 3d. & 4th acts. In Act 3d. Sc. 3d. Gloster says "there's part of a power (meaning the French) already footed"—i.e. landed & on the march, and Oswald speaks of "the army that was landed", in Act 4, scene 2d.—so there cannot be much time lost 'twixt 3rd. & 5th in which the battle ends with Scene 2d. According to this the entire action fills about 5 weeks—or less, as I figure it—but "I am ill at these *numbers*", and leave you to balance up.[36]—Your device for Lear is good; I do not quite catch the significance of that proposed for *Richelieu*. Did you get my last—in which I suggested emblems for these plays?

I saw your brother the other day—he thought you might be in N.Y., owing to some letter you mailed at that city. I hope no greater troubles may be added to those you already have to battle with, but at the worst be brave & fight the fight unflinchingly.

The day after I asked (in my last) after your poems I saw some notice of their approaching publication by the London folks.[37] I opened to a tremendous house; the two following good, not great: the cry is still for—Hamlet! D--n it! First chop attractions are too numerous here for any one to do greatly. My

---

[36] In the Appendix to *King Lear* Winter worked out the "time of action."
[37] Booth is referring to Winter's *Thistle-down*.

Othello & *Lear* are very bad & very good. My worst & my best, &c, &c, according to the papers.

Have had but a few hours as yet to look at the *Memoir*—will do my best; but do you think it will pay?

With the problems of *Lear* apparently solved by this last letter, Booth, despite a feeling of being "dead beat," turned to the publication of *Richelieu*.

36. The Brunswick, March 12, 1878

As you are frequently 'run to earth' by overstrain so I am often 'dead beat' by the incessant drain upon my vitality; my nervous system is on the rack continuously & the reaction is so depressing that it is impossible for me to find interest in anything, while my physical prostration unfits me for any exertion during the day. I might have rendered you more assistance in the 'prompt-book' business had it been otherwise with me. Regarding the matter of costume, I have told you how helpless I am without my authorities, for tho' I know just what is & what is not correct in such things I cannot describe (by their proper titles)—nor give dates &c—anything with clearness, such as is required for publication. So with the text—I cannot turn to several editions for my authority &c, although I *know* that there is warrant for whatever I may object to in this *Globe* edition. It is not altogether *bad*, for it retains the double negatives which Knight[38] discards, thereby weakening, I think, the force of words. Had I my books about me I could give you the information concerning the 'day of dupes', for when I produced *Richelieu* at *W. G.* I hunted up all the information to be found regarding the reign of *Louis 13.* —By-the-by, I remember when the play was running at W.G. the error of date, anent Cromwell & Charles,[39] was mentioned by one of the critics, but I really forget whether Bulwer anticipated or belated the incident.—This is all I can recall of the "Day of Dupes":—some political troubles occured in 1630, in which

[38] *The Pictorial Edition of the Works of Shakespeare*, ed. Charles Knight (London, 1839-1842).

[39] The reference is to *Richelieu*, V,iii, in which Louis XIII's secretaries try to convince him to aid Charles I in his fight against Cromwell. Richelieu advises that "Charles's cause is lost" and that Louis' help would be useless. This reference would be anachronistic since the conspiracy used in *Richelieu* occurred almost a decade earlier than Cromwell's defeat of Charles I.

Marie de Medici [sic] & Anne of Austria endeavored to oust Richelieu from his position of Minister, but he foiled them & established himself firmer than ever in his place.[40]—The books of *Richelieu* and *Hamlet*, published by Hinton, were my first (& last) feeble attempt to edit a series of plays as acted by *E.B.*, they have notes and cuts—you may have noticed. In the former I have the translation of Bulwer's French references, &c. I grew weary of the task & Hinton turned up with a terrible arrangement of *Richard* 3d. I intended to adapt this play for the 'Winter Garden', and fancying had found in *H.* just the man to do what I had found too tedious a job. I then told him to call the book "The prompt-book", & cut the several plays according to my method of acting them, &c. He chose to follow his own plan, & the result was a great expense, to him, and a useless lot of rubbish. I have frequently forbidden their sales in different cities, because purchasers naturally complained of their being unlike the acted play—but somehow they got into the hands of many 'hawkers' all over the country. In the present instance I did not prevent their being sold for this reason—Silverstone's pay comes from the sale of these books of those plays which we have not yet published; he takes no commission, as I understand it, from the sale of *ours*, but solely from the others & the photographs which *Horace* had sent to *him* from California. Of course when our books are all out the others will be stopped, but till then people will buy play-books of some kind, & so long as our *seller* can be paid without reducing our income— is it not best to let it be? Silverstone's account—some $19 & odd cents will be enclosed, if I don't forget it before sealing this; he is to render me a weekly statement. It appears the 'critics' did not receive copies—though McWatters sent some to the different papers, I believe—and McGlenan asked for *seven* to be distributed where they would 'do good', so he said. S. tells me he

---

[40] The "Day of Dupes" occurred on November 11, 1630, when Marie de Medicis and Gaston, Duc d'Orléans, extracted from Louis XIII the promise that he would dismiss Richelieu. The cardinal went immediately to Versailles, dissuaded the King, and became more powerful than ever. Marie and Gaston, the dupes, paid dearly for their short triumph. The conspiracy in *Richelieu* is actually an amalgamation of the "Day of Dupes" and the conspiracy of Henri Ruzé d'Effrat Marquis de Cinq Mars (1620-1642). See Bulwer-Lytton's Preface to *Richelieu*.

has them in several shops on sale, & the prospect is better than it was. I forget the *Picciola* man[41]—Edwina but lately read it & is now scratching her head after the authors name. I like your preface & appendix for Richd. very much, but some account of his early bravery in the *Wat Tyler* revolt would have helped him in our sympathy;[42] I wish Shakespeare had given him that advantage in the first act of the play. Where did you get that Property list for *Lear*? The bloody sponge for *Edmund* made me laugh a good old 'palmy day' laugh; when my Papa used to bloody his brow for *Richard* & *great actors* of the good old school used to 'chaw chalk' for frothy rages, &c.[43] To add such details the *blank* side of our leaves should be printed, but all promters prefer (and *will* do it) to write, in *very large letters*, the directions which are already printed; besides—some other actors may have different notions of many directions I might publish, & I think it best to leave such details as property-lists,

[41] *Picciola* (1836) was an historical romance by Joseph Xavier Boniface Saintine. Bulwer writes in his Preface to *Richelieu* that "for some portion of the intrigue connected with De Mauprat and Julie" and for "considerable if not entire reconstruction of character" he was indebted to *Picciola*.

[42] During the Peasants' Revolt of 1381, Richard II met with the rebel leader Wat Tyler. At the meeting, Tyler was killed by one of Richard's party, although the King had urged restraint. At the moment Tyler's men were poised to retaliate, Richard rode unaccompanied into their formation and asked whether they wished to shoot their King. His presence of mind in this situation prevented further bloodshed on both sides.

[43] Edmund in *King Lear*, II,i,44: "Look, sir, I bleed." A typescript promptbook of *King Lear* at the Walter Hampden Memorial Library [see Charles H. Shattuck, *The Shakespearean Promptbooks* (Urbana and London, 1965), *King Lear* 83] furnishes Edmund with "Blood on Sponge," but the direction is scored through with pencil. A Junius Brutus Booth promptbook of *Richard III* at the Folger Shakespeare Library (Shattuck, *The Shakespearean Promptbooks, Richard III* 10) contains a direction to apply "blood for scar" before the combat scene. John Wilkes Booth, according to the New York *Herald* (March 18, 1862), "blackened and smeared his face with blood" for the combat scene. Macready in his *Reminiscences*, I, 141, describes a piece of J. B. Booth's business as Sir Giles Overreach in which "one of the attendants who held him was furnished with a sponge filled with blood (rose pink), which he, unseen by the audience, squeezed into his mouth to convey the idea of having burst a blood vessel." Macready, however, noted that this business "was severely commented upon." I have found no references to "chawing chalk" for "frothy rages."

scene-plots & calls, music—& curtain-bells, lights up &c, &c, for the different 'stars' or managers who may use the book. Why did you send me De Vinne's receipt, you told me to have it sent to you—am I or you to keep it?—*Mar. 13th* was interrupted yesterday at this point. Have but a few moments before rehearsal to finish this. Have just recd. yours from N.Y. with enclosures—all right about the *Lear* & *Richelieu* title-pages. I just recd. a few leaves (author's proof) of *Richard* from De Vinne & sent them per post several days ago to Murray St.—that's all I've had since I left N.Y.—I forgot to say that I fancy the *dénoument* of 'day of dupes' refers to the episode of the last act—where the Secretaries lay before the King the condition of affairs in France, &c. Am I right? I remember something in 'Cinq Mars'[44] concerning that scene. I began a revision of the *Memoir*, but found it impossible to give it my full attention. I must talk it over with you & think it over by myself when the mood is on. I doubt if it will sell & I question the propriety of publishing such things during a fellow's life-time. As for England—heaven knows when, if ever, I shall go there. The poor sale of the books is rather discouraging, and as my season is drawing towards its close perhaps it would be as well (after Lear & Richelieu are out) to defer the rest—I mean their publication, but have them ready,—until the fall, what think you? We'll talk it over. I must close now. Hope this will catch you ere you leave N.Y.

Booth ended his engagement at the Boston Theatre on March 23. After a few days of rest in New York, he played in Pittsburgh from April 1 to April 13. At the end of this engagement, he wrote Winter about the financial returns of the Prompt Books and his difficulties with the Pittsburgh management.

37.   Monongahela House, Pittsburgh, April 14, 1878

I have lots of your letters to answer, but I've had so much dyspepsia to attend to during my miserable sojourn here that I've positively been unable to write—or attend to the corrections you desired me to make in *Richelieu*. I have now an idle week, five days at least, and I will mark what I think should be altered in each of the plays, & send the books to you.

I have run over your letters and made notes to guide me in this

[44] Alfred Victor Comte de Vigny's novel *Cinq Mars* (1826).

reply to all of them.—As to the expense & my reimbursement, I hardly know what to say—I hate to touch it, for I feel that I have not exactly carried out my part of the programme. I suppose it is just, in a business light, that the expenses should be first covered, and then the profits considered; but this was not to be a *business* matter at all—with me, and yet it seems that, by taking the proceeds as fast as they accrue, I am making a *business* of it. I decidedly decline to receive any of the proceeds from *L & S* sales,[45] or from any source outside the theatre, and I do not feel altogether comfortable in taking even those. So far I have received, from the Boston & Pittsburgh sales, *$146.92* (I have failed to keep acct. of what I've paid—but the bills will show)—if I am to be repaid before you get any of the 'swag' it will indeed, at this rate, be a long time before you get that little home from this source. No, I must stand by my original proposition & wait for the return of my investment; after the Baltimore sales I will turn over to you the receipts from the three cities. After the books get well under weigh, through L. & S.'s efforts, the dollars may come in fast enough to lift occasionally some part of the tax, & gradually remove it without depriving you of all immediate benefit; I can better afford to wait than you can. I want to avoid, however, any further outlay during my vacation— I mean after *Lear* is paid for, & the other "odds & ends" that require immediate attention. It is unfortunate that our venture occurs at a time when I must needs devote more time to rest than action; my vacation this year will be until October, and when I begin I shall *sandwich* my engagements with long intervals of rest. This, of course, will seriously lessen the theatre sales of the books, & I am, therefore, most anxious that *L. & S.* shall succeed in their plan of disposing of them. I do not like the idea of selling at *25* cts.—could that possibly pay? *35* perhaps— let's see what Baltimore does. As to Percy[46]—I can't see how

[45] The publishers Lee and Shepard in Boston, and Charles T. Dillingham in New York, seem to have acted as distributors of the Prompt Book outside of the theatre.

[46] A notation by Jefferson Winter on an earlier letter from Boston about March 17 (Folger Shakespeare Library, Y.c. 215 [268]) indicates that Percy Winter had thought of travelling with Booth to sell the Prompt Books in the theatre and elsewhere. Booth was reluctant to hire Percy, however, for the reasons given in this letter and because he had a previous commitment to Silverstone.

the commission could pay his expenses—it would be best to
'figure it up' & consider it well before risking it; besides it would
afford him no *occupation*—he would be merely an idler nine
tenths of the time; he should have some more active duties;
book & photograph pedlers, ticket-sellers, & theatrical agents
are in the main a loafing set of fellows; very few of the last named
are competent—for it requires a *detective* quality to deal with
doubtful managers & treasurers; he must be thoroughly
posted in all the "ways that are dark and the tricks that are
vain" to be of any service to his employer; he is tempted on all
sides by the genial reporter & the gay play-actor and at the best
his life is but one huge 'loaf.' Percy should have some steady
ennobling employment.

Now that I have acted *Richard* enough to feel its effect upon
the people I see how much better I could have arranged the play—
by dividing the acts differently, and a few omissions; much that
I have retained is wearisome to the audience—I mean in the first
two acts. It is impossible to see these defects until one has acted
a character a number of times & watched the effect. I think
with you that Irving should have been mentioned in connection
with the play, & also that *Richelieu* should have been dedicated
to "Owen Meredith",[47] but—'one can't think of all thinks to
oncet.' The new books of *Richard* you speak of & the circulars—
yes, get them out; I should have told you this several days ago.
I told *S.* to send you word if he wanted any more books, &
Henry to telegraph you at once (I was acting at the time) to send
Lear to Baltimore. My business here has been very *so-so*. I am
very much bothered about the "Merchant of Venice"—I must
have some time to think what changes I would like to make in its
stage-arrangements. The plays of Henry 8th, Taming the
Shrew, Don Caesar, are such mutilations & are acted by me just
as others do them that I question if it would be wise to add them
to our list, while *Ruy Blas* is such a monstrous mass of twaddle
that I'm rather ashamed of the version I have made up of several
other worse ones. Were it not that Shylock is so prominent in my
repertory I'd advise the omission of that play, but it is too
important a character & too often acted to be left out. When we

[47] "Owen Meredith" was the pen name of Edward Robert Bulwer-Lytton,
English poet and statesman and the only son of Edward George Bulwer-
Lytton.

meet we will talk over all these matters & decide just what we will do next season. I feel a little anxious about *Othello* & *Macbeth*, for I have been obliged frequently to change much in representation of them—according to the exigencies of the theatres & companies in different towns. I have found the actors so infernally *dumb* that I am half sorry I did not cut & slash *Hamlet* instead of adding to it. It's very discouraging.

Glad you thought of the insurance on plates. I enclose check for Hart's two bills, & also Silverstone's account for last week.— Edwina is looking eagerly forward, I should say *up*-ward, for Thistle-*down*; wife & she send kindly wishes. Had quite an event here t'other night. At 7 o'clk, as I was about going to the theatre, Henry Flohr came & told me to wait for an hour because the gas company had shut down on Gothold [sic] for last years debts. I waited—at 8 o'clk he & other messengers came to announce the closing of the house. A large number of people had collected & there would have been a fine house in spite of a heavy-rain storm. The papers went for Dr. Hostetters bitters (he being the head devil in this diabolical business) and sympathized with Gothold. The two following nights we lighted the house with some fluid called *Elaine*—not of Astalot,[48] but from some of the adjacent oil wells. Of course, this was an outrage on the dear public & on me as well, for it might have been prevented & I should have been consulted before any such step had been taken. I was naturally indignant, but I felt sorry for *G.* and said nothing *very* severe. He insisted that I should receive pay for that night at all events & we fixed an amount—the payment of which, however, was to be deferred until Saturday Matinée—yesterday. Although my engagement was with *G.* it appears that he was merely receiving a salary from the proprietor (some wealthy old cuss named Coleman), and could really do nothing although he *promised* much. After the *Matinée* word was sent me that the $ agreed upon for the vacant night would be paid in the evening, but during the performance (at the Matinée) Coleman came to see me & thought I ought to share the loss & endeavored to coax me off. I told him I had arranged everything satisfactorily with my manager, & could not recognize his (C's) right to interfere, &c. Before the play began at night I was told that Coleman had

[48] Elaine, the Lily Maid of Astalot, who died for love of Launcelot is the heroine of one of Tennyson's *Idylls of the King* (1859).

ordered the treasurer not to pay me. Gothold was limp & could do nothing. I dressed & would have gone on without bothering my head, or the audience about the matter, when *G.* in desperation begged me to wait & finding a son-in-law of the 'old one' in front told him that Mr. B. would not act until the money was paid, and that both he & Mrs. B would address the audience, relate our grievances & dismiss them.

The money came in just 2 3/4 minutes! I suppose the papers will have all sorts of versions of this affair; let me know if you see any of them. This is the crankiest letter! I doubt if you can read or understand half of it, I get so nervous & fidgetty when I write that it is really painful to me.

P.S. Have no stamp for check.

After his five-day vacation, Booth went to Baltimore where he acted at Ford's new Opera House from Easter Monday, April 22, until May 4. He then returned to New York for the summer, his apartments being at 68 Madison Avenue. Here he planned his engagements for next season and the publication of the remainder of the Prompt Books.

38.   68 Madison Avenue, Sunday Night, June 2, [1878]

I am positively tired out, but I must send you a few lines tonight, for my days are devoted (and will be for a week or two) to McEntee, who comes to town expressly on my account.[49] We are all ill here now—cook and all! I have just finished my revision of Oth. Brutus & Fool's R.—They are now in shape for print—so far as I can fix 'em. Exits & entrances are so frequently changed that they are not necessary—still put down all I've marked. The *Willow* song I have clumsily arranged—I mean its introduction. Yes, name the Cyprus town.[50] Wherever the text bothered me I turned to Knight; I think it all is correct. I'll go over the other books, but I think there is nothing except the "embrace" of Regan, in Lear,[51] and a needed transposition of a

[49] About this time Jervis McEntee began a series of full-length portraits of Booth in his various characters.

[50] The town was identified as "Famagusta, a Fortified Sea-Port Town on the Island of Cyprus." No other Shakespearian editor identified the town specifically.

[51] The reference is to II,iv,191 when Goneril enters to meet Regan and Lear says "O Regan, will you take her by the hand." In the Prompt Book

few lines in Richelieu. There is nothing I can find in print regarding my acting of the characters, but if you can suggest that my idea of Othello is not animal but poetic, and that I treat Macbeth more as a weak man, "full of the milk &c &c" rather than as a strong brute, it would not be amiss. For God's sake settle Preston![52] I thought his play would be good as a spectacular drama for Jarrett & Palmer, & think I told him so. I have to read so many that they are all run together in my memory. John & I did not mean to act together, but after a few weeks of my plays he was to revive "Coriolanus"—he to manage the theatre during both engagements;[53] " 'tis better as it is." Until I get rid of the *bothers* that now beset me I cannot think of the Memoir; if you have the copy you left with me a few days (while in Boston) send it [to] me, & I'll give it my earliest attention. Your emblems are good—I can't decide which is the better for Brutus though; would the "S.P.Q.R." serve with the *fa[s]ces*? How about the *Roman Eagle*?[54] All right about De Vinne—will go at them at once. What does he mean about a new style of binding? I think the books should be uniform in all outward *particulars*— don't you?

---

prior to this line is the direction "Regan advances to take Goneril by hand. Lear interposes." In an earlier letter to Winter dated Baltimore, March 5, 1878 (Folger Shakespeare Library, Y.c. 215 [277]), Booth wrote:

> As for Regan's *embrace*—if I ever knew its signification I certainly forgot it. Be sure there is no actor, & I doubt if one in a thousand readers, who would understand the stage direction so worded. Besides I do not permit Regan to touch her sister's hand—as she is about to do so I prevent the action, as if dreading such contamination, &c; ergo, the stage direction is incorrect & should be omitted.

[52] In the next letter Booth implies that this Preston wrote a play called *Columbus*. There were a number of playwrights in the late nineteenth century named Preston. Two wrote Columbus plays—Daniel S. Preston, *Columbus, or a Hero of the New World* (1887) and William Duncan Preston, *A Modern Columbus* (1890). Neither play seems ever to have been produced.

[53] Booth and McCullough had thought of an engagement at Booth's Theatre in the fall, 1878, but McCullough finally decided against it. See Booth to Winter, May 30, 1878 (Folger Shakespeare Library, Y.c. 215 [280]).

[54] The emblem for *Brutus* was an eagle rampant perched on a *fasces* on which was a banner initialed "S. P. Q. R." and encircling the whole was a wreath of laurel or oak leaves.

I've scratched so clumsily that I've mixed the pages, as you see. Will talk of Percy—can't write more at present. Have sent for my prompt-books of Shylock and Wolsey—you shall have them soon. It is impossible to do anything with the former more than is usually done in the way of cutting; I have tried it various ways. Several plays must be reserved for *double bills*—such as Wolsey & Petruchio; Shylock & Benedict; Ruy Blas & Don Caesar. Ruy Blas is such *bosh* that I dislike to publish it. The play of "Julius Caesar" (arranged by myself) at *Booth's*—is the same that Barrett, Bangs &c now claim as their own adaptation; shall we add it to the list? I sometimes play the 3 characters, not often. By-the-by, I notice that the 'notices' of the books don't give me any credit for the arrangements of the plays[55]—notwithstanding your "say so" in the prefaces, but 'nixy weedin'—I mean *n'importe*, that's French t'other's ["My-k-enic"?]

My back and head ache fit to split, as the old lady says. Good night!

Wife thanks you for yr. sympathy.

39.   [New York] June 4 [1878]

I'll take your letter line by line or rather—point by point. Imprimis: I don't doubt or worry in the least about the *credit* (if any be due) for my feeble part in the business, nor about the *money* neither. I have been frequently asked *who* cut & arranged the plays, and several papers said, in effect, that I but followed such and such plans, &c. That's why I spoke of that. My anxiety about the *outgo* was caused by the meagre *income* and with the cost being greater than I anticipated. But that feeling is past & gone; since DeVinne will wait I am serene, and doubt not of the future success and full return of all expenses; even if there be no return I shall grieve only on your account, not my own. So much for that.—

As for the scene in *Othello* you would like to have restored. My long and varied experience has taught me that the closer and the quicker tragedy can be acted the better is the audience pleased. As I told you, I think I once made so many restorations to Hamlet that nearly the entire play was acted. People gaped, slept and left before the final act—in squads. Dawison and I did

[55] Booth is probably referring to the "notices" in the *Tribune* which cited the Prompt Books as "William Winter's Prompt Books."

the scene in question and it seemed to drag.[56] The language
(with the exception of a line or so) cannot be spoken—in *English*
—"without offense to utter it."[57] The play is long enough with
all its cuts, and I think any additional matter would detract
from the enjoyment of its representation. I saw it done by
Fetcher and considered it a bore.[58] Do not let 'poetical justice' or
your love of Shakespeare win you into Hinton's error of restoring
scenes or speeches that I would be obliged, perhaps, to omit.
So much depends on the subordinate actors that I really have
regretted the restoration of such a simple bit as the "Truepenny"
&c in Hamlet;[59] Horatio and Marcellus cannot comprehend the
little required of them there, and invariably mar all I try to do.
I'm not particular about the reading of Iago's line, but you asked
for such 'notions' I may have, and that chanced to occur as I was
marking the book—just as the "misery" of Othello being given
without reference to himself; I think I am the first who ever
read it so—others give it "O misery! misery!" and become terribly

[56] Bogumil Dawison, Polish-German actor, toured America during the
season of 1866-67. On December 29, 1866, and January 2 and 4, 1867,
Othello was performed at the Winter Garden with Dawison in the title
role, Booth as Iago, and the German actress Marie Methua-Scheller as
Desdemona.

The scene in question was IV,i which is omitted in the Prompt Book. In
his preface to *Othello*, Winter says: "The hateful passages of Act Fourth,
in which Iago still further poisons the already jealous mind of Othello, by
making him overhear and misconstrue Cassio's talk of Bianca, were long
ago found, in the representation, to be needless and tedious; and they are,
accordingly, omitted."

[57] *Much Ado About Nothing*, IV,i,99-100: "There is not chastity
enough in language/ Without offence to utter them."

[58] Fechter first appeared in *Othello* at the Princess's Theatre in London
on October 23, 1861. Booth was in London at this time appearing at the
Haymarket Theatre and could have seen one of Fechter's performances
then. Fechter's Othello was not favorably received by discriminating critics.
George Henry Lewes thought it was "the very worst" he had ever seen
and Edmund Yates called it "a desperately poor performance, full of French
tricks and nonsense." I have found no evidence that Fechter ever appeared
again as Othello. See Kate Field, *Charles Albert Fechter*, Vol. VI of *The
American Actor Series*, ed. Laurence Hutton (Boston, 1882), pp. 51, 148.

[59] In 1870 Booth had restored the entire passage in *Hamlet*, I,v,150 in
which Hamlet calls the Ghost mocking names. In the 1878 edition of the
Prompt Book only "boy" and "Truepenny" are retained. See Shattuck, *The
Hamlet of Edwin Booth*, p. 156, fn. 61.

agonized with jealousy at the jump from this point;[60] you
perceive by the note further on that my jealousy does not begin
until during a pause, after Iago has spoken *slowly*—to give
Othello time to taste the poison—"Poor & content is rich &c &c."[61]
But all such little nothings I am reserving for my wife's
records;[62] I do not intend them for the prompt book. Some editions
give "peace" instead of "place"—the 1623 folio gives it so. I have
made my head ache searching for information concerning the
time of action of Othello; can find no clue to guide me. Let's
'figger': It's about 1400 miles from Venice to Cyprus, I think;
and although *Othello's* ship was "stoutly timbered" she must
have been a slow coach; at all events the tubs of 'them there days'
couldn't have been very fast sailers. I think it took my parents
six weeks to sail from England to Virginia—but that was not so
long ago as *Othello's* little excursion. Now calculate the time it
would take for him to scatter the Turks, outride a storm and get
to Cyprus. From that point I agree with you about 4 nights.[63]
And yet that seems a very brief time for such an officer to "rule in
Cyprus." He is not recalled for any more important business—
for "he goeth into *Mauritania* to his home" (in Algeria—is it

[60] H. H. Furness quotes Booth in *A New Variorum Edition of Shake-speare's Othello* (Philadelphia, 1886), III,iii,199 (p. 181): "Spoken without reference to himself. (I claim the credit of curing Othello's 'Misery! misery! misery!' as formerly given by actors. I directed my father's attention to it when I was a boy, and he approved.)"

[61] *Othello*, III,iii,200-204: Iago's speech is, "Poor and content is rich, and rich enough;/ But riches fineless is as poor as winter/ To him that ever fears he shall be poor:/—Good Heaven, the souls of all my tribe defend/ From jealousy." Furness quotes Booth on the reading of this speech of Iago's: "A pause. Spoken slowly and with significance; watch him curiously to observe the effect of your poison, suggest the 'evil eye.' Othello now, for the first time, begins to be conscious of a doubt—which, however, he immediately shakes off, and turns, as though from a trance, to Iago with a clear front."

[62] "My wife's records" refers to Mary McVicker's book about Booth's performances. In a letter to Horace Howard Furness on April 20, 1878, Booth wrote that Mary had been " 'takin notes' of how and why I do certain things in the course of my performances" ever since their marriage, with the intention of making a record of Booth's "stage tricks." The book unfortunately has been lost. See Edwina Booth Grossmann, *Edwin Booth: Recollections by his Daughter* (New York, 1894), p. 192.

[63] For Winter's "time of action," see Appendix to the Prompt Book, pp. 119-120.

not?) I give it up—I replaced the lines of *Bertuccio* simply because there is not one in ten thousand who have the least idea what 'horns' refer to; it occurs in the *tag* of 'Much Ado'[64] & it surely is not worse, or so bad as much that must be spoken in the Shakespearian plays; but, rip it out; 'tis not of any value whatever, and the less I have to say in that exhausting situation the better for me. Do you *really* think the change in that last act an improvement? I mean the action, not the *bosh*, of course. Hoyle[65] says (I'm told) "when in doubt take the trick"; how is it when you 'aint'? Take it any how, I suppose. Well, let's do so with the time of *Brutus* & *Bertuccio*; you feel no doubt—nor have I any means of information. . . . . (pause). I have just looked through Alfieri's Brutus,[66] but can find "no hinge to hang a" guess upon; as the play is now arranged the days and nights are terribly mixed, but the play acts better thus. The change was made by James Stark[67] (who was at Booth's, you may remember) & was suggested to me by McCullough. (By-the-by, I have set John at *Timon* for a revival;[68] hope he'll do it). Whatever appears in *Hinton*'s books in reference to costume came from *Booth's* Theatre—either from Joyce's books or mine; I have no recollection of the remarks you mention, nor have I a copy of the book; if they give a clear description of the Roman costume it wd. be well to use them. No, I do not think it well to quote from Gould[69]—I should have said 'yes', you are right; leave it out.

[64] *Much Ado About Nothing*, V,iv,126: "There is no staff more reverend than one tipped with horn."

[65] Edmund Hoyle (1672-1769), English authority on card games.

[66] Vittorio Alfieri's *The First Brutus* (1788). Alfieri's complete plays were available in English translation as early as 1815.

[67] James Stark, an American actor who came to prominence in California in 1849, is credited with being the first actor to tour a classic repertory across the American continent. His last appearance on the stage was at Booth's Theatre on February 2, 1873, in the role of Sextus in Payne's *Brutus* with Booth in the title role.

[68] McCullough, however, never acted *Timon of Athens*. See next letter, fn. 74.

[69] Thomas Ridgeway Gould, an American sculptor, was the author of *The Tragedian: An Essay on the Histrionic Genius of Junius Brutus Booth* (New York, 1868). In a copy of this book at the Walter Hampden Memorial Library inscribed to John E. Russell (see letter 65, fn. 37) Booth wrote that he thought Gould "suspect" in his description of certain points of his father's acting. Booth thought Gould remembered his acting

The emblem for Shylock should be typical of *Justice*—don't you think so? I've often fancied the old Jew with his 'balance,' as a sort of grotesquerie (is that the proper word?) of *Justice*, blind with malice and holding the scales. How about the bond, partly rolled, with the knife laid on it?—or the scales & knife?—or scales, knife, and bond, gracefully grouped? The heart is to be used so often—else I wd. suggest that, with a knife in it. Just as soon as I can do I will give my attention, *religiously*, to the five books already out; just now I am 'run to earth' almost.

I have already sent the 'Merchant' to you, and also 'Much ado'—so begin to brood; hatch away on 'em both. Let's get everything clean and clear in the more important plays before we touch the lesser ones. I will get at *Wolsey* as soon as Henry brings me a clean copy. I could not make you understand the cuts and transpositions as marked in the delapidated book I have. No *Lady*, nor *Stranger*—no, never! I mean, of course, each play to stand by itself; we can't give two plays, no matter how short they may be, for one price. I merely mean that when I play Shylock for instance I can do with it Petruchio, Benedict, or Don Caesar. All are very brief as they are cut. Don't blame me for butchering either *Shylock* or *Wolsey* (I mean the plays in which they appear) for Charles Kean treated the latter worse than I have done, and from Edmund Kean down all *stars* have done as badly by the former. You shall have a copy of my book (which was taken from *Booth's Theatre*) of *Julius Caesar*, its a poor affair but contains all the alterations, &c that I made for its revival. A year (perhaps two) after I read *Columbus* its author wrote me from 'Innsbruck'—I replied, after a pause, hoping the subject would end, but he still *Prest*-on (whew!) and I ceased to respond. He thinks me base, no doubt, but I am beset on every hand by authors and great actors of *every* sex; I cannot be eternally, unceasingly civil. The holy man who bored you for my address asked me to write an article for some paper with which he is connected!!! A jolly old article I'd make! I begged off, but my style and 'whatdyecall it' influenced his desire, and at me he came again—full tilt. I'm glad you played ['Classevia'?] and threw

---

better than his father's and throughout the book emended Gould's "recollections."

your body 'twixt' the assailant and 'poor Pillicoddy'[70]—thanks, therefore; I never answered him. Now there is a poet in Newark writing my verdict on some verses he has writ in honor of my mightiness. "Help me, Cassius, or I sink!"

My sick folks are bettering & McEntee is on the last portrait; *Lear*, Thank God! Now go to bed and sleep. Amen!

Just so soon as I have time I shall try the "Health Lift"—my circulation, my liver, my kidneys, my stomach, my spleen and, I verily believe my gizzard are all in commotion; physic does no good; nor liver pads; nor medicated waters—I'll see if I can't *lift* my ailments up by the roots.

40.   [New York] Sunday [ca. June 7, 1878][71]

Your day at 'Hohokus' was also the anniversary of my marriage—9 years!

I have added a line or two to correct the mistake regarding *Laertes* & the King. The difficulty is to make the actors speak 'new lines'; they have been accustomed to do thus & so, to speak this & that, and any introductions, no matter how essential to the sense of what they speak confuses them, & they will omit or mar by their bungling whatever they are unused to. That is one reason why so many plays are mutilated. The things we call actors are mere puppets—they do not study, they do not think; nine tenths of them have no idea who or what Shakespeare is; their sole object is to 'gobble' so many words, the fewer the better, for so many dollars, without any regard for propriety, art or sentiment. The strangest part is—they all are willing to be taught what to do, and how to speak in the flimsy modern plays, but they are beyond that in the Shakespearian drama; you can teach them nothing there.—

I think we can make Iago & Roderigo *fit*, by the omission, or transposition of a line or two.—*Theobald's* reference to *Herod* as being so applicable to *Othello's* case, and the smoother sound of *Judean* are my only reasons for preferring it to *Indian*, which seems to *hitch* a little. *Base* may mean ignorant, of course; but to a Christian (particularly to a convert—as Othello was) the

[70] John Maddison Morton's popular farce *Poor Pillicoddy* (1854).
[71] The letter is dated from Booth's reference to his wedding anniversary.

Judean was 'base,' as we understand the word. The "circumcised dog"—following the *Judean's* heels so closely leads me to infer that *Othello's* wits were a little mixed. (Were the Turks circumcised?) The *Herod* simile is to my mind more poetical (and quite as applicable) than the *Indian* one; both are more beautiful, however.

Tribe—whether taken as kindred or as a number, or nation, has no relation, that I can see, to either *cuss*—Jew or Injun. But use the word that you like best, & I'll repeat it.[72]

You have *Henry 8th*; I will send copies of *Petruchio* and *Don C. Ruy Blas* is such intolerable bosh that I feel ashamed to act it—yet it draws well, & seems as a relief now & then. I hesitate to send it [to] you. At first I studied the part from a printed book in blank verse; then I had a Ms. sent me & I blended the two. I became disgusted & for many years *shelved* it. In the meantime, however, Falconer[73] gave me a copy of his version— then I found a printed one in prose. I needed some short part for a *Matinée* performance, and set to work and concocted an arrangement whereby *Ruy & Don C.* were to be played by the same actor; for there is no earthly use for *Caesar*, except to show the likeness between him & *Ruy Blas*. Having cooked up this— hinging it here and there with a trifling plag[i]arism—I sent it to be printed. The proofs were never corrected, as you may see by this copy I have tried to work for you. Again becoming sick of the bungling bosh I destroyed all but a stray copy or two of the

---

[72] *Othello*, V,ii,420-422: "Of one, whose hand/ (Like the base Judean) threw a pearl away/ Richer than all his tribe." Booth, however, finally printed "Indian" in the Prompt Book. In a later letter to Winter dated Philadelphia, November 3, 1878 (Folger Shakespeare Library, Y.c. 215 [317]), he wrote:

> Since *Othello* has been acted from our book I am sorry I did not stick to my preference of *Judean*. Indian is no clearer & 'tis a *hitchy* word—less smooth in rhythm than *Judean*; Now we must say In-dian. Besides, I learn from my Shakespearian friends here that Knight is not always safe to follow—that *he* has changed some of his original views, &c.

Indian appears in F_2, F_3, F_4, and Qq. See *A New Variorium Edition of Shakespeare's Othello*, pp. 327 ff., for various readings of these lines including that of Lewis Theobald (1688-1744).

[73] Edmund Falconer, Irish actor and dramatist, best known for his numerous plays about Irish peasant life, adapted Hugo's *Ruy Blas* for Charles Albert Fechter who performed it at the Princess's Theatre in 1861.

worthless books, but in *Cal.*, I was induced to try the part—which I had not acted since the burning of Winter Garden. I found among my traps a copy of the play, & set to work cutting & slashing it; leaving out *Don Caesar* entirely, & this is the result (vide book of *Ruy Blas.*) If you can put it into decent shape let's use it—for it may sell at Matinées better than *Hamlet*; 'tis not much like Hugo—but no reference to the murdered author need be made. (I really am in doubt who was the original author—!)— No, not *Winter's Tale*; it is not in my repertory: I never heard of Brooke's *Timon*; I did not think it had been acted since Edmund Kean's time; was it ever done in America?[74]

Preston called here twice while I was at McEntee's; I sent him a "wash-down" for the *pill* you gave him; hope he is cured.

I horse-backed & row-boated for several summers at *L[ong] B[ranch]* & *C[os] C[ob]*, but derived no benefit from the exercise; I *rode* on land & water, and walked and drove 'till disgusted with my miserable fizzle. I halted permanently near a telegraph-pole by the roadside—*I ride no more*! I'd like to wander through Warwickshire, and loll in Lancashire, but to poke in the briny mere—not any, I thank you, dear. I'm a little off time there but the rhyme is not be sneezed at.

Guess I've gabbled eno' for Sunday, so bye-bye, I'll let you rest.

41. [New York] June 28 [1878]

First let me express my sincere regret for the "Thistle-down" business.[75] It is simply d--nable. You have my heartiest sympathy. The word 'heartiest' sounds *jolly* though—don't it? You know what I mean. I can't help you on that journey to Cyprus any

[74] *Timon of Athens* was rarely revived in the nineteenth century. Edmund Kean acted the role in 1816 and Samuel Phelps revived the play in 1851 and again in 1856 at Sadler's Wells, but I have found no evidence of the Irish tragedian Gustavus Vaughan Brooke ever playing Timon.

[75] Shortly after the publication of *Thistle-down*, Winter, according to Robert Young, received a letter of complaint from the attorney of an Esmeralda Boyle who had earlier published a book of poems entitled *Thistle-down* (Philadelphia, 1871). She claimed, of course, that Winter had stolen her title. Winter dismissed the matter saying he had never heard of Boyle or her book. She could not, at any rate, legally stop publication of Winter's book since it was published in London, and the matter was soon dropped.

further; some old Gloucester whaler (if there be any left) can give you all the *pints*. "Isaac" was the *front* name of the *Reed* you refer to.[76] I have not Payne's plays—but I send you his Life which may serve you.[77] The *prologue* in this work may be Croly's[78]—I don't know. The pages on costume you sent are from *Hope*,[79] it seems; I have Hope—but unfortunately, he is packed with other books, including *one* by Dana,[80] in boxes down stairs— my shelves being too few to hold all my books. I send you, how- ever, "Becker's Gallus" & "Carr's Antiquities"[81] which will give you—I think—all that Hope possesses. I prefer to let you read & select—for I am doing *nurse*-work now, besides—the Latin is too heavy for me. Please take good care of them; I vowed never to lend another book my library has been so scattered by this means. Fitz-Hugh Ludlow[82] & other friends raised Cain with my books. I like the suggestion John makes—also your ending of that parting scene; cut it, & so we need not discuss the *heard* & *served* question further. It is a better ending & I wonder it never oc-

[76] Isaac Reed was the English editor of the *First Variorum* of Shake- speare (1803).

[77] Probably Gabriel Harrison's, *The Life and Writings of John Howard Payne* (Albany, 1875).

[78] A prologue to *Brutus* was written by George Croly, Irish minister, author, and dramatic critic, and read by Henry Kemble at the first pro- duction of the play at Drury Lane on December 3, 1818.

[79] Thomas Hope, *Costume of the Ancients* (London, 1841). In the Appendix to the Prompt Book of *Brutus* Winter cites this book as his source for the essay on "Costume, Weapons, and Other Accessories for Brutus."

[80] Probably Richard Henry Dana, Sr. His *The Idle Man* (1823), or a later volume reprinting the same material entitled *Poems and Prose Writ- ings* (1833), contains an essay on "Kean's Acting." Dana gave a series of lectures on Shakespeare in New York from February 14 to March 9, 1849, but these were never published (see next letter).

[81] Wilhelm Adolf Becker's, *Gallus; or Roman scenes of the time of Augustus with notes and excurses illustrative of the manners and customs of the Romans*, first published in German in Leipzig in 1849 and subse- quently translated into English and published in London in 1866. It was republished in at least seven other English editions between 1866 and 1882.

Thomas Swinburne Carr's *A Manual of Roman Antiquities* (London, 1836).

[82] Fitz-Hugh Ludlow, an American writer, is best known for his auto- biographical novel *The Hasheesh Eater* (1856).

curred to me—for the lovers always bungle it.[83] You will see
that I have "thundered" throughout the *Macbeth* scene—as much
as I dare do.[84] I saw nothing to *axe* in the appendix to *Othello.*
Let Henry know where he can get the Cassell[85] copy of *Shylock*
& he will do so.

Yes, I sent the emblems to De Vinne—I told you so, I think,
in my New London letter. I think the one for *Macbeth* requires
something more than the *bat*—how would a pointed crown do—
[crown symbol]—just over or under the bat? As for the crescent—? it's a
mere fancy as is his costume & cimetar; I can't suggest anything
better; but I think the horns should be *up*, not *down* as in the
sketch; am I right? I really do not know anything about *numbers*
(I'm ill at 'em, as Hamlet says)—you must use your judgement
as to how many of each. No more tonight. Thanks for your good
advice about the papers; they have worried me considerably.

42. [New York] June 30, 1878

I hope I shall soon be rid of this *durned* tissue stuff[86]—don't
you? Your *P.S.* is on top, so I'll begin with that. "Passageroom"
is wrong[87]—evidently a Hintonism or some Prompter-ism. *Forres*
is the place of the King's residence, but as he first appears in
camp near *Forres*, I think it best for stage purposes to keep him
there; it saves a scene and whenever that can be done in a play so
crowded with shifting scenes the better 'tis for carpenter-work.
I thought of the "Thrice to thine" business, but forgot it again.
Put it thus: *They join hands, then turn and bow thrice,* or words
to that effect.[88] Banquo's *moon* has always induced me to omit the

[83] The reference is to V,i of the original play *Brutus* which was eliminated in the Prompt Book version. See Preface to the Prompt Book of
Brutus.

[84] The reference is to *Macbeth*, II,i (see next letter).

[85] *Cassell's Illustrated Shakespeare,* edited and annotated by Charles and
M. Cowden Clarke and illustrated by H. C. Selous (London, 1864).

[86] This and the preceding letter are written on a light weight onion skin
paper.

[87] Probably a direction for the locale of *Macbeth*, I,iv. Most editions,
based on F_1, locate I,iv at "Forres. The Palace." Booth, however, for
reasons which he states in this letter, located both I,ii and I,iv at "A Camp
near Forres."

[88] *Macbeth*, I,iii,35-36: "Thrice to thine, and thrice to mine,/ And thrice
again to make up nine." The direction as printed in the Prompt Book is as
follows: "Witches join hands, move round in circle, and bow thrice."

usual thunder and lightening business of that scene; however, I have—you will perceive—marked several directions for low & distant thunder. Too much of it would "take the horror from the time" &c.[89] If Lady M. does not mean her heart—what does she mean? here—in this place—or *now*? if the latter why not say so? I think it has reference solely to her feelings. Lady M. usually puts her hands to her two breasts—why? Genesee squaw [*Je ne sais quoi*] as the French say. Perhaps because she speaks immediately of her *milk*.[90] I also think Furness is right—I do not know if the reading is original with him. It is not possible for me to get at the Dana book without unpacking several large boxes stored in the cellar, and I doubt if 'tis the one you want. As I remember it contains poems, not lectures, with a few criticisms of *Kean's* and, if I mistake not, Cooke's—or Cooper's acting; could it be so far back? Henry says he cannot find *one* of the books you want—I think that was the reason he failed to bring it [to] you. As *Julius Caesar* is seldom acted—would it not be better to defer that, & get out all that are likely to be done first? the chances of sale are to be considered somewhat; but perhaps *all* can be ready by the time I begin—14th Octr.—? The *mistakes* in 'Hamlet' & 'Richelieu' should be corrected, cost or no cost. I cannot criticize, nor can I suggest more than I may have done already—if ever an *idea* does occur while reading your notes &c, I'll let you have it. I'm afraid I'm *out*—of *emblems*: how say you to a heart, or two, a *double one*, encircled by a crown of thorns, for 'Ruy Blas'? For the others—I am at a *non com*. I've seen, somewhere, a *cat* & *dog* for K.P.[91]—I'll *scratch* over it, but ask others, don't rely on me.

[89] *Macbeth*, II,i,1-2: Banquo asks, "How goes the night boy?" to which his son answers, "The moon is down; I have not heard the clock." At the end of Macbeth's "dagger" speech, Booth had "A low rumble of thunder." Throughout the remainder of the scene there are directions at specific lines for "distant thunder."

Macbeth, II,i,58-59: "The very stones prate of my whereabout/And take the present horror from the time."

[90] Lady Macbeth's "unsex me" speech, I,v,36-52. At the line "unsex me here" Booth wrote the direction, "Touching her heart." Ellen Tree crossed hands on her breast at "come to my woman's breast" and Madame Ristori clutched her breasts with her hands. See Arthur Colby Sprague, *Shakespeare and the Actors* (Cambridge, Mass., 1948), pp. 232-233.

[91] The emblem for *Katherine and Petruchio* was a yoke encircled by a wreath of roses.

I began to study 'Werner'[92] twenty-odd years agone—and at various epochs since have looked over the play, but failed to carry out the plan I had formed of acting it. I doubt if I shall ever do so — perhaps. An actor identifies with a certain set of characters [and] has little encouragement to depart from them: witness *Richard 2d.*! I like to act that character, but the play is *weak*, in the managerial sense, and I have to fight to have it done. If some odd mood should seize me during the next few months I may look again at 'Werner,' and will let you know how I feel about doing it. I'm afraid if Taylor[93] ever sees the book—that he[94]

43. [New York] July 1, 1878

I thought the 7 names were given in P[ayne]'s account of his 'Brutus.' All I can find regarding the other plays are *Downman, Cumberland, Lee,* and *Duncombe.* The others are not mentioned in the 'History of the Drama' &c, from which I have taken these four. The account is not very creditable to *Payne,* and I question if it would be worth while to use it; 'tis of no interest beyond the fact that he *took* much from each of the foregoing.[95] Regarding *our* version I told you that *John* [McCullough]

[92] Lord Byron's *Werner* (1823) was first acted by Macready at Drury Lane on December 15, 1830, and became part of his standard repertory.

[93] Tom Taylor, English dramatist, wrote *The Fool's Revenge* in 1859. Booth was concerned about Taylor's reaction to the cuts and alterations he had made in *The Fool's Revenge* for the Prompt Book version.

[94] The rest of this letter is missing.

[95] In his preface to *Brutus,* Payne wrote: "Seven plays upon the subject of Brutus are before the public. Only two have been thought capable of representation, and those two did not long retain possession of the stage. In the present play I have had no hesitation in adopting the conception and language of my predecessors, wherever they seemed likely to strengthen the plan which I had prescribed." Payne does not, however, identify the seven plays or their authors. Hugh Downman wrote *Lucius Junius Brutus* (1779) which was never performed. Richard Cumberland's play *The Sybil, or The Elder Brutus* was post-humously published as part of his complete works in 1813. It was never performed. Nathaniel Lee wrote *Lucius Junius Brutus* (1681), performed at the Duke's Theatre, 1681. William Duncombe wrote *Lucius Junius Brutus* (1734), performed at Drury Lane, 1734. Another source was Voltaire's *Brutus* (1730) and perhaps Alfieri's *The First Brutus.* The "seventh" play is obscure. See Arthur Hobson Quinn, *A History of the American Drama from the Beginning to the Civil War* (New York, 1923), pp. 171-174; also Grace Overmeyer, *America's First Hamlet* (New York, 1957), pp. 161-172.

suggested this arrangement which he said was made by *James Stark*—now dead; once a great actor (having *starred* some years ago in N.Y. & afterwards became famous in Australia & California—where he acquired & lost a large fortune) and finally one of my "sticks" at Booth's Theatre. He had served in the 'palmy days,' such men as *Cushman—Forrest—Macready* and my father in all the leading characters. *Joyce* was my *costumer*. Sich is fame! I have no old bills of *Brutus*, nor no copy of the play except the one I sent you, but Henry shall get you one of Hinton's if you will tell him whether you have one or not. I liked the emblem for Shylock, the *Fool's* cap didn't altogether suit me—what you say concerning it is just right, and I don't know but what you are correct about the *bond* & *scales*—as I recall them.

I have been looking over my costume books and find no particular mention of the Scotch dress of the period we are at. But the same style prevailed—all over Britain—Denmark and elsewhere for several centuries; therefore I dress *Hamlet, Macbeth* and *Lear* alike in general style—but of different colors & material; for *Lear* my dress is ruder than the others. For *M.* a simple tunic—reaching to the knees, with close-fitting sleeves, & trimmed with a "zig-zag" sort of pattern—the skirt of the tunic is opened a little at each side; the legs are bare & cross-gart[er]ed with leathern bands up to the knees; a short mantle, falling behind as low as the tunic—in front merely to the waist, like the Greek pallium; the hair long. Our modern trousers, or "truis" (I think it is spelled) were worn much earlier than the time of *Macbeth*. The female costume was similar except in length, of course. Chain armor-shields, battle-axes &c were their arms. This is about the gist of all that is spread over numerous pages on this subject. For the Kerne (yites) and Gallow (lager?) glasses[96] that supe-port me at *Dunsinane*—skins, clubs, shields of raw-hide, dirks et al, are used—I should say *were* used when I produced the play at Booth's long since; elsewhere my army is composed of *Romans, Greeks, Britons* and—*Faynians*.[97]

[96] *Macbeth*, I,ii,11-13: "The multiplying villanies of nature/ Do swarm upon him—from the western isles/ Of kerns and gallow glasses is supplied." A "kern" was a lightly armed Irish warrior; a "gallow glass" was heavily armed. See H. H. Furness, *A New Variorum Edition of Shakespeare's Macbeth* (Philadelphia, 1873), pp. 9-10.

[97] The Fenians were a semi-secret organization of Irish-Americans who advocated forceful separation of Ireland from England.

After tomorrow I hope to be free—my last tooth will be filled at *3:30* P.M.—*D.V.*, and if you'll send word when you can come be sure I shall be delighted to have you—often—here in my *loaferie*. I'm higher than the cedar's top: I dally with the wind (when there's a breeze) and scorn the roofs below me.[98] But we 'ave a helevator. When I get your reply to my last I'll write (or I won't) to *Harkin's*[99]—Good night.

Work on the Prompt Books continued through the summer and fall of 1878. On October 14, Booth began his theatrical season with a three-week engagement at the Broad Street Theatre in Philadelphia. This theatre was apparently owned or leased by John Sleeper Clarke and managed by Charles K. Burns. John T. Ford was also involved, probably as business manager, and was busy "puffing" the engagement, much to Booth's annoyance.

44.   St. George Hotel [Philadelphia], October 16, 1878

Since my performance of Monday night I have been in a state of utter helplessness—from nervous exhaustion; could do nothing all day yesterday, but lolled about—entirely "used up". I am now recovering, and do this on an empty stomach, waiting for my breakfast. I opened to a full & fashionable house & requested Chas. Ford to telegraph you, as you wished me to do, I had no other means than that of the manager's son, but I presume he neglected the matter. Last night the house was still greater— the largest by *$200* ever in the place. It's a very nice theatre & the location seems to be a favorable one; I shouldn't wonder if Clarke had again made a lucky stroke, it certainly was a bold one. The books, *100* each, arrived safely & are selling pretty well, so Burns tells me; don't know just how well. I presume *Macbeth* will arrive today—hope so—it was not here last night. Ford studies to annoy me. The shop windows are full of the d---dest caricatures—and the lesser papers have been (for a week & longer) puffing me with old articles revamped. Ford knows how I despise this thing, & yet he will persist. He takes credit for filling the houses by this d----d nigger-business. I hope I shall never act with

[98] *Richard III*, I,iii,264-265: "Our aery buideth in the cedar's top,/ And dallies with the wind and scorns the sun."

[99] Daniel H. Harkins, American actor and stage manager, was at this time the co-manager with Stephen Fiske of the Fifth Avenue Theatre where Booth was scheduled to appear in the fall, 1878.

him again. One of those "new idiots" you speak of says this is my first engt. here in five years, & that my present text of *Hamlet* is not the old version, but follows Shakespeare! Apropos of Holbein's Henry 8th—(which I hardly like for our purpose) we might have had *Richelieu's* and *Brutus'*—I think the block & axe, or some such significant affair better than a portrait.

Have just read your article on Louis XI. You are right—and for the very reasons you give I have dropped Sir Giles, Pescara & others—with a strong inclination to do the same by *Bertuccio*. I never acted *Louis*, though urged by Boucicault and others to do so.[100]

I am all right when in harness under the glare of gas—but during the day I feel weary & depressed; perhaps I'll brighten up when I get used to work. My long rest has spoiled me.

Will write you again in a few days.

[P.S.] Am glad the *Macbeth* article is found.[101] I really know nothing of *Ruy Blas*; hardly the silly stuff I have to recite in that character. Can't help you in the least—not even in costume, since my books are not with me; I do not adhere to *facts* in that particular in *Ruy Blas*. Perhaps the less said the better about it—as also with *Petruchio*—eh? Just *fix* the thing as easily as you can—'tis not of sufficient importance to work over—yet it may *sell* better than better books, I mean books of better plays. It is

[100] Jean François Casimir Delavigne's *Louis XI* (1832) was adapted by Dion Boucicault especially for Charles Kean, who acted it for the first time at the Princess's Theatre on January 13, 1849. The article referred to is Winter's review in the *Tribune* (October 16) of John W. Albaugh's performance of *Louis XI* in an adaptation by T. B. DeWalden at the Broadway Theatre on October 14. Winter wrote that "no permanent good can flow" from a play in which such a totally unsympathetic villain is portrayed: "No impression that is left by it can be rated as in any way more valuable than that which is given by a scorpion in a bottle of spirit." He then reiterated one of his major critical tenets: "The illustration or conveyance of a really great or lovely passage in human experience, real or ideal, is an affair of public importance—by virtue of the ennobling and salutary influences that proceed from such a work: and herein resides the value of acting. *Louis XI* on the contrary freezes feeling and limits mental vision to the mere analysis of technical skill."

[101] An earlier letter (Folger Shakespeare Library, Y.c. 215 [306]) indicates that the article referred to is Thomas DeQuincey's "The Knocking at the Gate in Macbeth," which was subsequently printed in the Appendix to the Prompt Book of *Macbeth*.

usually styled a drama—I don't know indeed what it is. Cassilda[102] is to the Queen what Nerissa is to Portia, I should say. As for the text—"any change will better its condition."

Booth's health did not improve during his engagement in Philadelphia.

45. Saint George Hotel, Philadelphia, November 6, 1878

Please forgive the pencil—I can't find either ink or pen fit to write with.

I send herewith the book returns. Keep 'em with the other accounts you have. I was in hopes that the unsold copies were now in New York, but I find they do not start until today. My illness has kept me indoors & my directions to send them on Monday last have been neglected.

I am sure that our books would have gone like 'hot cakes' (and may yet) if put for sale at all the stationary stands & shops about the country—say at 30 cts. just as I buy the handy volumes of essays, &c. We must make an effort to distribute them in that way.

My physician has made a thorough examination of my 'parts' and finds me organically sound—my stomach being the only functionary that is deranged, & that—oh Lord!—deprives me of mental peace. I must forgo the delights of life—one of the principal ones, I mean;—*wittles*. I may revel no more in my cordial curry—my comfortable coffee; spices and peppers and pickles—all such palatable kick-shaws, henceforth must be eschewed not chewed—nor even tasted by my blasé bowels. Do not *yourn* yearn for me? Like old 'Justice Greedy' I shall henceforth & for aye hereafter endure the agony of unglutted guts (Forgive my antiquated Saxon!)[103]—Farewell, a long farewell to all my gravies; my pies and *poodinks*; my turtle mock; my steak & inyuns, and such soul stirring et ceteras,—*Farwale!* I've eated the last of all my dinners.[104] Thin soups and lady-food must

[102] Cassilda is the confidante of Doña Maria, the Queen of Spain in Hugo's play *Ruy Blas*.

[103] "Justice Greedy" is a character in *A New Way to Pay Old Debts* whose single-minded devotion to food, its preparation, and especially its eating makes him an easy dupe for Sir Giles Overreach's schemes. Greedy constantly complains of his "empty guts."

[104] *Othello*, III,iii,403-413.

now *content* me. Mockery! Can a belly full of panada inspire the tragic muse? Fiddle! I must purge and leave such—simply to coddle that sick devil that croaks in Tom's belly; but not for a white herring.[105] By-the-by, that's the name of my doctor's father-in-law. He is Hanneman's [sic] successor—very old & good, à la vin, as I'm told. The son—Dr. Knerr—is Furness' physician—he sent him to me in my distress and I am satisfied (though you do turn up your proboscis at Homeopathics)—I am confident that he is on the right track & has got my disease—by the belly—well in hand.[106]

I fear the theatrical outlook for this season is gloomy; the fogs do not seem to rise in my quarter; 'tis what old salts term 'dirty weather' at every point. My kids are well thank the Lord! 'tis only the old he-goat, Me, that "grunts & sweats under a weary" ache.[107] I hope all yours & you are well & happy in the 'new birth.'

I start for home tomorrow noon. I dread my five weeks pull at the Fifth Avenue.

After a week of rest, Booth opened at the Fifth Avenue Theatre on November 11 for an engagement of five weeks through December 14. During this engagement, as in Philadelphia, he complained constantly to Winter of stomach aches, dyspepsia, diarrhea, even "yaller" tongue. On December 2, he wrote: "I was very *achey* both nights of Lear— nor did I feel quite well during the last scene of Shylock; on Saturday I griped through both performances" (Folger Shakespeare Library, Y.c. 215 [322]). His poor health undoubtedly affected the quality of his performances. The critic of the *Sun* complained on November 12 that Booth's impersonation of Hamlet "was not quite up to his best endeavors." The critic of the *Telegram* wrote that his performance of Richelieu on November 18 was "somewhat stiff and mechanical." Generally, however, Booth's performances were praised. The critic of the *Times* (November 12) wrote that as Hamlet "never, perhaps, was he seen to better advantage" and he also noted that many of Booth's

[105] *King Lear*, III,vi,32-33: "Hopdance cries in Tom's belly for two white herring."

[106] Constantine Hering, German-American physician, homeopath, and founder in 1867 of the Hahnemann Medical College in Philadelphia, named after the founder of homeopathy, Samuel Christian Friedrich Hahnemann. Calvin B. Knerr, American physician and homeopath, was Hering's associate and son-in-law.

[107] *Hamlet*, III,i,77: "Grunt and sweat under a weary life."

"most obtrusive mannerisms were toned down." The critic of the *Post* (November 12) wrote that Booth "probably never acted the part [of Hamlet] better." During this engagement, Booth and Winter completed the last of fourteen projected Prompt Books, and two weeks before his scheduled engagement at McVicker's Theatre in Chicago beginning April 7, 1879, copies of these were forwarded to his father-in-law.

About the time of Booth's arrival in Chicago, Winter bought the house on Fort Hill, New Brighton, Staten Island which was to become his home for the rest of his life. He did not, however, have the entire down-payment and he asked Booth to advance him the difference. With his usual generosity, Booth responded on April 14 with a check for $1000.[108]

46. Toronto, Canada, April 5, 1879

Dear Edwin:

I have closed the agreement as to the Staten Island house and I stand committed to buy it on or before the first of May. The owner, who is an old acquaintance, and friendly towards me, will make the terms easy. The price is $3500. I am to assume a mortgage of $1,500, at 7 percent interest which has two years yet to run, and is payable at 6 months notice. This is held by the Staten Island Savings Bank. I am to pay $2000 or as much of it as I can. I have a little money saved for this purpose, which with what you are willing to advance will enable me to complete the transaction.

I suppose you will let me have the money without interest. It will take about two years for me to pay it back to you—which I shall consider it a most sacred obligation to do. The mortgage can doubtless be extended should occasion require its payment to be deferred. The interest, taxes, and insurance, annually, amount to only about $140. I shall be enabled to save a considerable sum in two years, and within four years, shall have paid the whole. I am buying the place in my wife's name, and it will be settled on her and then on the children. I am quite satisfied that the investment is good. There are two and a half lots of land, valued at $2,500. The house is listed for $2,500 which is less than its value: and real estate must rise in value, as the time improves. The house is strongly built, and the older part of it is only 30

[108] The cancelled check is in the collection of Robert Young.

years old. The situation is good and the place is healthy. The owner undertakes to put it in perfect repairs and we are to take possession on May 1st, or sooner. I came here March 24th, and have been busily making arrangements for the removal. I shall probably return to New York, so as to be there on the 12th to look after the Easter theatricals, but this is not certain. Anyway, should you write, you might address me here till the 11th inst. I shall write again to apprize you of my departure from this place. I shall still, for the present, keep my lodging, at 221 East 18th St. N.Y. Percy is still there, but will come here, probably, on April 17th, to aid in the removal. I have many schemes of literary labor afoot for this summer to be carried out when I get settled in the new place. I hope all will go well with your season in Chicago. Has the new Richard arrived? The Boar's head is only to be had on the first hundred.[109] Twas the best we could get in, but a better one will replace it on the others. Should you want more of any of the books, please telegraph Lee & Shepard, 41 Franklin St. Boston, and they will send them. I should be glad to have a paper or two, with the notice of your first night in Chicago. Remember me kindly to Mrs. Booth and Edwina. All here are well. I have been reading Moore, whose centenary approaches (May 25th) and have re-read Bulwer's "Zanoni" and "Godolphin" and have written for the paper. Those are fine novels, by the way and in both there is talk of the stage and the principles of art. Take good care of your health, dear Edwin, and be all yourself. Those who love you wish for no more than that.

Ever yours
William Winter

During the third week of his Chicago engagement, on the night of April 23 during a performance of *Richard II*, a man by the name of Mark Gray attempted to assassinate Booth.

47. Chicago, April 27, 1879

Thanks for your good letter. I can tell no more than what the papers have told about this madman. From what has been extorted by reporters from him it appears he is a crushed genius

---

[109] The symbol for the Prompt Book of *Richard III* was originally a boar's head, but this was subsequently changed to two intertwining roses.

and because he never knew his father and looks about as much like a Booth as any dark eyed, black haired man might look, he imagines himself to be my offspring. His dates & mind don't tally, however, for at the time of his construction I was at the other end o' the world. I had first suspected it had something to do with Stuart's old plot against me in 1869, but it has not. The fellow is a lunatic, & a dangerous one. Have a letter from him today (from the jail) offering to let "the matter drop for $900, or he will follow me—*'till I dy*." I hope he will be sent for life to an Asylum; the case comes up this week.

Think of the horrible position I was in—made up for *Richard 2d.* in the dungeon scene, meditating upon the miseries of life with this fellow "blazing away" at me![110] My wife was prostrated by the shock! and though I preserved my self-control, for her and audience's sake, it oozed out at my toe tips in a day or two after, and all my thoughts have been a nervous yet fervent "Thank God!" ever since. What a series of narrow escapes I've had! and how tragical & sorrowful have been the events of my life. Think of my poor old mother, with all the horrible Past recalled so vividly by this newly averted horror! We are all well now, thank God! but not until we know that this maniac is secured for life can we feel safe.

I am sorry to find the cheap little second hand souls, are making light of it—as an advertising dodge; Robson & Crane[111] were about the first to utter such filth. Of course the moral poison this will exude will be nectar to Cauzauron, Byrne [sic],[112] Stuart & such like, but I am too profoundly grateful to God to care one straw for what such vermin may vomit forth. After the great

[110] The Chicago *Tribune* (April 24) reported that the first shot was fired at the line, "No thought is contented" (*Richard II*, V,v,11). On April 24 and 25, it reported in detail the incident and Mark Gray's—or Walter Lyon's (which seems to have been his real name)—statements and biography.

[111] Stuart Robson, and William H. Crane, popular American comedians, were engaged at Hooley's Theatre in Chicago, from April 14 through April 26, where they appeared as the two Dromio's in *A Comedy of Errors* to capacity audiences.

[112] Charles Alfred Byrnes, American drama critic, founded the *Dramatic News* in 1875. At various times he was the drama critic for *Truth*, the *Evening Standard*, and the *Morning Journal*.

mental strain relaxed a sort of nervous hilarity set in & I feel in the "well-said, old mole" mood.[113] I always appreciated (or fancied I did) the exact condition of *Hamlet's* mind after the ghost leaves him, but never so forcibly as now. I feel "the horrid joy men feel who scan with dizzy eye the height from which they leaped and live by miracle" (or words to that effect). My business has been injured by the author's carnival,[114] the first indifferent engagement I've ever played here, and yet the papers constantly speak of it as a natural weariness the public has for my plays (if not *me*). The last time I was here (as on all former occasions) my houses were crowded, and I have no doubt they would have been so now but for the fashionable cheap show at the Exposition Building.

I am writing in great haste (having to "tea" out this evening) and without reference to your letter. My next shall be an answer to any questions yours may contain.

I was obliged to leave you abruptly at this point and have just returned from my visit to Runion [sic], of the Tribune, he's an excellent fellow, and an old playmate of my wife in childhood.[115] He is to go with me tomorrow to see the State's Atty., Mills,[116] in relation to my affair.

Good night & kind regards to Mrs. W.

[113] *Hamlet*, I,v,162: "Well said, ole mole! canst work i' the earth so fast."

[114] The "Author's Carnival" was an annual charity extravaganza held in Chicago. It consisted mainly of a series of booths containing living tableaux of various scenes from literature. Scenes from *Hamlet, Othello, Ivanhoe, Faust, The Lady of the Lake*, and others were depicted, each dressed appropriately and with the appropriate music and sometimes dancing. Other booths were outfitted for games such as the "fish pond," the "rifle range," the "post office," and the "telegraph station." There were also numerous recitations, concerts, and parades. The 1879 carnival opened on April 15 and ran for two weeks through April 30.

The *Tribune* (April 29) reported that Booth's business was "light" and on May 1 wrote that "business during the preceding two weeks has been very poor, attributed by many to the draft of the carnival among the class who generally patronize Mr. Booth."

[115] James B. Runnion, reporter and the managing editor in 1873 of the Chicago *Tribune*, was a cousin of Mary McVicker.

[116] Luther Laflin Mills was the State Attorney of Illinois from 1876 to 1884.

[P.S.] Did you see what Harkins said of his intimacy with me, &c? He meant well, no doubt, but the aforesaid was longer ago than I can recollect. Ben Baker's report also was kindly intended, but not altogether "so."

I received your former letter & will answer when my brain & nerves are more at rest.

Upon receipt of Booth's letter, Winter penned the following reply.

48.   Fort Hill, New Brighton, Staten Island, May 3, 1879
Dear Edwin,

I received your letter of April 27th and am much relieved and rejoiced to see that your recent most trying and terrible experience has not shaken you in composure and purpose, nor injured your health. It was a dreadful trial. I have pictured it all to myself, many times: the prison scene, the dark cell, the moonlight, the distressed and darkly moralizing King, the deep wise words of the poet, the crowded, hushed theatre and then—the assassin at his fearful work! I don't wonder your wife was overwhelmed with the horror of such an experience. The man is, undoubtedly, mad—but he is dangerous, and should be confined for life. I have often mused on the strange dark vicissitudes of your life, as they are known to me. You have certainly passed through a very tragic and mournful ordeal. The wonder is that it has not embittered you. God grant that the worst of all trials is over now, and that the rest of the way will be smooth and bright. Remember me, with sympathy and kind regards, to Mrs. Booth and Edwina, and send me a line when you can to say how you are, and how the affair is settled. I saw McVicker last Wednesday night, at the Olympic, but we only exchanged a few words, about the play (of "The Assommoir"— very stupid and coarse),[117] as we met in the course of a wait, and I was pre-occupied with thinking what to write about the drama. Mac looked very well and was hearty and genial as ever.—I have taken some of the necessary steps to vest in you all the property rights in Prompt Books. The papers are all in readiness to deliver into your hands. The bill of sale of all my

[117] Augustin Daly's adaptation of Zola's *L'Assomoir* ran from April 30 to May 17, 1879, at the Olympic Theatre.

interest in the books will be dated April 25th. Notice of the assignment of the copyright has to be *recorded* in the office of the Librarian of Congress. I have given notice to Messrs. Lee and Shepard & Mr. Dillingham and shall give the same notice to Messrs. Hart & Co. (Theodore L. DeVinne is the head of the firm). I have renewed the policy of *Insurance* on the plates & caused it to be transferred to you. The amount is *$800.* My last account from L. & S. shows the number of the books on hand, in stock, exclusive of those in Chicago. The file of Hart & Co., for the new edition of Richard III has not yet been received. It will come a little later.

You will, of course, determine at your leisure, whether to go on with Lee & Shepard. They have 3749 vols. on hand, and, if sold, will pay over to you $937.25. Perhaps, the most immediate way to recover the original investment would be to sell the plates & copyright to some theatrical publisher, at about the cost— after a year or two of sales of the book new editions will not cost much. The plates, boxes, & all, at 63 Murray St. are yours, and subject to your order.—You will observe that the commission accounts of "The Trip to England" is stated with that of the Prompt Books in documents number 3 & 4.

The Prompt Books have, with L & S, done *nothing* this last quarter. The total received by me, from them, thus far, is $227.99 —to which is to be added what I received from the first *theatrical* sales, at Booth's Theatre, and the advance of one hundred from you. Altogether, in the present sale, &c.,—I have received about fifteen hundred dollars from the enterprise; and with this I am well satisfied. I shall improve every opportunity to push and help the sales of the books. L & S. will communicate with you, and send notices &c. My correspondence has been with Lee (William) and you will find him a good man—I have endorsed on assignment an [illegible word] certificate of copyright, and have written to A. R. Spofford, the Librarian, about the same.

I have bought the house here, paying $1500 down, & have to pay $2000 within one year. So you see I must work hard. Today I am almost disabled with muscular rheumatism and I can hardly write but I wanted to let you know how affairs are going on.

With love yours ever,
William Winter

[P.S.] I did not see that Harkin's talk nor that of Ben Baker. What was it? I have often heard Mr. Runnion—changes *pleasantly*.

Did you see my Stratford paper, in Harper's for May?

Booth ended his Chicago engagement on May 10 and immediately returned to New York for the summer. Here and later at Saratoga he recuperated from the chronic stomach ailments which had been annoying him since November. He was also suffering from another curious malady. A black fungus had appeared on his tongue. At first he thought the spot was cancerous, but Dr. Ghislani Durant, a New York cancer specialist, assured him it was not and began a painstaking but successful treatment. About July 5, Booth felt well enough to make the trip to Saratoga.

Chapter IV

American Tours

1879-1880

FOR several years, Booth's friends, among them Winter, had been urging him to visit England again. Booth was tempted, but because of Henry Irving's rise to power at the Lyceum Theatre he was hesitant. Booth did not want to be set up as the "leading American tragedian" in rivalry with the "leading English tragedian." He undoubtedly realized from such rivalries as that of his father and Edmund Kean, and of Forrest and Macready, that rivalries could in no way be profitable to either party or to the art of theatre. Thus Booth declined all offers to act in England. On February 3, 1878, he revealed in a letter to Winter (Folger Shakespeare Library, Y.c. 215 [329]) that he had declined an invitation to assist at the opening of the Stratford Memorial Theatre by acting Hamlet in a sort of tournament or jubilee with Irving and Barry Sullivan. "I am silly, no doubt," he wrote, "in throwing away such a splendid opportunity—finer than any actor has had offered, but I shall not go!" Less than a month later, however, this attitude towards an English engagement had changed.

This change was caused by rumors that Irving was planning an American tour for the next season. With Irving absent from London a rivalry would be eliminated and, not incidentally, the chances of Booth's success in London would be improved. Booth even hoped to effect something like an international theatrical exchange. On March 28, 1879, he wrote Irving to propose that during Irving's visit to America, he go to the Lyceum and perform "for a stated number of weeks and on such conditions as shall be mutually advantageous."[1] Booth acknowledged that he had no theatre to exchange for the Lyceum, but he was willing to do whatever he could to assure Irving's success in America. Booth told Irving that he would much rather avail himself of the advantages of Irving's company and scenery than "to risk the danger of having a 'scratch' company with its attendant cares." He advised Irving to appear at Booth's Theatre which, though no longer his, was still the finest theatre in New York City. To assure himself that Irving would not misinterpret his intentions, Booth asked

[1] Booth's letter to Irving is in the collection of the Walter Hampden Memorial Library. It is printed in Lockridge, pp. 246-248.

E. C. Stedman to write to his friend G. W. Smalley, who was on good terms with Irving, and direct him to assure Irving "that whatever Mr. B. shall write may be taken without reserve as the fresh and trustworthy expression of an honorable man."[2]

Booth kept his letter to Irving a secret between himself and Stedman. He took the position that he wanted to go to England as a tourist and would act there only with the greatest reluctance. He of course had to protect himself in case his negotiations failed. This is the tone of a letter to Winter in which he first hinted that a London engagement the next season was a possibility.

49. Saratoga, July 20, 1879

The delicious *laze* which over-comes one here renders the least exertion a labor to be deferred until "tomorrow, and tomorrow and &c"[3]—hence my neglect of you.

"Thou but rememberest me of my own conception"[4] by your remarks on my long contemplated visit to England. I have of late been encouraged by many—friends and partial strangers to go, not only to England & Germany, but—as you suggest—also to France. Since the Bernhardt furor in London there's no reason why an English actor should not risk a chance in Paris,—so Frenchmen tell me.[5] But, of course, Irving will have the advantage, being so near, so well known and having already had a

---

[2] Edmund Clarence Stedman, American poet, critic, editor, and stockbroker, was a close friend of both Booth and Winter. Stedman's literary interests had gained him many English friends and acquaintances.

George Washburn Smalley, American journalist, was the London correspondent for the New York *Tribune* from 1867 to 1890. His opinions of literary, theatrical, musical and social London were printed as "Letters from London" in the columns of the *Tribune* and were among the very few "signed" columns appearing in any American newspapers of the time. Some of his letters were collected and published as *London Letters and Some Others* (New York, 1890). Stedman's letter to Smalley is in the collection of the Walter Hampden Memorial Library. It is printed in Lockridge, pp. 244-246.

[3] *Macbeth*, V,v,19-21: "Tomorrow, and tomorrow, and tomorrow/ Creeps in this petty pace from day to day/ To the last syllable of recorded time."

[4] *King Lear*, I,iv,72.

[5] Sarah Bernhardt made her first London appearance with the Comédie Française which played at the Gaiety Theatre from June 2 through July 14, 1879.

French critic's favorable mention. I am glad that Bernhardt is received so well—socially, as well as professionally, but I wish she were other than she is (on the first count), and also that she were less of a charlatan—for despite her genius there's much Barnum about her.

I presume that Clarke is coming over; I wish he would come to stay and bring his family. Much as he would like to manage me in England (and I know he'd do his utmost in my behalf) I think it wd. be safer to deal with strangers in such business. Friendship & trade don't *gee* well—as both he & I know to our regret. I wish he had a small theatre of his own in N.Y.—with his two in Phila. (but *none* in England) he would do splendidly.[6]—

I hope you've had, or are having a good time with Joe & the trout. Here we have "hops" and races—the latter I attend occasionally, the former rarely; at both I am but a looker on. John McCullough is expected here; his agent, Conner,[7] looked for him last night—I have had the *mobled* line altered, and when the new (old) word, *inobled*, is used I want you to note it, and give my reasons for its use. I'm convinced that Shakespeare meant it.[8]

Apropos of the books—I wonder (if I ever go abroad) how I shall manage to take them with me—free of duty? They are all reprints of English works, of course, and would be taken there for sale. I doubt if it would be wise to risk the expense.

My ambition is so spasmodic and so feeble that when I think of Europe 'tis more as a tourist than as an actor that I contemplate the trip, and as such there is no need of haste—for my daughter has yet a year of schooling before her, & my wife is in constant dread of the journey; but—I suppose I *must* go, and go as a *mummer* too, some day.

Could we all see England with your eyes, we'd jump the life

[6] Clarke managed at this time the Holborn and Haymarket Theatres in London and the Broad Street and Walnut Street Theatres in Philadelphia.

[7] William M. Conner became McCullough's manager in the spring, 1878. According to Winter, Conner proved "an excellent manager and a wise friend." See *The Wallet of Time*, I, 253.

[8] Hamlet, II,ii,480-482. F_1 reads *inobled*, but other folios and early editors read *mobled*. H. H. Furness, *A New Variorum Edition of Shakespeare's Hamlet* (Philadelphia, 1877), p. 190. Booth gives some of the reasons for this change in the next letter. See also Shattuck, *The Hamlet of Edwin Booth*, pp. 174-175.

to come"[9]—"enjoy the present hour nor fear the last."[10] Wife and daughter send kind remembrances to you and yours as doth also yours always.

One week later, Booth disclosed to Winter that he had written to Irving four months ago and that Irving did not have the "common civility to acknowledge it!" This indignity was compounded when Booth discovered that, despite his request to the contrary, Irving had introduced an outside party—namely Augustin Daly—into their negotiations.

50. Saratoga, July 27, 1879

Charley Pope[11] is a man o' gab; I've know him these many years—in old California days. He can "talk high" and has impressed you with a golden opinion. He's a good enough fellow and I believe deserves success. Now, *mark!* So soon as *Mark Gray*, of St. Louis, made an ass (--assin) of himself the St. Louis managers made a desperate effort to get me there to "show off". Pope had no place for me at that time, but he wrote at once & asked me to open his new theatre next fall. I replied immediately and told him I could not do so. Despite of this he announced me as his opening attraction—so I am told by another manager, who felt hurt by my declining his offer (far better than Pope's), and asked if the advertisement was true—if not, would I not go to the old house. I again declined & at the same time assured him I shd. not act in St. Louis next season. I have arrived at that point where I find rest & a relief from travel are of more value to me than dollars—while I can earn sufficient for family requirements by

[9] *Macbeth*, I,vii,4-7: "That but this blow/ Might be the be-all and the end-all here,/ But here upon this bank and shoal of time,/ We'ld jump the life to come."

[10] Martial, *Epigrams*, x, 47, trans. Abraham Cowley (1668): "Enjoy the present hour, be thankful for the past,/ And neither fear nor wish th' approaches of the last." See *The Complete Works in Verse and Prose of Abraham Cowley*, ed. A. B. Grosart (Edinburgh, 1881), ii, 340.

[11] Charles R. Pope, American actor and theatre manager, was at various times the manager of theatres in Indianapolis, Kansas City, New Orleans, and St. Louis in the 1860's and 1870's. In the 1850's he toured California where Booth undoubtedly became acquainted with him. In 1879 he built Pope's Theatre in St. Louis, which he successfully managed until his retirement in 1888.

less wear & tear. I wish Pope the fullest success, but for the present I cannot aid him in accomplishing it.

And now for a *secret* (which I still wish guarded; for a few months, at least):—just four months ago I wrote a long and courteous letter to Mr. Irving, which he duly received but which he has not had the common civility to acknowledge. In it I proposed that I should fill certain weeks (to be named hereafter) at his theatre—*on his own terms*, regulated by his expenses & sense of equity, and advised him concerning his contemplated visit to America. Stedman was arranging to go to England & asked if he could do ought for me—advising me, as you & other friends do, to act in London. I then shewed him my letter which he approved of as a very generous and manly proposition, &c. I requested him to see Irving & he suggested a talk with Smalley, a friend of Irving—whom, doubtless, you know. My letter sailed for "England, ho!" on the 30th of last March. I requested Mr. Irving to keep the matter between ourselves until we had arrived at some conclusion, and above all I desired that no managers (hosts of whom were soliciting me to go) should get wind of what I suggested. In the course of a few weeks McVicker recd. a note from Daly (just after his return from Engd.) asking if I would entertain a project he had to propose relative to an exchange of Mr. Irving & myself. *Mac* sent him to me; I declined an interview & in reply to his note on the subject told him that Mr. Irving was fully aware of my wishes, and that I was not prepared to negociate. This brought an indignant reply from Daly— saying he was sure that Mr. Irving knew nothing of my views on the subject when he (Daly) left England.

A few days ago I recd. a letter from Stedman saying he had seen both Smalley & Irving; that the former said "Irving never answers letters," and the latter gave as his reason for not writing —Mr. Daly had undertaken to convey his reply, and to transact (or wished to do so) whatever business there might be in the matter (this is but the pith of it, not the exact way in which Stedman quotes Irving's words). Mr. Daly *knows* that Irving was ignorant of my views; Mr. Irving (notwithstanding my wish for privacy) leaves the matter in Mr. Daly's hands—as an answer to a very civil request from one who is his equal in position. To Stedman he showed the warmest desire to accommodate me; offering either to let me have the control of the theatre, while

he came to America, or—if he should not come—to manage it with me, on shares. All this is very generous—*in words*, but why not write it? Why offer the *boorish* excuse *I never answer letters*? I would to God I could recall the one I wrote to him, for it lowers me in my own esteem to be cooling my heels at the backdoor of Mr. Irving's playhouse, while other managers offer chances for far greater gain than he could afford to give me. There the matter rests. I have dropped him—unless his Lordship can humble himself to recognize a poor *muff* like me.[12] In the course of a few months I shall determine what to do—whether or not to go, and if I go whether to act or *loaf*. Don't speak of it yet— I'll tell you *when* the cat may jump. I cannot agree with you regarding Strakosh (I can't spell his name)[13]—he, like Pope, is a man of talk—"words, words, words." "Squab," I know nothing of.[14] I believe (I know not why though) that the English "Jarrett" (not our Black Crook one)[15] would be as good as any I could get to attend to my affairs abroad; he is to have Bernhardt, I believe. So much for England: the ice is broken.

I sent directions to Hart & the change in *Hamlet* is made. I merely meant that should the change be commented on, which I doubt—for nothing that I have ever done has been regarded as original with me,—if the word should be noted I'd like it to be known why I made the change. *Ignobled* (or *inobled*) expresses both the condition of *Hecuba* in her fallen state, and the Queen of Denmark's degradation. The word attracts Hamlet's notice as being applicable to his mother, and suggests the melancholy reflection wh. you noticed in my delivery of the line. Your note set me to thinking, & when I discovered the *true* word in the folio I saw at last some good reason for Hamlet's interruption of the player's speech. Being, doubtless, an unusual form of the word *ignobled* it receives the approval of Polonious [sic]: mobled

[12] Stedman's letter to Booth is at the Walter Hampden Memorial Library. It is printed in Lockridge, pp. 252-253.

[13] Probably Max Strakosch who was involved in the management of various theatrical and musical enterprises in the late nineteenth century. He was the brother of the Hungarian born pianist and opera impresario Maurice Strakosch.

[14] Probably Frederick A. Schwab of preceding letter 33.

[15] Edward Jarrett, English theatrical agent, was probably best known for his successful management of many of Sarah Bernhardt's world tours, including those of America.

was in common use and would have called forth no comment whatever. Why should *Hamlet* have caught at the Queen's *mob-cap*; it always strikes me as a silly, and a useless interruption; he might with a shrewder meaning have caught at the blanket about her loins; 'twould have been more in keeping with *Hamlet's* sarcastic treatment of his mother & his uncle.

If I am at all clear on this point use it—I'll be befogged presently, and had best stop just here. Adoo!

Winter apparently persuaded Booth that Irving meant no personal insult in ignoring his letter or in deferring the matter to his agents and that he should not "drop" the English tragedian. Booth was mollified, but not completely reconciled. He determined to seek "no further favor" from Irving.

51. Saratoga, August 9, 1879

I usually defer my correspondence until Sunday—therefore you have recd. no acknowledgement of your New London letter about the Irving matter; today I got your "8th." As my doctor is here and will doubtless *absorb* me tomorrow I steal this occasion (while my 'gals' are at the 'hop') to scratch a reply to both your epistles.—I acquit Mr. Irving of all professional pique or jealousy in the premises. I do not doubt the sincerity of his professions made to Stedman—who I may as well say at once, *has* my authority to negociate all *preliminary* arrangements tending towards an engagement with Irving & myself,—I have confidence in the man's generous impulses, but nothing (to my thinking) can excuse the indignity his neglect of my letters has put upon me. I have no sympathy with *mooney* actors who live in clouds. There is no excuse he can frame to palliate such a discourtesy. However, I am not angry any longer; but I shall certainly seek no further favor at his hands. I am sorry Stedman has spoken of the affair, but it appears both Irving & Smalley mentioned it to Aldrich—in an indirect way—some months ago, shortly after they rcd. Stedman's & my letters. I should not like to have anything said about it until I see Stedman: by no means would I have you, or anyone write Irving on the subject. It will shape itself, no doubt, and you shall have the first knowledge of my English affairs; at present I feel very little like going at all, but after my doctor has "set me up", which he vows he will do by the end of September, I may feel more ambitious.

Yes, I saw by the Herald, that I was at church looking "old & careworn"—while I thought myself "young & charming" here. I see the Tribune every day—and recd. your copy of it. You handled the Fetcher business splendidly, but—my dear Will— how about Palmer![16] Speak well of the dead &c—that's all right, but I'm afraid you laid on the *bonum* pretty thick when you treated the managerial part of him. I'm going to criticize you— hem! Why, my boy, of all the men who have striven to debase the actor's calling, and to make a brothel of the theatre Palmer & Jarrett bore away the bell! I aimed at just such speculating *showmen* in my "holy epistle"[17] which called forth the fury of their class upon me when it appeared in print. All that such fellows do in the *theatrical* way is but for their own sensual gratification. They all are free to extravagance—both with their money and their amours. Not a thought for the drama have they beyond dollars & debauchery. I knew very little of him as a *man*—I doubt not he had endearing qualities, but as a man— a—ger! Whew! God help the stage while such men direct it. I hope he is forgiven and is happy in Heaven! There now! Do you wince? My folks have just *hopped* in, so I must leave this for

[16] The reference is to Winter's obituaries of Charles Albert Fechter and Henry David Palmer appearing in the *Tribune* of August 6 and July 26, respectively. His obituary of Fechter was probably accurate but certainly uncomplimentary. He wrote that few would regret the death of Fechter, because he was "intensely selfish, devoured by self-esteem, arrogant and sensual in nature . . . and conscious of no moral purpose" in his acting. "There are some lives," he continued, "which create love and diffuse kindness. That of Mr. Fechter was constantly attended with strife, bitterness, and trouble." Considering Winter's attitude towards the moral purpose of the theatre in general, and *The Black Crook* in particular, one might have expected Palmer to be treated even less kindly than Fechter. Winter's obituary of Palmer was, however, laudatory. He praised Palmer as a leader of the public taste who thoroughly appreciated the "duties, responsibilities, and privileges of a director of the stage" and as "an agent in the best of good works for the benefit of the stage and community." He admitted that *The Black Crook* enterprise was not entirely a credit to Palmer, but he reminded his readers that "Harry Palmer . . . made no greater concession to the vulgar taste in theatrical management than were made by Garrick, and Rich, and Sheridan, and Hamblin, and Simpson, and Barry."

[17] The "holy epistle" was a letter Booth wrote to the editor of *The Christian Union* about Christmas, 1878. It is printed in *Life and Art*, p. 80.

the present—perhaps by tomorrow I may have another (!) idea. McCullough & I meet now & then at the races, but *he* wins all the money; my horse is always last! My doctor is very mysterious about my disease; is sure he has the key of it, and will tell me all about it *tomorrow*. Why tomorrow? The Revd. Mr. Booth's Pa & sister have the rooms adjoining ours & we are very chatty— on all but theological topics. May I never get rid of *D.D.'s*! I had one (Tyng), you remember, next door in my New York *flat*.[18] Good night!

Sunday

The Dr. has been with me nearly all the day and yet is he silent. He is quietly taking my bearings I suppose; watching me eat; getting at my *inards* through the medium of my *whittles*. He again defers the examination of my liver 'till tomorrow. I have worn my overcoat here—while you, no doubt, were sweltering at Staten Island; certainly this is a wonderful place—a world in itself, and I regret so many summers past without my knowing it. That's badly expressed, but you know what it means.

Bernhardt, I see, will not come. Her would-be manager is lucky. My sister, Mrs. Clarke, does not like her, but prefers Croisette [sic][19]—who, she says, is the greater favorite (or was, when the Clarkes were there) in Paris. *B.* she considers a ranting virago most of the time and at others an inanity. She likes Irving more the more she sees him act, but still wishes I were in London "to (as she expresses it) show them how to act." Perhaps, that young matron is prejudiced. Irving, I presume, is now on his cruise, on the "ebbless sea," for the vacation and I shall hear no more of the matter.[20] When Stedman returns I shall take some action, but I want to hear his full report of particulars first. Adieu! All join in kind remembrances to you & yours.

[18] Possibly William Agur Booth, prominent New York merchant and banker and the father of Henry M. Booth, notable New York Presbyterian clergyman.

Probably either Stephen Higginson Tyng or his son Stephen Higginson Tyng, Jr., both prominent New York Episcopal clergymen.

[19] Sophie Alexandrine Croizette, French actress, performed with the Comédie Française from 1870 to 1883 and was especially noted for her performances in *The Sphynx* (1874), *The Demi-Monde* (1877), and *The Princess of Bagdad* (1880).

[20] Irving was vacationing in the Mediterranean on the yacht of his patroness Baroness Angela Georgina Burdett-Coutts.

A week later Booth decided that since his negotiations with Irving would soon be in the papers anyway, Winter should print an authorized notice first.

52. Sarasota, August 17, 1879

Only a line or so—not in reply to yours, but merely to say that after thinking the matter over I do not see why you need delay publishing the fact that negociations are pending between Irving & myself; nothing definite, &c. The matter will surely soon be in print, as you surmise, and you had better have the first 'blow' of it. The subject might afford quite an article on the interchange of dramatic courtesy, international high art and all that sort of thing you know, with a touch of what Mr. Arnold speaks of regarding the "established theatre."[21] By-the-by, I see that both Smalley & the "Times" dig Irving about his poor support[22] (which Stedman says is "really superb"), just as the papers here have "dug" at me (and all successful tragic stars) about my companies. "Poor support" has ever been, will ever be, the cry at the heels of the "star", no matter how good his actors may be. There are several now in N.Y. lauded as excellent actors who were denounced when with me. Joe Wheelock & Sheridan for instance.[23] Irving I heard today—as I have [heard]

[21] In the August issue of *The Nineteenth Century*, Matthew Arnold published an essay entitled "The French Play in London" in which he compared the French Theatre, as represented by the Comédie Française then at the Gaiety Theatre, to the current English theatre. He found the French theatre well organized and irresistible and the English theatre chaotic and boring. He suggested that the English theatre take a lesson from the French theatre and found a national theatre and school of dramatic art supported by a governmental subsidy. In the *Tribune* (April 22) Winter announced the Booth/Irving exchange and observed that such an exchange would illustrate the "condition of American taste and scholarship with reference to the drama" and would establish "a warm and real correspondence of ideas and feelings between societies far distant from one another."

[22] Smalley commented in the *Tribune* (June 30) that if Irving "could pick up a few good actors" while he was yachting with the Baroness Burdett-Coutts "it would be a profitable voyage; for of such the Lyceum is in much need." The critic of the London *Times* (July 29) urged Irving to "improve the quality of his company." The New York *Times* picked up this criticism and printed it on August 10.

[23] Joseph Wheelock, Sr., American stock actor notably of "leading man" roles, was at various times during his career in the companies of Ford's

said of myself will not have young or good actors about him. The d---dest bosh! There are Stuarts still left in London.

I've heard nothing since I wrote you, nor do I know positively whether I will go next year, nor at which theatre I shall act. I'll post you duly. It may be I shall go to Newport next week—not yet decided.

Almost a week later, Winter, in poor health, answered Booth's letter.

53.   17 Third Ave., Fort Hill, New Brighton, Staten Island, August 26, 1879

My Dear Edwin: (I am so nervous that I must write by another hand).

Your letter of the 17th was received on the 19th, and I was very glad indeed to be authorized to announce your English expedition. I wrote an article, at once, but it did not get into the paper till the 22nd inst. (last week *Friday*) I inclose to you a copy of it, and hope that I have not muddled so important a matter. I thought it best to say but little about Mr. Irving and his share in the matter. I hope you will advise me instantly when the arrangements are definitely made. Have you heard yet from Stedman?—The other newspapers of the country are picking up the subject, and it will be widely discussed—both here and abroad. I look for the greatest success of your life when you act in London.

How about your health? Is it improving? My own, I regret to tell you, has again given way, under the pressure of hard work. I have a violent cough and sleepless nights.

Theatre in Baltimore, the Boston Theatre, and McVicker's Theatre in Chicago. In 1872 he joined the company of Booth's Theatre, playing such roles as Romeo, Laertes, Othello, and Iago.

William E. Sheridan, American actor notably of "heavy man" roles, was at various times during his career in the companies of the Chestnut Street Theatre in Philadelphia, the Boston Museum, the Globe Theatre in Boston, and the Olympic Theatre in St. Louis. From 1870 to 1873, he was a member of the company of Booth's Theatre, where he played such roles as Banquo, Baradas, and Lord Faenza in *The Fool's Revenge*. In 1879 he began a starring career, appearing most notably as Louis XI, his greatest "heavy" role.

Buckstone is reported dying, and I have just written his obituary.[24]

I notice what you say as to the "poor support" business, and when you come to town I should like to get a lot of positive points from you on that subject, with a view to an article which shall be a *settler*—a [illegible word].

New poems of mine will appear, very shortly, in Torney's *Progress*, the *Atlantic Monthly*, *Baldwin's Magazine*, the *Philadelphia Times* and the *Boston Gazette*.

Do you go to Newport—?

Kind regards to all in your home.

With love, yours always
William Winter

Stedman, writing to Booth from Paris on August 17, reported that Winter had been negotiating with Schwab and Strakosch for Booth. That Schwab, Strakosch, or even Winter were meddling in his affairs without authority annoyed Booth.[25]

54. 68 Madison Ave., September 9, 1879

Your last, written by another hand, reached me duly can't find it just now; we're all "higgeldy-piggeldy" as yet—tho' we've been here since Friday.

On my arrival I found a letter from Stedman in which he says—"now I have no grievance at all with you, unless one may be raised by Irving or Mr. Smalley—as if I had been unauthorized to act for you. This is quite likely to happen, especially in case they hear that Schwab & Strakosch have been negociating with you through Willy Winter. Schwab met me in London and told me that he had written Winter. I then said that I had been entrusted by you with certain matters relating to your coming abroad—but did not make known their nature. This was done to save Schwab trouble & he thanked me." And again: "If I again run against Schwab I shall tell him that I am not in his way."—

Now, what the d---l does all this mean? Has Mr. Schwab

[24] John Baldwin Buckstone, English actor, manager, and dramatist, died on October 31, 1879.

[25] Stedman's letter to Booth is in the collection of the Walter Hampden Memorial Library. It is printed in Lockridge, pp. 252-254.

been meddling without my authority in my affairs? Has he hinted at such a thing in his correspondence with you? If you write him again speak of this & assure him that I shall ratify no engagement, nor entertain any offer or suggestion from or through him. I could not honorably do so—under the circumstances. It places Stedman in an awful plight & makes me look like an ass or a nincompoop to the eyes of all concerned—I mean Stedman, Smalley & Irving. Of course I wrote Stedman at once —giving him full & sole authority. Telling him all that passed 'twixt you & me about Schwab & Strakosch—that I didn't know the former, & took no stock in the latter, or something to that effect. I hope no trouble will arise from the unlucky *muddle*, however it was caused. Stedman must curse me in his heart for placing him (as it, of course, seems) in such a position.

Hope you're well again. I am not—have had a deal of anxiety, & feel lost. Perhaps, someday, I'll go over to see you again. Hope your home is still delightful, & reposeful—which is better.

Wife & daughter send kind remembrances.

Adieu! Always my compliments to Mrs. W. whether I say so or don't

I failed to see your notice of Boucicault's opening—did you write one?[26]

A note by Jefferson Winter at the end of this letter reveals that Winter knew nothing about the Schwab/Strakosch negotiations. Winter, according to his son, was "annoyed . . . a good deal and there was some friction about the matter." Winter was "especially pungent in his remarks to Stedman" and apparently to Booth also.

55. 68 Madison Ave., September 16, 1879

Hold your horses! My letter put you into a *pheeze*—had I talked what I wrote 'twould not have sounded so serious as it looked. I suppose Stedman felt a little nettled (not that his letter

---

[26] Boucicault opened Booth's Theatre on September 4, 1879, with his melodrama *Rescued; or, a Girl's Romance*. Winter reviewed the opening of Booth's Theatre under Boucicault's management and *Rescued* in the *Tribune* on October 8. He said that he did not review Boucicault's activities sooner because he thought that the public was satiated with news of Boucicault from the pen of Boucicault himself. In his most scathing style, Winter went on to excoriate *Rescued* as the cheapest sort of old fashioned "blood and thunder" melodrama which could only appeal to those whose tastes ran to sensational "yellow covered literature."

indicated such a condition) at hearing a subject mentioned to which he alone was privy and, in his excitable mood, "flew off the handle," as it were, with the idea that I had trifled with him.

McVicker has just left me—he met Schwab today. If you've not already done so don't refer to the subject, for 'twill only grow out of a mole-hill to a mountain of bother; let it drop, and after I've seen Stedman (for in courtesy, I can do nothing until then) I'll hear what Mr. S. may have to say. My only objection to dealing with Clarke is the almost inevitable difficulties arising from business with friends & relatives; I've had so much of that sort of bickering,—seldom with strangers—that I wish, if possible to avoid such risks. Have been nowhere yet, but shall visit Boucicault's Booth's shortly; I see he does not advertise; is there a row in the camp? My doctor says I'm better; he ought to know. How are you? By-the-by, Boucicault (at Saratoga) suggested my going to Drury Lane, either with him—or under his *friendly* (!) "horse-pieces," & to act my last engagement on this side at Booth's. The latter O.K., the former—?

Irving continued to disregard Booth's letter. After Booth had talked to Stedman, who returned to New York about the first week in October, he must have learned that Irving would never answer it. He was advised to forget about Irving and continue his negotiations for a London engagement. Booth, thus encouraged, employed the London theatrical agent Morris Simmonds who began interviewing various theatre managers, such as John Hollingshead of the Gaiety, A. and G. Gatti of the Adelphi and Walter Gooch of the Princess's. Booth's desire to act in London continued, however, to be indefinite. On October 14, he wrote from Baltimore, where he was fulfilling an engagement at Ford's Opera House, to inform Winter that he had not yet determined what course he would take—"whether to engage or go free to London & look about me, as my sister suggests" (Folger Shakespeare Library, Y.c. 215 [347]). Four months later, Booth was no more definite about an engagement, although he had booked passage to England in June. He insisted that he was going more as a tourist than as an actor.

56. Mount Vernon, February 22, 1880

" 'Tis long since we have known each other"—so long that— save for the Tribune—I had lost all trace of you.[27] I'm the culprit,

[27] Booth's last letter to Winter had been written on Christmas Eve. (Folger Shakespeare Library, Y.c. 215 [354])

however, for I remember your last to me received no answer. I sent you a Mt. Vernon paper t'other day—the acct. of the Magonigle show up there, which it contained, I see you noticed. Last night he gave another & I took my folks to witness the performance. Harry's girls (3 of 'em) are really very clever—the youngest particularly, and Harry also did his little tragedy biz remarkably well. I always wished him to try the stage—as a light comedian he would be better than many we have, while his *serious* ability as shown last night surprised me. The girls are really talented & very pretty. Its a pity they cannot all go into the business—'twould pay 'em better than the labor of teaching kinder-garten & piano playing, which the girls do, while Harry drudges at bookkeeping in Stewarts shop.[28]—We went last week to MacKaye's theatre—once to see the play & again to inspect the building. It is certainly, by all odds, the perfection of a Comedy theatre. Beautiful to behold in every detail. I hope with all my heart that MacKaye will be amply repaid for this great improvement in theatrical decoration, & mechanism—for that (as far as it will go) is a wonderful advance. I don't see why the same method should not be applied to other than the "shiftless" scenes of comedies. By raising & lowering his stages a better effect in changing scenes would be given than that made in the old way; at *Booth's* the rising & sinking scenes often had a good effect & were applauded—especially when the two movements occurred simultaneously, as it did in *Hamlet* several times. Whether it could be successfully done on a larger scale is the question, if so—then the old split-scene style should be abandoned in every case where depth can be obtained.[29] I want to do something for "auld

[28] On Saturday, February 21, Magonigle acted David Murray in *Among the Breakers* at Steinway Hall, Mount Vernon. This was part of an entertainment given by the Teachers' Association to celebrate Washington's Birthday. See *Tribune*, February 18 and 21, 1880.

"Stewart's shop" could possibly be a reference to one of the numerous mills, factories, or retail stores founded by the New York merchant Alexander T. Stewart.

[29] Steele MacKaye, American actor, producer, dramatist, and inventor, opened the Madison Square Theatre on February 4, 1880, with his own play *Hazel Kirke*. The Madison Square, a small theatre on West Twenty-fourth Street, had been completely rebuilt in accordance with MacKaye's revolutionary ideas of theatre design. The most revolutionary innovation

Ireland" but have had so many calls of & on me that I felt unequal to much in the privy purse way, and therefore asked Abbey & Vincent to help me at the reading—as you may have seen by a card I sent to the Herald. After Abbey's Matinée, for the same cause, proved a failure he sent word that he wanted to talk with me about dropping the idea.[30] But I shall not—*cannot* do it, decently, and have been in a fume since he sent the message, for I've been unable to find him & the advertisement not appearing today, as I expected, annoys me exceedingly. I shall decide tomorrow what is to be done & let you know. My date in Brooklyn is May 3d.—one week.[31] It has taken me over a month to reply to your enquiry on that head, so I guess it is accurate.

My wife has been gradually wasting away for several months & is very feeble from nervous prostration. A severe cough, from last September—neglected, distresses her very much, & only since the last ten days has she had a doctor. I've been—at times— quite alarmed for her. Her will is something wonderful—she

---

was its double or elevator stage which preceded similar German innovations by twenty years.

The stage at Booth's Theatre was flat, not raked, and the old split-scene, wing-and-groove system of changing scenery was replaced by a new vertical system. Sets flew up or down from a ninety-six-foot fly-loft, or rose up or sank through the stage floor from a thirty-two-foot deep cellarage. The scene changes were "automated" by means of hydraulic rams powered by huge steam engines and were often made in full view of the audience. See Shattuck, *The Hamlet of Edwin Booth*, pp. 65-66, 122-123.

[30] Henry E. Abbey, American theatre manager, was at this time manager of the Park Theatre in Boston, and the New Park Theatre in New York.

Leon John Vincent, American stage manager born in England, was often associated with Abbey.

On February 20, Abbey gave a matinee of his current production, *Humpty-Dumpty*, for the benefit of "Ireland's sufferers." On the same day the New York *Herald*, which was sponsoring a fund-raising campaign for the victims of the Irish famine, printed a letter from Booth in which he proposed a benefit on March 4 at the New York Academy of Music. He also wrote that "Mr. Abbey of the Park Theatre has freely undertaken the management for me, and Mr. Vincent has as kindly volunteered his services as stage manager." On March 4 Booth performed the third act of *Hamlet* for the benefit of the Irish Relief Fund. On March 17, St. Patrick's Day, practically every theatre in New York held a similar benefit.

[31] Booth appeared at the Brooklyn Academy of Music from April 30 through May 8.

won't give up, but—tho' ill enough to be in bed—she insists on
going out as often as possible. This has kept me on the visiting
list pretty much all the time—in the evenings, either at the theatres
or at the houses of her acquaintances, none of which is altogether
agreeable to me; but then, you know, I have a daughter who
must have society, &c. Thank God, my boy, that your daughters
are boys & can go out without an escort, & that your boys can
look after *Ma*, when the old man wants to roost, as I do—often—
when I'm forced to become a swallow (tail) and flit "hither and
'yon" among the butterflies for wife & darter's sakes.—I hope
to sail by the Gallia in June, but I go without an engagement, &
may find all houses closed to me. I care very little whether or not
I act in London. Somehow I fancy that, like me as much as the
public might, the chance for any great success there is gone—
for the present, at all events. I have so little energy, less ambition,
& still less enthusiasm on the E. Booth subject that the bare idea
of acting there is irksome. Perhaps after a tour through Ireland
& Scotland I may feel more i' the vein; but I shall then arrive in
London too late for *the* season, & in midst of fogs & filth which
will depress us all so much that I'll lose heart again. I am going
more as a tourist, & chiefly because I believe the change will
benefit both wife & daughter—but of course, I shall, like old
Hackett, take my "fat" along: You know the old gentleman
always carried his Falstaff belly with him on all his hunting &
fishing tours—by mere chance, of course.[32] I may get an opportunity to act for some charity purpose, for some actor's benefit—to
"play with the people in" (or out), but unless I take a theatre—
which I won't do, I doubt if there'll *be* an opening for a crushed
tragedian[33] at any of the regular shops. Irving has evidently

[32] James Henry Hackett, noted American actor of "Yankee characters,"
was also known for his portrayal of Falstaff. He first acted this role at the
Park Theatre in New York on May 13, 1828, in *Henry IV, Part I*.

[33] *The Crushed Tragedian*, a play by the English dramatist Henry
James Byron, originally entitled *The Prompter's Box, A Story of the Footlights and the Fireside*, was first presented at the Adelphi Theatre in
1870. As *The Crushed Tragedian*, it was revived by the English actor
Edward Askew Sothern who made a great success with it during his
tour of America in 1879-80. The model for *The Crushed Tragedian* was
supposedly an eccentric actor named George Jones, who called himself
the Count Johannes, and who in all seriousness presented the most ridiculous impersonations of the great tragic roles for almost twenty years. He
was continually ridiculed and pelted with vegetables, yet he endured and

determined to keep on his beaten track, which is quite right, and
other managers are "Skeery" of Shakespeare. We shall see when
I get on the field. I wrote to Taylor & sent him a copy of the
F[ool's] R[evenge] apologizing for my change in his play, &
offering to pay the damage—if I act it in England. I've done more
for the play than it ever did for me, however; but that's "off", I'll
make it right with him.—I believe Abbey has secured Mrs.
Bowers & McCollum[34] to "support" me—but, as I have not seen
him since the idea was first suggested, I am not sure. I hope so at
all events—I mean as to Mrs. B.—the gentleman I know nothing
of—as an actor. My wife reads the distressing accts. of Ireland's
condition & dreads—more than ever—the idea of going; the
sickness in Ireland & the fogs of England keep her in a constant
scare. The mention of Bangs as one of Abbey's Company for me
is incorrect, I am happy to state. He's a good enough actor in
certain parts, but rather cranky to deal with & on the whole is too
"great". "Great actors" are very queer cusses to handle; besides,
there are so many of 'em; nearly every company counts a dozen
sich. Adieu! See me soon.

On March 30, following a three-week engagement at the Park
Theatre in Boston, Booth opened at Booth's Theatre, then under the
management of Henry E. Abbey, for another engagement of three
weeks. Several days after his opening, Booth wrote to Winter to thank
him for his kind review of his opening performance of *Macbeth*. In a
previous letter, Winter had apparently asked Booth if he might ac-
company him to England, probably at Booth's expense. This request
seems an unusual one for a critic, especially Winter, to make of an
actor. In later years, however, Winter made several trips to England

---

seems to have made a great deal of money. The resemblance between
Jones and Sothern's "Crushed Tragedian," Fitzaltamont, was reported
to be extraordinary. Jones even attempted to sue Sothern for slander, but
failed. See *Theatre Magazine* (February, 1880), pp. 99-100.

[34] Mrs. D. P. Bowers, née Elizabeth Crocker, American actress and
theatre manager, and James Clark McCollum, American actor, were a
popular starring team during the late 1860's and 1870's. Their last appearance together was in support of Booth at the Brooklyn Academy of
Music on May 30, 1880, after which they retired from the stage. On
January 28, 1883, Mrs. Bowers and McCollum were married, but he died
suddenly in November of the same year. In 1886, Mrs. Bowers once
again supported Booth and Tommaso Salvini when they appeared together
in *Othello* and *Hamlet*.

at Augustin Daly's invitation and expense; so perhaps his request was not as unusual then as it seems today. Booth refused the request, not because he did not want Winter's companionship, but because he was still undecided as to whether or not he would act in London. He continued to insist that acting would not be his "chief aim" but "recreation, health & a protracted loaf." Without a definite engagement, he could not afford to pay Winter's expenses.

57.   68 Madison Ave., April 2, 1880

My hands have been overfull since my return, or you should have had some word from me ere this. We are busy packing books, pictures, furniture &c, preparatory to our vacating these rooms (Monday, I hope) for others at the Brunswick.[35] After long rehearsals I've been up to my eyes in dust for several days, & am pretty well tired. Things are in such a bad way (among scenery, props &c at Booth's—to say nothing of the "dogans") that I have to attend rehearsals nearly every day—although the same people were with me in Boston & I have an experienced man at the helm.

I want to thank you—but don't know how—for your encouraging words about *Macbeth*.[36] I play the character so seldom that my infrequent performances are little better than rehearsals, and my own consciousness of my physical inappropriateness (if that's the word & the way to spell it) for the stalwart scot hampers me considerably. Old Tom Barry praised my performance of it years ago, but told me that I would never satisfy myself—as Macbeth was a high peg that even Kean (the elder) and Macready failed to hang on comfortably; that he never knew any Macbeth satisfy either his audience or himself. There's consolation in that! Immediately on receipt of yours I wrote to Abbey about the *ads* & he promised to do the right thing, but I suppose the Herald will always have the lion's share.—I now hear that Irving is engaged by Wallack—I don't believe it.—but do believe that Floyd is after him.

[35] The Brunswick Hotel on Madison Square.

[36] Winter's review appeared in the *Tribune* (May 31). It was as usual a complimentary criticism which noted that Booth's Macbeth was "one of the most truthful personations of that fiend-haunted soul which ever, in our time, have illustrated the dark and terrible spirit of that tremendous tragedy." Winter remarked, however, that Booth's slight figure limited his impersonation.

My dear Winter,
                    I am sorry I
have not seen you since the
first night — I know not how
to explain to you my unseemly
and crazy excitement of that
evening — when we meet I may
be able to do so. To the subject
I referred to in my last. What
I am about to do is in obedience
to a heart impulse — if I offend
blame the head. You know,
perhaps, that I am a member
of the "Century," and that is dev-
miration Society" is generally
well known by all its members. I ascertained that you
were desirous of joining that
Club, and were unanimously
elected — but that you were
obliged to decline on account
of pecuniary considerations.
Now, bearing a grateful
memory of your past ser-
vices and wishing to refute
what Stuart says of actors
in general & of myself in
particular, I venture — even
at the risk of wounding your
delicacy, to offer you a history
knowing you to be a thorough
gentleman, and professing

1. Manuscript page of Letter 2. Booth to Winter, February 7, 1869

2. Edwin Booth, his daughter Edwina, and his second wife Mary McVicker
Booth about 1875

3. Edwin Booth as Richelieu

4. Edwin Booth as Iago

5. Edwin Booth about 1880

the Prompt Book due, with L & S,
due within his last quarter. The
total received by me, for him,
this year, is $227 = 99 — to which
is to be added what I received
for the first theatrical sale, or
from the first theatrical sale, or
Booth's theatre, and the advance of one
hundred for so. Altogether, — in the
hundred sale, &c. — I have received
freight sale, &c. — I have received
about fifteen hundred Dollars from the
about fifteen hundred Dollars from the
entreprise; and with this I am well
satisfied. I shall inform every of-
keep the sales
for the books. L & S will communicate with you, and send notices &c.
My correspondence, as you will find
Lee (William), as you will find
him a good man. — I have obtained
an assignment a poor certificate of
copyright, and have written to A. R.
Shepherd, the Librarian, about the same. —
I have bought the Lowe Lee, paying
$1500. down, & have to pay $2000.
within one year. So you see I have
work ahead. To-day I am almost
disabled with muscular rheumatism, and
I can hardly write — but I wanted to
let you know how affairs are going on.
with love, from old William Winter.

Fort Hill. New Brighton.
Staten Island. May 3. 1879.

Dear Edwin: I received your letter of
April 27th and am much relieved &
rejoiced to see that your recent most
trying and terrible experience has not
shaken you in constitution and purpose,
shaken you in health. It was a
nor injured your health. I have pictured it all
dreadful trial. I have pictured it all
to myself, many times: the prison scene,
to myself, many times: the prison scene,
the dark cell, the moonlight, the
distressed and darkly moralizing king,
the deep, wise words of the poet, the
crowded, hushed theatre, and then — the
crowded, hushed theatre, and then — the
amazing at his people work! I dare
wonder your life was overwhelmed with
wonder your life was overwhelmed with
the horror of such an experience. The
man is, undoubtedly, mad — but he is
not to be caught for
danger, and should be caught for
daughter, and should be caught for
life. I have often mused on the
dark vicissitudes of your life, as they
are known to me. You have certainly
passed through a very tragic ordeal.
The world is the
more the world
it has not enlightened you. Sad great
that the worst of all trial is over now,

anxious ear, but it came not—(thank
God!) and in the early mornings, I
listened in vain for the larks & linnets
and each small deer, but in their
stead came "caw-caw" from my inky-
cloaked friends. Methought I so?
love old England solely because she
taught me to appreciate America—
which she has done, God bless her!
but having spent these few days
in Warwickshire I am prepared
to love her for herself—alone!
I have seen castles that interest me
—as castles—more than pretched at
Kenilworth does—which is after
all a monument to Walter Scott.
At Carnarvon (to me the most remarkable), at Edinboro' and Stirling
I felt myself carried—almost
bodily,—back to their early days—
but not so at Kenilworth or Warwick
—tho' the latter I think superb! While
at Chester we visited Toft Hall, at
Knutsford, where we were induced
to spend a day & night with Mr & Mrs
Lycester—whom we had met in New
York, and I had a pleasant ramble
over fields to the estate of Lord de
Tabley, a courteous old gentleman,
with whose lady we took tea on

Edwin Booth
Rec'd. August 30th.
Answered

RED HORSE HOTEL.
STRATFORD ON AVON
Aug. 17th 1880

My dear Will—
Here sit I at
last in Irving's chair, with
Geoffrey's scepter at my
pen—hand on the table. I
have had today, in this little sanctum, the first palatable, home-
like meal since I left "my own,
my native land". We arrived
this afternoon from Leamington (where we remained three
days) per carriage—by way of
Charlecote, through whose Park
I walked,—and after a light
lunch went at once to "the
house in Henly Street." Being
at once recognized there by some
American clergyman—one of a party
my name was buzzed about & the
dear old ladies who have charge
of the treasures at once over—

7. Manuscript pages of Letter 60. Booth to Winter, August 17, 1880

8. Edwin Booth in 1888   9. William Winter about 1890

10. Booth's Theatre, 1869

12. Edwin Booth as Hamlet

11. Manuscript page of Letter 123. Booth to Winter, March 17, 1890

13. Edwin Booth as Benedick in
Much Ado About Nothing

14. Edwin Booth as Pescara in
The Apostate

15. The Players, 16 Gramercy Park, New York City

16. Edwin Booth's bedroom at The Players

17. Edwin Booth's living room at The Players

18. William Winter in February, 1916   19. William Winter's home. New Brighton, Staten Island, New York

My dear boy, I was only joking (in my gloomy way) when I referred to your printing my letter.[37] I know you wouldn't do anything but for my good. My wife is somewhat better—but only a trifle somewhat, she eats nothing & is very thin & weak. I hope the voyage will build her up; if she has no sea-scares it certainly will do so.

Touching another point in your letter—that of your going over with me—I wish, my boy, that it could be so, for it would be a most delightful companionship for me—amid the scenes you love so well & have so gloriously described, and if I were differently circumstanced it should be so. But (I can't explain the why & wherefore) I cannot always do as I'd like to. It is still uncertain *when* I'll act in London; I shall go without engagement & travel for awhile—after seeing the London managers. It may not be 'till the following Spring—for I shall be slow & very cautious in all my movements. Acting will not be my chief aim this time, but recreation, health & a protracted loaf. Of course circumstances may change all this. The books will never pay. I saw the head cook at Lee & Shepard's & agreed with him to reduce the price to 25 cents. They don't sell—except in the theatre, and there very slowly. No matter, I can take only those of Shakespeare with me—the modern dramatists (reprints) are forbidden. Tom Taylor has not replied, but as I told him I wd. soon be in England I do not expect an answer. I regret to hear of your wife's misfortune— hope she has quite recovered; give my kindest regards to her. I saw Vedder's[38] pictures here & met him also. No word from Clarke—but my sister sends all sorts of fearful messages regarding my advent & Irving's great influence in London. There's a man named Clement Scot [sic]—who publishes ugly things

[37] Booth's last letter to Winter dated Boston, March 17, 1880 (Folger Shakespeare Library, Y.c. 215 [358]) mentioned that he had had tea with Ole Bull and James Russell Lowell and that he had met Longfellow, who came behind the scenes to talk with him during *Richelieu* on March 16. Booth, always dreading that such incidents would be "puffed" by the press, reminded Winter that this "ain't for printin'—moind!" Winter was apparently insulted by Booth's implication that he would print anything in the nature of "puffery."

[38] Elihu Vedder, American painter, muralist, and illustrator, exhibited some of his paintings at the Art Students League in New York during March. Later in the spring he had a more extensive exhibit in Boston.

about the Booth family constantly & is very anxious to know if & when I am going &c, &c.[39]

Wish I could hear your recital at the *Lotos*.[40] That reminds me—my wife has been collecting poems, & hopes to get sketches, from distinguished men, & wants a few lines from you. I promised to ask you. Longfellow, Holmes, Fields, Aldrich, Mrs. Howe & Ole Bull have contributed.[41] If you can spare so much brain & ink & patience I'll give you the sheet on which she wishes you to write—they are to be bound hereafter, when her collection is sufficiently large.

I want to get a sketch as well as a sentiment from Joe. The album is to be very select—only those of acknowledged eminence among writers, artists & actors—the latter numbers scarce ½ a dozen. I have scribbled away at lightning speed & have tried to answer your letter—line by line. Hope you can decipher it. Have yet a deal of packing to do, & must also go to the "show-shop." Sitting in front of Booth's I thought Boucicault had greatly improved the stage-part of the theatre, but acting on it I find he

[39] Clement Scott, English drama critic and author, was the critic of the *Daily Telegraph* from 1871 to 1898. From 1880 to 1889 he was also the editor of *Theatre Magazine* which in February published an anecdote regarding the elder Booth's drunkenness and violence during a performance of *Richard III* and in March a story about John Wilkes and the assassination of Lincoln. This was the extent of Scott's "ugly things about the Booth family" during 1880—at least in the *Theatre Magazine*.

[40] On March 27, at a dinner at the Lotos Club, Winter read his poem "The Lotos Flower" to celebrate the tenth anniversary of the club's founding. See *Poems*, pp. 123-124.

[41] Mrs. Booth's autograph album is in the collection of the Walter Hampden Memorial Library. In four consecutive issues of *Golden Book Magazine*, September to December, 1929, Otis Skinner wrote a description of it and reproduced some of its pages.

Julia Ward Howe, American authoress and social reformer, was an early friend of Booth and his first wife Mary Devlin. In 1858 she wrote a poem about his portrayal of Hamlet which was printed in *Atlantic Monthly* and occasionally reprinted in various papers during a Booth engagement.

Ole Bornemann Bull, Norwegian violinist and theatrical impresario, was on friendly terms with Booth. He came at Booth's request from Cambridge, Massachusetts to New York to play at the benefit for the Irish Relief Fund on March 4.

has destroyed it, for my pieces at all events.[42] I see Agnes[43] is after him. I am to have an interview with one of my old creditors today & expect a disagreeable time with him. My health is good—but a severe cold in the head & a settled cramp in the back for three days past, have kept me in agony. Hope soon to see you.

After his engagement at Booth's Theatre, which ended on April 24, Booth acted at the Brooklyn Academy of Music from April 30 to May 8. He then retired for the season and began to make preparations for his trip to England. On June 15, Booth was given a testimonial breakfast at Delmonico's, which was attended by about 150 persons. The *Tribune* (June 16) printed a detailed account of the festivities. Winter gave a speech in which he declared that when Booth "knocked the dust out of the cushions of Drury Lane there would be such a flight of owls as had not been seen since his illustrious father electrified the capital." Apparently Winter thought that Booth was to act at Drury Lane, although Booth had not yet made any definite commitment to act in any theatre in London. After his speech, Winter recited "Good-bye to Edwin," a poem which he had written especially for the occasion. (See *Poems*, pp. 136-138.) Several days later Booth wrote Winter the following note to complain about the ill effects he suffered from the "debauch" and to compliment him on his "glorious" recitation.

58. Friday, [June 18, 1880]

Ever since the breakfast I have been ill—not so much from the effects of my *debauch* (wasn't it glorious—weren't *you*!) as from

[42] Dion Boucicault managed Booth's Theatre from about September 4 to October 25, 1879. Before assuming management, he made several alterations in the theatre. He installed the first two rows of seats with large folding chairs, and extended the forestage over the orchestra pit, moving the orchestra underneath the stage. He also redecorated the auditorium. See New York *Tribune*, October 8, 1879.

[43] Agnes Robertson, English actress and singer, married Dion Boucicault in 1853. Although they had been separated for a number of years by mutual consent, she sued him for divorce in March, 1880, claiming infidelity. On March 27, she had him arrested because she claimed he was about to leave the state. Litigation continued, apparently, for almost the next ten years. In 1885, Boucicault, while on a tour of Australia, married Louisa Thorndyke. Four years later and less than a year before Boucicault's death, Agnes Robertson was finally granted her divorce in London on January 15, 1889. The divorce proceedings were reported in the New York *Times*, March 29, April 1, 13, May 22, and June 12, 1880.

a series of troubles that have been seething for some time &
culminated—all in a lump—the day before the *feed*. I was really
fitter for my bed than for my breakfast that day, & the loss of
sleep—the *hurrah* & the various conflicting emotions under which
I made my *debut*, prostrated me completely. I want very much
to see you in a few days, that I may explain & have a chat with
you before I sail. I intended to write this on Wednesday—but have
been so shaken till now, & even yet am in a whirl, that I cd. not
write. Hope you are not suffering. You dear old boy! How splendid
you were! I'm so glad I had an opportunity to see you "with
your back up" & all your feathers ruffled. It was superb! So all
whom I've seen since—who saw & heard you—say.

I'm interrupted now & must quit.

God bless you!

On June 28, Booth acted Petruchio at the Madison Square Theatre
for the benefit of the fund to erect a statue of Edgar Allan Poe in
Central Park. Two days later he sailed on the Cunard steamer *Gallia*.
He had still not contracted for an engagement in London, although
John S. Clarke was negotiating for him with Walter S. Gooch, the
manager of the Princess's Theatre. In dockside interviews Booth maintained that he was going abroad only for recreation and rest. He
planned to tour Ireland, Scotland, England, and he wanted to see the
Passion Play at Oberammergau, but he did not plan to act in London
or anywhere else.

Chapter V

England

1880-1881

ABOUT July 8, Booth arrived at Queenstown, in the south-east corner
of Ireland. From there he traveled west through Cork and Killarney,
then north to Dublin, Belfast, and Port Rush, visiting along the way
such scenic attractions as Lake Killarney, the Gap of Dunloe, and the
Giant's Causeway.

59. Port Rush (Antrim), July 18, 1880

Clarke forwarded your letter which caught me in Belfast—it
was evidently written before you received the one I sent you the
day before I sailed, in which, I am sure, I gave my address—
Brown, Shipley & Co., Bankers, London. I felt sorry to leave
without seeing you—I think Edwina caught a glimpse of you on
the pier as we sheered off, but I looked in vain. Hope the letter I
referred above reached you safely—it was important in more than
one respect. We had a canal-like voyage—not a ripple until the
last day & then but a ripple, nothing more. Had a pleasant set of
passengers & all the comforts, food &c, &c, were really home-like.
The trip did us all good—my wife particularly. Since we landed
at Queenstown we have traversed the length of the island as you
see. The weather has been what the folks here call charming
but to us Americans it is simply d--nable, fog & mist; but 2 clear
days—both passed in travel on the rail. Killarney in itself is a
poor apology for Lake George—but associations and the people
make it delightful. It's a kind of bliss to be be-be-ggared in the
gap of Dunlow, by the Colleen Bawns & the Kate Kearney's, &c.
Bless 'em—with a *D*! We have found all the folks civil 'till we
came to this cheerless, dull & deserted *fashionable* watering-place.
Here the servants are German & saucy, the lack of system in
European hotels is something marvelous; *Boots* has more to say
than the proprietor & the chamber-maid seems to be the only
creature possessed of a spark of intelligence. We have just returned from the *Giant Gouge-Way*, where you are waylaid &
made to deliver 5s. at every step you take. Am not sorry I've seen

the place, but if the guide was dead & the other ghouls that haunt the place were exorcised it wd. be more worth a visit. The hotel "beats" set the example to the other robbers, ostlers, guides, waiters, "Buttons" & "Boots", and—" 'fore God, they're all of a tail!"[1] I lost my luggage (baggage & buggage are constantly at war with each other on my tongue) en route from Dublin to Belfast, & have great doubt if I'll recover in a week—my other shirt is in one of the trunks!

Tomorrow early we return to Belfast & if my trunks are there will start at once for Glasgow. After a brief tour through Scotland we'll go through the English towns & reach London about the last of August—then for Ammergau & the German cities, perhaps to act, if not—through Italy &c, &c. McCullough, I hear, has got the Spring opening at Drury Lane & they tell me that the Princess' is now a sort of Bowery; Irving will not let me in, & consequently I may not act at all in England, unless some Yankee comes over & builds a theatre expressly for me. Where's Jarrett?[2] We have on our full winter clothing & yet are cold, but the ice-bergs here keep open doors & windows with no fires; You—I suppose—are roasting in New York.

Does it seem a month ago since we breakfasted together? Sometimes, to me, it seems long years ago, sometimes but a yesterday—owing perhaps, to the mood I'm in. It stands as a land-mark of joy & sorrow in my memory.

My kind remembrances to Mrs. Winter, the boys & such friends as may speak of me.

From Port Rush, the Booths returned to Belfast, then crossed the North Channel to Glasgow. They visited Stirling, Loch Lomond and the Trossachs, and Edinburgh. Leaving Scotland they traveled south into the Cumberland Lake District, then to Chester, Knutsford, and the ruins of Caernarvon Castle in Wales. After brief stops at Lichfield and Leamington in Staffordshire, they arrived at Stratford-on-Avon, where Booth reported his tourist activities to Winter and disclosed that he had closed with Gooch for an engagement in November. Gooch, who had been presenting "blood and thunder" melodrama at the Princess's, was rebuilding the theatre and wished to re-open it with

---

[1] *Much Ado About Nothing*, IV,ii,32: " 'Fore God, they are both in a tale."

[2] Henry Jarrett, the American manager.

Booth as a theatre dedicated to the classical drama, as it had been during Charles Kean's management from 1850 to 1859. This ambition undoubtedly appealed to Booth and to Clarke, his negotiator.

60.   Red Horse Hotel, Stratford-on-Avon, August 17, 1880

Here sit I at last in Irving's chair, with Geoffrey's sceptre at my pen-hand on the table.[3] I have had today, in this little sanctum, the first palatable, home-like meal since I left "my own, my native land." We arrived this afternoon from Leamington (where we remained three days) per carriage—by way of Charlecote, through whose Park I walked,—and after a light lunch went at once to "the house in Henly Street." Being at once recognized there by some American clergyman—one of a party—my name was buzzed about & the dear old ladies[4] who have charge of the treasures at once overwhelmed me with courtesies. I was *requested* to write my name on the wall—high up on the actor's pillar, although such scribbling is now forbidden. A Mr. Hodgson, of Clopton, at once made my acquaintance & invited me to dine tomorrow—where Shakespeare "smoked his pipe" as he put it. I doubt the pipe part of it—but at all events I shall smoke my cigarette there. As a *Gov*: of the Memorial Theatre &c, of course I am a distinguished body here. Mr. Flowers,[5] unfortunately is in Scotland—so I shall miss him. The old ladies referred to remembered yr. & Jefferson's visit and were full of *you* & *"Rip."*[6] We spent nearly all of daylight in the house & went to the church, but fatigue and the near approach of dinner time forced us to defer our inspection 'till tomorrow. A Mr. Seaver[7] (is it the publisher?) turned up; he had just heard from you—had a book, I think, & hoped you were coming over. He knows me very well &

[3] Washington Irving immortalized the parlor of the Red Horse Inn in *The Sketchbook of Geoffrey Crayon, Gentleman* (1820).

[4] According to Winter, *Life and Art*, p. 161, the "dear old ladies" were two spinster sisters named Chattaway.

[5] Charles Edward Flower, English brewer, was the mayor of Stratford and the founder of the Shakespeare Memorial Theatre. His brother Edgar was the chairman of the Shakespeare Memorial Trust.

[6] Winter and Joseph Jefferson had visited Stratford together in the summer of 1877.

[7] William A. Seaver was a close friend of Winter. In a note to his elegy to Seaver entitled "Good-night," Winter called him "the genial humorist of *Harper's Magazine*." See *Poems*, pp. 174-175.

I his face, but could not place him. I cannot of course describe my feelings—after all this flurry of travel is over & I get my wits straightened out into something like a *calm*—I shall live over & over again this charmed life by *bits*, as it were,—each bit will be as a dream of faerie. I confess that 'till I reached Warwickshire & caught my first glimpse of the grand old Castle (although it was from a rail-car window) I was hugely disappointed with dear old England. The absence of all my accustomed comforts; the insipid, coarse and badly cooked food, and the dreariness of the hotels disgusted me. The lovely scenery (but little of it equal to our own, however), of course, was a delight—but the awful stillness of the night, no sound—either buzz or chirp or croak, and the total absence of birds—save *crows*, which swarm all over those parts of the Kingdom that I have visited, prevented my *entire* enjoyment even of that. I passed lovely nights at the Trossacks; at lake Killarney; at Derwentwater & such charming spots—but not a frog, nor cricket, nor a buzz of any insect—why a mosquito would have sung sweet music to my anxious ear, but it came not—(thank God!) and in the early mornings I listened in vain for the larks & linnets and such small deer, but in their stead came "caw-caw" from my inky-cloaked friends. Methought I shd. love old England solely because she taught me to appreciate America, which she has done, God bless her, but having spent these few days in Warwickshire I am prepared to love her for herself—alone! I have seen castles that interest me—as *castles*—more than patched up Kenilworth does—which is after all a monument to Walter Scott. At Carnarvon (to me the most remarkable), at Edinboro' and Stirling I felt myself carried almost bodily—back to their early days—but not so at Kenilworth or Warwick—tho' the latter I think superb! While at Chester we visited Toft Hall, at Knutsford, where we were induced to spend a day & night with Mr. and Mrs. Leycester—whom we had met in New York, and I had a pleasant ramble over fields to the estate of Lord de Tabley,[8] a courteous old gentleman, with whose lady we took tea on the lawn in front of the grand *new* mansion—the old one being delapidated by time, and no longer tenable. But that same old Hall is one of the most delicious bits of antiquity you can imagine. Only a practical architect could describe it, and

---

[8] George Warren, the second Baron de Tabley and the father of John Byrne Warren, third and last Baron de Tabley, English statesman, poet, and dramatist.

as I *aint*—I *wont*. The records of Cheshire reek with Booths &
in the Leycester library (house of Elizabeth's time) I found in a
panel the Booth arms. I warned Mr. L. to look to his title-deeds
for I had read up the annals, while he was at church that blessed
day, & felt certain there was a chink somewhere in the pedigree.
We had a very delightful visit.—Here we may remain several
days—can't yet determine. I don't care to leave till I have taken
in *all*—to be digested at leisure at some future time. Your second
letter reached me the day after I mailed mine to you, your third
found me at Lichfield—looking at "old Sam"[9] in the market
place.—Since my last I have closed with Gooch to open his new
Princess' Theatre—when finished sometime in Octr: or 1st of
Novr. I shall open with *Hamlet*, as the safest for my *voice*—which
I know from experience is less liable to suffer from that as an
opening part after a long rest.[10] Somehow I feel little relish for
the work—but I saw no other opportunity for at least a year.
Irving evidently don't want me, he has not even recognized my
presence in his country; John has the Drury Lane in the Spring,
and by that time the *New* Princess's would be *down*, as was the
old, by force of "blood & thunder," unless kept up by the legitimate, & that was my only chance & choice; so I took it. This will
defer my long cherished plan for the Continent: if I succeed in
London—? if I *fail*, I know all about it; I shall be free to roam
about at will & loaf my fill, which I long to do. The news of poor
Neilson's death[11] has reached me—or rather it came yesterday.
What a miserable little game it is! The mere toss of a penny.

Good night!

Tomorrow night

We are just back from Clopton—had a very delightful evening
with the Squire and his family—dined in what was originally the
entrance hall but now a large dining room. The place is very
interesting. Tomorrow we lunch with Mr. Edgar Flowers; the
Mayor, his brother and president of Shak: society being absent

[9] A statue of Samuel Johnson was erected in Lichfield, his birthplace,
in 1838.

[10] Winter had advised Booth to open with Richelieu instead of Hamlet
in order to avoid comparison with Irving. Mary Booth, however, strongly
advised Hamlet and her advice prevailed.

[11] Adelaide Neilson, English actress, who had appeared at Booth's
Theatre during her first American tour in 1872, died from a stroke on
August 15, 1880, while driving in the Bois de Boulogne in Paris. She
was buried in Brompton Cemetery in London on August 20.

in Scotland. Went to Hathaway Cottage, through the church and theatre. Have found myself quite well known & perfectly at home here. After lunch tomorrow we go to Oxford for several days, then to Salisbury, then, perhaps, to isle of Wight & so to London. Clarke is looking after my London affairs & takes great interest in them. I recd. a copy of N.Y. Mirror with a letter from young Paulding in defense of me against the vile Dramatic News in which my character had been assailed. I shall have cart-loads of such filth hurled at me now. I wrote a letter of thanks to young P[aulding]—it was very generous in him; the paper also contained an editorial on the subject.[12] I see no American papers—impossible to get them, & the English ones give very little notice to our affairs. We are so unimportant. Queer folks these English! A young lady at dinner today told Edwina she was in New York in '67 (probably 17-67) & remembered the houses being mostly of *wood!* An English officer one day asked if our N.Y. houses were not *wooden*! Jolly bright, aren't they? Hope you will be over by the time I act—I hardly think it will be before November. I remember old Mr. Ludlow very well. He was for many years partners with Sol Smith (Mark's father) in the management of the St. Charles, N.O. and the St. Louis Theatres.[13]—Shall I get

[12] The day after he left New York, Booth was attacked by Charles A. Byrne, the editor of the *Dramatic News*. He was accused of drunkenness and of making arrangements with the agent of a steamboat proprietor to share the proceeds of the excursion boat *Grand Republic* which accompanied Booth's ship, the *Gallia*, out of New York harbor. Booth of course emphatically denied this last charge and was annoyed that his name was used in the advertising for this excursion. The New York *Dramatic Mirror* (July 10) defended Booth and demanded that the New York theatre managers stop supporting the "scandal sheet" with their advertising. The *Mirror* also printed a letter by the young actor Frederick Paulding, then appearing at the Fifth Avenue Theatre, which called for the American people to ignore, refuse to read or advertise in the *Dramatic News* and thus "quietly kill the viper that is already moribund from its own poison."

[13] Noah Miller Ludlow and Solomon Franklin Smith, pioneer American theatre managers, formed a partnership in 1835 which lasted until 1853. During this time they managed theatres in Mobile, St. Louis, Cincinnati, and New Orleans. This mention of Ludlow was undoubtedly motivated by the publication of his *Dramatic Life As I Found It* by G. I. Jones and Co. in St. Louis in 1880 (see letter 63).

Marcus "Mark" Smith, American comic actor, was best known as an interpreter of "good old English gentleman" in English comedies, most notably at Burton's Theatre in New York.

your new book in London? how jolly to have Joe illustrate the Trip![14] Wrote to him & Warren[15] before I left N.Y.—You shall hear often from me if it's but a wretched line or two. My sister tells me she is hard at her memoir of Father & wants me to give her some material &c, but I cannot possibly do so, in this unsettled state of mind and body. Haven't said half that I set out to say, but must perforce suspend—for 'tis late & before I *coucher* (that's Frinch, you Muggins) I must write an important note. So again goodnight!

On August 27, Booth arrived in London.

61. Bailey's Hotel, Gloucester Road, Kensington,
August 29, 1880

We arrived here day before yesterday & shall be very busy during our brief sojourn prior to the Ammergau trip—next week. I told you in my last that I had closed with Gooch. From all I hear expectation is on tip-toe. Wish I felt as confident as my friends do. Expect Clarke every moment—have not seen him yet. He is busy with rehearsals & I with many little nuisances, looking after luggage, &c, &c. I shall begin with *Hamlet*—for which Gooch is making great preparations. Saw Warner act *Coupeau*, in Drink, last night. An excellent performance. He goes to America with it, I believe.[16] Declined to act seconds with me & announced his early appearance as *Bertuccio*. Gooch bought the play for England & so cut Warner off from his evident intention to anticipate me in the part. This will occasion a split between Gooch & him— though there is an engagement of three years. He's a remarkably good actor & fine looking man. Did I tell you how delighted we

[14] A second edition of the *Trip To England* was published in 1880 with illustrations by Joe Jefferson.

[15] William Warren, American comic actor, was a member of the company of the Boston Museum for thirty-six years, playing 600 different roles in about 14,000 performances. On October 27, 1882, a gala benefit was given at the Museum to celebrate his fiftieth anniversary on the stage. His last appearance was as Old Eccles in T. W. Robertson's *Caste* at the Museum on May 12, 1883.

[16] Charles Warner, English actor, probably had his greatest success in the role of Coupeau in Charles Reade's *Drink*, an adaptation of Zola's novel *L'Assommoir*, in which he first appeared at the Princess's Theatre on June 2, 1879. He did not, however, make a trip to America until 1906.

all were with Miss Ward's "Forget-me-not"? I thought her performance very fine. I see there's trouble brewing between her &
Moss about the play.[17]

Dr. Durant sent me the enclosed. I think it a d--nable nuisance
to be advertised as an invalid. There's been already too much
said about my tongue—whose dark color proceeded from acidity &
is quite cured, thanks to Durant. And I hate to be "clap-trapped"
as a munificent donor of gifts, &c. It looks so shoppy. I have heard
a cricket! Yes, and several birds! Where? At Tunbridge Wells.
I sat by the open window late at night & was ravished with
homelike melodies of cricket & other bugs, and in the early
morning was awakened by the chirps of *real* birds! Since I was
here (17 years ago) they have introduced sun-light into London.
A great improvement too, you would not recognize the old town.
The food here is positively eatable, and I am beginning to feel
hearty and quite hopeful.

'Tis uncertain when the theatre will be ready for me—Gooch
says middle of October, Clarke says November at the earliest;
I hope the latter date. My engagement is for 3 months & I shall
endeavor to vary the bill often & avoid long runs. Have called on
no one yet—for I shall have no time for dinners & society 'till I am
permanently here. I fear for my chances at Ammergau; there
are flocks of people securing seats & beds—until I receive reply
to my dispatch, sent yesterday, I can't say when I shall be there,
but in a few days shall start haphazard.

September 2. Have just discovered to my surprise that this
letter did not go last Monday! We have secured the Ammergau
accommodations & shall start in four days—returning here
Octr. 1st. Went to the Haymarket last night—truly a most beauti-

---

[17] Genevieve Teresa Ward, American actress who spent most of her
life in England, produced and starred in *Forget-Me-Not* by Herman
Merivale and F. C. Grove, at the Lyceum Theatre on August 21, 1879.
She later toured this play all over the world.

Theodore Moss was Lester Wallack's business manager at Wallack's
Theatre in New York. Moss and Wallack produced *Forget-Me-Not*
on December 18, 1880. Miss Ward, however, claimed exclusive production rights. She crossed the Atlantic in late December and secured an
injunction which forced Wallack and Moss to close the play on January
13, 1881. For Winter's opinion of *Forget-Me-Not* see *The Wallet of Time*,
ii, 405-410.

ful theatre in its new dress. Empty benches. Business generally poor at the theatres. Clarke begins next week to help pull it up. Sothern[18] is still very ill at Brighton, hope to get a chance to see him before I leave. I think Florence[19] is a success—the papers those I've seen, speak kindly. The Post seems to be *the* paper—as to theatrical matters, the *Times* is not regarded as authority any longer. The weather still continues bright & London is—as I never knew it—charming, tho' a little dull; I like it all the better for being so—there's more elbow room. I must hurry off for "the city" now & so will leave you. All well. Hope you & yours are ditto. God bless you!

Returning from Oberammergau to London, Booth spent ten days in Paris where he wrote Winter to give him his opinion of the Passion Play, the Paris theatre, and the latest developments of his London engagement.

62. Hotel Continental, 3 Rue Castiglione, Paris,
    September 26, 1880

I really forget whether I've written to you since I left Ammergau or not. On my arrival here I was accidentally interviewed by a Herald man & doubtless my opinions of the play & players have been printed ere now.[20] As a whole it was really wonderful, but I discovered no approach whatever to any individual excellence. I'm glad I saw it but could not be hired to see it again. 'Tis fast becoming a worldly affair and if it really ever produced any religious effect, or inspired such sentiments among the spectators it certainly does not now. I'm too tired, & worried by various matters to enter into any description of it or indeed to write of

[18] Edward Askew Sothern died on January 20, 1881. See letter 68.

[19] William Jermyn "Billy" Florence and his wife Malvina Pray, American starring team, specializing in Irish comedies, filled an engagement at the Gaiety Theatre beginning on August 30.

[20] The *Herald* (September 18) reported Booth as saying that "as a theatrical performance the ensemble was marvelous—as perfect in all its details as if the peasants playing the parts were cultivated and experienced actors." He also liked the costumes and "the care and grace with which they were worn by the members of the troupe." For a complete description of the Passion Play see Clement Scott's articles in the *Theatre Magazine*, May and July, 1880.

anything worth the ink and paper. I merely mean to advise you of my whereabouts and post you as to my movements, so far as I know aught of them. A letter from Clarke, yesterday, informed me that he did not think the theatre would be ready for me before Novr. 15th—so I shall have at least six weeks loafing in foggy London, having leased apartments there dating from Octr. 1st. My wife requires medical treatment & perhaps the sooner we get there & the more idle time we have will be the better for her.[21]

We have not seen a tithe of what we expected to see in Paris— wearied with sightseeing & the general *hulabaloo*. The only excellent acting was by Coquelin,[22] who is really superb. Croisette was nothing in the role I saw, & the rest were pretty good—no more. The Porte St. Martin acting was quite as satisfactory as that of the Français. We passed most of the day at St. Cloud & saw much that reminded me of the old Bartholomew Fair shows. It was novel &, to me, highly interesting to see the "mumers" parading in front of their performances & to hear the crier calling out the excellences of the entertainments.

We shall leave here for London about Thursday next & as soon as I know when I am to appear I'll write you again. I wish I could feel the anxiety & enthusiasm that Clarke does about my advent. I am as dull & waxy about it as a "shotten herring."

Coquelin & Gôt[23] were invited to meet me here at dinner but they did not respond. The Yankee is said to be in bad odor with the Frenchman—why I know not—& perhaps we *barbarians* are not worth their notice. I hardly think we could hold our own upon the stage with them; but I'd like to see their tragedy. You'll have a chance at it with Bernhardt.[24]

I hear that the Corsican Brothers & Warner's Othello are not

[21] Mrs. Booth had been suffering for some time with a throat affliction, subsequently diagnosed as tuberculosis of the throat and lungs. Apparently at this time her condition had become acute.

[22] Constant-Benoît Coquelin, noted French actor, was at this time one of the principal actors of the Comédie Française. According to Winter, *Life and Art*, p. 145, Booth saw Coquelin and Sophie Croizette in a production of Augier's *L'Adventurière*.

[23] Edmond François Jules Gôt, principal actor of the Comédie Française since 1855, was considered one of the finest actors of his day.

[24] Sarah Bernhardt opened her first American tour at Booth's Theatre on November 8, 1880.

very great successes in London.[25] I'm very anxious to see Irving, but not in the present roles he is performing, in which there is nothing by which to fairly judge his ability as an actor.

I hope you are well—not having heard from you for—it seems so long a while—I've wondered if, like me, you are getting physically old. I feel very, very much so. Mentally *antique* we both are—or rather, I should say *spiritually* so, for your mind is ever young & full of "go". I get sadder and feel all over aged. I hope I shall be able to shake off this lethargy by the time my engagement begins.

God bless you, old boy. Write as often as you feel like—doing nothing else, and whenever I can get two or three sentences clearly arranged in my mud-box, I'll send you them.

Returning to London, Booth began rehearsing at the Haymarket Theatre, since construction at the Princess's was not yet complete.

63.   St. James's Hotel, Piccadilly, October 22, 1880

My intention was not to write you again until after my debut, but that seems so far off & it has been so long since we exchanged greetings that I shall wait no longer. For a while I seemed to get a batch of letters every week, but lately there has been a dearth—all appear to have ceased writing about the same time. My plan of alternating Richelieu & Hamlet could not be accomplished—new theatre, lack of time, great expense, &c, &c, but I am promised Richelieu for the 3d. week & *so* it is announced. Although work is carried on night & day the theatre looks very far from completion by the 30th—I fear it will not be ready by that date. I have begun rehearsals, however, at the Haymarket. A sad accident occurred there the first day to my Horatio (Harcourt) who fell down a dark opening (not a trap) used by the carpenters

[25] Henry Irving opened the 1880-81 season at the Lyceum Theatre with Boucicault's *The Corsican Brothers*, adapted from Dumas' novel *Les Frères Corses*, on September 18. This play was first presented by Charles Kean at the Princess's Theatre in 1852. The London critics were generally not enthusiastic about Irving's production. The public, however, was not deterred by the critics. The play ran until the end of the year and made a profit of £5000. See Laurence Irving, *Henry Irving: The Actor and his World* (London, 1951), p. 361.

Charles Warner appeared at the Sadler's Wells Theatre as Othello in September.

for stowing away scenery & left unguarded. The poor fellow fell some six feet & has been in the hospital for a week—he will not be able to act for some months I hear.[26] Clarke is ill & unable— he has suffered tortures with piles & is laid up. I've had a succession of coughs & colds in the head ever since I've been in the fog. Yesterday we had snow, today rain. Mrs. Booth is improving under Dr. Mackenzie's[27] treatment, although she is still coughing & has severe attacks of heart palpitation. Edwina breathes her native fog & is well. Clarke said he was willing to bet that Irving would announce Hamlet against me. I objected— did not think he would be guilty of such bad taste, when lo! his programme![28] If you hear that he proposed me as Honorary Member of the Garrick, don't swallow it—he seconded Bancroft's[29] proposition. He has not condescended yet to notice me. I'm sorry, but shall not allow it to influence my feelings in the least. From all I hear on the breezes there is a favorable sentiment for me, but I look for little in the face of such idolatry of the reigning favorite. If you shd. be near Carleton's shop ask him to send Mrs. Clarke (110 Haverstock Hill) a copy of your & Hennessey's Booth book—he had a copy in his window when I left which I intended to get for her—tell him to charge it up against me. There seems to be a dearth of good stuff here—we have far better in America, if we'd only learn to think so. Actresses particularly are scarce. Irving's company is weak, scenery superb. I could not judge of his ability in such a *role* as the Corsicans. Sothern is getting along remarkably well—his doctor lets him out, but I fear he is overtaxing his strength.

[26] Charles Harcourt, English actor, who since 1863 had been appearing in such roles as Claude Melnotte, Anthony Absolute, and Mercutio at various London theatres, died from the injuries received in this fall on October 27.

[27] Sir Morell Mackenzie was a famous English surgeon and pioneer in the treatment of diseases of the throat.

[28] Irving did not produce *Hamlet* at the Lyceum during the 1880-81 season, although it was advertised for production.

[29] Sir Squire Bancroft, English actor, and his actress-wife Marie Wilton Bancroft are probably best known for their management of the Prince of Wales's Theatre in London from 1865 to 1880. They elevated this theatre from a cheap, second-rate establishment to one of the most fashionable theatres in London and gained a special reputation for their presentations of the English comedies of T. W. Robertson.

Have been obliged to decline lots of invitations—for my health will not stand too many good things. Am sick of sightseeing & go only occassionally to the theatres—argal: it is rayther dullish. When I get to work I shall be gladder.

I told Bispham[30] to show you a letter I wrote him about the P[*assion*] P[*lay*] at Booth's[31]—have been too lazy since to write you about it as I intended. Hope it will not be attempted. Let me know what you think of Bernhardt. You did not send me the account of Ludlow's book—at least I never received it. It ought to be full of interesting theatrical matter—he was for many years with Sol Smith Co-manager of the Southern theatres.

My next to you will be after *Hamlet*. God help him! If I'm to suffer with these headaches & coughs I'll have a hard pull. Can you imagine me in a stove-pipe hat? Such is my cruel fate! 'Tis nigh on to twenty years since I committed such a folly & I am wretched! Pray for me. Adieu!

God bless you!

On November 6, Booth opened his London engagement as Hamlet at Walter Gooch's completely rebuilt Princess's Theatre. The reception of the London critics was generally polite and judicious, but not enthusiastic.[32] As Booth wrote they "damned him with faint praise." The critic of the *Times* (November 7) reported that Booth's performance was "scholarly" and "intelligent" and had the general effect

[30] William Bispham, a close friend of Booth since about 1862, was a partner in the firm of William H. Wallace, iron merchants. Booth's letter to Bispham is printed in Skinner, *The Last Tragedian*, p. 50.

[31] For the week of December 6 to 11, 1880, a production of a Passion Play as adapted by Salmi Morse with James O'Neill, Irish-American actor and father of Eugene O'Neill, as Christ, was advertised for Booth's Theatre. The play had been produced two years earlier at the Grand Opera House in San Francisco. Although in San Francisco O'Neill seems to have given an impressive performance and the production, as directed by David Belasco, was spectacular, the play was eventually withdrawn because of the religious furor it aroused. For the same reason, it was not produced at Booth's Theatre, much to Booth's satisfaction. See Booth's letter to David Anderson in Grossmann, p. 218.

[32] Booth's London engagement of 1880-81, as well as his London engagement and subsequent provincial tour of 1882, are well documented by a scrapbook of newspaper clippings at the Walter Hampden Memorial Library, which seems to have been prepared by Wynn E. Miller, English theatrical manager, who managed both Booth's provincial tour and his tour of Germany and Austria.

of being "beautifully musical and distinct." Charles Dunphie[33] of the *Morning Post* (November 8), which Booth had earlier characterized as *the* paper" of London, praised the performance for its evidence of "conscientious study" and instances of "true passion." He noted, however, that Booth occasionally fell into "artificial grooves" and that his scenes of tenderness and pathos were spoiled because of his exaggerated "vehemence" in other scenes. Some of the criticisms were contradictory. Clement Scott in the *Daily Telegraph* (November 8) complained that Booth looked as if he had stepped out of an old theatrical print. His acting style was old-fashioned and "elocutionary" rather than modern and "natural." The *Daily Chronicle* (November 8), on the other hand, found him to be too colloquial, too natural, especially in the advice to the players and the "To be or not to be" soliloquy. Booth's Hamlet and acting style were of course frequently compared with Irving's. The critic of the *Globe* (November 9) cautiously wrote that Booth's Hamlet was "on the level" with the Hamlets of Charles Kean, Samuel Phelps, Charles Albert Fechter, and Henry Irving. The critic of the *Morning Advertiser* (November 8) was perhaps the most laudatory in his criticism of Booth. He wrote that Booth's Hamlet was "picturesque and impassioned" and that Booth was "a fine actor" whose style had "power" and "careful finish." The critic of *Bell's Life in London* (November 7) was undoubtedly the most condemnatory. He wrote that Booth's Hamlet was from first to last "funereal and dispiriting." It was not the kind of critical reception Booth had hoped for, although he must have expected such, perhaps even worse.

64. St. James's Hotel, Piccadilly, November 13, 1880

The cable has told you all about my debut I presume. I have such a pile of letters waiting replies that I can barely give you a "scratch"—take it kindly.

Well, most of the papers "damn me with faint praise"—but all acknowledge that I speak good English & do not offend. Their attempts at *criticism*—even those that praise me are rather silly. I am told, by the Athenaeum, that I wear a "suit of sables" in the last act because I refer to such in the 3d.—forgetting that I wore the fur cloak in the 2d. Sc. of the first act. That raising the cross while following the Ghost is borrowed from the Germans—I never read or heard of anyone doing it;

[33] Charles James Dunphie, English drama critic, was the critic for the *Morning Post* from 1856 to 1895. He also contributed to the *Observer*, using the pen-name "Rambler."

'twas the result of an accident—years ago. That I do something else at the suggestion of that paper—I forget what, something I've done for the past 100 years (I am 47 today!) [34] From private sources—strangers, as well as friends, I have received great encouragement and there is no doubt of the public feeling for me. All will yet be well. They tell me Irving (who, by-the-by has not yet deigned to notice my existence) has a kitchen & French *chef* in his theatre & cooks a jolly repast while the play is on for the entertainment of his critics *et als* after the show. I can't afford such luxury *yet*. I don't think he has treated me altogether kindly, but no matter. I've had some of the big intellects to see my performances & have heard their very flattering opinions—but as yet no *Royalty* has visited the show. They'll come *poco tempo*. The first night's performance was not—by a league—my best, as you may well believe. A new theatre which gave me as much worry & anxiety as the opening of Booth's—a strange & unsympathetic audience (although they were kindly disposed & applauded—not "in the proper places")— the stalls full of adverse critics who, however well disposed towards me, were committed so far to Irving that I knew their verdict weeks before I appeared, and—altogether I labored under terrible disadvantage. Since the first night I have improved in my acting & the audiences have become warm & "homelike". Many of the "first nighters," too, have, I hear, altered their opinions of me, & the prevailing notion is that the critics have been unfair & went beyond their mark. Take it all in all we, the manager, my friends & I are satisfied. All speak of it as a "great sweep." *Richelieu* & other parts will, perhaps, meet with more

[34] In I,ii Booth wore a black furred cloak; in V,i and ii he wore a fur cloak and fur bonnet. The noted English critic and theatre historian Joseph Knight complained in *The Athenaeum* (November 13) that Booth's bringing Hamlet on stage in the last act in a suit of sables was the "strongest instance" of his over-cautious reading of certain lines. Booth's costume, he continued, suggests that he had read as "a direct declaration of purpose" the line "So long? Nay, then, let the devil wear black for I'll have a suit of sables." To take these words "seriously" was "a treatment altogether too matter of fact." Booth's sword-as-cross business in I,iv and v is described in Shattuck, *The Hamlet of Edwin Booth*, pp. 145-146, 151. The "something else" was the introduction of the actress in II,ii who would later become the Player Queen in the costume of a boy which, Knight noted, "was first recommended in these columns."

favor from the *Press*, for Hamlet has been so often done & has been so identified with Irving the past few years that I could not expect to be accepted in that character. Wishing to hurry up the changes of bill I am kept "on the go" at *240* speed all the time—for though the stage-manager gets the credit for what is done on the stage—he is simply the executor of the "Star" in almost all cases; invariably is it so in mine. This allows me barely time to eat & sleep 'twixt the rise & fall of the curtain. I'd like to have the London reports to the N.Y. papers, if you can send them do so.

On November 20 Booth appeared as Richelieu which the critics received more favorably than Hamlet. The *Standard* critic who had labeled Hamlet "disappointing" found Richelieu (November 22) "from the first to last . . . striking and effective!" The *Times* critic (November 22) lauded him as "a good, sound, conscientious and admirable artist, worthy of very high praise." His embodiment of Richelieu had "an intellectual breadth that takes admiration by storm." Even Clement Scott in the *Daily Telegraph* (November 22) wrote that it was "a carefully studied, even, and highly intelligent perform-ance." Dunphie in the *Morning Post* (November 22) wrote that Richelieu was "a performance with more beauties and fewer blem-ishes" than he had ever seen.

The papers had reported that the public was enthusiastic in its reception of *Hamlet*, but it was even more so for *Richelieu*. The gross receipts for the first week of *Hamlet* were £630.3.0 and for the second week £686.6.0. The first week of *Richelieu*, however, grossed £694.0.0 and the second grossed £845.11.6, the highest of Booth's entire engagement at the Princess's Theatre.[35] The engagement looked to be more of a success when Booth wrote Winter the following two letters.

65.   St. James's Hotel, Piccadilly, November 23, 1880

I have had two letters and as many papers together with some 'slips' from you since I last wrote you. Since then, too I have produced *Richelieu*, as you doubtless know. The well-fed Irvingites, who number among them nearly all the critics, I am told, reluctantly follow the popular lead. Except Scott—he gave Clarke to understand from the outset that my game would be

---

[35] From Booth's receipt books in the collection of the Walter Hampden Memorial Library.

blocked. He is the "Bolton Rowe" you mention.[36] Is it my dear old friend, John Russell,[37] of Connecticut you refer to? I really loved that man until—by lying most basely he succeeded in swindling me of *$3000*; he tried for *$7000* but my lawyer (the fellow that McV. sent from Chicago) discovered the fraud & took the balance, with nearly as much more, for himself. The last time I met Russell he squirmed & spluttered like a naughty boy—as he is. O, I have lots of villains plucking at my fame. The nearest, alas! were once the dearest. There was (I think it is well nigh squelched) an Irving set against me, but the Warner lot, if any, which I doubt is of no earthly account whatever. He's a mere nobody. From higher sources than the *Weakly* Press of London I have received splendid assurances of a full appreciation of my acting, and the veteran Ryder,[38] with tears in his eyes & choking voice, declared to me the first night of *Richelieu* that I had toppled down his idol (Macready)—after thirty years—&c—&c. The poor dear old man cd. say no more, but left my room abruptly. Now, that's worth a barrel of such nincompoops as Clement Scott—who, they say, owes his

[36] Clement Scott frequently used the pen-name "Bolton Rowe," the name of a street in London's Mayfair district. Scott's review of Booth as Hamlet in the *Daily Telegraph* was harsh and obviously prejudiced in favor of Irving. In the December issue of the *Theatre Magazine*, he printed an enthusiastic appraisal of Booth's Hamlet by J. Palgrave Simpson, although it was immediately followed by a reprint of Scott's review of Irving's Hamlet at the Lyceum on October 31, 1874. His review of Richelieu, while perhaps not enthusiastic, was warm and judicious. If Scott was trying to "block Booth's game," he was not trying very hard.

[37] Presumably John E. Russell, although he was from Massachusetts not Connecticut. Russell was the drama critic for the New York *Sun* in the 1860's when he first became acquainted with Booth. His subsequent career is largely obscure. He traveled extensively in Europe and South America and made "an ample fortune" during a "brief but brilliant business career." He served one term as a representative to Congress from Massachusetts and unsuccessfully campaigned twice for the office of Governor of Massachusetts. Russell alienated Booth in some financial matter concerned with Booth's Theatre and they were enemies for many years, but they became friendly again in 1881 (see letter 76). A number of Booth's letters to Russell are printed in Edward L. Partridge, "Edwin Booth to John E. Russell—Some Hitherto Unpublished Letters," *The Outlook*, CXXVII (April 20, 1921), 637-639.

[38] John Ryder, popular English actor, supported Booth at the Princess's Theatre as the Ghost in *Hamlet* and as the Capuchin in *Richelieu*.

Magazines to Irving's ducats.[39] But, this is not the strain of all, at all. I have not the least feeling—save that I came with [a spirit?] of co-operation not rivalry in a pursuit that should unite, not divide its followers.

I am glad of Irving's success—for he is a good worker in the right direction, & if I leave but a stone behind me fit for the temple I shall be satisfied. I am assured that I am doing a good job, teaching the English English & their actors how to act. And much of this comes from the actors—of other companies— who have seen me. All this is gratifying: Unfortunately my engagement ends just as the London season proper begins, & I fear I shall not have a fair lick at the *nobs*. Before I quit Irving I must tell you that after I had acted some ten nights he called. Referred to my letter of 18 months ago, but did not apologize or offer any excuse for not answering it. I shall return his call tomorrow. Was disappointed—prepared to like him, but he left a disagreeable impression on us all. Looked askant at me & mumbled 'tween compressed lips, &c. I've heard no more of his *Hamlet*. He told me he should next produce a play by Tennyson.[40]

Clarke is injudicious—he has talked *Me* too much, but I don't think he has said anything "American." He is not well & has been unlucky—dont see much of him now. He talks of going to New York next month. Heard from Sothern (at Brighton)—he is no better, no worse, he tells me. Dunphie has spoken well of Richelieu, but even he & Knight were too full of t'other *Hamlet* to swallow me whole. The latter was in my room last night & spoke very enthusiastically to me about *Richelieu*. The unfairness of judging an actor by his first night's work was never more manifest than in my case. A new theatre & company, with all things "hugger-mugger", and some dozen or more of adverse critics (who made no secret of their feelings) in the stalls—just under my eye-brows. Think of it! Every burst of applause was hushed

[39] The *Theatre Magazine* began publication in January, 1877. Irving acquired the magazine in August, 1878, and controlled it until December, 1879, when he turned it over to Scott for £1000, which was never paid, and for a quarter share in future profits, which were never realized. Scott sold the magazine in 1889. See Laurence Irving, pp. 349-353.

[40] Irving produced Tennyson's *The Cup* at the Lyceum on January 3, 1881. It was presented for four months until the Booth/Irving engagement interrupted the run in May.

down by these fellows & others in the gallery—at least they endeavored to kill all the applause & did succeed in marring most of it. I felt as much like *Hamlet* as a bull in a china-shop might be supposed to feel. I'd rather have *horned* the whole "Kerboodle" of 'em than bid for their favor that night. But let 'em go. Of course you heard of poor Harcourt's death. My next leading man, Redmund, fell ill the last rehearsal of *Richelieu* and a substitute has been doing Baradas—badly.[41] Indeed, the entire company is far inferior to any I've had at home. Julie wretched, yet her Ophelia was liked & praised almost as much as Miss Terry's.[42] How bad Miss T.'s must be! Of course you'll not say *my* say of the company, for they are all very kind and show me great courtesy. I seem to have won the hearts of nearly all the English actors—perhaps 'tis because they don't like t'other chap. Maybe. I hope the "Passion Play" will be prevented. Abbey is mad & will regret his obstinacy in this matter. I hope you'll let the people know that though the theatre bears my name I have no connection or authority—if I had I'd put a stop to the profanation. I am surprised that all the actors (except Jim O'Neil [sic]) do not refuse to take part in it. My wife is about the same—Edwina continues well & I also. I wish I could hear some good report of your health, my dear boy. God bless you! Goodnight—'tis now 1.A.M.—after *Richelieu*. Love to all yours

[41] William Redmund, English actor, made his first London appearance in 1875 at the Standard Theatre and during the next decade appeared in juvenile and leading-man parts at the principal London and provincial theatres. His substitute was Thomas Swinbourne, who made his first London appearance, after more than a decade touring the provinces, as Captain Randal Macgregor in Boucicault's spectacular drama *The Relief of Lucknow*. During the next decades he played leading-man roles in support of Samuel Phelps, Helen Faucit, and Henry Irving, among others.

[42] Julie was Florence Gerard, English actress, who after her stage debut in 1877 at the Theatre Royal, Portsmouth, gained prominence in the provinces and London as a brilliant comedienne. Her engagement at the Princess's Theatre in 1880 marked her return to the London stage after the absence of a year because of illness. Her performance of Ophelia was praised by the critics, but her Julie was generally found to be lacking in strength. Indeed the critics unanimously complained about Booth's "poor" and "indifferent" support in *Richelieu*.

Ellen Terry played Ophelia for the first time in Irving's production of *Hamlet* which opened at the Lyceum on December 30, 1878. Her performance in this role was generally acclaimed by the London critics.

66.   St. James's Hotel, Piccadilly, November 29, 1880

Your two books, Winter & Arnold,[43] came safely—for which much thanks. Edwina is at them now & when I get a moment's leisure will—I am sure—take great pleasure in them. I knew Arnold very slightly—he passed off during my early days in New York, but I have frequently chanced on his poetry. An article by *House* in the Herald, very kindly meant, says I was once a "negro minstrel" which he doubtless drew from the fact that during my apprenticeship as a stock actor I played "Dandy Cox", &c.[44] The *troubadour* business is not regarded as very reputable here & as the *Herald* article is copied by an unfriendly sheet in this little (minded) village it ought to [be] contradicted. I don't care much about it so little did I regard it that I answered House's letter & thanked him for the article, but my sister is hoppin' mad about it, for she says the Londoners will class me with those tramps who go about the streets with blackened faces, &c. The houses continue to grow & the enthusiasm also. The lack of scenery renders it very expensive to the manager to change the programme as often as I wished—there's not a stick or rag in the place that is not to be expressly made up into scenes & properties for my plays & therefore I must "let up" a little and will keep on "Richelieu" longer than I intended. *Hamlet* could have held its own & by this time, perhaps, been accepted by the mass (tho' not by the *critics*) as it was by the brains of London. But I want to bring it on again, later in the season & am satisfied to have drawn the foeman's fire. Now they're non-plussed, for *Richelieu* upset their pre-conceived notions of my lack of power &c, &c. 'Tis my fate to be interrupted at every sentence; now 'tis dinner. Called on Smalley

---

[43] The "two books" were Winter's new edition of *The Trip to England* and his edition of *The Poems of George Arnold* (Boston, 1880). Arnold, American poet and author, was a close friend of Winter during his Bohemian days before the Civil War.

[44] Edward Howard House, American journalist, drama critic, and musician, was at this time living in London with his close friend, the noted novelist and dramatist, Charles Reade and, according to his own report published in the *Century Magazine* (December, 1897), in close communication with Booth. His article reporting that Booth was once a "negro minstrel" appeared in the New York *Herald* (November 7, 1880). Booth's letter to House thanking him for the article is printed in the *Century* essay, p. 270.

today who had a letter from you. Will wind this up after dinner.

'Tis accomplished! Some noble-minded country woman of ours gave Mrs. Booth two pies! a mince and a pumpkin. The latter is no more—the former shall be kept till Xmas, if possible. This is the most important event that has occurred since I last wrote you.

If you see *Joe* give him my love & tell him all goes well. Shall write to him soonly.

Haven't sent you any notices—for even the favorable ones are *wishy-washy*. Knight is very enthusiastic in the "Sunday Times"—but I've no copy of the paper. Caricatures abound, Irving, Bernhardt and "Your humble [illegible words] all which denotes fame, I fancy. Mrs. Booth still coughs dreadfully—the rest of us well. Love to you and yours.

Booth continued *Richelieu* until Christmas Day when he appeared as Bertuccio in *The Fool's Revenge*. The critiques were generally favorable and the average receipts were larger than those for *Richelieu*, despite cold, snowy weather. The London engagement which had begun badly now appeared as if it might become a complete success and Booth's contract was extended six weeks. Booth was not, however, completely satisfied with the engagement. He complained of Gooch's "cheap Jew" managerial tactics, which had forced him to reduce his terms, and of his poor support. He was concerned, furthermore, about Mary, whose condition was not improving.

67. St. James's Hotel, Piccadilly, January 5, 1881

Your gift of books came timely and I also received the papers with bills you sent me. Thanks. How well Joe has illustrated the "Trip".—Well, I'm hammering away at my third role, Bertuccio, which seems to have stirred 'em up quite lively—my next will be Othello & Iago—in both characters I expect to be severely "gone for" by the critics. By-the-by one of the leading ones (Knight) was asked what he really thought of my Hamlet—"O, he looked too old" was his reply. A famous actress was similarly criticized by a Chicago lady of our acquaintance who disliked the actress because she "dressed her hair so badly". Scott is very full of gush over Tennyson's "Cup," which is a flat failure, I'm told.[45] I

---

[45] *The Cup* was hardly "a flat failure." It was an expensive production costing, according to Laurence Irving, p. 366, £2,370—about £300 more

believe the press generally praise it though. My business runs along very well & I have just agreed to extend the term of my engagement 'till March. I am very much fatigued and not at all pleased with my surroundings. My Managers are cheap jews and the company very poor. My wife's health does not improve, as it once showed signs of doing; she loses flesh rapidly—though her cough is better. Edwina has the best time of it—having formed some very pleasant acquaintances who dine and dance her to her heart's content, while I am kept at the mill grinding life out— without any relaxation whatever. Fourteen weeks more of it & then for a rest I hope. Smalley is very genial—but I've seen no important mention of my engagement in his letters; he'll go heavy on the "Cup," I presume.[46] Have not seen Irving since he called—some weeks ago. Many of his ardent admirers are becoming very *Boothy* & speak of him as "poor, dear Irving!" He told me he should never go to America. I wish he would & let me have his theatre during his absence. I assured him that he would receive great kindness there—he is led to suspect that the American press would pitch into him on my account, & such rot. Of course, he did not say so, but I've heard it from his friends. Did Hatton's letter to the "Times" appear?[47] He read it to me before sending it & I expected to see the paper ere this. Clarke is still quivering on the "go"—but has not yet set a day for sailing. I received the book you ordered from Carleton's & gave it to my sister on Xmas. I spent the day with her, the first time since our childhood. I hope you all at "Fort Hill" had a happy time of it.

---

than Irving's splendid production of *The Merchant of Venice*. The critics were generally disappointed by the quality of Tennyson's play but praised Irving's and Terry's acting and the lavish settings and costumes. *The Cup* was not as popular as *The Corsican Brothers*, but Irving still netted £8,500 at the end of his fourth season on April 9, 1881.

[46] Smalley's "Letter from London" column concerning *The Cup* appeared in the New York *Tribune* (January 16). He reported that he did not attend the opening night of the production and proceeded to summarize the reviews from the London papers.

[47] Joseph Hatton, English novelist, journalist, and editor, was the London correspondent of the New York *Times*. He dispatched a letter to the *Times*, regarding the reception of Booth's Richelieu, and reported an interview with Booth on December 17 which was printed on January 2, 1881.

Browning, Martin, Huxley, Froude and McCarthy[48] are the only interesting men I've met thus far—but have had kindly treatment from many of the "effete" aristocrats. The weather is called lovely here, I presume it is & should be grateful that I've had no worse. I shall now have a new set of audiences—as parliament opens tomorrow & *the* season begins, much earlier than usual—all for my sake. Prince Leopold[49] is the only blood royal that has yet *honored* my performances with his presence, but I've had lords & ladies *galore* in front. Love to all. Happy New Year!

On January 17, Booth opened in the role of Othello, alternating Iago with Henry Forrester.[50] The critics were almost unanimous in their condemnation of his Othello. Generally it was found to be "artificial," "over-elaborate," "too elocutionary," and "old-fashioned." Booth's Iago, on the other hand, was declared a "masterpiece," "brilliant," "artistic," "intelligent," and "scholarly." The financial success of this production was also seriously affected by a terrific snow storm and cold wave which blanketed London during the week. Booth's attitude towards Gooch, moreover, was even more vehement than in his last letter to Winter. Generally, the reception of his Othello, Gooch's management, the weather, and the death of E. A. Sothern had thrown Booth into a despondent mood.

68. St. James's Hotel, Piccadilly, January 22, 1881

After I had finished Othello last evening I was told of poor Sothern's death—I had spent an hour with him on Monday & really thought he might die while I was by his bedside, so weak and emaciated was he. They had taken him to Bournemouth, at

[48] Robert Browning, noted English poet and playwright.

Sir Theodore Martin, English author and editor, and the husband of the popular English actress Helena Faucit, Lady Martin.

Thomas Henry Huxley, famous English scientist.

James Anthony Froude, English editor and historian.

Justin M'Carthy, Irish politician, journalist, and historian.

[49] Prince Leopold, the Duke of Albany, was Queen Victoria's fourth and youngest son. According to a letter from Booth to E. C. Stedman dated December 24, 1880, Prince Leopold saw Booth perform Richelieu on December 23. See Grossmann, p. 221.

[50] Henry Forrester, English actor, first played Iago at the Lyceum Theatre on February 14, 1876, in support of Irving's Othello. He was acclaimed for his performance and was generally considered to be one of the finest Iago's of the time. In the season following his debut as Iago, Forrester appeared at the St. James's Theatre as Othello with equal success.

the Dr.'s suggestion, but he suffered so much there he was taken back to London. The journey there and returning—was enough to prostrate a man in his condition. Poor fellow! What a full company of players has made its exit this twelve-month past! I think I have acknowledged all the books you sent me—as well as bills & papers—some of the latter came yesterday. Ryder remembers you with apparent delight & wants to see your "Trip" —can it be got here? He's a dear old boy—full of good.

I'm after the *Era* for you; one of the almanacks ('79, I think) is out of print—but Ledger[51] says he will hunt up a copy and let me have it—if he can possibly get one; soon as I get them they shall be mailed. Clarke sails on Tuesday and perhaps I can send them by him. I have sent no criticisms home to anyone—don't know why; have lost my interest in such matters—tho' some of them have been very . . . . Forget what I was about to say, twenty-four hours have elapsed since I said "very"—but you may guess how *very* some of them are. *Othello* has been d----d by the severe snow storm, the worst in 50 years! and by the faintest of faint praise from the few critics who ventured out last week. My *Iago* is *clever*; my *Othello* is feeble. Both compared with Vezin's & Forrester's.[52] The former's I've never seen, the latter's I wish I never had, for his acting is about as direful as you'll find in Ypsilanti or Oskosh. Comparisons, too, with frumpy old Phelps[53]— whose tragedy was dreary, are still indulged in & not very complimentary. The fact is I'm not sociable. Dunphie has at last become tired of fishing for me and has dropped me—flat as a flounder. The public should not be judged by the critics—in

[51] Edward Ledger was the proprietor and editor of *The Era Almanack*, an annual publication concerned with drama and music in England founded by his father Frederic Ledger in 1837. Winter, since 1873, had occasionally contributed articles to the magazine.

[52] Hermann Vezin, English actor born in America, was generally considered an outstanding actor, especially in such roles as Macbeth, Othello, Jacques, and Sir Peter Teazle. Vezin's wife, Jane Elizabeth Thompson, formerly the wife of the American actor Charles Young, was a member of the Princess's Theatre company supporting Booth.

[53] Samuel Phelps, English actor-manager, is probably best known for his management of the Sadler's Wells Theatre from 1843-1862, during which time he produced all but six of Shakespeare's plays. He was a versatile actor and was generally considered excellent as Othello, Lear, Sir Pertinax McSycophant, and Bottom.

London, at all events. The applause has not flagged since the first night of my engagement—the houses have been well-filled, and commendations from private sources continue—verbally as well as by letter. I never felt more warmth and genuine appreciation, and have seldom seen such demonstrations of satisfaction at home, yet the critics say I fail to touch the hearts of my audience, and attribute the applause I receive simply to the kindly feeling the Englishman bears his guest.

Isn't it encouraging to read such rot? True, I have not felt much like acting since I began—for everything connected with the management is distasteful to me. Gooch, a cheap-John Jew, has taken a vile advantage of me & coerced me into a filthy bargain on his own terms—closing my engagement at a time when it should be in its bloom (next March), and hampering me every way—with a wretched low company, and in many other things making my spirit ill-at-ease. He is about the vilest cur I ever dealt with—not excepting Robertson, Clarke Bell, Barnett or John E. Russell. The story is too tedious and too "madding" to tell here & now. No actor ever had more odds to fight against, on the one hand, while—I must confess—none has had more favorable auspices on the other. My lack of judgement & foresight, & my idiotic credulity prevented me from securing the advantages that were really in my hand; I failed to close my fingers tightly & the chances slipped from my fist. What opportunity may be offered in the Spring I know not—at present I see none. I can hire Irving's place in the fall, but I've had my swing at that season and want the spring-time *nobs*—who come just as I must go. Irving has been very pleasant of late—and I can't blame him from guarding his own interests which are very important & not to be neglected. I hope the American press will treat him kindly. His friends think the papers' treatment of me will influence the American critics: I hope not. I've already seen several such hints in print, and many "blatherskite" compatriots of mine have openly threatened to roast the English actors, &c, &c. People I don't know, thank God! but their remarks have come 2d. hand to me, & I'm told they have been used as coming from my *friends*. I've no doubt the "spread-eagle" has created a sour feeling for me in certain quarters.—Yes, it is strange that *scalliwags* are permitted to remain while the good ones are so frequently called away. And how sad it seems that when the good fellow, who has had

such hosts of friends, goes to his grave (very often of late, at all events) so poorly attended. Witness Brougham's & Floyd's[54] cases; I don't know what is to be done in Sothern's—I've heard nothing, and on account of a severe cold, which I fear to aggravate while acting, I shall not be able to take any part in the affair—beyond visiting the house of mourning; I doubt very much if there'll be many of the hundreds who flocked about him in life will follow his corpse—or think of him after today. My acquaintance with House is very slight—I occassionally met him, years ago, but never encouraged his advances towards intimacy, but he always seemed to be familiar with me; he told Clarke he was coming here—I hope not till I am gone at all events. I'm sorry that Percy has been so unfortunate, but hope his accident has left no bad results as mine did; my stiff arm is a nuisance & renders me—in some respects—helpless, at least greatly dependant upon others' help. I do not dread the Russell & "Rowe" clique— the latter has greatly moderated his tone; the former is an ingrate & a falsehood to the core. Have lunched with Lady Coutts & Irving lately—she saw Bertuccio & Othello, but is too much taken up with her approaching "change of life" to think of theatres just now—or I'd ask her to build me one. Supper is being served and I must again defer this most interesting epistle.

X (Supper's done gone.)

What have I gained by acting here? I haven't knocked the dust out of old Drury's Cushions (I think you prophesied I could) nor scattered the old owls from their roosts—"or words to them effex". It seems to me a loss of time & labor. I shall leave no impression—there seem to be few minds here worth impressing. The actor's art is judged by his costumes and the scenery if they are not "esthetic" (God save the mark!) he makes no stir. I believe Clarke is the most disappointed of my set—for he had built high hopes on the trad[it]ions of Kean's and others' great successes here, which I never did—I did expect, however, more interest for the Shakespearian drama than is manifested. Chas. Kean, Fechter, & Irving have feasted the Londoners so richly they cannot relish undecorated dishes. I was at the same ill work at home—but was fortunately checked, by fire first and afterwards

---

[54] John Brougham died on June 7, 1880. Booth and Winter were among his pallbearers. William R. Floyd died on November 25, 1880.

by bankruptcy. I do not regret my losses now—since I've seen the evil results of "grand revivals."

January 24.

[P.S.] I failed to seal & post my letter last night & finding yours of the 10th on my breakfast table this morning I add this scrap by way of acknowledgement. The fog is dense—my head aches & I've had a severe cough. The business is what is called good in London—for this season, at home I'd think *queer*; all London is one huge *dead-head*, particularly the wealthy & "nobby" part of it. Palmer is misinformed as you will have seen by the forepart of this letter. Gooch is anxious to get rid of *stars* & likes the cheap & nasty drama better than Shakespeare. We are still freezing here—it looks as though a real American winter of the hardest sort had set in here for the season. 'Tis unfortunate that it should have struck the very first week of *Othello*.

Am now making engagements for a provincial tour next fall. Hope something may open here for the Spring.

On February 14 Booth appeared as King Lear. The critics commended his performance without reservation. Charles Dunphie in the *Morning Post* (February 16) wrote that Booth's performance of Lear "has touches of beauty and grandeur as pure and as lofty as any within the reach of any actor." The critic of the *Sunday Times*, probably Joseph Knight, wrote (February 20) that it was a "remarkable performance," "a brilliant illustration of method and capacity" which had "seldom, if ever, been seen on our stage." Even Clement Scott in the March number of the *Theatre Magazine* called Booth's Lear his "greatest intellectual success."

69.   St. James's Hotel, Piccadilly, February 15, 1881

I send with this two Era *Almanacks*—the 3d. cannot be obtained, although I have a scout still on the search for it; with them will be mailed a Sat: Review anent Othello—my conception & performance of which part are so rarely understood or commended by my critics.[55] A Shakespearian (!) friend—who gloats

[55] The article concerning Othello appeared in *The Saturday Review* (February 5). It was written by the English author Walter Herries Pollock, who was at this time a sub-editor, later the editor (1883-1894), of *The Saturday Review*. Pollock insisted in his essay that the "poetical view" of Othello was the only "true one." Othello, to his mind, was a

over my Iago—fails to discover in Othello—either read or acted (by me or any one)—anything but a "beast"—as Sh: intended him to be! Did you ever?!!! I cannot possibly see the least animalism in him—to my mind he is poetically pure & noble; even in his rage, which wrings from him occasional *curse*'ry remarks, not over-refined, perhaps, I perceive no beastiality—such as my learned friend discovers. I've just received a Tribune (Clara Morris—article) and two packages of bills from you—t'other day from Brougham arrived.[56] I've read & like it. What a pity the poor old fellow hadn't completed his autobiography! Last night I produced *Lear*, & from the wild demonstrations that greeted my every scene from first to last—I judge it to be a "go"—I've seen but two articles this morning. Most of the papers wait a day or two before publishing their notices of important plays & players. All the critics were there in force, however, & I daresay tomorrow's press will record their opinions. I'm glad I've been able to "hit 'em again" with a test part—Shakespearian, before my too brief engagement closes. I've had a cold in the head & a distressing cough the past three weeks & the anxiety about my wife's suffering rather disturbs my artistic moods, but latterly I've done some of my best work—I think. Day after tomorrow I play at the Lyceum (4th act of Richelieu) for Chatterton's benefit: don't know why—but "everybody" said I ought to—so I consented; shortly I shall act again (Drury Lane) for the Dramatic Fund (Petruchio)[57]—this is all right, of course; but 'tis pretty hard to

romantic, noble character, not "a violent savage," nor even "a mere savage overlaid with a thin veneer of acquired civilization." He thought Booth captured this concept almost ideally. Even Booth's occasional "thrilling bursts of passion" were not those of "a demi-savage carried away by the inflamed temper of his blood, but of a noble nature led astray by the diabolical wiles that work upon its freedom and openness to believe that it has been wronged to the uttermost and must exact, not blind vengeance, but the uttermost penalty."

[56] Clara Morris, American actress, was then appearing in *Camille* at the Union Square Theatre. Winter's generally laudatory review of her performance was printed in the *Tribune* (February 3).

About January 25 *The Life, Stories, and Poems of John Brougham* edited by Winter was published by James R. Osgood in Boston.

[57] Booth appeared at the Lyceum Theatre on February 17 in the fourth act of *Richelieu* at a matinee benefit for F. B. Chatterton, the manager of Drury Lane from 1863 to 1878, whose last season had ended in financial

be acting in the day, at other theatres—and other parts while working at *Lear*. I have not yet presented my letter to Mr. Lord—though I have met his wife & have accepted an invitation to dine & lunch. I really have little chance to go out during the day; shall call tomorrow. I think I thanked Mr. Butler[58] for the letter, but whether I did as much for your kindness in getting it for me or not is beyond my memory. I do so now at all events. I am in doubt as to what I shall wind up my term with—Richelieu is asked for, but I somehow feel as if I'd like to give a week or two of Richard 2d. The company is vile—the management mean & very disagreeable, & I rather rejoice at the prospect of getting rid of such embarrassments, but where else to go?—If my wife's health does not improve I'd rather give up acting though it should deprive me of the best of the London season. The weather at present is beastly! wet & foggy—though the late nights are frequently clear & star lit. I hope Clarke arrived safely—my love to him. There is a lot of money for him in America & I hope he'll get it.

Love to all your folks.

Shortly after his last letter to Winter, Booth made an agreement with Irving to play in *Othello* at the Lyceum, in which they would alternate Iago and Othello. It would be presented only three, or at the most four, times a week, with productions of *The Cup*, *The Corsican Brothers*, and *The Belle's Stratagem* on the other days. The opening was scheduled for May 2. As early as February 24, a notice of this agreement was released to the press, but Booth did not inform Winter until almost two weeks later. His delay, he wrote, was caused by his concern for Mary. Her condition was critical and her physicians did not expect her recovery. Mary's illness apparently had also affected her mind. She seems to have suffered from delusions of persecution, accusing Booth of not loving her and of neglecting and mistreating her. Booth traced the cause of her mental anxiety to a relatively obscure incident involving Mrs. McVicker and a Mrs. Young which occurred at Saratoga the previous summer. They first seem to have planted the idea in Mary's mind that her husband did not love her, that he married her for her money, and that he neglected and abused her. On June 29, the day before he left for England, Booth wrote Winter about his

---

disaster. On February 28 he played Petruchio at a matinee at Drury Lane for the benefit of the Royal Theatrical Fund.

[58] Probably Richard Butler, English dramatic author, who was for many years editor of *The Referee*.

domestic difficulty (Folger Shakespeare Library, Y.c. 215 [370]). Most of this letter was destroyed but a few fragments survived which provide a partial insight into Booth's "secret sorrow" as he called it. "You could not credit, Will," he wrote, "the vile & wicked things I've been told by husbands of women to whom my crazy wife & her wicked mother have 'confided' their sorrows. I certainly deserve a worse death than my married life has been—for the wicked deeds I have committed!" Booth, however, did not accuse J. H. McVicker: "Remember whatever I may say about the misery I have endured these eleven years past, that the *he* McVicker is not included in my list of devils— he is simply a good old man that will be—I won't say *talking* for he does not talk, at least I think not, but he will believe the lie & not the truth." The rift between Booth and the McVickers widened as Mary's death became more imminent.

70.   St. James's Hotel, Piccadilly, March 6, 1881

It seems a long while since I wrote you—surely not since the arrival of Fitz-James O'Brien,[59] only your preface to whose memoirs have I had a chance to read. I have rather a confused recollection of him. I hope to have a clear chance at you tonight, but an invitation to a late reception must not be neglected, although neither my wife nor I are in a fit condition to go out. I am merely run down by nervous strain—owing to anxiety, though, on her account, as much as to the nightly drain upon my system. This run of *Lear* is the heaviest pull I've yet had. Mary's case is a more serious one than mine. I should have told you earlier—but I've been unfit for writing much of late, & have barely managed to answer long delayed communications. The doctors, MacKenzie and Sir Wm. Jenner[60] consider her case an alarming one, and yet propose no change of place or treatment. They do not think it well that she should know the worst, for it would hasten the end & render the little remnant of her life miserable. She may, with her indomitable will keep up much longer here,

[59] Fitz-James O'Brien, American journalist and author born in Ireland, was one of Winter's Bohemian friends. *The Poems and Stories of Fitz-James O'Brien*, edited by Winter, who also contributed a biography of the author, was published about February 1 by James R. Osgood in Boston.

[60] Sir William Jenner, distinguished English physician to Queen Victoria and the Royal family, was considered a specialist in the treatment of tuberculosis.

with the social excitement she has—in spite of her daily increasing weakness—than under other conditions, & so I have to humor her every whim. Her mind is so affected that everybody she talks with notices it—she is conscious of it, & 'tis pitiful to see her break down, like a baby, at every mistake she makes & at her frequent loss of memory. She is visibly wasting away, and must soon wear out entirely. The doctor has written her parents, without her knowledge, for she is sure of getting well & does not wish them to know how ill she is. She coughs all night & sleeps from exhaustion all day. 'Tis well not to let the true condition of affairs be known publicly yet, for should it reach her it would increase her wretchedness. Strange! All this dates from the d--nable break in my domestic circle, caused by her mother's & Mrs. Young's deviltry. Since the unfortunate experience at Saratoga, of which—I think—I told you, Mary has been less sane than ever & has had severe coughs and loss of flesh. It is a terrible strain upon me, as you must know, and acting the madman everynight seems to keep me in a semi-craze all the while. The 21st will terminate my "job" with Gooch and after a month's rest I shall begin a series of performances (alternating Othello & Iago) during May, perhaps longer, with Irving—who very kindly met my request for an occassional Matinée during the *season*, with the proposition to act three nights per week with me instead. This is very good in him & proceeds, I am sure, from a genuine artistic feeling. If you can give him a lift & let him know that I appreciate his feelings I'm sure 'twill make him happier than anything I can say. He had a doubt of how he wd. be received in America, on my account, for many ugly things have reached him from the lesser press of America, though I assured him he would receive a cordial welcome there, and that no ill-feeling whatever existed concerning us "poor players"—except, perhaps, among a few feeble minded, little-brained bummers of the press. Forrest set all such d----d nonsense at rest by the bloody farce of 1849. "Lear", to my surprise has toppled over even Richelieu & Iago. 'Tis the one character which even Hazlitt admitted my father to have acted better than Kean. I have received many flattering congratulations on the performance, while not a few want to see Hamlet again. But I see no opportunity at present to do that or other characters I'd like to act here. If I could only have a swing

at a theatre for a year—untrammeled by cheap-John managers—
I could do something worth working for.[61]

On March 21, Booth began the last week of his engagement at the
Princess's Theatre with a double bill of Petruchio and Shylock. The
London critics admired his versatility. They found Shylock, however,
old-fashioned, unsympathetic, and generally much inferior to that of
Henry Irving. Financially, it was the worst week of Booth's entire en-
gagement. On March 26, he closed and in a letter to E. C. Stedman
sighed "Thank God a thousand times, again and again repeated."[62]
After securing less expensive quarters in an apartment house on
Weymouth Street, Booth began rehearsals with Irving.

71. 23 Weymouth St., Portland Place, April 28, 1881

Various worries prevented my writing you—even a line, as you
requested in yours of March 30th, until the advent of
McCullough[63]—so I thought I'd wait a day or two longer to tell
you the result. So far as one can judge of these Englishers he has
made a favorable impression. The papers treat him well & the
"first night" audience seemed mightily pleased. 'Twas the second
time only that I have seen him from the front—once at the Opera
House, as Coriolanus. On both occassions I was impressed by
his repose & freedom from that fault we tragedy-boys indulge
so freely in—over gesticulation, both qualities are most difficult
to acquire—but I think they are natural with him. If he had a
longer engagement in order to keep up the paper-talk about him
it could be better for his popularity, as it is the one play may not
give him the means of riveting the nail. I heartily wish him the
fullest success. I have been rehearsing Othello with Irving the
past two weeks. The play will be finely set & well acted by the
subordinates. Whatever may be the opinions concerning his
histrionic ability (wh., after all is but a matter of *opinion*) he
certainly has the element of greatness, largely developed, for
the very difficult art of stage-management. His patience & untiring
energy—his good taste & superior judgement in all pertaining
to stage-craft are marvellous! He is very gentle & courteous—

[61] The rest of this letter is missing.

[62] Grossmann, p. 224.

[63] John McCullough opened at the Drury Lane as Virginius on April 18,
1881. His engagement lasted through May 21, during which time he
appeared also as Othello.

though tried to the extremity of endurance by the "awkward squad." To be on his stage & find all one's own ideas perfectly carried out is delightful! He imparts to the humblest member of his corps a somewhat of the true artistic feeling that animates himself. Rehearsals, which were always an exhaustion & a bore to me, are, on his stage, with him as director, positively a pleasure; it revives all my old interest in my profession. He deserves all the honors heaped upon him & wears them modestly.

Poor Mary lingers still in lunacy, but is physically stronger than she was; yet there is no hope of her recovery. Edwina bears up bravely & superintends the household & social duties wonderfully well. My health is very good—in spite of the anxiety I suffer. Expect McV. soon, his wife has been two weeks here. Love to all your folks & you.

After months of publicity and several weeks rehearsal, *Othello* was brought out as scheduled on May 2 with Booth in the title role and Irving as Iago. The London critics hailed the production as a complete triumph. Booth's impersonation of Othello, which five months earlier had been condemned almost unanimously, was now almost unanimously commended. Charles Dunphie wrote in the *Morning Post* (May 3) that Booth's Othello still possessed all its former "beauty and finish" but displayed "more concentration, more force, and more balance." The critic of the *Globe* (May 3) echoed Dunphie, praising Booth's performance as "more sustained, riper, and more mature than before, and more passionate also." Even Walter Pollock, who had been the only critic with whom Booth's Othello had found complete acceptance earlier, now noted that in the more favorable circumstances afforded by the Lyceum production the "truth of his conception was deepened." Irving's Iago and Ellen Terry's Desdemona were acclaimed as their finest performances.

On May 10 Booth and Irving exchanged roles for the first time. Once again, as before, the critics endorsed Booth's Iago, but they were more reserved about Irving's Othello. The critic of the *World* (May 18), probably Dutton Cook,[64] found Irving "not wholly satisfying," but more so than when he first acted the Moor in 1876. The *Observer* (May 18) characterized Irving's Othello as "unequal" and "difficult to place at its proper value." The consensus was that both actors were more effective as Iago than Othello and Booth's Othello perhaps had a slight edge on Irving's. All the critics agreed, however,

[64] Edward Dutton Cook, English drama critic and author, was the critic for the *World* from 1875 to 1883.

that never before had London witnessed such a generally excellent production of the play. It was as Joseph Knight wrote in the *Sunday Times* (May 15) "a credit to English art."

The engagement was prevented from being one of the happiest theatrical experiences in Booth's career, as he wrote Winter, only by his "domestic misery." His relationship with the McVickers, who were now in London, was badly strained. To Booth they became the "McWickeds" or the "McDevils." At their suggestion he agreed to postpone a proposed tour of the English provinces and go with them to New York. In August he planned to return to England for his tour and leave Mary in the care of her parents, if she was still living.

72. Brunswick Hotel, May 22, 1881

After six weeks of gloomy apartment experience I have shifted my quarters to a more cheerful and comfortable locality. Mary's partial return to reason caused me so much trouble on account of her antipathy to the lodging house that Edwina & I bundled up in haste and left half our things behind—unpacked. I have concluded to go with the family (McVickers) to New York and leave Mary there with them—returning here in Augst. to fulfill my fall engagements in the Provinces. Her increased strength surprises all & her partial sanity also, although her doctors expect her death at any time—so they have done for some weeks, indeed *months*, past. She may linger a year—if so I do not know how to act. I can do her no good & yet to leave her in such an uncertain condition is simply awful to think of!

My engagement continues with Irving 'till June 12th & is very great in all respects—only my domestic misery prevents it from being the happiest theatrical experience I've ever had. I wish I could do so much for him in America as he has done here for me. A suggestion has reached me (yet indefinite) that I should again take the reins at Booth's—but be relieved of all financial cares, &c. I cannot *scrape* up even the scraps of my old-time ambition to *decorate* the plays & act as unceasingly as I would be obliged to do were I at the head of affairs as once I was on that stage.

John closed his engt. last night & this evening I shall meet him at dinner, at a Mr. Bigelow's (of Philadela.). Nearly all the papers speak highly of him & many call him, as they did me, "old school" &c, &c. I wonder what the devil is the new school

of tragedy. My voyage home & back will be very sad—but alas! My experience during the past two years has been wretched enough. The *McWickeds* have been very decent & act as if nothing ever came between us; while they know that I know *she* [Mrs. McVicker] has slandered me most vilely, & they also know my sentiments regarding her—from interrupted letters from me to Horace—who caught h-ll on my account. They doubtless see now how they have wronged me in supporting poor Mary's madness against me. I've had quite a hurt letter from Gill[65] because I said his sketch of me (sent to the [Tinsleys?]) was inaccurate—he says I've injured him com-mercially—at least $250. A gentle way of putting it after boreing my life out for the purpose of a sketch I had no wish that he should write. Gill ought to have a fine funeral; can't you get up something for him in that line? A farewell benefit, or something that would immortalize him. Adieu! My love to your folks & you.

On June 16, Booth and Irving acted together for the last time in *Othello* for the benefit of Ellen Terry. Several days later the Booths and the McVickers sailed from Liverpool to New York aboard the *Bothnia*.

[65] Probably William Fearing Gill, American journalist and author who was connected with leading New York papers from 1878.

Chapter VI

The American Tour

1881-1882

ON JUNE 29, Booth arrived in New York and proceeded immediately to apartments at the Windsor Hotel. Mary's physical condition had improved, superficially at least, during the Atlantic crossing, but there was no hope for her complete recovery. She still denounced Booth and, moreover, seems to have barely tolerated his presence. The McVickers of course sided with their daughter against Booth. The entire situation was becoming extremely tense, and Booth was driven to drastic measures.

73. Windsor Hotel, New York, Wednesday Night
    [June 29, 1881][1]

Only a line to say—Forgive me for not notifying you of my arrival. But my life of late has been as near d--nation as I care to be & my poor weak brain's a whirl all the while. Today I have been beset & run to death by interviews (with a splitting headache) and other callers & this is the first chance I've had to drop you a line since my arrival at this hotel at 10 this morning. Hope soon to see you. Must go to Long Branch to see Mother[2] as soon as I can get things settled—all is now at sixes & sevens. On all sides I hear nothing but McVicker slanders—even in London she began her devilish work. Were it not for Mary's weak condition & the scandal that would arise I'd prosecute the she-devil for defamation of character & apply for a divorce from her lunatic daughter. I have sacrificed everything for this poor creature's whims and—but enough of this. I want to see you & have a good long chat; when shall it be? God bless you! Love to all your dear ones.

I'm tired out.

[1] Winter appended the following note to the end of this letter: "This first letter to me, after his arrival home from England—Wednesday June 29th 1881. I dined with him & Edwina on the 30th at the Windsor, & did not receive this letter till afterwards. A very sad letter."

[2] Booth's mother, who at this time was seventy-nine, was in poor health.

The papers soon picked up Booth's "domestic misery" and it became a matter of notoriety. According to the *Dramatic Mirror*, which published a two-column defense of Booth on July 9, he was accused of "monstrous cruelty, selfishness, and wicked design," of refusing to take Mary to the South of France when her physicians had ordered such a trip, of forcing her to perform "certain onerous offices" in his dressing room every night at the theatre although she was gravely ill, and of returning her to America at great peril so that he could fulfill certain remunerative engagements. The *Mirror* identified the author of these obvious calumnies as "a woman" who "manufactures her insidious weapons of rank scandal in her own imagination, barbs them with hatred, points them with malice and assisted by her husband, a poor, weak, miserable fool, who is a mere tool in her hands, sends them with merciless aim to the spots where she believes they will harm Edwin Booth the most." Although no names were mentioned, the "woman" and the "miserable fool" would be understood by a wide circle of knowing readers as the McVickers.

"Friends" of the McVickers also printed libelous attacks against Booth. The Chicago *Tribune* of October 16, for example, printed an especially vicious letter dated October 15 and signed by "Hogang," probably an anagram for one Grace Logan, a friend of the McVickers. Hogang, who claimed "to know something of the Booth family," accused Booth of "heartless neglect and outrageous abuse of his heartbroken wife." Mary was reported to have left her sick bed to follow Booth to Europe "to save him from becoming a complete moral wreck." Booth then "dragged her through the rain and chilly mists of Ireland and Scotland" in the most uncomfortable manner as though he "designed to put the finishing stroke to a life spent in watching over his every interest, personal, pecuniary and moral and in training and educating his daughter." Moreover, when Booth was advised by physicians in Edinburgh that Mary's condition was serious and that he should take her immediately to a more genial climate in the South of France he refused and dragged her instead to "chilly foggy London." After the London specialists had pronounced Mary hopeless, Booth committed her to the care of "hired nurses" and had "a rollicking time at the clubs, varying these amusements by returning in the small hours of the morning and cursing his sick wife." From such treatment from an unfeeling husband whom she nevertheless loved, Mary's "reason quitted her throne for a season" and in the state of temporary insanity she divulged Booth's abuse to her physicians who immediately sent for her parents. Her mother "nursed, soothed, consoled and brought her back to reason" but her physical condition was still hopeless. After her parents forced Booth to bring Mary back to New York, he seemed

apparently unconcerned with his dying wife or with the fact that she refused to see him. He sported at Newport and Long Branch "in a natty jockey suit of clothes and a slim walking stick" and took quantities of " 'fluid' morphine, as it is now called." "Hogang" also implied that Booth had married Mary for her money and had forced her to will him all her property, including that which her father gave her.

Such attacks continued throughout the summer and early fall as Booth visited his mother at Long Branch, his friends Benedict at Greenwich and Magonigle at Mount Vernon, and made a business trip to Newport to inspect a lot he had recently purchased.

During this time he sent Winter only brief notes or telegrams which tersely reported his activities and rarely mentioned his tribulations. On July 23 he reported that he had stopped the scandal-mongering *Dramatic News* by sending his lawyer, apparently with a threat of a libel suit. He did this, he wrote, not "for the purpose of fighting the paper but to scare McV. wh. it did" (Folger Shakespeare Library, Y.c. 215 [389]). For the most part, however, Booth took no action, kept silent, and patiently waited for a turn of events, even when Mary drove him from her room and refused to see him. On August 10 he wrote that he had not seen her for a week but that the doctor reported her "full of fun" and "quite chippy" (Folger Shakespeare Library, Y.c. 215 [390]). On August 23, he reported that he had not seen her for four weeks but that he "heard every day or two from the Dr. of her condition" (Folger Shakespeare Library, Y.c. 215 [393]). On October 2 he announced that Edwina was engaged to Downing Vaux, the son of Calvert Vaux, the landscape architect,[3] and that the "McWickeds" would find "fresh food for gossip" when they heard the news. "Poor Mary," he continued, "I have not seen her for eight weeks, but hear she's doing well" (Folger Shakespeare Library, Y.c. 215 [397]).

On October 3, Booth opened a four-week engagement at Booth's Theatre. This was the first engagement in a tour of the United States, which had been arranged for him by Henry E. Abbey, after Booth had decided to postpone his contracted English tour until next season because of Mary. His reception by the New York critics and audience was enthusiastic, despite the fact that during his first week Booth suffered from a cold and hoarseness. During a performance of *Othello*

---

[3] Calvert Vaux emigrated from England to America in 1852 to work with the noted American landscape architect Andrew Jackson Downing. Vaux is principally known for his design, in collaboration with Frederick Law Olmstead, of Central Park. In 1854 he married Mary Swan McEntee, the sister of Jervis McEntee. Edwina's engagement to Downing Vaux was eventually broken.

on October 7, he was too ill to finish the play. Nevertheless, the consensus of the critics was that he had never appeared in better form.[4] On October 14, Booth and Lawrence Barrett, then appearing in what was described as a "friendly rivalry" at the Fifth Avenue Theatre, appeared together as Iago and Othello in a benefit performance for the sufferers of a disastrous fire in Detroit, Michigan. On October 29, Booth closed his engagement at Booth's Theatre, and, on the following Monday, he opened at Haverly's Theatre in Brooklyn for one week.

The Sunday before his opening in Brooklyn, Booth wrote Winter to apologize for his neglect and to report the latest of his conflicts with the McVickers and his wife. Mary still refused to see him; and about October 10, the McVickers moved her from the Windsor to a house on West Fifty-third Street. Only if Booth subscribed to a set of conditions which she proposed would she see him. In good conscience, he could not subscribe to them.

74. Sunday [received November 1, 1881]

I am really ashamed of my neglect but indeed, my dear boy, I have been in such a miserable condition—mentally—that I've been unable to collect my wits although for the past week I've sat up—wide awake—'till day-light nearly every night. (Here am I interrupted again for at least the 'dozeneth' time this morning, and must defer this 'till tonight.—It is now after 9 P.M.—just back from a visit—and I'll endeavor to make an end o'this:—I am sorry you troubled about the remittance, there was no need to send it, but 'tis yours at call—any time. The last news from Mrs. B. reports her better, having rallied last Thursday & is fairly well today, tho' the end may come at any moment. She is willing to see me provided I will subscribe myself a villain and give a written apology & acknowledgment of my guilt to her parents. People who have used me for their purposes and slandered me most vilely in return for all the slavelike devotion to their daughter these many years. On such conditions Mary will never see me in the flesh & I hope not in the spirit—for God knows I've had enough of the McVicker tribe. Horace is the only decent one of the lot & he is consequently debarred their favor. I think it best for me to write a full & free statement of my relations with that

---

[4] Booth's national tour of 1881-82 is chronicled in detail by a scrapbook of press clippings and playbills at the Walter Hampden Memorial Library. Dates of engagements and quotations from reviews are taken from this scrapbook.

family from the very beginning to the present time & if necessary swear to it before a justice for use in case of my death. Their sole object is to destroy my reputation for sheer malice—in order to shield their daughter. Of course my self-defence places me in the cowardly position of fighting women, one of whom (and that one my wife) may be dead when the time comes for my defense. The parents are doing all they can to prevent her from sending for me—except on the conditions named. They have succeeded, I think in blinding Dr. Collyer,[5] an old Chicago friend of *Mac's*, & he, of course, thinks hardly of me because I will not agree to the plan proposed. This is surely the acme of ingratitude, the vilest case of wickedness ever practised by bad women against a fool of a man who has for twelve years submitted to injustice, disgrace and incessant mental torture. My position is irksome chiefly because I cannot be at the Death-bed of my wife, and when I do go to the house & to her funeral as I must for decency's sake, I shall be surrounded by a set of their friends who do not know the truth.—The Tribune libel & others quite as bad come from McV. through a couple named Angel, I'm told, and Grace Logan.[6] I hardly think it best to make any advance whatever— but to wait patiently the turn of events. McV. in two blackguard letters threatens to expose me in the Courts, when Mary dies, as a devil of the d---dest dye; vile & filthy in every respect, &c, &c. My lawyer advises entire silence on my part. Edwina's engagement was no secret—no time is yet fixed. I had an interview with Dr. Collyer & offered to "forget & forgive"—his report next day, after seeing the McV.'s. gave me hope that peace would be restored & that further slander would be avoided—that Mary wd. express a wish to be buried in Boston by her child & so prevent scandal after death, but he says the Mc's yielded (pretense, I'm sure) but that Mary refused to see me 'till I had written the letter

[5] Robert Collyer, prominent Unitarian minister, became the pastor of Unity Church in Chicago in 1859. In 1879 he assumed the pastorate of the Church of the Messiah in New York, which post he retained until his death.

[6] This letter was subsequently refuted as "wholly false" by "M.H.F.," a New York correspondent of the *Tribune*, in a dispatch printed on October 30. M.H.F. noted also that Booth was "a good husband" and that anyone familiar with the Booth family knew that Mary was the real authoritarian figure and not Edwin.

she required. Mrs. Pateman[7] first told her of my Collyer experience, & the next day Mrs. McV. *excused* her from calling any more on Mary. What the devil am I to do! 'Tis only propriety and public decency that urges me to even approach Mary again— alive or dead, for she has killed the very germs of affection in my heart for her, and I know full well that I shall be insulted & perhaps denied admittance to the house when I go.—You may judge how near madness I've been at times; passing sleepless nights after heavy work and not daring to stimulate for fear of falling off my perch, wh. but a pint of mild beer can accomplish for me. I wish I could have a quiet set of hours with you on this business—though 'tis not fair to bore my friends with such matters. God bless you!

On November 13, while he was engaged at the Lyceum Theatre in Philadelphia, Booth received a telegram which notified him that Mary had died at five o'clock. He immediately wired this news to Winter at Staten Island (Folger Shakespeare Library, Y.c. 215 [400]). Winter and Lawrence Hutton[8] accompanied Booth to Chicago where Mary was buried on November 18. On November 21 Booth resumed his tour with an engagement of one week at the Academy of Music in Baltimore. He then played a series of one-night stands in Wilmington, New Haven, Waterbury, Hartford, Springfield, and Worcester. At Waterbury the company's baggage car, by mistake, had never been unhitched and was carried on to the next town, so that the company was without costumes or scenery. According to Mrs. Charles Calvert, who reports the incident in her autobiography, pp. 208-210, Booth suggested that the audience be appealed to and if they wanted to remain and see *Hamlet* in "modern tourist costumes" he would perform the first four acts, if not, their money should be refunded. Under the circumstances Booth would not consider the fifth act. The audience clamored for *Hamlet* and so it was performed in "modern dress." Mrs. Calvert says that the company played "earnestly and conscientiously"

[7] Isabella "Bella" Pateman, English actress, and her husband Robert Pateman, English actor, supported Booth as Katherine and Grumio in his benefit performance of *Katherine and Petruchio* for the Royal Theatrical Fund on February 28, 1880. From about 1869 to 1876 the Patemans had lived in America, during which time they were members of the company at Booth's Theatre. They were also members of Booth's company during his London and provincial tour in 1882.

[8] Lawrence Hutton, American drama critic, theatre historian and editor, was a close friend of both Booth and Winter.

but that "it required a somewhat powerful effort at self-control when the Play Scene came, and the murdered Player King in coat and trousers was sent to his doom sitting upright on a kitchen chair."

75. ~~Springfield,~~ No! Worcester, December 3, 1881

I'm so dazed that I hardly know where "*I'm at*", as Jno. Ford, the oldest manager on earth, would say. Flitting thus from town to town, in a hand-bag, disconcerts me, & you'll pardon my aberration & not attribute it, as McWicked would, to *drunk*, when you N.B. the heading of this epistle. As he who croaked "I'm saddest when I sing"—so you will, I trust, consider me—"not merry, but beguiling the thing I am by *seeming* otherwise."[9] Not sad, neither: but—thoughtful; or regretful—which implies a condition of unhealthy thoughtfulness. I had not the heart—nor the energy to write you a description of my work-a-day *Hamlet* adventures. I reminded the Wart-er-bury-ians that Shakespeare's actors did it, but they didn't see the point of my refined wit—they yelled "go on," however, and I took it for granted that they preferred the text of Shakespeare in its *naked truth* to the [filthy?] lucre they were offered as a substitute. The play went well, but it required all my iciness of look & manner to keep my playmates serious, a wink from me (& I was strongly tempted to give it) would have made a farce of the whole performance. Therein, only I felt *my* strength; the respectful attention of the audience was due to our lord & master, W[illiam] S[hakespeare].

Today I read the Tribune's article anent our friend Booth—I suspect Larry Hutton wrote that letter.[10] I'm glad you know him & I'm sure you'll *love* him. He's solid to the core. The enclosed is from a "gent" who asked a pass.—By Hercules! he deserved a dozen! Voila! le cochon! *il Porco*! Robert! d--me! what's fame? Who's who! and his success in *this* city! Porbleu! Morbleu! Scré papier &c!

Bless thee, Will! I'm not answering your good letter—that reserved for Boston, where I hope to have some respite from

[9] *Othello*, II,i,123-124: "I am not merry; but I do beguile/ The thing I am, by seeming otherwise."

[10] On December 3, the New York *Tribune* reprinted a "private letter" sent to the Boston *Gazette* describing Booth's "great dignity and propriety and the most perfect taste" through the "trying ordeal" of his wife's funeral in Chicago. Winter rather than Hutton, however, seems to have written the letter (see letter 77).

handbags & railways for at least a month. Edwina & I write love-letters to each other—she late at night, before going to bed, & I put my replies where she can find them in early morning—I've just found a most beautiful one from her. O, my boy, if God, in his wiseness should take *her* from me—what's left here for me? I think I should go head-long hellwards! She's all I've lived for through the dreary years since Mollie died! And to think that my intermediate struggles with the demon of my life should have been so wasted!!—I want to get at my autobiographical account of those sad years of *ache*—for you; I will, some day. Good night. Love to you all.

From Worcester Booth proceeded to Boston, where he was scheduled to open on December 5 at the Park Theatre for an engagement of four weeks. On the morning of his arrival, he was greeted by an article in the Boston *Sunday Herald* written by Junius Henri Browne defending him against the McVickers.[11] Booth was not pleased. He wished everyone would just "let the matter rest." Such an article, moreover, appearing the day before his opening looked to Booth like the sort of muckraking, theatrical "puffery" that he despised.

76.   The Vendome, Boston, December 4, 1881

I was sorry to find the enclosed when I reached Boston this morning. It looks like an advertisement coming as it does on the eve of my engagement. I was in hopes that both my friends & enemies would let the matter rest. This article contains much of fact & a quantity of error, the latter may provoke the McVickerites to fume again—for, of course, they will attribute it to my influence—while God knows I'd rather no more shd. be said of poor Mary's mistakes. I want to forget all that was wrong & remember only the right—the true & good of the poor girl. Hutton was here this evening for a few moments. I had letters at Worcester from both Mr. & Mrs. Russell—they seem anxious to renew our old friendship. Somehow I can't hate longer than it takes to "cuss" a round oath or two, then I'm serene; so I've frankly told Russell

[11] Junius Henri Browne, American journalist, was an acquaintance of Booth and Winter. He was connected with the editorial staffs of the New York *Times* and the *Tribune*, and he acted as correspondent for many other newspapers throughout the country. Lockridge, pp. 277-279, prints excerpts and a synopsis of the article.

"the cause of my distemper" and left it to his conscience to reciprocate. I'd rather shake hands than fists anytime.

Edwina joins me in love for *il porco*.[12]

On the day of his opening the *Herald* carried an article concerned with John Wilkes. Following the Sunday article about the McVickers, this definitely appeared to be a planned scheme of advertising of the type that drove Booth "half-wild."

77. The Vendome, December 6, 1881, After *Richelieu*

In my last I sent you a copy of the *Domestic* troubles, published in the Sunday Herald; today I have read—in *Monday's* Herald a 3 column article about *John*. Now this looks to me as a set thing in the way of advertising, & though I believe Abbey free from such a spirit, as he invariably has shown respect for my feelings in everything, I suspect that some of his men—Schoffle [sic][13] or Byrne (who, by the by, is connected with the Boston Herald)—has something to do with it. Of course it will be copied far & near & I shall be credited with the appearance of the articles—coming, as they do, just as I begin my engt.—It is enough to drive me to hell at once! I talked with Parker[14] & Hutton about it, & they feel as I do. Can it be stopped? Wd. it be well for you to write Parker again (by-the-by, I made a great mistake about that good letter you wrote—seeing only an extract & Hutton being in Boston at the time)—a letter deprecating such ill-timed, unfeeling—though, perhaps, well-meant publications? I don't know. Perhaps it would make it worse, but by the Jumping Jingo! I feel like throwing up the sponge & not acting at all. If Abbey has any knowledge of the fact I am much deceived in him, & don't believe it. If it is continued God knows what I'll do—for these two articles have kept me awake o'nights & were it not for my girl I believe I should quit the town & let the d--ned theatre go to pot! I'll be guided by what you advise in regard to any comment on the indelicacy of the d----d business. If I find that Abbey's

---

[12] This was a "joke" of Edwina's, according to letter 79. It means "the pig" and might possibly refer to Winter's last child and only girl, Viola Rosamund, who was born on July 13, 1881.

[13] John B. Schoeffel was a Rochester theatrical manager who in 1876 joined Henry Abbey as a partner in the management of his various theatres.

[14] Henry G. Parker was the editor and publisher of the Boston *Gazette*.

rabble have anything to do with this filthy work (I'm sure *he* has not) I'll never act again except under my own management. Excuse this. I am half-wild! When will you come? I'm glad & sorry you had the ring inscribed—I intended to do it. I'm glad it's done. Love to all of yours.

Although Mary was dead and buried, the McVickers continued to irritate Booth by sending him clippings from biased Chicago newspapers.

78.   Hotel Vendome, December 27, 1881

I send with this a paper—just recd. with a kind & sympathetic letter, from Lady Martin. The phraseology seems quaint to me, & I send it for yr. edification; I cdnt. read *his* article, on the inner side.[15]

I've recd. a clipping from Chicago Inter-Ocean, describing a beautiful Xmas gift from McV. to his wife—an ebony casket, with portrait & mottoes & labels anent poor Mary—all bearing deep meaning to the afflicted parents, &c. Today I got two of the old Chicago Tribunes relating the circumstances of the funeral at Chicago—sent by McV. either in a drunken or an attempted disguised hand. The impotent donkey uses every means he can devise to irritate me, as he thinks. I expect a *blast* when I act in that city.—but if I should be forced to tell my story—wh. I can prove—they wd. feel more inclined to drive the McV's. from Chicago than scowl at me. Isn't it strange that but *one* paper (the N.Y. Times) has yet thought of the fact that the son of McV. & Mary's favorite brother, Horace, should be in sympathy with me & sides with me in all particulars relative to this business? If a chance ever occurs to ventilate this fact it wd. be well to do so. It is sad for me to think that Runyon [sic] shd. be warped in his feelings by the McV's.—but *Mac* has become so great a liar under his wife's influence that I'm not surprised at Runyon, who regards *Mac*, of course, with a sort of veneration as his father's old friend. I'll have a heavy old time when I get there, next April, I guess. All I fear is that I may be forced to speak & tell the truth in my own defence, & as all my feelings for the poor dead girl are those of tender pity I would rather be kicked out of town than denounce the mother—by doing which I must tell things of Mary

[15] Undoubtedly "*his* article" was an article by Sir Theodore Martin.

that are better buried. I have not yet 'set to music' the story of my 2d. marriage—I am so tired after the play, & during the day I have so many interruptions; but I will do so, & let you have it. O, I just remember what I wanted to say about the stereotyped plates of our *P.B.*—I find so much to add & to omit as my studies progress that the expense of alteration prevents my changing the text, as I would like to—very often.

I find a word or a sentence here & there, as *I act* (and this I *study*, you know) that seems to me essential to be added or, in accordance with the curtailment, to be omitted; to do which would cost ten or fifteen dollars now & then—and it seems like throwing $ in the gutter, for the books *wont* sell—even at the cheap rate I've now set them. I hoped, at the start, that you'd get Wallack's, Barrett's & McCullough's plays, & establish a standard prompt-book, but I came into the field too late with mine. *All* my household *is* asleep. The dear girl has had a happy Xmas, thank God! Have you read the "Mystery of Hamlet"?[16] aside from the absurdity of his being a woman—there is much in it that accords with my idea of the character. I'm now playing with Darwin's monkey,[17] which interests me to that degree that I fancy I feel an "os cocyx" (If that's the way to spell it) at the end of my vertebrae (diphthong)[18]—but I guess its piles! Good night—before I say more on the subject.

Booth closed his Boston engagement on December 31. He then headed west to Pittsburgh, stopping on the way at Fall River, Providence, Newark, Easton, and Scranton. From Pittsburgh, where he opened at Library Hall on January 11 for an engagement of four days, he wrote the following letter listing various injuries that he had received thus far on his tour and denying rumors that he was to lease the Grand Opera House in New York and return to theatrical management in association with John Sleeper Clarke.

[16] Edward P. Vining's *The Mystery of Hamlet: An Attempt to Solve an Old Problem* (Philadelphia, 1881) was a monograph which postulated that Hamlet's motivations are feminine rather than masculine, that Hamlet is essentially a woman's mind in a man's body (see letter 80).

[17] Booth's personal copy of Charles Darwin's *The Descent of Man, and Selection in Relation to Sex* (New York, 1880) is preserved at the Walter Hampden Memorial Library.

[18] In an attempt to join the "a" with the "e" Booth had made an almost illegible smear.

79. Pittsburgh, January 12, 1882

'Tis so long since I heard from you that I wonder where you *is*!
I hoped to see you in Boston but you came not as you promised.
The last note I sent you was about Xmas time—it contained a
couple of wax impressions & a "porco," a little joke of Edwina's,
who now thinks it may have offended you & that she was too
forward; she's sensitive as a sore spot. That reminds me I have
had a swelled arm for ten days, the result of vaccination, & not a
little pain from it. Also quite a severe burn on my ankle causes
me great inconvenience. I was bathing aboard my portable hotel
& the lurch of it sent me naked against the hot air pipe—there
I was pinned for a few seconds & received an ugly wound, the
ankle is swollen, & to cap the whole I have a big *jaw*, all these ills
are on my larboard side, so I act "right" as much as possible.

You can give a full & flat denial to all the stuff about the new
theatre that Clarke & I are to build & the Opera House lease, &c.
I can't imagine who started such reports. Clarke has never suggested such an idea to me—nor do I believe that he had any such.
All I know of the business is what I've seen in the papers. Have
already received offers of eligible sites, &c. Clarke's manager is
Harry Wall & he may have some object in the way of advertising
relative to it—but I hardly think so. There is certainly not the
least shadow of truth in it so far as I am concerned. Barton Hill
now fills poor Piercy's place.[19] Business is good. We two are good,
too, but mighty dull in smoky Pittsburgh. Where we stop longer
than one night we abandon the car for hotels—as here.

Thought I had more to write about but 'tis gone. Let me know,
dear Will, if I can ever serve you & 'twill give me pleasure.

Edwina joins me in love to you all.

Following his Pittsburgh engagement, Booth traveled west to
Louisville and Memphis, then south to Galveston and New Orleans,
then north again to Montgomery, Atlanta, Chattanooga, and Nashville.

[19] Samuel Piercy, American actor who had been playing second leads
in support of Booth, died from smallpox on January 9.

Barton Hill, American actor born in England, replaced Piercy at Booth's
Fall River engagement and continued with the company. Hill was a capable
and versatile supporting actor, especially noted for his portrayals of
Malvolio, Sir Charles Surface, Iago, and Othello.

80. Chattanooga, February 9, 1882

Since I last wrote you I've rcd. 2 letters but have been so tired & out of sorts—though perfectly well, but nervously used up—that I've not been able to write again. Being in a car, bounced from end to end all day & acting at night unfits one for such pastime, then, when we've stopped over a few days we have generally had numerous callers & other distracting causes to prevent one's attention to correspondence. I have just finished Hamlet & am in my car—to start at 5 A.M. tomorrow, therefore this will not be mailed 'till I reach Nashville. The horrid weather has kept me confined all day—but the house was well-filled, in spite of mud & rain. The leader of Orchestra was drunk, so we did à la Française —without music. I have *pic-nic'd* my company several times— for Edwina's sake—and had jolly times. Once at Galveston, by the Mexican Gulf, & again 'neath the pines of Alabama—where we had a set of darkies to sing, dance & 'cut up' for us.[20] I intended another *go* here, on Lookout Mt., above the clouds, you know, but the weather prevented.

Since I left Boston I've heard nothing from the McV's, & I am heartily sick of that whole business. Edwina met some ladies of Chicago in New Orleans who assured her that I have more friends there than I suppose & that Mrs. McV. is too well known to excite any sympathy among decent people. I shall certainly visit Mary's grave, but not as you surmise—[to] place a stone *there*, some day I shall put a slab at Mt. Auburn, where her baby lies, to her memory. I have no right to meddle with the McV. grave-yard & he wd. object & raise a stink at once. Thank God! I am on the return voyage—if He spares us in our travels over these in-fernal roads I shall never cease my thanks & will never take such a trip again.—No, I've never read Collins,[21] indeed I have read very little of the novelists—Thackeray, whom I can't remember, Dickens, whom I've half forgotten, are all that I can recall just now. When I've not been mentally acting, my life has drifted away in regrets for the Past or building airy castles for the future. It has been a waste!

I had a letter from Miss Ward, at the time Piercy was called upon to act the Corsican, cautioning me against Nelson Decker—

[20] See Mrs. Calvert's, *Sixty-Eight Years on the Stage*, pp. 205-206.

[21] William Wilkie Collins, noted Victorian novelist and dramatist, was a close friend of Winter. See Winter's *Old Friends*, pp. 203-222.

who had played some trick, as Barrett's brother did with me, and lately I've seen the death of Nelson Decker by drowning—can it be the same? both had a wife & child, that is if there are *both* & they be not one & the same.[22] Poor Waller's[23] gone! I liked the man for many things, but—he is dead: God rest him! Edwina is happy that 'il porco' did not annoy you & because you speak so sweetly of her. She is well & enjoys the travel very much—that is, the *variety* part of it, not the 'go' & the bumps & the thumps we get. I'll leave this 'till tomorrow, for 'tis very late. Good night!

At Nashville, February 10, 1882

All the clippings about E. B. have reached me but not *your* articles of wh. you speak. It is seldom I can get the Tribune out this way—the Herald seems to be the only N.Y. paper to be found regularly. The *"Mystery of Hamlet"* is an argument by Mr. Vining, dedicated to H. H. Furness,[24] attempting to prove that *Hamlet* is really a woman. (Good for Anna!)[25] It is very ingenious & aside from the absurdity of the writer's theory I agree with much that he urges in support of it. I have always endeavored to make prominent the femininity of Hamlet's character and therein lies the secret of my success—I think. I doubt if ever a robust and masculine treatment of the character will be accepted so generally as the more womanly and refined interpretation. I know that frequently I fall into effeminancy, but we can't always hit the proper key-note. You must see the book—it will amuse &

[22] Apparently it was not the same, since Nelson Decker, a minor American actor who had been in the Booth's Theatre company, is cited in Odell's *Annals*, xii, as still theatrically active in the spring of 1884.

[23] Daniel Wilmarth Waller, American actor and stage manager and the husband of the tragedienne Emma Waller, was Booth's stage manager at Booth's Theatre. He died on January 30.

[24] Horace Howard Furness, noted American Shakespearian scholar and editor of the *New Variorum Edition of Shakespeare*, was a close friend of Booth.

[25] Anna Elizabeth Dickenson, American authoress, actress, and social reformer, on January 19, 1882, appeared at the Grand Opera House in Rochester, New York for the first time as Hamlet. She subsequently played the role during an engagement beginning on March 20 at Haverly's Fifth Avenue Theatre in New York. She was not well-received as Hamlet, however. She retired from the stage after this engagement and returned to her former career as a lecturer for woman suffrage and other social reforms.

I think it will interest you too.—I am glad of John's success[26] & sorry for your ails—mental & physical. God bless you!

We were sadly disappointed anent our long-hoped for visit to Joe. He telegraphed me not to attempt the journey; rains had rendered the roads impassable & so we had to lose that delightful experience. We stopped twice at Iberia, but only long enough for water—I found a note there from Joe on my return from Texas. Edwina met some young girls in N.O. who were to have met her at Joe's invitation. This has been one of my chief regrets of this trip. We are homeward bound, but 2 months yet must elapse before we meet. Will it not be odd if my engagement at Booth's shd. close that house! I think it will. After I end with Abbey in April (or May) I shall be busy getting the 'will' probated & its conditions settled: I am called upon to deliver to the McV's. a lot of things that really were not Mary's by rights—but of no value to me, & they are packed in various places & numerous boxes. Already they have many thousands worth of my property which the poor demented creature gave them—at her request I sent silver, linen, & some furniture to the 53d. St. house. The last will made (in my favor) is the 3d., I exploded the 2d. wh. caused McV's. bitterness—that left all my Chicago property to them. It became public, & to wriggle out of the scrape they got her to sign a will leaving the property (at *their request*) to me![27] I'll explain all when we meet. I must now close, as I have but an hour or two of sunlight & I wish Edwina to have a drive; tomorrow

[20] John McCullough had successful engagements of six weeks at Haverly's Fifth Avenue Theatre from November 14 through December 24 and of one week at Haverly's Theatre in Brooklyn from January 23 through January 28.

[27] In a brief letter to Winter dated Boston, December 11, 1881 (Folger Shakespeare Library, Y.c. 215 [408]) Booth wrote:

My suspicions anent the *will* were right & have been confirmed by the appearance of another (a 3d)—copy of wh. was sent to me. I am co-executor with a Mr. Goodwin of the Vanderpoel firm. To me is left (at the request of loving and beloved parents) all the estate except her jewels, dresses, trinkets, &c. This is a secret—only Hutton who was with me when it came, knows of it, except—of course—my lawyer. This will proves that they were afraid to use the one made before going to England of which she raved while crazy in London!—Now what's the next trick! Accepting the trust of co-extr.ship I admit that she was not insane—thus the poor girl's memory is deprived of that excuse for her misdemeanors. The wicked *iditos* (aint I an idiot to spell the word so?).

I shall be busy at the Matinée. She joins me in loving regards to you & yours.

[P.S.] Just returned to our car, after a pleasant drive, in time for dinner—then a *nap*—then *Richelieu*.

After engagements in St. Louis, Evansville, Terre Haute, Lafayette, Fort Wayne, Indianapolis, Cincinnati, Akron, Youngstown, Milwaukee, Madison, Davenport, Burlington, Rock Island, and Rockford during the remainder of February and most of March, Booth opened in Chicago on March 27 for an engagement of two weeks at Haverly's Theatre. Contrary to the "blast" he expected, Booth was enthusiastically received by the public; and the press was, with a few exceptions, cordial or, at the least, indifferent. The critic of the *Inter-Ocean* (March 28) was perhaps the most cordial, writing of his opening night performance of Richelieu that he never gave "a more consistent intelligent, and generally admirable portraiture of the Cardinal." But W. D. Eaton,[28] the critic for the *Herald* and, according to Booth, "a strong McVickerite," characterized Booth's Richelieu (March 28) as "mechanical and stilted," "labored," and "destitute of spirit"; nevertheless, he could still admire "the network of movements and gestures— the mere mechanism of the personation." Eaton, however, was more commendatory in his review of *Hamlet* (March 30). He wrote that although Booth's impersonation of Hamlet was substantially the same that he had been presenting for almost twenty years, he had "infused more life, spirit, animation into the part than has characterized his performance for years." He was equally commendatory of Booth's performances of Othello, Bertuccio, Macbeth, and Shylock. Undoubtedly the critic of the *Tribune* was the most indifferent. He was of the opinion that practically nothing new could be written about Booth's acting, and so he wrote practically nothing. On March 30, he wrote that the Chicago public had witnessed innumerable Hamlets and of them all Booth's was "perhaps the most acceptable." "Everybody knows the story," he continued, "and pretty much everybody has a different idea as to the mental characteristic of the young Prince of Denmark. Mr. Booth presents the character as that of a mind overburdened with woe, but does not give way under strain, and his conception is both a

[28] The career of W. D. Eaton is largely obscure. He was the drama critic for the Chicago *Herald* during the 1880's. On January 14, 1882, he delivered an address on "The Press and the Stage" to the members of the Chicago Press Club. The following year he was elected vice-president of the club. See William H. Freeman, *The Press Club of Chicago: 1880-1894* (Chicago, 1894), p. 33.

pleasing and a dignified one. The support is good." And that was the extent of his critical remarks for two weeks.

81. The Grand Pacific Hotel, Chicago, April 2, 1882

As you predicted all has gone well here, even the Herald-man, Eaton, a strong McVickerite & a fellow *spiritist*, has gently touched me & the Miln breakfast, which cd. not be foregone, is over; it went off with modesty & discretion, no mention having been made of it. It brought me into acquaintance with some excellent people & really—socially—I have found more friends than I ever dreamed of having here. The wives & daughters of some of the best men—several at the breakfast—have called on Edwina. So, after all, I feel I am indebted to Mr. Miln. I believe him to be earnest and sincere, & that's what very few of his profession are.[29] I endeavored to get Dowden's book[30] at Carleton's some time ago, but they failed to obtain a copy for me. Runion, I grieve to say, is an insincere man, he has completely gone over to McV. and believes all the evil he hears from that quarter; Horace[31] had to shut him up t'other day. There is some suspicion that he is after McV.'s theatre. The old man is getting rid of all his Chicago ties & intends to leave the city for good—I hear he goes to Europe in May—to be close on my heels I suppose, to bark me down there as he has tried to do here. A letter for Orlando Tomkin's tells me that Stuart had written him to get at certain points about me, if I did vote for Lincoln &c; that he is to "re-write" (review, I presume he meant) the *Booth* book for Dana.[32] On Easter Sunday I start for Buffalo—arriving there

[29] Reverend George C. Miln was a Chicago clergyman. He had succeeded Robert Collyer as pastor of Unity Church. During the week of March 27 he gave Booth a luncheon to which were invited a number of noted Chicagoans. On February 19, Booth wrote Winter from St. Louis (Folger Shakespeare Library, Y.c. 215 [413]) that the purpose of this luncheon was "to express sympathy &c., all in a *quiet* way, of course." Booth, as usual, was reluctant to accept this invitation. Reverend Miln later left the pulpit for a career as an actor (see letter 86, footnote 21).

[30] Booth's personal copy of Edward Dowden's *A Critical Study of Shakespeare's Mind and Art* (London, 1875) is preserved at the Walter Hampden Memorial Library.

[31] Horace McVicker was Booth's business manager in Chicago.

[32] The "*Booth* book" is Asia Booth Clarke's *The Elder and the Younger Booth*.

Charles Anderson Dana was the editor of the New York *Sun* from 1868 to 1897.

Monday about noon. I shall stop, I hope, at the Brunswick when in New York. Good night. God bless you. Love to all.

[P.S.] Old McV. is here, arrived Saturday. Perhaps we'll collide!

Booth closed his Chicago engagement on April 8 and started toward New York, stopping for one- or two-night engagements at Buffalo, Rochester, Utica, and Albany. On April 17 he opened at Booth's Theatre for a two-week "farewell engagement" prior to his tour of England and Germany. The month of May was spent preparing for his trip abroad, visiting his mother, and caring for Edwina who about May 15 had been put to bed with a severe case of pleurisy. By June 4, however, Edwina, whose illness Booth feared would cause him to postpone his trip, was "rapidly improving," as he wrote Winter (Folger Shakespeare Library, Y.c. 215 [421]). On June 14, again on the Cunard steamer *Gallia*, Booth left America for Europe. He was accompanied by Edwina and Julia Vaux, Downing Vaux's sister.

Chapter VII

England

1882

ON JUNE 26, Booth opened in *Richelieu* at the Adelphi Theatre in London for an engagement of six weeks through August 5. The critical reception of Booth's acting in this play was as cordial as in the preceding year. A number of critics, such as Joseph Knight, found his performance to be "far more interesting, expressive, and dignified than last year" (*Sunday Times*, July 1). The audiences were equally enthusiastic, although reportedly "scanty" because of the warm weather. After four weeks of *Richelieu*, Booth presented Bertuccio in *The Fool's Revenge*, which was also generally well-received. On August 3, he gave a special matinee performance of *Don Caesar de Bazan*, the only other character in which he appeared during this engagement. Booth's impersonation of this character was new to the London audience and critics. Most of the critics found the role to be, of course, less important than Richelieu, but generally effective in its performance. The *Observer* (August 4) characterized it as "admirable and delightful" and in the *Morning Post* (August 4) Charles Dunphie wrote that the performance "abounded in fun and fancy, not unrelieved at times with impressive touches of tenderness and sensibility."

During this engagement there was no need to write Winter, since he had arrived in London the week after Booth. Winter was not the only one of Booth's friends in London. Thomas Bailey Aldrich, Lawrence Barrett, and Lawrence Hutton were there also. Winter recorded some of their meetings in his journals. On July 8, he wrote that Booth and he attended a gala midnight supper given in Booth's honor by Henry Irving at the Lyceum Theatre. It was attended by numerous English and American friends of both Booth and Irving, including Barrett and Aldrich, but "no ladies." Winter noted that Booth and he did not get away until morning, and that he "conveyed Edwin to his hotel." That evening Booth, on his way to another party, dispatched the following gay note to Winter.

82. Sunday 7 P.M.

Left bed at *2.30*—feeling quite decent. Have a rehearsal of *Fool's Revenge* tomorrow but if possible must see you during

the day. I want yr. advice & help maybe. If you are engd. during the day see me at night. I wish you had stayed here with me. *Betty*[1] says she wanted you to do so, but you wdnt. Have not the faintest shadow of recollection anent last evening beyond the fact that the table was luxurious, the lager cold, and the spirits warm.

Did I make a very great 'hass' of E.B.? I fear so: I usually do. Hereafter, when I can afford it, I shall take a page with me to shut me up & take me away at a certain cue.

Bless thee, my boy! I am now *starting* for *Lady Martin's*— to dine & wine again. Pray for your tragedy boy!

Adoo!

[P.S.] Love to Lawrence. O, I wish his wife had been there last night to start as ere the sun rose!

Ta-ta

About the middle of July, Winter accompanied Lawrence Barrett on a tour of Holland and Germany. They returned to London about August 1, and Winter spent several days in Booth's company before he sailed from Liverpool for New York on August 7. Before he left, Winter wrote Booth to advise him to guard his health and to warn him against excessive drinking. On the day of Winter's departure Booth wrote to reassure him that he would try to guard his health. "But if you knew," he continued, "how exhausted is my nervous system and what a strain I have to endure you would not wonder at my stimulating a wee bit now and then to help me up." Besides he tippled "less than many a prudent dame daily drinks" because his poor brain, like Cassio's, was so feeble that "two glasses of beer will often stupefy me" (Folger Shakespeare Library, Y.c. 215 [425]).

On August 8, Booth left for a month of sightseeing in Holland, Germany, and Switzerland. On September 11 he began his provincial tour at Sheffield. Two weeks later he wrote a long letter in reply to two of Winter's in which Booth was apparently once again advised about his drinking. Booth's reply was much less casual than his last and he expressed his sincere regrets for his inebriety at Irving's stag supper.

83.   Harrogate, September 24, 1882

I've had two letters from you since I wrote—two dear good letters, and only sheer laziness, or fatigue, or lack of opportunity

---

[1] Presumably Booth's Irish maid who had been in his employ since the 1870's.

has prevented an earlier acknowledgement. Do not think that I take your kindly advice amiss—I value it, dear Will, and believe me I *do* strive my very best to follow it. My poor brain is so weak that if I attempt the least jollity I am overthrown. The day I was at Morley's with you I must have been unusually *feeble-minded*, for previous to the glass of beer I drank in your room I had had but one that day & yet I lost consciousness before I reached home.[2] It was unfortunate that you mentioned to Edwa. the circumstances of the Irving night, for she knew nothing & after you left she spoke to Betty, who is close-mouthed on the subject; so it came to me. That night looms up in my memory like an ugly dream. It was one of my most unlucky slips. The persons who were there Genl. Porter,[3] Barrett & I know not who else, are not likely to keep the silence imposed on all who carouse in that hallowed place. I daresay my enemies have their maws full of my "infamy" & are devouring it with great relish. Tell me all you can recall of the infernal mess, & destroy this as I shall your letters that relate to either's folly. I see Hutton is arrived— I intended to send a letter after him by the next mail, but really I have been wearied with travel & acting every night since I left London. While there, only a week, I had long rehearsals every day & could go nowhere, indeed, I had to give up a promised trip to Windsor with McHenry and could not possibly visit Burgess,[4] or I should certainly have done so. On our tour in Holland, Germany & Switzerland we had much quiet enjoyment, but the dyspepsia assailed me & the care of two girls prevented much pleasure I might have had; still it made me happy to see my charges so, & the health of all was improved by the trip, hurried tho' it was. *This* tour is irksome because of my having to work yet it is interesting. Thus far the country has been very beautiful & I yet hope to see the places you mention. Today we visited 'Eugene Aram's' locality & had a delightful drive & ramble

[2] An entry in Winter's journal dated July 10 undoubtedly refers to this incident at Morley's Hotel: "Booth called. Went home with him to dinner. Met Edwina & Miss Vaux. Went with Edwin to the Theatre & remained in his dressing-room all the evening—to take care of him. (Drink)."

[3] Horace Porter, distinguished Union general, businessman and statesman, was an acquaintance of Booth and Irving.

[4] I have been unable to identify "McHenry."

Frederick Burgess was for many years the manager of the Moore and Burgess Ministrels at St. James Hall in London.

through some woods & curious rocks.[5] At York I made the most of the one short day I had—I cd. have passed a week there with profit & great pleasure. This is a quiet, lovely place, & Scarborough from whence we came hither is superb! But, you know, we have the dismal Autumn weather with us now & that keeps us indoors very often when we should be going about. Your last letter brought me your charming 'Rambles'—that of 'Stoke Pogis' is delicious, but I hardly know which I liked most—let me have the rest of them.[6] I wish, indeed, that I had urged you to remain & go on the Continent—I am sure it would have been well for us both. I was strongly tempted to coax Hutton, but a fellow does not know exactly what to do for his own sake when he has two girls to consider: I was somewhat lonely at times—when they were ill, or tired & "didn't feel like it"; besides they retired at nine usually & I was left 'till midnight *solus* & wide awake. I couldn't fix my mind to either read or write—so I smoked & ruminated & got *blue*. I have already declined offers to act in N.Y. and 'tour' the states while Irving is there—but what am I to do? I can't be loafing for six months for courtesy. Tell Percy if I had anything superfluous here in the way of 'traps' he shd. have it, but I brought just as few things as I could possibly get on with. I hope he will have the best success & am glad that he's with *L.B.*[7] I have one of Clarke's boys with me—but I fear he is dull.[8] Miller says he sent for my route & some notices

[5] Eugene Aram (1704-1759) was an English scholar who, while a schoolmaster at Knaresborough, murdered his friend Daniel Clarke after discovering an illicit liaison between Clarke and his wife. Aram buried Clarke's body in St. Robert's Cave, a well-known shrine near Knaresborough where the thirteenth-century hermit lived. The murder was discovered and Aram was brought to trial, condemned, and executed. Thomas Good romanticized this incident in his poem "Eugene Aram's Dream" (1831), Bulwer-Lytton in his novel *Eugene Aram* (1832), and Henry Irving and William Gorman Wills in their play *Eugene Aram* (1873).

[6] Winter, as on his previous visit to England, dispatched weekly letters to the *Tribune* describing his sightseeing. These were printed under the heading "English Rambles." In 1884 they were collected and published in a book form along with some of Winter's poems as *English Rambles: and Other Fugitive Places* by James R. Osgood, Boston.

[7] Percy Winter joined Barrett's company in the fall of 1882 and played minor roles with him for several seasons.

[8] Creston Clarke appeared with Booth's company as Francisco and

t'other day; the German tour is not yet complete—I have signed only for Berlin, Jan: 16th, and four other places Hamburg, Bohn, & I think Hanover is one of the other two, there are twelve cities to be *bored*. How the deuce will I ever get through with the job! The more I think of it & the nearer I get to it—the more I don't know how I shall do it! I shall surely visit Newstead[9] if possible, but I am not free, every day is tied up by my engagements 'till near Xmas & during the few weeks interval—prior to the German tour—I have promised to take the girls to France: but I hope for a full idle month in England before I go home. Hope the 'hay-fever' is allayed by the Autumn breezes & that all are well at Fort Hill. I will write as often as possible—I seldom feel like it nowadays & have done more for you in this tonight than I thought myself capable of. Love to your folks, to *L[awrence]* *H[utton]* and to *L.B.* when *you* see or write to them—not forgetting dear old Joe & McCullough. Edwina sends love & Miss V. desires her kind remembrances—I forget to mention a fine ruin at Knaresboro' (where we were today) in wh. Richd. 2d., & the murderers of Thos. A'Becket, were held for awhile[10]—very picturesque. St. Robert's chapel—cut in the rock & curious series of apartments, one above the other, cut in rocks near by, in which the descendants live—very cozy & jolly—I mean descendants of the man who made the curious house. Good night, my boy! Write often & direct to American Exchange Strand.

[P.S.] A beautiful 'collie' has just walked into my room & placed his nose on my lap, it's the sheep-dog, you know, that's why he "cuddles up" to me—maybe.

During the final scene of *Hamlet* at Dundee on October 3, Booth was accidentally stabbed in the right arm by his Laertes, Henry

---

Marcellus in *Hamlet* and Leonardo in *The Merchant of Venice*. He was billed as "Mr. H. Creston." He later became a relatively successful supporting actor, although he was never of the stature of his father or uncle. Creston also seems to have suffered from a "lisp" (see letter 118).

[9] Newstead Abbey was the ancestral home of Lord Byron. He lived there from 1806 to 1816 and many of his manuscripts and mementoes were preserved there.

[10] The reference is to the keep at Knaresborough Castle. Hugh de Morville, one of the murderers of Thomas à Becket, owned the castle and was said to have hid there at one time for almost a year.

George. The wound was poorly treated and as a result Booth developed erysipelas. Apparently Winter had recently written Booth regarding a lecture which McVicker was scheduled to present. Rumor had it that the lecture was to be about his relationship with Booth. On November 28 at the Central Music Hall in Chicago, McVicker delivered a harmless lecture entitled "The Press, the Pulpit, and the Stage" which was a defense of the theatre against attacks by Chicago clergymen. The speech was reported by the Chicago *Tribune* on December 3 and published the next year by The Western News Company, Chicago. It contained no references to Booth.

84. Edinburgh, October 22, 1882

Your last (Sept. 26th) reached me the day after I had been slightly stabbed by a 'dogan' who stood in defence of the King in the final scene of Hamlet, and owing to my own treatment of the wound I have suffered the torments of erysipelas (in *both* arms) ever since—even now the itching is intolerable & I am bandaged o' both sides with ointment & lint. I have not been able to write more than a few lines to mother for two weeks. With both arms swollen & itching all the while I have not ceased acting, though you can well believe my heart was not in my work. Had to fence left-handed, which arm you know is so badly mashed that I can barely use it, and the exertion caused that member to swell, so the next night I had both fins in limbo. I presume that I am getting better—but I don't feel so I assure you.

You speak of the dog McV. & his *lecture*. I hardly know what to say about it, I could tell so much & prove so much against that family—were it not so hard on the poor lunatic girl who, really, in her lucid intervals tried to do & indeed *did* so much that was worthy. I think his threatened *lecture* is a "scare" for me as he imagined that his frequent announcements of his visit to England would be. If the donkey knew how I laugh at him he would keep quiet. It surprises me that he should do anything to provoke me—for he knows that I can tell a d---ning tale of him, his wife & daughter. But he always loved controversy & is happy in print; he likes to squabble with editors and preachers.[11] Mrs. Pateman says she had a letter from him—but did not

[11] For a brief description of some of McVicker's "squabbles" with "editors and preachers" see Jay Ferris Ludwig, "McVicker's Theatre, 1857-1896," pp. 117-127.

mention me. I think the Pateman's & the McV's. have been in communication all the while. If the old ass forces me to the wall, I suppose I'll have to say something—but all I can say is just what a man should *not* say, what a husband should forget of his dead wife. God knows I try to remember only the good of her & forgive the wrongs she did me—even her devilish mother, who rides old donkey Mac, has but a part of a d--n when I chance to think of her. The worst I wish them both is that they each had one of my arms at present. If we were together at times when the talking mood is on me I could give you rough material to shape for my defence—if it be ever necessary to explain matters to the public. And this leads to your reference to the German tour—why not spend those few weeks (from Jan 16th to end of March) with me? In April I shall have young Vaux over & he will return with us in June. See if you can arrange with Reid for a German job & if so I will be mighty glad to have you with me. My week here has been very satisfactory (of course the financial part of my success is a farce) but the press & the audiences have been very enthusiastic. In all my English work I have barely cleared expenses. Houses crowded with shillings & sixpences; & I daresay the Kreutzers & Pfennigs of Germany will amount to nothing more than beer & baccy money.

I wrote Warren a day or so before your reminder came & I shall *cable* him also on the 28th.[12] Bless him! I believe that Joe with Rip, wd. be Abbey's best card at the Lyceum during Irving's absence—I intended saying as much to both of them, but failed to do so. Tell Joe & suggest the same to Abbey. I hear Joe talked of a great deal—all retain an affectionate remembrance of him. The play would cost little & could be splendidly cast & mounted. There is a recent operatic success of *Rip* in London which would add to the desire to see Joe.[13] I fear that the Barrett scheme wd. fail—for good as his Cassius is it cannot carry the play, even with great scenes & numbers, *three* strong names are necessary

---

[12] On October 28, 1882, William Warren was honored with a benefit and banquet at the Boston Museum to celebrate his fiftieth anniversary on the stage. Booth's letter to Warren is printed in *Life and Art*, p. 208 (see next letter).

[13] *Rip Van Winkle*, a comic opera, by Robert Planquette opened at the Comedy Theatre on October 14 with Fred Leslie in the title role.

for a 'run' of Caesar.[14]—I wish you were here now; the sun is
flooding the old castle, opposite my windows, with a light seldom
seen at this season in Scotland, and to the left I see Scott's
monument, & the quaint old houses of the Canongate are in sight;
Holyrood is but a short walk and "Arthur's seat" with the
Saulsbury Crags near by. Have you ever been here? The most
charming city I have ever visited! We would have 'revelled' on
the Rhine, I mean if we had been alone & unencumbered; girls'
comforts must be considered & they are not always interested in
what other fellows enjoy, & are not always equal to the *climb*
& 'poke about' requisite for a thorough satisfaction of such visits.
When they were 'used up' I was left to my *stoopid* guide or still
more stupid self—which wasn't funny.

Give my benediction to *little* Mary Anderson, my love to
McCullough & Barrett, Joe & Warren & Hutton when you see
or write to any of them. Haven't heard from Clarke—the opening
of the Strand is deferred, I fancy, the papers do not mention
him or it. I hear that that fellow, Richardson, of Dramatic News
or Times, is to have an office like Gillig's in the Strand.[15] He's
one of McV's minions. Horace writes me that his Pa still ignores
him, and that he is the father of a girl. My lawyer is Luke
A. Lockwood 59 Liberty St., perhaps it would be well to keep
him *posted* if McV. opens fire. Lockwood has lots of letters &
other scurrilous effusions of the Mc party. I don't think anything
of importance will be attempted; but they will keep up a ceaseless
fusilade of filth for awhile. Let 'em rot.—We start at 10 P.M.
& ride all night to Hull—for six nights, but you have a list of
my route. All the German engagements are not yet signed—but
they are in train; six—I think—are closed. All the while I write
my arms are itching & burning dreadfully; I'd rather had solid
pain than this infuriating torment! The temptation to relieve it

[14] Apparently Barrett was proposing that he lease the Lyceum during
Irving's absence and produce *Julius Caesar*. In the spring of 1884 he did
appear at the Lyceum while Irving toured America.

[15] Probably Leander Pease Richardson, American drama critic, playwright, and novelist, who was the editor of *The Dramatic News* from
1891 to 1896.

Probably Henry F. Gillig, London correspondent for the Chicago *Tribune*, who was also an official of the American Exchange, then located at
249 The Strand.

by scratching is almost irresistable. I have written longer than I thought possible under the circumstances & must now yield to my agony & 'cuss' a little.

Adieu!

Love to yours all. Edwina sends kind regards to "il Porco" and Miss V. desires to be remembered also.

[P.S.] Miller tells that he sends you clippings, &c.

Booth kept Winter au courant as he continued his tour with the following letters from Hull and Leeds.

85. Hull, October 29, 1882

Your 2d. letter to this point, with enclosures reached me this morning—I leave for Leeds tomorrow, 9 A.M. So, you see, you just hit it. I hope last night's *jubilee* was a happy one.[16] Dear old William! I cabled him the previous day—fearing lest some hitch might delay my message, but no doubt it was read, if any were, on the occasion: I had also written him a few lines some days ago. Of course, *yes*, let me place a leaf on his laurel wreath. Good for Barrett in suggesting it! Burgess was with me an hour yesterday. He looks badly, poor fellow! but was chatty enough. He told me he thought of writing to Joe about Mrs. Farjeon[17] who is seriously ill, *enceinte* (3d. time) and he fears

---

[16] A full account of Warren's golden jubilee is given in *Life and Memoirs of William Warren* (Boston, 1880). Booth's telegram, printed in this book, read: "Cordial congratulations. Love and best wishes." Apparently, Barrett, Winter, Jefferson, McCullough, Mary Anderson, and Booth had planned to present Warren with a testimonial laurel wreath, but later decided to present him with a loving cup instead (see letter 87). The cup was made of beaten gold and silver and was inscribed "To William Warren/On the Completion of His Fiftieth Year on the Stage/October 27, 1882/From Joseph Jefferson, Edwin Booth, Mary Anderson, John McCullough, and Lawrence Barrett." At the testimonial, which was held at the house of Warren's niece Amelia Fisher, with whom he had lived for twenty-nine years, Winter also read a poetical tribute he had written. It is printed in *Vagrant Memories*, pp. 18-22.

[17] Margaret "Maggie" Jefferson Farjeon was the daughter of Joseph Jefferson and the wife of the English novelist Benjamin Leopold Farjeon. She was pregnant with Joseph Jefferson Farjeon who was born on June 4, 1883. A delightful biography of the Farjeon family was written by the noted authoress of children's books Eleanor Farjeon entitled *Portrait of a Family* (New York, 1936).

she will not get through safely: I hope it may not be so bad as he thinks. Yes, I have all your "Rambles" and Edwina—who thanks you for your kind words of her & sends love to you—is keeping them as her own property. I shall be glad to get your Warren poem. Wish I could have been at Miss Fisher's last night; I can imagine how the dear old lady trembled & wept & was "O, so happy" at William's coronation—for I hope the wreath was ready for his venerable pate last night. I wanted Clarke to cable—but I doubt if he did for he is in bed sick & has been there since his return. His boy, "Ted,"[18] has resigned his position at the Walnut St. & is coming back—this will distress John very much, I know, for he had hopes of the boy's becoming settling permanently in the business there. While it is in my mind I'll tell you what I've meant to for a long while past. When I declined to act with Ford & went to the Academy in Balto. he shewed his spite by having ugly things published about me. 'Tis but a trifle, I know, but it proves the man to be unworthy of the feeling you, Joe, Clarke & even I have entertained for him. Let it pass: it just came into my mind at the moment. He's a Stuart. By-the-by, what is *he* up to nowadays? I daresay he'll be after Irving & Stoker[19]—his fellow countryman. The latter has been honored with a medal for bravery in a grand attempt he made to save a suicide on the Thames. Stoker is not only a fine fellow, but a very clever man & I think quite as agreeable, personally, as Irving: two men were never better fitted. Edwina's marriage is set for the October that Irving opens in New York—so, of course, I shall *have to act* the role of 'heavy Pa' which will suffice for a month at least, but I don't see how I can be idle longer. Am glad at Percy's success—wish I could report so well of Creston Clarke; I fear the boy will not advance so rapidly; he may do better in comedy & I have asked his father to take him in his Strand company after he ends with me in December. Wonder if "L.B."

[18] Probably Wilfred Booth Clarke who at the age of thirteen could have been employed as a call-boy at the Walnut Street Theatre. He later had a successful career as an actor, manager, and producer.

[19] Abraham "Bram" Stoker, English author, most notably of *Dracula* (1897), was a close friend of Irving and his stage manager at the Lyceum Theatre. He later wrote *The Personal Reminiscences of Henry Irving* (London, 1906).

will try the Lyceum venture? Burgess was in Nuremburg when I was last in London. Of all the German towns that is the one I longed most to see & B. says it is worth a week at least. I cd. not reach it on my tour, but hope to do so in the Winter—if but for a day. Love to all friends & to your immediate dear ones, of course.

86.   Leeds, November 2, 1882

We've had a shock in the news of Downing Vaux's narrow escape from death by suffocation. His father found him unconscious on his bed & the room filled with gas—since when he has been just hovering over the precipice. Edwa. had not heard from him for several weeks & was, of course, very much distressed all the while; his parents wisely kept the facts from us until he was out of danger when they wrote to me at this place. I cabled & recd. an encouraging reply but to my second dispatch asking them to send him here to us if possible I've had no answer tho' it was sent two days ago, & it renews our anxiety. His sister, Julia, bravely subdues her own fears in order to sustain & encourage Edwina, tho' I can see how distressed she is. This bad news reaching us in this miserable place—where we've had not a ray of sunshine, but rain & fog every day, has made us all wretched enough—but I hope for better news tonight & better weather too when we leave Leeds.

In many respects the theatre here is the best I've ever seen; too large for anything but Opera & Pantomime, the 'shows' on which they chiefly depend, but in all the arrangements 'fore & aft' it is perfect—thousands of pounds have been spent in needless decoration, of course, & on useless conveniences—such as baths & marble tileing, Roman pavements—even at the back door, &c, but it is superb![20] It should be in London, Paris or New York. The Opera last week took all the money & I am acting to vacancy, of course.

Hutton sent me a Chicago paper—from which I send you the enclosed; isn't it funny? Note my "ignobled Queen"—cool, to say the least. Perhaps you've seen it & the account of his performance too; Horace McV writes that he saw his Iago wh.

---

[20] The Grand Theatre in Leeds was built, leased, and managed by the English actor-manager Wilson Barrett. It was purportedly the finest theatre in Great Britain in the 1880's.

was terrible. Let me know what you hear of his acting, &c.[21]

The "Referree" said that Mrs. Langtry would probably see a 'ruined Abbey' in America if there were no old castles, & the Park disaster must have made the joke seem to her prophetic.[22]

I long to hear particulars of the Warren Jubillee (I find that I spell queerly today—I've already *spilt* a *har* & a *hell* on Referee & Jubilee. Why?)—you will send me the full accounts I hope.

Glad to hear that John is doing so well—hope he will lay by some ducats for that rainy season which must come to us all. Love to him, to Joe, Barrett & all our boys.

After his engagement at Leeds which ended on November 3, Booth crossed to Dublin where on November 6 he opened for an engagement of two weeks at the Gaiety Theatre, presenting, with the exception of Iago, his entire repertoire. It was undoubtedly the most successful engagement of his tour. The audiences were friendly, receptive, and large. The gross receipts for the first week amounted to £834.18 and for the second week £829.12.6, the largest of his tour. Booth was also well-received by the Dublin critics. They recognized that Booth's acting was not altogether faultless, disagreed about some of his interpretations, and complained, as countless other critics had, that he was occasionally "spiritless" and without force and passion, but

[21] George C. Miln made his debut in the character of Hamlet on October 23 at the Grand Opera House in Chicago. Later in the same week he appeared as Iago. Miln was advertised as having restored and revised certain portions of the text. Apparently one of his "restorations" was "*inobled* queen" which, of course, Booth had restored. The critic of the Chicago *Tribune* (October 22) wrote that Miln's Hamlet was marred by "ranting," "offensive contortions" and "want of dignity," but still admitted that it was "spirited, scholarly, well-conceived, and on the whole an effective impersonation." He condemned Iago, however, as an "incongrous mixture of the coarse diabolism of a penny-show *Mephistopheles* and the eel-like workings of the boneless contortionist of a traveling circus." After his one-week engagement in Chicago, Miln made a tour of the Midwest and West.

[22] *The Referee* was a London journal which began publication in August, 1877.

The Park Theatre in New York was destroyed by fire on the afternoon of October 30. That evening Lillie Langtry, English actress renowned for her beauty, was scheduled to make her American stage debut, under the management of Henry Abbey, as Rosalind in *As You Like It*. Miss Langtry's debut was postponed until November 6, when she opened at Wallack's Theatre as Hester Grazebrook in Tom Taylor's *An Unequal Match*.

generally they acclaimed his brilliant talent, originality, faultless elocution, scholarship, and sensitivity. Unlike other critics on his tour, they regarded Booth's Bertuccio and, much to his surprise, Othello, more highly than his Hamlet or Richelieu. The Dublin *Telegraph*, which on November 8 had complained that his Hamlet was "too tame," on November 9 praised Bertuccio unreservedly as "a triumph and a masterpiece." The *Irish Times* (November 7) also lauded Bertuccio as "his highest achievement," and "a representation of surpassing power." The *Freeman's Journal* (November 9) wrote that Booth's performance of Bertuccio was better than either Richelieu or Hamlet and, on November 10, commended Othello as "one of the most perfectly artistic representations that he has given," a "fervid, intense, and singularly impressive" performance. The consensus was, as the *Freeman's Journal* (November 17) observed, that "however opinions may differ as to his special fitness for this or that particular character, there would seem to be almost universal recognition amongst the Dublin public of the fact that Mr. Booth is a great actor."

87. Shelbourne Hotel, Dublin, November 12, 1882

I send you a London Times (to Tribune office) containing news from Berlin of Haäse's [sic][23] Richelieu, lest the paper fails to reach you I enclose a slip from the same journal. Haäse advised me by all means to *do* Richelieu there (now he has done it)—it was 'one of my best'—'new to Germany' &c. I lately sent over a lot of the books translated for me by Bayard Taylor's daughter[24] & it has been known some time in Berlin that I am to open in that part at the very theatre Haase has selected for his unprofessional movement. He has merely succeeded in rubbing the gilt off my one little bit of gingerbread, no more. But let it

[23] Friedrich Haase, German actor, especially noted for his portrayals of Mephistopheles and Richard III, made a tour of the United States from October, 1881, to May, 1882. Upon his return to Germany, he made his first appearance on November 3, 1882, at the Victoria Theatre in Berlin as Richelieu. The London *Times* (November 6) reported that "the piece was coldly received and pronounced by all the critics to be of poor dramatic achievement." Haase "gushed" at a testimonial luncheon given for Booth at Delmonico's on May 1, 1882. See Haase's autobiography *Was ich erlebte: 1846-1896* (Berlin, 1896), p. 124.

[24] Bayard Taylor, American translator and man-of-letters, had only one daughter, Lillian Bayard Taylor Kiliani. The New York *Dramatic Mirror* (May 6, 1882) reported that a Miss Taylor, "a niece of Bayard Taylor," had translated Booth's versions of *Hamlet* and *Richelieu* for use on his German tour. I have been unable to locate any copies, however.

pass away among the many theatrical deceits that I have experienced. After his 'gush' at Delmonico's he will no doubt be full of affection when we meet & I must receive it & submit. But isn't it funny. Miln has sent a copy of his paper 'The Alliance' full of his success, &c.[25] He'll make a deal of money in the West. Your charming poem (Warren) is before me. Glad our offering was in the Cup & not a wreath—I wish I could have had a sip, just a wee sip, from it that night at Miss Fisher's: I know how happy you were & indeed all that took part in the affair. I have been very successful here, but strange to say *Othello*, not Richelieu (as expected), or Hamlet, has been *the* hit, & it has been so much talked of & asked for at the box office that I shall repeat it this week. This is my first experience of the kind—for either Richelieu or Hamlet has, everywhere, for years, been the preferred play—never before has Othello been called for, though Iago has been frequently as strong a card as the other two. The press has been kind, the audiences fashionable. I have received many kind offers here but declined them all. I keep very close—being hard worked & very tired.

My girls keep well, but Edwina is still anxious about Downing —of whose illness I think I told you in my last; we expect him here (in England) very soon. It looks now as though I shall return home in April & miss my Italy again.

Love to all. Expect a letter from you in a day or two. Tomorrow I will be 49!

"Darling, I am growing old!" &c.

After his Dublin engagement which ended on November 18, Booth crossed back to England where he filled engagements of one week in Manchester, two weeks in Liverpool, and one week in Birmingham. He then returned to London for the Christmas holidays.

88. Birmingham, December 14, 1882

I expected to find a letter from you at this point giving an account of McV's. lecture, which—I presume—'eventuated' on

[25] *The Alliance*, a weekly journal devoted to religion, literature, and government, was founded in 1873 by David Swing, a Presbyterian minister and writer from Chicago. He edited this paper until 1881 when it was taken over by George C. Miln. See Joseph Fort Newton, *David Swing: Poet-Preacher* (Chicago, 1909), p. 145.

the 28th, I have waited to hear from you before writing. A package of paper slips came here last week & was forwarded to Liverpool—I suppose you forgot I had two weeks in that city. I close here Saturday night & hurry on to Berlin—expecting to arrive there day before Xmas; I shall stop but a day or perhaps two in London. If you knew England as I know it—in fall & Winter—I fear your account of it would be less ardent than what you have hitherto given. Such beastly, god-forsaken places as these provincial towns can only be equalled by miserable London —which, after the Summer sets, is the embodiment of wretchedness & the horrors! I sent my cheque for $70 to Barrett's address, as you suggested. I have not heard from him, or Warren, nor have I seen the latter's letter in print that you spoke of in your last.[26] I am in good health & take the best of care of it, but my poor girl is suffering with a very severe cold which worries me. I was so prostrated by nervous exhaustion & consequent loss of sleep that I had a doctor see me in Liverpool, who lifted me up in the course of a few days and I now get my rest o'nights & 'nap,' but the gloom & dampness & the reaction from my night's labors depress me very much. I seem to be acting with strength & the papers are very much 'enthused' over me. Business averages about the same—pit & gallery full, sparse elsewhere. Will leave this open till after the play (Othello).

December 16

Didn't write after the play as I intended, nor yesterday—from shear weariness. Yours of Decr. 2d. (from Gillig's) came yesterday. A cablegram from Vaux tell us that Downing is on the Servia—she is due the 22d., so this changes again our plans. We cannot go 'till after he arrives & a day or two of rest will be necessary for him—therefore we will not start from London till the 27th & then go direct to Berlin. Italy must wait 'till the Spring.—I hope to be there at Easter. I will endeavor to get the book & Almanacs for you. 'Tis strange Mc's lecture was not reported next day, perhaps it was a failure. You say Miln has fizzled—I saw accounts of him in some Western towns. I enclose

[26] Warren wrote Winter on October 30 thanking him, Booth, Barrett, McCullough, and Mary Anderson for the "loving cup." The letter was printed in the *Tribune* on November 12. Booth's share of the expense of the cup was probably seventy dollars.

a slip I found somewhere—the writer evidently has not visited the Harp but wrote from hearsay. I doubt if I'll finish this 'till I get to London—for I am constantly interrupted & write *spasmodically*, besides, I have two letters of yours to answer. Flower has invited me to Stratford for a few days' rest, but I can't go, nor do I think it would be very delightful there at this season—amid fogs & snow & slush. Business has improved. The audiences are excellent to act for & the papers are very enthusiastic over my acting; Miller tells me that he sends you the notices else I would do so. Edwina is somewhat better, but the dreariness of the place is very depressing & no one can feel entirely well here. I am now interrupted & must put you off again 'till—when?

London, December 18

Now I am at Morley's—in rooms just under Barrett's. Went first to Charing X, but was so uncomfortable that I came this noon. Find here your Monday Tribune[27] (thanks for the notice) and a letter from *Horace* who says his *Dad's* lecture was fair, drew about $250, dealt principally with the pulpit—no personalities. Have seen nothing about it in print, so concluded 'twas not worth reporting. Shall see 'Much Ado' tonight.[28] Expect Clarke every moment—he is rehearsing. Burgess promised (at Hull) to send me a book of Barton Booth,[29] but it has not come. Sent Miller after the Almanacs &c this morning. Will post this tonight or tomorrow *sure*. Dreary weather—wet & warm.

Midnight, December 18

I have returned from the Lyceum. 'Much Ado' is superbly 'staged' and very finely acted. Irving's conception & treatment of the hero is excellent & were it not for his unfortunate voice & manner it wd. be as good as Miss Terry's Beatrice—which is

[27] In "Dramatic Notes" in the *Tribune* (November 11), a brief notice appeared that Booth was shortly to complete his provincial tour.

[28] *Much Ado About Nothing* was revived by Irving in a spectacular production which opened on October 11 at the Lyceum Theatre.

[29] Barton Booth, English actor and manager, was not related to Booth, although he liked to believe so. The reference is probably either to the *Memoirs of the Life of Barton Booth* (London, 1733) or Theophilus Cibber's *The Life and Character of that Excellent Actor Barton Booth, Esq.* (London, 1753).

perfect save for a little lack of power in the great scene with Benedict: Edwina & I were charmed with all we saw & heard tonight, despite the little exceptions I have noted. Terry will set 'em wild in America if she does this part. The scenery & sets are the finest I ever saw! I read all you sent me of Langtry: I don't see how else you could have treated so tender a subject. I have read some very unkind things of her, her person as well as her personations.[30] I regret sincerely the loss you sustained by the fire—I enclose a couple of rails for the new fence![31] Glad of Percy's success—with Barrett he is in good hands, but he must not have all serious work, as much comedy & character bits as he can carry will do him a world of good; my principal experience was in a variety of quaint & comic parts & I am endeavoring to get Clarke's boy, Creston, with that sort of work; he was with me on the tour & did very well for his 1st. season. I fear, my dear boy, the time is far distant when I shall settle down to reading the several useful & charming things you suggest—of each of them I have had the merest taste, a page or two, here & there at odd times in my past vacations, but now it seems that when I have time I have not energy, or the ability to concentrate my thoughts

[30] The "tender subject" was undoubtedly the split between Langtry and her acting teacher, former English actress Henrietta Hodson Labouchère, which developed because of Langtry's affair with Frederick Gebhard, a wealthy New York socialite. The split and its cause were reported in the drama section of the *Tribune* (December 2). On December 12, the *Tribune* printed a column on the Langtry/Labouchère quarrel and noted that a "*Tribune* reporter" had called on Mrs. Labouchère but she had refused to answer any questions. This same article also reported an interview with John Stetson, the manager of the Globe Theatre in Boston where Langtry was then playing, concerning Stetson's having Gebhard thrown out of a box at the theatre during Langtry's performance. Perhaps Winter was the "*Tribune* reporter" and the author of this article. He had been very friendly in his reviews of Langtry's performances in New York and had also made her personal acquaintance. For more information concerning Langtry, Labouchère, and Gebhard see Pierre Sichel, *The Jersey Lily: The Story of the Fabulous Mrs. Langtry* (Englewood, New Jersey, 1958), pp. 153 ff. Winter prints some of his reviews of Langtry in *The Wallet of Time*, I, 516-597.

[31] On December 16, the Pavilion Hotel at New Brighton, Staten Island, was almost completely destroyed by a fire attributed to an incendiary. The *Tribune* (December 17) reported two previous attempts to burn the hotel which, however, inflicted only slight damage. Perhaps Winter's "fence" was also among the incendiary's victims.

on anything either useful or agreeable: I am always "tired out" when not acting. I hope your adaptation & your tragedy will both be great successes.[32] I shall not breathe a word of either. Edwina's mother used to read Longfellow to me, years ago, & that first gave me the desire to visit Nuremburgh[33]—I missed it on my Spring tour, but I was at Heidelberg 3 days. We expect Downing here at Xmas, and we four will dine under the holly & misteltoe in our snug little quarters here—a *select* & cozy party. Have not yet heard from either Barrett or Warren—hope they will write. Saw Clarke but a moment or two today—he looks well & is busy getting up 'Comedy of Errors.'[34] Shall call on Irving in a day or so. I forget if I told you that Edwina reached her majority on the ninth: Percy is a few weeks ahead of her. Many happy years for them both. She is still coughing & I shall have a doctor for her tomorrow. Goodnight, old friend, & may God bless you! Kind regards & best wishes of the season for your dear ones.

On December 27, Booth, Edwina, and Julia and Downing Vaux left London for Berlin.

[32] Winter was apparently working on a dramatic adaptation and a play of his own. He published and produced only one adaptation during his life —Paul Heyse's *Mary of Magdala*—which was first produced in Milwaukee on October 23, 1902, with the American actress Minnie Maddern Fiske in the title role. The play was subsequently toured across the United States. In the fall of 1903 it was published by the Macmillan Company in New York.

[33] Longfellow's poem "Nuremberg," written in the spring of 1844, is a reminiscence of his visit there in 1842. See *Longfellow's Works* (The Craigie Edition), pp. 219-223.

[34] *The Comedy of Errors* opened January 18, 1883, at the Strand Theatre, which Clarke had leased and was managing.

Chapter VIII

Germany and Austria

1883

SHORTLY after Booth's arrival in Berlin, his German representative and stage manager for the tour, Emanuel Lederer, informed him that Gustav Scherenberg, the manager of the Victoria Theater where Booth was scheduled to appear, had canceled his contract.[1] Scherenberg had a success running, a spectacular entitled *Frau Venus*, and refused to interrupt the run. For a few days it appeared as if Booth would never be seen on a Berlin stage, but then Gustav Sanftleben, "a well known theatrical agent" according to Lederer, arranged an engagement for Booth at the smaller Residenz Theater. On Thursday, January 11, Booth made his first stage appearance in Germany as Hamlet.

Booth played in English of course and his supporting actors played in German. The company at the Residenz Theater had never before played *Hamlet*. This bilingual acting was not always successful. Karl Frenzel, a noted German critic, complained that the German actors, because they had learned their movements and facial expressions artificially, acted more like puppets than human beings.[2] An American who witnessed the performance reported to the New York *Tribune* (December 6, 1885) that whenever Booth was not on the stage "one would have supposed that the audience was listening to a capital farce" by their laughter and "mock applause." When Booth was on the stage, however, everyone was raptly attentive. Despite the inadequacy of his support, Booth's reception by the audience, which included prominent men of art and letters, the American and English ambassadors, and the Crown Prince and Princess, was enthusiastic. He was called before

[1] Gustav Otto Scherenberg was the manager of the Victoria Theater from 1882 to 1889. In the 1860's he managed Fanny Janauschek's tour of the United States and was active in theatres in Baltimore and New York.

Emanuel Lederer was an Austrian actor and stage manager who had met Booth in America. Lederer preceded Booth on the tour and rehearsed the companies with which Booth was to appear. Booth seems normally to have had only one rehearsal with a company prior to performance and even then only his cues were given. In the Theatre Collection of the New York Public Library is a manuscript by Lederer in English describing Booth's German tour.

[2] Karl Frenzel, "Die Berliner Theater," *Deutsche Rundschau*, xxxv (April-June, 1883), p. 466.

the curtain after every change of scenery. The Berlin critics were also generally commendatory, praising his naturalness, his complete lack of theatrical exaggeration, artificiality, and trickery, and his ability to make Hamlet an entirely believable, *living* character by his complete immersion in the role and the unity of his performance.

The critic of the *Norddeutsche Allgemeine Zeitung* (January 12) wrote that Booth's Hamlet was not like the German Hamlets he had seen, not *"halb Sauertopf, halb Narr"* but rather a Hamlet of *"Fleisch und Blut."* *"Jeder empfindet,"* he continued, *"bei seiner Darstellung zum ersten Male, dass eine Persönlichkeit, wie Shakespeare den Hamlet zeichnet, möglich ist."* The critic of the *Deutscher Reichs-Anzeiger* (January 12) lauded him as a *"grossartigen Tragöden von hinreissender Begabung."* Of course not all of the critics were so impressed by Booth's Hamlet. The young critic, Otto Brahm, writing in the *Vossische Zeitung* (January 13) conceded Booth's rich histrionic ability, his vitality, and splendid virtuosity; nevertheless, his impression of the performance as a whole was that it was not much above *"ein mittleres Niveau."*[3]

89.   Berlin, January 11, 1883

I have just accomplished the one great object of my professional aspiration. 'Tis after one o'clk, & I am very weary but cannot go to bed without a line to *you*. I have not written you for some weeks (it seems), nor have I heard from you lately, my reason—I wished to send you something of interest; what's yours?—I cannot tell you of my triumph tonight without a gush of egotism & you know how difficult that is for me, therefore I'll let Miller tell you the particulars. O, I wish you had been present tonight.— When I am cooler I will try to give you a full account of the nights work. The actors as well as the audience were very enthusiastic, many of the former "kissing" my hand & thanking me over & over again—for what I know not unless it was because they recognized in me a sincere disciple of their idol Shakespeare. I am at a different house from the one I expected to perform in; the manager of the Victoria had a successful running play & took dirty advantage of a quibble to break my engt.—at the eleventh hour [a] star at the Residenz (a smaller house) was taken ill & I had to hurry up & jump in hap-hazard—without scenery at all

---

[3] Brahm reviewed all of Booth's performances in Berlin for the *Vossische Zeitung*. The reviews are reprinted in Otto Brahm, *Theater, Dramatiker, Schauspieler* (Berlin, 1961), pp. 76-85.

appropriate & costumes of the queerest description; but the company is really excellent—good in every particular. Crowded & enthusiastic audience—very distinguished—the Crown Prince in a box & all the big ones represented. I scratch in great haste, with a wretched pen & hardly ink enough to keep it going. In a future letter I'll be more explicit. I have only 3 weeks here & I'll do Hamlet & Lear only, Hamlet most of the time. Well, this is the realization of my 20 years' dream! What shall I do now? Act in Italy & France?—No, I care as little for them as I did & do for flimsy London. Poor London! (I wont say England) what a cheap, little minded village it is—dramatically! Good night, dear boy! I thought to have written you a long letter & several to several other friends tonight, but I am too tired. Good night. God bless you!

During his four-week engagement at the Residenz Theater, Booth was also seen as King Lear, Iago, and Othello. The critics were more divided in their opinions of his Lear than of his Hamlet. Some, like Brahm and the critic of the Berlin *Börsen Zeitung* (January 25), thought that Booth lacked that elemental passion, wrath, and force which Lear should display in the first part of the play. But most critics agreed that from the mad scenes to the end of the play Booth's performance was a histrionic triumph. Some, like the critic of the *Deutscher Reichs-Anzeiger* (January 24), thought that Booth's Lear was even better than his Hamlet—"*Hier steht der Künstler offenbar auf der Höhe seines künstlerischen Schaffens.*" Some critics thought Booth lacked power and passion as Othello too, but most were highly impressed by the uniqueness of Booth's poetic, noble interpretation of Othello as compared to the more bestial, "African" interpretations of Tommaso Salvini and Ernesto Rossi. Generally Othello was Booth's greatest critical and popular success in Berlin. On February 3 and 4 he gave his only performances of Iago in Berlin. Most of the critics agreed that Booth's Iago was not Shakespeare's Iago, yet they still admired his performance. The critic of the *Börsen Zeitung* (February 4) characterized Booth's Iago as a fine, elegant Venetian with all the mobility, cunning, and perfidy of an Italian Tartuffe. As such it was a thoroughly admirable performance which, for the most part, kept the audience spellbound, but which also estranged them by its radical departure from tradition. Iago, after all, is not a "*Lieutenant*" but a "*Feldwebel.*" Otto Brahm was perhaps more satisfied by Booth's Iago than any of his other characterizations. Although he could not agree with the dominance the role assumed when played by Booth, he praised

it as "*eine Leistung von bewunderungswürdiger Lebendigkeit, Klarheit und Noblesse.*"

The reservations of the critics had little effect, however, on the attitudes of the audience. The public continued to receive Booth with enthusiasm. A dispatch from Berlin to a New York newspaper reported that during the performance of King Lear, Booth was called before the curtain eighteen times. A week after his opening as Lear, however, Booth wrote Winter that "all this *glory*" was marred by the continued illness of Downing Vaux, which was distressing to him and, more importantly, to Edwina.

90. Hotel de Rome, Berlin, January 29, 1883

Yours of the 12th reached me yesterday. I have not written you since 'after the play' Jan: 11th, and could then scribble you only a few lines. I have been suffering all the while with a racking cough & headache, and have had wearisome rehearsals every day. Not a moment's leisure & utterly 'fagged out.' My success has continued and I am told that fine critiques have been numerous but my interpreter is not able to give me any clear idea of what is published. *Lear* seems to have taken even a deeper hold on Shakespearian Germans than *Hamlet*. It (the language) is *Greek* to me, of course, & consequently my head is not turned by the praises showered upon me. As an offset to all this *glory* I have sadness in my little family circle. Young Vaux has been with us about six weeks and his condition is very distressing, especially to Edwina; his mind is improved in many respects, but he seems to have lost all his interest in Edwina, & his profession. Is in a dazed condition—yet talks sensibly on indifferent topics. This has so affected Edwina that after mature consideration and several consultations with Profr. Leyden,[4] the highest authority on brain & nerve diseases, we have concluded to send him & his sister home. They will sail from Bremen in the *Donau*, Sunday next. The boy should have been kept at home under careful treatment for several months, but we all thought for the best. It may be *two* years, certainly *one*—the Dr. says, before he can pursue his profession. This apathetic state has rendered him an object of pity & Edwina feels as though he were not the man she knew & loved at home, for though bodily here his spirit seems

[4] Ernst von Leyden was an eminent German physician specializing in neurological disorders.

absent. You can imagine the poor child's wretchedness & my anxiety. It is affecting her health, but I have hope that the separation will restore her former feelings; while [he] is here in this condition she is hourly reminded of his sad change. Why should two young lives be thus suddenly darkened? It is very cruel! Of course my spirits are at low water mark & it is a double labor for me to act with such a weight oppressing me.—I shall play Iago next week for a few nights, then *Othello* after which a repeat of *Lear* & *Hamlet* to close the engagement. Haäse has been acting *Richelieu* again with the same result—failure. This has so prejudiced the critics & managers against the play that I am begged not to do it. Tag & Brachvogel[5] (of the Delmonico party, you remember) are 'hopping' mad at Haäse's conduct, but it has not injured me in the least. He was in town t'other day, I was told; have not seen him. He speaks well of me in his book, it seems, and otherwise, so perhaps he did not mean to block my way, but he has d----d the play for Germany.—We have had lovely weather till three days ago when a splendid snow fell which melted yesterday & all today we've had rain & wind but no *slush*, thanks to the wonderful management of street affairs here. It may be that I shall return in April or May, but my movements will be controlled by Edwina's health and wishes. All looks dark to the poor girl now. I was too ill to rehearse this evening (I have one night off) after a tiresome tug this morning, & thus I have an opportunity to write a few letters. The death of the Emperor's brother[6] has put a stop to my "Royal patronage" as the Court will be in mourning till after my engagement ends. I learned, at the only dinner I have accepted, that it was intended by some of the 'Nobs' here to have a grand fete of some kind to which I wd. be invited & presented to the Imperial folks, but that's upset. It is wonderful how well the plays go in "polyglot"! When a word or sentence sounds very English, as is frequently the case, it rather confuses me for tho' I have not acquired a word of the language I seem to think it & comprehend it while I am acting with the Germans. Isn't it

[5] Possibly Casimir Tag, prominent New York banker of German origin. Probably Udo Brachvogel, New York journalist of German origin, who contributed articles on the stage to the New York *Staatszeitung*.

[6] Prince Karl, the brother of Kaiser Wilhelm I, died on January 21.

curious? at rehearsal I've had to ask my interpreter to give me the last few sentences of a speech in German when he has endeavored to assist me in English. Judge Daly is right regarding the Residenz Theatre being small, it is very small, but it is not near the palace, it is a twenty-five minutes drive. An elevated road (a splendid structure) takes me quicker. What a pity that Miln has made such a donkey of himself! Glad of John's success & sorry that you think him in ill health. Edwina wrote Miss Barrett & I had forgotten her address—so she sent it to your care. Have not written to Hutton for some weeks—indeed I've had neither time nor energy to write at all since I began acting here.

Love to all your dear ones & to thee.

[P.S.] Have declined several fine 'invites' but am booked for one next Monday at the house of a Shakespearian.

On February 11, Booth closed his Berlin engagement with a performance of Othello and a few days later he departed for Hamburg where he was scheduled for four performances beginning February 15. As Booth wrote in his last letter to Winter, he had "had neither time nor energy to write at all" since he began his German tour. Even when he did write, he refrained from giving much of an account of his acting or of his reception by the German audience, critics, and his fellow actors. Edwina, however, felt no such restraint. From Hamburg and Bremen she wrote Winter a long letter to give him "*some* idea of Papa's great and genuine successes."

91.   Hotel de l'Europe, Hamburg, February 20, 1883
My dear Mr. Winter.

Many times since my arrival in Germany, I have been tempted to write you, but somehow, it has always seemed to me rather a bold venture, and so I have even allowed your pretty Xmas card to go unacknowledged. I am willing to make this an excuse, (if necessary) for writing *now*, though my real motive, is to speak of Papa's success, and of the wonderful impression he has made, not only *generally*, but *individually*, speaking. In the latter instance, I can mention Professors Werder and Leo[7]—men of

---

[7] Karl Friedrich Werder, professor of philosophy at Berlin University, was noted for his lectures on *Hamlet* and *Macbeth*, which were published as *Vorlesungen über Shakespeare's Hamlet* (Berlin, 1875) and *Vorle-*

learning and Shakespearian students. The former was tutor to the Crown Prince and his enthusiasm was most delightful to see. Unable to speak English his conversation was an intermixture of German and French, though the latter tongue, like a true Teuton, he held in contempt, and so, unmindful of our ignorance, would ramble off into his native language, with an unconsciousness quite comical to behold. This little old man addressed most of his conversation toward me and gave me to understand that Papa was the greatest actor he had ever seen during his seventy years' experience—that he saw in the genius of the actor, the nobility of the man—not only his *mind*, but his *heart*—was touched. Prof. Leo declares Papa to be, not only the greatest actor, but also the best *commentator* of Shakespeare, and intends writing an article to that effect. If you but could witness Papa's various performances, I am sure you would marvel at their continued vigor, and at the remarkable smoothness and ease with which he follows his fellow-actors. The audiences you would find more appreciative, though slow to give applause *during* a performance. "Othello" made such a profound impression in Berlin, that "Hamlet" was withdrawn, and indeed *I* begin to feel that it is one of Papa's strongest characters. Urgent requests, for his return to Berlin, came from all sides, and the actors thanked him for coming to teach them "how to act" indeed the *tragic* art seems in rather a primitive stage amongst them. How different are they from the English, who are so slow to accept any ideas save their own! The Germans appear to me, less narrow, and certainly more *Shakespearian*, and I am proud to have Papa win their approval so readily. I speak, perhaps a trifle enthusiastically, on this subject, but to you, who are interested, I can do so, without reserve—for I doubt if Papa ever writes you of the honors he receives. His last performance in Berlin was a fitting close to his preceding triumphs, being the occasion for several flattering presentations from the press, the actors and the public. The

---

sungen über Shakespeare's Macbeth (Berlin, 1885). The lectures on *Hamlet* were translated into English by Elizabeth Wilder as *The Heart of Hamlet's Mystery* (New York and London, 1905).

Friedrich August Leo, a prominent German Shakespearian scholar, was, from 1880 to 1898, the editor of *Shakespeare Jahrbuch*. In the 1883 volume of this publication he wrote a highly commendatory essay on Booth's acting. See *Shakespeare Jahrbuch*, XVIII, 270-272.

former offered him a handsome address, the actors a beautiful wreath in silver,[8] and there were many wreaths of genuine bay leaves from the public: The members of the company bade him quite a touching farewell, many of them embracing him! his carriage was surrounded by enthusiastic men and women, waving their hats and handkerchiefs—some calling out "Auf Wiedersehen". The same scene was repeated on our departure from the railway station.

Here he has also been the recipient of great enthusiasm, and last night, six large wreaths were handed him across the footlights. His "Lear" seems to have excited more enthusiasm here than in Berlin, where "Othello" was the favorite. His support here is mediocre, with the exception of Carl Formes,[9] who, Papa says, is a fine actor, playing the Fool in "Lear", and the First Grave-digger most admirably. The Germans are vile stage-dressers, their costumes being all out of character, as to time and place. It was amusing to see "Horatio" in military boots, and armed with a sword, and Ophelia in satins and gold lace. As Papa remarked, the costumes of several centuries are seen on the stage at once. I am sure we should find more care and more knowledge in these matters, even amongst our poorest companies in America. The Residenz Theatre in Berlin has a company much better adapted to *tragedy* than this one. We saw a comedy here, however, exquisitely performed in every detail—in this respect quite equal to any French performance we have seen. You can fancy how many difficulties Papa has to contend against—most of his actors knowing no English whatever, and thus speaking before their "cue," interrupting Papa's lines. Yet he follows them with great accuracy, never saying the wrong thing, even when more than the usual text is spoken. It is curious how, even the oldest actors, depend, almost entirely upon the aid of the prompter, who sits

[8] Booth was also given a silver crown of laurel in Bremen, a silver goblet in Hanover, a silver wreath of laurel in Leipzig, and a silver spray of laurel in Hamburg. These trophies are preserved at The Players.

[9] Edwina undoubtedly meant Ernst Formes, the son of Carl Formes, a German opera singer who emigrated to America about 1864. Ernst was a leading comic actor of the Thalia Theater in Hamburg from 1878 to 1892. About his portrayal of the Fool the critic of the Hamburg *Nachrichten* (February 19) complained that he "*war aber mehr ein gutmüthiger lustiger Geselle, als ein sarkastischer und melancholischer wie er fast allgemein zur Darstellung kommt.*"

in his little box facing the stage, giving them every word in an audible whisper,—this too, whilst Papa is reciting his lines, which is most annoying as well as confusing.

Papa says, that though accustomed to applause from his child-hood almost, this demonstration from his fellow-actors, is a new experience and has given him a fresh incentive and ambition. The director of this Thalia theatre, who remembers Talma and Devrient,[10] awards the palm to Papa, and after watching his rehearsals, bestowed on him the title of "Meister," in which light the actors also regard him. They crowd about the "wings" and "flies" watching his performances, and at his exits, men and women alike embrace him warmly, the above mentioned director too, having actually kissed him most affectionately before the actors. All this would seem amusing, and a trifle *overdone*, but for its very genuineness, and is therefore quite touching to Papa, who is scarcely able to respond to so many marks of favor.

Now I trust that my long letter has not wearied you. I could not resist, the desire to tell you, what I feel sure you would never hear from Papa or even the newspapers. I hope I have given you *some* idea of Papa's great and genuine successes. He joins me in best remembrances to yourself and your wife, to whom, please give my warm thanks for her pretty souvenir.

Very Sincerely
your young friend
Edwina Booth

Bremen, February 23, 1883

P.S. I left this open to tell you somewhat of Papa's closing night in Hamburgh, which was such a complete triumph—so unlike anything he has ever before experienced, that I am positive it will interest you. The piece was "Lear" and went extremely well and smoothly throughout—Papa having quite outdone himself, and as

---

[10] Charles "Cheri" Maurice, German theatre manager, was the director of the Hamburg Thalia Theater from 1843 to 1893.

François-Joseph Talma, French actor, was at his finest in roles such as Voltaire's Mahomet, Corneille's Le Cid, and Shakespeare's Macbeth, Shylock, and Hamlet in the Ducis translations.

Ludwig Devrient, German actor of Dutch origin and the first of a famous family of German actors, was best known for his Falstaff, although he preferred tragic roles such as Franz Moor, Shylock, and King Lear.

I glanced around at the audience I saw very few dry eyes. He was recalled before the audience many times, as many as eight consecutive calls at the end of the play. Miller had confided to me that a presentation was to be made, so he escorted me behind the scenes, where were assembled nearly a hundred gentle-men and ladies in evening dress—many of them being actors & employes attached to the theatre. After Papa's last call and the curtain descended the people clustered about him, and Herr Formes, made him a most flattering address in excellent English. Then some ladies came forward bearing a satin cushion on which rested a delicate sprig of silver bay leaves—a match to his Berlin presentation. All this while the people in front who had not yet left the theatre were howling for Papa's reappearance—and this he acknowledged by a neat little speech surrounded by his little crowd of admiring friends. On this occasion, let me add, I made my first appearance on "any stage," and distinguished myself— though well in the background—by joining in the cheers of the others. But this act of demonstration did not close this never-to-be forgotten scene. What *do* you guess follows! Well, it fairly took my breath away! A whole bevy of pretty young girls— actresses & otherwise—rushed up with open arms and actually *embraced* and **kissed** Papa as he stood on the stage still in Lear's robes!! Then the men too, many with tears in their eyes, embraced him warmly pronouncing him the "greatest artist" they ever saw. The young girls appeared so modest—and their demonstration so genuine, not in the least silly or bold as one might imagine, that I stood speechless—not with displeasure but delight. I must not omit to mention either, that some handsome bay wreaths were thrown to Papa during the performance. The old director of the theatre who had seen Talma—told me in French that Papa was the greatest artist he knew. Well, the enthusiasm did not subside until we were fairly settled in our carriage. For as we passed out, its doors were surrounded by well dressed ladies and gentlemen— through whose midst we had to pass, as many of them gave three cheers lifting their hats as they did so. For a moment I fancied I must be a member of the Royal Family. Oh! it was very exciting and delightful, I assure you, and an event to be proud of too. I only wish you and some of Papa's other warm friends could have witnessed it.—They tell us, and so the papers state, that his success is unrivalled—not even native artists having called

forth such genuine enthusiasm in many years. It quite touched me to see the aged stage director—(a pupil of Devrient's) throw his arms about Papa, and kiss his hand with profoundest respect. The morning of our departure a chorus of male voices serenaded Papa in the corridor of our hotel—just outside his door, and we were awakened by their exquisite strains and vocal harmonies—for they sang without instruments. Papa arose and thanked them for their charming farewell. In a few hours we had left the pretty town, whither Papa was entreated by both managers and actors—to return again soon. He wants me to tell you how delighted he was with Herr Formes' performance of the "Fool"—really the first that has ever satisfied *his* conception of that part. Formes acted with true pathos and helped Papa immensely by his quick intelligence and sympathy. Yet in spite of this, the critics found fault with him, however, Papa expressed *his* opinion publicly, and I am sure it will have some effect. Papa is now rehearsing "Hamlet," "Lear" and "Othello" here. For the purpose of seeing the latter many are coming from Hamburgh. I believe, every box and seat is gone for the entire five performances. I had a charming evening walk about the town today. It is very quaint and picturesque, and the streets old and narrow. There is a fine old windmill in this vicinity, and as I saw it this evening with a sunset-sky for a background, and the surrounding old lindens dipping their branches into the stream beneath, I think I never saw a more inviting picture. Will you send me a photo of your baby-girl "Viola"? I can see you going to sleep over this long postscript—so now goodnight with Papa's love and mine for yourself and family.

Edwina

After Hamburg Booth filled brief engagements in Bremen, Cologne, Bonn, Hanover, and Leipzig. He then traveled south towards Vienna, via Dresden and Prague. In Dresden he stopped for a day of sight-seeing and relaxation.

92. Dresden, Good Friday [March 23], 1883

I closed my brief but brilliant engt. at Leipsic Wednesday & came here to see the sights en route for Vienna, but the holiness of the day has shut up all places of interest, even the shops—so we must be content with a general view of the city. I should have

acted here, but for some difficulty about dates, I believe. The exterior of the theatre, just opposite this hotel, is the grandest I've yet seen & the splendid old Cathedral across the way (with communication by means of a covered bridge with the Palace) is also superb—the Palace is a *stupid* looking structure. We have just returned from a two hours drive about town & the suburbs, very interesting; a charming day, but very cold, an attempt to walk after our drive failed on this account—consequently we must sit indoors, before a delicious wood fire, & scribble & read the guide book—for we take no interest in the papers of this country. Tomorrow we go as far as Prague & the following day push to Vienna. Miller & my interpreter & valet with the traps left Leipsic this A.M. by through train. News from Vaux & his family is encouraging & Edwina is well & hopeful. The stories from Berlin correspondents to our papers are grossly exaggerated; they have brought many letters of condolence to Edwina & she, as well as Vaux's folks are greatly worried about them—'tis bad enough as it really is without exaggeration. It could not be kept quiet, for interpreters were required to describe the particulars to the Doctor; at first I had mine—then he brought his, & so it got talked about in Berlin. I shall make no arrangement for America 'till after I get there & have had a good rest—on my (silver) laurels, &c. I recd. another token from the Company at Leipsic & an engraving & plaster copy of death mask of Ludwig Devrient— one of the greatest of German actors, you know: his grand- nephew, the Director of the Leipsic Theatre gave them to me— whose old mother told him that she saw in my acting his uncle Emil Devrient as he appeared in his prime.[11] Zo, goot! Ya.—that's the limit of my Deutsch thus far. I think I have recd. all your letters & papers. I must not only thank you, but Mr. Reid also for the Tribune column of translations—through them I've learned what the critics of Berlin said of me—though I was told there, as elsewhere, that the Press was enthusiastic over my acting. I was

[11] Emil Devrient was a nephew of Ludwig. He had two older brothers who were also actors, Karl and Eduard. Emil, for forty years a leading actor at the Dresden Court Theatre, excelled in such roles as Hamlet, Egmont, and Tasso.

Max Staegemann, whose mother was the sister of Emil, Karl, and Eduard, became the director of the Leipzig Neues Stadt Theater in the summer of 1882. He was still active in this capacity as late as 1906.

invited to Coburg & Weimar & recd. hints of *decorations* in both cities, but I cd. not arrange the time so as to secure a little rest, and declined.[12] I shall have a week yet before Vienna. Sorry I did not know of the Barnay[13] breakfast in time to send my greetings. Gillig sent dispatch to Drencker[14] [sic] at Berlin & the latter *mailed* it to me—the consequence was I did not get it till long after the wittles was dispozed of. I hope it was successful. I have directed Miller to send you papers & bills for I never see either, but he keeps files of the latter for me. I can hardly realize that my dream of twenty years ago is so near its end. In a week or so I shall terminate at Vienna the series of engagements towards wh. I have looked with hope, fear and despair—for I hardly believed I should ever accomplish them, or if—not so successfully.

My health is excellent, a little fatigued & occassionally a cough depresses me, but on the whole I was never better. What shall I do after I end my work at Vienna?—I've never been ambitious to act in Italy, France, Russia or Spain—yet I have been urged, at various times to visit each country, but 'tis too late now for me to be wandering about the world in harness—'tis time to sit down & begin to be old, as some other chap remarked. Don't worry about me & the wines—the beverage I indulge in is the mildest lager beer, and that only at night—just to keep my 'mad' up. I see that Irving is to have a grand banquet before going to America—I may

[12] According to Emanuel Lederer, Booth was called on after the third act of *Hamlet* on his opening night at Leipzig by a Baron von Loen, an official of the Grand Ducal Theatre at Weimar. The Baron invited Booth to perform before the Court. Booth would have gladly accepted the invitation but he was already scheduled for Vienna and could not postpone that engagement. Lederer reports that Booth told the Baron: "I never had a will of my own; I always found myself, as now, under the yoke of obligations which others assumed for me."

[13] Ludwig Barnay, German tragedian who was a leading actor with the Meininger Company from 1871 to 1882, made his New York stage debut as Coriolanus on January 3, 1883, at the Thalia Theatre. During the next three months of his engagement he appeared with great success in a variety of roles, including William Tell, Hamlet, King Lear, Antony, Kean, Wallenstein, Essex and Narciss. On March 19 he was honored with a testimonial breakfast at Delmonico's given by the members of the New York theatrical profession.

[14] Emil Drenker, German theatrical agent, seems to have been the booking agent for Booth's tour, for which he was paid 2½ percent of the gross receipts of each performance.

be in London at the time; I presume there'll be lots of *gush* about me & Henry's magnanimity towards me! Had Henry acted properly I might have had his theatre during his absence—he told Miller he would do nothing 'till he had consulted me, just a day or two before Abbey announced the arrangement between himself & Irving. But, doubtless, 'tis for the best: I could never be made fashionable in London & without that aid there is no chance for positive success there.—It is now uncertain what course we will take after my Vienna engagement. The distances are so great & my time so limited (and the season so far advanced, too) that we may not visit Italy, but go direct to Paris thence to London for a few weeks in each city, rather than rush from place to place for a day in each & wear & tear ourselves to bits instead of resting. I regret this very deeply—for I want to see Rome, Florence & Venice; perhaps I may come abroad again some day solely for recreation—I hope so. I presume Edwina's letter has reached you ere this—she sends her love with mine to thee & thine. Regards to all who speak kindly of me. Thank Mr. Reid for me. God bless you!

[P.S.] Would it be well to contradict the reports of Vaux's hopeless condition, &c?

On March 31, Booth opened at the Stadttheater, Vienna, as Hamlet. He subsequently apeared as Othello on April 4 and King Lear on April 7, after which his engagement was extended for another five performances during which he repeated Lear and Hamlet. The Viennese critics generally commended him, but were much more reserved in their assessments of his Hamlet and Othello than had been the Berlin critics. Most of them complained of Booth's lack of passion, of force, of *"elementare Gewalt"* as Hamlet, especially in the scene at Ophelia's grave. But they admired his technical skill, especially his gestures and, on the whole, thought the performance *"interessant."* Othello also was criticized for its lack of force and passion, especially in the first two acts, which the critic of the Vienna *Presse* (April 5) characterized as *"ruhig, fast farblos."* The third act impressed most critics, but still the consensus was, in the words of the critic of the *Neue Freie Presse* (April 5), that Booth's Othello was more of a *"Katze"* than a *"Löwe."* Booth's King Lear was perhaps his most successful impersonation in Vienna. The critic of the Vienna *Allgemeine Zeitung* (April 8) praised the clarity and unity of Booth's delineation of the character of Lear; it was, he continued, a *"Meisterstück"* of acting. Not every critic

concurred. The critic of the *Neue Freie Presse* (April 8) thought that for Booth the character of Lear was *"unzugänglich."* He admired Booth's technical virtuosity and thought the performance was generally effective, but Booth's Lear was not Shakespeare's Lear, not King Lear, but *"ein alter Hausvater des bürgerlichen Trauerspieles"* or, at best, *"ein rührender alter Mann."*

The Viennese public was of course much more enthusiastic than the critics, although perhaps less exuberant than the Berlin audiences. Booth was called before the curtain after every act, often every scene, and at the end of the performance was presented with garlands of flowers and wreaths of laurel.

93. Vienna, April 5, 1883

Yours of March 17th has just reached me and as I am *in* for the day, on account of weather & fatigue, I will answer it at once—for Sunday, my regular letter-day, will be spent at a friendly dinner and the Opera after, and therefore I take Time by the 'padlock.' I'm sincerely sorry for your severe illness & happy at your recovery. Your enclosure from the *Mail* is very gratifying—so were the others wh. I had seen before. Edwina is much pleased that you liked her letter—she feared it wd. bore you. She is just recovering from a severe cold & I am very tired—otherwise both are well. I shall look with interest for your report of the *Festival* at Cincinnati.[15] As for the books—I found the *derned* things cost so much for printing & transportation that I got tired of the whole affair: I have a large case of a cheaper edition locked up which Henry Flohr's stupidity failed to send with my theatrical things, though they were prepared expressly for England. There shall be no change in the plates whatever, so don't annoy yourself on that account any more—it doesn't worry me. I only regret that they were stereotyped for there are so many changes I'd like to make but cannot without great & useless expense. I began here last

[15] The Dramatic Festival was held from April 29 to May 4, 1883, at the Cincinnati Academy of Music. During this week such prominent American actors and actresses as John McCullough, Lawrence Barrett, Mary Anderson, James E. Murdoch, Clara Morris, Nat C. Goodwin, and John Ellsler appeared as one company in Shakespeare's plays and other plays from the standard repertory. Winter dispatched reports which appeared daily in the *Tribune* from May 1 to 6. See Otis Skinner, *Footlights and Spotlights* (Indianapolis, 1924), pp. 120-122 for several anecdotes about this festival.

Saturday but was dull & heavy as the spritely Dane. Monday I did better, but failed to draw very much; as Othello, last night the audience (a fine one) received me with great enthusiasm, but I am told the critics think my quiet fifth act spoiled the effect of my third & fourth wh. they commend highly. They want the *howler* here &, I presume, they think it best not to endorse fully the Berlin verdict. I don't care. They can't make me *yank* Desdemona about the floor like a clothes bag. I shall close my work here Monday next—then for an easy journey to Paris, stopping at Nuremburg—the place of all Germany I long to see, at Frankfort & somewhere else to break the trip. When will I act again? Maze Edwards[16] is a good fellow & very capable—I wish him well, but fear that he is in a rut of mistake. I forget how often I have told you that I revelled at Heidelberg last August (I think it was) but I know that you've several times told me to go there. *I HAVE Bin DAR, Honey.* As the Barrett's go to Stuttgart I will miss them, I fear.[17] Sorry I was not in time for a cable message to the Barnay breakfast. This is a beautiful city, but a horrible place for throat & lung diseases. Edwina joins me in love to you all.

April 9, 1883.

I just discovered this letter stowed away in my table drawer—since writing it I've had a solid success here with Lear: all the papers, I'm told, are strong in my favor & my engagement is prolonged—the manager wants an indefinite renewal, but I'm tired & have not sufficient time to see the places, leisurely, that I've set my heart on visiting & get the rest I so much require. There is even a call for a repeat of Hamlet—as though the impression it made had set folks thinking. Miller says all the seats were gone early this A.M. & the crowd was great about the box-office. Have seen some excellent Comedy & domestic acting on my "off" nights at the Bourg Theatre—in the Palace, about 150 years old. Lucca is singing and on three occassions Edwina & I have heard her with great delight[18]—The operas are splendidly

[16] Maze Edwards was an American theatrical agent. During the late 1870's and early 1880's he was Henry Abbey's business manager.

[17] Barrett's daughter Mary Agnes was married to Baron Hermann von Roder in Stüttgart during the summer of 1883.

[18] Pauline Lucca, Viennese soprano, was from 1874 until her retirement in 1889 a permanent member of the Hofoper, although she made

mounted & acted & I presume the singing & music is excellent—
everybody says so. I shall act three or four nights longer & then
start for quaint old Nuremburg. I see that Burgess is to marry a
Miss Meyrick who travelled with her aunt Mrs. Billington,[19] both
in my company through the Provinces—a very worthy lady. I
thought of him in connection with Nuremburgh—the last time
I saw him he had just returned from a visit there. I'm sorry to
know that John has been ill & hope the dear fellow is all right,
he ought to visit Joe for a few weeks—wish I could. Profr. Leo of
Berlin has written an essay on my Hamlet for the Shakespeare
book, wh. appears every year on the 23d. of April—if I can get
a copy I'll send it [to] you, or a translation maybe. The people
here seem to think their endorsement of all artistic matters of
greater value than that of the Berliners or Parisians: they do not
accept all the great actors or singers of Germany (Barnay, for
example). I regret that you were not at his breakfast & that I cd.
not have sent a word by cable: I saw him but once, as Antony,
at Drury Lane.[20] The director of the Court (Bourg) Theatre &
the opera is an Excellence Baron & High Cock-a-lorum & [illegible
word][21]—a fine old gentlemen, sent me his box to see a translation

---

numerous starring tours, including three to America in 1872-73, 1882, and
1884. According to Lederer, Lucca was the first to congratulate Booth at
his triumph in Vienna. She "rushed to his dressing room between the acts
and was detained at the door by an attendant until Booth was able to do
the honors himself. The great songstress realized that she was too im-
petuous, and after an awkward pause, a cordial greeting took place."

[19] Ellen Meyrick, minor English actress, played the Player Queen,
Francesca Bentivoglio in *The Fool's Revenge*, Lady Anne, and Marion de
Lorme in *Richelieu* in Booth's company during his 1882 tour of the English
provinces. Adeline Billington, another minor English actress, played
Gertrude, and occasionally appeared in farcical curtain-raisers or after-
pieces during Booth's 1882 provincial tour. Both Miss Meyrick and Mrs.
Billington were frequently commended by the critics for the excellence
of their support.

[20] Barnay appeared as Antony in the Meininger Company's production
of *Julius Caesar* which opened at the Drury Lane on May 30, 1881.

[21] Adolph Wilbrandt, Austrian poet, dramatist, and novelist, was the
director of the Burgtheater from 1881 to 1887. Booth, however, is probably referring to Heinrich Laube who was the director of the Stadttheater
in Vienna from 1875 to 1880.

of Calderon's "Judge of Salamica"[22]—excellently acted,
particularly the Judge—by Barmeister [sic],[23] one of the best of
their actors; also met him at dinner yesterday & in the box at the
Opera last night—he is full of praise of my Lear. Frau Wolter
& Herr Sonnenthal[24]—the leading ones of the company are on
leaves of absence & therefore I shall not see a tragedy: their
comedy & domestic drama is very fine—equal to the French in
every respect. Adieu! I'll be sure to post this now. Love to all.

On April 17, Booth closed his engagement in Vienna with a benefit
performance of Iago. On the next day a number of critics, such as
the critic of the *Neue Freie Presse* (April 18), complained that
Booth's Iago was more of a devil than a soldier. Yet the same critic
praised him for his *"runde Leistung"* and the critic of the Vienna
Allgemeine Zeitung (April 18) also lauded his *"erstaunliche Deu-
tungskunst."* After his performance, the audience and his fellow
performers reportedly presented him with a *"ganzen Garten"* of flowers
and wreaths. Booth rested in Vienna for a few days and then started
back to London, making leisurely stops at places of interest along
the way, such as Frankfort on the Main, Nuremberg, Metz and
Paris.

94.   Frankfort on the Main, April 23, 1883

I wrote you last from Vienna at the close of my German work
but expect no letters from any one till I reach Paris—thither I
now wend my way leisurely. It seems strange to have time all
to myself, yet—such is the force of habit, and advancing age,
that I regularly take my nap after dinner & thus lose much time

[22] As *Der Richter von Zalamea*, Calderón's *El alcalde de Zalamea* was
then being produced at the Burgtheater.

[23] Bernhard Baumeister, leading actor with the Vienna Burgtheater
from 1852 to 1912, was a versatile actor who, at the close of his career,
had played over 500 different roles but was probably best known for his
Falstaff.

[24] Charlotte Wolter, Austrian actress born in Germany, was a leading
actress at the Vienna Burgtheater for over thirty-five years, excelling in
roles such as Antigone, Lady Macbeth, Phaedra, and Fedora.

Adolph von Sonnenthal, Austrian actor born in Czechoslovakia, was
a principal actor at the Burgtheater for more than forty years, excelling
in such roles as King Lear, Nathan, and Wallenstein. He made numerous
starring tours, including three to America in 1885, 1899, and 1902.

that might be profitably employed. Friday & Saturday I passed in dear old Nuremburg—where Edwina read me Longfellow's poem, wh. her mother twenty-two years ago! used to read for me & gave me the desire to visit the place. How odd, & weird almost, it seems to be there with her daughter, her counterpart in form and feature, reading the same poem to me! I confess to a sense of disappointment, however—in the perfect comforts and modern, conveniences I found there; in spite of the quaint streets and houses and atmosphere of antiquity that pervades them, I felt the lack of harmony in the people. They are not picturesque in the least, not even *dirty* enough to interest one—they seem out of place. By-the-by the cleanliness of nearly all the German towns is very remarkable; with us that people are regarded as quite otherwise. I mean the servants &c, but here at home the poorer classes are neat & the meanest little byways of streets are scrupulously clean. I've seen a great deal of the country and nearly all of it is very beautiful. Today we went first to Geothe's house—where he was born, & spent a very interesting hour with a queer old woman who told us all about the Poet, his parents &c, &c in Deutsche which bored us intensely—but made us laugh. After we left the house Edwina reminded me 'twas Shakespeare's birthday. A most appropriate occasion for such a visit. Immediately we purchased Lewes' life of Geothe,[25] & Edwina is now deep in it—up to her eyebrows, while I write. I am about finishing the journal of Caroline Fox,[26] very full of interest & information. I was much encouraged to find in her quoted remarks of Hartley Coleridge the very idea expressed regarding Genius, which I had conceived as long ago as boyhood when contemplating my father in the character of Hamlet: that it is a disease which prevents its possessors from sympathetic communion with ordinary men; isolates them, as it were, in the

[25] George Henry Lewes, *The Life and Works of Goethe* (London, 1855).

[26] Caroline Fox, daughter of the English scientist Robert Were Fox, is principally noted for her journals in which she recorded the conversations of such famous Victorians as John Stuart Mill, Thomas Carlyle, John Sterling, and Hartley Coleridge, who frequently visited her father at their home in Cornwall. These were published in London in 1882 as *Memories of Old Friends Being Extracts from the Journals and Letters of Caroline Fox of Penjerrick, Cornwall*, edited by Horace N. Pym. The remarks of Hartley Coleridge to which Booth refers are on pp. 21-22.

very midst of the masses that they attract about them, and makes them *think* differently from the rest of mankind. It's a *hellish* condition. Perhaps a *perfect* Genius, like Shakespeare's, is more equable and less self-torturing, but he must have been "less alone when alone" than with the most congenial companions.

This is a charming city—with gay shops, quite Parisian, and a delightful hotel with excellent cooking (the best I've had in Germany—in fact in Europe)—cozy and comfortable rooms & the best bath I've ever had—anywhere: a large marble-lined opening in the floor. Now, how the devil can I enjoy such luxuries where I looked with eagerness for dinginess & musty reminders of old Shylock? You [know that] here he bought the diamond (for two thousand ducats) that Jessica stole some years ago. Even after my 'set back' at Nuremberg I anticipated a little sniff of ancientry. To cap the climax of my misery Edwina has just interrupted me with a passage from Lewes which says the present Geothe house was built when Master G. was a boy at school. "Lord, lord, how this world is given to lying!"[27] Why, dern it! I uncovered & stepped tip-toe over the floor of the room in which he was *not* born. In Prague I had a gorgeous dish of eld, in the Jews' quarter & other nasty little bits about the town, but for "all in all" Ober-Ammergau is the one only spot I've visited that is out of the modern atmosphere; yet even there, at the time I went, the place was crowded by the riff-raff of today, tourists & fashionable folks from all parts of earth, but they were not of the place & did not prevent my appreciation of it. Tomorrow I shall look for the 'Ghetto' & whatever else there may be of interest here.

The trip hence to Paris is so fatiguing that I shall break it at Metz to-morrow night & proceed next day at noon—expecting to reach Hotel Meurice, Rue Rivoli, about midnight Wednesday. I sent Miller & my traps ahead & he should now be in London. I told him to give you a full account of my wind up at *Wien*; it was a large *Hurrah*! I could scarcely get away from the crowd that beset me after the play. Well, 'tis over. The experiment proved successful and I've no doubt—if I had a year to delve a little deeper here—I could do some good and lasting work. But 'tis too late: I must go home & think of 'roosting' now. I wish

[27] *Henry IV*, *Part I*, V,iv,148.

I had a practical head, like Irving's, I could devote the 'fall' of life to the work I foolishly began at Booth's, too early, and which he (Irving) and the Germans have accomplished. But I lack executive *nous*,[28] and can only act my part in the scene— no more. But I must not complain; indeed, I do not—but am content and happy in the knowledge that, for the time—at least, my successes in Germany are regarded as a National triumph & sometimes feel that "the Summer of my Fate seems fruitless beside the Autumn,"[29] then I fancy this to be the most egregious vanity & get 'blue' over the emptiness, the nothingness of it all.—

Paris has no charm for me now—twenty years ago it had; I must agonize through a fortnight there & dine & Opera most of the time, I suppose. Then for London—where I hope to see Irving in other roles than Benedict & others I don't like him in, and expect to get a peep at Canterbury, also to revisit Stratford & dine again at Clopton Hall. I shall have four weeks for England before sailing June 9th per Gallia—I expect for New York.— We get very encouraging news from home about Downing—his letters show an anxiety about his profession, the best evidence of his restoration yet given. I hope John is entirely well by this time & at his work again, my love to him & to Joe—when you see or write to them—Edwina joins me in dearest rememberances & best wishes for you & yours. God bless you!

95.   Hotel Meurice, Paris, May 1, 1883

On my arrival here from Vienna I found yours of April 6th with the papers you sent—I sent you one a day or two after to shew how I am compelled to taste the bitter dregs of my cup of nectar: The first mention of me in Germany made the same reference and scarcely a month has passed since that d--nable event without my seeing some allusion to it in connection with my name.[30] It's a delicious 'top-off' to my recent successes!— To break the journey I stopped at Nuremburgh (from which place I think I wrote you) at Frankfort & at Mertz—arriving here Wednesday night. I now remember it was at Frankfort that I wrote & told you of my visit to Göethe's house on

[28] *Nous* is Greek for "mind."
[29] *Richelieu*, II,i.
[30] John Wilkes and the assassination were mentioned in many bio-graphical sketches of Booth published in Germany during the season.

Shakespeare's birthday. The next morning, before I left there, I drove about town & had with me a very interesting old man who had lived many years in America—he acted as guide. He remembered Göethe, & also some of my "ancestors"—Forrest, J. R. Scott,[31] & Chas. Kean; saw my father act in the old Bowery & had a bill of the play &c. He gave me his address & was proud that we bore the same name—tho' his is spelt differently from mine—"Buhs"—but sounded the same—they called me *Boos* in Germany & *Boot* in Austria. I am now reading Lewes' life of Göethe—Abbey wanted me to fill Irving's place at the Lyceum & was sure of me, my refusal, I think, rather bothered him. I don't care to waste my efforts on Londoners—*fashion* alone secures success there & nothing could make me fashionable; it would be labor lost—I've got all I can possibly get out of the English & am satisfied. Don't know what move I shall make in America—feel as though I'd rather not act any more. It was kind in Browne & kind in you to oppose his proposition—I hope there'll be no *reception* of any kind whatever. I have already told you, I think, that our Italian trip is given up & that we will sail for New York, per Gallia, June 9th. A letter just recd. from Hutton tells me he will come by the same vessel May 23d— will barely have a chance to shake hands with him. I wish I could stay this side another year. Met "Oscar" twice & he called today—but didn't see him. He has cut his hair, dresses quietly & looks quite the gentleman.[32] Have been but once to a theatre— missed Sara's [sic] Fedora,[33] she ceased just before I came;

[31] John R. Scott, American actor, made his first New York appearance on July 2, 1829, at the Park Theatre as Malcolm in support of Junius Brutus Booth's Macbeth. Scott subsequently became one of the favorite actors at the Old Bowery and Chatham Theatres in New York.

[32] The reference is probably to Oscar Wilde, who was in Paris from January to the middle of May, 1883. During these months Wilde met Hugo, Mallarmé, Zola, and Degas, and his esthetic attitudes changed considerably. He had his hitherto shoulder-length hair cut and curled in imitation of a bust of Nero in the Louvre and began to dress in the height of fashion. Reportedly, he said, "the Oscar of the first period is dead." Wilde had met Booth at least once on a previous occasion long enough to get his autograph. See *The Letters of Oscar Wilde*, ed. Rupert Hart-Davis (London, 1962), p. 135.

[33] Sarah Bernhardt appeared in Victorien Sardou's *Fédora* for the first time at the Vaudeville Theatre in Paris on December 12, 1882.

nothing of note is going on. Paris is full & very gay—weather fine. No, my *profit* on the German tour did not amount to much, better than England, but not more than I'd make in a fortnight at home;[34] its about gone already & before I get through 'shopping' here & in London I'll have to draw on Benedict, I fear. I'm mighty glad that I've only one girl to dress. Edwina is quite well & very busy with the milliners &c, when she gets through with them we'll go sight-seeing.

Love to all your dear ones & to your own self. Adieu!

96.   London, May 23, 1883

It has been a very long time since I had a letter from you and I fear you are ill. I wrote you soon after my arrival in Paris, which place I left four days ago & am now at Morley's—the rooms that Barrett had when you were here & which Mrs. B. & her girls vacated just before my arrival; I did not see them—I presume Lawrence will be shortly here.[35] Well, you know, of course, all about the new trouble that has fallen upon us. Just as Downing was improving he was allowed to work & the result is overtaxed brain and his relapse. I have managed thus far to keep the fact from Edwina's knowledge, but the absence of home-letters and my tell-tale face & manner cause her to suspect some ill. It must be known in a day or two. I see that she is very anxious and her appearance indicates loss of sleep and debility. What an up & down my life has been! I fear the boy will never be strong again—and what a prospect for her! Our pretty plans for a home are scattered & we return to America without an idea what we shall do.—After Paris London seems very dull despite the "season" and the Derby—I presume it's chiefly owning to the depressing news from home that London is unlovely just now.—The weather is superb and the city full—of course. Did I tell you that an effort was made in Paris to have me act there? The time was too short to get a proper company together for the many rehearsals necessary for a French version of Shakespeare. Have yet seen no one but Clarke—he takes Miller

[34] A list of the nightly receipts of Booth's tour through Germany is in the collection of the Walter Hampden Memorial Library.

[35] Mrs. Barrett and her daughters proceeded to Stuttgart without Lawrence Barrett, who had to fill an engagement in San Francisco. Presumably he arrived in Stuttgart in time for the wedding.

to the Provinces in the Fall & Winter. We are up to our ears packing & shall keep to ourselves 'till all is finished, then I shall try to get Edwina out as much as possible. I have heard only from McEntee and Benedict since Downing's departure & return and they both write discouragingly. "What new sorrow claims acquaintance at my hands?"[36] Almost every day, for years, I've asked this question. At the height of each career I make, a break-neck fall appears to check me. So be it! God bless you!

About June 1, Booth left England for the last time.

[36] *Romeo and Juliet*, III,iii,5-6: "What sorrow craves acquaintance at my hand/ That I yet know not."

Chapter IX

American Tours

1883-1886

Booth spent the summer of 1883 arranging a house near Newport which he called Boothden. About the middle of August, he and Edwina—whose engagement to Downing Vaux had been broken because of his continued mental illness—were settled in their new quarters, and Booth began to plan his next season. The planning was complicated since Henry Irving had finally decided to tour America and was to open at the Star Theatre in New York on October 29. Booth was careful to avoid an engagement which would appear to put him into competition with Irving. On November 5, while Irving was in New York, Booth opened at the Globe Theatre in Boston for two weeks. On December 10, after Irving moved to Philadelphia, Booth opened for six weeks at the Star. The critics in Boston had been unusually severe about Booth's support at the Globe. The New York critics continued this attack, contrasting the general inadequacies of Booth's company to the general excellencies of Irving's. Even Winter joined the chorus of complainers. The cast of *Richelieu*, he wrote in his *Tribune* review (December 11), "cannot be extolled." In his review (December 13) of *King Lear* he complained that it was "scarcely possible to speak with patience" of the supporting company. He was even more caustic in a letter to Booth.

97. (Saturday Afternoon, [December 15, 1883], Laid up in Bromide and Quinine)

Dear Edwin,

I am glad to receive a line from you. It cheers me much. And I'm glad that you and Edwina like my Rambler.[1] There is still no news of John that I can think credible. He is, I am sure, a very sick man: but of course it is best not to say so.[2] I have heard

---

[1] About December 6 Winter sent Booth a copy of his *English Rambles: and Other Fugitive Pieces in Prose and Verse*, recently published by James R. Osgood and Company, Boston. Booth thanked him for the book in a brief note dated December 13, 1883 (Folger Shakespeare Library, Y.c. 215 [456]).

[2] In 1883 John McCullough's health began to decline. Winter, in *Other*

nothing from Irving, since he left New York. I did not go over
to Philadelphia to see him in Hamlet and that performance is
unknown to me. One of the adverse judges in that city wrote
that it would be a splendid Hamlet if "the execution" were only
as good as "the conception." In Boston Irving seems to have met
with a frost. The mental habit there is too much that of judging
by *"comparison."* Not one of the writers, thus far has shown the
capacity to form an independent judgement. "Puzzled" is the
word used to describe the critical attitude of the auditors of his
Louis XI which is a performance that no expert, surely, could
hesitate to recognize. He will do best to appear but little in
Shakespeare. I understand that Stedman has written an *adverse*
article about him for the *Atlantic*. This will be gratuitous and
unwise for Stedman knows nothing about acting. Indeed literary
men seldom do know anything of the art of acting, viewed
abstractly and apart from *"the intention of the* AUTHOR."—I
have sent to you my "notices" of Richelieu & Lear with corrections
of typographical errors. If you find them full of more serious
blunders, it won't be strange, for I'm sick and tired, and all that
I write is forced out of my mind by effort of the will. The wish
to do some sort of justice to your ripe and beautiful work, will,
however, be seen in those articles, and that may atone for their
dullness. I dread *Hamlet*—it is so hard to write *anything*, fit to
print, on that subject![3] The Star people have not given you
such scenery as they ought to have given you, for the plays
thus far seen. As to the *Company*—I received a letter from one of
your oldest admirers in Boston, bewailing that it was so
incompetent &c.; & I wrote a little article for the Tribune (which
got crowded out & I suppose, killed), the week before last,
stating it had always been the fashion to decry your companies
and giving the names of some of the players that used to act
with you; I mentioned the names of the chief members of the
present company, showing that on the face of it, the organization
seemed to be a good and sufficient one.—But I see no talent

Days, p. 215, writes that he was suffering from "softening of the brain."
In the summer of 1885 he was finally placed in a mental institution in
Bloomingdale, New York. In the fall he was removed to Philadelphia
where he died on November 8.

[3] Booth introduced *Hamlet* on Monday, December 17.

whatever in Miss Weaver or Mr. Bock.⁴ On the contrary, they
spoil whatever they touch. Last night I saw *Lear* again; and
Miss Weaver nearly ruined your exquisite and most pathetic and
beautiful scene by her precipitation and shrill noise, and rushing
upon you before the line was out of your lips and by her extreme
loveliness. The latter cannot be helped; but nobody with such a
face and such a voice,—not to speak of the acting, which is
conventional and clumsy,—could ever seem like Lear's Cordelia.
She spoke of the King, too, as "crowned with rank fumiter and
furrow seeds" instead of furrow weeds.⁵ In fact she knows
nothing about Cordelia, nor ever will know. Mr. Bock always
was a bad actor, but, generally, I hear, is a very good fellow.
Plympton,⁶ of course, does very well, but Edgar is "too many
guns" for him. In the first scene he is simply silly; & in the
raving scene he shows no purpose—no *mind* as an artist. Still he
gets through creditably. But De Mauprat is his high-water mark
& through even that his vanity & self-consciousness are painfully
conspicuous.—Couldn't *Edmund* and *Kent* get along without
that Broadway greeting and hand-shake in the first scene? And
in the great imprecation scenes, couldn't the people who are
[illegible word] with the king be got to manifest some sign or
interest in what is going on?—I touch on these things to explain
my own somewhat cold manner of referring to the subsidiary
players, & also why it is that (as I suppose) the press people
are so caustic. With Plympton, Fawcett, Lane, Anderson, Ida
Vernon, and Mrs. Eldridge,⁷ of course you have got the substance

⁴ Affie Weaver was the wife of Horace McVicker and the daughter of
Henry A. Weaver, Sr., American actor born in England, who had supported Booth at various times in the 1870's and 1880's.

Frederick Bock was a minor American actor.

⁵ *King Lear*, V,iv,3.

⁶ Eben Plympton, American actor, had a long career on the stage
playing various leading roles in support of Booth, Barrett, Modjeska,
and McCullough.

⁷ Owen Fawcett and John A. Lane were popular American supporting
actors. Fawcett was probably best known for his portrayal of the First
Gravedigger, while Lane played such roles as Cassius, Macduff and
Horatio.

Louisa Eldridge, affectionately known as "Aunt Louisa," excelled in the
portrayal of "eccentric old women," although she also acted such roles as
Lady Capulet, Emelia and Gertrude.

David C. Anderson had been an intimate friend of Booth ever since they

of a good company; but you have not got a competent leading lady and such players as Bock & Dulcis ought not to be upon the stage at all. These are my impressions, and I should be unjust to you if I did not give them. Everything of course depends upon the point of view. What *I* see is—the best tragic actor in the English language acting most nobly in the greatest of plays & surrounded with old "fakements" for scenery & with persons who, in others are out of their depth, and in still others are painful sticks. The man who plays Gloster is a laughing-stock—, I know, of course, that, under ordinary circumstances, these people would do well enough. The existing circumstances, however, are *not* ordinary. Irving's plays were got up very carefully. The mounting was not extraordinary; it has often been surpassed; but the painting was fine: the people had been carefully drilled and there was a sense of completeness about everything. And Miss Terry was charming. The multitude will compare that effect with this, and many will suppose that *you* are responsible for all imperfections.—I don't know where you could get an Othello, but W. E. Sheridan cannot act it, and Plympton in either Othello or Iago would be ghastly. He might do Cassio. I hear that you have Ida Vernon for Lady Macbeth & I look forward with natural apprehension to that. She, at least, however, has experience.—Well, "all shall finish well at last."—I will remember your message to Miss Anderson, when next I write. She would be greatly gratified to get a letter from you and Edwina.—no name was mentioned by M. A. but I have heard somebody; on this side of the ocean (whom I don't remember) speak of the sister of Mrs. Aldrich (one of the Woodman girls), but, *Mum* is the only safe word. I'm devoutly thankful the story is unfounded.[8]

---

toured Australia and Hawaii together in 1854. He acted at Booth's Theatre during Booth's management, retiring from the stage in 1872. In 1881, however, he returned and acted, for the most part, in support of Booth until his death.

Ida Vernon, American actress, excelled in such roles as Ophelia, Desdemona, Pauline, Marguerite Gautier, and Mary Stuart.

[8] A rumor was at this time circulating that Booth was contemplating remarriage. On December 6 (Folger Shakespeare Library, Y.c. 215 [455]) he wrote to Winter: "Marry again?—Whew!—No, sir! Though I had a taste of its bliss in my early venture. I had a hell of it for nearly 13 years

God bless you, dear Edwin. I hope to be able to come up and have a quiet hour at your fireside soon.

Ever yours,
Will

Almost a week later, Booth replied to Winter's letter.

98.   42 East 25th St., December 20, 1883

All you say anent my 'dogans' is gospel true, I can't gainsay it, but what affects me is the constant, 'hammer-and-tongs' clatter that the papers keep up abt. *my* people while the same quality of 'sticks' used by others is approved. Like the speculator nuisance it must exist—*I* cannot cure it. Apropos: I have recd. lots of unpleasant letters calling my attention to, & some containing slips from the papers that blame me for the sidewalk speculators. I have not read any of the articles, for of late years I have carefully avoided all unpleasant paper talk. But what can I do? I have protested in my feeble way & managers declare they are innocent & helpless, so long as these fellows are licensed & have political influence. I've known several instances (one at the Lyceum—under McVicker) when the papers sided with the pedlars against the managers & I also have heard several persons (but very recently too) declare that they'd rather buy of speculators than have to wait in line for tickets, &c. I am sure it is a detriment to business as it is a curse to the public—but how prevent it?[9]—You can form no idea of the miserable condition of the *Star* theatre! Filthy dirty & so crowded with rubbish that I am in dread every night lest some accident may occur. It is impossible to get anything decent in the way of scenery or *props* & I will not be able to produce all the plays in consequence of this neglect—so far as I can ascertain the fault is with Moss who agreed to furnish all such things. Such a hole I've not been

---

in my second marriage and that has scared me to death." Mattie Woodman, whose sister Lillian was married to Thomas Bailey Aldrich, was one of the candidates to become Mrs. Booth. Ida Vernon was another (see next letter).

[9] Laurence Irving writes in *Henry Irving*, p. 423, that during his grandfather's engagement at the Star Theatre from October 29 to December 8, 1883, a ticket speculator named T. J. McBride parlayed $7,200 worth of tickets into $30,000.

in for many years—I mean in this country. The rats carry off everything that is lying about loose & raise merry 'h-ll & Tommy' in my dressing-room. I think the old rattle trap ought to have been condemned years ago—it never could have been safe; for the actors, at all events, there is little chance for escape. The report anent Miss W[oodman] was rife about three months (or less) after the death of my first wife—twenty years ago. On the heels of this I have just heard that some Dramatic sheet couples my name with Ida Vernon.[10] Although I do not see these papers yet a 'dear friend' is always ready to inform me of their scandals. I can't imagine who starts such things—there is not the least foundation for them. I suppose the best way is not to notice them. I fancy (perhaps I am too suspicious, though) that they are designed to worry Edwina. Fiske of the Mirror once had an article similar to the last report & as he & *Marie*[11] are friends & she envious of Edwa. (God forgive me if I wrong her!) I suspected her influence in that business; she more than once gave me cause to think so meanly of her. Is it not better to ignore such reports? or would it be well for you to 'item' a general denial stating that I am "fancy free?" Thank you, a thousand times, dear Will, for your splendid notices—I have them all. I can appreciate the great difficulty you have to say *anything* of these old performances—'tis even harder than my constant repetition of them. Regarding the little matters of *stage*

[10] An article entitled "Love Story of Ida Vernon and Edwin Booth" by the pseudonymous "Madame Qui Vive," printed in the Chicago *Record Herald* (November 17, 1917) states that Ida Vernon and Edwin Booth were lovers in the mid-1860's after the death of Mary Devlin. A lover's quarrel occurred and they went their separate ways. Miss Vernon, however, was in love with Booth for the rest of her life. After the accidental death in 1867 of her husband, an actor named I. V. Taylor, to whom she had been married for less than a year, she once again sought Booth's affection. According to Mme. Qui Vive, after the death of Mary McVicker "the old love story began a new chapter," but Booth and Miss Vernon were prevented from marrying by Edwina's jealousy. Yet "when Booth knew that for him the play of human existence was nearly finished he sent for Miss Vernon, and it was she who made happy his last hours." Although the authoress seems to have gotten her information from Ida Vernon, there is little in the story which is really credible.

[11] The reference is probably to Marion Booth who was the daughter of Junius Brutus Booth, Jr. and his second wife Harriet Mace. Booth called her "Marie."

business you refer to—at rehearsals I have had such matters noted & my instructions followed satisfactorily, yet so d--nably dull are the wretched idiots that compose the rank & file of my profession that none of the details are remembered or comprehended. It would surprise & amuse you to hear the arguments advanced in defense of some senseless action or reading by the conceited asses that I have to deal with. 'Tis very disheartening & takes the *act* out of me by the roots. God grant there be no actors in the other life!

Apropos of Weaver (& of Davey too) there's where my foolish weakness interferes with my business. I promised Horace to let her have a chance without really knowing ought of her; & I somehow couldn't decline Uncle Dave, who thinks he can act Shakespearian characters, wh. he can't. It was the same with McCullough at the 'California'—he had a lot of fossils & incompetents on his salary list because they were his friends & therefore could not dismiss them. The same weakness has hampered me always. "Save me from my friends!"

This is about the longest letter I've written in many months— I'm surprised at it for my nerves are at an ebb & have been all the week.

Edwina joins me in loving remembrances with best wishes for the season. Adieu!

After his engagement at the Star Theatre, which closed on January 19, 1884, Booth filled engagements in Philadelphia and Baltimore. On February 18 he returned to the Globe Theatre in Boston for an engagement of two weeks.

99. Parker House, Boston, February 24, 1884

I have reproached myself often during the many weeks that have passed since I last wrote you & threatened you frequently with one of my prosy epistles, but laziness has prevented me from writing. I sent you a paper from Baltimore—not for what is said of my performance of Shylock but for what I regard as the true Shakespearian portrait of the Jew. I believe you hold a different estimate of the character, as many do, but I have searched in vain for the slightest hint of anything resembling dignity or worthiness in the part. I thought the article an

excellent analysis of it. I believe it was the view taken of it by Burbage, despite his red wig (I never heard of his false nose), and revived by Macklin after an interval of 'low comedy' and doubtless the dignified notion of the beastly old miser.[12]—The constant rain injured my first three nights here but when it ceased Thursday the houses were packed—altho' it rained badly yesterday, my Matinée house was crowded. The indications are that my business will not suffer by Irving's return, but I am sorry that circumstances have placed us in opposition & will do so again in New York. I expect to meet him at Luncheon Friday. The Atlantic & Harpers, no—the Century have articles about him.[13]—I shall act for a charity benefit Saturday night & start for Newport Monday (having an idle week) en route for New York, playing in Brooklyn the following Monday two weeks & then two weeks at the 14th St. House, being the only place open to me, and there terminate my season. The same 'strong' company will assist me in my slaughterous work. By-the-by, the critics are now praising the poor devils—gently. Mother is improving, but her first nurse retarded her recovery by her insolence & positive brutality in handling the poor helpless body: she was discharged & a new one is now in attendance.—I think with you that *dust choked souls* is rather a mixed metaphor, but if certain spirits have to "fast in fire" it would seem that they have 'gullets.'[14]—'Tis a pity that Joe is so remiss in correspondence—until I wrote to Irving I never knew a man so negligent in that respect as Joe. I fear that I am falling into that bad habit of late. I thought Joe was older. I fear that Barrett has made a mistake in declining the offer—Mary should have a full swing & Barrett will gain nothing by interrupting

[12] Although Winter for many years had disagreed with Booth's impersonation of Shylock, he later accepted Booth's ideal as "correct." See Winter's *Shakespeare on the Stage*, 1st ser. (New York, 1911), p. 156.

[13] Irving returned to the Star Theatre on March 31 and played through April 26.

Henry Austin Clapp, "Henry Irving," *Atlantic Monthly*, LIII (March, 1884), 413-422.

John Ranken Towse, "Henry Irving," *Century Magazine*, N.S. V (March, 1884), 660-669.

[14] Presumably Booth is referring to an image in one of Winter's poems. I have been unable, however, to identify the poem.

her *run*.[15] I have scribbled awkwardly & told you all I know quite hastily, for I've a cold hand & bad pen with numerous interrupters; very little leisure have I—especially in Boston.

Edwina is not so well today—looks pale & haggard, she can't stand much excitement & there are so many calls upon her here, all kindly meant, that she gets little rest. Next week we both will relax a little. She joins me in loving thoughts of you. Adieu!

Booth filled engagements at Haverley's Theatre in Brooklyn from March 10 to March 22 and at the Fourteenth Street Theatre in New York from March 24 to April 5. During the week of March 31, Booth's engagement at the Fourteenth Street Theatre overlapped Irving's engagement at the Star Theatre. Much to Booth's relief no rivalry ensued, nor did either actor suffer at the box-office. The relationship between Booth and Irving continued to be cordial. On April 14 Booth gave Irving a breakfast at Delmonico's, to which Winter and McCullough were also invited, and they seem to have met frequently while both were in New York. After he ended his season on April 5, Booth remained in New York for several weeks. About the beginning of May he moved to a new house at 29 Chestnut Street in Boston which he had recently bought and which he hoped to make his permanent residence.

During the summer of 1884, while Booth was busy entertaining his friends at Boothden, Winter made his third trip to England. He accompanied Augustin Daly's company, probably at Daly's expense, to report the debut of the first American company to appear in London. About October 7, Winter returned to New York.

100.   Newport, October 9, 1884

Am very glad of your safe return & that the trip did you so much good. Until I heard of you from Hutton I was really in doubt as to your whereabouts. I've had neuralgia lately wh. has prevented my writing or you should have heard earlier from me. The season has been a busy one at *Boothden*, so many guests that the strain was severe on Edwa. but now she is having a *rest* and a week or two of quiet before we leave for Boston will be beneficial to us both. I see by the papers today that I am to

[15] Mary Anderson appeared at the Lyceum from September 1, 1883, to April 5, 1884, when her engagement was interrupted by Lawrence Barrett who appeared as Richelieu and as Yorick in William Dean Howell's *Yorick's Love*, an adaptation of Joaquin Estebanez's *Un drama nuevo*.

appear in Berlin in February—not the least foundation for the
report. Poor McCullough's case is indeed sad, but I hope 'tis
exaggerated & that rest will restore him. As for the Lyceum
school[16]—I know not what to advise. I sincerely hope for the
success of such an Institution, yet cannot, for the life of me,
feel any faith in the present enterprise. It may be the opening
wedge, however; the foundation for the future "Conservatoire,"
as it were—but are the 'heads' of it sufficiently responsible to push
through the thousand & one difficulties that will beset it? I cannot
conscientiously advise you regarding the subject. My dear old
friend, Dave Anderson, has been lying at death's door for some
weeks past & every day I expect to be summoned to his bedside
the second time—I was there a week (I mean in N.Y.—visiting
him every day) and as he seemed better & the doctors gave hope I
returned. I wish you could have been with us this Summer—all
our guests, except the Jeffersons, had a good time. They promised
a week & after several disappointments they came for a day
& that a beastly foggy one. I took them for a sail in a gale which
disgusted them, of course, and they were glad enough to get
away. I anticipated a delightful week with Joe, but the fates
were against us. I hope he will return to London—he said nothing
of going although I spoke of his doing so. I may be in New York
a few days before going to Boston. Thanks for the thought of
me at Stratford. I enclose the autograph. I'm afraid that I shall
be obliged to 'tour' this country—even to California next season,
too late to arrange it now as all my time is filled. I've made two
big mistakes—last season (wh. should have been a great one)
and this—which I have limited to sixteen weeks. My plan for a
royal loaf cannot yet be carried out.—[17]

[16] The Lyceum Acting School was founded in the late summer of
1884 by Steele MacKaye and Franklin Sargent. Charles Frohman was
the business manager and David Belasco was one of the principal in-
structors. As the American Academy of Dramatic Arts, it has survived to
the present day.

[17] Shortly before Winter sailed to England on July 5, 1884, Booth
wrote him a letter, of which only a fragment survives (Folger Shakespeare
Library, Y.c. 215 [465]), in which he was more explicit about his
financial status:

The two years in Europe were simply a large expense, and my last
season yielded less than my Boston & this place required to arrange
[i.e. his Chestnut Street house and Boothden.] Of course the fault was

With a thousand welcomes & affectionate regards from us both—

After his engagement at the Boston Museum, November 17 to December 13, 1884, Booth rested for four weeks prior to his scheduled engagement at the Fifth Avenue Theatre on January 19.

101. Boston, 29 Chestnut St., January 1, 1885

When not tired out by rehearsals I've been so 'low down' with nervous depression that the bare thought of writing a letter has distressed me. I've not had sufficient energy to thank you for the interesting 'Tobacco Talk' you gave me or to wish you the compliments of the season. I don't know what the trouble is—unless it be malaria, contracted some years ago when "on the road" with Ford. I am well enough & take good care of myself, but now & then my nerves collapse & I can do nothing but *glare*—unable to think or read; & smoke mechanically—a stolid idiot. This has prevented me from acknowledging your kindness & the favors of other friends, except Miss A—who sent Edwina & me two quaint cards at Xmas. Daughter got two for her & I wrote a few words on one & sent it yesterday. That's all I've done, unless I except a reluctant visit to the Opera as escort to Edwina who had none other.

When Barrett was in Washington I wrote to him, c/o Opera House, relative to the daughter of "Lucy" Hooper,[18] but he has not replied—when you see him ask, please, if he received the

---

my own, I should have *tour-d* the country & taken my slice of New York before Irving & the Operas gobbled the cake, & then (on my daughter's account) I refused to travel, & so lost many thousands. I have repeated the folly for this coming season by declining engagements except in Boston & New York—while all my expenses have increased. I hoped to reach a point where I could act for pleasure & stand by my less fortunate friends, and had I not made several severe mistaken investments I think I could have carried out the plan but I find that I will have to 'buckle' up and slave away some years longer.

[18] Probably Annie Hooper, minor American actress, who was the daughter of Lucy Hamilton Hooper, American journalist and authoress. In 1884 Lucy Hooper wrote a play, *Helen's Inheritance*, which was first produced in America at the Madison Square Theatre on December 23, 1889, with Annie in the leading role.

letter. I am sorry that he and I must play against each other in New York,[19] but I guess he is safe: I have not so much confidence in my own engt. at the 5th Ave:—my first one there yielded me within a fraction of *$1000 per* night, the 2d didn't pay at all—& the theatre closed, you may remember, after I finished & in the midst of Joe Jefferson's engt.; neither of us could make it go, after Daly gave it up. Miller writes me good news of Mary Anderson's hold on the people of England, & of W. Barrett's increasing strength, but fears that Irving is losing ground there—I hope not for the Drama's sake. It wd. be sad to see his labor lost. If what I hear be true (that Terry is to *star* it here alone) he will lose one of—if not the chiefest pillar of his Temple.—I don't know what to do next year; whether to make an extended tour in order to replenish my coffers or to retrench & risk the chances by long intervals of rest, as I am doing this Winter. A dozen managers wait my reply & I'm like the donkey twixt the bundles of hay. "I stand in doubt where I shall first begin and both neglect."[20] Adieu! God be with you!

During his engagement at the Fifth Avenue, Booth revived two roles—Sir Edward Mortimer and Pescara—in which he had not appeared for over ten years. On February 14 he closed at the Fifth Avenue and for the next two weeks he toured New England, making one-night engagements in New Haven, Hartford, Holyoke, Haverhill, Providence (two nights), Worcester, New Bedford, Lowell, Lawrence, and Brockton. On March 2 he returned to the Boston Museum for an engagement of four weeks. On March 4 he revived Sir Giles Overreach, another role he had not played for over ten years. On the Sunday before his last week at the Boston Museum Booth wrote Winter to express his sympathy for his poor health and to remove fears that his critical support of another actor (probably Henry Irving) had jeopardized their friendship.

[19] Lawrence Barrett was at the Star Theatre from January 5 to February 14. During this engagement he produced spectacular revivals of Boker's *Francesca da Rimini* and *Julius Caesar*. The supernumeraries for *Caesar* were students from the Lyceum School of Acting, and with them, in imitation of the Meininger Company, the realistic ensemble crowd scenes were effected. Barrett also produced, for the first time in New York, Browning's *A Blot in the 'Scutcheon*.

[20] *Hamlet*, III,iii,42-43: "I stand in pause where I shall first begin,/ And both neglect."

102. Boston, March 22, 1885

The tone of your letter is very sad & pains me. I hope, however, that better health has already dissipated the woeful clouds and that the sunny side of your nature brightens the sky for you. I know, my dear boy, what a hell of suffering one endures when the nervous system is strained by overwork & anxiety, and 'tis a settled fact, I think, that such as we are doomed to agonize through life; there's no escape—no help for us, and we must fight it out. Do not fear for a moment that the expression of yr. sentiments regarding any other actor can, in the least degree, affect my feeling towards you; nor can I understand why anyone should reproach you for finding in others besides myself something worthy of your commendation. Thank God my successes have not blinded me to the merits of others. Don't let any such stuff annoy you. Sonenthal [sic] & I cannot act together, but there is little doubt (tho' 'tis not definitely settled) that Ristori & I will give *Macbeth* in N.Y., Phila. & Boston—one night each—sometime in May, the 1st week—probably. I saw Sonenthal but once—in Vienna; he is highly esteemed there socially & professionally; I understand that he rarely acts tragedy.[21]

Tomorrow I begin my last week here, I wish it was the last of my season, for I am very tired. I shall be a few days—nearly a week in N.Y., at the Albermarle, after this week & we must meet then. Send me a few words for the Poe affair;[22] I got about as far as "My lords & gentlemen:"—or words to that effect & made a protracted pause which has not yet been interrupted, so

[21] Adolph von Sonnenthal was engaged at the Thalia Theatre in New York from March 9 to April 4, 1885. While Sonnenthal was a versatile actor who seems to have been most effective in melodrama and light comedy, he often, contrary to Booth's remark, acted tragedy. During his engagement at the Thalia he was seen in a variety of plays among which were *Hamlet*, Dumas fils' *La Père Prodique*, Karl Gutzkow's *Uriel Acosta*, and Henri Meilhac's *L'Attaché d'Ambassade*.

Adelaide Ristori, Italian tragedienne, celebrated for her impersonations of Maria Stuart, Antigone, Phaedra, and Adrienne Lecouvreur, appeared as Lady Macbeth to Booth's Macbeth at the New York Academy of Music on May 7, 1885.

[22] On May 4, Booth gave a speech at the dedication of the Actor's Poet Memorial at the Metropolitan Museum of Art. It is printed in *Life and Art*, pp. 406-408. Winter seems to have written the speech for Booth (see letter 104).

help me out, as gently as you can. The Bartoldi scheme was entirely Horace's—at least I knew nothing of [it] 'till he suggested it—I do not favor it.[23] I offered to share the profits with Sonenthal or act free with him for a German Charity. This between our agents, Sonenthal doubtless knows nothing of it, but Horace says the report is that I asked 75 pct. of gross receipts or to give the Actor's Fund a benefit. Nothing can be done for lack of time for rehearsals. Unless the 75 pct. report gets into print 'tis best not [to] notice it; I hope Sonenthal does not believe it. Edwina is well & always joins me in love to you. "New Way" seems to have hit 'em here, but not Apostate.[24]

After Edwina's engagement to Downing Vaux had been broken, she became her "dear Papa's" housekeeper and hostess when he was at home in Boston or at Boothden. Booth, as he wrote in an earlier letter, refused to travel extensively only that he might not be separated from Edwina for long periods of time. Except that Booth "lost many thousands," the arrangement seems to have been very satisfactory for both him and Edwina. He was a doting father and she was an equally possessive daughter. Yet about the fall of 1884, Edwina became engaged again, this time to Ignatius Grossmann, an Hungarian emigrant who, after a brief career as a language teacher, had established himself as a relatively successful stockbroker. The wedding, the "all-important event" of his life as Booth called it, was to be on May 16. At the end of a one-week engagement in Philadelphia, Booth wrote to plead for "some brief remarks" for the Poe Memorial dedication, if, indeed, he could not escape this duty.

[23] Presumably Horace McVicker suggested that Booth give a benefit for the Bartholdi Statue Pedestal Fund, a fund being raised by public subscription to finance the pedestal upon which the Statue of Liberty was to be erected. The fund was named after the sculptor of the statue Frederic Auguste Bartholdi. Although numerous theatrical benefits were given for the fund between 1883 and 1886, Booth, adhering to his policy of not giving benefits, did not participate.

[24] In a later essay entitled "Edwin Booth in Some Non-Shakespearean Parts," *Outing*, XI (June, 1885), 343-349, Henry Austin Clapp observed that *A New Way to Pay Old Debts* had "some obvious faults and defects" but it was still " a wonderful play, full of force, and fire and virile imagination." He praised Booth's Sir Giles Overreach as "brilliant," "strongly and freshly realistic," and "unfaltering, elastic and vigorous." He condemned *The Apostate*, however, as an exasperating and disgusting piece, which succeeded only by Booth's "masterly" and "thoroughly artistic and immensely effective" Pescara.

103. Hotel Bellevue, Philadelphia, Sunday [April 13], 1885

Your dispatch recd. last night. I can give you but a hasty scrawl just now as I rose late & must dine with Furness (& spend the evening at his house) as soon as I can get off a few letters that have been too long deferred. Is there no possible escape for me from that *Poe* affair? Won't Lester Wallack do it? He is a veteran & a poet and an actor—I am but a *babe* in such affairs & will "mewl & puke"[25] all over the business. If there's no escape for me then 'help me up' by writing for me some brief remarks & I'll endeavor to speak 'ex tempore.' Business very fine & the papers (I've seen but one—the Times) are, I'm told, full of all that's kind for me. This is the longest engagement I've ever had in Phila. This is my last week, thank God! and when my Ristori week is ended & the all-important event of my life is passed I shall hope for a long unbroken season of rest. My head is full of "quick-coming fancies," as you may readily believe, and the future seems a seething mass of impossibilities. But I daresay matters will shape themselves in good time. I do not permit myself to attempt to straighten them out. I did think I'd like to have you here with me, but I can't see what time we'd be together— 'twixt my rising & napping & acting & bed-time there's little leisure for sociability. I hope, however, you will be with me at 'Boothden' this Summer (if you don't go to Europe) wh. I hope you won't & I hope you will. Let me know *sure* if & when you sail.

About a week later Winter answered Booth's plea and mailed him the Poe speech (the "doc").

104. Boston, 29 Chestnut St., April 24, 1885

Mr. McWatters kindly sent me the 'doc' that you were good enough to prepare for me—it is just the thing, not beyond my capacity & quite long enough for my endurance which is not great under such circumstances. The intolerable vanity of the bashful man ought to be pounded out of him and I wish that mine had been in early life. Such a trifling affair as this (I mean my part in it) tortures me to think of, & I daresay I'll blunder it worse than any school-boy. However, I'm in for it, and must take the consequences of my jackass-icality in consenting to make a donkey of myself.—I will be at the Albemarle the evening of Saturday—

[25] *As You Like It*, II,vii,144.

tomorrow week, May 2d. & hope to see you Sunday & we can then arrange our movements for the morrow.

I find that beast Joe Howard[26] has been interviewing Stuart & had in the Boston Herald (I did not see it) an abusive article about me. If I were not so indifferent to praise or blame, so indolent & forgetful of injuries (not virtuously but cowardly so) I'd let out the truth about those curs, but after I've d---d a little my 'goody-goody' soothes my temper. 'Tis not manly & I despise myself for not hating better. That reminds me of a surprise I had yesterday—3 packages of my old pictures (minus the most rare, & not above half the set) came from McVicker's—tumbled into the boxes &, of course, very thoroughly smashed. What has induced the old frump to send them, I wonder! I suspect that he gave Runion his pick of them as Mrs. McV. gave Mrs. R. of Mary's "Worth" dresses—for which I paid 'much' hundreds & with wh. the little lady R. garnishes herself.[27]—I'm sorry I did not see Hutton before he sailed; I fear his bride is not a very well woman—but I hope the trip will recover her.—The daughters & I have been discussing flowers and other marriage matters these two days.—the wedding will be in our parlor May 16th at noon. That's an ordeal that 'dazes' me to think of—I dread it—and yet I am sure that the man she has chosen will be a tender, devoted and fatherly husband. His record is of the purest—what strength he has I cannot tell, but all who know him endorse him in the highest terms; Edwina & he have many mutual friends here & in New York & all sorts of "golden opinions" of him pour in upon her with enthusiastic congratulations. 'Till after the event I can give no thought to myself—what I shall do, where [to] go during her absence abroad. If you remain this side you must be with me a while at Boothden! I enclose a *souvenir*[28] wh. I hope will interest you.

[26] Joseph Howard, Jr., American drama critic, was, along with Stuart and Charles A. Byrne, one of Booth's worst critical enemies. During his long career he worked for various papers, among which were the New York *Herald*, *Times*, *Democrat*, and *Star*, the *Brooklyn Eagle*, and the *Leader*. In 1880 he became a freelance journalist, but his main contributions were to the Boston *Globe*.

[27] Charles Frederick Worth, English couturier, was mainly responsible for developing the Paris dressmaking business into a national industry.

[28] A note at the end of the letter by Winter indicates that the "souvenir" was an engraving of Junius Brutus Booth as Richard III.

During the summer of 1884, Booth, while visiting Lawrence Barrett at his summer home at Cohasset, agreed to let him manage a national tour for the 1886-87 season. Booth wanted their arrangement to be kept secret until the end of his 1885-86 season. The news, however, soon leaked out and it was—as is usually the case—distorted in the process. Booth wrote Winter to give him the "*facts*" of the arrangement.

105. Boston, October 18, 1885

For the past two weeks I've felt as you doubtless felt the day that you *didn't* write your last letter to me & even now I am merely edging along as it were, against the tide.—When I can summon sufficient energy to do so I'll write my cordial congratulations to 'Mary,'[29] and attend to the other matter too; meanwhile tell her, when you see her, how glad I am at her success & how sincerely I hope for a life full of it for her. Now for the *facts* of the Barrett arrangement. While at Cohasset with him we talked of our work, of course, & I of my weariness & wishes to cease travelling &c, "but, I supposed that I'd have to make a tour of the South & West pretty soon for a final visit, and to replenish a purse which of late has been pretty well drained." He spoke of his own great outlay & expenses & said "by Jove! I'd like to manage a tour for you; it would help me very much and I'd feel a great pride in it" &c, &c. Although I had *partly* promised another to give him the 'job'—provided I determined to make the tour, I reasoned thus—'if I am to have a manager who will receive a per-centage of my profits, why not give it to a brother actor, rather than to an agent who has no claim on my consideration whatever'? "Upon this hint—I spoke"[30] & said I would do it—with the understanding that I should have (as he suggested) his scenery, costumes for company, stage manager & *company*—whose prestige was good, although I did not consider it as good as any that I have had—which he frankly acquiesed in—"No matter whom I engage the company will be condemned, but as the leading members of yours are praised I may not be abused, as usual for my 'sticks.' " Such was the gist of our talk & we put the thing in shape (I giving him a larger interest than I'd give a mere agent). All is perfectly

[29] Mary Anderson filled a successful engagement at the Star Theatre in New York from October 12 to November 21, 1885.

[30] *Othello*, I,iii,166.

satisfactory to me & I believe that he feels a sincere pride in the project & will do all in his power to make 'honors easy.' But, it seems—today—a report, or—perhaps—an interview, I have not seen it (Boston Herald) nor do I care to, is published to the effect that 'tis purely a matter of friendship on Barrett's part to help me get good support, &c, &c, &c.—all rot! I don't believe he said a word to authorize this—intentionally, but he *does* talk a little 'off' when excited & incautiously may have said too much, or too little, to the purpose. This is the whole of it. The strongest member (Louis James) has left him & applied to me (since our arrangement was made) but Field's cast is full & I cannot use him.[31] There is no reason why all this should go to the public, for I wish all good for Barrett & shall say nothing to him of the report. The whole matter was to be kept quiet—but of course, all those with whom he has been negociating for me must know something of it & thus it has leaked out & swollen. As for the friendship part of it, although there is more of it than was ever dreamed of in Irving's case, it is very like my experience with the generous Henry.—After I had humbled myself to *ask* him for a chance to act at the Lyceum, he was lauded for his magnanimity in *inviting* me [to] act there (which put more shekels into the house than ever 'till Miss Anderson 'beat the score.') But all this is idle. I wouldn't, for the world, have any break with Lawrence, & am willing to let him, as I did Henry, have the glory of *goodness* in the business. I have the warmest feeling for him & all his family. I am sure if he sees the printed report & reflects on it he will be sorry.—At first I thought it best to keep the engagement secret; then I fancied it would have a good 'moral' effect to let the world know of it—as far as it had a right to know anything of such affairs, but he thought secrecy, so far as it could be kept, was better; & to this end he has denied rumors that have reached him &, as I say, has probably said too much or too little. It is awkward for me to face the man to whom I had partly committed

[31] Louis James, American actor, was from about 1880 to 1885 the foremost member of Lawrence Barrett's company. From 1886 until his death he made starring tours, achieving some success as Virginius, Othello, Richelieu, Falstaff, and Joseph Surface.

In the seasons of 1884-85 and 1885-86, Booth was supported by the company of Richard Montgomery Field's Boston Museum.

myself—for the article mentions Mr. Chase,[32] I think it is, as my agent. He is Barrett's man of business whom I have never seen—to my knowledge, nor even held any kind of communication with; I must now tell *my* man the entire facts rather than let him think that I have thrown him over for another agent, & he may incautiously let the cat entirely out of the bag. I hope nothing more unpleasant will come of the matter.

My folks are well & the daughter very happy. I shall have my hands full now 'till I begin to act. Am weary & apathetic to a degree, & have had a task to get through this. God bless you! I wish I could have had you with me at "Boothden" some of the time that I had no visitors & when I felt in the mood to be talked to—as *you* talk; I had many nights of loneliness—when I could not read, or write or think. I have promised to try a slight sketch of my father for Hutton & 'tis far more irksome that I imagine, for the subject (with all reverence, I say it) is very worn & tedious.[33] I fear that I cannot keep my promise.

Since his return from Europe, Booth had felt less and less like writing letters and his correspondence with Winter diminished. On Christmas Day, a few days before his opening at the Boston Museum on December 28, he managed to send Winter a hasty greeting.

106. Christmas, 1885

Dear Will—

I can't write a letter—I owe you one, I know, but—I simply *Karnt*!

Health good—Memory too old to study new parts; am struggling now with the 2 Brutes—*Lucius* "physics *Payne*", but *Marcus* is a pill—so are *Cassius* & *Anthony*.[34] I've promised

[32] Arthur Branscomb Chase was the manager of the Booth/Barrett tours. Charles H. Thayer was Booth's business manager during the 1884-85 and 1885-86 seasons.

[33] The reference is to the biographical sketches of Junius Brutus Booth and Edmund Kean which Booth contributed to *Actors and Actresses of Great Britain and The United States*, edited by Laurence Hutton and Brander Matthews (New York, 1886).

[34] *Macbeth*, II,iii,55: "The labor we delight in physics pain."

Booth opened his engagement at the Boston Museum in Payne's *Brutus* which he had not played in for almost five years. During the last week of his engagement, January 25 through January 30, he played Brutus in *Julius Caesar*, another role which he had not played for many years.

to do the piece—and Richd. 2d, if possible; don't think the latter is.[35]—Got your *cherries* and relished 'em at breakfast yestreen.— Here's my Xmas greeting in return therefore.[36] 'Tis sent in sincere brotherly feeling; & my blessing goes with it.—Begin work Monday—for 5 weeks; the 'take' is (*they say*) very large.

The household is asleep—"not a mouse stirring"—I am re-covering the *Caesar* 'Brutus' & reading the life of Young—just finished Siddons & ½ of Crabbe [sic] Robinson.[37] I'm glad the itch for reading has come o'er me again—after a lapse of many years—the *Scribendi* business, though, is not at all to my liking. I've tried to 'reminiscence' but cannot fashion a sentence that does not disgust me. God help my poor sketch! Some one here is filling three columns of the Herald with just such matter; Booth & Kean—doubtless all the great actors will be dealt with before Hutton's book is ready. It looks like a trick to forestall him—or will it advertize his work? Adieu!—All well.

[P.S.]—Am afraid that I must go back to the Cibber Richd.— Shakespeare's is not popular.

On January 16, 1886, Winter's favorite son Arthur was severely injured when his head struck a rock in a sledding accident. Nine days later the boy died. Winter was desolated by this loss. Shortly after his opening at the Fifth Avenue Theatre on February 1, Booth sent Winter his condolences.

107. Albemarle Hotel, Madison Square, New York, Thursday
      [postmarked February 4, 1886]

Your last sad letter has been before me whenever I have taken the pen to answer my many correspondents but I could find no words for you, dear boy. In vain I have tried to say something in

---

[35] Booth did not play *Richard II* during this engagement.

[36] According to a note by Jefferson Winter at the end of this letter, Booth sent Winter a pansy stick pin with a diamond in the center as a Christmas gift.

[37] Booth's personal copies of the following books are preserved at the Walter Hampden Memorial Library:

Julian Charles Young, *Charles Mayne Young: A Memoir with Extracts from His Son's Journal* (London, 1871).

Thomas Campbell, *Life of Mrs. Sarah Siddons* (New York, 1834).

Thomas Sadler (ed.), *Diary, Reminiscences, and Correspondence of Henry Crabb Robinson* (Boston, 1880).

response to your deep heart-wail, but after reading & rereading the touching words I have sighed—"the rest is silence"—and put your letter by 'till another time. How good of Curtis[38] to speak as he did of the sweet boy! I remember him as a most lovely and fascinating child—on the occassion of my one visit to you at Fort Hill.

After the hard experiences of life, such as we have had—tho' of a different sort, we should rather rejoice at the escape our early dead have made from the agony of living. But if no philosopher can bear the *toothache* patiently how can one be expected to endure the heart-ache without a groan?[39] "With meditating that she must die once, I have the patience to endure it now"[40]—is the most comforting of our great Philosopher's reflections, yet I never speak the words without a pang for my past losses and the probable loss of that fragile one to whom the very tendrils of my being cling. Philosophy, Hope and Faith have but little weight in the balance against the load of an over-burthened soul—which kicks the beam despite all reasoning. We must suffer! It is the *Great* decree and we must endeavor to endure with patience. How often do we wish for death without a thought for those to whom our absence would be pain. To us it seems that our absence would be but temporary; feeling assured that they must come to us bye-&-bye. But when *they* go from us we sink in the trial—hopeless! They are gone! that's all we know, and we are desolate!—But be assured, my friend, that your sweet boy will be the first to meet you, as of old, and his little arms will clasp your very neck and welcome you, as but a few days ago he did. I, who have received a succession of soul shocks & have learned to "suffer and be strong" in sorrow, am more than hopeful—I am confident of our reunion by whichever path (so it tends upward) we may travel, in our Father's House.

God bless you! Convey to your dear wife my profoundest sympathy and dearest condolences.

[38] George William Curtis, author, orator, and editor of "The Easy Chair" in *Harper's Magazine*, was, from 1854 until his death, a close friend of Winter. For Winter's reminiscences of their friendship see *Old Friends*, pp. 223-274.

[39] *Much Ado About Nothing*, V,i,35-36: "Never yet philosopher/ That could endure the toothache patiently."

[40] *Julius Caesar*, IV,iii,191-192.

Do not attempt to write a line for me, or work at anything,—yet, on second thoughts,—perhaps it were better that you should hammer at the anvil awhile for relief, but let it be on something more congenial than the monotonous drudgery of criticism. To record the woes of mimic life would now be more distasteful than ever it was. Yet in that mimic world I almost nightly find the actual sufferings of my real life rehearsed. Think, my dear boy, how much heavier than our own is the cross our neighbors bear & that "the worst is not, so long as we can say:—'This is the worst.'"

Only God can comfort you both, and He will.

The day after he closed his engagement at the Fifth Avenue Theatre Booth wrote Winter to again convey his sympathy and to express his "utter dismay" at forgetting his lines during a performance of *Julius Caesar* on January 26.

108.   New York, Sunday [postmarked February 28, 1886]

Tomorrow at 11 a.m. I shall go to Phila.—to be absent two weeks; on my return I hope to see you and to find you recovered from the terrible wrench you've had. So far recovered, I mean, that you will be able to resume your work and divert your thoughts from the great shadow that has fallen on your life. After a while, my dear boy, the shadows will give way to sunlight—you will see the 'silver lining' of the cloud. Be patient; be strong.—

My sojourn here has not been a pleasant one & I shall be glad when the next fortnight ends and my connection with Field (outside of Boston, at least) shall terminate. I have worked hard & am weary. With the exception of one misstep (Iago)[41] I have done my duty well—until Friday when, to my utter dismay I found that Brutus has slipped from my memory! I recently acted the part, 7 times, to my own & the public's satisfaction & felt so sure of myself that I did not rehearse or read the part—but when I entered the scene my wits flew off & I made a bungle of the part. At the Matinée, next day, I was but a trifle better—although I read & re-read the lines; at night—last night—I felt my way cautiously & consequently my acting was indifferent, yes—*bad*! I never had anything of the sort happen to me 'till now, & it sets

[41] Booth gave one performance of Iago on February 6, which, according to a note by Winter at the end of this letter, he did not play to suit himself.

me thinking of poor John.[42] But don't let this worry you—I have but two more weeks of work & then comes a good long respite. Of course I am somewhat anxious for Edwina[43] & it may be my brain has been too much distracted by this & other matters foreign to my work.—I hope the poor mother keeps her health—in such trials women are stronger than men. God comfort you both!

During the spring of 1886, Booth and the great Italian actor Tommaso Salvini, then making his fourth and last tour of the United States, agreed to act together in bilingual performances of *Othello* and *Hamlet*. At the New York Academy of Music on April 26, 28, and at the matinee on May 1, Booth acted Iago to Salvini's Othello; on the evening of May 1 Booth acted Hamlet and Salvini the Ghost. Booth had wanted Salvini to play Claudius, as indicated by the letter he wrote Winter a few days before the opening.

109. Boston, April 23, 1886

Forgive my neglect of your last note, I've tried in vain to write to you, since the receipt of it, several times, but the horrible depression succeeding my nervous strain during the past month has rendered me unfit for such a 'labor'—though it is one that 'physics pain'; even my darling vice of smoking has been more of a task than pleasure to me.—I hope I shall be able to visit you during my stay in New York, but as I shall have rehearsals while there, on the 'off' days, I may be prevented. If you *can*, come to the 'polyglot' (it will amuse you) & stay over-night with me—at the Albermarle. I am sorry that I consented to play Hamlet— I did so under the impression that Salvini wd. act *Claudius* & thus give a precedent for other actors to perform that neglected character in preference to the *Ghost*, which leading actors—in their blindness choose; but S. concluded that it was too long a part for him to study, and at the last decided not to act *Lear*, which prevents me from playing a subordinate role with him in return for his compliment to me.[44] Of course he will be 'wasted' as the

[42] Booth apparently was afraid that he might be losing his mind as John McCullough did.

[43] Edwina was pregnant with her first child, Mildred, who was born on March 24, 1886.

[44] Salvini acted King Lear in America for the first time at the Fifth Avenue Theatre on February 21, 1883. Presumably Booth would have played Edgar or perhaps Kent.

Ghost, and I must drag through the heavy play of Hamlet unaided by him. With some great name for the *King* that character wd. always after be accepted by the donkeys who decline it as unworthy of them. I am writing this on the eve of my departure for New York ('tis Shakespeare's birthday, by-the-by)—I shall arrive there about seven o'clk & rehearse Sunday noon and night— again Monday morning. When we meet we will discuss the biography business; I have little faith in the pecuniary result of it—or of any 'actor-book,' but we will talk it over.[45] I hope, my dear Will, that you & your wife will let the dead past bury its dead and look forward to the near future in which life is sure and ever-lasting. Think of our sorrows and rejoice that yr. dear one has escaped the common lot of suffering. Think as I have thought in all my many trials that—"the worst is not so long as we can say 'this is the worst!' and consider how brief is the very longest sojourn here; how soon you will have your boy again. As I sat by what all believed to be Edwina's death-bed the thought of her dear mother was always present and I thanked God for her early death which spared her the sufferings she wd. have endured in the misfortunes that so frequently have befallen me. So let yr. minds reflect, and grow strong in hope and faith—Edwina and little *Mildred*, thank God! are now doing well—but it was a very narrow escape for the mother. She wishes to pass the Summer at 'Boothden' and I have countermanded the order to sell the place—perhaps she will again change her mind, but doting fathers must humor their only daughters—especially when health if not life is in question. This and other changes of whim have de- ranged me so that I can form no idea what my movements will be after I get through with Salvini—three weeks hence. By-the-by, I must study a garbled useless scene for him[46]—in wh. according to his arrangement, nothing is to be gained by either. 'Tis morning and I must to bed although the intense heat does not promise me much rest.

Drop me a line at the Albermarle and tell me that sweet resignation has settled in your house, and when I may see you.

God bless and comfort you both!

[45] Winter had asked Booth to write a book of reminiscences as early as the summer of 1884, but Booth had refused (Folger Shakespeare Library, Y.c. 215 [465]). Now Winter wanted to write a biography of Booth.

[46] *Othello*, IV,i which Booth omitted from his acting version.

Salvini and Booth acted in *Hamlet* in Philadelphia on May 3. At the Boston Theatre on May 10 and 12 and at the matinee on May 15 they acted *Othello* and on May 14 they acted *Hamlet*. These bilingual performances were highly successful for both actors, except for one accident to Booth. During *Othello* on April 28, Booth suffered from an attack of vertigo, undoubtedly caused by his incessant smoking. He wrote to Edwina that he was "dizzy from the effects of dyspepsia and being *jerked* up from the stage by Salvini, who let me go before I had regained my footing, I stumbled on my heels, and a rent in the carpet laid me flat on my back."[47] Many papers, however, attributed his fall to drunkenness, and even Salvini thought that Booth was intoxicated. He was publicly defended by several of his friends, including Barrett and Bispham, and after his impressive performance as *Hamlet* on May 1, the attacks lost force and stopped. After he closed at the Boston Theatre, Booth retired for the season.

On July 7 Winter wrote an article for the *Tribune* refuting a report that Junius Brutus Booth was a slave-holder. From his Boston home Booth wrote Winter the following long letter (unfortunately partially destroyed) to thank him for the article, to arouse him from his continued grief at the loss of Arthur, and to complain about his own domestic and financial difficulties.

110.   Boston, 29 Chestnut St., July 17, 1886

Thanks, my dear Will, for your refutation of the silly & malicious report. It is so d--nably absurd that I gave it no notice, yet I asked my sister, Rose, the other day—at L. Branch, if father ever owned a slave & if ever a boy named Alexander was employed on the farm. She said "No—father bought a woman servant because he could not hire an obedient one, but as she threatened to beat Mother's brains out with a fence-rail he gave her free papers three days after purchase & sent her away." That he afterwards 'leased' for ten years a black girl named Hagar who asked him at the end of 3 to set her free from the lease, wh. he did—on her promise to return to her owner, Judge Bond of Harford Co. Of course she didn't go back & father had to pay for her. A rabid Republican, or Democrat rather, my father could not hold slaves. 'Old Joe Booth' (or Hall—his master's name) lived on the farm from early boyhood until he died of old age—a full-blooded African. Henry (colored) & Geo Brown (of 'Green-room' fame) were hired boys on the farm—the latter to look after the

[47] Grossmann, p. 71.

horses; a great favorite of father & a companion to Junius. These were the only 'help' my father ever had on the farm. Possibly the dirty dog who wrote the article may have met a darky named Booth & concocted his filth from that circumstance. A negro barber, shaving me, told me (years ago) that his name was *Booth*, & wondered if we were related! He was a d----d handsome fellow,—but he had the razor in hand & I was at his mercy. This was in slavery days & in a slave state, too.—

I intended to pass a day with you last week but lumbago laid me *flat* on my back in Bispham's *flat* in 18th St. & I could not go. Nor could I write—even if I had words to comfort you withal. My dear Will—you know that all that I could do I would do for you in your distress, but I look at life & its d--nable impositions in a different light from that in wh. you see them. I cannot grieve at Death, wh. to me seems to be the greatest boon the Almighty has granted us. Consequently I cannot appreciate the grief of those who wail the loss of loved ones, particularly if they go early from this hell of misery to wh. we have been doomed. A lady friend of mine who has never known sorrow told me once that she had no doubt of Heaven being a very delightful place but for the present her house on 5th Ave: was good enough for her. She goes to church 2ce. every Sunday, too! The shadow of Death is now over her house & if it falls I've no doubt her grief will be excessive—as her happiness has been. But in one of such solid sense, religious sentiment and large experience of earthly suffering I cannot comprehend your despair. Great God! Will, have you no faith? Have you no Philosophy?—"Man's life's but a span,"[48] tho' it be a Century long, and who knows how soon you & the dear mother may go to join the 'avant-courier,' the angel-guide?— When I last saw my mother alive she had just entered her 84th year[49]—after a weary battle of certainly 60 years of sorrow, her face was scored with wrinkles; in every one of which could be plainly seen the ravages of suffering. No son ever loved his parent dearer than I & yet, for years, I prayed—silently, deeply in my soul, for her release, & when it came & I was hastily summoned to her death-bed I found the weary old woman transformed into a most beautiful object—so beautiful that I would not have believed it to be my poor old Mother's corpse had I seen it by mere chance.

[48] *Othello*, II,iii,73-74: "A soldier's a man;/A life's but a span."
[49] Booth's mother died from pneumonia on October 22, 1885.

The natural grief that possessed me, from the moment I was summoned until I raised the cloth from her dear face, ceased at once & my soul said—'God be thanked!' & I was happy in her happiness which the good God revealed to me in the exquisite loveliness of her dead features. Do you mean to say this change is but Natural? because it so frequently occurs? *Bosh*! if so our dead dogs & other pets would be beautiful—wh. they are never— if it were a mere process of Nature. It is God's sign manual of Immortality! In it I see more than all the books and all the dullards of the pulpit teach. "I have looked upon the world for (nearly double) four times seven years, and since I could distinguish"[50] good from bad I have regarded what men term misfortune as the best tonic an apathetic spirit can receive. Horatio as Hamlet depicts him as—one in suffering that suffers nothing; one that Fortune's buffets & rewards hast ta'en with equal thanks,[51] (your selection of those lines for your early sketch of my life was most appropriate; I hope they will be my epitaph, if I have one, for they are *true*—) I have always regarded as the type of the true belief. He (Horatio) is not a stoic, but a calm & firm believer of what you my friend, seem to doubt. "With mediating that she must die one day I have the patience to endure it now" are words that bear more meaning to me than they convey to him who condemns philosophy as 'cold comfort.' All my life has been passed on 'picket duty,' as it were, I have been 'en guarde'—on the 'lookout' for disasters for wh. when they come, I am prepared. Therefore I've seemed to those who do not really know me, callous to the many blows that have been dealt me. Now, why do not you, my dear boy, look at this miserable little life, with all its ups & downs, as I do? At the very worst 'tis but a scratch, a temporary ill to be soon cured by that dear old doctor Death—who gives us a life more healthful & enduring than all the physicians, temporal or spiritual, can give. You have advised me well & wisely in my perplexities and I am therefore bold in admonishing you, my dear fellow:—[about half a page cut out] this Summer: I have been living on my friends at various places. Today I received your book on Mary Anderson have not yet read it.[52] In your [about one

---

[50] *Othello*, I,iii,313-317.

[51] *Hamlet*, III,ii,72-73: "A man that fortune's buffets and rewards/ Hast ta'en with equal thanks."

[52] *The Stage Life of Mary Anderson* (New York, 1886).

sentence cut out] your wife's sea-voyage.[53] Why did she not go?
You seemed to be hopeful. [about half a page cut out] have been
[illegible word] considerably in financial affairs & therefore have
I consented to take the tour with Barrett next season [about one
sentence cut out] years I have him living on my capital, [not]
my interest, & with several families of poor relations to maintain
I am awakened to the fact that I am not so rich as I thought & as
the newspapers make me out to be. I must sell this house & get
rid of 'Boothden' as soon as possible, in order to support my
son-(of a b---h) in-law, who does nothing but spend money, smile,
& live on his wife's father. A jolly old home is mine—would you
not rather have your child in heaven—? I would. Particularly if
I knew that her husband was in *Hell* where he belongs![54]

Be at peace, my friend; your griefs are but shadows of the
sorrows others bear.[55]

What a superb picture is that you gave Mary! Why could
not *I* have a copy of it?

I am in the midst of trouble & have written hurriedly,—I hope
I have said nothing that will pain you but something that will
encourage you.

God bless you![56]

Despite Booth's desire not to cause Winter any pain by the preceding letter, he nevertheless seems to have hurt him deeply by his
detached, almost callous tone. Almost two weeks later he wrote to
explain the motive behind his "cruel" words.

[53] In either the late summer or early fall Mrs. Winter *did* go to England,
taking Jefferson and Viola with her. Winter and his eldest son Percy
remained at Fort Hill.

[54] Booth's attitude toward his son-in-law had changed considerably since
the wedding (see letter 115).

[55] *Richard II*, IV,i,296-297: "These external manners of laments/ Are
merely shadows to the unseen grief"; and II,ii,14: "Each substance of a
grief has twenty shadows."

[56] Jefferson Winter wrote at the end of this letter: "A long section (8,
if not 12, pages) cut out and destroyed by my father.—The whole of
passages about Barrett; Clarke; Daly—and I think something else. Also, by
me, the worst part about his son-in-law. 'What the devil does he mean' &c.
(memo)—J. W." A close examination of the manuscript leads me to
believe, however, that Jefferson exaggerated the number of destroyed
pages. Only about half a page and about two complete sentences seem to
be missing.

111.   Cohasset, July 29, 1886

Indeed I meant no reproof, my dear friend—if what I wrote seemed such it was the result of my awkwardness in striving to rouse you from the crushing despondency beneath which you suffer, and you must forgive the cruel surgeon for cutting beyond the wound, or rather for probing so deeply. Would to God I had it in my power to lift you both out of the slough and set you high up in the sunshine of hope & comfort—but I am helpless in others' griefs, but will not be hopeless of your ultimate relief in the blessed resignation which God, in his own good time, will yield to you.

I cannot write much—I am mentally depressed and physically distressed at present; my health is good, but the heat is oppressive and the flies distract me to such a degree that 'tis not possible to sit quiet long enough to write more than one sentence at a time; mentally I continually revolve the 'might have been' and the 'what is'—of my domestic matters until my head nearly splits.

I shall remain here a few days longer & then to Boston & Newport until I start for my long tour—God bless and comfort you both.

[P.S.] Barrett joins me in loving regards.

Chapter X

The Booth-Barrett Tours

1886-1890

ON SEPTEMBER 13, 1886, at Buffalo, New York, Booth began an extensive tour of the United States under the management of Lawrence Barrett. Barrett had selected Booth's supporting company, arranged the schedule of engagements and furnished the scenery and costumes. Barrett did not, however, act or even travel with the company this season, but rather filled a number of independent starring engagements. During this tour Booth traveled over 14,000 miles and gave 233 performances.[1]

112. Detroit, September 19, 1886

Your last letter, August 27th, has been before me every day, or night, since I recd. it and I have made futile efforts to acknowledge it & several others that claim my consideration, but my idle spirit masters me most of the time, as you know, and I procrastinate without heed of time which runs into months before I can accomplish the simplest duty. I am glad to know that your wife is gone abroad for I am confident the change of scene & climate & novel surroundings will do wonders for her in every respect—I wish that you also could have such a change. Let the hope of her restoration to health stimulate you, my dear boy—I am sure it will.

I have little to tell you of myself. My first week of the many I must travel ended last night very satisfactorily. I failed to send you a list of my dates and stopping-places but will enclose it now. 'Twill be a hard pull—especially hard to work on plain, un-adulterated water, to which I restrict myself; it lends but feeble inspiration to one who so much requires stimulant. But I'll try to 'stick it out!' The company is liked & they do very well—A Miss

[1] A chronology of this tour is given in *Life and Art*, pp. 287-288. With the exceptions that Booth seems to have played in Lancaster on March 31 and Newark on January 1 and not to have played in Wilmington, it is otherwise correct. For further information on this and subsequent Booth-Barrett tours see Katherine Goodale "Kitty Molony," *Behind the Scenes with Edwin Booth* (Boston and New York, 1931) and Edwin Milton Royle, *Edwin Booth as I Knew Him* (New York, 1933).

Vaders (whose elder sister is with Joe) bids fair to become an excellent actress.[2] By-the-by, I shall see Joe in Chicago—he will be there during my engagement. Edwina went to Franconia in the White Mountains for change of climate two weeks ago, but I believe she is in Boston now; she is evidently very happy & perhaps the strange companionship is for the best. A woman's love is incomprehensible. I shall sell the Boston house & return to New York, while she will have her home in a small house or apartments in Boston. I shall never again attempt to settle down outside a graveyard. God bless you, old boy, I hope to find you in health & the comfort of resignation when I reach New York.

Five weeks and nine engagements later Booth sent Winter another brief progress report.

113.   Cleveland, October 28, 1886

Although I have had no letter from you for many a day, not since I began my *tour*, yet I recd. your charming little book[3]— while I was in Chicago, I think. I know, my dear boy, how little like writing you must feel and have not looked for any direct news from you, but I've heard of you often through others. I have been so fatigued by work & travel, & have still a heavy cold, contracted several weeks since, at Minnehaha Falls,[4] that letter-writing has been an irksome labor & only an occassional note to Edwina have I been able to scribble.—I trust you have good reports from England of your wife's health and that your plans for a trip thither will be accomplished—if you still hold that purpose—My engagement thus far has been remarkably success-ful and my health (but for this stupid cold) unusually good. The

[2] Emma Vaders, minor American actress, was Booth's leading actress during the Booth-Barrett tours playing Ophelia, Desdemona, Julie, Kathe-rine, Cordelia, etc. She subsequently appeared as the leading actress to Joseph Jefferson, Stuart Robson and other starring actors. Her sister Henrietta Vaders was Joe Jefferson's leading actress in the 1880's.

[3] The "little book" was probably *Shakespeare's England*, a one volume, revised edition of *The Trip To England* and *English Rambles*. It was pub-lished in 1886 by David Douglas, Edinburgh. Between 1886 and 1916 *Shakespeare's England* was issued in a number of editions and it became Winter's most popular book.

[4] See Goodale, pp. 44-47, for a description of Booth's excursion to Min-nehaha Falls with the actresses of his company and Arthur Chase.

travel is better than I thought it would be—but the easiest car-travel is to me a torture; the hotels are very comfortable and the food (in most of them) equals anything Delmonico affords. The improvement in these matters since my last trip is marvellous. The constant rain this week prevents my enjoyment of this most lovely city—where I anticipated pleasant walks & drives—I must leave here after the play Saturday night & expect to be in New York Sunday—have not yet determined where I shall reside, am tired of the Albermarle. I fear the long journey will weary me considerably—for I cannot possibly sleep on the rail, no matter how comfortably the car may be arranged for me.

Barrett is somewhere in this neighborhood & Joe follows me here; I saw him several times in Chicago & he let me read some of his reminiscences (the rough draft merely) and I was delighted with his simple & chatty style of telling many anecdotes of deep interest, some of his experiences, as related, are novel, amusing and pathetic; I think it will be the "boss" book of the kind.[5] I hope to see you during my five weeks sojourn in N.Y.—one of the weeks I will act in Brooklyn but shall 'bunk' in Gotham.

God bless you, Will! I hope your heart is lighter than it was erewhile & that your physical condition is good.

On November 1 Booth opened at the Star Theatre for an engagement of four weeks. During the second week of their engagement, he caught a cold which then developed into a slight case of pneumonia. In a week, however, Booth had recovered sufficiently to return to the theatre where he played until November 27.

114. Victoria Hotel [New York], Sunday Eve. [November 13, 1886]

Your affectionate attention during my illness was not wasted—I felt it sensible, only I wished you had come in person. That beastly Saturday night you might have passed with me & thus avoided the wet journey home & the cold you took. Stedman has just left me, he told me of a beautiful letter about me that he had received from you; Bless you! I've had a short, sharp turn wh.

[5] Jefferson began writing his autobiography about 1877 at the suggestion of William Dean Howells. It was serialized in the *Century Magazine* in 1889 and published in book form in 1890 by The Century Company, New York.

might have unhorsed me quite, and had not my *53* winters weighed in the scale against me I doubtless would have risked the worst. It is lucky I stopped when I did; I am all right now, but obey the doctor who bids me keep still until the last hour; I shall not leave the room 'till I go to the theatre at 7 o'clk tomorrow. All pain is gone & I feel in good condition—except a weakness, which will soon wear off. I believe with you that the cold and dampness (to say nothing of the putrid odors) of the Star Charnel house caused the sudden illness, which, however, I felt approaching several days in advance.

I had much difficulty in preventing Edwina coming; I telegraphed her not to do so before the announcement was made to the public; but she begged & not 'till the Dr. and several friends told her my true condition would she abandon the desire to come. It would have worried me to have her take the journey & be helpless & anxious here while her baby would be an additional care to her.

My room has been filled with flowers in memory of my 13th[6]—yesterday, and lots of kind messages and "goodies" (none of wh. may I eat) have come from various friends, some anonymous. So—my hospital life has not been altogether dismal—I hope you will let me see you soon & that I shall find you well and cheerful over good news from abroad.

I've had two hours interruption & am pretty well wearied—Must therefore close with best wishes, 'goodnight' and God bless you!

On a one-day business trip to New York, following a two-week tour of one-night stands, Booth sent Winter the following note.

115. Everett House, Sunday Night [postmarked January 3, 1887]

Having some business-matters to talk over with Bispham & Benedict I came from Newark—where I played last night instead of Wilmington as per list—and shall start for Philadelphia tomorrow, where I will be two weeks—if during that time you can get there we can have that Hamlet chat you refer to; I shall be mightily pleased to have you, I shall stop at the Lafa[y]ette

---

[6] On November 13, 1886, Booth celebrated his 53rd birthday.

Hotel.—So Stewart [sic] is gone at last![7] I wonder that he held out so long. I have seen no reference to myself in connection with him, but heard there were some Joe Howard & Crinkle[8] mud thrown. Was there anything worth noticing?

I'm glad the books were welcome; I am having a set of the Prompt Books bound for the library,[9] also; Aldrich is attending the matter for me.

Barrett is delighted with the success of Rienzi[10]—which I hope will hold out for him. I saw him for a few hours in Boston, two weeks ago.—'Tis very sad to read your report of your dear wife, but I hope a turn for the better has come & that she will find relief & solace in the dear ones still left to her. God comfort you both!—[11]

Hard work seems to agree with me this season, but I must confess the constant travel & irregular meals do not—I have suffered somewhat on that account—but I have agreed to 'do it again' next season.—Give me Mary Anderson's address—Edwina wants it & as she sent me a Xmas card I must acknowledge it.—

[7] William Stuart died in New York at age 65 on December 27, 1886.

[8] Andrew Carpenter Wheeler, American drama critic who wrote under several pen-names, the most famous of which was "Nym Crinkle," was for many years the drama critic of the New York *World*. He also contributed to other papers, notably the *Sun* and the *Dramatic Mirror*.

[9] About March, 1885, Winter established the Arthur Winter Memorial Library at his late son's school, The Staten Island Academy. Booth, Irving, Jefferson, Augustin Daly, Mary Anderson, Ellen Terry and others contributed books to the library. For a brief description of the library and its collection see "The Arthur Winter Memorial," New York *Dramatic Mirror*, XLI (January 7, 1899), 2.

[10] During the season of 1886-87, Barrett revived Mary Russell Mitford's tragedy, *Rienzi*. He presented it at the Boston Theatre for two weeks beginning January 17, and for three weeks at Niblo's Garden beginning May 2, and in other principal American cities.

[11] Mrs. Winter and the children returned from England about January 1. On January 6, Winter wrote to Laurence Hutton: "The anniversary of his [Arthur's] death is close by us now. I need not speak of the horror and grief of these days. I have been foolish ever to speak of our misery at all, for no one understands it. My wife came back from England the other day, calmer, more patient, but crushed under the weight of despair that has broken her heart. There is no more to be said about the matter. The effort to live and to work and to endure has worn upon me till I am more dead than alive." Quoted from Richard Ludwig, "The Career of William Winter," p. 394. The original is in the Harvard Theatre Collection.

Business has been very fine everywhere & from all I hear the outlook is excellent.

Edwina was ill-abed when I left her, but she is better & has been down stairs of late; she is *on the way* again! The heavy-hipped Hungarian hog, who, "like a full-acorned boar cries 'ugh!' and mounts,"[12] will soon kill the poor fragile girl, or beget a breed of puny weaklings—following too fast to permit the mother to gain strength to nourish them. My anxiety for her is constant & very great.—I must see Charly Jefferson about the Couldock benefit—a sort of crazy-quilt programme has been in print which is false, so far as Barrett & Booth are concerned. I told Joe I would do either 3d. act of Hamlet, or 4th of Richelieu, & have arranged to keep my company together for that & the subsequent benefit to Wallack.[13] God bless you!

On March 7 Booth opened at the San Francisco Theatre for an engagement of four weeks. At the end of his third week, he wrote Winter to tell of his "utter fatigue" and other matters.

116. Palace Hotel, March 26 [1887]

I know not how many eons are gone by since we held each other's fist foreninst us, or who writ last. Utter fatigue—a 'goneness' of all energy (save for my three hours pull at night) has been the cause of my long silence. If it were not for this chronic 'tire' I shd. enjoy this dreaded tour very much—everything has been done to make it comfortable for me & if my bones were looser I should have had a 'picnic' all this while; but, alas! I'm

[12] *Cymbeline*, II,v,16-17: "Like a full-acorn'd boar, a German one,/ Cried 'O'!"

[13] Charles W. Couldock, American actor born in England, made his greatest success as Dunstan Kirke in Steele Mackaye's *Hazel Kirke*. He purportedly played this role over 1500 times. On May 10, 1887, a benefit was given for Couldock at the Star Theatre in celebration of his fiftieth anniversary on the stage. Booth appeared in the third act of *Hamlet*. For a description of the benefit see *Annals*, XIII, 233.

Charles Burke Jefferson, Joe Jefferson's eldest son, was for many years his father's manager.

On May 10, Lester Wallack retired from management and transferred the lease of Wallack's Theatre to Theodore Moss. I find no reference, however, to a benefit for Wallack until May 21, 1888. For this benefit, which was one of the most lavish in the history of the American theatre, Booth gave *Hamlet* in its entirety. See *Annals*, XIII, 507-508.

growing rusty—I feel it in every joint, & besides—I seem to need
mental or spiritual rest—I have been kept up to a high nervous
pitch by the constant thought of Edwina; her condition causes
me great anxiety. Her letters—six days old, of course—are full of
hope & cheerfulness, but frequent telegrams are requisite to allay
my fears for her. Acting while under such a nervous pressure, as
it were, pulls one all to pieces—I hope, my dear boy, that you
are well & in a calmer frame of mind than when I last heard from
you, & that your wife is more resigned & in better health. A *squib*
went the round of the press to the effect that I had donated 700
vols. to your boy's library—did the check I sent to the Principal
of the school procure so many? What books are they?—A *squib*
of another sort is now & then 'picked' up by some scoundrel who
hates Barrett—I mean the report of my engt. to marry his
daughter. Poor little Gertie! To drag that dear little sufferer into
publicity is the work of a filthy cur![14] I can't help suspecting that
fellow Thayer, who hoped to have the control of my engagements,
but—I may wrong him.—Gillig is giving Barrett annoyance, but
in the end *B.* must triumph. G. is a bloodsucker & has bled
Lawrence for several years of many thousands—in the way of
usury.[15]—The *tour* has thus far been a *boom*; in every place the
interest has reached fever point, and *Hamlet* despite the gray hair
& wrinkles, is in greater demand than other parts; there seems to
be here an increasing appetite for it. I think I've acted fairly well
most of the time—The growth of this city since '76 is marvellous!
greater, it seems to me, than that of the 20 years previous; and
the streets—all of them—are crowded—I never saw such bustle

[14] Gertrude Barrett was Lawrence Barrett's youngest daughter. Why
Booth should refer to her as a "sufferer" is obscure. In January, 1889, she
married Joseph Anderson, minor American actor and the brother of Mary
Anderson.

[15] In the San Francisco *Argonaut* (March 26, 1887) "Betsy B." (the
pen name of Mary Therese Austin) wrote:

> The legal persecution to which Lawrence Barrett is being subjected by
> Henry Gillig, of the American Exchange in London, is remarkable in
> two things—the first (which was strongly suspected before), that
> Barrett lost a considerable sum by his London engagement two years
> ago; & the second, that the American Exchange extorted thirty-five per
> cent interest when the actor was in a pinch, & tried to enforce the terms
> in the hope that he would rather pay up than to acknowledge that his
> London season had been a failure.

here 'till now.—The McVinegar's are under this roof with me; they were at Los Angeles while I was there &—I hear they go to Monterey when I expect to be there. They are anxious for reconciliation & hope that some chance may bring us together. 'Tis better as it is.—Now I'm off for the Matinée; not yet fully awake & am fitter for a lounge than for a tussle with Bertuccio; think of my playing the Fool today & Richard tonight! I am surely nearing my dotage to undertake such hard work—even for a young man. Another week here then Salt Lake; give me a line there. Adieu! God bless you both.

On May 14, Booth ended his national tour at New Bedford and retired for the season. During the summer he visited various of his friends such as Benedict, on whose yacht, the *Oneida*, he took a cruise about the middle of July. Barrett, Hutton, Aldrich, and Bispham were also aboard and it was during their conversations that the idea for The Players was first discussed. During this summer Booth also sold his Chestnut Street home and his summer house at Newport. On May 18, he wrote Winter that his home "will be where my *hat* may be, not where my heart is" (Folger Shakespeare Library, Y.c. 215 [496]).

On September 12 at Buffalo, Booth and Barrett began their first tour together. During this tour they traveled over 16,000 miles and gave 255 performances. *Julius Caesar*, because it gave both Booth and Barrett an almost equal opportunity as Brutus and Cassius to display their talent, became the most frequently performed play in their repertory.[16] From Chicago, where he opened for an engagement of three weeks, Booth wrote Winter to describe his tour with Barrett thus far.

117. Chicago, October 12, 1887

My impression is that I've not heard from you since I last wrote to you—from what point I know not—but I've not forgotten you for a' that. Should have written long ago but for the lack of aught to tell you. Fine business and very pleasant time finally, always excepting the travelling; that under any conditions, is a torture to me. Barrett & I get along cosily; keeping very much to ourselves & working together harmoniously in every respect. I think I will, hereafter, so long as I continue to act, play short engagements with Barrett—he can fill his regular seasons with

[16] For a chronology of this tour see *Life and Art*, pp. 288-290. With the exception of performances in Erie, Toledo, and Youngstown, on November 17, 18, and 19, respectively, it otherwise seems to be correct.

new productions when I am idle—and so I may glide gently &
gradually into the shade, instead of making an abrupt 'flop' of my
last act. He is an excellent stage manager & not only executes his
own ideas but carries out whatever I suggest but which, for many
years, I have lacked energy to put into practice. Our *Caesar*
(here) was admirably done & the *Othello* scenes, also, recalled
the Winter Garden & Booth Theatre revivals: I believe that both
plays, as done here, would—in as good a theatre—make long
'runs' in New York. The company is remarkably good.[17]—

Tell me more of your work; what is the *Memoir* (of myself) to
be like? Your excellent one already published covered all of
me—I fancy. That book was killed by *the caricature illustrations*
& the large price of it. It is a pity &—in view of the great interest
in theatricals—it is strange, too, that dramatic books are so
little read. One would think that, at least, all theatrical folk would
read such books, but a very, very few of them do so. Cheap
novels & Ledger[18] stories seem to be the favorite literature of the
majority of players.

I hope all your dear ones are well. Mine are in their Beacon St.
house[19]—not yet settled in household affairs & without a nurse—
wh., of course, distracts Edwina & I fear will affect her health;
her husband (who is not one of my *dear* ones—in affection tho'
much so financially) is ill, and altogether the poor girl's troubles
are well begun already.

Now I must draw to a close with you—with affectionate regards,
in wh. Barrett joins me.

On March 5 Booth and Barrett opened at the Baldwin Theatre in
San Francisco for an engagement of three weeks. Three days after
their opening, Booth wrote Winter a long overdue letter.

118. San Francisco, March 8, 1888

As the years pile up I find the lack of energy increases & it has
become a real labor to write—even a few occasional lines to

---

[17] The Booth-Barrett *Julius Caesar* was similar in many respects to the
1875 Jarrett and Palmer production. See Benson, pp. 194-197, for a brief
account of this production.

[18] The reference is undoubtedly to stories about the theatre published
in Edward Ledger's *The Era Almanack*.

[19] After Booth sold the Chestnut Street house, Edwina moved to a house
on Beacon Street formerly owned by William Dean Howells.

Edwina. Letters from friends are heaped before me all the while & I sit dazedly gazing at them fully determined to answer them *all*, tomorrow. So it goes to the last nick of recording Time;[20] or words to that effect. You sent me a lot of papers which kept me awake during a long car-trip & for wh. I never thanked you: now I do; with an old man's blessing. I have had lumbago & a bilious turn the past week—the journey thro' Texas was *evil*; bad weather, long & rough jumps with consequent fatigue; otherwise all well. What an escape Irving had from the fire at Union Square.[21] I hope it did not affect his engagement; sorry he is to leave the country without my meeting him—for the last time; 'tis not at all likely I shall ever go again to London—though, of course, he may come here. I've heard nothing definite or reliable about my nephew Creston's success; have you? The last I heard of him he had finished Hamlet & had taken up Claude. I hope he will be able to carry on the tragedy business another generation.[22] June's second boy (Sidney) now at school seems to be full of the 'horrid stuff' & looks it too; the first boy, *J. B. B.*, affects the Society walk;[23] Clarke's youngest boy[24] is successful in comedy, I believe, and all I hear of *Cres* is good, but I fear his lisp: "Theemis! nay, it ith; I know not Theemis."[25] is a little rough for the "even road of

[20] *Macbeth*, V,v,21.

[21] Irving was engaged at the Star Theatre from February 20 to March 24. On the evening of February 28, the Union Square Theatre, which adjoined the Star, was gutted by a fire which also threatened to destroy the Star. The *Dramatic Mirror* (March 3) reported that Irving's costumes and any scenery which was not in use were moved to a place of safety, but the performance was not interrupted.

[22] Creston Clarke made his American stage debut at Wallack's Theatre as Roy Marston in Henry Hamilton's *Harvest* on October 13, 1886. On May 29, 1890, he appeared as *Hamlet* in the closet scene from the play at the Vaudeville Theatre in London. I have found no reference to his playing Claude Melnotte.

[23] Sidney Barton Booth, youngest son of Junius Brutus, Jr., and his third wife Agnes Land Rookes Perry, made his stage debut at Wallack's Theatre in 1892. He subsequently was a successful leading man to such actresses as Maud Adams, Lillian Russell, Grace George, and Alice Brady. Junius Brutus Booth III, the elder son, also became an actor. In 1887 he was in the company of the Lyceum Theatre in New York. In 1912 he shot himself and his wife in a London hotel.

[24] John S. Clarke's youngest boy was Wilfred Booth Clarke.

[25] *Hamlet*, I,ii,76: "Seems, madam! Nay; it is. I know not 'seems.' "

blank verse" but I hope he can o'ercome that obstacle. Barrett & I keep well up to the mark & our business continues wonderfully fine—despite foul weather & high rates. After next season I hope to settle down & act seldom—if I last so long. I see Barnay has come back:[26] I don't know how you regard him, but I'm aware of your objection to the foreign Shakespearians. Whatever you may feel I hope you will be kind to him—the Germans 'did me proud,' you know, and they are sensitive and I suspect somewhat suspicious; they might think it cruel if their courtesy to the Yankee was not reciprocated. I have never met him—nor do I know his acting; but I hear nothing but good of him in every respect. Have not heard anything of Mary recently; presume she is still in full tide; am sorry she will be with Abbey—she should have better management. Edwina's letters are always full of her wonderful babies & her own happiness—for this I am thankful, in spite of my disappointment otherwise. God bless you. Love to you all. Give me a line now & then & forgive my laziness.

On May 19 at the Amphion Academy in Williamsburg (Brooklyn), New York, Booth and Barrett ended their first tour. Booth then retired for the summer and devoted his energies to the organization of The Players which had been legally incorporated on January 7, 1888. Booth wrote Winter, who had been in England since April 29, to explain the aims of his club.

119. Boston, June 8, 1888

I have two letters from you of early date, but I shall not attempt to reply to either. I am so 'fagged out' that I can barely scribble you a cranky line or so in response to your 'goodbye' greeting (isn't that a bull?), merely to assure you that his "heart is true to Poll &c, &c." I hope you are having a Royal time—taking it for granted that Royalty's time is 'O, so gay & jolly!'—The long tour ended gladly & now all but I—are taking their ease for the Summer; I am homeless & hang up at hotels, here & there, above with my pipe & reflections—which there are many & not over cheerful. No matter: the play will soon be ended. I shall remain here a

[26] Ludwig Barnay filled engagements at the Academy of Music from March 15 to March 24 and at the Thalia Theatre from April 2 to May 8. For Winter's opinion of European actors in Shakespeare, see *Shakespeare on the Stage*, 1st ser., pp. 30-34.

couple of weeks to select certain books & pictures, from the dusty mass I have stored in this town for presentation to the *Players Club* & to destroy lots of letters & old papers that have accumulated for many years.[27]

The Players will be the best institution yet organized for the ultimate good of the Actor—but I can't explain its object now, I shd. rather say its *methods*—the object is simply to excite ambition & self respect in those who waste the better part of their nature in happy-go-lucky, Bohemian habits. All I hear of my young ones is good—they are at Stockbridge now. My sister, Asia, was placed beside our parents in Baltimore last Friday—I have but just returned from her funeral. Clarke brought her remains from Bournemouth, where she died, and will soon return to London.[28] Give my best regards to Mary, her Mamma & the beauty boy, my namesake;[29] to Irving, Toole,[30] & others who may kindly remember.

From his London residence Winter penned the following warm reply.

120. The Grange, Brook Green, London, Kensington, W.,
    June 18, 1888

My dear Edwin:

I was not surprised today, although much pleased, when a letter from you was brought to me: for you have been much in my thoughts for many days past. Yesterday afternoon I was at Laleham, sitting beside the grave of Matthew Arnold, and I thought of you *there*—for those who are especially dear to my heart always seem present with me whenever I am in a *sacred* place. Your letter is dearly welcome. I hear the voice of sadness in

---

[27] Undoubtedly many of Winter's letters to Booth were among those destroyed.

[28] Asia died in London on May 16, 1888.

[29] Mary Anderson's father, Charles Joseph Anderson, died in 1863 in the Civil War in which he served as a Confederate officer. Her mother, Marie Antoinette Leugers, subsequently married a Dr. Hamilton Griffin. They had one child who was named Edwin Booth Griffin.

[30] John Laurence Toole, popular English comic actor and theatre manager, was a close friend of Irving. From 1879 until his retirement in 1895, Toole successfully managed the Charring Cross Theatre, giving it his own name in 1882.

it—but how could that be otherwise? I very well understand the loneliness of your life; how much you are constrained to live among memories and regrets; how widely your spirit is withdrawn from the world that is around you; how little comfort you can find, or are likely to find, in society or in the commonplace current of ordinary affairs. It is my deliberate judgment that if it were not for the refuge that is afforded by *Art*, no man of exalted mind and true sensibility could retain his reason or find patience to live! I can only reiterate the precept of Goethe, as phrased by Arnold: "Art still has hope—take refuge there!"[31]

I heard, by letter from Charles Dunphie, of the death of your sister Asia some time after it occurred. Clarke had already left England with her remains. This sad event must have affected you deeply and strangely—with, I dare say, a new realization that the spiritual world is a more real world than this one; certainly that this one is most evanescent and shadowy. I hear nothing of the nature of her illness or the cause of her death. It seemed shockingly sudden. I deeply sympathize with you and with her husband in the sorrow that has thus come upon you. It comes to all. People have only to stay *long enough* in this world to make *sure* that its burdens will be laid upon their hearts.

Asia must have been a woman of very strong character and of fine intellect, if I may judge from what I have seen of her writings. You told me once that she had written poetry. A little book of her literary remains might perhaps afford you a solace, and a congenial, if sad, occupation for a while. I wish that you would give my love to Clarke. I answered, to the Gilsey House, a very kind letter received from him in New York last spring, but have never heard from him or seen him since. That was just before we dedicated the monument at Evergreen. Should he return to England before I leave it I trust to meet him again. Mr. Dunphie proposed a dinner, but we found that Clarke had gone.

I shall convey your message of kind remembrance to Mrs. Griffin, Miss Anderson and Edwin. I have seen them only two or three times and had but little talk with them. They live at Hampstead, a long distance from here, and I am rather more than ever reclusive and silent. I have devoted most of the time to

[31] "The end is everywhere/Art still has truth, take refuge there!" Matthew Arnold, "Memorial Verses" (1850).

prowling around London and its suburbs with Willy, to show him the sights that he ought to see. We have been to Kew, Richmond, Bushes Park, Hampton Court, Strawberry Hill, Twickenham, the Crystal Palace, the Tower, the Abbey, the Temple, many churches in the city, many theatres, up and down the Thames in steamboats; up and down the city in busses and on foot; the National Gallery, the Kensington Museum, the several exhibitions, Hampstead Heath, Finchley (with Fred Burgess) Barnet, Henley, Totteridge, &c., &c.,—Stoke Poges and the Churchyard of the Elegy, Laleham; in fact, almost everywhere. It will be a great education for the boy, and it has been a great help to me—in absorbing my thoughts in new channels, away from personal grief, at least for a time.

We sailed from New York on April 21, arrived in London at midnight of April 29, and it will be two months next Thursday since we left home. For my part, I should like to return to this country for a year, and devote myself to writing in a thoughtful way, and this I certainly should do if I were able to make any practicable arrangement to that end. There is a magnificent field, especially in Scotland, for the kind of work that I have, in time past, done well; and I might perhaps do as well again. Meanwhile the unsettled feeling and the anxiety distract me somewhat, and defeat my industry. Still I manage to do a little every day.

I have been to the Lyceum Theatre four times; the Gaiety twice, the Haymarket, the Vaudeville, the Empire, Toole's—we went to Oxford with Toole and saw him play The Don[32] there—the Adelphia, and one or two other places. But I am weary of playgoing: I feel that I have done my work in that direction; and I dread returning to that wearisome routing and most afflicting field of strife.

I observe what you say about The Players Club, and I hear, through the papers, that you have given a most magnificent property to them. Irving spoke of this to me with the greatest sympathy and admiration. No man, at home or abroad, let me say here, since the fact suggests itself now, could show a feeling of deeper kindness, respect, comradeship and sweet appreciation than is evident in every word that Irving speaks whenever your name

[32] Toole first appeared as Mr. Milliken, M.A. in Herman Merivale's *The Don* on March 7, 1888, at the King William Street Theatre.

is mentioned. He will be sincerely glad to receive your message of regard, and so will Toole.

Reverting to The Players: Various intimations, not distinct, have been conveyed to me in a particularly kind way by Daly, Palmer and Lawrence Barrett, as to my being in some way associated with the club. "Dark hints to seize my person, in this palace. His Highness trembled while he spoke!"[33] I don't know what it all means—but I should like to know; because if there is anything that I could do to serve the institution the subject ought to be taken at once into my consideration and thought of carefully. My feeling is that the time has fully come for a radical change in my literary life. That is why I so much wish it were possible for me to remain here, at least for a time—so that I might get into other grooves of artistic labor and quit the drudgery of night work and uncongenial criticism of evanescent and usually trivial dramatic performances.

I shall send this letter to the care of Benedict, as I know not where else it will certainly find you. I am at present at a house owned by Irving in the Suburbs, and otherwise unoccupied. He himself lives at his lodgings, in Bond Street, Grafton Street. I purpose going to Stratford-upon-Avon within a few days; and afterward to York and Edinburgh, if I can manage it. You will have received before this a paper that I lately sent to you containing my speech at the Green Room Club dinner on June 3. It went very well, and seems to have attracted some attention.

The news that I have from home is not bad news, and that in itself is a comfort. The place is safe and in good order. The children are well. Miss Campbell[34] remains about the same. My wife is in better health, though completely alienated from any interest in this world. Percy is at home. The place of Artie's rest has been made beautiful with roses. I wish you could see it. Write to me whenever you can; it is always a comfort to get a letter from you. Give my dear love to Lawrence and his wife and children, and to Edwina. I saw Gertrude and Joe[35] just before they started for America. Your namesake, little Edwin, is a very pretty child, and very interesting, I have passed a week at the Hotel Victoria, and a week with Fred Burgess at Finchley.

[33] The source of this quotation is obscure.
[34] Presumably a sister of Lizzie Winter.
[35] Joseph Anderson and Gertrude Barrett.

There is nothing of essentially great value on the London stage at present. Ellen Terry is beautiful and touching in The Amber Heart, which is a poor play, saved by her. Irving gives a strong farcical twist to Robert Macaire.[36] Tree is picturesque but unsympathetic as Narcisse.[37] Tom Thorne is good, but not good enough, as Parson Adams.[38] Toole is funny in The Don. Sweet Lavender, which I have not yet seen, is, I hear, the best play now afloat here.[39] There is a raft of indifferent actors visible. Irving will entertain the Daly Company on July 8;[40] and Tom Thorne has asked me to a garden party on July 13.

The enclosed leaf I took from the wall of Laleham Church tower, close to the spot where Matthew Arnold is buried. God bless you, dear Edwin.

Ever faithfully your affectionate friend,

William Winter

[P.S.] The weather here is so cold that we must have fires almost all the time, and today is as dark as November.

Shortly after the founding of The Players, Booth suggested that Winter become the club's librarian and historian at an annual salary of $2000. If Winter accepted the position, however, he would have to resign as the Tribune's dramatic critic, since The Players excluded practising critics from their membership. After consideration, Winter declined the offer. In a letter to A. M. Palmer, dated January 15, 1889, now in the New York Public Library Theatre Collection, he notified the Board of Directors of his decision:

[36] In May, 1888 Irving presented at the Lyceum revivals of A. C. Calmour's *The Amber Heart* with Ellen Terry as Ellaline and his own adaptation of Frederic Lemaitre's *Robert Macaire* in which he appeared in the title role.

[37] Sir Herbert Beerbohm Tree, noted English actor-manager, appeared in the spring of 1888 at the Haymarket Theatre as Narcisse in *Pompadour*, an adaptation by W. G. Wills and Sydney Grundy of A. E. Brachvogel's *Narcisse*.

[38] Thomas Thorne, English actor and theatre manager, was for years the manager of the Vaudeville Theatre in London.

[39] Arthur Wing Pinero's farce *Sweet Lavender* was first produced in London in the spring of 1888.

[40] Daly's company made their third trip to London and the continent in 1888. The "hit" of this tour was *The Taming of the Shrew* starring Ada Rehan as Kate and John Drew as Petruchio.

I have decided, after thoughtful consideration of the subject, in all its bearing, that it is my duty to decline the proposal that I shd. become Librarian & Historian for the Players. Will you be so kind as to communicate the decision of mine to the Directors of the Club. I have apprised Mr. Booth & Mr. Barrett. My chief reason for declining this offer is my conviction that I can be of greater service to the stage, to my own art, and to society by remaining as I am—absolutely free. There are other reasons, but it is not material to state them. I need not answer you, because you know already, that the Club has my respect & sympathy, & cordial good wishes: and therewithal I remain.

From Philadelphia, where he and Barrett were filling a three-week engagement, Booth wrote Winter to express his regrets at this decision.

121. The Stratford, Philadelphia, February 24 [1889]

I sent you the copy of 'Richard'—with a hurried line or two, from Boston—I think. The 'Bohemian Club', of Frisco,[41] want a set of the Prompt Books (wh. I have ordered & will have bound) for their library & they wish your, Barrett's & my autographs inserted; when I get the books & we meet in New York, I'll ask you & Barrett to grant their request. Pure weariness (some laziness, mayhap) is the sole cause of my neglect of correspond-ence—I have a dozen unanswered letters (some three months old) before me every day but all efforts at writing are vain, & I put it off 'till "tomorrow, and tomorrow, and tomorrow" with the et ceteras that follow. I am positively fagged out, but not by work, which has been comparatively light the past two seasons; I think it is the result of some liver derangement, and yet—with the exception of this chronic fatigue—my health seems better than ever it was. I deeply regret your decision regarding the Club; so does Barrett, & I know that "many of our Players do," but, of course, you know better than we what may be involved by any change of your position.—All I heard of the Langtry-Booth 'combine' was a sort of second hand suggestion that she & I should act *Macbeth* (not Henry 8th) sometime next season; nothing

[41] The Bohemian Club of San Francisco was founded in 1872 as a social club for journalists, writers, and businessmen. Winter was a member of the club.

more. I'd rather not.[42] It is pretty well settled that Modjeska will act with me & that Barrett & I will branch off from each other 'till near the close of our season—he with a new play & I with several that I have omitted the past two seasons.[43] The Club house cost went $20,000 beyond the estimated sum aside from the art additions to the gift,[44] but I will be amply repaid if the supreme purpose be accomplished. The house requires one or two more rooms wh. I presume I shall have constructed. Business here is very fine & the prospect is good. Hope you & yours are well. Affectionately

Booth's letters to Winter became less frequent near the end of the 1880's. From The Players where he had established quarters in rooms on the third floor, he wrote Winter the following letter concerning his "Memoirs" on which Winter had been working for the past several years.

122.   The Players, July 10, 1889

Believe my neglect is due entirely to a sort of spiritual and physical flabbiness—solely that and nothing else; I seem to have no energy or strength sufficient to scribble half a dozen words or to stir from the house for exercise: I am saturated with fatigue, as you are; *tired* as a wheel—all round.

[42] Apparently there were rumors that Lillie Langtry and Booth would act together in *Macbeth* during the 1889-90 season. They did not ever act together.

[43] Helena Modjeska and Booth made a national starring tour together in the 1889-90 season. It began in Pittsburgh on September 30 and ended in Buffalo on May 10. Modjeska played Ophelia, Lady Macbeth, Portia, Julie de Mortemar, and Beatrice. The "new plays" that Booth introduced were *Much Ado About Nothing* and *Don Caesar de Bazan*. Modjeska also appeared in either *Maria Stuart* or Westland Marston's comedy *Dona Diana* in a double bill on the nights Booth played in either *Don Caesar* or *The Fool's Revenge*, in which plays she did not appear. Although the tour was under Barrett's management, he did not tour with Booth and Modjeska nor did he join them at the close of the season. He spent most of the 1889-90 season in Europe trying to cure himself of a peculiar swelling of the face. For a chronology of the Booth/Modjeska tour see *Life and Art*, pp. 291-292.

[44] Booth had originally estimated that The Players club house might cost $150,000 but the house at 16 Gramercy Park when finally purchased, redecorated by the noted architect Stanford White, and furnished cost over $200,000. The entire sum was paid by Booth.

I hoped to see you here before you sailed—you sent word you would call—and therefore I did not write. The book[45] to S. F. came here promptly & I recd. the memoir also, I want to go over it with you, it is at Edwina's house Narragansett, where I shall be next week: I will again endeavor to annotate it, tried once in vain, but could do better in a chat with you. My health seems to be as good as usual, but I feel no interest in anything whatever— consequently the memoir is as tedious as a twice-told tale,[46] but I will try it again next week.

I have to answer a letter from Mr. Partington,[47] recd. a month ago, but the great drain upon me lately & the increased number of demands on my purse, prevent me from doing what I'd like to do for the Library—which, I'm glad to know has been successful. I'll write Mr. P. in a day or two. I hope you will obtain rest and return full of vigor. Barrett wrote me from Southampton— disgusted with all things & folk on the German steamer; a letter from Hutton, full of jollity also has come—my love to him & to Osgood when you meet. Aldrich sailed without my knowledge of his going—till I got his note of "ta-ta"; wish I were with you all— If I could stand the 'racket.' I've been with Benedict on his yacht for a few days, but declined a cruise with the squadron: I prefer the quiet outlook from these windows & find my rooms quite as cool as the cottage at Narragansett or the house at Greenwich—I shall pass most of my vacation here.

Let me hear from you soon & as often as you can but don't be annoyed if I do not respond promptly; be sure that I always want to do so. God bless you

During his engagement at the Chicago Opera House with Helena Modjeska from March 10 to March 29, 1890, Booth had the opportunity to inspect the new Auditorium Theatre. It had been completed

[45] The "book" is probably one volume of the three which comprise Winter's *Brief Chronicles* published in March and April, 1889, by The Dunlap Society in New York. Booth had earlier received one part of the *Chronicles* in San Francisco (Folger Shakespeare Library, Y.c. 215 [503]). Winter had probably sent another volume there also, but Booth left the city before he received it. Winter seems to have sent him another copy of the *Chronicles* eight months later (see next letter).

[46] *King John*, III,iv,108: "Life is as tedious as a twice told tale."

[47] Frederick E. Partington was the principal of the Staten Island Academy.

in December, 1889, having been built over a three-year period by the Chicago firm of Adler and Sullivan. Dankmar Adler was primarily responsible for the engineering of the building which resulted in an acoustically perfect auditorium, and Louis Sullivan was responsible for the overall architectural design and decoration. It was the largest permanent theatre in America and one of the largest in the world, with a maximum seating capacity of 4,237. The seating capacity could, however, be reduced to 2,574 by the lowering of huge panels which closed off the first and second galleries and the rear third of the balcony. The stage was also one of the largest in the world, measuring 62 feet from the curtain line to back wall and 98 feet across with the side walls 110 feet apart. The size of the proscenium opening could be reduced, like the auditorium, by means of a huge false proscenium which operated automatically. The stage was equipped with numerous mechanisms among which were 26 hydraulically operated sections which could be variously elevated or depressed as much as 18 feet and a panoramic horizon, an endless canvas roll which ran on a steel-linked belt and track and carried around the three sides of the stage, on which was painted in various sections the sky of every season of the year in every weather condition. Booth thought it was a "perfect" theatre.

123. Chicago, March 17, 1890

Thanks for the "Chronicles"—as I am a member of the Dunlap I presume I shall receive another copy.

I have nothing to write about—all goes well, so far as health & business are concerned. Utter collapse during the day—but I brighten up at night & get through my work with (for me) great vigor. Cheery news from Barrett—at Pau.[48]

A letter (two in fact) from Miss Robbins [sic][49] asking my consent to publish some account she has written regarding a tour with me as a member of my company. She said she had sent the sketch for my approval, &c. It never reached me. A second letter from her stated she had sent you a copy of it. Do you know of it? I have lost her address—can you enlighten me?

[48] Barrett had become ill in the fall of 1889 and had gone to Germany to try and regain his health at some of the noted spas.

[49] Elizabeth Robins, American actress and dramatic author, was a member of the Boston Museum Company which supported Booth during his 1884-85 tour. She appeared as Marion de Lorme, Jessica, Casilda, Blanche in *The Iron Chest*, and Margaret in *A New Way to Pay Old Debts*. She later became a successful supporting actress. I have been unable to find her article, if it was ever published.

But for the smoke the sky would be blue, as it is one's spirit is instead.

The "Auditorium" here—a grand affair comprising Hotel, Theatre, shops, offices, &c, &c, on a large scale, is doubtless the grandest & most complete affair of the kind in the world. I saw the model of the *theatre* part of it when I was in Vienna. The Opera House at Pesth was then in course of construction on the plan I saw, & this one is built by the same party, I believe, with improvements. The mechanical contrivances for all kinds of stage effects are wonderful & the scenery, painted in Vienna, is superb. The stage, and the auditorium too, can be compressed—made smaller for dramatic performances, enlarged for Opera. The electrical storm effect, which I had for Lear in Leipsic, and every conceivable trick for theatrical illusion is here in perfection.[50] Everything about the place, dressing-rooms, store & property-rooms, all conveniences—superb! It is worth a trip here to see it. New York should have such a house. *Booth's* in all its glory, sinks into insignificance beside this perfect theatre. Hope all your folk are well & that Mrs. Winter & the children are home again. My love to all. Edwina is in Florida still; her chicks have just recovered from whooping-cough. I send you a bit of my travels; isn't it killing? No more of such!

In the spring of 1890, Booth restored the tomb of the actor George Frederick Cooke in the cemetery of St. Paul's Chapel, New York. On May 31, Booth and Winter made a formal visit to the cemetery. Probably shortly after that visit, and prior to Winter's departure to England for a seventh visit on June 10, Booth wrote him the following note.

[50] During the construction of the Auditorium, Adler went to Europe to study the latest improvements in stage design, especially in the opera houses at Halle, Prague, and Budapest. All of these used moveable stage floors and the houses at Halle and Budapest had panoramic horizons. These mechanisms, and those for the Auditorium, were designed by the *Asphalia Gesellschaft* in Vienna. Adler and Sullivan did not design any European theatres, nor could Booth have seen a model of the Auditorium since the idea for the building did not originate until 1886. The Auditorium Theatre was closed in 1941 but it was restored to its original 1889 condition between 1960 and 1967. See Hugh Morrison, *Louis Sullivan: Prophet of Modern Architecture* (New York, 1935), pp. 85-109; also Montgomery Schuyler, "A Critique of the Works of Adler and Sullivan," *The Architectural Record* (December, 1895), pp. 4-25.

124. The Players, Tuesday

3 notes from you. Sorry & glad you're going—would I could
[go] with thee! I am down-a-down-a; weary unto death, well-nigh.
Can't shake off my *tire* tho' I be a broken wheel. "O, how the
wheel becomes it!"[51]—I mean that tired pun.[52]

Thanks for the precious girdle. Did the Cooke article appear?
In Harper's? & when? Send word when you'll come & I'll await
you. I rarely go out—only to the 'Swedish cure' & not every day:
that's about all the exercise I can endure. Premature antiquity,
"with stealing step, hath clawed me in his clutch,"[53] and I am older
in aches than age.—I hope your boy will recuperate & lie
fallow for awhile; hold him back: a season of loaf will do his
brain more good than book-study can. Educate his body for
awhile & let his head rest.[54] Do you know aught of Miss Marlow's
[sic] ability?[55] Now that Mary has left us she seems to be the only
hope we have—I mean from what I hear of her. I advised my
nephew, Creston Clarke, to accept an offer to support her & the
papers state that he has done so. I hope the best for both.
Edwina—but I've told you already that she has rented her Boston
house & will reside in New York. I miss her very much of late,
and (strange it seems) I miss my mother, as I go nearer to her,
more than I ever did. (Just here I was interrupted by a notice for
the S. Island Academy, of a meeting of stockholders—odd coincidence!)—Dr. Smith does not approve of the *pill* for my case: he
is *pilling* me for it.—Should you see Irving, the Kendals,[56] Walter

---

[51] *Hamlet*, IV,v,172.

[52] One paragraph has been inked out here.

[53] *Hamlet*, V,i,80-81: "Age with his stealing steps,/ Hath claw'd me in
his clutch."

[54] Winter's son Louis, at this time a student at Brown University,
suffered from tuberculosis, from which disease he died in 1905. Louis
accompanied Winter to England in the summer of 1890 and again in
1891.

[55] Julia Marlowe, American actress born in England, after a few years
of touring and a period of intense study, appeared at the Bijou Theatre in
New York on October 20, 1887, as Parthenia in *Ingomar*. She was immediately successful and began a long career as a leading actress, excelling
in such roles as Rosalind, Galatea, Beatrice, Imogen, and Pauline. In 1904
Miss Marlowe and Edward H. Sothern formed a combination and successfully toured the United States and England for the next ten years.

[56] The English acting team of William Hunter Kendal and his wife

Pollock & others who may chance to remember me, give 'em good greetings on my behalf. Toole, I presume, is in Australia. I had a letter from Edwards[57] & recd. the paper you sent me from him. He'll soon be here &, I hope, will stay—for the Club's sake. Try to stir up the Shakespearians to buy the Hathaway & Mary Arden cottages—or I'll bribe Barnum to secure them for the 'World's Fair'; 'twas he saved the Henley House & New Place, by his offer to purchase the former. God bless you! Come, if you can, & loaf a day in my den.

On November 3, 1890, Booth and Barrett began their last tour together at Albaugh's Theatre in Baltimore. During their two-week engagement there, Booth's fifty-seventh birthday was celebrated by the citizens of his birthplace, Bel Air, Maryland. A portrait of him by Louis Dietrich, a Baltimore artist, was hung in the Bel Air court house and he was given a testimonial banquet in Baltimore at which he was presented with a painting on a panel of wood made from a cherry tree which once grew on his father's farm. According to Winter, *Life and Art*, p. 297, the painting showed a cherry branch, garnished with cherries and two birds flying near it.

After the engagement in Baltimore, Booth and Barrett filled an engagement of two weeks in Philadelphia and then on December 1 opened for another engagement of two weeks at the Park Theatre, Boston.

125.   Boston, December 3, 1890

Thanks, dear Will, for yr. kind birthday remembrance. I've been tardy in saying so much—but I've felt it all along since it's receipt.

'Tis rather hard for me to write, to do anything, in fact, that requires half a thought or the least exertion.

I recd. about the same time with the Arnold your Dunlap— Gilbert—both gratified me much, but was the Gilbert bust placed in the Player's house? I do not remember it there, but saw it at

---

Dame Madge Robertson Kendal managed the St. James's Theatre in London from 1879 to 1891.

[57] Presumably Henry "Harry" Edwards, American comic actor born in Wales, seems to have made his first American appearances in California between 1849 and 1853, where Booth probably first made his acquaintance. In 1888 he made a second trip to Australia, returning to the United States in the summer of 1890.

the artist's studio & have (or had) a photograph of it.[58] The Balto. & Bel-Air papers had accounts of my birthday & the portrait I sent to the latter place to be hung among my legal townsmen in the Court House—which stands on the site of the old one, in which Clarke & I gave our Shakespearian & Ethiopian entertainment years agone! Several of the accounts were cut from those papers for you, but my laziness o'erlooked them & they were lost.

We've had good old fashioned Winter these three days past: 'tis bitter cold, and the business is correspondingly frosty. J. J. & Billy [Florence] closed last week at the same theatre & with similar patronage—can it be that the *legitimate* is fading in classic 'bean-town'! For two weeks in Phila. recently Shakespeare & the standard plays were rampant at three theatres—here the *Soudan* triumphs.[59] I am at the Thorndike opposite the public garden where the boys & girls are skating on the lake; 'tis a pleasing sight, but the room is cold, the chimney smokes & the radiator don't radiate worth a dern, I have all the gas-jets ablaze but they barely take off the chill. I am gradually gaining vigor & my tongue doesn't 'lapse' so often in the text as it has been doing all this season.[60] Adieu! old boy. God bless you.

No letters of later dates have been preserved, but there was little need for either Booth or Winter to write. After his last performance on April 4, 1891, at the Brooklyn Academy of Music, Booth retired to his rooms at The Players. Except for occasional visits to his friends such as Benedict and Jefferson and to his daughter during the sum-

---

[58] According to an entry in Winter's journal dated November 26, the "Arnold" was a book of poems by Matthew Arnold. Winter's *A Sketch of the Life of John Gilbert* was published by the Dunlap Society, New York, in November, 1890. Gilbert, popular American comic actor excelling in such roles as Sir Peter Teazle, Sir Anthony Absolute, Justice Greedy, and Dominie Sampson, was a close friend of both Booth and Winter.

[59] *The Soudan* was a spectacular melodrama by Henry Pettitt and Augustus Harris. It was first presented at the Boston Theatre on September 16, 1890, and ran until January 10, 1891—seventeen weeks. It was revived on April 20 and ran four weeks more. Booth and Barrett were originally booked for the Boston Theatre, but their contract was cancelled because of the successful run of *The Soudan*. They were paid $1500 and opened at the Park Theatre instead. See Tompkins, pp. 377-381.

[60] Booth had a slight stroke on April 3, 1889, during an engagement in Rochester, which caused him to lose his voice for several days. He seems never to have completely recovered his former vigor.

mer, he remained there until his death. Since Winter and he were almost always in New York, they could meet at any time. Winter's journals record their frequent visits, during which they would spend hours, often entire days, in conversation. Usually the entries are terse, such as the following of October 30, 1891: "Passed the afternoon & evening with Edwin Booth at the Players Club.—We passed nine hours in talk had luncheon and supper." Occasionally an entry is more detailed:

June 27, 1892: Passed the afternoon with Edwin Booth at the Players Club. He is in better health, tho' feeble. We talked of many things. He spoke much of his father, & of family matters. Did not care to read old plays but liked to read of the old dramatists &c. Wondered what could have been the claim of Garrick's performance of Abel Drugger. . . . Edwin spoke of acting. Had never cared much for it: lately not at all. Probably wd. never act again. Had much enjoyed acting with Irving in London, & all thro the season in Germany. We indulged in reminiscences of Barrett &c, & spoke of Clarke, & other absent friends. . . . Edwin said, speaking of himself, "I was always of a boyish spirit and if my physical health were good I should still be very boyish. But there was always an air of melancholy about me that made me seem much more serious than I ever really was."—Spoke of his enjoyment of Paris, when he first went there, under the Empire, in company with Boughton.— Said that Lawrence Barrett & he attended the funeral of E. P. Whipple & described the contortions of Dr. Bartol. Said it was intolerable, and that he quickly remarked to Barrett "I knew Mr. Whipple slightly, but I never thought I should be so sorry for his death." And both left the church, in laughter.

Winter's last visit to Booth was on February 11, 1893. They had a "long talk" and Booth told him some anecdotes about his Sandwich Island experiences and autographed some photographs of himself. Shortly after this visit Winter left New York for a brief vacation at a home he had recently bought in Mentone, California. Here on June 7 he received the news of Booth's death. In his journal, Winter recorded the date and place of death and then surrounded it with a border of thick black lines. Almost three months later, on August 25, Winter recorded that he had finished the *Life and Art of Edwin Booth* on which he had worked, consciously and unconsciously, for thirty-five years.

BIOGRAPHICAL NOTE
SELECTED BIBLIOGRAPHY
INDEX

BIOGRAPHICAL NOTE

Abbey, Henry E. 1846-1896
Adams, Edwin 1834-1877
Aldrich, Thomas Bailey 1836-1907
Alfieri, Vittorio 1749-1802
Anderson, David C. 1813-1884
Anderson, Mary 1859-1940
Aram, Eugene 1704-1759
Arnold, George 1834-1865

Baker, Benjamin A. 1818-1890
Bancroft, Marie Wilton 1839-1921
Bancroft, Sir Squire 1841-1926
Bangs, Frank C. 1833-1908
Barnay, Ludwig 1842-1924
Barrett, Lawrence 1838-1891
Barrett, Wilson 1846-1904
Barry, Thomas 1798-1876
Bateman, Kate 1842-1917
Baumeister, Bernhard 1828-1917
Bell, Clarke 1832-1918
Benedict, Elias Cornelius 1834-1907
Bernhardt, Sarah 1844-1923
Booth, Barton 1681-1733
Booth, Junius Brutus, Jr. 1821-1883
Booth, Sidney Barton 1873-1937
Booth, William Agur 1805-1895
Boucicault, Dion 1822-1890
Bowers, Mrs. D. P. 1830-1895
Brooke, Gustavus Vaughan 1818-1866
Brougham, John 1810-1880
Browne, Junius Henri 1833-1902
Browning, Robert 1812-1889
Buckstone, John Baldwin 1802-1879
Bull, Ole Bornemann 1810-1880
Bulwer-Lytton, Edward George 1803-1873
Bulwer-Lytton, Edward Robert 1831-1891
Burdett-Coutts, Baroness Angela Georgina 1814-1906
Butler, Richard 1844-1928
Byrnes, Charles Alfred 1848-1909
Byron, Henry James 1834-1884

Calvert, Mrs. Charles 1837-1921
Clapp, Austin 1841-1904
Clarke, Creston 1865-1910
Clarke, John Sleeper 1833-1899
Clarke, Wilfred Booth 1867-1945
Collins, William Wilkie 1824-1889
Collyer, Robert 1823-1912
Conrad, Robert Taylor 1810-1858
Cook, Edward Dutton 1829-1883
Coquelin, Constant-Benoît 1841-1909

Couldock, Charles W. 1815-1898
Crabtree, Charlotte "Lotta" 1847-1924
Crane, William H. 1845-1928
Croizette, Sophie Alexandrine 1847-1901
Croly, George 1780-1860
Cumberland, Richard 1732-1811
Curtis, George William 1824-1892

Daly, Augustin 1838-1899
Dana, Charles Anderson 1819-1897
Dana, Richard Henry, Sr. 1787-1879
Dawison, Bogumil 1818-1872
De Quincey, Thomas 1785-1859
deVigny, Alfred Victor Comte 1797-1863
Devrient, Emil 1803-1872
Devrient, Ludwig 1784-1832
Downman, Hugh 1740-1809
Duff, James C. 1855-1928
Duff, John 1787-1831
Duff, Mary Ann 1794-1857
Dumas père, Alexandre 1802-1870
Dunphie, Charles James 1820-1908

Edwards, Henry "Harry" 1824-1891
Eldridge, Louisa 1829-1905
Eytinge, Rose 1835-1911

Falconer, Edmund 1814-1879
Farjeon, Benjamin Leopold 1838-1903
Farjeon, Eleanor 1881-1965
Faucit, Helena 1817-1898
Fawcett, Owen 1839-1904
Fechter, Charles Albert 1824-1879
Field, Richard Montgomery 1832-1902
Fiske, Stephen Ryder 1840-1916
Flohr, William Henry 1836-1901
Florence, William Jermyn "Billy" 1831-1891
Flower, Charles Edward 1829-1892
Floyd, William A. "Billy" 1832-1880
Ford, John Thompson 1829-1894
Formes, Carl 1816-1889
Formes, Ernst 1841-1898
Forrester, Henry 1827-1885
Fox, Caroline 1819-1871
Fox, George Washington Lafayette 1825-1877
Fox, Robert Were 1789-1877
Froude, James Anthony 1818-1894
Furness, Horace Howard 1833-1912

Gautier, Theophile 1811-1872
Gilbert, John 1810-1889
Gill, William Fearing 1844-1917
Gôt, Edmond François Jules 1822-1901
Gould, Thomas Ridgeway 1818-1881
Greeley, Horace 1811-1872

Haase, Friedrich 1825-1911
Hackett, James Henry 1800-1871
Hahnemann, Samuel Christian Friedrich 1755-1843
Harcourt, Charles 1838-1880
Harkins, David H. 1836-1902
Hatton, Joseph 1839-1907
Hennessy, William John 1839-1917
Herbé, Charles Auguste 1801-1884
Hering, Constantine 1800-1880
Hill, Barton 1831-1911
Holland, George 1791-1870
Holmes, Oliver Wendell 1809-1894
Hooper, Lucy Hamilton 1835-1893
House, Edward Howard 1836-1901
Howard, Joseph, Jr. 1833-1908
Howe, Julia Ward 1819-1910
Hutton, Lawrence 1843-1904
Huxley, Thomas Henry 1825-1895

James, Louis 1842-1910
Janauschek, Fanny 1830-1904
Jarrett, Henry C. 1827-1903
Jefferson, Charles Burke 1851-1908
Jefferson, Joseph 1828-1905
Jenner, Sir William 1815-1898

Kellogg, Clara Louise 1812-1912
Kendal, Dame Madge Robertson 1848-1935
Kendal, William Hunter 1843-1917
Knerr, Calvin B. 1847-1940
Knight, Joseph 1829-1907

Labouchère, Henrietta Hodson 1841-1910
Lander, Jean Margaret Davenport 1829-1903
Langtry, Lillie 1852-1929
Laube, Heinrich 1806-1884
Leo, Friedrich August 1820-1898
Leyden, Ernst von 1832-1910
Linton, William James 1812-1897
Lucca, Pauline 1841-1908
Ludlow, Fitz-Hugh 1836-1870
Ludlow, Noah Miller 1795-1886

Mackaye, Steele 1842-1894
Mackenzie, Sir Morell 1837-1892

Marlowe, Julia 1867-1950
Martin, Sir Theodore 1816-1909
Maurice, Charles "Cheri" 1805-1896
M'Carthy, Justin 1830-1912
McCollom, James Clark 1838-1883
McCullough, John 1832-1885
McEntee, Jervis 1828-1891
McVicker, Horace 1838-1913
McVicker, James Hubert 1822-1896
Mestayer, Maria Pray 1826-1911
Miller, Wynn E. 1847-1932
Mills, Luther Laflin 1848-1909
Modjeska, Helena 1840-1909
Moody, Dwight Lyman 1837-1899
Moore, Thomas 1779-1852
Morris, Clara 1849-1925

Neilson, Adelaide 1847-1880

O'Brien, Fitz-James 1828-1862
O'Neill, James 1847-1920

Palmer, Albert Marshall 1838-1905
Pateman, Isabella "Bella" 1843-1908
Pateman, Robert 1841-1924
Paulding, Frederick 1859-1937
Perry, Harry 1826-1862
Phelps, Samuel 1804-1878
Piercy, Samuel 1849-1882
Planché, James Robinson 1796-1880
Plympton, Eben 1853-1915
Pollock, Walter Herries 1850-1926
Porter, Horace 1837-1921

Reade, Charles 1814-1884
Reed, Isaac 1762-1804
Reid, Whitelaw 1837-1912
Richardson, Leander Pease 1856-1918
Ristori, Adelaide 1822-1906
Robertson, Agnes 1833-1916
Robins, Elizabeth 1856-1952
Robson, Stuart 1836-1903
Rookes, Agnes Land 1845-1910
Rowe, George Fawcett 1834-1889
Russell, Benjamin F. 1828-1880
Ryder, John 1814-1885

Saintine, Joseph Xavier Boniface 1798-1865
Sankey, Ira D. 1840-1908
Schwab, Frederick A. 1844-1927
Scott, Clement 1841-1904
Sheridan, William E. 1839-1887
Smalley, George Washburn 1833-1895
Smith, Marcus "Mark" 1829-1874
Smith, Solomon Franklin 1801-1869

Sonnenthal, Adolph von 1834-1909
Sothern, Edward Askew 1826-1881
Stark, James 1819-1875
Stedman, Edmund Clarence 1833-1908
Stewart, Alexander T. 1803-1876
Stoker, Abraham "Bram" 1848-1912
Stuart, William 1821-1886
Swing, David 1830-1894

Tag, Casimir 1847-1913
Talma, François-Joseph 1763-1826
Taylor, Bayard 1825-1878
Taylor, Tom 1817-1880
Terry, Ellen 1847-1928
Thompson, Jane Elizabeth 1827-1902
Thompson, Launt 1833-1894
Thorne, Thomas 1841-1918
Tompkins, Eugene 1841-1909
Toole, John Lawrence 1830-1906
Towse, John Ranken 1845-1933
Tree, Sir Herbert Beerbohm 1853-1917
Tyng, Stephen Higginson, Jr. 1839-1898

Vaux, Calvert 1824-1895
Vecellio, Cesare 1521-1601
Vedder, Elihu 1863-1923
Vernon, Ida 1843-1923

Vezin, Hermann 1829-1910
Vincent, Leon John 1833-1925

Wallack, Lester 1820-1888
Waller, David Wilmarth 1823-1881
Waller, Emma 1820-1889
Ward, Genevieve Teresa 1838-1922
Warde, Frederick Barkham 1851-1935
Warner, Charles 1846-1909
Warren, George 1811-1887
Warren, John Byrne 1835-1895
Warren, William 1812-1888
Watterson, Henry 1840-1921
Weaver, Affie 1855-1940
Weaver, Henry A., Sr. 1832-1903
Werder, Karl Friedrich 1806-1893
Wheatley, William 1816-1876
Wheeler, Andrew Carpenter 1835-1903
Wheelock, Joseph, Sr. 1839-1908
Wilbrandt, Adolf 1837-1911
Williams, Barney 1823-1876
Winter, Captain Charles 1800-1878
Winter, Charles, Jr. 1833-1903
Winter, Elizabeth 1840-1922
Witham, Charles 1842-1926
Wolter, Charlotte 1834-1897
Worth, Charles Frederick 1825-1895

SELECTED BIBLIOGRAPHY

SOURCES OF MANUSCRIPTS

The original manuscript of Letter 12 (Winter to Booth, ca. fall, 1872) has not survived. This letter and missing parts of Letter 120 (Winter to Booth, June 18, 1888) were taken from an article by Jefferson Winter, "As I Remember: Glimpses of Old Actors—Edwin Booth," *The Saturday Evening Post* (October 30, 1920), pp. 34 ff. The other letters are taken from original manuscripts in the following collections:

The Folger Shakespeare Library, Washington, D.C.: Letters 1-5, 7-9, 11, 13-45, 47, 49-52, 54-96, 98-117, 120-125.

The Walter Hampden Memorial Library at The Players, New York: Letters 46, 48, 53, 97, 119.

Manuscript Division, The New York Public Library, Astor, Lenox and Tilden Foundations: Letter 118.

Robert Young, Montpelier, Vermont: Letters 6, 10.

BOOKS AND ARTICLES

Allgemeine Deutsche Biographie. 56 vols. Leipzig, 1875-1912.

Appleton's Cyclopedia of American Biography. ed. James Grant Wilson and John Fiske, rev. ed. 7 vols., New York, 1900.

Arnold, Matthew, "The French Play in London," *The Nineteenth Century*, vi (August, 1879), 228-243.

Bartlett, John. *A New and Complete Concordance to the Dramatic Works of Shakespeare*. London and New York, 1894. Globe edition.

Boucicault, Dion. "The Decline of the Drama," *North American Review*, cxxv (September, 1877), 235-245.

Brahm, Otto. *Theater, Dramatiker, Schauspieler*. Berlin, 1961.

Brown, T. Allston. *A History of the New York Stage, From the First Performance in 1732 to 1901*. 3 vols. New York, 1903.

Calvert, Mrs. Charles. *Sixty-Eight Years on the Stage*. London, 1911.

Clapp, Henry Austin. "Edwin Booth in Some Non-Shakespearean Parts," *Outing*, vi (June, 1885), 343-349.

Clarke, Asia Booth. *The Elder and the Younger Booth*. New York, 1882.
————. *Passages, Incidents and Anecdotes in the Life of Junius Brutus Booth, The Elder*. New York, 1866.

Coad, Oral Sumner and Edwin Mims, Jr. *The American Stage*. New Haven, 1929.

Coleman, Marion Moore. *Fair Rosalind: The American Career of Helena Modjeska*. Cheshire, Conn., 1969.

Daly, Joseph Francis. *The Life of Augustin Daly*. New York, 1917.

Dictionary of American Biography. ed. Allen Johnson, 20 vols., New York, 1928-1958.

Dictionary of National Biography. ed. Leslie Stephen and Sidney Lee. 27 vols. London, 1908-1959.

Dramatic Compositions Copyrighted in the United States 1870-1916. 2 vols., Washington, 1918.

Dramatic Notes, a Chronicle of the London Stage: 1879-1882. London, 1883.

Eisenberg, Ludwig. *Grosses Biographisches Lexicon der Deutschen Bühne.* Leipzig, 1903.

Elderkin, John. *A Brief History of The Lotos Club.* New York, 1895.

Enciclopedia dello Spettacolo. ed. Silvio D'Amico. 10 vols., Rome, 1954-1962.

Encyclopaedia Britannica. 24 vols., Chicago and London, 1968.

Farjeon, Eleanor. *Portrait of a Family.* New York, 1936.

Field, Kate. *Charles Albert Fechter.* Vol. VI of *The American Actor Series.* ed. Laurence Hutton. Boston, 1882.

Fielding, Mantle. *Dictionary of American Painters, Sculptors and Engravers with Addendum by James Carr.* New York, 1965.

Freeman, William H. *The Press Club of Chicago: 1880-1894.* Chicago, 1894.

Frenz, Horst. "Edwin Booth in Polyglot Shakespeare Performances," *Germanic Review,* XVIII (December, 1943), 280-285.

Frenzel, Karl. "Die Berliner Theater," *Deutsche Rundschau,* XXXV (April-June, 1883), 466.

Furness, Horace Howard. *A New Variorium Edition of Shakespeare's Hamlet.* Philadelphia, 1877.

————. *A New Variorium Edition of Shakespeare's Macbeth.* Philadelphia, 1878.

————. *A New Variorium Edition of Shakespeare's The Merchant of Venice.* Philadelphia, 1888.

————. *A New Variorium Edition of Shakespeare's Othello.* Philadelphia, 1886.

Gaye, Freda (ed.). *Who's Who in the Theatre.* 14th ed. London, 1967.

Genest, John. *Some Account of the English Stage From 1660 to 1830.* 10 vols. Bath, 1832.

Goodale, Katherine. *Behind the Scenes with Edwin Booth.* Boston and New York, 1931.

Gould, Thomas Ridgeway. *The Tragedian: An Essay on the Histrionic Genius of Junius Brutus Booth.* New York, 1868.

Grossmann, Edwina Booth. *Edwin Booth: Recollections by his daughter and letters to her and to his friends.* New York, 1894.

Grove, Sir George (ed.). *Grove's Dictionary of Music and Musicians.* 5th ed. 10 vols. London, 1954.

Haase, Friedrich. *Was ich erlebte: 1846-1896.* Berlin, 1896.

Hall, Lillian Arvilla. *Catalogue of Dramatic Portraits in the Theatre Collection of the Harvard College Library.* 4 vols. Cambridge, Mass., 1930-1934.

Hart-Davis, Rupert. *The Letters of Oscar Wilde.* London, 1962.

Hartnoll, Phyllis (ed.). *The Oxford Companion to the Theatre.* 3rd ed. London, 1967.

Hennessy, W. J., W. J. Linton, and William Winter. *Edwin Booth in Twelve Dramatic Characters.* Boston, 1871.

Hewitt, Barnard. *Theatre U.S.A.: 1668-1957.* New York, 1959.

Hirsch, August, et al. (eds.). *Biographisches Lexicon der hervorragenden Ärzte aller Zeiten und Völker vor 1880.* 6 vols. Munich and Berlin, 1962.

House, Edward Howard, "Edwin Booth in London," *Century Magazine,* LV (December, 1897), 269-79.

Ireland, Joseph Norton. *Records of the New York Stage, From 1750 to 1860.* 2 vols. New York, 1866-1867.

Irving, Laurence. *Henry Irving, The Actor and His World.* London, 1951.

Johnson, Albert E. and W. H. Crain, Jr. "A Dictionary of American Drama Critics," *Theatre Annual* (1955), pp. 65-89.

Kimmel, Stanley. *The Mad Booths of Maryland.* Indianapolis, 1940.

Kosch, Wilhelm. *Deutsches Literatur-Lexicon.* 4 vols. Bern, 1947.

————. *Deutsches Theater-Lexicon.* 2 vols. Klagenfurt and Vienna, 1953.

Ledger, Edward, (ed.). *The Era Almanack.* London, 1880-1883.

Leo, Friedrich August, "Edwin Booth," *Shakespeare Jahrbuch,* XVIII (1883), 270-272.

Life and Memoirs of William Warren. Boston, 1889.

Lockridge, Richard. *Darling of Misfortune.* New York and London, 1932.

Ludwig, Richard M. "William Winter and the New Drama," *Theatre Research/Recherches Théâtrales,* I (December, 1959), 5-19.

MacDougall, Sally, "Edwin Booth Counted His Ducats," *Century Magazine,* CVII (December, 1928), 198-204.

McVicker, J. H. *The Press, The Pulpit and The Stage.* Chicago, 1883.

Matthews, Brander and Laurence Hutton (eds.). *Actors and Actresses of Great Britain and the United States.* Vol. V, *The Present Time.* New York, 1886.

Modjeska, Helena. *The Memories and Impression of Helena Modjeska.* New York, 1910.

Morrison, Hugh. *Louis Sullivan: Prophet of Modern Architecture.* New York, 1935.

Mumford, Lewis. *The Brown Decades: A Study of the Arts in America, 1865-1895.* 2nd rev. ed. New York, 1955.

The National Cyclopedia of American Biography. 49 vols. New York, 1898-1966.

Newton, Joseph Fort. *David Swing: Poet-Preacher.* Chicago, 1909.

Odell, George C. D. *Annals of the New York Stage.* 15 vols. New York, 1927-1949.

Overmyer, Grace. *America's First Hamlet.* New York, 1957.

Parker, John (ed.). *The Green Room Book, or Who's Who on the Stage.* London, 1908.

Partridge, Edward L. "Edwin Booth to John E. Russell—Some Hitherto Unpublished Letters." *The Outlook,* CXXVII (April 20, 1921). 637-639.

Pascoe, Charles E. (ed.). *The Dramatic List*. London, 1880.

Pollock, Sir Frederick (ed.). *Macready's Reminiscences and Selections From His Diaries and Letters*. New York, 1875.

Pym, Horace N. (ed.). *Memories of Old Friends, Being Extracts From The Journals and Letters of Caroline Fox*. London, 1882.

Ruggles, Eleanor. *Prince of Players*. New York, 1953.

Schuyler, Montgomery. "A Critique of The Works of Adler and Sullivan," *Architectural Record* (December, 1895), pp. 4-25.

Shattuck, Charles H. "Edwin Booth's Hamlet: A New Promptbook," *Harvard Library Bulletin*, xv (January, 1967), 20-48.

————. *The Hamlet of Edwin Booth*. Urbana and London, 1969.

————. *The Shakespeare Promptbooks*. Urbana and London, 1965.

Sichel, Pierre. *The Jersey Lily: The Story of the Fabulous Mrs. Langtry*. Englewood, New Jersey, 1958.

Skinner, Otis. *Footlights and Spotlights*. Indianapolis, 1924.

————. *The Last Tragedian*. New York, 1939.

Sprague, Arthur Colby. *Shakespeare and The Actors*. Cambridge, Mass., 1948.

Thieme, Ulrich and Felix Becker. *Allgemeines Lexicon der Bildenden Künstler*. 36 vols. Leipzig, 1907-1947.

Tompkins, Eugene and Quincy Kilby. *The History of the Boston Theatre: 1854-1901*. Boston and New York, 1908.

Vining, Edward P. *The Mystery of Hamlet: An Attempt to Solve an Old Problem*. Philadelphia, 1881.

Warde, Frederick Barkham. *Fifty Years of Make-Believe*. New York, 1920.

Who Was Who in America, 1607-1896. Chicago, 1963.

Who Was Who in America, 1897-1942. Chicago, 1943.

Wilson, Francis. *Joseph Jefferson: Reminiscences of a Fellow Player*. New York, 1906.

Winter, William. *Brief Chronicles*. New York, 1889.

————. "Edwin Booth," *Harper's Magazine*, lxii (June, 1881), 61-68.

———— (ed.). *The Edwin Booth Prompt Books*. New York, 1877-1878. *Brutus, or The Fall of Tarquin*, July 4, 1878; *Don Caesar de Bazan*, November 2, 1878; *The Fool's Revenge*, October 27, 1878; *Hamlet*, February 7, 1878; *Henry VIII*, November 2, 1878; *Julius Caesar*, sent to the press in 1887 according to Winter, although it had been edited, cut, and arranged for publication much earlier; *Katherine and Petruchio*, October 15, 1878; *King Lear*, March 25, 1878; *Macbeth*, September 27, 1878; *The Merchant of Venice*, October 30, 1878; *Much Ado About Nothing*, October 24, 1878; *Othello*, June 27, 1878; *Richard II*, February 12, 1878; *Richard III*, December 25, 1877; *Richelieu, or The Conspiracy*, 1878; *Ruy Blas*, 1878.

————. *Life and Art of Edwin Booth*. New York, 1894.

————. *Life and Art of Joseph Jefferson*. New York, 1894.

————. *Old Friends*. New York, 1909.

————. *Other Days*. New York, 1908.

Winter, William. *The Poems of William Winter.* New York, 1909.
————. *The Press and The Stage.* New York, 1889.
————. *Shakespeare on The Stage.* 1st ser., New York, 1911; 2nd ser., 1915; 3rd ser., 1916.
————. *Vagrant Memories.* New York, 1915.
————. *The Wallet of Time.* 2 vols. New York, 1913.

NEWSPAPER AND MAGAZINE REVIEWS

The Athenaeum. November 13, 1880.
Bell's Life in London. November 6, 1880.
Berlin *Börsen Zeitung.* January 25, February 4, 1883.
Berlin *Deutscher Reichs-Anzeiger.* January 12, 24, 1883.
Berlin *Norddeutsche Allgemeine Zeitung.* January 12, 1883.
Boston *Advertiser.* March 7, April 31, 1878.
Boston *Evening Transcript.* October 30, 31, 1871.
Boston *Herald.* December 4, 5, 1881.
Chicago *Evening Journal.* February 12, 1874.
Chicago *Herald.* March 28, 30, 1882.
Chicago *Inter-Ocean.* February 12, 1874; March 28, 1882.
Chicago *Record Herald.* November 17, 1917.
Chicago *Tribune.* April 11, 12, 14, 15, 18, 21, 1876; April 16, 24, 25, 29, May 1, 1879; October 16, 30, 1881; March 30, October 22, 1882.
Dublin *Freeman's Journal.* November 9, 10, 17, 1882.
Dublin *Irish Times.* November 7, 1882.
Dublin *Telegraph.* November 8, 9, 1882.
Hamburg *Nachmittags-Ausgabe.* February 20, 1883.
Hamburg *Nachrichten.* February 19, 1883.
London *Daily Chronicle.* November 8, 1880.
London *Daily Telegraph.* November 8, 21, 1880.
London *Globe.* November 9, 1880; May 3, 1881.
London *Morning Advertiser.* November 8, 1880.
London *Morning Post.* November 8, 1880; February 15, May 3, August 4, 1881.
London *Observer.* May 11, August 4, 1881.
London *Standard.* November 21, 1880.
London *Times.* July 29, 1879; November 8, 1880; February 15, 1881; July 1, November 6, 1882.
London *World.* May 11, 1881.
Louisville *Courier-Journal.* March 12, 1876.
New York *Daily News.* February 5, 1869.
New York *Dramatic Mirror.* July 10, 1880; July 9, 1881; May 6, 1882; March 3, 1888.
New York *Evening Post.* October 26, 1875.
New York *Herald.* March 18, 1862; November 9, 1875; August 15, 1876; January 8, 1878; February 20, September 18, November 7, 1880.

New York *Telegram*. January 8, 1878.

New York *Times*. January 27, 1876; February 24, 1878; February 22, 1880; January 2, 1881; December 3, 1882; December 29, 1886.

New York *Tribune*. April 18, December 26, 1871; October 26, November 9, 12, 17, 1875; July 21, August 16, 1876; October 16, 1878; June 30, July 26, August 6, October 8, 1879; February 8, 21, March 27, May 31, June 16, 1880; January 16, December 3, 1881; November 12, December 2, 12, 17, 1882; December 11, 13, 18, 1883; December 6, 1885; July 7, 1886.

"Our Dramatic Stars," an unidentified newspaper clipping ca. 1878 in the Harvard Theatre Collection.

Philadelphia *Inquirer*. February 20, 26, March 6, 1877.

San Francisco *Chronicle*. September 5, 1876.

The Saturday Review. February 5, May 14, 1881.

The Season. January 14, 1871.

Spirit of the Times. February 6, 1869; November 13, 1875.

Theatre Magazine. February, March, December, 1880; March, 1881.

Vienna *Allgemeine Zeitung*. April 8, 18, 1883.

Vienna *Neue Freie Presse*. April 1, 5, 8, 18, 1883.

Vienna *Presse*. April 5, 8, 1883.

UNPUBLISHED MATERIAL

Benson, Richard Lee. "Jarrett and Palmer's 1875 Production of *Julius Caesar*: A Reconstruction." Unpublished doctoral dissertation, University of Illinois, 1968.

Booth, Edwin.

A collection of autographed letters at the Walter Hampden Memorial Library.

A collection of letters and papers concerned with Booth's bankruptcy and loss of Booth's Theatre. The Folger Shakespeare Library. Box T.b.5.

Edwin Booth's letters to Henry L. Hinton. The Folger Shakespeare Library. Y.c. 215, letters 118 to 133.

"Edwin Booth Season 1881-1882 Under the Management of Henry E. Abbey." A scrapbook of press clippings and programs at the Walter Hampden Memorial Library.

Financial records of Booth's German tour in 1883 at the Walter Hampden Memorial Library.

Financial records of Booth's London engagements and Provincial tour in 1880-1881 and 1882 at the Walter Hampden Memorial Library.

A scrapbook of Booth press clippings and programs from 1861 to 1873 at the Walter Hampden Memorial Library.

A scrapbook of press clippings and programs of Booth's London engagements and Provincial tour in 1880, 1881, and 1882 at the Walter Hampden Memorial Library.

A scrapbook of Booth press clippings, programs, and photographs from 1881 to 1883 at the Walter Hampden Memorial Library.

A scrapbook of Booth press clippings and programs from 1888 to 1893 at the Walter Hampden Memorial Library.

Lederer, Emanuel. "Edwin Booth in Germany." An unpublished manuscript essay in the New York Public Library Theatre Collection.

Ludwig, Jay Ferris. "McVicker's Theatre, 1857-1896." Unpublished doctoral dissertation, University of Illinois, 1958.

Ludwig, Richard M. "The Career of William Winter, American Drama Critic: 1836-1917." Unpublished doctoral dissertation, Harvard University, 1950.

McGaw, Charles J. "An Analysis of the Theatrical Criticism of William Winter." Unpublished doctoral dissertation, University of Michigan, 1940.

Rubenstein, Gilbert M. "The Shakespearean Criticism of William Winter: An Analysis." Unpublished doctoral dissertation, Indiana University, 1951.

Sollers, John Ford. "The Theatrical Career of John T. Ford." Unpublished doctoral dissertation, Stanford University, 1962.

Young, Robert. "Frosty, But Kindly." An unpublished typescript biography of William Winter.

INDEX

Abbey, Henry E., manages New Park Theatre, 62n; manages Booth's national tour (1881-82), 192; leases Lyceum Theatre (London), 214; mentioned, 48n, 147, 173, 198, 289

Adams, Edwin, at Booth's Theatre, 31; death of, 80, 85-86; mentioned, 41n

Adler, Dankmar, 298, 299

Albaugh, John W., 122n

Aldrich, Lillian Woodman, 253, 254n

Aldrich, Thomas Bailey, 13, 37, 44, 152, 208, 297

Alfieri, Vittorio, *The First Brutus*, 66, 119n

Ames, Oakes A., 43n, 69, 71, 97

Ames, Oliver, 43n

Anderson, David C., 252, 253n, 256, 259

Anderson, Joseph, 293

Anderson, Mary, at Lyceum Theatre (London), 257-58; mentioned, 80, 215, 260, 261, 266, 267, 276, 283, 289, 290, 291, 300

Aram, Eugene, 210-11

Arnold, George, 13, 48n, 174

Arnold, Matthew, 141, 290, 294

Astor Place Riot, 185

Auditorium Theatre, praised by Booth, 297-99

Badeau, Adam, 8

Baker, Benjamin A., 30, 129

Bancroft, Marie Wilton, 166n

Bancroft, Sir Squire, 166

Bangs, Frank C., 75, 149

Barnay, Ludwig, 238, 241, 242, 289

Barnett, T. J. ("Judge"), 43, 44, 50, 63, 179

Barnum, Phineas Taylor, 301

Barrett, Gertrude, 285, 293

Barrett, Lawrence, manages California Theatre, 57n; produces *Julius Caesar*, 71n, 261n; produces *King Lear*, 75; in London (1882), 208; travels in Holland and Germany with Winter, 209; at Lyceum Theatre (1884), 257-58; produces *A Blot in the 'Scutcheon*, 261n; produces *Francesca da Rimini*, 261n; Booth-Barrett tours, 266-68, 279, 286-87; produces *Rienzi*, 283; travels to Germany for health, 297, 298; mentioned, 4, 15, 41, 60, 200, 210, 214-15, 217-18, 222, 241, 248, 260-61, 281, 285, 286-87, 293, 303

Barrett, Mary Agnes, marries Baron Hermann von Roder, 241n

Barrett, Wilson, 218n, 261

Barry, Thomas, 30, 150

Bartholdi, Frederic Auguste, 263

Bateman, Kate, 24n

Bates, Ernest Sutherland, vii

Baumeister, Bernhard, 243

Becker, Wilhelm Adolf, *Gallus*, 116

Belasco, David, 15, 167, 259

Bell, Clarke, 43, 44, 50, 63, 82, 179

Benedict, Elias Cornelius, 75, 192, 249, 282, 286, 293

Bernhardt, Sarah, 133, 137, 140, 164, 167, 175, 247

Billington, Adeline, 242

Bispham, William, 167, 275, 282

Bock, Frederick, 252, 253

Bohemian Club, 295

Booth, Barton, 223

Booth, Edwin,
    Career, relations with William Stuart, 18-19, 20-21; failure as theatre manager and bankruptcy,

Booth, Edwin (*cont.*)
32-33, 42-46, 67, 69-70, 82, 83n; carriage accident, 50, 52, 59-60; at Fifth Avenue Theatre, 51-52; Southern tour (1876), 54-55; in San Francisco (1876), 72-75; New England tour (1877), 78; at Booth's Theatre (1878), 91-92; assassination attempt, 126-27; relations with Henry Irving, 132, 136, 138, 165, 172, 179, 183, 185, 186-87, 208-09, 223-24, 238-39, 258, 261, 267: travels in Ireland, 155-56; travels in Scotland, 156; travels in England, 156-60; in Paris (1880), 163-64; conflict with McVickers, 183-85, 188-89, 190-94, 199-200, 202, 204, 213-14, 285, 286; in London (1882), 208; travels in Holland, Germany, and Switzerland, 209-10; in English provinces, 209-21; in Edinburgh, 215; in Dublin, 219-21; in Berlin, 226-31; in Vienna, 239-40; in Paris (1883), 246-48; writes biographical essay on his father and Edmund Kean, 268, 269; acts with Tommaso Salvini, 272, 274; in San Francisco (1888), 287-88; founds The Players, 286, 289, 290; acts with Helena Modjeska, 296; last visits with Winter, 303; death, 303
  Personal Characteristics, chronic lethargy and fatigue, 85, 98, 122, 134, 165, 268, 279, 280, 284-85, 287-88, 296, 300; drinking habits, 7, 195, 209-10, 238; dyspepsia, 73, 124, 131, 210; generosity, 10, 19-23, 34, 35, 60, 62, 69, 125, 147n, 154, 283; intellectual limitations, 7-9, 202; interest in homeopathy, 61, 113, 124; interest in spiritualism, 10, 47-48; melancholy, 10-11, 159, 179-80, 196, 202, 211, 246, 249, 262, 270-71, 275-76, 278; personal modesty, 9, 22, 66, 76-78, 113, 264; sense of humor, 11, 26, 36, 38, 76, 78, 108, 113, 123-24, 160, 200; sharp temper, 7, 143-44, 198-99
  Attitudes and Opinions, actors, 9, 46, 57, 72, 105, 109, 113, 138, 149, 173, 254, 255-56, 287; artistic tastes, 8, 31, 49, 78-79; critics, 30, 66, 74, 127, 168-69, 172-73, 179; European acting, 5, 164, 166, 241-42; friendship, 24; his own acting, 6, 74, 79, 90, 98, 104, 108, 114, 119, 122-23; marriage, 253-54n; scenic spectacle, 5-6, 180-81; theatre art, 22, 31-32, 46, 139; theatrical benefits, 9-10, 26-27; theatre managers, 75, 105-06, 176; theatrical puffery, 9, 45, 121-22, 197; Winter's poetry, 31, 48-49, 53, 55, 78-79
  See also Plays and dramatic characters

Booth, Edwina. *See* Grossmann, Edwina Booth
Booth, Harriet Mace, 255n
Booth, Rev. Henry M., 140n
Booth, John Wilkes, 7, 58, 101n, 152n, 198, 246
Booth, Junius Brutus, 52-53, 58-59, 61n, 101n, 152n, 274-75
Booth, Junius Brutus, Jr., 33, 42, 49
Booth, Junius Brutus, III, 288
Booth, Marion, 255n
Booth, Mary Ann Holmes, 190, 257
Booth, Mary ("Mollie") Devlin, 28n, 197, 225, 244
Booth, Mary McVicker Runnion, her record of Booth's stage business, 110; her autograph album, 152; final illness and death, 164, 166, 175, 176, 182,

183-85, 187, 188, 190-94, 195; mentioned, 46, 56, 63, 72, 127-48, 151

Booth, Rose, 274

Booth, Sidney Barton, 288

Booth, William Agur, 140

Booth's Theatre, failure of, 4, 42-43, 67, 85; mentioned, 27n, 60, 146

Boucicault, Agnes Robertson, 153

Boucicault, Dion, feuds with Winter, 86, 88; manages Booth's Theatre, 144, 152-53; marries Louisa Thorndyke, 153n; mentioned, 19n, 122, 145.
    See also Plays and dramatic characters

Bowers, Elizabeth Crocker, 149n

Boyle, Esmeralda, 115n

Brachvogel, Udo, 230

Brady, John R., 29

Brahm, Otto, reviews Booth's Berlin engagement, 227, 228-29

Brathwaite, Richard, 48n

Brooke, Gustavus Vaughan, 115n

Brooklyn Theatre, fire at, 76

Brougham, John, death of, 180; mentioned, 49

Browne, Junius Henri, 197, 247

Browning, Robert, 177

Bruorton, William, 69, 71

Buckstone, John Baldwin, death of, 143

Bull, Ole Bornemann, 8, 151n, 152

Bulwer-Lytton, Edward George, 90, 126

Bulwer-Lytton, Edward Robert, 104

Burbage, Richard, 257

Burdett-Coutts, Baroness Angela Georgina, 140n, 180

Burgess, Frederick, 210, 216, 218, 223, 242, 293

Burns, Charles K., 121

Butler, Richard, 183

Byrnes, Charles Alfred, 127, 198

Byron, George Gordon, Lord, 3, 23

Byron, Henry James, *The Crushed Tragedian*, 148n

Calvert, Adelaide Helen Biddles (Mrs. Charles), 52n, 195

Carr, Thomas Swinburne, *A Manual of Roman Antiquities*, 116

Cassell's Illustrated Shakespeare, 117

Cazauran, Augustus R., criticizes *Richelieu*, 28; mentioned, 29, 47, 127

Century Club, 19, 20

Chase, Arthur Branscomb, 268

Chatterton, F. B., 182, 183n

Chicago Fire, 4, 32

Church, Frederick, 8

Cincinnati Dramatic Festival, 240

Clapp, Henry, Jr., 13

Clapp, Henry Austin, praises Booth's Richard ii, 82n; praises Booth's Sir Giles Overreach and Pescara, 263n

Clare, Ada, death of, 48; mentioned, 13

Clarke, Asia Booth, biography of her father, 161; death of, 291; mentioned, 51, 54, 61, 140, 151, 166

Clarke, Creston, acts with Booth, 211, 212n, 224; his American debut, 288; mentioned, 217, 300

Clarke, John Sleeper, manages theatres in London, 25n, 215, 225; and George Holland benefit, 27, 28; portrait as Zekiel Homespun, 78; negotiates Booth's London engagement, 154; illness of, 166, 217; mentioned, 19n, 25, 26, 80, 81, 91, 97, 121, 134, 145, 151, 162, 163, 164, 172, 176, 178, 201, 248

Clarke, Wilfred Booth, 51, 217, 288

Coleridge, Hartley, 244

Coleridge, Samuel Taylor, 15

Collins, William Wilkie, 16, 202

Collyer, Rev. Robert, 194, 195

Conner, William M., 134

Conrad, Robert Taylor, *Jack Cade*, 96n

Cook, Edward Dutton, reviews Booth-Irving *Othello*, 187

Cooke, George Frederick, 118, 299

Cooke, Jay, 42

Cooper, Thomas Abthorpe, 118

Coquelin, Constant-Benoît, 5, 164

Couldock, Charles W., 284

Crabtree, Charlotte ("Lotta"), at Booth's Theatre, 31-32

Crane, William H., 127

Croizette, Sophie Alexandrine, 5, 140, 164

Croly, George, 116

Cumberland, Richard, *The Sybil, or The Elder Brutus*, 119

Curtis, George William, 270

Cushman, Charlotte, 120

Daly, Augustin, and George Holland benefit, 27; engages Booth for Fifth Avenue Theatre, 51; withdraws from management of Fifth Avenue Theatre, 82; and proposed Booth-Irving exchange, 135, 136; mentioned, 4, 14, 15, 57, 72, 261, 293

Daly, Joseph Francis, 51, 231

Dana, Charles Anderson, 19n, 206

Dana, Richard Henry, Sr., 116, 118

Darwin, Charles, *The Descent of Man*, 8, 200

d'Avrigny, Charles-Joseph Loeillard, Count, *Le médecin des enfants*, 48n

Dawison, Bogumil, acts with Booth, 108-09

Decker, Nelson, 202, 203n

Delavigne, Jean François Casimir, *Louis* XI, 122n

de Vigny, Alfred Victor, Comte, *Cinq Mars*, 102

De Vinne, Theodore S., 95, 107, 117, 130

Devrient, Emil, 237

Devrient, Ludwig, 234, 237

Dickens, Charles, 8, 16, 202

Dickenson, Anna Elizabeth, acts Hamlet, 203

Dietrich, Louis, paints Booth's portrait, 301

Dowden, Edward, *A Critical Study of Shakespeare's Mind and Art*, 206

Downman, Hugh, *Lucius Junius Brutus*, 119

Drenker, Emil, 238

Drew, John, 15

Duff, James C., 82

Duff, John A., 82n

Duff, Mary Ann, 61

Dumas père, Alexander, 66

Duncombe, William, *Lucius Junius Brutus*, 119

Dunphie, Charles James, reviews Booth's London engagements, 168, 170, 181, 187, 208; mentioned, 172, 178, 291

Durant, Dr. Ghislani, treats Booth, 131, 162

Eaton, W. D., 205-06

Eaves, Henry J., 89

Edwards, Henry ("Harry"), 301

Edwards, Maze, 241

Eldridge, Louisa, 252

Eytinge, Rose, 89-90

Falconer, Edmund, 114

Farjeon, Benjamin Leopold, 216n

Farjeon, Eleanor, 216n

Farjeon, Margaret ("Maggie") Jefferson, 216

Faucit, Helena. *See* Martin, Lady Helena Faucit

Fawcett, Owen, 252

Fechter, Charles Albert, manages New Park Theatre, 43, 48; as

Othello, 109; death of, 139; mentioned, 5-6, 26, 180

Field, Richard Montgomery, 96, 267, 271

Fields, James T., 152

Fisher, Amelia, 216, 217, 221

Fisk, Jim, 20

Fiske, Stephen Ryder, 97

Flohr, William Henry, 86, 105, 240

Florence, Malvina Pray, 163

Florence, William Jermyn "Billy," 33, 163, 302

Flower, Charles Edward, 157, 223

Flower, Edgar, 157, 159

Floyd, William A. ("Billy"), death of, 180; mentioned, 47, 150

Folger, Henry Clay, vii

Ford, Charles, 121

Ford, John Thompson, manages Booth's Southern tour (1876), 54, 55, 56, 57, 59; mentioned, 80, 96, 121, 196, 217

Formes, Carl, 233

Formes, Ernst, as Fool in *King Lear*, 233, 236; mentioned, 235

Forrest, Edwin, 19, 120, 132, 185, 247

Forrester, Henry, acts with Booth, 177; mentioned, 178

Fox, Caroline, *Memories of Old Friends*, 8, 244

Fox, George Washington Lafayette, death of, 63

Fremont, General John C., 13

Frenzel, Karl, reviews Booth's Berlin engagement, 226

Frohman, Charles, 259n

Froude, James Anthony, 177

Furness, Horace Howard, 124, 203, 264

Gatti, A., 145

Gatti, G., 145

Gautier, Theophile, 41, 42n

Gebhard, Frederick, 224n

George, Henry, 212-13

Gerard, Florence, 173n

Gifford, Sanford, 8

Gilbert, John, 301-02

Gill, William Fearing, 189

Gillig, Henry F., 215, 238, 285

Goethe, Johann Wolfgang von, 8, 244, 245, 246, 247, 291

Gooch, Walter S., engages Booth for Princess's Theatre, 156-57, 159; mentioned, 145, 154, 161, 162, 167, 175, 177, 179

Goodale, Katherine, 279n

Gôt, Edmond Françoise Jules, 164

Gotthold, J. Newton, 4, 96, 105

Gould, Thomas Ridgeway, *The Tragedian*, 111, 112n

Grand Theatre, praised by Booth, 218

Grant, Ulysses S., 38

Gray, Mark, attempts to assassinate Booth, 127, 135

Greely, Horace, 14, 37-38

Griffin, Edwin Booth, 290, 291, 293

Griffin, Marie Antoinette Anderson, 290, 291

Grossmann, Edwina Booth, engagement to Downing Vaux, 192, 217, 218, 221, 229-30, 248; illness of, 207, 222-23, 258; describes father's reception in Germany, 231-36; marries Ignatius Grossmann, 263; birth of daughter Mildred, 272, 273; mentioned, 61, 72, 96, 101, 176, 197, 203, 204, 206, 225, 280, 282, 284, 285

Grossmann, Ignatius, marries Edwina Booth, 263; Booth's opinion of, 277, 280, 284, 287

Haase, Friedrich, 220, 230

Hackett, James Henry, 148

Hahnemann, Dr. Samuel Christian Friedrich, 124

Hall, A. Oakey, 62n

Hanley, J. G., 27n

Harcourt, Charles, death of, 165-66, 173

Harkins, Daniel H., 121, 129, 131

Hart, Francis, 95

Hatton, Joseph, 176

Hazlitt, William, 15, 185

Hennessy, William John, 61-62, 65, 67, 68, 96, 166

Herbé, Charles Auguste, *Costumes François*, 93

Hering, Dr. Constantine, 124

Hill, Barton, 201

Hillyard, Henry, 71n

Hinton, Henry L., edits Booth's acting versions, 64, 65, 67, 96, 100; Booth's letters to, 65; mentioned, 71n, 109, 120

Holland, George C., benefit for, 25, 26, 27; death of, 27

Hollingshead, John, 145

Holmes, Oliver Wendell, 77, 152

Hooper, Annie, 260n

Hooper, Lucy Hamilton, 260

Hope, Thomas, *Costume of the Ancients*, 116

House, Edward Howard, 14, 174

Howard, Bronson, *The Banker's Daughter*, 28n

Howard, Joseph, Jr., 265, 283

Howe, Julia Ward, 152

Howells, William Dean, 13

Hoyle, Edmund, 111

Hugo, Victor, 115

Hutton, Lawrence, accompanies Booth to Mary's funeral, 195; mentioned, 197, 208, 210, 211, 215, 231, 258, 265, 268, 297

Huxley, Thomas Henry, 177

Ireland, Joseph Norton, *Records of The New York Stage*, 30

Irving, Henry, his acting version of *Richard* III, 80; and proposed Booth-Irving exchange, 132, 136, 143, 144; produces *The Corsican Brothers*, 164-65; advertises production of *Hamlet*, 166; produces *The Cup*, 172, 175-76; produces *Much Ado About Nothing*, 223-24; American tour (1883-84), 250, 251, 258; produces *Robert Macaire*, 294; mentioned, 5, 6, 9, 15, 138, 140, 141, 142, 150, 151, 169, 170, 172, 175, 176, 179, 180, 208, 214, 225, 238-39, 253, 288, 290

Irving, Washington, 157

Jackson, T. B. ("Black"), 19n, 38, 39

James, Henry, 8

James, Louis, 267

Janauschek, Fanny, 26

Jarrett, Edward, 137

Jarrett, Henry C., 49n, 69, 71n, 75, 80, 107, 139, 156

Jefferson, Charles Burke, 284

Jefferson, Joseph, III, and George Holland benefit, 28; as painter, 31, 66; interest in spiritualism, 47-48; in London (1875), 53, 54; death of his son Henry, 54; portrait as Dr. Pangloss, 78; travels in England with Winter, 84, 157; at Booth's Theatre (1877), 88; illustrates *The Trip to England*, 161, 175; writes *Autobiography*, 281; mentioned, 3, 4, 25, 26, 28, 33, 47, 55, 82, 91, 134, 152, 204, 214, 215, 242, 257, 259, 261, 280, 281, 284, 302

Jefferson, William Winter, birth of, 56

Jenner, Sir William, treats Mary McVicker Booth, 184

Jennings, Clara, 81

Jennings, John W., 81n

Johnson, Eastman, 8, 13

Johnson, Samuel, 3, 15, 159

Jones, George ("Count Johannes"), 148-49n

Joyce, Thomas, 36, 120

Karl, Prince (brother of Kaiser Wilhelm Friedrich I), death of, 230

Kean, Charles, 5-6, 64, 71n, 89, 112, 122n, 180, 247, 248n

Kean, Edmund, 112, 115, 118, 132, 150, 185, 269

Kellogg, Clara Louise, 8, 53, 54

Kendal, Dame Madge Robertson, 300-01n

Kendal, William Hunter, 300-01n

Kiliani, Lillian Bayard Taylor, 220

Knerr, Dr. Calvin B., 124

Knight, Charles, *The Pictorial Edition of the Works of Shakespeare*, 95, 99, 106

Knight, Joseph, reviews Booth's London engagement, 169n, 175, 181, 188, 208; mentioned, 15, 172

Labouchère, Henrietta Hodson, 224n

Lander, Jean Margaret Davenport, 38-39

Lane, John A., 252

Langtry, Lillie, 219, 224, 295-96

Laube, Heinrich, 242n

Lederer, Emanuel, 226

Ledger, Edward, 178

Ledger, Frederic, 178n

Lee, Nathaniel, *Lucius Junius Brutus*, 119

Leo, Friedrich August, 231-32n, 242

Leopold, Prince, Duke of Albany (son of Queen Victoria), 177

Leslie, Fred, 214n

Leutze, Emanuel, 8

Lewes, George Henry, *Life and Works of Goethe*, 244-45, 247

Leyden, Dr. Ernst von, treats Downing Vaux, 229

Lincoln, Abraham, assassination of, 7. *See also* Booth, John Wilkes

Linton, William James, 55, 60n

Lockwood, Luke A., 215

Logan, Grace, 191-92, 194

Longfellow, Henry Wadsworth, 8, 12-13, 16, 152, 225, 244

Lowell, James Russell, 151n

Lucca, Pauline, 5, 8, 241-42n

Ludlow, Fitz-Hugh, 13, 116

Ludlow, Noah Miller, 160, 167

Lyceum Acting School, 259, 261n

Lyons, Walter. *See* Gray, Mark

Macauley, Bernard ("Barney"), 57

Macdonald, Alexander E., 76n

MacKaye, Steele, 146, 259n

Mackenzie, Sir Morell, treats Mary McVicker Booth, 166, 184

Macklin, Charles, 257

Macready, William Charles, 3, 52, 53n, 120, 132, 150, 171

Madison Square Theatre, praised by Booth, 146

Magonigle, John Henry, relations with Booth, 28-29; and failure of Booth's Theatre, 67; as actor, 146; mentioned, 29, 35, 37, 50, 82, 117, 120, 192

Mansfield, Richard, 14, 15

Marlowe, Julia, 300

Martial, Marcus Valerius, 8, 135n

Martin, Lady Helena Faucit, 199, 209

Martin, Sir Theodore, 177, 199

Mason, Lyman, 13

Maurice, Charles ("Cheri"), 234

McBride, T. J., 254n

M'Carthy, Justin, 177

McCollom, James Clark, 149n

McCullough, John, manages California Theatre, 57, 58, 72, 75; in Boston, 96; in London, 156, 186; in New York, 204; final illness and death of, 242, 250, 251n, 259; mentioned, 15, 73, 74, 80, 81-82, 93, 94, 107, 111, 116, 119-20, 134, 140, 200, 215, 219, 231, 272

McDonough, Thomas B., 66

McEntee, Catherine Devlin, 28n

McEntee, Jervis, paints portraits of Booth, 106, 113; mentioned, 8, 18, 115, 249

McVicker, Horace, 92, 100, 189, 193, 199, 206, 218, 223, 263

McVicker, James Hubert, loses theatre in Chicago fire, 32; and Booth's bankruptcy, 50, 63, 67, 68, 69-70; defends theatre against clerical attacks, 213; mentioned, 32-33, 43n, 47, 65, 66, 74, 76, 129, 136, 145, 171, 206, 223, 265

McVicker's Theatre, 4, 32

McWatters, Thomas, 90, 92, 94, 95, 100, 264

Methua-Scheller, Marie, 109n

Meyrick, Ellen, 242

Miller, Wynn E., 167n, 211-12, 223, 227, 235, 239, 248-49, 261

Mills, Luther Laflin, 128

Miln, George C., 206, 218, 219n, 221, 231

Modjeska, Helena, in Boston, 96, 97; acts with Booth, 296; mentioned, 14, 15, 40n

Molony, Kitty. *See* Goodale, Katherine

Moody, Dwight Lyman, 53, 54

Moore, Thomas, 61

Moray, John S. *See* Cazauran, Augustus R.

Morris, Clara, 182

Morse, Salmi, *Passion Play*, 167n

Morton, John Madison, *Poor Pillicoddy*, 113n

Moss, Theodore, 162, 284n

Mumford, Lewis, *The Brown Decades*, 11

Murray, Gaston, 68

Neilson, Adelaide, death of, 159; mentioned, 15

Nicholson, Paul F., 29

O'Brien, Fitz-James, 13, 184

O'Neill, Eugene, 167n

O'Neill, James, 167n, 173

Osgood, James R., 60, 68, 88, 297

Palmer, Albert Marshall, 181, 293, 294

Palmer, Henry David, manages Booth's Theatre, 49n; death of, 139; mentioned, 69, 71n, 75, 80, 107

Panic of 1873, 4, 42, 79

Parker, Aurelius D., 13

Parker, Henry G., 198

Partington, Frederick E., 297

Pateman, Isabella ("Bella"), 195, 213-14

Paulding, Frederick, 160

Payne, John Howard, 116, 119. *See also Plays and dramatic characters*

Perry, Harry, 80

Phelps, Samuel, 178

Piercy, Samuel, death of, 201; mentioned, 4, 202

Planché, James Robinson, *The History of British Costumes*, 89, 92, 93

Platt, John H., 69

The Players, 289, 292-93, 295, 296

Plays and dramatic characters,

The Amber Heart, A. C. Calmour, 294

The Apostate, Richard Lalor Shiel (Pescara), Prompt Book edition of, 72; revival of, 261; mentioned, 129

L'Assomoir, Augustin Daly, 129

The Black Crook, Charles Barras, Winter's opinion of, 49n

A Blot in the 'Scutcheon, Robert Browning, 261n

Brutus; or, the Fall of Tarquin, John Howard Payne (Brutus), Prompt Book edition of,

72, 74, 79, 107, 116-17, 122; costume, 111; properties, 116n; sources, 119; mentioned, 11, 268

Commodus, Lew Wallace, 82

Coriolanus, Shakespeare (Coriolanus), 94, 107

The Corsican Brothers, Dion Boucicault, 164-65n

The Crucible, A. Oakey Hall, 62n

The Cup, Alfred Tennyson, 172, 175-76n

Don Caesar de Bazan, Dion Boucicault (Don Caesar), Prompt Book edition of, 72; review of, 208; mentioned, 11, 104, 114

Enoch Arden, Alfred Tennyson, 31n

The Exiles, George Fawcett Rowe, 94n

The Fool's Revenge, Tom Taylor (Bertuccio), Prompt Book edition of, 72, 74, 79, 107; reviews of, 175, 208, 220; mentioned, 11, 286

Forget-Me-Not, Herman Merivale and F. C. Grove, 162

Francesca da Rimini, George Henry Boker, 261n

Frou-Frou, Augustin Daly, 27

Hamlet, Shakespeare (Hamlet), interpretation, 6, 128, 137-38, 200, 203-204, 244-45; reviews of, 25, 51, 73, 79n, 124-25, 167-68, 220, 226-27, 239; Prompt Book edition of, 68, 71, 73, 74, 93n, 109n, 134, 137-38; costume, 92, 98, 120, 233; stage business, 168-69n; in modern dress, 195-96; mentioned, 6, 11-12, 38, 44, 90, 105, 109, 165, 170, 221, 285

Henry VIII, Shakespeare (Wolsey), 104, 108, 112

The Iron Chest, George Colman (Sir Edward Mortimer), revival of, 261

Jack Cade, Robert Taylor Conrad, 96

Julius Caesar, Shakespeare (Brutus), revival of, 33n; review of, 34n; Prompt Book edition of, 71, 73, 112; mentioned, 261n, 268

King Lear, Shakespeare (Lear), reviews of, 54n, 66n, 73, 181, 228, 239-40; Prompt Book edition of, 66, 67-68, 71, 96, 98; costume, 97-98, 120; time of action, 98; properties, 101; stage business, 101, 106, 107n, 252; mentioned, 53, 54, 68, 93

The Lady of Lyons, Edward Bulwer-Lytton (Claude Melnotte), 66n, 72

Leah the Forsaken, Augustin Daly, 24

Louis XI, Dion Boucicault (Louis XI), 122

Love's Penance, Charles Albert Fechter, 48n

Lucie d'Arville, Mary Provost, 38

Macbeth, Shakespeare (Macbeth), Prompt Book edition of, 71, 73, 93n, 117; costume, 98, 120; interpretation, 107, 150; stage business, 117-18; sound effects, 118; mentioned, 105

The Marble Heart, Charles Selby (Raphael), 41

Maternus, Edward Spencer, 82

The Merchant of Venice, Shakespeare (Shylock), Prompt Book edition of, 71, 74, 104, 108, 112, 120; interpretation, 39-40, 256-57; reviews of, 66n, 73, 186; mentioned, 90

Mesalliance, Jean Davenport Lander, 38-39

Much Ado About Nothing, Shakespeare (Benedict), Prompt

Plays and dramatic characters
(*cont.*)

Book edition of, 72; mentioned, 5, 68

A New Way to Pay Old Debts, Philip Massinger (Sir Giles Overreach), Prompt Book edition of, 72; revival of, 261; mentioned, 18, 122

Othello, Shakespeare (Othello, Iago), interpretation, 6-7, 107, 109, 110, 181-82, 241; reviews of, 66n, 73, 177, 228, 239, 243; Prompt Book edition of, 71, 96, 106, 108, 110, 113-14; time of action, 110; Booth-Irving production, 187-88; mentioned, 6, 105, 221

Oberammergau Passion Play, 163

Passion Play, Salmi Morse, 167, 173

Pompadour, W. G. Wills and Sydney Grundy, 294

Poor Pillicoddy, John Madison Morton, 113n

Richard II, Shakespeare (Richard), Booth acts for first time, 51; Prompt Book edition of, 66, 67-68, 71, 95; reviews of, 54n, 62n; mentioned, 6, 52, 57n, 62, 93, 97, 101, 119

Richard III, Shakespeare (Richard), stage business, 52, 101; Prompt Book edition of, 72, 81, 87, 89, 91, 93, 95, 126; reviews of, 73, 79n, 91; costume, 91; mentioned, 6, 33, 54, 90, 97, 286

Richelieu, Edward Bulwer-Lytton (Richelieu), reviews of, 28, 73, 79n, 124, 170, 205; Prompt Book edition of, 71, 74, 81, 96, 98, 99-100, 122; mentioned, 11, 93, 97, 107, 165, 169, 208, 221

Rienzi, Mary Russell Mitford, 283

Romeo and Juliet, Shakespeare (Romeo), 21n

Ruy Blas, Victor Hugo (Ruy Blas), Prompt Book edition of, 85, 114, 118; mentioned, 6, 104, 122

The Soudan, Henry Pettitt and Augustus Harris, 302

The Stranger, August von Kotzebue, 72

Sweet Lavender, Arthur Wing Pinero, 294

The Taming of the Shrew, Shakespeare (Petruchio), Prompt Book edition of, 118; reviews of, 186; mentioned, 11, 27, 104, 114

Timon of Athens, Shakespeare (Timon), 111, 115

Werner, Lord Byron (Werner), 119

The Winter's Tale, Shakespeare, 71, 73-74, 115

Yorick's Love, William Dean Howells, 258n

Plympton, Eben, 252

Poe, Edgar Allan, 41n, 53n, 154

Pollock, Walter Herries, praises Booth's Othello, 181-82n; reviews Booth-Irving *Othello*, 187; mentioned, 300, 301

Pope, Charles R., 135, 136, 137

Porter, Horace, 210

The Prompt Books, profits from, 103, 108, 130; listed, 70n. *See also Plays and dramatic characters*

Provost, Mary, 39

Randall, James Ryder, 60n

Redmund, William, 173

Reed, Isaac, 116

Rehan, Ada, 15

Reid, Whitelaw, 29, 214, 239

Richardson, Leander Pease, 215

Ristori, Adelaide, 262, 264

Robertson, Richard A., 31n, 32, 50, 179

Robins, Elizabeth, 298

Robinson, Henry Crabb, 269

Robson, Stuart, 127

Rochefoucauld, François Duc de La, 8, 35

Rowe, George Fawcett, 53n, 94n

Runnion, James B., 128, 199, 206

Russell, Benjamin F., 85

Russell, John E., 171, 179, 180, 197-98

Ryder, Albert Pinkham, 12

Ryder, John, 171, 178

Saintine, Joseph Xavier Boniface, *Picciola*, 101

Salvini, Tommaso, acts with Booth, 272, 274

Sanftleben, Gustav, 226

Sankey, Ira D., 53, 54

Sargent, Franklin, 259n

Scherenberg, Gustav Otto, 226

Schoeffel, John B., 198

Schultz, Carl H., 26

Schwab, Frederick A., and proposed Booth-Irving exchange, 143, 144, 145; mentioned, 94, 137

Scott, Clement, reviews Booth's London engagement, 168, 170, 181; and *Theatre Magazine*, 171-72; mentioned, 15, 151-52, 175

Scott, John R., 247

Seaver, William A., 157

Shanks, William F. G., 47

Sheridan, William E., 141, 142n, 253

Shook, Sheridan, 80

Siddons, Sarah, 269

Simmonds, Morris, 145

Simpson, J. Palgrave, reviews Booth's Hamlet, 171n

Sinclair, Catherine, 41n

Smalley, George Washburn, and proposed Booth-Irving exchange, 133, 136, 138, 143, 144; mentioned, 141, 174-75, 176

Smith, Marcus ("Mark"), 160

Smith, Solomon Franklin, 160, 167

Sonnenthal, Adolph von, 5, 243, 262

Southern, Edward Askew, final illness and death of, 163, 166, 172, 177-78, 180; mentioned, 148n

Spofford, A. R., 130

Staegemann, Max, 237

Star Theatre, poor working conditions in, 254-55, 282

Stark, James, 111, 120

Stedman, Edmund Clarence, and proposed Booth-Irving exchange, 133, 138, 142, 143, 144-45; mentioned, 13, 136, 140, 141, 186, 251, 281

Sterne, Laurence, 3

Stetson, John, 224n

Stewart, Alexander T., 146n

Stoddard, Richard Henry, 13

Stoker, Abraham ("Bram"), 217

Stowe, Harriet Beecher, *Lady Byron Vindicated*, 23

Strakosch, Maurice, 137n

Strakosch Max, and proposed Booth-Irving exchange, 137, 143, 144

Stuart, William, relations with Booth, 19; manages New Park Theatre, 43, 48, 62; death of, 283; mentioned, 21, 44, 127-28, 206, 265

Sullivan, Louis, 298, 299

Swinbourne, Thomas, 173n

Swing, David, 221n

Tag, Casimir, 230

Talma, François-Joseph, 234

Taylor, Bayard, 13, 220

Taylor, Douglas, 93

Taylor, Tom, 119, 149, 151. *See also Plays and dramatic characters*

Terry, Ellen, 5, 173, 223-24, 253, 261, 294

Thackeray, William Makepeace, 8, 202

Thayer, Charles H., 268n, 285

Theobald, Lewis, 114n

Thompson, Launt, 8, 13, 18

Thorndyke, Louisa, 153n

Thorne, Thomas, 294

Tompkins, Eugene, 94n

Tompkins, Orlando, 94, 206

Toole, John Laurence, 290, 292-93, 294, 301

Towse, John Ranken, 15, 51

Tree, Herbert Beerbohm, 294

Trelawney, Edward John, 23, 24

Tyng, Rev. Stephen Higginson, 140

Tyng, Rev. Stephen Higginson, Jr., 140

Vaders, Emma, 279-80

Vaders, Henrietta, 280

Vaux, Calvert, 192

Vaux, Downing, engagement to Edwina Booth, 192, 250; illness of, 218, 225, 229; mentioned, 214

Vaux, Julia, 207, 218

Vecellio, Cesare, *Habiti antichi et moderni di tutto il mundo*, 93

Vedder, Elihu, 151

Vernon, Ida, 252, 253n, 255

Vezin, Hermann, 178

Vezin, Jane Elizabeth Thompson, 178n

Vider, Louis, 4, 47

Vincent, Leon John, 147

Vining, Edward P., *The Mystery of Hamlet*, 200, 203

Voltaire, François Marie Arouet, *Brutus*, 119n

Wall, Harry, 201

Wallack, Lester, and George Holland benefit, 25, 26, 28; retires from management, 284; men-

tioned, 14, 57, 60, 67, 72, 73, 74, 162, 200, 264

Waller, Daniel Wilmarth, death of, 203

Waller, Emma, 203n

Ward, Artemus, 13

Ward, Genevieve Teresa, 5, 162, 202

Warde, Frederick Barkham, 56n

Warker, Thomas G. H., 26

Warner, Charles, 161, 164-65, 171

Warren, George de Tabley Baron, 158

Warren, John Byrne, 158n

Warren, William, celebrates fiftieth anniversary on stage, 214, 216; mentioned, 161, 215, 221, 222

Watterson, Henry, 58

Weaver, Affie, 252

Weaver, Henry A., Sr., 252n, 256

Werder, Karl Friedrich, 231-32

Wheatley, William, 63

Wheeler, Andrew Carpenter ("Nym Crinkle"), 283

Wheelock, Joseph, Sr., 141, 142n

Whitman, Walt, 8, 13

Wilbrandt, Adolph, 242n

Wilde, Oscar, 247

Williams, Barney, death of, 62-63

Williams, Maria Pray Mestayer, 62

Winter, Arthur Elliot, birth of, 14; killed in sledding accident, 269; mentioned, 91

Winter, Charles, death of, 94; mentioned, 12, 34, 37

Winter, Charles, Jr., 44, 81, 98

Winter, Elizabeth ("Lizzie") Campbell, as authoress, 14; as actress, 14, 18, 24, 90; mentioned, 279, 283, 293

Winter, Louis Victor, birth of, 14; mentioned, 300

Winter, Percy Curtis, birth of, 14; acts in Barrett's company, 211, 224; mentioned, 91, 103-04, 108, 180, 217, 293

Winter, Viola Rosamund, birth of, 14

Winter, William, characterizes Booth's letters, 3; career and personality, 12-17; critical principles, 15; reviews Booth's Romeo, 21n; arranges George Holland benefit, 25, 26, 27; moves to Staten Island, 37; advises Booth, 38-42; opinion of Shylock, 39; opinion of Booth's Hamlet, 40-41; opinion of *The Marble Heart*, 41-42; praises Booth's Hamlet, 51-52; illness of, 59, 142, 251; moves to New York, 73-74; first trip to England (1877), 81, 84, 86, 157; moves family to Toronto, 84; feuds with Dion Boucicault, 86, 88; writes biographical sketch of Booth, 93, 95, 99, 102; reviews *Louis* xi, 122n; buys house on Staten Island, 125-26; and attempted assassination of Booth, 129; his obituaries of Palmer and Fechter, 139; reviews Macbeth, 150n; accompanies Booth to Mary's funeral, 195; in England (1882), 208; tours Holland and Germany with Barrett, 209; advises Booth about drinking, 209-10; writes play, 225; reviews Cincinnati Dramatic Festival, 240; criticizes Booth's supporting company, 250, 251-52; in England (1884), 258; *Shakespeare's England*, 280; death of son Arthur, 269, 283; in England (1888), 289, 290-94; biography of Booth, 273, 287, 303; declines Librarianship of The Players, 294-95; last visits with Booth, 303

Winter, William Jefferson, birth of, 14; mentioned, vii, 33, 292

Witham, Charles H., 36, 64

Wolter, Charlotte, 243

Woodman, Mattie, 253, 254n, 255

Worth, Charles Frederick, 265n

Young, Charles Mayne, 269

DATE DUE			